The Basic Design of Two-Stroke Engines

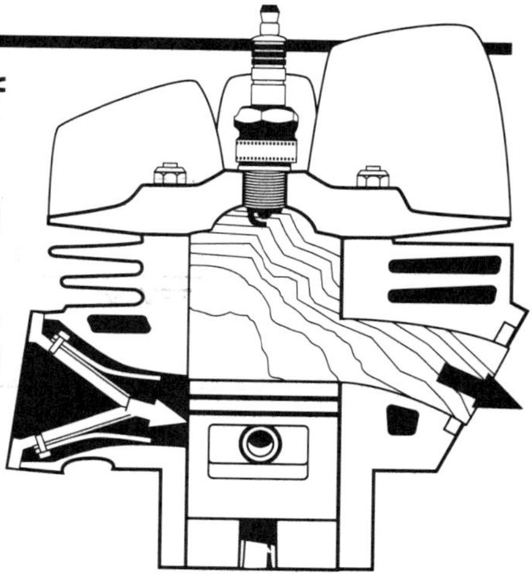

by
GORDON P. BLAIR
Professor of Mechanical Engineering
The Queen's University of Belfast

Published by:
Society of Automotive Engineers, Inc.
400 Commonwealth Drive
Warrendale, PA 15096-0001

**Library of Congress
Cataloging-in-Publication Data**

Blair, Gordon P.
The basic design of two-stroke
engines / Gordon P. Blair.
p. cm.
"R-104."
Includes bibliographical references.
ISBN 1-56091-008-9
1. Two-stroke cycle engines--Design and
construction. I. Title.
TJ790.B57 1990 89-48422
621.43--dc20 CIP

The Mulled Toast

To Sir Dugald Clerk I raise my glass,
his vision places him first in class.
For Alfred Scott let's have your plaudits,
his Squirrels drove the four-strokes nuts.

To Motorradwerk Zschopau I doff my cap,
their Walter Kaaden deserves some clap.
Señores Bulto and Giro need a mention,
their flair for design got my attention.

Brian Stonebridge I allege,
initialled my thirst for two-stroke knowledge.
In academe I found inspiration
in Crossland and Timoney and Rowland Benson.

All of my students are accorded a bow,
their doctoral slavery stokes our know-how.
The craftsmen at Queen's are awarded a medal,
their precision wrought engines from paper to metal.

For support from industry I proffer thanks,
their research funds educate many Ulster cranks,
but the friendships forged I value more
as they span from Iwata to Winnebago's shore.

To the great road-racers I lift my hat,
they make the adrenalin pump pitter-pat,
for the Irish at that are always tough,
like Dunlop and Steenson and Ray McCullough.

In case you think, as you peruse this tome,
that a computer terminal is my mental home,
I've motorcycled at trials with the occasional crash
and relieved fellow-golfers of some of their cash.

Gordon Blair
May 1989

Foreword

This book is intended to be an information source for those who are involved in the design of two-stroke engines. In particular, it is a book for those who are already somewhat knowledgeable on the subject, but who have sometimes found themselves with a narrow perspective on the subject, perhaps due to specialization in one branch of the industry. For example, the author is familiar with many who are expert in tuning racing motorcycle engines, but who would freely admit to being quite unable to design for good fuel economy or emission characteristics, should their industry demand it. It is the author's experience that the literature on the spark-ignition two-stroke engine is rich in descriptive material but is rather sparse in those areas where a designer needs specific guidance. As the two-stroke engine is currently under scrutiny as a future automobile engine, this book will help to reorient the thoughts of those who are more expert in designing camshafts than scavenge ports. Also, this book is intended as a textbook on design for university students in the latter stages of their undergraduate studies, or for those undertaking postgraduate research or course study.

Perhaps more important, the book is a design aid in the areas of gas dynamics, fluid mechanics, thermodynamics and combustion. To stop the reader from instantly putting the book down in terror at this point, rest assured that the whole purpose of this book is to provide design assistance with the actual mechanical design of the engine, in which the gas dynamics, fluid mechanics, thermodynamics and combustion have been optimized so as to provide the required performance characteristics of power or torque or fuel consumption.

Therefore, the book will attempt to explain, inasmuch as the author understands, the intricacies of, for example, scavenging, and then provide the reader with computer programs written in Basic which will assist with the mechanical design to produce, to use the same example, better scavenging in any engine design. These are the very programs which the author has written as his own mechanical design tools, as he spends a large fraction of his time designing engines for test at The Queen's University of Belfast (QUB) or for prototype or production development in industry. Many of the design programs which have been developed at QUB over the last twenty-five years have become so complex, or require such detailed input data, that the operator cannot see the design wood for the data trees. In consequence, these simpler, often empirical, programs have been developed to guide the author as to the data set before applying a complex unsteady gas dynamic or computational fluid dynamic analysis package. On many occasions that complex package merely confirms that the empirical program, containing as it does the distilled experience of several generations, was sufficiently correct in the first place.

At the same time, as understanding unsteady gas dynamics is the first major step to becoming a competent designer of reciprocating ic engines, the book contains a major section dealing with that subject and the reader is provided with an engine design program of the complete pressure wave motion form, which is clearly not an empirical analytical package.

The majority of the book is devoted to the design of the two-stroke spark-ignition (si) engine, but there will be remarks passed from time to time regarding the two-stroke diesel or compression-ignition (ci) engine; these remarks will be clearly identified as such. The totality of the book is just as applicable to the design of the diesel as it is to the gasoline engine, for the only real difference is the methodology of the combustion process.

It is to be hoped that the reader derives as much use from the analytic packages as does the author. The author has always been somewhat lazy of mind and so has found the accurate repetitive nature of the computer solution to be a great saviour of mental perspiration. At the same time, and since his schooldays, he has been fascinated with the two-stroke cycle engine and its development and improvement. In those far-off days in the late 1950's, the racing two-stroke motorcycle was a music-hall joke, whereas a two-stroke engined car won the Monte Carlo Rally. Today, there are no two-stroke engined cars and four-stroke engines are no longer competitive in Grand Prix motorcycle racing! Is tomorrow, or the twenty-first century, going to produce yet another volte-face?

The author has also had the inestimable privilege of being around at precisely that point in history when it became possible to unravel the technology of engine design from the unscientific black art which had surrounded it since the time of Otto and Clerk. That unravelling occurred because the digital computer permitted the programming of the fundamental unsteady gas-dynamic theory which had been in existence since the time of Rayleigh, Kelvin, Stokes and Taylor.

The marriage of these two interests, computers and two-stroke engines, has produced this book and the material within it. For those in this world who are of a like mind, this book should prove to be useful.

Acknowledgements

The first acknowledgement is to those who enthused me during my schooldays on the subject of internal combustion engines in general, and motorcycles in particular. They set me on the road to a thoroughly satisfying research career which has never seen a hint of boredom. The two individuals were, my father who had enthusiastically owned many motorcycles in his youth, and Mr. Rupert Cameron, who had owned but one and had ridden it everywhere—a 1925 350 cc Rover. Of the two, Rupert Cameron was the greater influence, for he was a walking library of the Grand Prix races of the twenties and thirties and would talk of engine design, and engineering design, in the most knowledgeable manner. He was actually the senior naval architect at Harland and Wolff's shipyard in Belfast and was responsible for the design of some of the grandest liners ever to sail the oceans.

I have to acknowledge that this book would not be written today but for the good fortune that brought Dr. Frank Wallace (Professor at Bath University since 1965) to Belfast in the very year that I wished to do postgraduate research. At that time, Frank Wallace was one of perhaps a dozen people in the world who comprehended unsteady gas dynamics, which was the subject area I already knew I had to understand if I was ever to be a competent engine designer. However, Frank Wallace taught me something else as well by example, and that is academic integrity. Others will judge how well I learned either lesson.

Professor Bernard Crossland deserves a special mention, for he became the Head of the Department of Mechanical Engineering at QUB in the same year I started as a doctoral research student. His drive and initiative set the tone for the engineering research which has continued at QUB until the present day. The word engineering in the previous sentence is underlined because he instilled in me, and a complete generation, that real "know-how" comes from using the best theoretical science available, at the same time as conducting related experiments of a product design, manufacture, build and test nature. That he became, in latter years, a Fellow of the Royal Society, a Fellow of the Fellowship of Engineering and a President of the Institution of Mechanical Engineers seems no more than justice.

I have been very fortunate in my early education to have had teachers of mathematics who taught me the subject not only with enthusiasm but, much more importantly, from the point of view of application. I refer particularly to Mr. T. H. Benson at Larne Grammar School and to Mr. Scott during my undergraduate studies at The Queen's University of Belfast. They gave me a lifelong interest in the application of mathematics to problem solving which has never faded.

The next acknowledgement is to those who conceived and produced the Macintosh computer. Without that machine, on which I have typed this entire manuscript, drawn every figure which is not from SAE archives, and developed all of the computer programs, there would be no book. In short, the entire book, and the theoretical base for much of it, is there because the Macintosh has such superbly integrated hardware and software so that huge workloads can be tackled rapidly and efficiently.

It would be remiss of me not to thank many of the present generation of doctoral research students, and my four colleagues involved in reciprocating engine research and development at QUB, Drs. Douglas, Fleck, Goulburn and Kenny, who read this book during its formation and offered many helpful suggestions on improving the text. To one of our technicians at QUB, David Holland, a special mention must be made for his expert production of many of the photographs which illustrate this book.

As the flyleaf poem says, gentlemen, take a bow.................

Gordon P. Blair,
The Queen's University of Belfast,
May 1989.

Table of Contents

CHAPTER 7

CHAPTER 8

CHAPTER 1

Introduction to the Two-Stroke Engine

1.0 Introduction to the two-stroke cycle engine

It is generally accepted that the two-stroke cycle engine was invented by Sir Dugald Clerk in England at the end of the 19th century. The form of the engine using crankcase compression for the induction process, including the control of the timing and area of the exhaust, transfer and intake ports by the piston, was patented by Joseph Day in England in 1891. His engine was the original "three-port" engine and is the forerunner of the simple two-stroke engine which has been in common usage since that time.

Some of the early applications were in motorcycle form and are well recorded by Caunter(1.5). The first engines were produced by Edward Butler in 1887 and by J. D. Roots, in the form of the Day crankcase compression type, in 1892; both of these designs were for powered tricycles. Considerable experimentation and development was conducted by Alfred Scott, and his Flying Squirrel machines competed very successfully in Tourist Trophy races in the first quarter of the twentieth century. They were designed quite beautifully in both the engineering and in the aesthetic sense. After that, two-stroke engines faded somewhat as competitive units in racing for some years until the supercharged DKW machines of the thirties temporarily revived their fortunes. With the banning of supercharging for motorcycle racing after the Second World War, the two-stroke engine lapsed again until 1959 when the MZ machines, with their tuned exhaust expansion chambers and disc valve induction systems, introduced a winning engine design which has basically lasted to the present day. A machine typical of this design approach is shown in Plate 1.1, a 250 cc twin-cylinder engine with a rotary sleeve induction system which was built at QUB about 1969. Today, two-stroke engined motorcycles, scooters and mopeds are still produced in very large numbers for general transport and for recreational purposes, although the legislative pressure on exhaust emissions in some countries has produced a swing to a four-stroke engine replacement in some cases. Whether the two-stroke engine will return as a mass production motorcycle engine will depend on the result of research and development being conducted by all of the manufacturers at the present time. There are some other applications with engines which are very similar in design terms to those used for motorcycles, and the sports of go-kart and hydroplane racing would fall into this category.

The two-stroke engine is used for lightweight power units which can be employed in various attitudes as hand-held power tools. Such tools are chainsaws,

Plate 1.1 A QUB engined 250 cc racing motorcycle showing the tuned exhaust pipes.

brushcutters and concrete saws, to name but a few, and these are manufactured with a view to lightness and high specific power performance. One such device is shown in Plate 1.2. The manufacturing numbers involved are in millions per annum worldwide.

The earliest outboard motors were pioneered by Evinrude in the United States about 1909, with a 1.5 hp unit, and two-stroke engines have dominated this application until the present day. Some of the current machines are very sophisticated designs, such as 300 hp V6 and V8 engined outboards with remarkably efficient engines considering that the basic simplicity of the two-stroke crankcase compression engine has been retained. Although the image of the outboard motor is that it is for sporting and recreational purposes, the facts are that the product is used just as heavily for serious employment in commercial fishing and for everyday water transport in many parts of the world. The racing of outboard motors is a particularly exciting form of automotive sport, as seen in Plate 1.3.

Some of the new recreational products which have appeared in recent times are snowmobiles and water scooters, and the engine type almost always employed for such machines is the two-stroke engine. The use of this engine in a snowmobile is almost an ideal application, as the simple lubrication system of a two-stroke engine is perfectly suited for sub-zero temperature conditions. Although the snowmobile has been described as a recreational vehicle, it is actually a very practical means of everyday transport for many people in an Arctic environment.

Plate 1.2 A Homelite chainsaw engine illustrating the two-stroke powered tool (courtesy of Homelite Textron).

The use of the two-stroke engine in automobiles has had an interesting history, and some quite sophisticated machines were produced in the 1960's, such as the Auto-Union vehicle from West Germany and the simpler Wartburg from East Germany. The Saab car from Sweden actually won the Monte Carlo Rally with Eric Carlson driving it. Until recent times, Suzuki built a small two-stroke engined car in Japan. With increasing ecological emphasis on fuel consumption rate and exhaust emissions, the simple two-stroke engined car disappeared, but interest in the design has seen a resurgence in recent times as the legislative pressure intensifies on exhaust acid emissions. Almost all car manufacturers are experimenting with various forms of two-stroke engined vehicles equipped with direct fuel injection, or some variation of that concept in terms of stratified charging or combustion.

The two-stroke engine has been used in light aircraft, and today is most frequently employed in the recreational microlite machines. There are numerous other applications for the spark-ignition engine, such as small electricity generating sets or engines for remotely piloted vehicles, i.e., aircraft for meteorological data gathering or military purposes. These are but two of a long list of multifarious examples.

Plate 1.3 A high-performance multi-cylinder outboard motor in racing trim (courtesy of Mercury Marine).

The use of the two-stroke engine in compression ignition, or diesel, form deserves special mention, even though it will not figure specifically in terms of design discussion within this book. The engine type has been used for trucks and locomotives, such as the designs from General Motors in America or Rootes-Tilling-Stevens in Britain. Both of these have been very successful engines in mass production. The engine type, producing a high specific power output, has also been a favorite for military installations in tanks and fast naval patrol boats. Some of the most remarkable aircraft engines ever built have been two-stroke diesel units, such as the Junkers Jumo and the turbo-compounded Napier Nomad. There is no doubt that the most successful of all of the applications is that of the marine diesel main propulsion unit, referred to in my student days in Harland and Wolff's shipyard in Belfast as a "cathedral" engine. The complete engine is usually some 12 m tall, so the description is rather apt. Such engines, the principal exponents of which were Burmeister and Wain in Copenhagen and Sulzer in Winterthur, were typically of 900 mm bore and 1800 mm stroke and ran at 60-100 rpm, producing some 4000 hp per cylinder. They had thermal efficiencies in excess of 50%, making them the most efficient prime movers ever made. These engines are very different from the rest of the two-stroke engine species in terms of scale but not in design concept, as Plate 1.4 illustrates.

Plate 1.4 A Harland and Wolff uniflow scavenged two-stroke diesel ship propulsion engine of 21,000 bhp (courtesy of Harland and Wolff plc).

It is probably true to say that the two-stroke engine has produced the most diverse opinions on the part of either the users or the engineers. These opinions vary from fanatical enthusiasm to thinly veiled dislike. Whatever the view of the reader, at this early juncture in reading this book, no other engine type has ever fascinated the engineering world to quite the same extent. This is probably because the engine seems so deceptively simple to design, develop and manufacture. That the very opposite is the case may well be the reason that some spend a lifetime investigating this engineering curiosity. The potential rewards are great, for no other engine cycle has produced, in one constructional form or another, such high thermal efficiency or such low specific fuel consumption, such high specific power criteria referred to either swept volume, bulk or weight, nor such low acid exhaust emissions.

1.1 The fundamental method of operation of a simple two-stroke engine

The simple two-stroke engine is shown in Fig. 1.1, with the various phases of the filling and emptying of the cylinder illustrated in (a)-(d). The simplicity of the engine is obvious, with all of the processes controlled by the upper and lower edges of the piston. A photograph of a simple two-stroke engine is provided in Plate 1.5. It is actually a small chainsaw engine, giving some further explanation of its construction.

In Fig. 1.1(a), above the piston, the trapped air and fuel charge is being ignited by the spark plug, producing a rapid rise in pressure and temperature which will drive the piston down on the power stroke. Below the piston, the opened inlet port is inducing air from the atmosphere into the crankcase due to the increasing volume of the crankcase lowering the pressure below the atmospheric value. The crankcase is sealed around the crankshaft to ensure the maximum depression within it. To induce fuel into the engine, the various options exist of either placing a carburetor in the inlet tract, injecting fuel into the inlet tract, injecting fuel into the crankcase or transfer ducts, or injecting fuel directly into the cylinder before or after the closure of the exhaust port. Clearly, if it is desired to operate the engine as a diesel power unit, the latter is the only option, with the spark plug possibly being replaced by a glow plug as an initial starting aid and the fuel injector placed in the cylinder head area.

In Fig. 1.1(b), above the piston, the exhaust port has been opened. It is often called the "release" point in the cycle, and this allows the transmission into the exhaust duct of a pulse of hot, high-pressure exhaust gas from the combustion process. As the area of the port is increasing with crankshaft angle, and the cylinder pressure is falling with time, it is clear that the exhaust duct pressure profile with time is one which increases to a maximum value and then decays. Such a flow process is described as unsteady gas flow and such a pulse can be reflected from all pipe area changes, or at the pipe end termination to the atmosphere. These reflections have a dramatic influence on the engine performance, as later chapters of this book describe. Below the piston, the compression of the fresh charge is taking place. The pressure and temperature achieved will be a function of the proportionate reduction of the crankcase volume, i.e., the crankcase compression ratio.

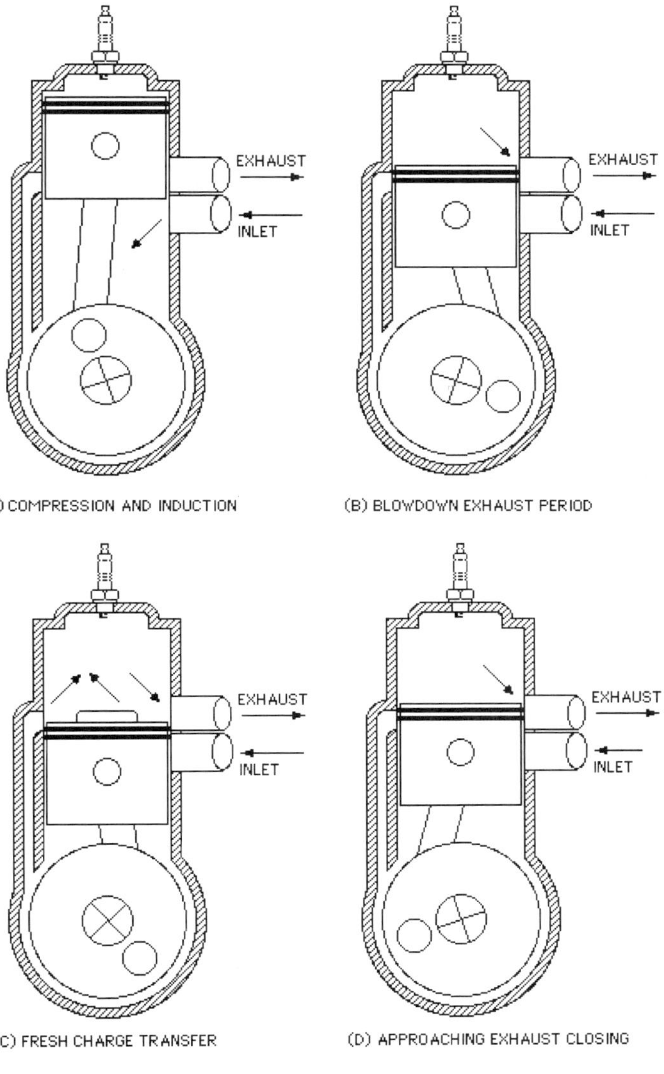

Fig. 1.1 Various stages in the operation of the two-stroke cycle engine.

In Fig. 1.1(c), above the piston, the initial exhaust process, referred to as "blowdown", is nearing completion and, with the piston having uncovered the transfer ports, this connects the cylinder directly to the crankcase through the transfer ducts. If the crankcase pressure exceeds the cylinder pressure then the fresh charge enters the cylinder in what is known as the scavenge process. Clearly, if the

Plate 1.5 An exploded view of a simple two-stroke engine.

transfer ports are badly directed then the fresh charge can exit directly out of the exhaust port and be totally lost from the cylinder. Such a process, referred to as "short circuiting," would result in the cylinder being filled only with exhaust gas at the onset of the next combustion process, and no pressure rise or power output would ensue. Worse, all of the fuel in a carburetted configuration would be lost to the exhaust with a consequential monstrous emission rate of unburned hydrocarbons. Therefore, the directioning of the fresh charge by the orientation of the transfer ports should be conducted in such a manner as to maximize the retention of it within the cylinder. This is just as true for the diesel engine, for the highest trapped air mass can be burned with an appropriate fuel quantity to attain the optimum power output. It is obvious that the scavenge process is one which needs to be optimized to the best of the designer's ability. Later chapters of this book will concentrate heavily on the scavenge process and on the most detailed aspects of the mechanical design to improve it as much as possible. It should be clear that it is not possible to have such a process proceed perfectly, as some fresh charge will always find a way through the exhaust port. Equally, no scavenge process, however extensive or thorough, will ever leach out the last molecule of exhaust gas.

In Fig. 1.1(d), in the cylinder, the piston is approaching what is known as the "trapping" point, or exhaust closure. The scavenge process has been completed and the cylinder is now filled with a mix of air, fuel if a carburetted design, and exhaust gas. As the piston rises, the cylinder pressure should also rise, but the exhaust port is still open and, barring the intervention of some unsteady gas-dynamic effect generated in the exhaust pipe, the piston will spill fresh charge into the exhaust duct to the detriment of the resulting power output and fuel consumption. Should it be feasible to gas-dynamically plug the exhaust port during this trapping phase, then it is possible to greatly increase the performance characteristics of the engine. In a single-cylinder racing engine, it is possible to double the mass of the trapped air charge using a tuned pipe, which means doubling the power output; such effects are discussed in Chapter 2. After the exhaust port is finally closed, the true compression process begins until the combustion process is commenced by ignition. Not surprisingly, therefore, the compression ratio of a two-stroke engine is characterized by the cylinder volume after exhaust port closure and is called the Trapped Compression Ratio to distinguish it from the value commonly quoted for the four-stroke engine. That value is termed here as the Geometric Compression Ratio and is based on the full swept volume.

In summary, the simple two-stroke engine is a double-acting device. Above the piston, the combustion and power processes take place, whereas below the piston in the crankcase, the fresh charge is induced and prepared for transfer to the upper cylinder. What could be simpler to design than this device?

1.2 Methods of scavenging the cylinder

1.2.1 Loop scavenging

In Chapter 3, there is a comprehensive discussion of the fluid mechanics and the gas dynamics of the scavenging process. However, it is important that this preliminary descriptive section introduces the present technological position, from a historical perspective. The scavenging process depicted in Fig. 1.1 is described as Loop Scavenging, the invention of which is credited to Schnurle in Germany about 1926. The objective was to produce a scavenge process in a ported cylinder with two or more scavenge ports directed towards that side of the cylinder away from the exhaust port, but across a piston with essentially a flat top. The invention was designed to eliminate the hot-running characteristics of the piston crown used in the original method of scavenging devised by Sir Dugald Clerk, namely the deflector piston employed in the Cross Scavenging method discussed in Sect. 1.2.2. Although Schnurle was credited with the invention of loop scavenging, there is no doubt that patents taken out by Schmidt and by Kind fifteen years earlier look uncannily similar, and it is the author's understanding that considerable litigation regarding patent ownership ensued in Germany in the 1920's and 1930's. In Fig. 1.2, the various layouts observed in loop scavenged engines are shown. These are plan sections through the scavenge ports and the exhaust port, and the selection shown is far from an exhaustive sample of the infinite variety of designs seen in two-stroke engines. The common element is the sweep back angle for the "main" transfer port, away from the exhaust port; the "main" transfer port is the port next to the exhaust port. Another advantage of the loop scavenge design is the availability of a compact combustion chamber above the flat-topped piston which permits a rapid and efficient combustion process. A piston for such an engine is shown in Plate 1.6. The type of scavenging used in the engine in Plate 1.5 is loop scavenging.

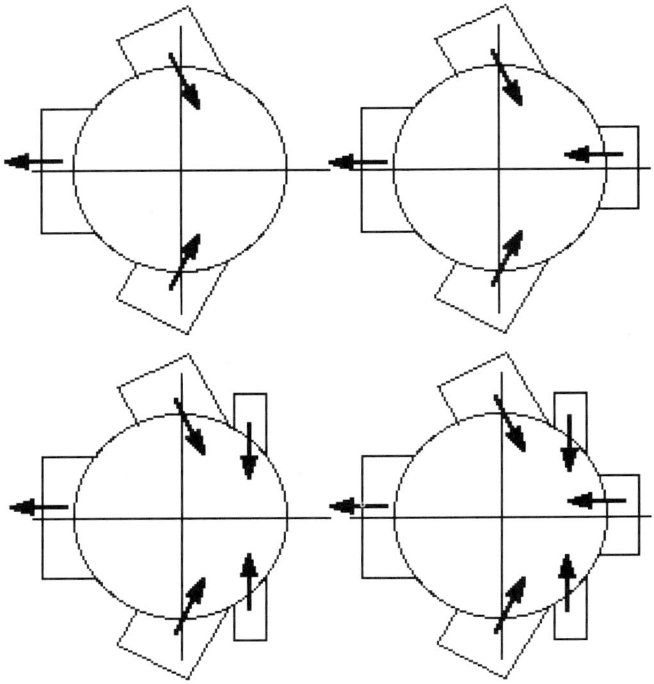

Fig. 1.2 Various scavenge port plan layouts found in loop scavenging.

1.2.2 Cross scavenging

This is the original method of scavenging proposed by Sir Dugald Clerk and is widely used for outboard motors to this very day. The modern deflector design is illustrated in Fig. 1.3 and emanates from the Scott engines of the early 1900's, whereas the original deflector was a simple wall or barrier on the piston crown. To further illustrate that, a photograph of this type of piston appears in Plate. 1.6. In Sect. 3.2.4 it will be shown that this has good scavenging characteristics at low throttle openings and this tends to give good low-speed and low-power characteristics, making it ideal for, for example, small outboard motors employed in sport fishing. At higher throttle openings the scavenging efficiency is not particularly good and, combined with a non-compact combustion chamber filled with an exposed protuberant deflector, the engine has rather unimpressive specific power and fuel economy characteristics (see Plate 4.2). The potential for detonation and for pre-ignition, from the high surface-to-volume ratio combustion chamber and the hot deflector edges, respectively, is rather high and so the compression ratio which can be employed in this engine tends to be somewhat lower than for the equivalent

Plate 1.6 The pistons, from L to R, for a QUB type cross scavenged, a conventional cross scavenged and a loop scavenged engine.

loop scavenged power unit. The engine type has some considerable packaging and manufacturing advantages over the loop engine. In Fig. 1.3 it can be seen from the port plan layout that the cylinder-to-cylinder spacing in a multi-cylinder configuration could be as close as is practical for inter-cylinder cooling considerations. If one looks at the equivalent situation for the loop scavenged engine in Fig. 1.2 it can be seen that the transfer ports on the side of the cylinder prohibit such close cylinder spacing; while it is possible to twist the cylinders to alleviate this effect to some extent, the end result has further packaging, gas-dynamic and scavenging disadvantages(1.12). Further, it is possible to drill the scavenge and the exhaust ports directly, in-situ and in one operation, from the exhaust port side, and thereby reduce the manufacturing costs of the cross scavenged engine by comparison with an equivalent loop or uniflow scavenged power unit.

One design of cross scavenged engines, which does not have the disadvantages of poor wide-open throttle scavenging and a non-compact combustion chamber, is the type designed at QUB(1.9) and sketched in Fig. 1.4. A piston for this design is shown in Plate 1.6. However, the cylinder does not have the same manufacturing simplicity as that of the conventional deflector piston engine. It has been shown by Blair(1.10) and in Sect. 3.2.4, that the scavenging is as effective as a loop scavenged power unit and that the highly squished and turbulent combustion chamber leads to

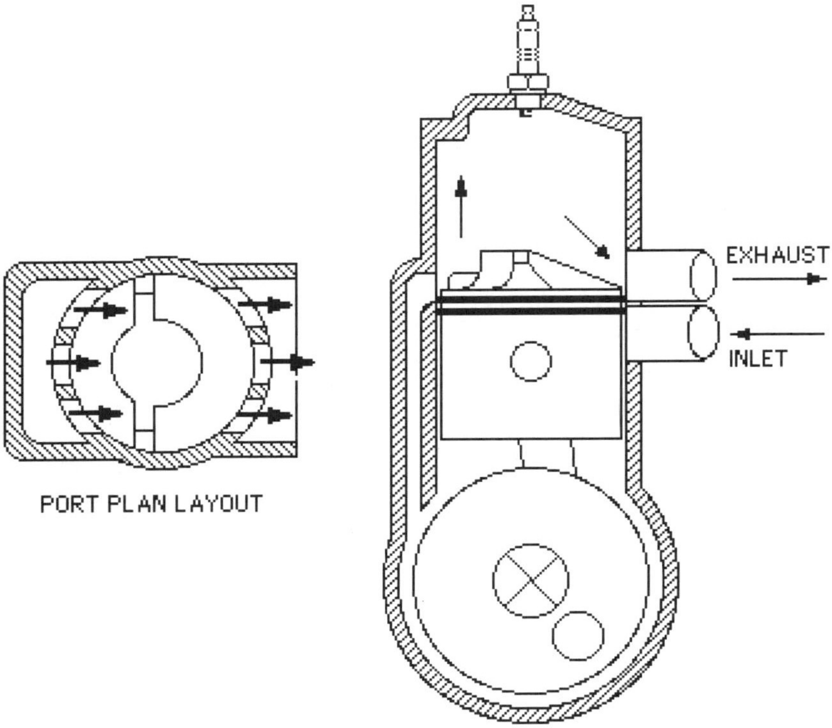

Fig. 1.3 Deflector piston or cross scavenged engine.

good power and good fuel economy characteristics, allied to cool cylinder head running conditions at high loads, speeds and compression ratios(1.9) (see Plate 4.3). Several models of this QUB type are in series production at the time of writing.

1.2.3 Uniflow scavenging

Uniflow Scavenging has long been held to be the most efficient method of scavenging the two-stroke engine. The basic scheme is illustrated in Fig. 1.5 and, fundamentally, the methodology is to start filling the cylinder with fresh charge at one end and remove the exhaust gas from the other. Often the charge is swirled at both the charge entry level and the exhaust exit level by either suitably directing the porting angular directions or by masking a poppet valve. The swirling air motion is particularly effective in promoting good combustion in a diesel configuration. Indeed, the most efficient prime movers ever made are the low-speed marine diesels of the uniflow scavenged two-stroke variety with thermal efficiencies in excess of 50%. However, these low-speed engines are ideally suited to uniflow scavenging,

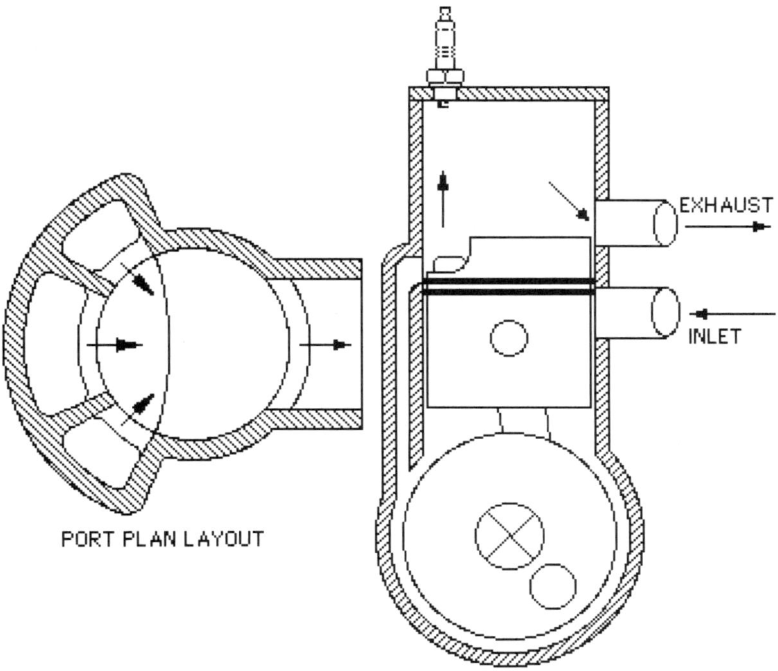

Fig. 1.4 QUB type of deflector piston or cross scavenged engine.

with cylinder bores about 1000 mm, a cylinder stroke about 2500 mm, and a bore-stroke ratio of 0.4. For most engines used in today's motorcycles and outboards, or tomorrow's automobiles, bore-stroke ratios are typically between 0.9 and 1.3. For such engines, there is some evidence (presented in Sect. 3.2.4) that uniflow scavenging, while still very good, is not significantly better than the best of loop scavenged designs(1.11). For spark-ignition engines, as uniflow scavenging usually entails some considerable mechanical complexity over simpler methods and there is not in reality the imagined performance enhancement from uniflow scavenging, this virtually rules out this method of scavenging on the grounds of increased engine bulk and cost for an insignificant power or efficiency advantage.

1.2.4 Scavenging without employing the crankcase as an air pump

The essential element of the original Clerk invention, or perhaps more properly the variation of the Clerk principle by Day, was the use of the crankcase as the air-pumping device of the engine; all simple designs use this concept. The lubrication of such engines has traditionally been conducted on a total-loss basis by whatever

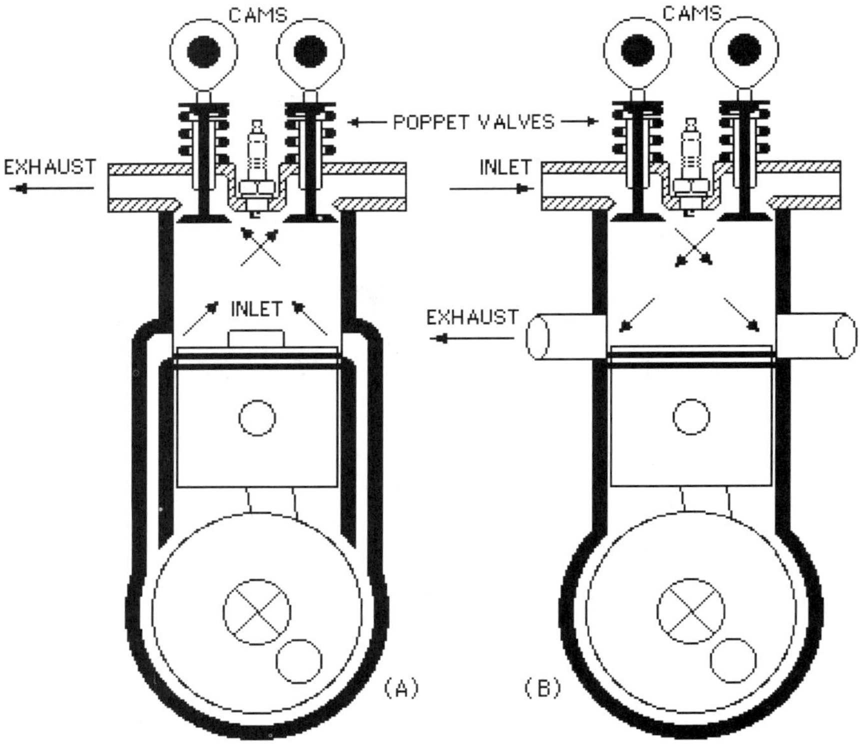

Fig. 1.5 Two methods of uniflow scavenging the two-stroke engine.

means employed. The conventional method has been to mix the lubricant with the petrol (gasoline) and supply it through the carburetor in ratios of lubricant to petrol varying from 25:1 to 100:1, depending on the application, the skill of the designers and/or the choice of bearing type employed as big-ends or as main crankshaft bearings. The British term for this type of lubrication is called "petroil" lubrication. As the lubrication is of the total-loss type, and some 10-30% of the fuel charge is short-circuited to the exhaust duct along with the air, the resulting exhaust plume is rich in unburned hydrocarbons and lubricant, some partially burned and some totally unburned, and is consequently visible as smoke. This is ecologically unacceptable in the latter part of the twentieth century and so the manufacturers of motorcycles and outboards have introduced separate oil-pumping devices to reduce the oil consumption rate, and hence the oil deposition rate to the atmosphere, be it directly to the air or via water. Such systems can reduce the effective oil-to-petrol ratio to as little as 200 or 300 and approach the oil consumption rate of four-stroke cycle engines. Even so, any visible exhaust smoke is always unacceptable and so, for future designs, as has always been the case for the marine and automotive two-stroke diesel engine, a crankshaft lubrication system based on pressure-fed plain bearings with a wet or dry sump may be employed. One of the successful compression ignition engine designs of this type is the Detroit Diesel engine shown in Plate 1.7.

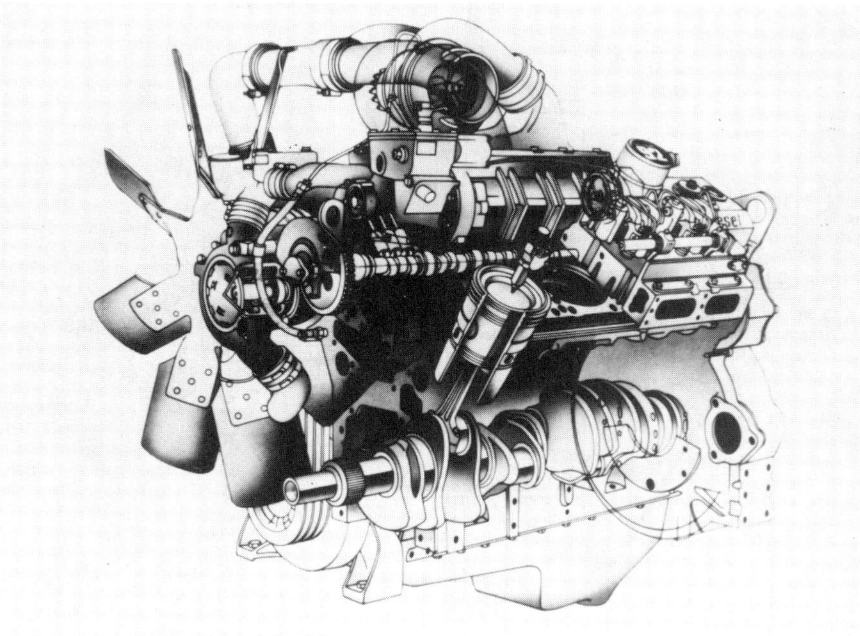

Plate 1.7 The Detroit Diesel Allison Series 92 uniflow scavenged, supercharged and turbocharged diesel engine for truck applications (courtesy of General Motors).

By definition, this means that the crankcase can no longer be used as the air-pumping device and so an external air pump will be utilized. This can be either a positive displacement blower of the Roots type, or a centrifugal blower driven from the crankshaft. Clearly, it would be more efficient thermodynamically to employ a turbocharger, where the exhaust energy to the exhaust turbine is available to drive the air compressor. Such an arrangement is shown in Fig. 1.6 where the engine has both a blower and a turbocharger. The blower would be used as a starting aid and as an air supplementary device at low loads and speeds, with the turbocharger employed as the main air supply unit at the higher torque and power levels at any engine speed. To prevent short-circuiting fuel to the exhaust, a fuel injector would be used to supply petrol directly to the cylinder, hopefully after the exhaust port is closed and not in the position sketched, at bottom dead center (bdc). Such an engine type has already demonstrated excellent fuel economy behavior, good exhaust emission characteristics of unburned hydrocarbons and carbon monoxide and superior emission characteristics of oxides of nitrogen, by comparison with an equivalent four-stroke engine. This subject will be elaborated on in Chapter 7. The diesel engine shown in Plate 1.7 is just such a power unit, but employing compression ignition.

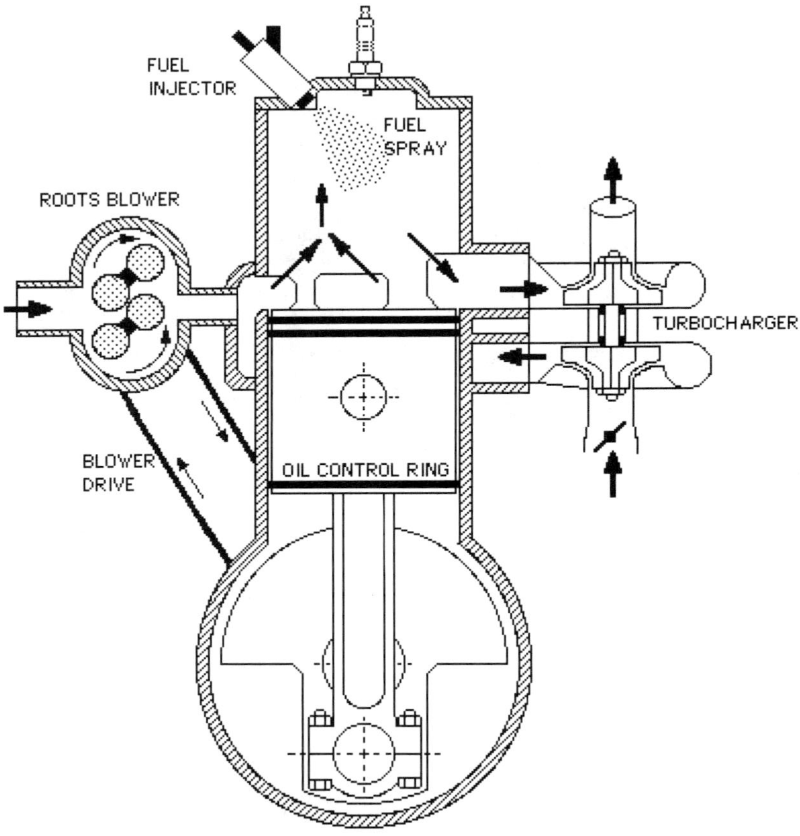

Fig. 1.6 A supercharged and turbocharged fuel-injected two-stroke engine.

Nevertheless, in case the impression is left that the two-stroke engine with a "petroil" lubrication method and a crankcase air pump is an anachronism, it should be pointed out that this provides a simple, lightweight, high specific output powerplant for many purposes, for which there is no effective alternative engine. Such applications range from the agricultural for chainsaws and brushcutters, where the engine can easily run in an inverted mode, to small outboards where the alternative would be a four-stroke engine resulting in a considerable weight, bulk, and manufacturing cost increase.

1.3 Valving and porting control of the exhaust, scavenge and inlet processes

The simplest method of allowing fresh charge access into, and exhaust gas discharge from, the two-stroke engine is by the movement of the piston exposing ports in the cylinder wall. In the case of the simple engine illustrated in Fig. 1.1 and Plate 1.5, this means that all port timing events are symmetrical with respect to top dead center (tdc) and bdc. It is possible to change this behavior slightly by offsetting the crankshaft center-line to the cylinder center-line, but this is rarely carried out in practice as the resulting improvement is hardly worth the manufacturing complication involved. It is possible to produce asymmetrical inlet and exhaust timing events by the use of disc valves, reed valves and poppet valves. This permits the phasing of the porting to correspond more precisely with the pressure events in the cylinder or the crankcase, and so gives the designer more control over the optimization of the exhaust or intake system. The use of poppet valves for both inlet and exhaust timing control is sketched, in the case of uniflow scavenging, in Fig. 1.5. Fig. 1.7 illustrates the use of disc and reed valves for the asymmetrical timing control of the inlet process into the engine crankcase. It is virtually unknown to attempt to produce asymmetrical timing control of the scavenge process from the crankcase through the transfer ports.

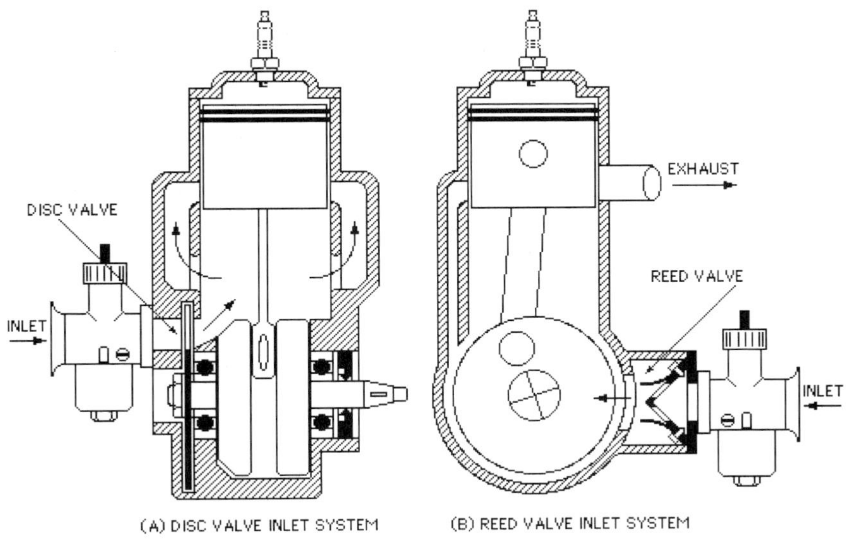

Fig. 1.7 Disc valve and reed valve control of the inlet system.

1.3.1 Poppet valves

The use and design of poppet valves is thoroughly covered in texts and papers dealing with four-stroke engines(1.3), so it will not be discussed here, except to say that the flow area-time characteristics of poppet valves are, as a generality, considerably less than are easily attainable for the same geometrical access area posed by a port in a cylinder wall. Put in simpler form, it is difficult to design poppet valves so as to adequately flow sufficient charge into a two-stroke engine. It should be remembered that the actual time available for any given inlet or exhaust process, at the same engine rotational speed, is about one half of that possible in a four-stroke cycle design.

1.3.2 Disc valves

The disc valve design is thought to have emanated from East Germany in the 1950's in connection with the MZ racing motorcycles from Zchopau, the same machines which introduced the expansion chamber exhaust system for high specific output racing engines. A twin-cylinder racing motorcycle engine which uses this method of induction control is shown in Plate 1.8. Irving(1.1) attributes the design to Zimmerman. Most disc valves have timing characteristics of the values shown in Fig. 1.8 and are usually fabricated from a spring steel, although discs made from composite materials are also common. To assist with comprehension of disc valve operation and design, the reader should find useful Fig. 6.13 and the discussion in Sect. 6.4.

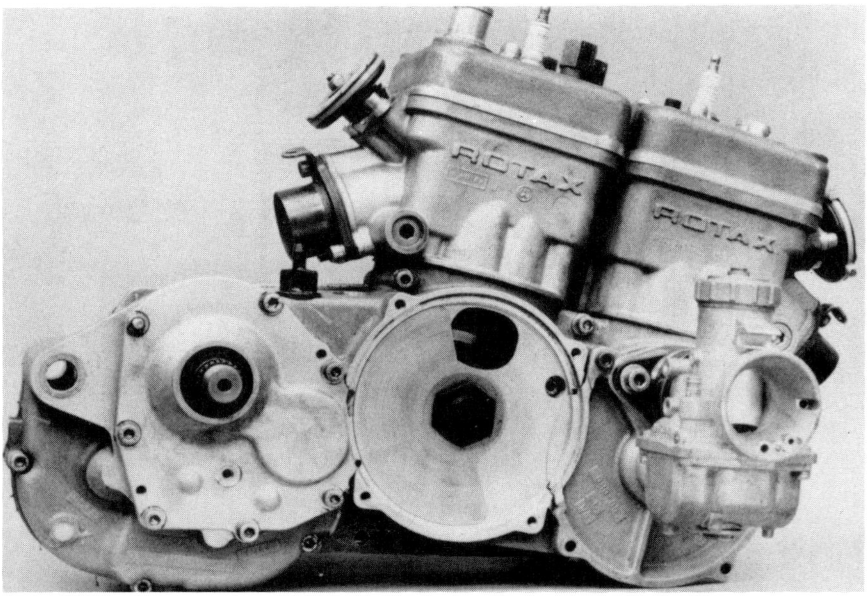

Plate 1.8 A Rotax disc valve racing motorcycle engine with one valve cover removed exposing the disc valve.

1.3.3 Reed valves

Reed valves have always been popular in outboard motors, as they provide an effective automatic valve whose timings vary with both engine load and engine speed. In recent times, they have also been designed for motorcycle racing engines, succeeding the disc valve. In part, this technical argument has been settled by the inherent difficulty of easily designing multi-cylinder racing engines with disc valves, as a disc valve design demands a free crankshaft end for each cylinder. The high-performance outboard racing engines demonstrated that high specific power output was possible with reed valves(1.12) and the racing motorcycle organizations developed the technology further, first for motocross engines and then for Grand Prix power units. Today, most reed valves are designed as V-blocks (see Fig. 1.7 and Plates 1.9 and 6.1) and the materials used for the reed petals are either spring steel or a fiber-reinforced composite material. The composite material is particularly useful in highly stressed racing engines, as any reed petal failure is not mechanically catastrophic as far as the rest of the engine is concerned. Further explanatory figures and detailed design discussions regarding all such valves and ports will be found in Section 6.3.

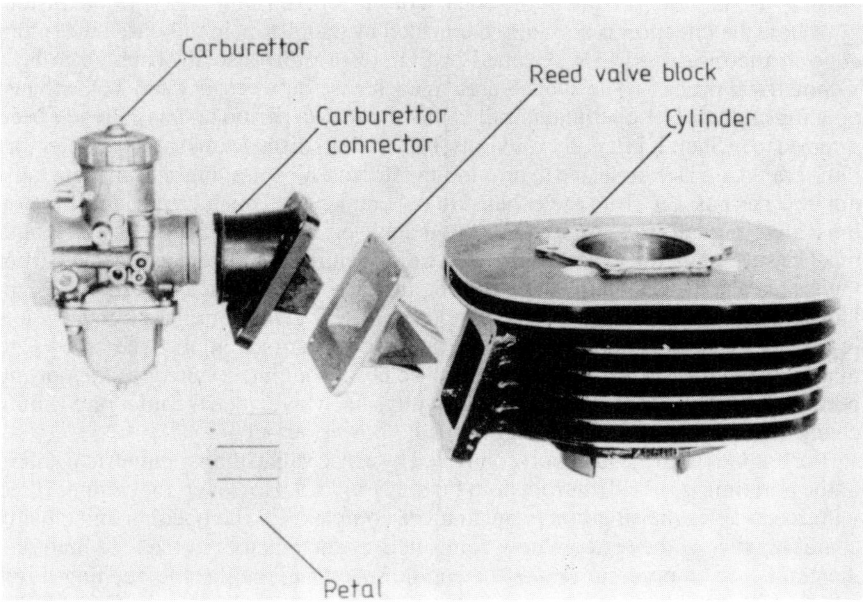

Plate 1.9 An exploded view of a reed valve cylinder for a motorcycle.

Fig. 1.7 shows the reed valve being given access directly to the crankcase, and this would be the design most prevalent for outboard motors where the crankcase bottom is accessible (see Plate 5.2). However, for motorcycles or chainsaws, where the crankcase is normally "buried" in a transmission system, this is somewhat impractical and so the reed valve feeds the fresh air charge to the crankcase through the cylinder. An example of this is illustrated in Plate 4.1, showing a 1988 model 250 cc Grand Prix motorcycle racing engine. This can be effected(1.13) by placing the reed valve housing at the cylinder level so that it is connected to the transfer ducts into the crankcase.

1.3.4 Port timing events

As has already been mentioned in Sect. 1.3.1, the port timing events in a simple two-stroke engine are symmetrical around tdc and bdc. This is defined by the connecting rod-crank relationship. Typical port timing events, for piston port control of the exhaust, transfer or scavenge, and inlet processes, disc valve control of the inlet process, and reed valve control of the inlet process, are illustrated in Fig. 1.8. The symmetrical nature of the exhaust and scavenge processes is evident, where the exhaust port opening and closing, EO and EC, and transfer port opening and closing, TO and TC, are under the control of the top, or timing, edge of the piston.

Where the inlet port is similarly controlled by the piston, in this case the bottom edge of the piston skirt, is sketched in Fig. 1.8(a); this also is observed to be a symmetrical process. The shaded area in each case between EO and TO, exhaust opening and transfer opening, is called the blowdown period and has already been referred to in Sect. 1.1. It is also obvious from various discussions in this chapter that if the crankcase is to be sealed to provide an effective air-pumping action, there must not be a gas passage from the exhaust to the crankcase. This means that the piston must always totally cover the exhaust port at tdc or, to be specific, the piston length must be sufficiently in excess of the stroke of the engine to prevent gas leakage from the crankcase. In Chapter 6 there will be detailed discussions on porting design. However, to set the scene for that chapter, Fig. 1.9 gives some preliminary facts regarding the typical port timings seen in some two-stroke engines. It can be seen that as the demand rises in terms of specific power output, so too does the porting periods. Should the engine be designed with a disc valve, then the inlet port timing changes are not so dramatic with increasing power output.

For engines with the inlet port controlled by a disc valve, the asymmetrical nature of the port timing is evident from both Figs. 1.8 and 1.9. However, for engines fitted with reed valves the situation is much more complex, for the opening and closing characteristics of the reed are now controlled by such factors as the reed material, the crankcase compression ratio, the engine speed and the throttle opening. Figs. 1.8(c) and 1.8(d) illustrate the typical situation as recorded in practice by Fleck(1.13). It is interesting to note that the reed valve opening and closing points, marked as RVO and RVC, respectively, are quite similar to a disc valve engine at low engine speeds and to a piston controlled port at higher engine speeds. For racing engines, the designer would have wished those characteristics to be reversed! The transition

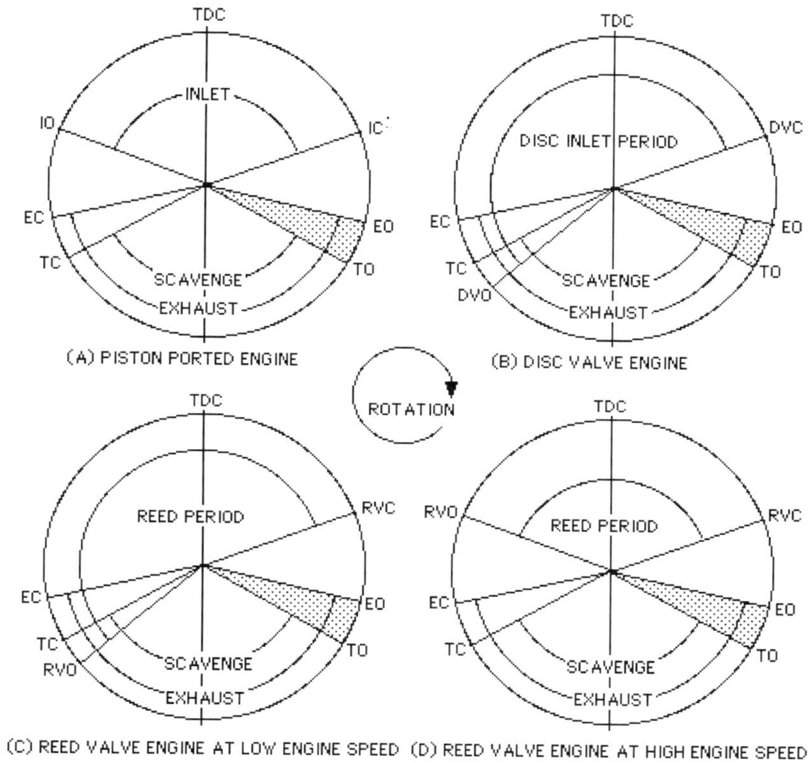

Fig. 1.8 Typical port timing characteristics for piston ported, reed and disc valve engines.

in the RVO and the RVC points is almost, but not quite, linear with speed, with the total opening period remaining somewhat constant. Detailed discussion of matters relating specifically to the design of reed valves is found in Sect. 6.3.

The reader should examine Fig. 6.1, which shows the port areas in an engine where all of the porting events are controlled by the piston. The actual engine data used to create Fig. 6.1 is that for the chainsaw engine design discussed in Chapter 5 and the geometrical data displayed in Fig. 5.3.

1.4 Engine and porting geometry

The point has been reached in the text where some mathematical treatment of design will commence. This will be conducted in a manner which can be followed by anyone with a mathematics education of university entrance level. The fundamental principle of this book is not to confuse, but to illuminate, and to arrive as

ENGINE TYPE	PISTON PORT CONTROL EXHAUST OPENS,ºATDC	 TRANSFER OPENS,ºATDC	 INLET OPENS,ºBTDC	INLET DISC VALVE CONTROL OPENS,ºBTDC	 CLOSES,ºATDC
INDUSTRIAL, MOPED, CHAINSAW, SMALL OUTBOARD	110	122	65	130	60
ENDURO, LARGE OUTBOARD, RPV	97	120	75	120	70
MOTOCROSS, GRAND PRIX M/C	81	113	100	140	80

Fig. 1.9 Typical port timings for various types of two-stroke engine applications.

quickly as is sensible to a working computer program for the design of the particular component under discussion.

(a) Units used throughout the book

Before embarking on this section, a word about units is essential. This book is written in SI units, and all mathematical equations are formulated in those units. Thus, all subsequent equations are intended to be used with the arithmetic values inserted for the symbols of the SI unit listed in the Notation section at the end of this, or the appropriate, chapter. If this practice is adhered to, then the value computed from any equation will appear also as the strict SI unit listed for that variable on the left-hand side of the equation. Should the user desire to change the unit of the ensuing arithmetic answer to one of the other units listed in the Notation section, a simple arithmetic conversion process can be easily accomplished. One of the virtues of the SI system is that strict adherence to those units, in mathematical or computational procedures, greatly reduces the potential of arithmetic errors. The author writes this with some feeling as one who was educated with great difficulty, as one of his American friends once expressed it so well, in the British "furlong, hundredweight, fortnight" system of units!

(b) Computer programs presented throughout the book

The intention is that the user can type in the program emanating from any of the design discussions in any chapter directly into a Macintosh® II computer, and be able to carry out the calculations which illustrate this book. In short, the reader has available directly the programs which the author uses to design two-stroke engines.

The listing of all computer programs connected with this book is contained in the Computer Program Appendix. Logically, programs coming from, say, Chapter 5, will appear in the Computer Program Appendix as ProgList.5. In the case of the first programs introduced below, they are to be found as ProgList.1.1, ProgList.1.2, and ProgList.1.3. As is common with computer programs, they also have names, in this case, PISTON POSITION, LOOP ENGINE DRAW, and QUB CROSS ENGINE DRAW, respectively. All of the computer programs have been written in Microsoft®

Macintosh is a registered trademark of Apple Computers, Inc.
Microsoft is a registered trademark of Microsoft Corporation.

QuickBASIC for the Apple Macintosh and this is the same language prepared by Microsoft Corp. for the IBM® PC's and their many clones. Only some of the graphics statements are slightly different for various IBM-like machines.

The author uses a Macintosh II, but the programs will run on any Macintosh computer, albeit some of the graphics may not fit so well on the smaller screen of the earlier Macintosh variants. A hard disk addition to any computer is recommended for speed and ease of use.

Almost all of the programs are written in, and are intended to be used in, the interpreted QuickBASIC mode. It is preferable to run some of the programs in the unsteady gas flow sections in Chapters 4 and 5 in the compiled mode, as the speed advantage in the compiled mode makes for more effective use of the software. In Microsoft QuickBASIC, a "user-friendly" computer language and system, it is merely a flick of a mouse to obtain a compiled version of any program listing.

The software is intended to be available in disk form for direct use on either Macintosh II or IBM PC computers, thus saving the operator from a lengthy typing procedure. For further details of the programs and the methodology of data handling at the input and output stages, the reader should consult the Computer Program Appendix.

1.4.1 Swept volume

If the cylinder of an engine has a bore, BO, and a stroke, ST, as sketched in Fig. 4.2, then the Swept Volume, SV, of that cylinder is given by:

$$SV=\pi*BO*BO*ST/4 \qquad (1.4.1)$$

If the engine has a number of cylinders, Ncy, of that same swept volume then the total swept volume of the engine is given by Ncy*SV.

If the exhaust port closes some distance called the Trapped Stroke, TS, before tdc, then the Trapped Swept Volume, TSV, is given by:

$$TSV=\pi*BO*BO*TS/4 \qquad (1.4.2)$$

The piston is connected to the crankshaft by a connecting rod of length, CRL. The throw of the crank (see Fig. 1.10) is one half of the stroke, and is designated as length, CT. As with four-stroke engines, the connecting rod-crank ratios are typically in the range of 3.5 to 4.

1.4.2 Compression ratio

All compression ratio values are the ratio of the maximum volume in any chamber of an engine to the minimum volume in that chamber. In the crankcase that is known as the Crankcase Compression Ratio, CCR, and is the ratio defined by:

$$CCR=(CCV+SV)/CCV \qquad (1.4.3)$$

IBM is a registered trademark of International Business Machines Corp.

where CCV is the crankcase clearance volume, or the crankcase volume at bdc.

While it is true that the higher this value becomes, the stronger is the crankcase pumping action, the actual numerical value is greatly fixed by the engine geometry of bore, stroke, con-rod length and the interconnected value of flywheel diameter. In practical terms, it is rather difficult to organize the CCR value for a 50cc engine cylinder above 1.4 and almost physically impossible to design a 500cc engine cylinder to have a value less than 1.55. Therefore, for any given engine design the CCR characteristic is more heavily influenced by the choice of cylinder swept volume than by the designer. It then behooves the designer to tailor the engine air-flow behavior around the crankcase pumping action, defined by the inherent CCR value emanating from the cylinder size in question. There is some freedom of design action, and it is necessary for it to be taken in the correct direction.

In the cylinder shown in Fig. 4.2, if the clearance volume, CV, above the piston at top dead center, TDC, is known, then the Geometric Compression Ratio, GCR, is given by:

$$GCR=(SV+CV)/CV \qquad (1.4.4)$$

Theoretically, the actual compression process occurs after the exhaust port is closed, and the compression ratio after that point becomes the most important one in design terms. This is called the Trapped Compression Ratio. Because this is the case, in the literature for two-stroke engines the words "compression ratio" are sometimes carelessly applied when the precise term "trapped compression ratio" should be used. This is even more confusing because the literature for four-stroke engines refers to the geometric compression ratio, but describes it simply as the "compression ratio." The trapped compression ratio, TCR, is then calculated from:

$$TCR=(TSV+CV)/CV \qquad (1.4.5)$$

1.4.3 Piston position with respect to crankshaft angle

At any given crankshaft angle, B, after tdc, the connecting rod center-line assumes an angle, A, to the cylinder center-line. This angle is often referred to in the literature as the "angle of obliquity" of the connecting rod. This is illustrated in Fig. 1.10 and the piston position of any point, X, on the piston from the tdc point is given by length H. The controlling trigonometric equations are shown on Fig. 1.10. Clearly, it is essential for the designer to know the position of the piston at salient points such as exhaust, transfer, and inlet port opening, closing and the fully open points as well, should the latter not coincide with either tdc or bdc. These piston positions define the port heights and the mechanical drafting of any design requires these facts as precise numbers. In later chapters, design advice will be presented for the detailed porting design, but often connected with general, rather than detailed, piston and rod geometry. Consequently, the equations shown in Fig. 1.10 are programmed into three computer programs, Prog.1.1, Prog.1.2 and Prog.1.3, for use in specific circumstances.

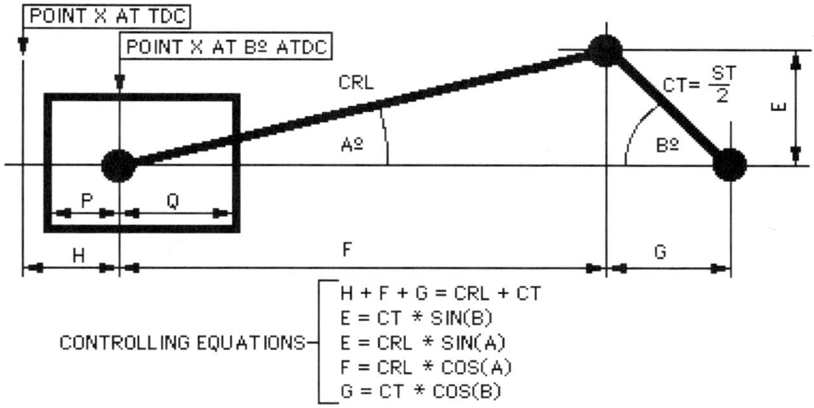

Fig. 1.10 Position of a point on a piston at a crankshaft angle from tdc.

1.4.4 Computer program, Prog.1.1, PISTON POSITION

This is a quite unsophisticated program without graphics. The program should be typed in from the listing, and upon a RUN command the prompts for the input data are self-evident in nature. An example of the simplistic nature of the input and output data is shown in Fig. 1.11. If the reader is new to the ways of the Macintosh or IBM system of operation then this straightforward program will provide a useful introduction. The printout which appears on the line printer contains further calculated values of use to the designer. The program allows for the calculation of piston position from bdc or tdc for any engine geometry between any two crankshaft angles at any interval of step between them. The output shown in Fig. 1.11 is exactly what the operator sees on the Macintosh computer screen.

1.4.5 Computer program, Prog.1.2, LOOP ENGINE DRAW

This program is written in a much more sophisticated manner, using the facilities of the Macintosh software to speed the process of program operation, data handling and decision making by the user. The program listing, ProgList.1.2, details the data handling features and so it will not be repeated here. Fig. 1.12 illustrates the operator's view of a completed calculation which will be printed in that form on demand by the user.

The data input values are written on the left-hand side of the figure from "BORE" down to "WRIST PIN TO SKIRT." All other values on the picture are output values. "Wrist pin" is the American term for a gudgeon pin and the latter two data input values correspond to the dimensions P and Q on Fig. 1.10. It is also assumed in the calculation that all ports are opened by their respective control edges to the top or bottom dead center positions. The basic geometry of the sketch is precisely as Fig. 1.1; indeed that sketch was created using this particular program halted at specific crankshaft angular positions.

```
RUNNING THE PROGRAM(Y OR N?)?  Y
enter BORE in mm 60
enter STROKE in mm 60
enter CON-ROD LENGTH  in mm 110
enter  CRANK-ANGLE  after tdc at  start  of calculation 100
enter  CRANK-ANGLE  after tdc at end of calculation 120
enter  CRANK-ANGLE interval for the calculation step from start to finish  5

Swept volume,CM3,=  169.6
crank-angle after tdc       height from tdc        height from bdc
              100.00              39.25                 20.75
              105.00              41.65                 18.35
              110.00              43.93                 16.07
              115.00              46.09                 13.91
              120.00              48.11                 11.89
WANT A PRINT-OUT(Y OR N?)?  N
```

Fig. 1.11 Example of a calculation from Prog.1.1, Program "PISTON POSITION."

```
BORE,mm=  60
STROKE,mm=  60
CON-ROD,mm=  110
EXHAUST OPENS,ºatdc=  100
TRANSFER OPENS,ºatdc=  120
INLET OPENS,ºbtdc=  65
TRAP COMPRESSION RATIO=  7
SQUISH CLEARANCE,mm=  1.5
WRIST PIN TO CROWN,mm=  30
WRIST PIN TO SKIRT,mm=  32
SWEPT VOLUME,cm3=169.6
TRAP SWEPT VOLUME,cm3=111.0
CLEARANCE VOLUME,cm3=  18.5
```

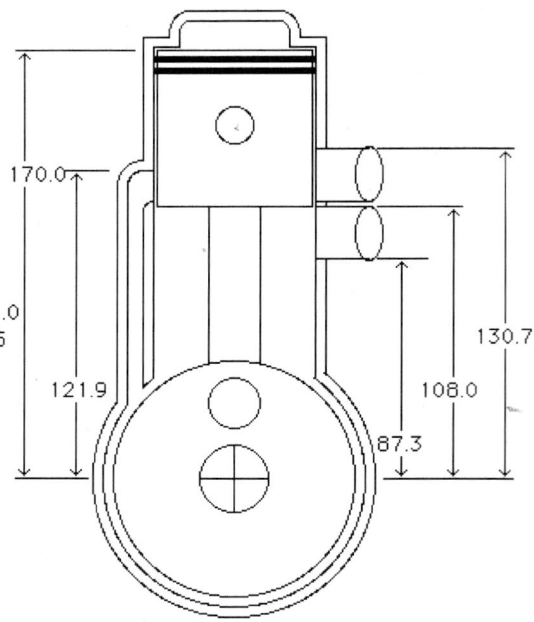

Fig. 1.12 Example of a calculation from Prog.1.2, Program "LOOP ENGINE DRAW."

When the user runs this program he will discover that the engine on the screen rotates for one complete cycle, from tdc to tdc. When the piston comes to rest at tdc, the linear dimensions of all porting positions from the crankshaft center-line are drawn, as illustrated. By this means, as the drawing on the screen is exactly to scale, the designer can be visually assured that the engine has no unusual problems in geometrical terms. It can be observed, for example, that the piston is sufficiently long to seal the exhaust port and preserve an effective crankcase pumping action!

The inlet port is shown on Figs. 1.1 and 1.12 as being underneath the exhaust port for reasons of diagrammatic simplicity. There have been engines produced this way, but they are not as common as those with the inlet port at the rear of the cylinder, opposite to the cylinder wall holding the exhaust port. However, the simple two-stroke engine shown in Plate 1.5 is just such an engine. It is a Canadian-built chainsaw engine with the engine cylinder and crankcase components produced as high pressure aluminium die castings with the cylinder bore surface being hard chromium plated. The open sided transfer ports are known as "finger ports."

1.4.6 Computer program, Prog.1.3, QUB CROSS ENGINE DRAW

This program is exactly similar, in data input and operational terms, to Prog.1.2. However, it designs the basic geometry of the QUB type of cross scavenged engine, as shown in Fig. 1.4 and discussed in Sect. 1.2.2.

The reader might well inquire as to the design of the conventional cross scavenged unit as sketched in Fig. 1.3 and also discussed in the same section. The timing edges for the control of both the exhaust and transfer ports are at the same height in the conventional deflector piston engine. This means that the same geometry of design applies to it as for the loop scavenged engine. Consequently, program Prog.1.2 applies equally well.

Because the exhaust and transfer port timing edges are at different heights in the QUB type of engine, separated by the height of the deflector, a different program is required and is given here as Prog.1.3 and in the Computer Program Appendix as ProgList.1.3. An example of the calculation is presented in Fig. 1.13. As with Prog.1.2, the engine rotates for one complete cycle. One data input value deserves an explanation: the "wrist pin to crown" value, shown in the output Fig. 1.13 as 25 mm, is that value from the gudgeon pin to the crown on the scavenge side of the piston, and is not the value to the top of the deflector. As the engine rotates, one of the interesting features of this type of engine appears: the piston rings are below the bottom edge of the exhaust ports as the exhaust flow is released by the deflector top edge. Consequently, the exhaust flame does not partially burn the oil on the piston rings, as it does on a loop scavenged design. As mentioned earlier, the burning of oil on the piston rings and within the ring grooves eventually causes the rings to stick in their grooves and deteriorates the sealing effect of the rings during the compression and expansion strokes, reducing both power output and fuel economy. Therefore, the QUB type engine has enhanced engine reliability and efficiency in this regard over the life span of the power unit.

The data values used for the basic engine geometry in Figs. 1.10-1.13 are common for all three program examples, so it is useful to compare the actual data

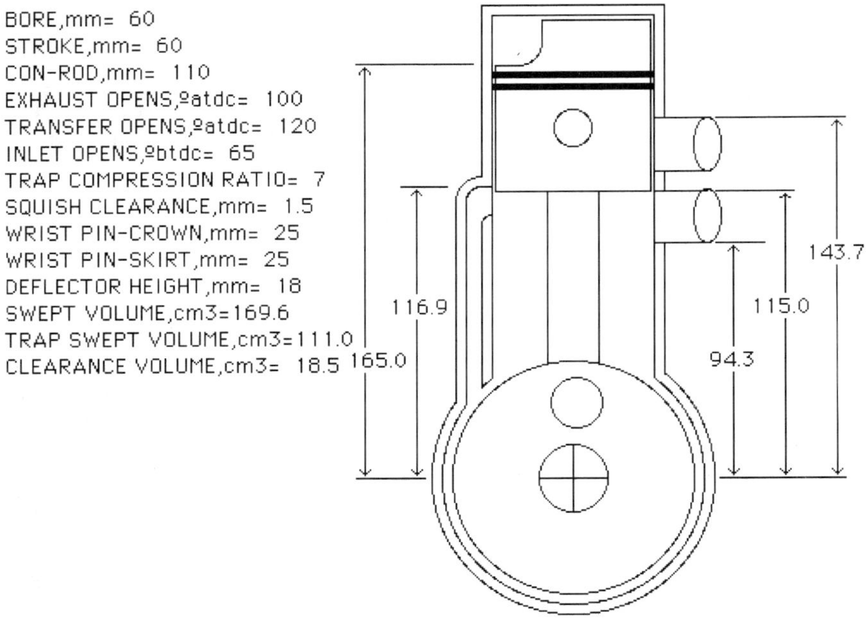

Fig. 1.13 Example of a calculation from Prog.1.3, "QUB CROSS ENGINE DRAW."

output values for similarities and differences. For example, it can be seen that the QUB cross scavenged engine is taller to the top of the deflector, yet is the same height to the top of the combustion chamber as the loop scavenged power unit. In a later chapter, Fig. 4.14 shows a series of engines which are drawn to scale and the view expressed above can be seen to be accurate from that comparative sketch. Indeed, it could be argued that a QUB deflector engine can be designed to be a shorter engine overall than an equivalent loop scavenged unit with the same bore, stroke and rod lengths.

1.5 Definitions of thermodynamic terms used in connection with engine design and testing

1.5.1 Scavenge ratio and delivery ratio

In Fig. 1.1(c), the cylinder has just experienced a scavenge process, in which a mass of fresh charge, Mas, has been supplied through the crankcase from the atmosphere. By measuring the atmospheric, i.e., the ambient pressure and temperature, Pat and Tat, the air density will be given by Dat from the thermodynamic equation of state, where Rair is the gas constant for air:

$$\text{Dat}=\text{Pat}/(\text{Rair}*\text{Tat}) \qquad (1.5.1)$$

The Delivery Ratio, DR, of the engine defines the mass of air supplied during the scavenge period as a function of a reference mass, Mdr, which is that mass required to fill the swept volume under the prevailing atmospheric conditions, i.e.:

$$Mdr=SV*Dat \tag{1.5.2}$$
$$DR=Mas/Mdr \tag{1.5.3}$$

The Scavenge Ratio, SR, of a naturally aspirated engine defines the mass of air supplied during the scavenge period as a function of a reference mass, Msr, which is the mass which could fill the entire cylinder volume under the prevailing atmospheric conditions, i.e.:

$$Msr=(SV+CV)*Dat \tag{1.5.4}$$
$$SR=Mas/Msr \tag{1.5.5}$$

The SAE Standard J604d(1.24) refers to, and defines, delivery ratio. For two-stroke engines the more common nomenclature in the literature is scavenge ratio, but it should be remembered that the definitions of these air-flow ratios are mathematically different.

Should the engine be supercharged or turbocharged, then the new reference mass, Msr, for the estimation of scavenge ratio, is calculated from the state conditions of pressure and temperature of the scavenge air supply, Ps and Ts.

$$Ds=Ps/(Rair*Ts) \tag{1.5.6}$$
$$SR=Mas/((SV+CV)*Ds) \tag{1.5.7}$$

The above theory has been discussed in terms of the air flow referred to the swept volume of a cylinder as if the engine is a single cylinder unit. However, if the engine is a multi-cylinder device, it is the total swept volume of the engine which is under consideration.

1.5.2 Scavenging efficiency and purity

In Chapter 3 it will be shown that for a perfect scavenge process, the very best which could be hoped for is that the Scavenging Efficiency, SE, would be equal to the scavenge ratio, SR. The scavenging efficiency is defined as the mass of delivered air which has been trapped, Mtas, by comparison with the total mass of charge, Mtr, which is retained at exhaust closure. The trapped charge is composed only of fresh charge trapped, Mtas, and exhaust gas, Mex, and any air remaining unburned from the previous cycle, Mar, where:

$$Mtr=Mtas+Mex+Mar \tag{1.5.8}$$

Hence, scavenging efficiency, SE, defines the effectiveness of the scavenging process, as can be seen from the following statement:

$$SE=Mtas/Mtr=Mtas/(Mtas+Mex+Mar) \tag{1.5.9}$$

However, the ensuing combustion process will take place between all of the air in the cylinder with all of the fuel supplied to that cylinder, and it is important to define the purity of the trapped charge in its entirety. The purity of the trapped charge, PUR, is defined as the ratio of air trapped in the cylinder before combustion, Mta, to the total mass of cylinder charge, where:

Mta=Mtas+Mar (1.5.10)

PUR=Mta/Mtr (1.5.11)

In many technical papers and textbooks on two-stroke engines, the words "scavenging efficiency" and "purity" are somewhat carelessly interchanged by the authors, assuming prior knowledge by the readers. They assume that the value of Mar is zero, which is generally true for most spark ignition engines, but it would not be true for two-stroke diesel engines where the air is seldom totally consumed in the combustion process, and it would not be true for similar reasons for a stratified combustion process in a gasoline fueled spark-ignition engine.

1.5.3 Trapping efficiency

Definitions are also to be found in the literature(1.24) for Trapping Efficiency, TE. Trapping efficiency is the capture ratio of mass of delivered air which has been trapped, Mtas, to that supplied, Mas, or:

TE=Mtas/Mas (1.5.12)

It will be seen that expansion of Eq. 1.5.12 gives:

TE=(SE*Mtr)/(SR*Msr) (1.5.13)

It will be seen under ideal conditions, in Chapter 3, that Mtr can be considered to be equal to Msr and that Eq. 1.5.13 can be simplified in an interesting manner, i.e., TE=SE/SR. In Sect. 1.6.3, a means of measuring trapping efficiency in a firing engine from exhaust gas analysis will be described.

1.5.4 Charging efficiency

Charging Efficiency, CE, expresses the ratio of the filling of the cylinder with air, by comparison with filling that same cylinder perfectly with air at the onset of the compression stroke. After all, the object of the design exercise is to fill the cylinder with the maximum quantity of air in order to burn a maximum quantity of fuel with that same air. Hence, charging efficiency, CE, is given by:

CE=Mtas/Msr (1.5.14)

It is also the product of trapping efficiency and scavenge ratio, as shown here:

CE=(Mtas/Mas)*(Mas/Msr)=TE*SR (1.5.15)

It should be made quite clear that this definition is not precisely as defined in SAE J604d(1.24). In that SAE nomenclature Standard, the reference mass is declared to be Mdr, from Eq. 1.5.2, and not Msr as used from Eq. 1.5.4. The author's defense for this is "custom and practice in two-stroke engines," the convenience of charging efficiency assessment by the relatively straightforward experimental acquisition of trapping efficiency and scavenge ratio, and the opinion of Benson(1.4, Vol. 2).

1.5.5 Air-to-fuel ratio

It is important to realize that there are narrow limits of acceptability for the combustion of air and fuel, such as gasoline or diesel. In the case of gasoline, the ideal fuel is octane, C_8H_{18}, which burns "perfectly" with air in a balanced equation, called the stoichiometric equation. Most students will recall that air is composed, volumetrically and molecularly, of 21 parts oxygen and 79 parts nitrogen. Hence, the chemical equation for complete combustion becomes:

$$2C_8H_{18} + 25O_2 + (25*79/21)N_2 = 16CO_2 + 18H_2O + (25*79/21)N_2 \qquad (1.5.16)$$

This produces the information that the ideal stoichiometric Air-to-Fuel Ratio, AF, is such that for every two molecules of octane, one needs twenty-five molecules of air. In the case where one needs the information in mass terms, then:

$$AF=(25*32+25*79*28/21)/(16*12+36*1)=15.06 \qquad (1.5.17)$$

As the equation is balanced, with the exact amount of oxygen being supplied to burn all of the carbon to carbon dioxide and all of the hydrogen to steam, such a burning process yields the minimum values of carbon monoxide emission, CO, and unburned hydrocarbons, HC. Mathematically speaking they are zero, and in practice they are at a minimum level. As this equation would also produce the maximum temperature at the conclusion of combustion, this gives the highest value of emissions of NO_x, the various oxides of nitrogen. Nitrogen and oxygen combine at high temperatures to give such gases as N_2O, NO, etc. Such statements, although based in theory, are almost exactly true in practice as illustrated by the expanded discussion in Chapters 4 and 7.

As far as combustion limits are concerned, although Chapter 4 will delve into this area more thoroughly, it may be helpful to the reader to point out at this stage that the rich misfire limit of gasoline-air combustion probably occurs at an air-fuel ratio of about 9, peak power output at an air-fuel ratio of about 13, peak thermal efficiency (or minimum specific fuel consumption) at an air-fuel ratio of about 14, and the lean misfire limit at an air-fuel ratio of about 18. The air-fuel ratios quoted are those in the combustion chamber at the time of combustion of a homogeneous charge, and are referred to as the Trapped Air-Fuel Ratio, TAF. The air-fuel ratio derived in Eq. 1.5.17 is, more properly, the trapped air-fuel ratio, TAF, needed for stoichiometric combustion.

To briefly illustrate that point, in the engine shown in Fig. 1.6 it would be quite possible to scavenge the engine thoroughly with fresh air and then supply the appropriate quantity of fuel by direct injection into the cylinder to provide a TAF of, say, 13. Due to a generous oversupply of scavenge air the overall AF could be in excess of, say, 20.

1.5.6 Cylinder trapping conditions

The point of the foregoing discussion is to make the reader aware that the net effect of the cylinder scavenge process is to fill the cylinder with a mass of air, Mta, within a total mass of charge, Mtr, at the trapping point. This total mass is highly dependent on the trapping pressure, as the equation of state shows:

$$Mtr=Ptr*Vtr/(Rtr*Ttr) \quad (1.5.18)$$
$$Vtr=TSV+CV \quad (1.5.19)$$

In any given case, the trapping volume, Vtr, is a constant. This is also true of the gas constant, Rtr, for gas at the prevailing gas composition at the trapping point.The gas constant for exhaust gas, Rex, is almost identical to the value for air, Rair. Because the cylinder gas composition is usually mostly air, the treatment of Rtr as being equal to Rair invokes little error. For any one trapping process, over a wide variety of scavenging behavior, the value of trapping temperature, Ttr, would rarely change by 5%. Therefore, it is the value of trapping pressure, Ptr, which is the significant variable. As stated earlier, the value of trapping pressure is directly controlled by the pressure wave dynamics of the exhaust system, be it a single-cylinder engine with, or without, a tuned exhaust system, or a multi-cylinder power unit with a branched exhaust manifold. The methods of design and analysis for such complex systems are discussed in Chapters 2 and 5. The value of the trapped fuel quantity, Mtf, can be determined from:

$$Mtf=Mta/TAF \quad (1.5.20)$$

1.5.7 Heat released during the burning process

The total value of the heat, which will be released from the combustion of this quantity of fuel, will be Qc:

$$Qc=Mtf*CAL*BE \quad (1.5.21)$$

where BE is the Combustion Efficiency and CAL is the Calorific Value of the fuel in question.

A further discussion of this analysis, in terms of an actual experimental example, is given in Sect. 4.2.

1.5.8 The thermodynamic cycle for the two-stroke engine

This is often referred to as a derivative of the Otto Cycle, and a full discussion can be found in many undergraduate textbooks on ic engines or thermodynamics, e.g., Taylor(1.3). The result of the calculation of a theoretical cycle can be observed in Figs. 1.14 and 1.15, by comparison with measured pressure-volume data from an engine of the same compression ratios, both trapped and geometric. In the measured case, the cylinder pressure data is taken from a 400 cc single-cylinder two-stroke engine running at 3000 rev/min at wide-open throttle. In the theoretical case, and this is clearly visible on the log P-log V plot in Fig. 1.15, the following assumptions are made: (a) compression begins at trapping, (b) all heat release (combustion) takes place at tdc at constant volume, (c) the exhaust process is considered as a heat rejection process at release, (d) the compression and expansion processes occur under ideal, or isentropic, conditions with air as the working fluid, and so those processes are calculated as $PV^g = k$, where k is a constant. For air, the ratio of specific heats, g, has a value of 1.4. A fundamental theoretical analysis would show(1.3) that the Thermal Efficiency, ThE, of the cycle is given by:

$$ThE=1-(TCR)^{(1-g)} \qquad (1.5.22)$$

Thermal efficiency is defined as:

$$ThE=(\text{Work produced per cycle})/(\text{Heat available as input per cycle}) \qquad (1.5.23)$$

As the actual 400 cc engine has a trapped compression value of 7, and from Eq. 1.5.22 the theoretical value of ThE is readily calculated as 0.541, the considerable disparity between fundamental theory and experimentation becomes apparent, for the measured value is about one half of that calculated, at 27%.

Upon closer examination of Figs. 1.14 and 1.15, the theoretical and measured pressure traces look somewhat similar and the experimental facts do approach the theoretical presumptions. However, the measured expansion and compression indices are at 1.33 and 1.17, respectively, which is rather different from the ideal value of 1.4 for air. On the other hand, the actual compression process clearly begins before the official trapping point at exhaust port closure, and this in an engine with no tuned exhaust pipe. The theoretical assumption of a constant volume process for the combustion and exhaust processes is clearly in error when the experimental pressure trace is examined. The peak cycle pressures of 54 bar calculated and 36 bar measured are demonstrably different. In Chapters 4 and 5 a more advanced theoretical analysis will be seen to approach the measurements more exactly.

The work on the piston during the cycle is ultimately and ideally the work delivered to the crankshaft by the connecting rod. The word "ideal" in thermodynamic terms means that the friction or other losses, like leakage past the piston, are not taken into consideration in the statement made above. Therefore, the ideal work produced per cycle (see Eq. 1.5.24) is that work carried out on the piston by the force, F, created from the gas pressure, P. Work is always the product of force and distance, X, moved by that force, so, where A is the piston area:

$$\text{Work produced per cycle}= \int(F*dX)=\int(P*A*dX)=\int(P*dV) \qquad (1.5.24)$$

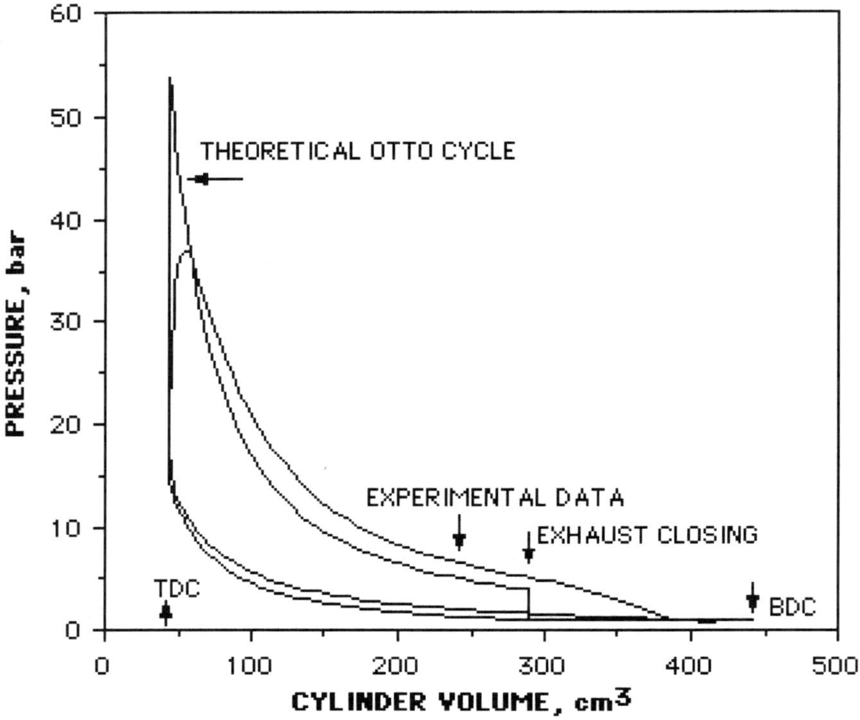

Fig. 1.14 Otto cycle comparison with experimental data.

Therefore, the work produced for any given engine cycle, in the case of a two-stroke engine for one crankshaft revolution from tdc to tdc, is the cyclic integral of the pressure-volume diagram in the cylinder above the piston. By the same logic, the pumping work required in the crankcase is the cyclic integral of the pressure-volume diagram in the crankcase. In both cases, this work value is the enclosed area on the pressure-volume diagram, be it a theoretical cycle or the actual cycle as illustrated in Fig. 1.14. The above statements are illustrated in Fig. 1.16 for the actual data shown previously in Figs. 1.14 and 1.15.

1.5.9 The concept of mean effective pressure

As stated above, the enclosed P-V diagram area is the work produced on the piston, in either the real or the ideal cycle. Fig. 1.16 shows a second rectangular shaded area, equal in area to the enclosed cylinder P-V diagram. This rectangle is of height IMEP and of length SV, where IMEP is known as the Indicated Mean

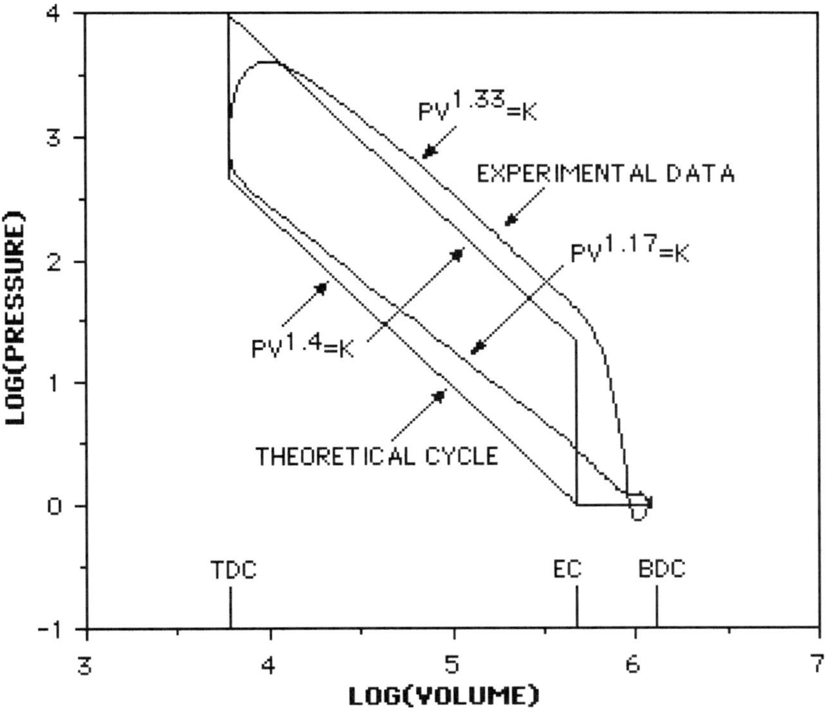

Fig. 1.15 Logarithmic plot of pressure and volume.

Effective Pressure and SV is the swept volume. The word "indicated" stems from the historical fact that pressure transducers for engines used to be called "indicators" and the P-V diagram, of a steam engine traditionally, was recorded on an "indicator card." The concept of mean effective pressure is extremely useful in relating one engine development to another for, while the units of IMEP are obviously that of pressure, the value is almost dimensionless. That remark is sufficiently illogical as to require careful explanation. The point is, any two engines of equal development or performance status will have identical values of mean effective pressure, even though they may be of totally dissimilar swept volume. In other words, Figs. 1.14, 1.15 and 1.16 could have equally well been plotted as pressure-compression (or volume) ratio plots and the values of IMEP would be identical for two engines of differing swept volume, if the diagrammatic profiles in the pressure direction were also identical.

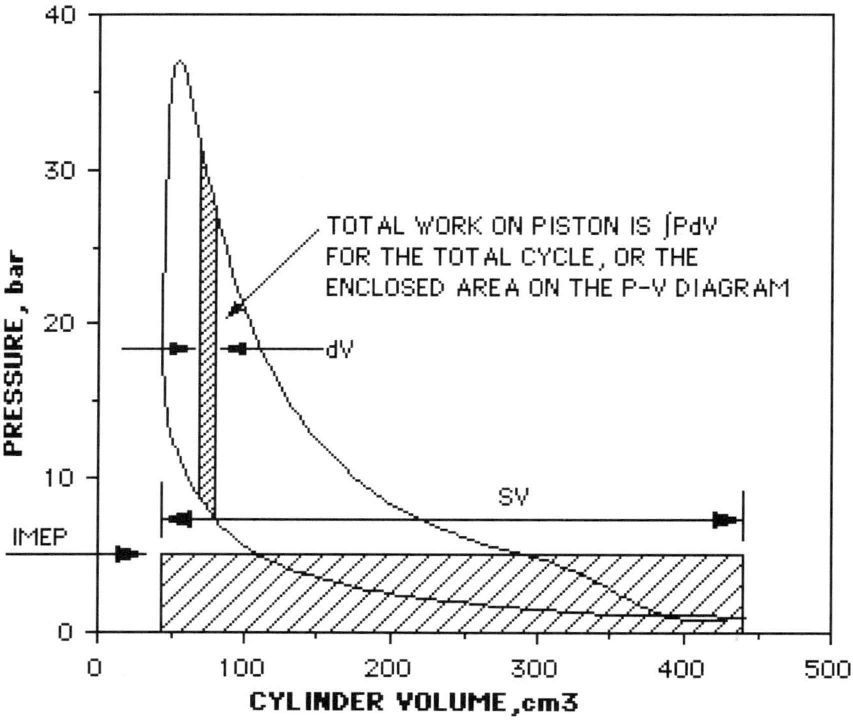

Fig. 1.16 Determination of IMEP from the cylinder P-V diagram.

1.5.10 Power and torque and fuel consumption

Power is defined as the rate of doing work. If the engine rotation rate is RPS revolutions per second, and the two-stroke engine has a working cycle per crankshaft revolution, then the power delivered to the piston crown by the gas force is called the Indicated Power Output, IPR, where:

IPR=IMEP*SV*(work cycles/s)
=IMEP*SV*RPS - for a two-stroke engine (1.5.25)
=IMEP*SV*RPS/2 - for a four-stroke engine

For a four-stroke cycle engine, which has a working cycle lasting two crankshaft revolutions, the working cycle rate is RPS/2 and this should be inserted into Eq. 1.5.25 rather than RPS. In other words, a four-stroke cycle engine of equal power output and equal swept volume has an IMEP value which is double that of the two-stroke engine. Such is the actual, if somewhat illogical, convention used in everyday engineering practice.

The Indicated Torque, ITRQ, is the turning moment on the crankshaft and is related to power output by the following equation:

$$IPR=2*\pi*ITRQ*RPS \qquad (1.5.26)$$

Should the engine actually consume fuel of calorific value CAL at the measured or theoretically calculated mass flow rate of FC, then the Indicated Thermal Efficiency, IThE, of the engine can be predicted from an extension of Eq. 1.5.23:

$$IThE=(\text{power output})/(\text{rate of heat input}) = IPR/(FC*CAL) \qquad (1.5.27)$$

Of much use in engineering practice is the concept of specific fuel consumption, the fuel consumption rate per unit power output. Hence, to continue the discussion on indicated values, Indicated Specific Fuel Consumption, ISFC, is given by:

$$ISFC=(\text{fuel consumption rate})/(\text{power output}) = FC/IPR \qquad (1.5.28)$$

It will be observed from a comparison of Eqs. 1.5.25 and 1.5.26 that thermal efficiency and specific fuel consumption are reciprocally related to each other, without the employment of the calorific value of the fuel. As most petroleum based fuels have almost identical values of calorific value, then the use of specific fuel consumption as a comparator from one engine to another, rather than thermal efficiency, is quite logical and is more immediately useful to the designer and the developer.

1.6 Laboratory testing of two-stroke engines

1.6.1 Laboratory testing for power, torque, mean effective pressure and specific fuel consumption

Most of the testing of engines for their performance characteristics takes place under laboratory conditions. The engine is connected to a power absorbing device, called a dynamometer, and the performance characteristics of power, torque, fuel consumption rate, and air consumption rate, at various engine speeds, are recorded. Many texts and papers describe this process and the Society of Automotive Engineers provides a Test Code J1349 for this very purpose(1.14). There is an equivalent test code from the International Standards Organization in ISO 3046 (1.16). For the measurement of exhaust emissions there is a SAE Code J1088 (1.15) which deals with exhaust emission measurements for small utility engines, and many two-stroke engines fall into this category. The measurement of air flow rate into the engine is often best conducted using meters designed to British Standard BS 1042(1.17).

Several interesting technical papers have been published in recent times questioning some of the correction factors used within such test codes for the prevailing atmospheric conditions. One of these by Sher(1.18) deserves further study.

There is little point in writing at length on the subject of engine testing and of the correction of the measured performance characteristics to standard reference pressure and temperature conditions, for these are covered in the many standards and codes already referred to. However, some basic facts are relevant to the further discussion and, as the testing of exhaust emissions is a relatively new subject, a simple analytical computer program on that subject, presented in Sect. 1.6.2, should prove to be useful to quite a few readers.

A laboratory engine testing facility is diagrammatically presented in Fig. 1.17. The engine power output is absorbed in the dynamometer, for which the slang word is a "dyno" or a "brake". The latter word is particularly apt as the original dynamometers were, literally, friction brakes. The principle of any dynamometer operation is to allow the casing to swing freely. The reaction torque on the casing, which is exactly equal to the engine torque, is measured on a lever of length, L, from the center-line of the dynamometer as a force, F. This restrains the outside casing from revolving, or the torque and power would not be absorbed. Consequently, the reaction torque measured is the Brake Torque, BTRQ, and is calculated by:

$$BTRQ=F*L \tag{1.6.1}$$

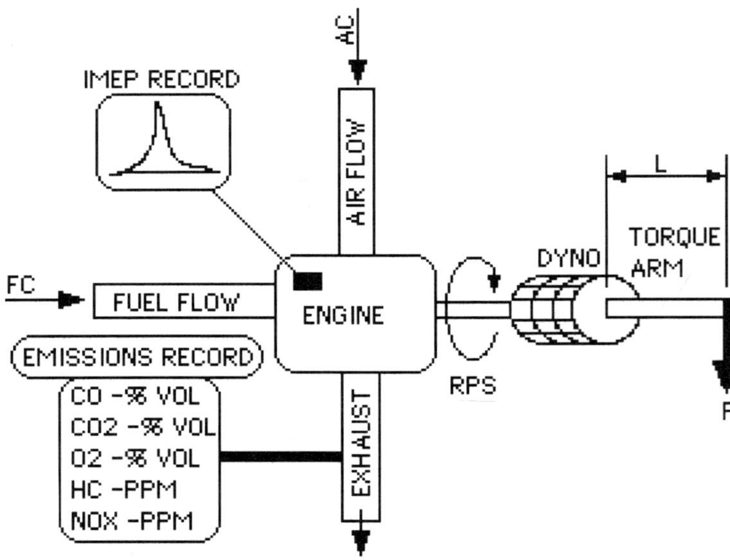

Fig. 1.17 Dynamometer test stand recording of performance parameters.

Therefore, the work output from the engine per engine revolution is the distance "travelled" by the force, F, on a circle of radius, L:

$$\text{Work/revolution}=2*\pi*F*L=2*\pi*BTRQ \qquad (1.6.2)$$

The measured power output, the Brake Power, BPR, is the work rate, and at RPS rotational speed, is clearly:

$$BPR=(\text{Work/revolution})*(\text{rev/s})= 2*\pi*BTRQ*RPS \qquad (1.6.3)$$

To some readers, this equation may clear up the mystery of the use of the operator π in the similar theoretical equation, Eq. 1.5.26, when considering the indicated power output, IPR.

The Brake Thermal Efficiency, BThE, is then given by the corresponding equation to Eq. 1.5.23:

$$BThE=(\text{power output})/(\text{rate of heat input}) = BPR/(FC*CAL) \qquad (1.6.4)$$

A similar situation holds for Brake Specific Fuel Consumption, BSFC, and Eq. 1.5.28:

$$BSFC=(\text{fuel consumption rate})/(\text{power output}) = FC/BPR \qquad (1.6.5)$$

However, it is also possible to compute a mean effective pressure corresponding to the measured power output. This is called the Brake Mean Effective Pressure, BMEP, and is calculated from a manipulation of Eq. 1.5.25 in terms of measured values:

$$BMEP=BPR/(SV*RPS) \qquad (1.6.6)$$

It is obvious that the brake power output and the brake mean effective pressure are the residue of the indicated power output and the indicated mean effective pressure, after the engine has lost power to internal friction and air pumping effects. These friction and pumping losses deteriorate what is known as the engine's Mechanical Efficiency, Meff. Friction and pumping losses are related simply by:

$$IPR=BPR+(\text{friction and pumping power}) \qquad (1.6.7)$$
$$Meff=BPR/IPR=BMEP/IMEP \qquad (1.6.8)$$

This raises the concept of the Friction and the Pumping Mean Effective Pressures, FMEP and PMEP, which can be related together as:

$$IMEP=BMEP+FMEP+PMEP \qquad (1.6.9)$$

It is often very difficult to segregate the separate contributions of friction and pumping in measurements taken in a laboratory except by recording crankcase pressure diagrams and by measuring friction power using a motoring methodology which eliminates all pumping action at the same time. It is very easy to write the last fifteen words but it is much more difficult to accomplish them in practice.

Lastly, the recording of the overall air-fuel ratio, AF, is relatively straightforward as:

$$AF=AC/FC \tag{1.6.10}$$

Continuing the discussion commenced in Sect. 1.5.5, this overall air-fuel ratio, AF, is also the trapped air-fuel ratio, TAF, if the engine is charged with a homogeneous supply of air and fuel, i.e., as in a carburetted design for a simple two-stroke engine. If the total fuel supply to the engine is, in any sense, stratified from the total air supply, this will not be the case.

1.6.2 Laboratory testing for exhaust emissions from two-stroke engines

There have been quite a few technical contributions in this area (1.7) (1.15)(1.19)(1.25) as the situation for the two-stroke engine is subtly different from the four-stroke engine case. Much of the instrumentation available has been developed for, and specifically oriented towards, four-stroke cycle engine measurement and analysis. As has been pointed out in Sects. 1.1 and 1.2, a significant portion of the scavenge air ends up in the exhaust pipe without enduring a combustion process. Whereas a change from a lean to a rich combustion process, in relation to the stoichiometric air-fuel ratio, for a four-stroke engine might change the exhaust oxygen concentration from 2% to almost 0% by volume, in a two-stroke engine that might produce an equivalent shift from 10% to 8%. Thus, in a two-stroke engine the exhaust oxygen concentration is always high. Equally, if the engine has a simple carburetted fueling device, then the bypassed fuel along with that short-circuited air produces a very large count of unburned hydrocarbon emission. Often, this count is so high that instruments designed for use with four-stroke cycle engines will not record it!

In today's legislative conscious world, it is important that exhaust emissions are recorded on a mass basis. Most exhaust gas analytical devices measure on a volumetric or molecular basis. It is necessary to convert such numbers to permit comparison of engines on their effectiveness in reducing exhaust emissions at equal power levels. From this logic appears the concept of deriving measured, or brake specific, emission values for such pollutants as carbon monoxide, unburned hydrocarbons, and oxides of nitrogen. The first of these pollutants is toxic, the second is blamed for "smog" formation and the last is regarded as a major contributor to "acid rain." These and other facets of pollution are discussed more extensively in Chapter 7.

As an example, consider a pollutant gas, PG, with molecular weight, MWpg, and a volumetric concentration in the exhaust gas of proportion, VCpg. In consequence, the numerical value in ppm would be 10^6*VCpg and as % by volume it would be

100*VCpg. The average molecular weight of the exhaust gas is MWex. The power output is BPR and the fuel consumption rate is FC. The total mass flow rate of exhaust gas is M'ex:

M'ex=(1+AF)*FC kg/s =(1+AF)*FC/MWex kgmol/s (1.6.11)
Pollutant gas flow rate =(1+AF)*FC*VCpg*MWpg/MWex kg/s (1.6.12)

Brake specific pollutant gas flow rate
=(1+AF)*FC*VCpg*MWpg/(BPR*MWex) kg/Ws (1.6.13)
=(1+AF)*BSFC*VCpg*MWpg/MWex kg/Ws (1.6.14)

By quoting an actual example, this last equation is readily transferred into the usual units for the reference of any exhaust pollutant. If BSFC is employed in the conventional units of kg/kWh and the pollutant measurement of, say, carbon monoxide, is in % by volume, the brake specific carbon monoxide emission rate, BSCO, in g/kWh units is given by:

BSCO=10*(1+AF)*BSFC*(%CO)*16/29 g/kWh (1.6.15)

where the average molecular weight of exhaust gas is assumed to be 29 and that for CO is known to be 16.

The actual mass flow rate of carbon monoxide in the exhaust pipe is
=BSCO*BPR g/h (1.6.16)

These equations are programmed into Prog.1.4, called "EXHAUST GAS ANALYSIS," and should be useful to those who are involved in this form of measurement in connection with engine research and development. The pollutants covered by Prog.1.4 are carbon monoxide, carbon dioxide, hydrocarbons and oxides of nitrogen. Brake specific values for air and oxygen are also produced. An example of the use of this calculation, an encapsulation of the computer screen during a running of the program, is illustrated in Fig. 1.18.

It should be pointed out that there are various ways of recording exhaust emissions as values "equivalent to a reference gas." In the measurement of hydrocarbons, either by a NDIR (non-dispersive infrared) device or by a FID (flame ionization detector), the readings are quoted as ppm hexane, or ppm methane, respectively. Therefore Prog.1.4 contains, in the appropriate equations, the molecular weights for hexane, C_6H_{14}, or methane, CH_4. Some FID meters use a $CH_{1.85}$ equivalent(1.15) and in that case, in the relevant equation in Prog.1.4, the molecular weight for HC equivalent would have to be replaced by 13.85. The same holds true for nitrogen oxide emission, and in Prog.1.4 it is assumed that it is to be NO and so the molecular weight for NO, 30, is employed. Should the meter used in a particular laboratory be different, then, as for the HC example quoted above, the correct molecular weight of the reference gas should be employed.

```
enter air to fuel ratio, AF?  20
enter brake specific fuel consumption, BSFC, as kg/kWh?  .315
enter oxygen concentration in the exhaust gas as % vol, O2VOL?  7.1
enter carbon monoxide emission as % vol, COVOL?  .12
enter carbon dioxide emission as % vol, CO2VOL?  6.9
enter oxides of nitrogen emission as ppm, NOX?  236
the unburned hydrocarbon emission values will have been measured as-
either HC ppm by a NDIR system as hexane equivalent, C6H14,
or as HC ppm by a FID system as methane equivalent, CH4
type in the name of the type of measurement system, either 'NDIR' or 'FID'?  NDIR
enter the HC as ppm measured by NDIR system?  356
OUTPUT DATA
The brake specific air consumption, BSAC, in kg/kWh units, is      6.3
The brake specific emission values printed are in g/kWh units
The brake specific carbon monoxide value, BSCO, is     7.7
The brake specific nitrogen oxide value, BSNOX, is     1.6
The brake specific carbon dioxide value, BSCO2, is  692.5
The brake specific oxygen value, BSO2, is  518.3
The brake specific hydrocarbon value by NDIR system, BSHC, is      7.0
The Trapping Efficiency, TE , as %, is    64.5
WANT A PRINT-OUT(Y OR N?)?  N
```

Fig. 1.18 Example of the use of Prog.1.4, program "EXHAUST GAS ANALYSIS."

A further complication arises for the comparison of HC values recorded by NDIR and FID instrumentation. This is discussed by Tsuchiya and Hirano(1.19) and they point out that the NDIR reading should be multiplied by a sensitivity factor "K" to obtain equality with that recorded by a FID system. They illustrate this in graphical form (Fig. 1.19). In Prog.1.4, this sensitivity factor "K" is taken as unity.

1.6.3 Trapping efficiency from exhaust gas analysis

In an engine where the combustion is sufficiently rich, or is balanced as in the stoichiometric equation, Eq. 1.5.16, it is a logical presumption that any oxygen in the exhaust gas must come from scavenge air which has been lost to the exhaust system. Actually, a stoichiometric air-fuel ratio in practice would still have some residual oxygen in the exhaust gas. So, such an experimental test would be conducted with a sufficiently rich mixture during the combustion process as to ensure that no free oxygen remained within the cylinder after the combustion period. If the actual combustion process was deliberately stratified, as in a diesel engine, then that would be a very difficult condition to satisfy. However, on the assumption

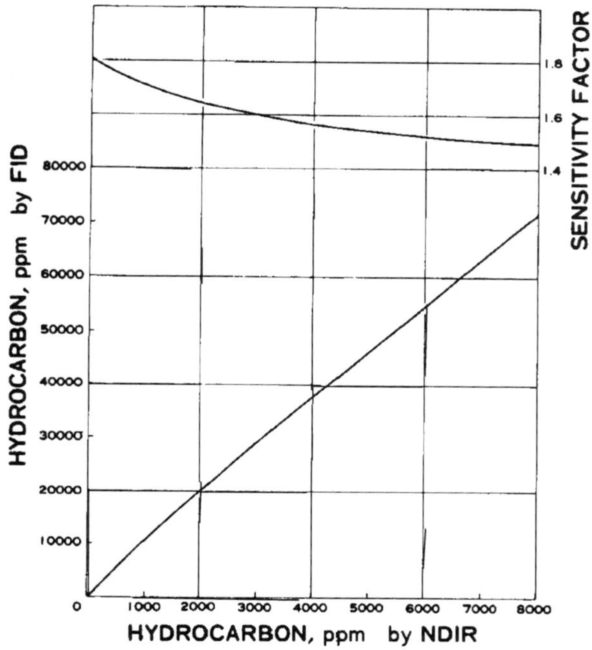

Fig. 1.19 HC concentration from NDIR and FID analysis (from Tsuchiya and Hirano(1.19)).

that this condition can be met, and it is possible, as most two-stroke engines are homogeneously charged, the trapping efficiency, TE, can be calculated from the exhaust gas analysis as follows:

Total mass flow out of the exhaust=(1+AF)*FC/MWex kgmol/s (1.6.17)

Oxygen flow rate out of the exhaust
=(1+AF)*FC*(O_2%)/(MWex*100) kgmol/s (1.6.18)

Oxygen flow rate into engine=0.2314*AF*FC*/32 kgmol/s (1.6.19)

The numerical value of 0.2314 is the mass fraction of oxygen in air and 32 is the molecular weight of oxygen. The value noted as O_2% is the percentage volumetric concentration of oxygen in the exhaust gas.

Trapping efficiency, TE, is given by:
TE=(air trapped in cylinder)/(air supplied)
=1-(air lost to exhaust)/(air supplied)
=1-(Eq.1.6.18/Eq.1.6.19)
=1-((1+AF)*O_2%*32)/(AF*MWex*23.14) (1.6.20)

Assuming the average molecular weight of exhaust gas to be 29, this becomes:

$$TE=1-((1+AF)*O_2\%)/(21*AF) \quad (1.6.21)$$

This equation, produced by Kee(1.20), is programmed together with the other parameters in Prog.1.4.

The methodology emanates from history in a paper by Watson(1.21) in 1908. Huber(1.22) basically uses the Watson approach, but provides an analytical solution for trapping efficiency, particularly for conditions where the combustion process yields some free oxygen. The use of this analytical technique, to measurements taken in a two-stroke engine under firing conditions, is described by Blair and Kenny(1.23); they provide further data on the in-cylinder conditions at the same time using the experimental device shown in Plate 3.3.

1.7 Potential power output of two-stroke engines

At this stage of the book, it will be useful to be able to assess the potential power output of two-stroke engines. From Eq. 1.6.6, the power output of an engine delivered at the crankshaft is seen to be:

$$BPR=BMEP*SV*RPS \quad (1.7.1)$$

From experimental work for various types of two-stroke engines the potential levels of attainment of brake mean effective pressure are well known within quite narrow limits. The literature is full of experimental data for this parameter, and the succeeding chapters of this book provide further direct information on the matter, often predicted directly by engine modelling computer programs. Some typical levels of BMEP for a brief selection of engine types are given in Fig. 1.20.

The engines, listed as A-H, can be related to types which are familiar as production devices. For example, type A could be a chainsaw engine or a small outboard motor of less than 5 hp. The type B engine would appear in a motorcycle for both on- or off-road applications. The type C engine would be used for competition purposes, such as motocross or road-racing. The type D engine could also be a motorcycle, but is more likely to be an outboard motor or a snowmobile engine. The type E engine is almost certain to be an outboard motor. The type F engine is possibly an electricity-generating set engine, whereas type G is a truck power unit, and type H could be either a truck engine or a marine propulsion unit. Naturally, this table contains only the broadest of classifications and could be expanded into many sub-sets, each with a known band of attainment of brake mean effective pressure.

Therefore, it is possible to insert this data into Eq. 1.7.1, and for a given engine total swept volume, SV, at a rotation rate, RPS, determine the power output, BPR. It is quite clear that this might produce some optimistic predictions of engine performance, say, by assuming a BMEP of 10 bar for a single-cylinder spark-ignition engine of 500 cc capacity running at an improbable speed of 20,000 rpm. However, if that engine had ten cylinders, each of 50 cc capacity, it might be mechanically

ENGINE TYPE	BMEP, bar	PISTON SPEED, m/s	BORE/STROKE RATIO
SPARK-IGNITION ENGINES			
SINGLE CYLINDER UNITS			
A-UNTUNED AND SILENCED EXHAUST PIPE	4.5-6.0	12-14	1.0-1.3
B-TUNED AND SILENCED EXHAUST PIPE	8.0-9.0	12-16	1.0-1.2
C-TUNED AND UNSILENCED EXHAUST PIPE	10.0-11.0	16-22	1.0-1.2
MULTI-CYLINDER UNITS			
D-TWO-CYLINDER WITH EXHAUST TUNING	6.0-7.0	12-14	1.0-1.2
E-THREE/FOUR CYLINDER WITH EXHAUST TUNING	7.0-9.0	12-20	1.0-1.25
COMPRESSION IGNITION ENGINES			
F-NATURALLY ASPIRATED ENGINE	3.5-4.5	10-13	0.85-1.0
G-SUPERCHARGED ENGINE	6.5-10.5	10-13	0.85-1.0
H-TURBOCHARGED ENGINE	8.0-14.0	10-13	0.5-0.9

Fig. 1.20 Potential performance criteria for some types of two-stroke engines.

possible to rotate it safely at that speed! Thus, for any prediction of power output to be realistic, it becomes necessary to accurately assess the possible speed of rotation of an engine, based on criteria related to its physical dimensions.

1.7.1 Influence of piston speed on the engine rate of rotation

The maximum speed of rotation of an engine depends on several factors, but the principal one, as demonstrated by any statistical analysis of known engine behavior, is the mean piston speed, PS. This is not surprising as a major limiting factor in the operation of any engine is the lubrication of the main cylinder components, the piston and the piston rings. In any given design the oil film between those components and the cylinder liner will deteriorate at some particular rubbing velocity, and failure by piston seizure will result. The mean piston speed, PS, is given by:

$$PS=2*ST*RPS \qquad (1.7.2)$$

As one can vary the bore and stroke for any design within a number of cylinders, Ncy, to produce a given total swept volume, the bore-stroke ratio is determined as follows:

$$BS=BO/ST \qquad (1.7.3)$$

The total swept volume of the engine can now be written as:

$$SV=Ncy*((\pi/4)*BO*BO*ST) \quad =Ncy*((\pi/4)*BS^2*ST^3) \qquad (1.7.4)$$

Substitution of Eqs. 1.7.2 and 1.7.4 into Eq. 1.7.1 reveals:

$$BPR=(BMEP*PS/2)*((BS*SV)^{0.666})*(\pi*Ncy/4)^{0.333} \quad (1.7.5)$$

This equation is strictly in SI units. Perhaps a more immediately useful equation in familiar working units, where the power output is in kW, BPRk, the BMEP is in bar, BMEPb, and the total swept volume is in cm^3 units, SVc, is:

$$BPRk=(BMEPb*PS/200)*((BS*SVc)^{0.666})*(\pi*Ncy/4)^{0.333}$$
$$=(BMEPb*PS/216.77)*((BS*SVc)^{0.666})*(Ncy^{0.333}) \quad (1.7.6)$$

The values for bore-stroke ratio and piston speed, which are typical of the engines listed as types A-H, are shown in Fig. 1.20. It will be observed that the values of piston speed are normally in a common band from 12-14 m/s for most spark-ignition engines, and those with values about 20 m/s are for engines for racing or competition purposes which would have a relatively short lifespan. The values typical of diesel engines are slightly lower, reflecting not only the heavier cylinder components required to withstand the greater cylinder pressures but also the reducing combustion efficiency of the diesel cycle at higher engine speeds and the longer lifespan expected of this type of power unit. It will be observed that the bore-stroke ratios for petrol engines vary from "square" at 1.0 to "oversquare" at 1.3. The diesel engine, on the other hand, has bore-stroke ratios which range in the opposite direction to "undersquare," reflecting the necessity for suitable proportioning of the smaller combustion chamber of that higher compression ratio power unit.

1.7.2 Influence of engine type on power output

With the theory developed in Eqs. 1.7.5 or 1.7.6, it becomes possible by the application of the BMEP, bore-stroke ratio and piston speed criteria to predict the potential power output of various types of engines. This type of calculation would be the opening gambit of theoretical consideration by a designer attempting to meet a required target. Naturally, the statistical information available would be of a more extensive nature than the broad bands indicated in Fig. 1.20, and would form what would be termed today as an "expert system." As an example of the use of such a calculation, three engines are examined by the application of this theory and the results shown in Fig. 1.21. The engines are very diverse in character such as a small chainsaw, a racing motorcycle engine, and a truck diesel powerplant. The input and output data for the calculation are declared in Fig. 1.21 and are culled from those applicable to the type of engine postulated in Fig. 1.20. The target power output in the data table are in kW units, but in horsepower values are 7 bhp for the chainsaw, 62 bhp for the racing motorcycle engine and 250 bhp for the truck diesel engine. The physical dimensions predicted for the three engines are seen to be very realistic, and from a later discussion in Chapters 5 and 6, the reported behavior of engines such as the chainsaw and the racing motorcycle engine will confirm that statement. The varied nature of the specific power performance from these very different engines is observed from the 95 bhp/liter for the chainsaw, 247 bhp/liter for the racing

ENGINE TYPE	TYPE A	TYPE C	TYPE G
INPUT DATA	CHAINSAW	RACING MOTOR	TRUCK DIESEL
POWER, kW	5.2	46.2	186
PISTON SPEED, m/s	12	20	10
BORE-STROKE RATIO	1.3	1	0.9
BMEP, bar	4.5	10	7
NUMBER OF CYLINDERS	1	2	6
OUTPUT DATA			
BORE, mm	49.5	54.1	106
STROKE, mm	38.0	54.1	118
SWEPT VOLUME, cc	73.8	250.9	6300
ENGINE SPEED, rpm	9440	11080	2545

Fig. 1.21 Output from calculations predicting potential engine performance.

motorcycle engine and 39.7 bhp/liter for the truck diesel power unit. The most useful part of this method of initial prediction of the potential power performance of an engine is that some necessary pragmatism is injected into the selection of the data for the speed of rotation of the engine.

NOTATION FOR CHAPTER 1

NAME	SYMBOL	SI UNIT	OTHER UNITS
Air-Fuel Ratio	AF		
Bore-Stroke Ratio	BS		
Brake Thermal Efficiency	BThE		
Charging Efficiency	CE		
Crankcase Compression Ratio	CCR		
Geometric Compression Ratio	GCR		
Indicated Thermal Efficiency	IThE		
Number of engine cylinders	Ncy		
Mechanical Efficiency	Meff		
Ratio of specific heats	g		
Scavenge Ratio	SR		
Scavenging Efficiency	SE		
Trapped Air-Fuel Ratio	TAF		
Trapped Compression Ratio	TCR		
Trapping Efficiency	TE		
Air Consumption Rate	AC	kg/s	kg/h

NAME	SYMBOL	SI UNIT	OTHER UNITS
Airflow Quantity	Mas	kg	
Density	D	kg/m^3	
Fuel Flow Quantity	Mtf	kg	
Gas Constant (air)	Rair	J/kgK	
Heat released during combustion	Qc	J	
Indicated Power Output	IPR	W	kW, hp
Brake Power Output	BPR	W	kW, hp
Brake Mean Effective Pressure	BMEP	Pa	bar, atm, psi
Brake Specific Fuel Consumption	BSFC	kg/Ws	kg/kWh,lb/hp.hr
Brake Torque	BTRQ	Nm	ft.lb
Clearance Volume	CV	m^3	cc or cm^3
Cylinder Bore	BO	m	mm, in
Cylinder Stroke	ST	m	mm, in
Dyno Torque Arm Force	F	N	
Dyno Torque Arm Length	L	m	
Engine Rotation Rate	RPS	rev/s	rev/min
Fuel Calorific Value (lower)	CAL	J/kg	kJ/kg
Fuel Consumption Rate	FC	kg/s	kg/h
Indicated Mean Effective Pressure	IMEP	Pa	bar, atm, psi
Indicated Specific Fuel Consumption	ISFC	kg/Ws	kg/kWh,lb/hp.hr
Indicated Torque	ITRQ	Nm	ft.lb
Piston Speed	PS	m/s	ft/min
Pressure	P	Pa	bar, atm, psi
Specific Carbon Monoxide Emission	BSCO	kg/Ws	g/kWh, g/hp.hr
Specific Hydrocarbon Emission	BSHC	kg/Ws	g/kWh, g/hp.hr
Specific Nitrogen Oxides Emission	BSNOX	kg/Ws	g/kWh, g/hp.hr
Swept Volume	SV	m^3	cc or cm^3, in^3
Temperature	T	K	ºC, ºF
Trapped Swept Volume	TSV	m^3	cc or cm^3

REFERENCES FOR CHAPTER 1

1.1 P.E. Irving, Two-Stroke Power Units, their Design and Application, Temple Press Books for Newnes, London, 1967.

1.2 K.G. Draper, The Two-Stroke Engine, Design and Tuning, Foulis, Oxford, 1968.

1.3 C.F. Taylor, E.S. Taylor, The Internal Combustion Engine, International Textbook Company, Scranton, Pa., 1962.

1.4 R.S. Benson, N.D. Whitehouse, Internal Combustion Engines, Volumes 1 and 2, Pergamon, Oxford, 1979.

1.5 C.F. Caunter, Motor Cycles, a Technical History, Science Museum, London, HMSO, 1970.

1.6 P.H. Schweitzer, Scavenging of Two-Stroke Cycle Diesel Engines, Macmillan, New York, 1949.

1.7 G.P. Blair, R.R. Booy, B.L. Sheaffer, Editors, "Technology Pertaining to Two-Stroke Cycle Spark-Ignition Engines," Society of Automotive Engineers, SAE/PT-82/26, Progress in Technology Series No. 26, 1982.

1.8 W.J.D. Annand, G.E. Roe, Gas Flow in the Internal Combustion Engine, Foulis, Yeovil, Somerset, 1974.

1.9 G.P. Blair, R.A.R. Houston, R.K. McMullan, N. Steele, S.J. Williamson, "A New Piston Design for a Cross-Scavenged Two-Stroke Cycle Engine with Improved Scavenging and Combustion Characteristics," SAE Intnl. Off-Highway Vehicle Meeting, Milwaukee, WI, September, 1984, SAE Paper 841096.

1.10 G.P. Blair, R.G. Kenny, J.G. Smyth, M.E.G. Sweeney, G.B. Swann, "An Experimental Comparison of Loop and Cross Scavenging of the Two-Stroke Cycle Engine," SAE Intnl. Off-Highway Vehicle Meeting, Milwaukee, WI, September, 1986, SAE Paper No. 861240.

1.11 G.P. Blair, "Correlation of Theory and Experiment for Scavenging Flow in Two-Stroke Cycle Engines," Society of Automotive Engineers International Off-Highway & Powerplant Congress, Milwaukee, WI, September 12-15, 1988, SAE Paper No. 881265.

1.12 J.D. Flaig, G.L. Broughton, "The Design and Development of the OMC V-8 Outboard Motor," Society of Automotive Engineers International Off-Highway & Powerplant Congress, Milwaukee, WI, September 9-12, 1985, SAE Paper No. 851517.

1.13 R. Fleck, G.P. Blair, R.A.R. Houston, "An Improved Model for Predicting Reed Valve Behavior in Two-Stroke Cycle Engines," Society of Automotive Engineers International Off-Highway & Powerplant Congress, Milwaukee, WI, September 14-17,1987, SAE Paper No. 871654.

1.14 SAE J1349, Engine Power Test Code, Spark Ignition and Diesel, June 1985.

1.15 SAE J1088, Test Procedure for the Measurement of Exhaust Emissions from Small Utility Engines, June 1983.

1.16 ISO 3046, Reciprocating Internal Combustion Engines: Performance-Parts 1, 2, and 3, International Standards Organization, 1981.

1.17 BS 1042, Fluid Flow in Closed Conduits, British Standards Institution, 1981.

1.18 E. Sher, "The Effect of Atmospheric Conditions on the Performance of an Air-Borne Two-Stroke Spark-Ignition Engine," *Proc. I. Mech. E.*, Vol. 198D, No. 15, pp. 239-251 and as SAE No. 844962.

1.19 K. Tsuchiya, S. Hirano, "Characteristics of 2-Stroke Motorcycle Exhaust Emission and Effects of Air-Fuel Ratio and Ignition Timing," Society of Automotive Engineers, Automobile Engineering and Manufacturing Meeting, Detroit, MI, October, 1975, SAE No. 750908.

1.20 R.J. Kee, "Stratified Charging of a Cross-Scavenged Two-Stroke Cycle Engine," Doctoral Thesis, The Queen's University of Belfast, October, 1988.

1.21 W. Watson, "On the Thermal and Combustion Efficiency of a Four-Cylinder Petrol Motor," *Proc. I. Auto. E.*, Vol. 2, p. 387, 1908-1909.

1.22 E.W. Huber, "Measuring the Trapping Efficiency of Internal Combustion Engines Through Continuous Exhaust Gas Analysis," SAE International Congress, Detroit, MI, February, 1971, SAE Paper No. 710144.

1.23 G.P. Blair, R.G. Kenny, "Further Developments in Scavenging Analysis for Two-Cycle Engines," SAE International Congress, Detroit, MI, February, 1980, SAE Paper No. 800038.

1.24 SAE J604d "Engine Terminology and Nomenclature", June, 1979.

1.25 G.P. Blair, B.L. Sheaffer, G.G. Lassanske, Editors, Advances in Two-Stroke Cycle Engine Technology, Society of Automotive Engineers, SAE/PT-89/33, Progress in Technology Series No. 33, 1989.

Chapter 2

Gas Flow Through Two-Stroke Engines

2.0 Introduction

In Chapter 1, the basic method of operation of the two-stroke engine was described. The gas flow processes into, through, and out of the engine are all conducted in an unsteady manner. The definition of unsteady gas flow is where the pressure, temperature and gas particle velocity in a duct are variable with time. In the case of exhaust flow, the unsteady gas flow behavior is produced because the cylinder pressure falls with the rapid opening of the exhaust port by the piston. This gives an exhaust pipe pressure that changes with time. In the case of induction flow into the crankcase through an intake port whose area changes with time, the intake pipe pressure alters because the crankcase pressure is affected by the piston motion, causing volumetric change in that crankcase.

To illustrate the dramatic variations of pressure wave and particle motion caused by unsteady flow, by comparison with steady flow, the series of photographs taken by Coates(8.2) are shown in Plates 2.1-2.4. These photographs were obtained using the Schlieren method(3.14), which is an optical means of using the variation of the refractive index of a gas with its density. Each photograph was taken with an electronic flash duration of 1.5 μs and the view observed is around the termination of a 28 mm diameter exhaust pipe to the atmosphere. The exhaust pulsations occurred at a frequency of 1000 per minute. The first photograph, Plate 2.1, shows the front of an exhaust pulse about to enter the atmosphere. Note that it is a plane front, and its propagation within the pipe up to the pipe termination is clearly one-dimensional. The next picture, Plate 2.2, shows the propagation of the pressure wave into the atmosphere in a three-dimensional fashion with a spherical front being formed. The beginning of rotational movement of the gas particles at the pipe edges is now evident. The third picture, Plate 2.3, shows the spherical wave front fully formed and the particles being impelled into the atmosphere in the form of a toroidal vortex, or a spinning doughnut of gas particles. Some readers may be more familiar with the term "smoke ring." That propagating pressure wave front arrives at the human eardrum, deflects it, and the nervous system reports it as "noise" to the brain. The final picture of the series, Plate 2.4, shows that the propagation of the pressure wave front has now passed beyond the frame of the photograph, but the toroidal vortex of gas particles is proceeding downstream with considerable turbulence. Indeed, the flow through the eye of the vortex is so violent that a new acoustic pressure wave front is forming in front of that vortex. The noise which emanates from these pressure pulsations is composed of the basic pressure front propagation and also from the turbulence of the fluid motion in the vortex. Further discussion on the noise aspects of this flow is given in Chapter 8. The author has always found this series of photographs, obtained during research at QUB, to be particularly illuminating. When a schoolboy, on a farm in Co. Antrim too many years ago, the milking machines were driven by a single-cylinder diesel engine with a long straight exhaust pipe and, on a frosty winter's morning, it would blow a "smoke ring" from the exhaust on start-up. That schoolboy used to wonder how it was possible; he now knows.

As the resulting performance characteristics of an engine are significantly controlled by this unsteady gas motion, it behooves the designer of engines to understand this flow mechanism thoroughly. This is true for all engines, whether they are destined to be a 2 hp outboard motor or a 150 hp Grand Prix power unit. A simple example will suffice to illustrate the point. If one were to remove the tuned exhaust pipe from a single-cylinder racing engine while it is running at peak power output, the pipe being an "empty" piece of fabricated sheet metal, the engine power output would fall by at least 50% at that engine speed. The tuned exhaust pipe harnesses the pressure wave motion of the exhaust process to retain a greater mass of fresh charge within the cylinder. Without it, the engine will only trap about half

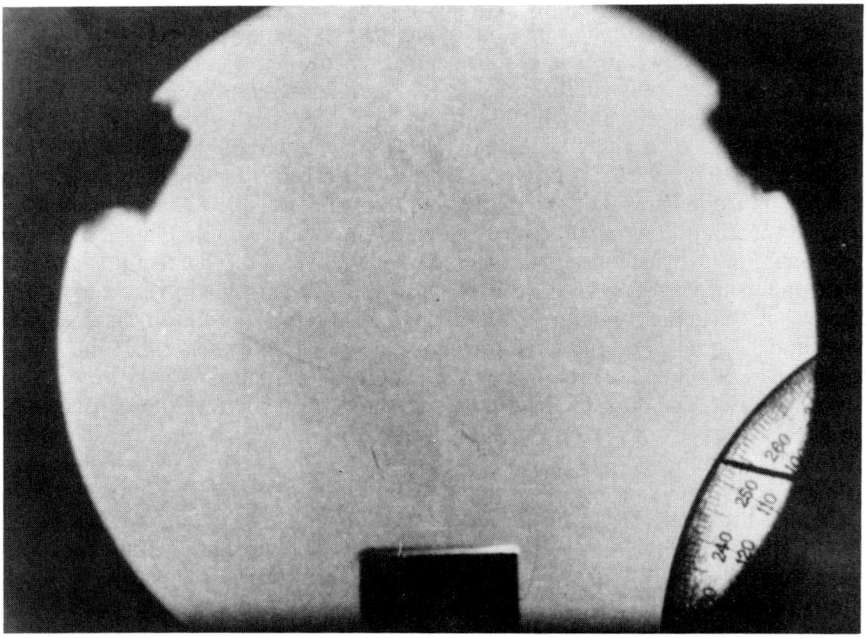

Plate 2.1 Schlieren picture of an exhaust pulse at the termination of a pipe.

Plate 2.2 The exhaust pulse front propagates into the atmosphere.

as much fresh air and fuel in the cylinder. To design such exhaust systems, and the engine which will take advantage of them, it is necessary to have a good understanding of the mechanism of unsteady gas flow. For the more serious student interested in the subject of unsteady gas dynamics in depth, the series of lectures(2.2) given by the late Prof. F. K. Bannister of the University of Birmingham is an excellent introduction to the topic; so too is the book by Annand and Roe(1.8), and the books by Rudinger(2.3) and Benson(2.4). The references cited in this chapter will give even greater depth to that study.

Therefore, this chapter will explain the fundamental characteristics of unsteady gas flow in the intake and exhaust ducts of reciprocating engines. Such fundamental theory is just as applicable to four-stroke engines as it is to two-stroke engines, although the bias of the discussion will naturally be for the latter type of power unit. As with other chapters of this book, computer programs will be provided for the reader to use for his/her own education and experience. Indeed, these programs are used here as a means of explaining the behavior of pressure waves and their effect on the filling and emptying of engine cylinders. In the later sections of the chapter, there will be explanations of the operation of tuned exhaust and intake systems for two-stroke engines.

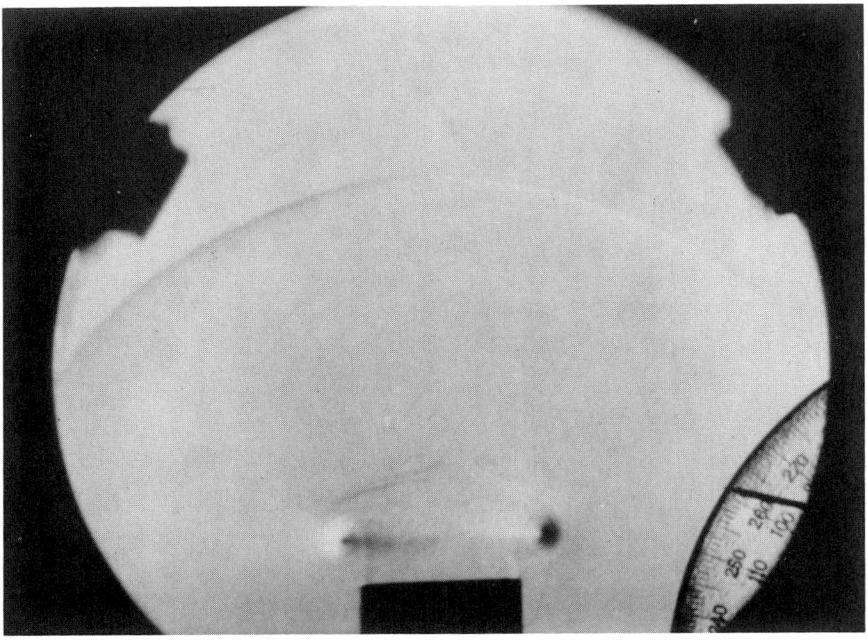

Plate 2.3 Further pulse propagation followed by the toroidal vortex of gas particles.

2.1 Motion of pressure waves in a pipe

2.1.1 Nomenclature for pressure waves

The reader is already familiar with the motion of pressure waves of small amplitude, for they are acoustic waves, or sound. Some of our personal experience with sound waves is helpful in understanding the fundamental nature of the flow of the much larger amplitude waves to be found in engine ducts. Pressure waves, and sound waves, are of two types: (1) compression waves or (2) expansion waves. This is illustrated in Fig. 2.1. In both Fig. 2.1(a) and (b), the undisturbed pressure and temperature in the pipe ahead of the pressure wave is Po and To, respectively.

The compression wave in the pipe is shown in Fig. 2.1(a) and the expansion wave in Fig. 2.1(b), and both waves are propagating towards the right in the diagram. At a point on the compression wave the pressure is Pe, where Pe>Po, and it is being propagated at a velocity, CPe. It is also moving gas particles at a gas particle velocity of CGe and in the same direction as the wave is being propagated. At a point on the expansion wave in the pipe the pressure is Pi, where Pi<Po, and it is being propagated at a velocity, CPi. It is also moving gas particles at a gas particle velocity of CGi, but in the opposite direction to that which the wave is being propagated.

At this point the reader can draw on his personal experience of sound waves to help understand the physical nature of the statements made in the preceding

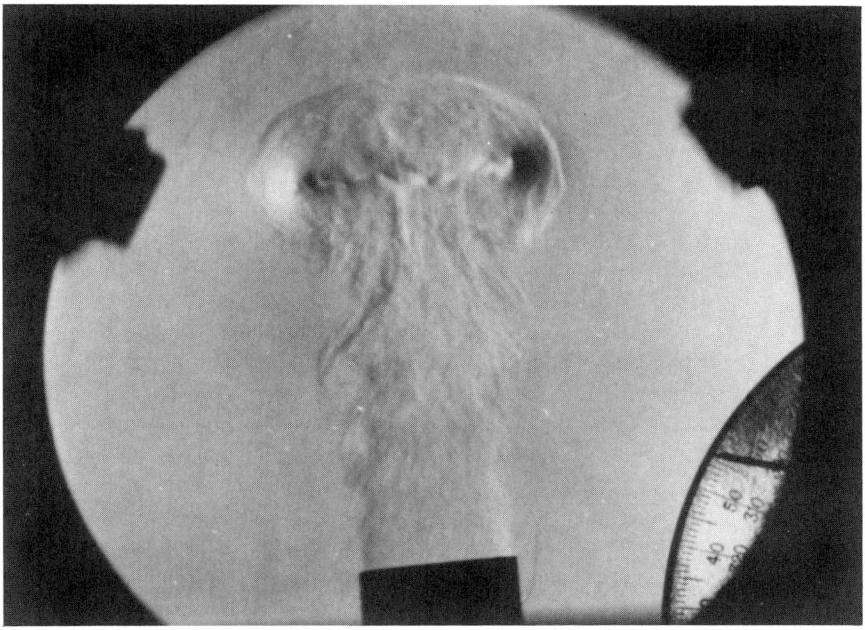

Plate 2.4 The toroidal vortex of gas particles proceeds into the atmosphere.

paragraph. Imagine standing several meters away from another person, Fred. Fred produces a sharp exhalation of breath, for example, he says "boo" somewhat loudly. He does this by raising his lung pressure above the atmospheric pressure due to a muscular reduction of his lung volume. The compression pressure wave produced, albeit of small amplitude, leaves his mouth and is propagated at the local acoustic velocity, or speed of sound, to the ear of the reader. The speed of sound involved is on the order of 350 m/s. The gas particles involved with the "boo" leaving Fred's mouth have a much lower velocity, probably on the order of 1 m/s. However, that gas particle velocity is in the same direction as the propagation of the compression pressure wave, i.e., towards the ear of the reader. Contrast this simple experiment with a second test. Imagine that Fred now produces a sharp inhalation of breath. This he accomplishes by expanding his lung volume so that his lung pressure falls sharply below the atmospheric pressure. The resulting "u....uh" heard by the reader is caused by the expansion pressure wave leaving Fred's mouth and propagating towards the reader's ear at the local acoustic velocity. In short, the direction of propagation is the same as before with the compression wave "boo", and the propagation velocity is, to all intents and purposes, identical. However, as the gas particles manifestly entered Fred's mouth with the creation of this expansion wave, the gas particle velocity is clearly opposite to the direction of the expansion wave propagation.

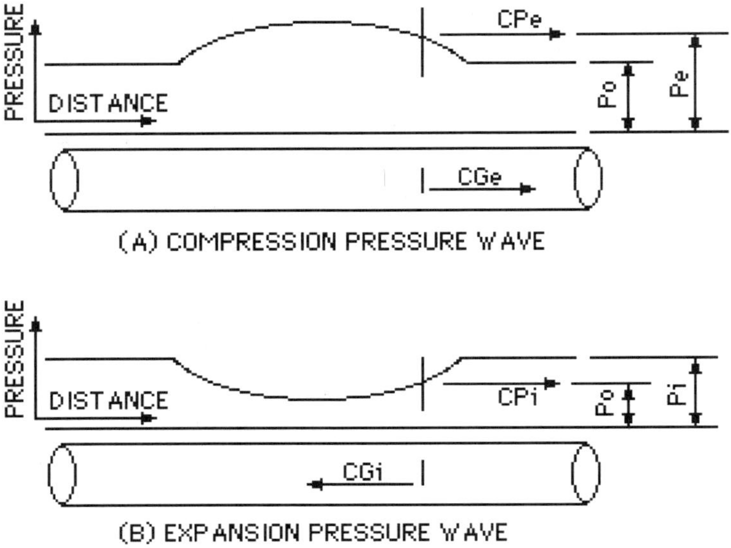

Fig. 2.1 Pressure wave nomenclature.

It is obvious that exhaust pulses resulting from cylinder blowdown come under the heading of compression waves, and expansion waves are generated by the rapidly falling crankcase pressure during induction. However, as in most technologies, other expressions are used in the literature as jargon to describe compression and expansion waves. Compression waves are variously called "exhaust pulses," "compression pulses" or "ramming waves." Expansion waves are often described as "suction pulses," "sub-atmospheric pulses" or "intake pulses." However, as will be seen from the following sections, expansion and compression waves do appear in both inlet and exhaust systems.

2.1.2 Acoustic pressure waves and their propagation velocity

As already pointed out, acoustic pressure waves are pressure waves where the pressure amplitudes are small. Let dP be the pressure difference from atmospheric pressure, Pe-Po or Po-Pi, for the compression or expansion wave, respectively. The value of dP for Fred's "boo" would be on the order of 0.2 Pa. The pressure ratio, PR, for any pressure wave is defined as the pressure, P, at any point on the wave under consideration divided by the undisturbed presssure, Po. The undisturbed pressure is more commonly called the reference pressure. Here, the value for Fred's "boo" would be:

$$
\begin{aligned}
PR &= P/Po \\
&= 101325.2/101325 \\
&= 1.000002
\end{aligned}
$$

For the loudest of acoustic sounds, say a rifle shot at about 0.2 m away from the human ear, dP could be 2000 Pa and the pressure ratio would be 1.02. That such very loud sounds are still small in pressure wave terms can be gauged from the fact that a typical exhaust pulse in an engine exhaust pipe has a pressure ratio of about 1.5.

According to Earnshaw(2.1), the velocity of a sound wave in air is given by Ao, where:

$$Ao=\sqrt{(g*Ra*To)} \qquad (2.1.1)$$
$$=\sqrt{(g*Po/Do)} \qquad (2.1.2)$$

The value denoted by g is the ratio of specific heats for air. In the above equations, To is the reference temperature and Do is the reference density, which are related to the reference pressure, Po, by the state equation:

$$Po=Do*Ra*To \qquad (2.1.3)$$

For sound waves in air, Po, To and Do are the values of the atmospheric pressure, temperature and density, respectively.

2.1.3 Finite amplitude waves

Any pressure wave with a pressure ratio greater than an acoustic wave is called a wave of finite amplitude. Earnshaw(2.1) showed that the gas particle velocity associated with a wave of finite amplitude was given by CG, where:

$$CG=m*Ao*(X-1) \qquad (2.1.4)$$

Bannister's(2.2) derivation of this equation is explained with great clarity. Within this equation the parameters m and X are expanded to reveal:

$$m=2/(g-1)$$
$$X=PR^{n} \qquad (2.1.5)$$
$$PR=P/Po \qquad (2.1.6)$$
$$n=(g-1)/(2*g)$$
$$Ao=\sqrt{(g*R*To)}$$

As stated above, PR is the pressure ratio of a point on the wave of absolute pressure, P. The notation of X is known as the pressure amplitude ratio and R is the gas constant for the particular gas involved.

Incorporation of the above shorthand notation reveals the complete equation for Eq. 2.1.4 as:

$$CG=[2/(g-1)]*[\sqrt{(g*R*To)}]*[(P/Po)^{((g-1)/(2*g))}-1]$$

If the gas in which this pressure wave is propagating has the properties of air, then the several shorthand parameters and constants become:

g=1.4 R=Ra=287 J/kgK m=5 n=1/7

If the gas properties are assumed to be as for air, then Eq. 2.1.4 for the gas particle velocity reduces to the following:

$$CG=5*Ao*(X-1) \quad (2.1.7)$$

$$\text{where } X=PR^{(1/7)} \quad (2.1.8)$$

The propagation velocity at any point on a wave where the pressure is P and the temperature is T is like a small acoustic wave moving at the local acoustic velocity at those conditions, but on top of gas particles which are already moving. Therefore, the absolute propagation velocity of any wave point is the sum of the local acoustic velocity and the local gas particle velocity. Thus, the propagation velocity of any point on a finite amplitude wave is given by CP:

$$CP=Ap+CG \quad (2.1.9)$$

where Ap is the local acoustic velocity at the elevated pressure and temperature of the wave point, P and T.

However, Ap is given by Earnshaw from Eq. 2.1.1 as:

$$Ap=\sqrt{(g*R*T)} \quad (2.1.10)$$

Assuming that the state conditions change from Po and To to P and T to be isentropic, then for such a change:

$$T/To=(P/Po)^{((g-1)/g)}=(D/Do) \quad (2.1.11)$$

$$Ap/Ao=\sqrt{(T/To)} \quad (2.1.12)$$

$$=(P/Po)^{[(g-1)/(2*g)]} \quad (2.1.13)$$

$$=PR^{n}$$

$$=X \quad (2.1.14)$$

Hence, the absolute propagation velocity, CP, defined by Eq. 2.1.9, is given by the addition of Eqs. 2.1.4 and 2.1.14:

$$CP=m*Ao*(X-1)+X*Ao$$

$$=Ao*[((g+1)/(g-1))*X-m]$$

If the properties of air are assumed for the gas, then this reduces to:

$$CP=Ao*(6*X-5) \quad (2.1.15)$$

The density, D, at any point on a wave of pressure, P, is found from Eqs. 2.1.11 and 2.1.14 as follows:

$$D/Do=(P/Po)^{(1/g)} =X^{(2/(g-1))} \tag{2.1.16}$$

For air, where g is 1.4, this translates to:

$$D=Do*X^5 \tag{2.1.17}$$

2.1.4 Propagation and particle velocities of finite amplitude waves in air

From Eqs. 2.1.4 and 2.1.15, the propagation velocities of finite amplitude waves in air in a pipe are calculated by the following equations:

Propagation velocity, $CP=Ao*(6*X-5)$ (2.1.18)
Particle velocity, $CG=5*Ao*(X-1)$ (2.1.19)
Pressure amplitude ratio, $X=(P/Po)^{(1/7)}= (PR)^{(1/7)}$ (2.1.20)
Acoustic velocity at the reference gas state conditions,
$Ao=\sqrt{(1.4*287*To)}$
Density of the reference gas, $Do=Po/(287*To)$

It is interesting that these equations corroborate the experiment which the reader conducted with his imagination regarding Fred's lung-generated compression and expansion waves.

Fig. 2.1 shows compression and expansion waves. Let us assume that the undisturbed pressure and temperature in both cases are at standard atmospheric conditions. In other words, Po and To are 101325 Pa and 20ºC, or 293K, respectively. The reference acoustic velocity, Ao, and reference density, Do, are, from Eqs. 2.1.1 and 2.1.3:

$Ao=\sqrt{(1.4*287*293)}=343.11$ m/s
$Do=101325/(287*293)=1.2049$ kg/m^3

Let us assume that the pressure ratio, PRe, of a point on the compression wave is 1.2 and that of a point on the expansion wave is PRi with a value of 0.8. In other words, the compression wave has a pressure differential as much above the reference pressure as the expansion wave is below it. Let us also assume that the pipe has a diameter, ØP, of 25 mm.

(a) The compression wave

First, consider the compression wave of pressure, Pe. This means that Pe is:

$Pe=PRe*Po=1.2*101325=121590$ Pa

The pressure amplitude ratio, Xe, is calculated as:

$Xe=(1.2)^{(1/7)}=1.02639$

Therefore, the propagation and particle velocities, CPe and CGe, are found from:

CPe=343.11*(6*1.02639-5)=397.44 m/s
CGe=5*343.11*(1.02639-1)=45.27 m/s

From this it is clear that the propagation of the compression wave is faster than the reference acoustic velocity, Ao, and that the air particles move at a considerably slower rate. The compression wave is moving rightwards along the pipe at 397.4 m/s and, as it passes from particle to particle, it propels each particle in turn in a rightwards direction at 45.27 m/s. This is deduced from the fact that the sign of the numerical values of CPe and CGe are identical.

The local particle Mach number, MNe, is defined as the ratio of the particle velocity to the local acoustic velocity, Ae, where:

MNe=CGe/Ae

From Eq. 2.1.14, Ae=Ao*Xe.

Hence, Ae=343.11*1.02639 =352.16 m/s,
and the local particle Mach number, MNe=45.27/352.16 =0.1285.

The mass rate of gas flow, MGe, caused by the passage of this point of the compression wave in a pipe of area, FPe, is calculated from the thermodynamic equation of continuity:

Mass flow rate=density*area*particle velocity
MGe=De*FPe*CGe

From Eq. 2.1.17,
$De=Do*Xe^5=1.2049*1.02639^5=1.2049*1.1391=1.3725\ kg/m^3$.

The pipe area, $FPe=\pi*ØP^2/4=3.14159*0.025*0.025/4=0.000491\ m^2$.

Therefore, the mass rate of flow is in the same direction as the wave propagation as:

MGe=1.3725*0.000491*45.27 =0.0305 kg/s

(b) The expansion wave

Second, consider the expansion wave of pressure, Pi. This means that Pi is given by:

Pi=PRi*Po=0.8*101325 =81060 Pa

The pressure amplitude ratio, Xi, is calculated as:

$Xi=(0.8)^{(1/7)}=0.9686$

Therefore, the propagation and particle velocities, CPi and CGi, are found from:

CPi=343.11*(6*0.9686-5)=278.47 m/s
CGi=5*343.11*(0.9686-1)=-53.87 m/s

From this it is clear that the propagation of the expansion wave is slower than the reference acoustic velocity but the air particles move somewhat faster. The expansion wave is moving rightwards along the pipe at 278.47 m/s and, as it passes from particle to particle, it propels each particle in turn in a leftwards direction at 53.87 m/s. This is deduced from the fact that the numerical values of CPe and CGe are of opposite sign.

The local particle Mach number, MNi, is defined as the ratio of the particle velocity to the local acoustic velocity, Ai, where:

MNi=CGi/Ai

From Eq. 2.1.14, Ai=Ao*Xi.

Hence, Ai=343.11*0.9686=332.3 m/s.

The local particle Mach number, MNi=(-53.87)/332.3=-0.1621.

The mass rate of gas flow, MGi, caused by the passage of this point of the expansion wave in a pipe of area, FPi, is calculated by:

MGi=Di*FPi*CGi

From Eq. 2.1.17:

$Di=Do*Xi^5=1.2049*0.9686^5 =1.2049*0.8525 =1.0272\ kg/m^3$

The pipe area, $FPi=FPe=0.000491\ m^2$.

Therefore, the mass rate of flow is in the opposite direction to the wave propagation as:

MGi=1.0272*0.000491*(-53.87)=-0.0272 kg/s

2.1.5 Distortion of the wave profile

It is clear from the foregoing that the value of propagation velocity is a function of the wave pressure and wave temperature at any point on that pressure wave. It should also be evident that, as all of the points on a wave are propagating at different velocities, the wave must change its shape in its passage along any duct. To illustrate this the calculations conducted in the previous section are displayed in Fig. 2.2. In Fig. 2.2(a) it can be seen that both the front and tail of the wave travel at the reference acoustic velocity, Ao, which is 53 m/s slower than the peak wave velocity. In their travel along the pipe, the front and the tail will keep station with each other in both time and distance. However, at the front of the wave, all of the pressure points between it and the peak are travelling faster and will inevitably catch up with it. Whether that will actually happen before some other event intrudes (for instance, the wave front could reach the end of the pipe) will depend on the length of the pipe and the time interval between the peak and the wave front. Nevertheless, there will always be the tendency for the wave peak to get closer to the wave front and further away from the wave tail. This is known as "steep-fronting." The wave peak could, in theory, try to pass the wave front, which is what happens to a water wave in the ocean when it "crests." In gas flow, "cresting" is impossible and the reality is that a shock wave would be formed. This can be analyzed theoretically and Bannister(2.1) gives an excellent account of the mathematical solution for the particle velocity and the propagation velocity, Csh, of a shock wave of pressure ratio, PRsh, in a gas medium of reference acoustic velocity, Ao. The theoretically derived expressions for propagation velocity, Csh, and particle velocity, CGsh, of a shock front are:

Csh=Ao*√{[(g+1)*PRsh+(g-1)]/(2*g)} (2.1.21)

CGsh=[2/(g+1)]*[Csh-(Ao*Ao/Csh)] (2.1.22)

The situation for the expansion wave, shown in Fig. 2.2(b), is the reverse, in that the peak is travelling 64.6 m/s slower than either the wave front or the wave tail. Thus, any shock formation taking place will be at the tail of an expansion wave, and any wave distortion will be where the tail of the wave attempts to overrun the peak.

2.2 Motion of oppositely moving pressure waves in a pipe

In the previous section, the reader was asked to conduct his imaginary experiment with Fred, who produced compression and expansion waves by exhaling or inhaling sharply, producing a "boo" or a "u...uh," respectively. Once again, the reader is asked to conduct another experiment so as to draw on his experience of sound waves to illustrate a principle, in this case the behavior of oppositely moving pressure waves. In this second experiment, the reader and his friend Fred are going

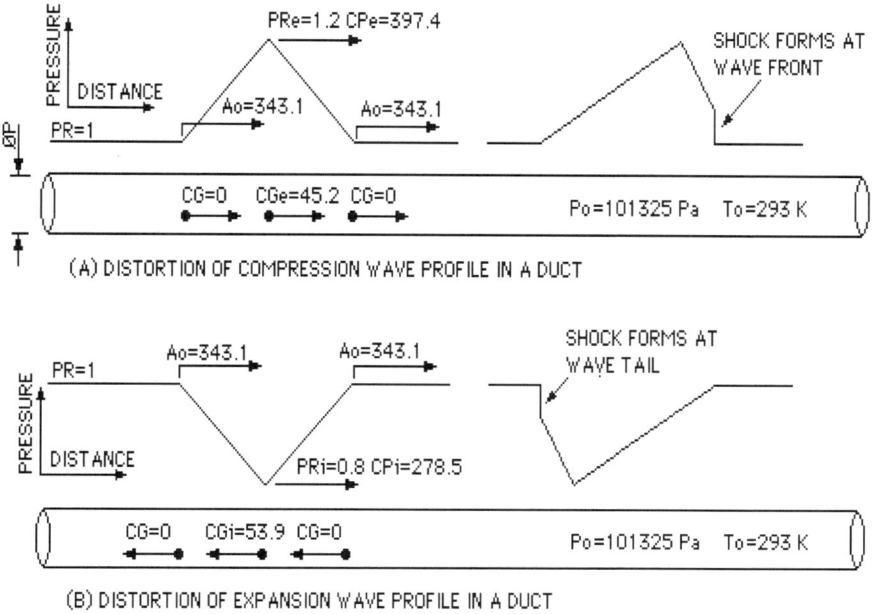

Fig. 2.2 Distortion of wave profiles and possible shock formation.

to say "boo" at each other from some distance apart, and at the same time. Their ears, being rather accurate pressure transducers, will record their own "boo" first, followed a fraction of time later by the "boo" from the other party. Obviously, the "boo" from each passed through the "boo" from the other and arrived at both Fred's ear and the reader's ear with no distortion caused by their passage through each other. If distortion had taken place, then the sensitive human ear would have detected it. At the point of meeting, when the waves were passing through each other, the process is called "superposition."

2.2.1 Superposition of oppositely moving waves

Fig. 2.3 illustrates two oppositely moving pressure waves in air in a pipe. They are shown as compression waves, ABCD and EFGH, and are sketched as being square in profile, which is physically impossible but it makes the task of mathematical explanation somewhat easier. In Fig. 2.3(a) they are about to meet. In Fig. 2.3(b) the process of superposition is taking place for the front EF on wave top BC, and for the front CD on wave top FG. The result is the creation of a superposition pressure, Ps, from the separate wave pressures, P1 and P2. Assume that the reference acoustic

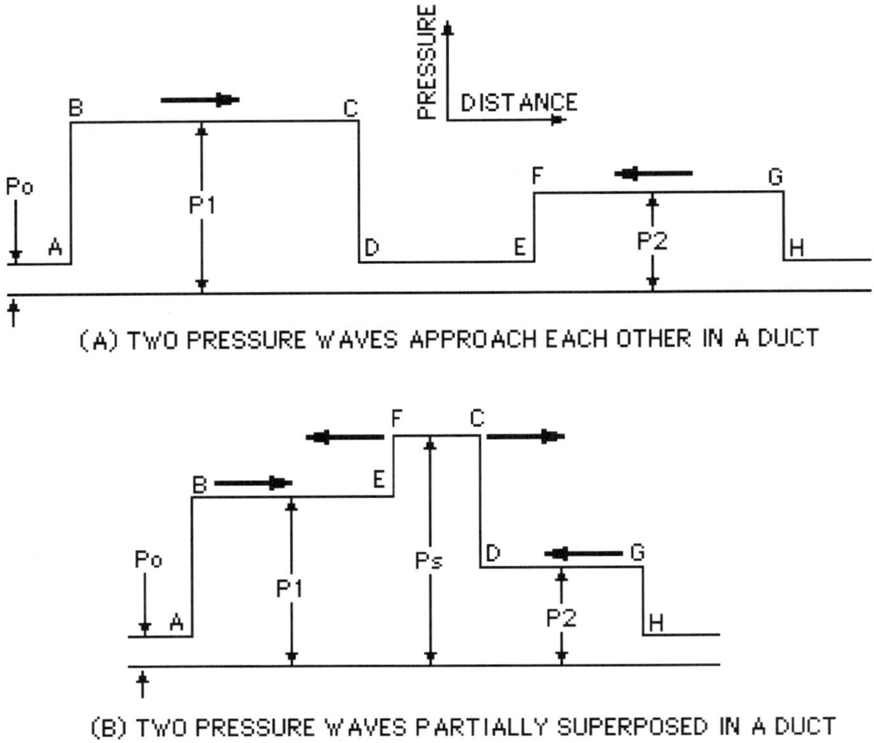

Fig. 2.3 Superposition of pressure waves in a pipe.

velocity is Ao. Assuming also that the rightwards direction is mathematically positive, the particle and the propagation velocity of any point on the wave top, BC, will be CP1 and CG1. From Eqs. 2.1.18-20:

CG1=5*Ao*(X1-1)
CP1=Ao*(6*X1-5)

Similarly, the values for the wave top FG will be (with rightwards regarded as the positive direction):

CG2=-5*Ao*(X2-1)
CP2=-Ao*(6*X2-5)

From Eq. 2.1.14, the local acoustic velocities in the gas columns BE and DG during superposition will be:

Ap1=Ao*X1
Ap2=Ao*X2

During superposition, the wave top F is now moving into a gas with a new reference pressure level at P1. The particle velocity of F relative to the gas in BE will be:

$$\begin{aligned} CGFrelBE &= -5*Ap1*[(Ps/P1)^{(1/7)}-1] \\ &= -5*Ao*X1*[(Ps/P1)^{(1/7)}-1] \\ &= -5*Ao*(Xs-X1) \end{aligned}$$

The absolute particle velocity of F, CGs, will be given by the sum of CGFrelBE and CG1, as follows:

CGs=CGFrelBE+CG1
=-5*Ao*(Xs-X1)+ 5*Ao*(X1-1)
=5*Ao*(2*X1-Xs-1)

Applying the same logic to wave top C proceeding into wave top DG gives another expression for CGs, as F and C are at precisely the same state conditions:

CGs=CGCrelDG+CG2
=5*Ao*(Xs-X2)- 5*Ao*(X2-1)
=-5*Ao*(2*X2-Xs-1)

Equating the two expressions for CGs for the same wave top FC gives two important equations as a conclusion, one for the pressure of superposition, Ps, and the other for the particle velocity of superposition, CGs:

Xs=X1+X2-1 (2.2.1)
CGs=5*Ao*(X1-1)- 5*Ao*(X2-1) (2.2.2)
=CG1+CG2 (2.2.3)

Note that the latter expression for superposition particle velocity reserves the need for a declaration of a positive direction, whereas Eq. 2.2.1 is independent of direction.

In the pipes of an engine, the superposition process is the norm as pressure waves continually pass to and fro. Further, if one places a pressure transducer in the wall of a pipe, it is the superposition pressure-time history which is recorded, not the individual pressures of the rightwards and the leftwards moving pressure waves. This makes it very difficult to interpret recorded pressure-time data in engine ducting. A simple example will make the point. In Sect. 2.1.4 and in Fig. 2.2, there is presented an example of two pressure waves, Pe and Pi, with pressure ratio values of 1.2 and 0.8, respectively. Suppose that they are in a pipe but are the oppositely moving waves just discussed and are in the position of precise superposition. There is a pressure transducer at the point where the wave peaks coincide and it records a superposition pressure, Ps. What will it be and what is the value of the superposition particle velocity, CGs?

The values of Xe, Xi, Ao, CGe, and CGi were 1.0264, 0.9686, 343.1, 45.3, and -53.9, respectively, in terms of a positive direction for the transmission of each wave.

If Pe is regarded as moving rightwards and that is defined as the positive direction, and consequently, Pi as moving leftwards in a negative direction, then Eqs. 2.2.1 and 2.2.2 show:

$$Xs=1.0264-.9686-1=0.995$$

Hence,

$$PRs=Xs^7=0.965 \text{ and } Ps=PRs*Po=0.965*101325=97831 \text{ Pa}$$
$$CGs=45.3+(-(-53.9))=99.2 \text{ m/s}$$

Thus, the pressure transducer in the wall of the pipe would show little of this process, as the summation of the two waves would reveal a trace virtually indistinguishable from the atmospheric line, and exhibit nothing of the virtual doubling of the particle velocity.

The opposite effect takes place when two waves of the same type undergo a superposition process. For example, if wave Pe met another wave Pe going in the other direction and in the plane of the pressure transducer, then:

$$Xs=1.0264+1.0264-1=1.0528 \text{ and } PR=Xs^7=1.0528^7=1.434$$
$$CGs=45.3+(-(+45.3))=0$$

The pressure transducer would show a large compression wave with a pressure ratio of 1.434 and tell nothing of the zero particle velocity at the same spot and time.

This makes the interpretation of exhaust and intake pressure records a difficult business if it is based on experimentation alone. As this is a most important topic, it will be returned to in later sections of this chapter.

2.2.2 Reflection of pressure waves

Oppositely moving pressure waves arise from many sources, but principally from reflections of pressure waves at the boundaries of the inlet or the exhaust duct. Such boundaries include junctions at the open or closed end of a pipe and also from any change in area, gradual or sudden, within a pipe. The reader is already aware of sound reflections, as an echo is a classic example of a closed-ended reflection. All reflections are, by definition, a superposition process, as the reflection proceeds to move oppositely to the incident pressure wave causing that reflection. Fig. 2.4 shows some of the possibilities for reflections in a twin-cylinder engine, where at

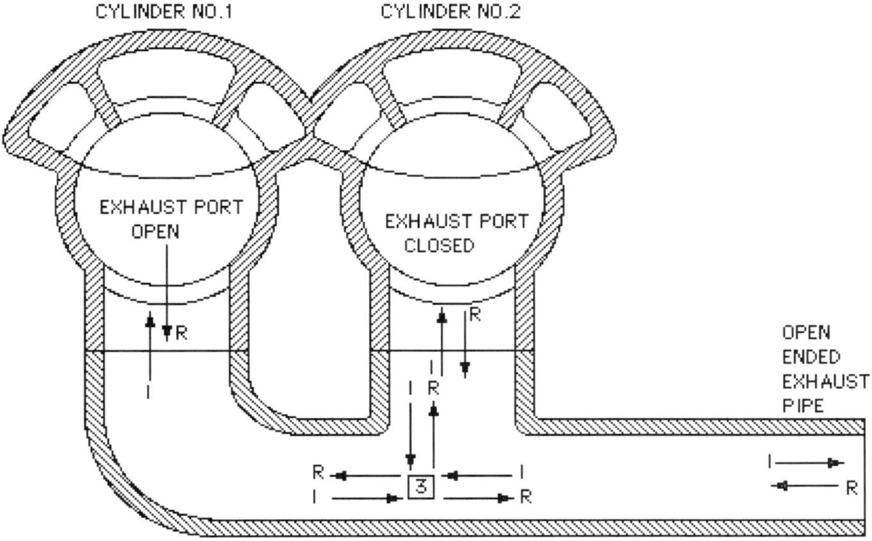

Fig. 2.4 Some reflection possibilities in the exhaust manifold of an engine.

a particular point in time one cylinder is producing an exhaust pulse through an open port and the other cylinder, timed to operate 180º crankshaft later, has a closed exhaust port. All incident and reflected waves are indicated by arrows marked I and R, respectively. At the exhaust port of cylinder no.1, the new exhaust pulse, shown as R, will have an amplitude which will be determined by the size of the reflection incident upon the exhaust port at that time, the area ratio from the port to the pipe, and the cylinder pressure. At the other exhaust port, an echo is taking place because the pipe end is closed. When a pressure wave arrives at the three-way junction, the values transmitted into each of the other branches will be a function of the areas of the pipes and all of the reflections present. When the pressure waves arrive at the open pipe end, a reflection will take place which will be immediately superposed upon the incident pressure pulse. In short, as stated above, all reflection processes are superposition processes.

The resulting pressure-time history is complicated and beyond the memory tracking capability of the human mind. Computers are the type of methodical calculation tool ideal for this pedantic exercise, and the reader will be introduced to their use for this purpose. Before that juncture, it is essential to comprehend the basic effect of each of these reflection mechanisms, as the mathematics of their behavior must be programmed in order to track the progress of all incident and reflected waves.

2.3 Reflections of pressure waves in pipes

A wave arriving at a position where it can be reflected is called the incident wave. In the paragraphs which follow, all incident pressure waves, whether they be compression or expansion waves, will be designated as I, i.e., pressure Pi, pressure ratio PRi, pressure amplitude ratio Xi, particle velocity CGi, acoustic velocity Ai, and propagation velocity CPi. All reflections will be designated as R, i.e., pressure Pr, pressure ratio PRr, pressure amplitude ratio Xr, particle velocity CGr, acoustic velocity Ar, and propagation velocity CPr. All superposition characteristics will be designated as S, i.e., pressure Ps, pressure ratio PRs, pressure amplitude ratio Xs, particle velocity CGs, acoustic velocity As, and propagation velocity CPs.

2.3.1 Reflection of a pressure wave at a closed end in a pipe

When a pressure wave arrives at the plane of a closed end in a pipe, a reflection takes place. This is the classic echo situation, so it is no surprise to discover that the mathematics dictate that the reflected pressure wave is an exact image of the incident wave, but travelling in the opposite direction. The one certain fact available, physically speaking, is that the superposition particle velocity is zero in the plane of the closed end, as is shown in Fig. 2.5(a).

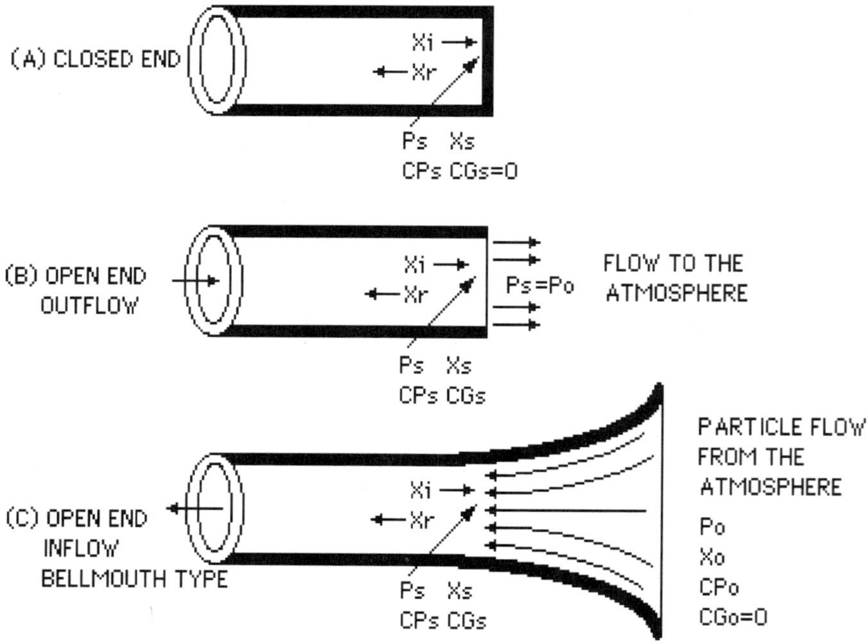

Fig. 2.5 Wave reflection criteria at some typical pipe ends.

From Eq. 2.2.3:

$$CGs=CGi+CGr=0 \tag{2.3.1}$$
$$CGr=-CGi \tag{2.3.2}$$

From Eq. 2.2.2, as CGs=0:

$$Xr=Xi \tag{2.3.3}$$

Hence, Pr=Pi. (2.3.4)

From Eq. 2.2.1, as Xr=Xi:

$$Xs=2*Xi-1 \tag{2.3.5}$$

2.3.2 Reflection of a pressure wave at an open end in a pipe

Here the situation is slightly more complex, in that the fluid flow behavior is different for compression and expansion waves. The first one to be dealt with is that for compression waves, which, by definition, must produce gas particle outflow from the pipe into the atmosphere.

(a) Compression waves

In this case, illustrated in Fig. 2.5(b), the first logical assumption which can be made is that the superposition pressure in the plane of the open end is the atmospheric pressure. The reference pressure, Po, in such a case is the atmospheric pressure. From Eq. 2.2.1, this gives:

$$Xs=Xi+Xr-1=1$$

Hence, Xr=2-Xi. (2.3.6)

For compression waves, as Xi>1 and Xr<1, the reflection is an expansion wave.

From Eq. 2.2.2:

$$CGs=5*Ao*(Xi-1)- 5*Ao*(Xr-1) =5*Ao*(Xi-1-2+Xi+1)$$
$$=2*5*Ao*(Xi-1) =2*CGi \tag{2.3.7}$$
$$=CGi+CGr \tag{2.3.8}$$

This equation is applicable as long as the superposition particle velocity, CGs, does not reach the local sonic velocity, when the flow regime must be analyzed by further equations. The sonic regime at an open pipe end is an unlikely event in most engines, but the student will find the analysis described clearly by Bannister(2.2).

Equally it can be shown that, as CGr=CGi, and as that equation is sign dependent, so the particle flow is in the same direction. This conclusion can also be reached from the earlier conclusion that the reflection is an expansion wave which will impel particles opposite to its direction of propagation. This means that an exhaust pulse arriving at an open end is sending suction reflections back towards the engine which will help to extract exhaust gas particles further down the pipe and away from the engine cylinder. Clearly this is a reflection to be used by the designer so as to aid the scavenging of the cylinder of its exhaust gas.

(b) Expansion waves

This reflection process is connected with inflow and therefore it is necessary to consider the fluid mechanics of the flow into a pipe. Inflow of air in an intake system, which is the normal place to find expansion waves, is usually conducted through a bellmouth ended pipe of the type illustrated in Fig. 2.5(c). This form of pipe end will be discussed in the first instance.

The analysis of gas flow to and from a thermodynamic system, which may also be experiencing heat transfer and work transfer processes, is analyzed by the first law of thermodynamics. The theoretical approach is to be found in any standard textbook on thermodynamics(5.11). In general this is expressed as:

Δ(heat transfer) + Δ(energy entering)
= Δ(system energy) + Δ(energy leaving) + Δ(work transfer)

The first law of thermodynamics for an open system flow from the atmosphere to the superposition station at the full pipe area in Fig. 2.5(c) is as follows:

$$\Delta Q + \Delta Ms*(hs + 0.5*CGs^2) = \Delta E + \Delta Mo*(ho + 0.5*CGo^2) + \Delta W \qquad (2.3.9)$$

If the flow at the instant in question can be presumed to be quasi-steady and steady-state flow without heat transfer, and also to be isentropic, then ΔQ, ΔW, and ΔE are all zero. The mass flow increments must be equal to satisfy the continuity equation. The difference in the enthalpy terms can be evaluated as:

$$hs-ho=cP*(Ts-To) =[g*R/(g-1)]*(Ts-To) =[g*R/(g-1)]*[As^2-Ao^2]/(g*R)$$

Although CGo is virtually zero, Eq. 2.3.9 for air reduces to:

$$CGo^2+5*Ao^2=CGs^2+5*As^2 \qquad (2.3.10)$$

As the flow is isentropic, from Eq. 2.1.14:

$$As/Ao=Xs \qquad (2.3.11)$$

Substituting Eq. 2.3.11 into 2.3.10 and regarding CGo as zero:

$$CGs^2=5*Ao^2*(1-Xs^2) \qquad (2.3.12)$$

From Eq. 2.2.2:

$$CGs=5*Ao*(Xi-Xr) \tag{2.3.13}$$

From Eq. 2.2.1:

$$Xs=Xi+Xr-1 \tag{2.3.14}$$

Bringing these latter three equations together:

$$5*(Xi-Xr)^2=1-(Xi+Xr-1)^2$$

This becomes:

$$3*Xr^2 - Xr*(4*Xi+1) + (3*Xi^2-Xi) = 0 \tag{2.3.15}$$

This is a quadratic equation solution for Xr. Neglecting the negative answer from the square root yields:

$$Xr=\{(4*Xi+1)+\sqrt{[1+20*Xi-20*Xi^2]}\}/6 \tag{2.3.16}$$

This equation shows that the reflection of an expansion wave at a bellmouth end of an intake pipe will be a compression wave. Take as an example the induction Pi pressure wave of Sect. 2.1.4(a), where the expansion wave has a pressure ratio, PRi, of 0.8 and a pressure amplitude ratio, Xi, of 0.9686. Substituting these numbers into Eq. 2.3.16 to determine the magnitude of its reflection at a bellmouth open end gives:

$$Xr=\{(4*0.9686+1)+\sqrt{[1+20*0.9686-20*0.9686*0.9686]}\}/6 =1.0238$$

Thus, $PRr=Xr^7=1.178$.

As predicted, the reflection is a compression wave and, if allowed by the designer to arrive at the intake port while the intake process is still in progress, will push further air charge into the cylinder or crankcase. In the jargon of engine design this effect is called "ramming" and will be discussed thoroughly later in this chapter.

The reflection of expansion waves at the end of plain pipes can be dealt with in a similar theoretical manner, but is more easily handled together with other pipe boundary conditions dealt with in the next section. Should the reader wish to examine the theory separately, Bannister(2.1) describes it in some detail.

2.3.3 Reflection of pressure waves in pipes at a cylinder boundary

This situation is fundamental to all unsteady gas flow generated in the intake or exhaust ducts of a reciprocating ic engine. Fig. 2.6 shows an exhaust port and pipe, but the theory to be generated is equally applicable to an intake port and pipe into

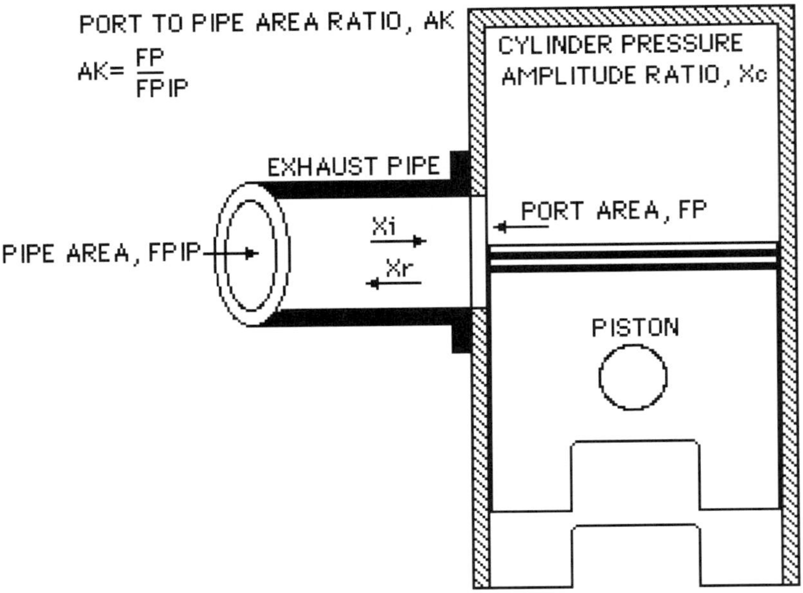

Fig. 2.6 Criteria for unsteady gas inflow or outflow.

a crankcase, a throttled end of an exhaust pipe leading into the atmosphere or silencer box, or the throttled end of an intake pipe from the atmosphere or a silencer box. Anywhere in an unsteady flow regime where a pressure wave in a pipe is incident on a pressure filled space, box, or cylinder, the following method is applicable to determine the magnitude of the reflected pressure wave.

The solution of the gas dynamics of the flow must include separate treatments for subsonic inflow, sonic inflow, subsonic outflow and sonic outflow. The first presentation of the solution for this type of flow is by Wallace and Nassif(2.5) and their basic theory was used in a computer-oriented presentation by Blair and Cahoon(2.6). Probably the most detailed exposition of the derivation of the equations involved is that by McConnell(2.7). A complete description of the solution is out of place in this book, as the theoretical solution would occupy almost a book in itself, but the result of that solution is presented in graphical form in Fig. 2.7 for all subsonic and sonic inflow and outflow for the case of a gas with the properties of air, where g is 1.4. Since exhaust gas has properties not dissimilar to air, and inflow is air by definition, this is the most commonly used representation and solution.

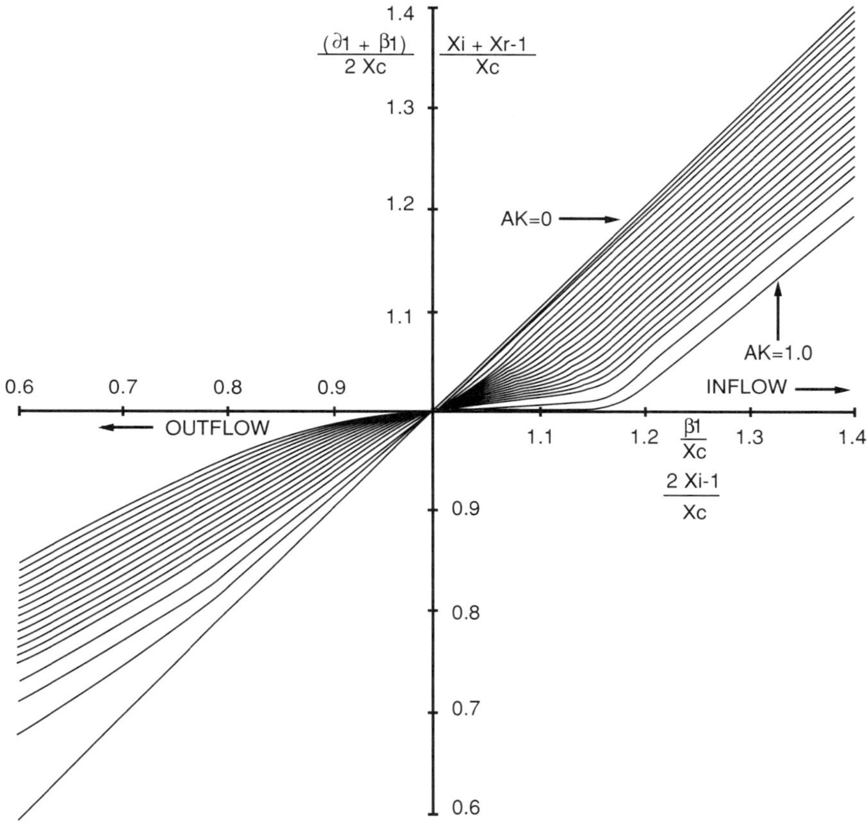

Fig. 2.7 Boundary condition chart for unsteady gas inflow and outflow.

Fig. 2.7 has lines of constant AK expressed as:

AK=(effective port area)/(pipe area)
=CC*FP/FPIP (2.3.17)

where CC is a contraction coefficient of the flow.

In reality, the effective port area is the area of the vena contracta of the flow. On certain occasions the value of CC will be as high as unity and in others it will be as low as 0.6. A discussion regarding typical CC values is presented in Chapter 5.

The horizontal axis has scale values of (2*Xi-1)/Xc, where Xi and Xc are the pressure amplitude ratios of the incident wave in the pipe and the cylinder, respectively. The vertical axis has scale values of (Xi+Xr-1)/Xc, where Xr is the reflected pressure amplitude ratio. The value of Xi+Xr-1 is, of course, Xs, the superposition pressure amplitude ratio at the port, but is the pressure over the full pipe area at the port.

For any given calculation, the values of AK and (2*Xi-1)/Xc will be known, so the value on the Y-axis can be determined for (Xi+Xr-1)/Xc. As the remaining unknown is Xr, that too is capable of being calculated. This is relatively difficult to carry out with precision from a graphical representation, but is rapidly and accurately effected by interpolation from a data bank of numbers stored in a computer file. For the computer calculations which are carried out to illustrate this book, the relevant file is called UFLO and is a 21 by 41 matrix of six-digit numbers. This file is presented in number format in the Computer Program Appendix as Data Bank 2.0. As stated earlier, this approach was presented by Blair and Cahoon(2.6) and the mechanism of solution for the UFLO file, and of its use by interpolation in a computer program, are both described in that paper.

So that the reader may gain some understanding of the nature of outflow and inflow in an unsteady gas flow regime, a computer program has been prepared and presented as Prog.2.3, CYLINDER-PIPE FLOW. This program indexes the UFLO data bank and interpolates the reflected pressure wave amplitude and properties in the manner described above. The data used to illustrate the flow behavior is shown in Fig. 2.8 for six calculations of inflow and outflow. The pressure ratio of the incident pressure wave is PRi and the resulting calculated pressure ratio of the pulse produced in the pipe is PRr. The superposition pressure wave in the pipe at the port, and acting across the full pipe area, is PRs. The common data for all of the calculations are given in Fig. 2.8.

(a) Outflow

For the three outflow cases the cylinder has a pressure of 2.0 bar, or 1.97 atm. The pipe reference temperature is calculated from an isentropic pressure drop from the cylinder pressure to the pipe reference temperature, in this case 364°C. The first outflow case shows the pipe to be undisturbed by any incident reflection as PRi=1, and a superposition pressure ratio of 1.738 is created. The mass flow rate is 0.1208 kg/s and the particle velocity is 207.7 m/s.

The second case has an opposing compression pulse as PRi=1.2, and although this produces a higher superposition pressure ratio of 1.907, the outgoing reflected pulse pressure ratio is lower at 1.608. The result is a reduced mass flow of 0.0688 kg/s and particle velocity of 110.7 m/s. Remember that if there is a pressure transducer in the pipe at that point, it would be the superposition pressure which it would record and the interpreter of the pressure trace could be duped into thinking that an apparently larger amplitude exhaust pulse implied increased mass flow.

The last case of outflow shows the presence of a suction reflection pulse of PRi=0.8. This enhances the mass flow rate to 0.1479 kg/s and the gas particle velocity to 291.9 m/s. The superposition pressure ratio is 1.432, the lowest value of the three cases discussed, reinforcing the viewpoint regarding pressure record interpretation in the previous paragraph and section. The actual exhaust pulse

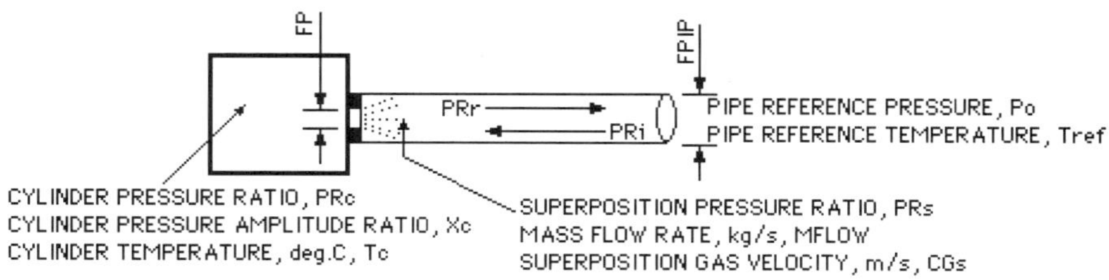

COMMON DATA: Po=1 atm, FP=25 mm dia., FPIP=30 mm dia., AIR g=1.4								
PRc	Tc	Tref	PRi	PRr	PRs	MFLOW	CGs	FLOW TYPE
1.97	500	364	1.0	1.738	1.738	0.1208	207.7	OUTFLOW
1.97	500	364	1.2	1.608	1.907	0.0688	110.7	OUTFLOW
1.97	500	364	0.8	1.759	1.432	0.1479	291.9	OUTFLOW
0.59	50	30	1.0	0.653	0.653	-0.0626	-103.0	INFLOW
0.59	50	30	0.8	0.750	0.594	-0.0088	-15.5	INFLOW
0.59	50	30	1.2	0.661	0.802	-0.1029	-146.3	INFLOW

Fig. 2.8 Calculation of cylinder to pipe boundary conditions with Prog.2.3.

pressure ratio is 1.759, the highest of the cases discussed, but pressure transducers do not provide information on leftwards or rightwards moving waves only!

In Sect. 2.3.2 there was a comment that suction reflections created by exhaust pressure wave reflection at the open end of a pipe could be used to assist with extraction of exhaust gas from the engine cylinder. The last example illustrates that this mechanism is indeed effective.

(b) Inflow

In this set of three examples the physical port and pipe geometry is as before but the cylinder pressure is set at 0.6 bar or 0.59 atm. The pipe reference temperature, Tref, is now an input parameter because the flow is into the cylinder and the cylinder temperature has no influence on the instantaneous pipe flow situation.

The first case has an "undisturbed" incident pressure wave of PRi=1.0. Here a superposition, and reflected, pressure wave is created with pressure ratios, PRs and PRr, equal to 0.653. The particle velocity is inwards at -103 m/s and the mass flow rate is -0.0626 kg/s.

The second case has an opposing suction wave present as an incident pulse at PRi=0.8 and this reduces the mass flow rate considerably to -0.0088 kg/s and the particle velocity to -15.5 m/s. The superposition pressure ratio is deeper at PRs=0.594, although the actual intake pulse created is weaker by being closer to the reference pressure at PRr=0.750.

The last of the three inflow examples shows what happens when a "ramming" wave is present during inflow. The incident wave pressure ratio is set at PRi=1.2 and the result is an increased inflow mass flow rate of -0.1029 kg/s and a higher particle velocity of -146.3 m/s. The superposition pressure is almost up to the reference pressure at PRs=0.802, but the actual intake pulse created is stronger at PRr=0.661. The potential effectiveness of ramming during an induction process was indicated in Sect. 2.3.2 and these examples give an indication of the order of magnitude of the effect which can be expected.

Once again, the difficulty of interpreting pressure transducer records is highlighted.

2.3.4 Reflection of pressure waves in a pipe at a sudden area change

It is quite common to find a sudden area change within a pipe or duct attached to an engine. This will produce a change in amplitude of the transmitted pulse beyond the area change and also cause a wave reflection from it. Such sudden area changes are sketched in Fig. 2.9, and it can be seen that the pipe area can contract or expand at the junction. In each case, the incident wave at the sudden area change is depicted as propagating rightwards, with the pipe nomenclature as 1 for the wave arrival pipe, with "i" signifying the incident pulse, "r" the reflected pulse and "s" the superposition condition. The X value shown is the conventional symbolism for pressure amplitude ratio. For example, at any instant, the incident pressure pulses at the junction are Xi1 and Xi2 which, depending on the areas F1 and F2, will give rise to reflected pulses, Xr1 and Xr2.

It was Benson(2.4) who suggested a simple theoretical solution for such junctions. He assumed that the superposition pressure at the plane of the junction was the same in both pipes at the instant of superposition. Such a simple junction model will clearly have its limitations, but it is the author's experience that it is remarkably effective in practice, particularly if the area ratio changes, FR, are in the band, $0.17<FR<6$.

From Benson, Xs1=Xs2. (2.3.18)

Consequently, from Eq. 2.2.1:

Xi1+Xr1-1= Xi2+Xr2-1 (2.3.19)

From the continuity equation, equating the mass flow rate on either side of the junction:

mass flow rate=density*area*particle velocity
Ds1*F1*CGs1=Ds2*F2*CGs2 (2.3.20)

Using the theory of Eqs. 2.1.17 and 2.2.2, where the reference conditions are Po, To and Do, Eq. 2.3.20 becomes:

$Do*Xs1^5*F1*5*Ao*(Xi1-Xr1)=Do*Xs2^5*F2*5*Ao*(Xr2-Xi2)$

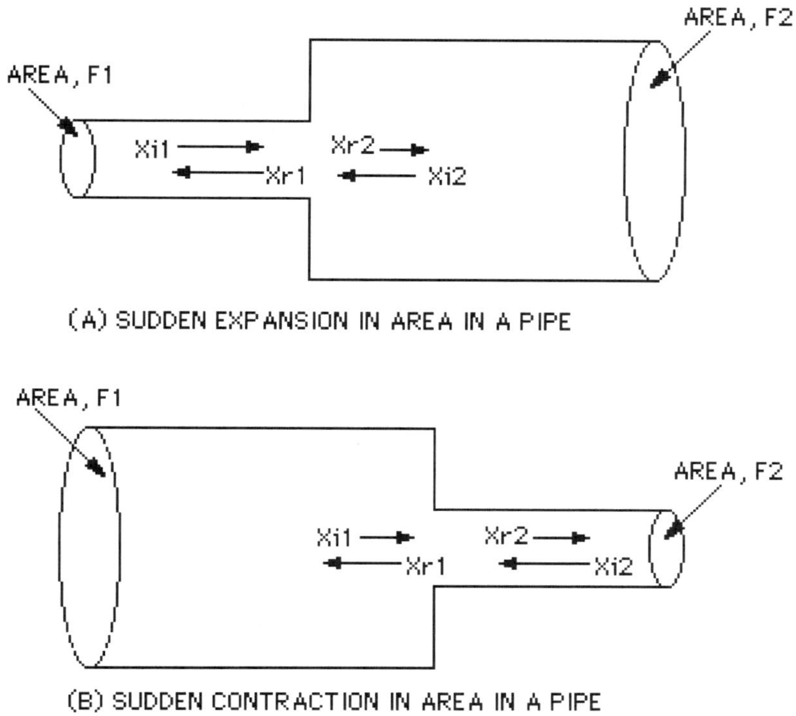

Fig. 2.9 Sudden contractions and expansions in area in a pipe.

As Xs1 equals Xs2, this reduces to:

F1*(Xi1-Xr1)=F2*(Xr2-Xi2) (2.3.21)

Joining Eqs. 2.3.21 and 2.3.19, and eliminating each of the unknowns in turn, Xr1 or Xr2, and letting FR=F2/F1:

Xr1=[(1-FR)*Xi1+2*FR*Xi2]/(1+FR) (2.3.22)
Xr2=[2*Xi1-Xi2*(1-FR)]/(1+FR) (2.3.23)

To get a basic understanding of the results of reflections of compression and expansion waves at sudden enlargements and contractions in pipe area, consider an example using the two pressure waves, PRe and PRi, previously used in Sect. 2.1.4.

PRe is a compression wave of pressure ratio 1.2 and PRi is an expansion wave of pressure ratio 0.8. Such pressure ratios are shown to give pressure amplitude ratios of 1.02639 and 0.9686, respectively. Each of these waves in turn will be used as data for Xi1 arriving in pipe 1 at a junction with pipe 2, where the area ratio will be either halved for a contraction or doubled for an enlargement to the pipe area. In each case the incident pressure amplitude ratio in pipe 2, Xi2, will be taken as unity, which means that the incident pressure wave in pipe 1 is facing undisturbed conditions in pipe 2.

(a) An enlargement, FR=2, for an incident compression wave Xi1=1.02639

From Eqs. 2.3.22 and 2.3.23, Xr1=0.9912 and Xr2=1.01759.

Hence, the pressure ratios, PRr1 and PRr2, of the reflected waves are:

PRr1=0.940 and PRr2=1.130

The sudden enlargement behaves like a slightly less effective "open end," as a completely open ended pipe from Eq. 2.3.6 would have given a reflected pressure ratio of 0.829 instead of 0.940. The onward transmitted pressure wave in pipe 2 is one of compression, of reduced pressure ratio 1.13.

(b) An enlargement, FR=2, for an incident expansion wave Xi1=0.9686

From Eqs. 2.3.22 and 2.3.23, Xr1=1.0105 and Xr2=0.97908.

Hence, the pressure ratios, PRr1 and PRr2, of the reflected waves are:

PRr1=1.076 and PRr2=0.862

As above, the sudden enlargement behaves as a slightly less effective "open end" because a completely bellmouth open end to a pipe in Sect. 2.3.2 was shown to produce a reflected pressure ratio of 1.178, instead of 1.076 found here. The onward transmitted pressure wave in pipe 2 is one of expansion, of diminished pressure ratio 0.862.

(c) A contraction, FR=0.5, for an incident compression wave Xi1=1.02639

From Eqs. 2.3.22 and 2.3.23, Xr1=1.0088 and Xr2=1.0352.

Hence, the pressure ratios, PRr1 and PRr2, of the reflected waves are:

PRr1=1.063 and PRr2=1.274

The sudden contraction behaves like a partially closed end, sending back a partial "echo" of the incident pulse. The onward transmitted pressure wave is also one of compression, but of increased pressure ratio 1.274.

(d) A contraction, FR=0.5, for an incident expansion wave Xi1=0.9686

From Eqs. 2.3.22 and 2.3.23, Xr1=0.9895 and Xr2=0.9582.

Hence, the pressure ratios, PRr1 and PRr2, of the reflected waves are:

PRr1=0.929 and PRr2=0.741

As in (c), the sudden contraction behaves like a partially closed end, sending back a partial "echo" of the incident pulse. The onward transmitted pressure wave is also one of expansion, with an increased expansion pressure ratio of 0.741.

The theoretical presentation here, due to Benson(2.4), is clearly too simple to be completely accurate in all circumstances, and the subject was studied recently by Richardson(2.20), who produced some alternative theoretical treatments for these sudden area changes. While the new treatment by Richardson(2.20) is more accurate, it is also more complex.

2.3.5 Reflections of pressure waves at branches in a pipe

The simple theoretical treatment for this situation was also suggested by Benson(2.4) and in precisely the same form as for the sudden area changes found in the previous section. A sketch of a typical branch is shown in Fig. 2.10. The sign convention for the branch theory is that inwards propagation of a pressure wave towards the branch will be regarded as "positive." Benson(2.4) postulates that the superposition pressure at the junction, at any instant of wave incidence and reflection, can be regarded as a constant. Therefore, the theoretical solution involves expansion of Eqs. 2.3.19 and 2.3.21 to deal with the superposition state and mass flow rate of the extra pipe 3 at the junction, thus:

$$Xs1=Xi1+Xr1-1=Xs2=Xi2+Xr2-1=Xs3=Xi3+Xr3-1 \qquad (2.3.24)$$

The net mass flow rate at the junction is zero:

$$F1*(Xi1-Xr1)+F2*(Xi2-Xr2)+F3*(Xi3-Xr3)=0 \qquad (2.3.25)$$

There are three equations to solve for the three unknowns, Xr1, Xr2 and Xr3. It is presumed that in the course of any computer calculation one knows the values of all incident pressure waves from one calculation time step to another.

The solution of the above simultaneous equations gives:

$$FT=F1+F2+F3$$
$$Xr1=[(2*F2*Xi2)+(2*F3*Xi3)+((F1-F2-F3)*Xi1)]/FT \qquad (2.3.26)$$
$$Xr2=[(2*F1*Xi1)+(2*F3*Xi3)+((F2-F3-F1)*Xi2)]/FT \qquad (2.3.27)$$
$$Xr3=[(2*F1*Xi1)+(2*F2*Xi2)+((F3-F2-F1)*Xi3)]/FT \qquad (2.3.28)$$

where FT is the sum of the pipe areas.

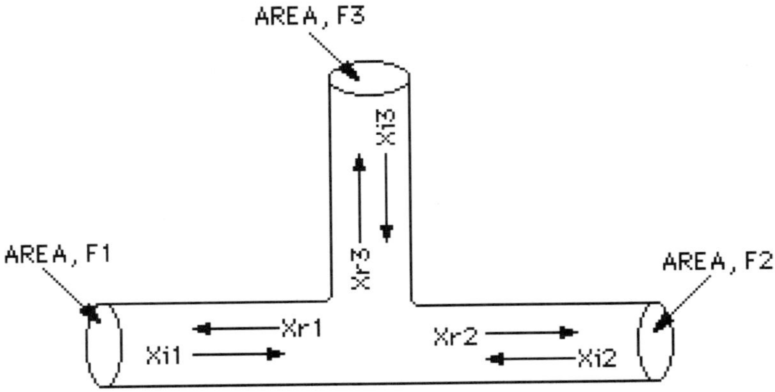

Fig. 2.10 Pressure waves at a branch in a pipe.

Perhaps not surprisingly, the branched pipe can act as either a contraction of area to the flow or an enlargement of area to the gas flow. Consider these two cases where all of the pipes are of equal area:

(a) A compression wave is coming down to the branch in pipe 1 and all other conditions in the other branches are "undisturbed." The compression wave has a pressure ratio of PRi1=1.2, or Xi1=1.02639. The result of the calculation is:

Xr1=0.9911, Xr2=Xr3=1.01759, PRr1=0.940 and PRr2=PRr3=1.13

As far as pipe 1 is concerned the result is exactly the same as that for the 2:1 enlargement in area in the previous section. In the branch, the incident wave divides evenly between the other two pipes.

(b) Compression waves of pressure ratio 1.2 are arriving as incident pulses in pipes 1 and 2. Pipe 3 is undisturbed as PRi3=1.0. Now the branch behaves as a 2:1 contraction to this general flow, for the solution of Eqs. 2.3.26-28 shows:

Xr1=Xr2=1.0088, Xr3=1.0352 and PRr1=PRr2=1.0632 while PRr3=1.274

These numbers are already familiar as computed data for pressure waves of identical amplitude at the 2:1 contraction discussed in the previous section.

When one has dissimilar areas of pipes and a mixture of compression and expansion waves incident upon the branch, the situation becomes much more difficult to interpret by the human mind. At that point the programming of the mathematics into a computer will leave the designer's mind free to concentrate more upon the relevance of the information calculated and less on the arithmetic tedium of acquiring that data.

It is also physically obvious that the angle between the several branches must play some role in determining the transmitted and reflected wave amplitudes. This subject was studied most recently by Bingham(2.18) and Blair(2.19) at QUB. While the branch angles do have an influence on wave amplitudes, it is not as great as might be imagined and can be neglected in most calculations. The discussions of Bingham(2.19) and Blair(2.19) deserve further study for those who wish to achieve greater accuracy for all such calculations.

2.4 Computational methods for unsteady gas flow

The original calculation method for unsteady gas flow is known as the method of characteristics and is usually attributed to De Haller(2.8). As this was in an era before the advent of the digital computer, this was a graphical method. It is not well suited for computational methods, although Jones(2.9) did succeed in transferring the mathematical theory of the graphical solution to a digital computer. Since the 1960's the most popular computational method has been that based on Riemann variables. It was first suggested as applicable for unsteady gas flow in ic engine ducting in a paper by Benson(2.10), and is based on earlier work by Hartree(2.11) and Rudinger(2.3). Since that time there have been further mathematical methods pursued in the search for a more rapid or more accurate solution of the basic theory. Such techniques are variations of finite-difference theory(2.12) and the Lax-Wendroff approach(2.13).

2.4.1 Riemann variable calculation of flow in a pipe

(a) ∂ and ß characteristics

As already discussed in Sect. 2.1.3, a pressure wave is propagated along a pipe in a relatively "pure" manner with little deterioration of its signal due to either friction or other viscosity effects. The Riemann variable technique is based on this knowledge and establishes dimensionless values of ∂ and ß to denote the rightward and leftward characteristics, respectively. The values of ∂ and ß are called the Riemann variables and appear from the mathematical solution of the equations of continuity and momentum applied to the unsteady flow regime. The result of this analysis is that the collection of thermodynamic terms, represented by ∂ or ß, are constants for a point on a propagating pressure wave in a parallel pipe, as illustrated in Fig. 2.11. They are discussed more fully by Rudinger(2.3) and by Benson(2.4). The values of ∂ and ß are as follows:

$$\partial = a + ((g-1)/2)*u \quad \text{-for air } \partial = a + u/5 \qquad (2.4.1)$$
$$ß = a - ((g-1)/2)*u \quad \text{-for air } ß = a - u/5 \qquad (2.4.2)$$

The values of a and u are the dimensionless values of the local acoustic and particle velocities of a wave point, where:

$$a = A/Ao \qquad (2.4.3)$$
$$u = CGs/Ao \qquad (2.4.4)$$

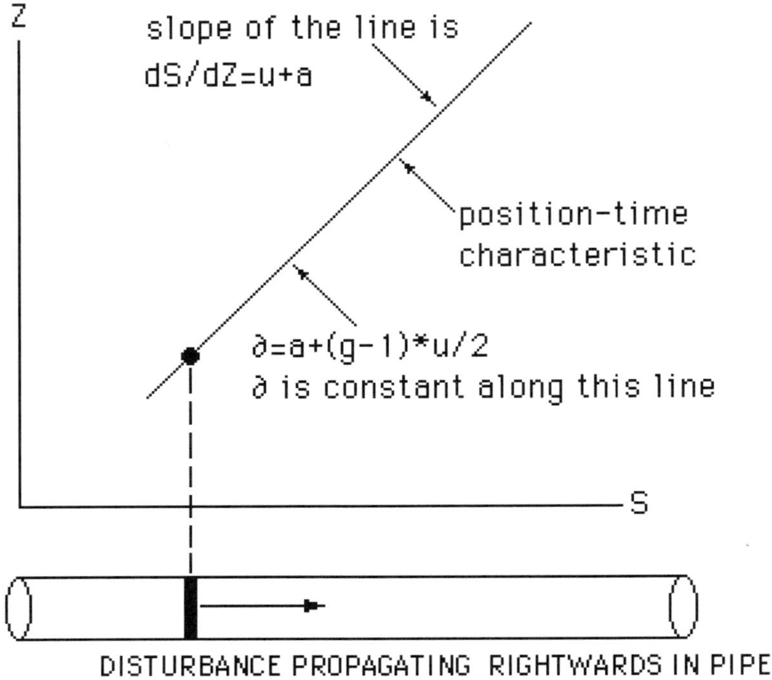

Fig. 2.11 Position-time characteristic.

As the pressure amplitude ratio on a wave point is evaluated as,

Xs=A/Ao,

and the particle velocity as,

CGs=(2/(g-1))*Ao*(Xr-Xl), or for air CGs=5*Ao*(Xr-Xl).

Then Eqs. 2.4.1 and 2.4.2 become:

∂=Xr+Xr-1=2*Xr-1 (2.4.5)
ß=Xl+Xl-1=2*Xl-1 (2.4.6)

The values of Xr and Xl are the rightward and leftward values of pressure amplitude ratio which are superposed at the same instant in time and space as the

∂ and ß characteristics. Consequently, the superposition pressure and particle velocity can be deduced from the superposition pressure amplitude ratio Xs:

$$Xs=Xr+Xl-1=(\partial+ß)/2 \quad (2.4.7)$$
$$CGs=(2/(g-1))*Ao*(Xr-Xl)=Ao*(\partial-ß)/(g-1) \quad (2.4.8)$$

For air, where the value of g is 1.4, this reduces to:

$$CGs=5*Ao*(Xr-Xl)=2.5*Ao*(\partial-ß)$$

The density during superposition, Ds, originally derived in Eq. 2.1.16, can be evaluated by the Riemann variables as follows:

$$Ds=Do*Xs^{(2/g-1)}$$
$$=Do*[(\partial+ß)/2]^{(2/g-1)} \quad (2.4.9)$$

For air, where the value of g is 1.4, this reduces to:

$$Ds=Do*[(\partial+ß)/2]^{5}$$

Consequently, the mass flow rate, MGs, at any point in a pipe of area, FPs, becomes:

$$MGs=Ds*CGs*Fs=\{Do*[(\partial+ß)/2]^{(2/g-1)}\}*\{Ao*(\partial-ß)/(g-1)\}*FPs \quad (2.4.10)$$

For air, where the value of g is 1.4, this reduces to:

$$MGs=\{Do*[(\partial+ß)/2]^{5}\}*\{2.5*Ao*(\partial-ß)\}*FPs$$

The characteristics can be plotted in the time-distance plane, just as for a graphical solution to the method of characteristics, and this is illustrated in Fig. 2.11. The distance and time values are also declared as dimensionless parameters, S and Z, respectively, where:

$$S=s/L \quad (2.4.11)$$
$$Z=Ao*t/L \quad (2.4.12)$$

The values of s and t are distance and time, while L is the length of a calculation segment in the distance plane, and is called the mesh length. As can be seen in Fig. 2.11, the slope of the ∂ characteristic is given by:

$$\Delta Z/\Delta S=1/(a+|u|) \quad (2.4.13)$$

The use of the absolute value of u, i.e., |u|, allows for the calculation of the slope of the ß characteristic in the leftwards direction.

(b) The mesh layout in a pipe

The pipe is that segment of a duct in an engine between two boundaries where a reflection process is taking place at both ends. For example, it could be straight, tapered, or partially straight and partially tapered, between an exhaust port and an open end to the atmosphere. At each end reflections take place, but in the pipe the ∂ and ß characteristics are tracked from mesh to mesh in individual time steps. The layout in a pipe for such a mesh calculation is shown in Fig. 2.12.

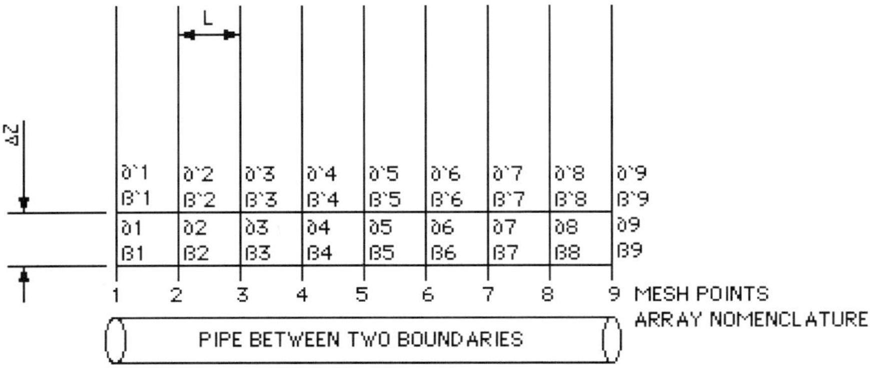

Fig. 2.12 Mesh calculation of Riemann variables between boundaries.

For pictorial reasons, nine meshes are displayed. Obviously, the number of meshes is linked to the mesh length, L, and the total pipe length. To put a number on a typical mesh length found to give acceptable accuracy in two-stroke engines, a value of L between 20 and 40 mm would be sensible. For four-stroke engines or medium-speed two-stroke diesel engines this value could be between 30 and 60 mm. In fact, the mesh length is usually adjusted to give a time step of between 1 and 2 crankshaft degrees in an application for any engine.

In a computer calculation, the ∂ and ß values are stored in arrays. For the example cited in Fig. 2.12, this would give ∂(j) and ß(j), for j=1 to 9. At the start of any time step, all of these values would be known. The objective is to calculate the new values of ∂ and ß at each mesh point by the end of the time step, ΔZ. The new values of ∂ and ß at each mesh point at the end of the time step are shown in Fig. 2.12 as ∂‘(j) and ß‘(j), for j=1 to 9.

(c) Reflection of characteristics at the pipe ends

However, not all values are given simply from the mesh calculation. There are two which emanate from the reflection of the appropriate characteristic at each end of the pipe. For example, ∂`9 is the new value of the ∂ characteristic arriving at the extreme right-hand boundary. Therefore, the value of ß`9 will be dependent on the type of boundary involved. For instance, assume that the right-hand end of the pipe is a plain open end to the atmosphere. If the value of the ∂`9 characteristic is greater than unity it means that it is a wave of compression and the solution for ß`9 is based on Eq. 2.3.6:

Xs=1

From Eq. 2.4.7, Xs=(∂+ß)/2.

Hence, the solution for ß`9 is given by:

Xs=1=(∂`9+ß`9)/2
ß`9=2-∂`9

At the left-hand end of the pipe, the value of ß`1 is known but the value of ∂`1 will depend on the type of boundary encountered there. Suppose that it is an exhaust port open to a cylinder. In this case the boundary condition chart UFLO given in Fig. 2.7, and contained in the computer file of the same name, must be used for evaluation of ∂`1. It will be observed in Fig. 2.7 that the axes are also labelled in terms of ∂ and ß values. The x-axis has values of ß1/Xc, where ß1 is the value of ß`1 in question. The methodology of solution is exactly as discussed in Sect. 2.3.3, in terms of the port-to-pipe AK value and the cylinder pressure amplitude ratio, Xc. The y-axis is also plotted in terms of (∂`1+ß`1)/2*Xc so that the parameter can be evaluated by interpolation, and ∂`1 determined.

For any particular type of boundary (all of the relevant ones have been discussed in Sect. 2.3), the equations have already been presented for reduction to ∂ and ß parameters instead of the X values previously employed.

(d) Interpolation of ∂ and ß values at each mesh point

The selection of the time step, ΔZ, is a vital factor at each stage of the calculation. It is not possible to select a time step which will allow all ∂ and ß characteristics to perfectly intersect the mesh-time point. Therefore, as illustrated in Fig. 2.13 for a ∂ characteristic between the j and the j+1 mesh points, it will be necessary to linearly interpolate a value of ∂w between the ∂(j) and ∂(j+1) values at time Z1, which would have the precise slope to intersect the j+1 position at time Z2. In this case, for a "perfect" transmission of ∂ characteristics, ∂`(j+1) would equal ∂w.

First, the time step ΔZ must be determined. This is done by searching through all of the mesh points for the fastest wave propagation characteristic. In this manner the value of ΔZ/ΔS, or 1/(a+|u|), is determined at each mesh position and the minimum value found for ΔZ is used, as that is the fastest wave propagation characteristic. This is known as the "stability criterion" in the paper by Courant, et al(2.13). This allows all wave point values to be determined by interpolation, for it cannot be done by extrapolation.

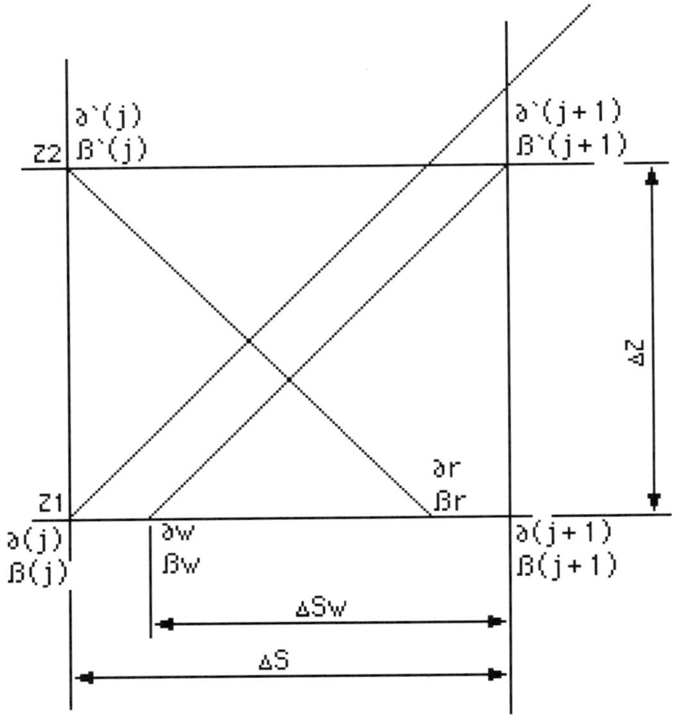

Fig. 2.13 Interpolation procedure between meshes.

After the setting of ΔZ at each time step, the interpolation procedure already outlined is followed for all ∂ and ß characteristics at each mesh point. The values of ∂w and ßw are shown below.

$$∂w=∂(j+1)-\Delta Sw*[∂(j+1)-∂(j)]/\Delta S \quad (2.4.14)$$
$$ßw=ß(j+1)-\Delta Sw*[ß(j+1)-ß(j)]/\Delta S \quad (2.4.15)$$

The $\Delta Sw/\Delta Z$ value is deduced from the prevailing "stability criterion" giving ΔZ:

$$\Delta Sw/\Delta Z=[(g+1)*∂w-(3-g)*ßw]/[2*(g-1)] \quad (2.4.16)$$

The equivalent values for all ß characteristics are then determined for all such leftward running characteristics.

At this point the new value of ∂`(j+1) can be found as:

∂`(j+1)= ∂w+d∂ (2.4.17)

The value of d∂ is zero if the flow is frictionless, without heat transfer, without any entropy changes, or without area change from the j to the j+1 mesh point. Such a flow would be described as homentropic. In non-homentropic flow, and all real gas flows, some or all of these effects would be present. To predict such changes another mesh procedure, called "path lines," is necessary to track particles. It is possible to evaluate, with reasonable accuracy, the unsteady flow behavior in most engines using a homentropic calculation approach. This will be used in this book, for it would take a book by itself to explain the non-homentropic method, as an examination of Benson's excellent treatise(2.4) will demonstrate. That the homentropic method is sufficiently accurate for most engineering purposes is demonstrated in many technical papers(2.6)(2.7) and (2.14)-(2.17).

Returning to the determination of d∂ in homentropic flow with friction and area change, the following expression for d∂ can be derived(2.10):

d∂=-(g-1)*dZ*{[(f*L/Ø)*(u^2)*(u/|u|)*(1-(g-1)*u/a)]
-[(a*u*dF/F)/dS]}/2 (2.4.18)

The a and u values are those for the ∂w and ßw characteristics and can be determined using Eqs. 2.4.1-2.4.4.

For the leftward running ß characteristics, the value of dß is given by(2.10):

dß=(g-1)*dZ*{[(f*L/Ø)*(u^2)*(u/|u|)*(1+(g-1)*u/a)]
-[(a*u*dF/F)/dS]}/2 (2.4.19)

Consequently, ß'(j) can be found from the equivalent exercise for the ß characteristics from the determination of the ßr values at each mesh point:

ß'(j)=ßr+dß (2.4.20)

In both of the above expressions for d∂ and dß, dF and F are the change of area and the mean area, respectively, between mesh points j and j+1. The value denoted by f in the same equations is a friction factor. Blair and Goulburn(2.14) showed that a value of 0.004, a value consistent with experimental data from the steady flow of air in smooth pipes, would be applicable to unsteady gas flow. The value denoted by Ø is the mean pipe diameter between the j and the j+1 mesh points.

2.4.2 The computation of "cylinder" state conditions at each time step

During the previous discussion of reflection at cylinder boundaries, the prediction of the reflected Xr or ∂ characteristic was shown to depend on the value of port-to-pipe area ratio, AK, and the cylinder pressure amplitude ratio, Xc. The word "cylinder" is used broadly, for it could just as well be the crankcase of the two-stroke engine during the induction phase of the operation, with the pipe in question being the intake pipe. Due to gas inflow or outflow from time step to time step, it is clear that the cylinder pressure will alter, and it is imperative that the calculation method be capable of predicting the new cylinder state conditions at the beginning of any calculation step. That this is a complex situation is made clear in the sketch of a two-stroke engine in Fig. 2.14. On the left is a sketch of the engine during the scavenge phase. The sketch on the right depicts the two thermodynamic systems involved which are separated by the piston, with the three pipes in which unsteady gas flow takes place.

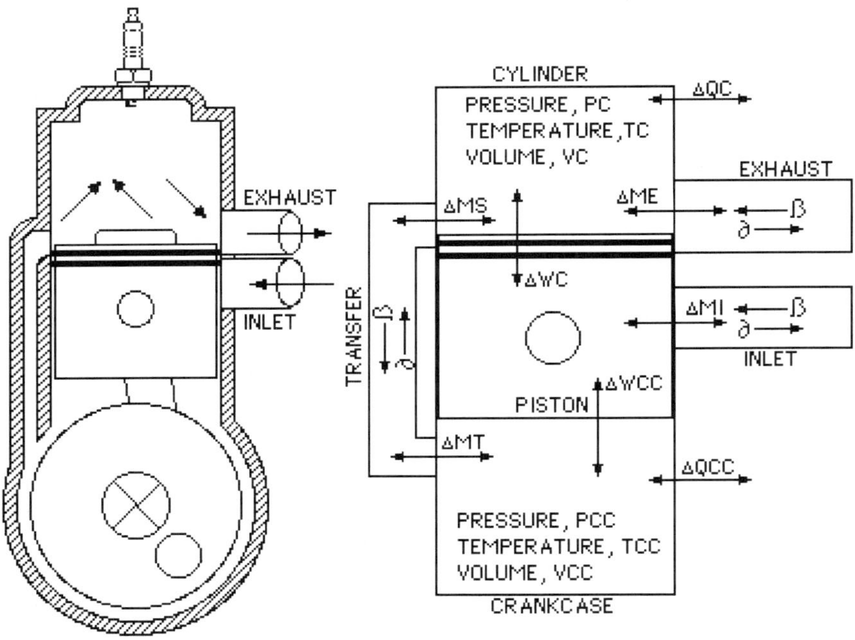

Fig. 2.14 Gas-dynamic and thermodynamic systems in the two-stroke engine.

The first thermodynamic system is the cylinder, which has two possibilities for thermodynamic analysis: (1) a closed system period from exhaust closing through compression, combustion, and expansion to exhaust opening, and (2) an open system period from exhaust opening to exhaust closing. The open system period is the one which will be analyzed here, for the closed system period will be dealt with specifically in the chapters on combustion and engine modelling, Chapters 4 and 5.

The second thermodynamic system is the crankcase, which is an open system at all times because the transfer duct is (almost) always open to the crankcase.

The mathematical analysis of these systems is carried out by using the first law of thermodynamics on the assumption that the flow is quasi-steady, i.e., it behaves as steady flow during the period of time for the analysis, which will be the time step, ΔZ, controlled by the stability criterion of the unsteady gas flow calculation. From Eq. 2.4.13, the actual value of time involved in the step, Δt, can be calculated.

The cylinder, as with any open system, is analyzed to find the effect of inflow and outflow, heat transfer, and work upon the cylinder state conditions during the time interval involved. Applying the first law of thermodynamics to the cylinder during the open cycle period reveals:

$$\Delta QC+\Delta HS=\Delta EC+\Delta HE+\Delta WC$$
$$\Delta QC+\Delta MS*(hS+0.5*CS^2)=\Delta EC+\Delta ME*(hE+0.5*CE^2)+\Delta WC \qquad (2.4.21)$$

where the system energy term, ΔEC, is given by:

$$\Delta EC=\Delta[UC]=\Delta[MC*cV*TC] \qquad (2.4.22)$$

Each of these terms in turn deserves some explanation.

The heat transfer, ΔQC, will be heat convected into, or away from, the cylinder. This is given by a separate equation, where HC is a convection heat transfer coefficient, FC is the instantaneous cylinder and piston surface area and Tcw is a representative cylinder wall temperature:

$$\Delta QC=HC*FC*(Tcw-TC1)*\Delta t \qquad (2.4.23)$$

This value has little influence on the calculation of the open cycle behavior, but becomes much more significant during the closed cycle phase.

Because the cylinder state conditions are going to change from known to yet uncalculated values, i.e., from PC1 to PC2, TC1 to TC2, the volume is changing from VC1 to VC2 and the cylinder mass from MC1 to MC2, the following thermodynamic equations of state and continuity will apply:

$$\text{State: } PC1*VC1=MC1*R*TC1 \qquad (2.4.24)$$
$$PC2*VC2=MC2*R*TC2 \qquad (2.4.25)$$

where R is the gas constant.

During any calculation step the cylinder volumes, VC1 and VC2, are known from the engine geometry.

$$\text{Continuity: } \Delta MC=MC2-MC1=\Delta MS-\Delta ME \qquad (2.4.26)$$

Now, the values of ΔMS, ΔME, TS, TE, hS, hE, CS, CE, the mass flow increments, temperature, specific enthalpy and particle velocity for the scavenge flow and exhaust flow, respectively, will be available from the unsteady gas dynamic calculation at the first mesh points in the transfer and exhaust ducts. These are the superposition values at these meshes and are determined in the same manner as described in Sect. 2.3.3 and preceding sections on superposition behavior. As MC1 and ΔMS and ΔME are already evaluated, the new cylinder mass, MC2, can be calculated from Eq. 2.4.26. Only the enthalpy values have not been discussed previously, and these are evaluated thermodynamically, where cV and cP are the specific heats at constant volume and constant pressure, respectively.

$$hS=cP*TS \qquad (2.4.27)$$
$$hE=cP*TE \qquad (2.4.28)$$

The change of system energy term, given in Eq. 2.4.22, contains the value of UC, the internal energy of the system and this is evaluated in the usual way:

$$UC=MC*cV*TC \qquad (2.4.29)$$

The change of this term during the period of the process, Δt, gives:

$$\Delta EC=MC2*cV*TC2-MC1*cV*TC1 \qquad (2.4.30)$$

The work term on, or by, the piston, ΔWC, can be found by standard theory:

$$\Delta WC=\int P*dV=PC1*(VC2-VC1) \qquad (2.4.31)$$

Upon substituting the information obtained from the unsteady gas flow calculation and that contained in Eqs. 2.4.23-31 into Eq. 2.4.21, it is possible to solve for either of the remaining two unknown values, PC2 or TC2, thus:

$$TC2=[\Delta QC+\Delta HS-\Delta HE+MC1*cV*TC1-PC1*(VC2-VC1)]/(MC2*cV) \qquad (2.4.32)$$
$$PC2=MC2*R*TC2/VC2 \qquad (2.4.33)$$

where enthalpy of outflow, $\Delta HE=\Delta ME*(cP*TE+0.5*CE^2)$ (2.4.34)
where enthalpy of inflow, $\Delta HS= \Delta MS*(cP*TE+0.5*CS^2)$ (2.4.35)

For the second thermodynamic system in the crankcase the same analytic approach is taken, with the inlet system regarded as inflow and the crankcase end of the transfer duct viewed as outflow. The same reasoning used above for the cylinder is followed for the crankcase.The result of that analysis is a similar set of equations with the appropriate symbolism for the crankcase as given in Fig. 2.14, as follows:

$$MCC2-MCC1=\Delta MI-\Delta MT \qquad (2.4.36)$$
$$TCC2=[\Delta QCC+\Delta HI-\Delta HT+MCC1*cV*TCC1-PCC1*(VCC2-VCC1)]/(MCC2*cV) \qquad (2.4.37)$$
$$PCC2=MCC2*R*TCC2/VCC2 \qquad (2.4.38)$$

where enthalpy of outflow, $\Delta HT=\Delta MT*(cP*TT+0.5*CT^2)$ (2.4.39)
where enthalpy of inflow, $\Delta HI= \Delta MI*(cP*TI+0.5*CI^2)$ (2.4.40)

In any crankcase calculation the properties of the crankcase, inlet and transfer gas are those of air, where the values of g, cV, cP and R, found in any thermodynamic textbook, are 1.4, 718, 1005 and 287 J/kgK, respectively. For the open cycle cylinder calculation, where the cylinder gas is being replenished by air and, as exhaust gas properties are closely that of air, it is deemed sensible to use the same values for the thermodynamic constants for that calculation. For those who wish to model the properties of cylinder gas precisely, its properties are given in Chapter 5.

The inlet system, shown in Fig. 2.14 as being piston controlled, can, of course, be replaced by any other form of induction control as long as the unsteady gas dynamic flow can be predicted for its behavior. Disc valve and poppet valve time-area characteristics are straightforward geometrical calculations, but reed valves pose some difficulty in that regard. That reed valves can be handled by such calculations has been demonstrated by Fleck et al(1.13).

2.5 Illustration of unsteady gas flow into and out of a cylinder

Two almost identical computer programs have been specially written for this book to illustrate unsteady gas flow in the pipes of engines. All of the transmission and reflection characteristics discussed in previous sections have been incorporated into these programs, with the exception of sudden area changes and branched flow. The program is written in the Riemann variable format discussed in Sect. 2.4 and is designed to provide the reader of this book with a physical understanding of unsteady gas flow in a manner which the mere study of the mathematics will not provide. Both of these computer programs refer to the physical geometry shown in Fig. 2.15.

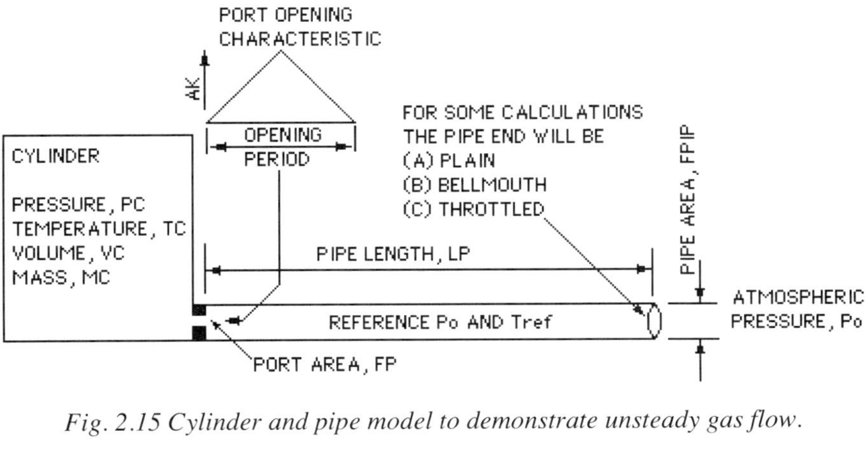

Fig. 2.15 Cylinder and pipe model to demonstrate unsteady gas flow.

The cylinder shown in Fig. 2.15 is filled with air and is connected to the parallel pipe of length, LP, and area, FPIP, to the atmosphere. The pipe feeds from the cylinder through a valve of maximum area, FP. This valve has a triangular area profile commencing at zero area at port opening, rising linearly for half of the total port opening period to area FP, and closing linearly over the latter half of the period to zero area. The total port opening period, in crank-angle terms, is a program input data value.

The cylinder pressure can be set to levels which are either above or below atmospheric pressure so that both unsteady gas inflow and outflow can be simulated. The cyclic speed of operation of the valve can be altered, in cycles per minute, RPM, and at the conclusion of each cycle of 360 crankshaft degrees, the original data values for the cylinder state conditions are restored and a second cycle of gas flow is commenced. The cylinder state conditions mentioned are for pressure, temperature and volume, PC, TC, and VC, respectively. The cylinder does not have a piston, which would induce a variable cylinder volume in the calculation, so that the reader may observe only the effect of the unsteady gas flow upon the emptying or filling of the cylinder.

The pipe length can be set to any value up to 1500 mm and the mesh length of the calculation is set at 30 mm. While almost all of the calculations are for a plain open ended pipe, for the inflow case the operator is invited to select either a bellmouth pipe end or a plain pipe end. For certain calculations, the author has programmed the effect of throttling the end of the pipe, and of a pipe with no reflections at all, so that further understanding may be gained by the reader from the explanations which follow.

The two computer programs mentioned are contained in the Computer Program Appendix as Prog.2.1, called EXHAUST, and Prog.2.2, called INDUCTION. All of the major subroutines are common and they are also common with many of those used for the major engine design program given in Chapter 5. All of these programs operate with dynamic graphics, so that the propagation of the pressure waves are visually observed on the computer screen.

An example of the output screen is shown in Fig. 2.16. The top four graphs have the x-axis as the length along the pipe. The graph at top-left is the superposition pressure ratio profile along the pipe. The graph at top-right is the pressure ratio profile along the pipe of the outward moving pressure wave. The graph at bottom-right is the pressure ratio profile along the pipe of the inward moving pressure wave. The graph at bottom-left is that of the Mach number of the gas particles and the arrow at each mesh point indicates the direction in which they are moving. The pressure-time graph at the bottom of the screen is the superposition pressure ratio in the pipe at the port, mesh position 1, as a function of crank angle. In this lower graph, each new cycle is overwritten upon the history of the previous cycle.

The user is invited by the program to select a cycle and an angle for a printout of the screen. That is how Fig. 2.16 was produced, and all subsequent figures referred to in these sections.

It will be observed that another parameter is also printed at each crank angle, Cylinder Mass Ratio, CMR. This is a definition for this application only, and is

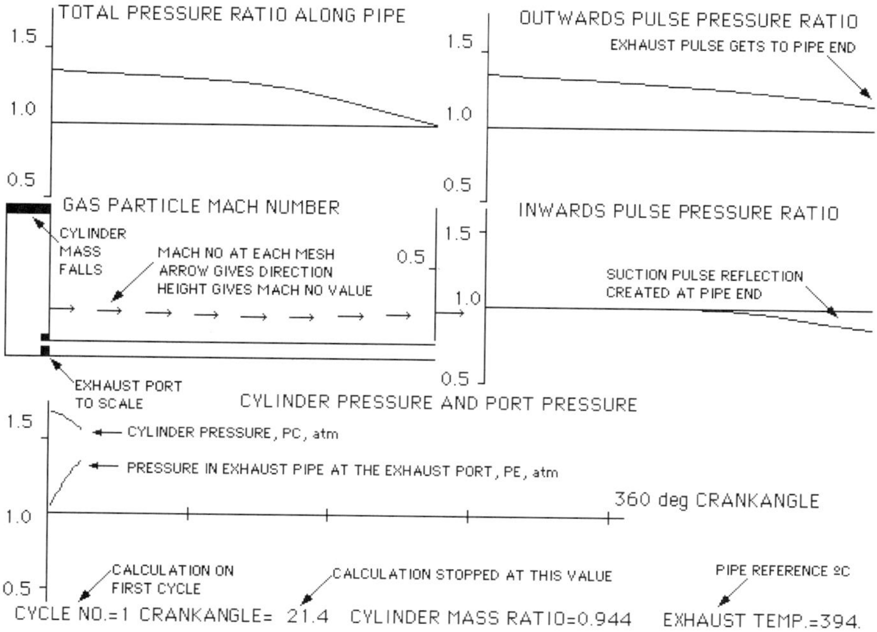

Fig. 2.16 Creation of exhaust pulse and its reflection at a plain open ended pipe.

defined differently for the EXHAUST and the INDUCTION programs, so that the effect of the unsteady gas flow on the cylinder mass can be assessed. For the EXHAUST program, where the pipe effects should assist with the emptying of the cylinder, it is defined as:

CMR=(instantaneous cylinder mass)/(original cylinder mass)

The cylinder which is drawn on the bottom-left diagram is then shown as if it were being emptied and the ratio of the blackened area to the total area is also the CMR value. This visual effect, along with the numerical value of CMR, informs the user of the effectiveness of the exhaust process. The calculation commences at CMR equal to unity.

For the INDUCTION program, where the pipe effects should assist with the filling of the cylinder, it is defined as:

CMR=(instantaneous cylinder mass)/(mass to fill cylinder volume at STP)

The STP conditions are 101325 Pa and 20ºC. The cylinder which is drawn on the bottom-left diagram is then shown as if it were being filled and the ratio of the blackened area to the total area is also the CMR value.The calculation commences at CMR equal to the initial conditions and, should the CMR value ever reach unity, the cylinder shown would be completely clear and "unblackened."

Lastly, at the port between the cylinder and the pipe, the area ratio between port and pipe, AK, is used to show visually the extent to which the port is open at any crank angle and also when it is functioning as a closed end.

2.5.1 Simulation of exhaust outflow with Prog.2.1, EXHAUST

For most of the simulations which are described the following common data values are used:

cylinder pressure, 1.7 bar; cylinder temperature, 500ºC;
cylinder volume, 300 cc; cyclic speed, 6000 rev/min;
maximum port diameter, 22.4 mm; pipe diameter, 25 mm;
pipe length, 250 mm; total port opening period, 100º crank angle.

Any change of data value, from any of the numbers above, will be indicated in the text at that juncture.

As with the example given in Sect. 2.3.3, it is the cylinder temperature and the isentropic pressure ratio between the cylinder and the reference pressure which control the pipe reference temperature, Tref. It can be seen in Fig. 2.16 that this is predicted as 394ºC.

The several effects which can be observed from the EXHAUST program will be discussed below.

(a) Calculation stopped on cycle number 1 at 21.4º crank angle

This is shown in Fig. 2.16. The calculation has been stopped at 21.4º crank angle, and the cylinder pressure is observed to have fallen, the port has commenced opening and an exhaust pulse has been created. The front of the exhaust pulse has reached the open end and an expansion wave reflection has been created. The superposition pressure along the pipe dips sharply towards the open end, for the criterion is that the superposition pressure ratio in the plane of the open end is unity (see Eq. 2.3.6).

The gas particles are being impelled towards the open end and further moved in that direction by the suction reflection. The cylinder mass has already dropped by 4.6%.

(b) Calculation stopped on cycle number 1 at 34.5º crank angle

This is shown in Fig. 2.17. At this point the cylinder pressure is still falling rapidly and the cylinder now contains 88.1% of its original contents after the passage of 34.5º crank angle at 6000 cycles per minute. The time which has passed is just (34.5*60)/(360*6000) seconds or 0.96 ms. The exhaust pulse is past its peak amplitude and the front of the suction reflection has just arrived at the exhaust port. The suction reflection is due to get stronger at the exhaust port as the expansion wave deepens with time. The gas particle velocity is seen to be increasing towards the pipe

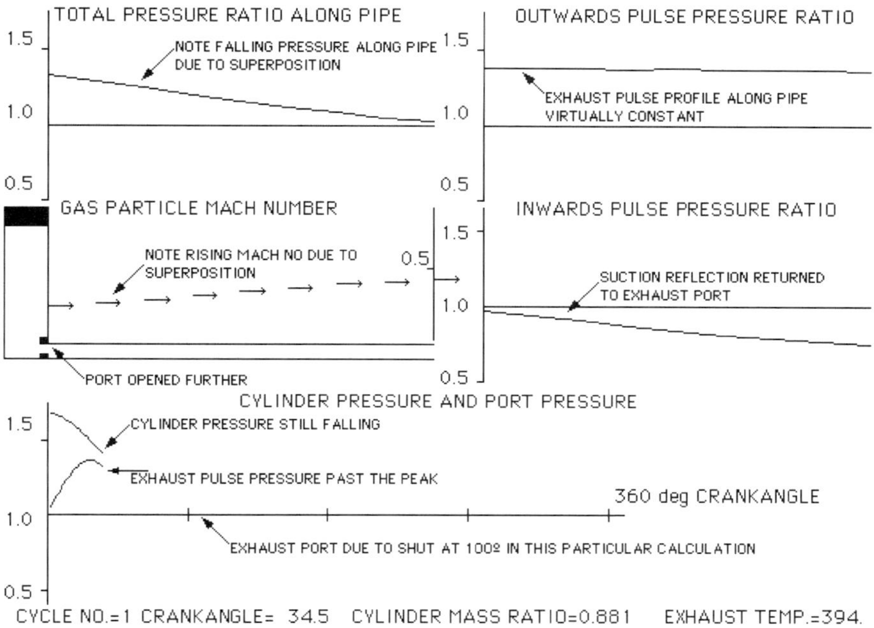

Fig. 2.17 The expansion wave reflection returns to the exhaust port to assist scavenging.

exit where the strong suction pulse is accelerating the flow. The exhaust pulse profile along the pipe is virtually constant, but the superposition pressure profile is seen to fall towards the open end. The exhaust port is still not yet fully open; that will occur in this calculation data set at 50º crank angle.

(c) Calculation stopped on cycle number 1 at 88.3º crank angle

This is shown in Fig. 2.18. The effect of the suction reflection is now evident. The cylinder pressure has been dragged below the atmospheric pressure and the CMR value is 0.625. The cylinder has been emptied of 37.5% of the original mass contents. The port is due to shut in 11.7º crank angle, so the suction reflection peak at the exhaust port is being reflected off this partially closed end as another expansion wave. It can be seen being propagated outward and following on the tail of the original exhaust compression pulse. The gas particle velocity profile along the pipe is still outward.

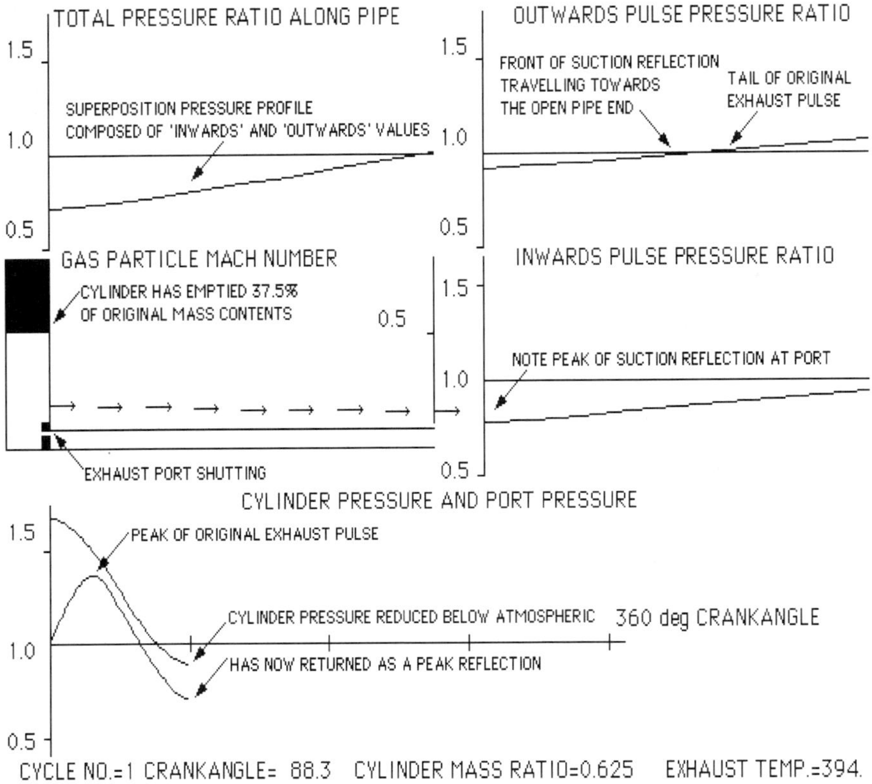

Fig. 2.18 Expansion wave reflection has pulled cylinder pressure below atmospheric pressure.

(d) Calculation stopped on cycle number 2 at 360º crank angle

This is shown in Fig. 2.20, for the pipe length of 250 mm, where the pressure crank-angle diagram has been extracted and plotted with others of differing pipe length, to be discussed later. To continue with the sequence for the 250 mm pipe length, it can be seen that the final CMR value of 0.62 was lower than at 88.3º in (c). After the port has shut, the pressure wave is reflected to and fro between the open and closed ends of the pipe, changing its type from compression to expansion, or vice-versa, at each open ended reflection. Indeed, the pipe is traversed eight times by such reflections before the next cycle commences.The cylinder pressure remains constant at its "trapping" condition until the next cycle commences, when its mass contents and state conditions are "miraculously" restored mathematically.

(e) Effect of residual pressure waves on the exhaust process

The first cycle allowed the exhaust pulse to proceed into a pipe with no residual pressure waves present, i.e., the ∂ and ß values at the eight meshes in the pipe were all unity at the start of cycle number 1. This will not be the case at the start of cycle number 2, when all of the ∂ and ß values at the mesh points will have values other than unity. That this has an effect on the exhaust process can be observed in Fig. 2.19, for a calculation for a pipe length of 400 mm, all other data being common with the original data set. The first cycle leaves a residual compression wave at the exhaust port as it opens on the second cycle, and this changes the exhaust port pressure-crank angle history on cycle number 2. It is not a major change, but the CMR value is altered from 0.677 to 0.672 on cycle number 2 and to 0.671 on cycle number 3. Thus, all cycles after the second are more or less the same. In all subsequent calculations, to demonstrate the effects of changing various parameters, the results of the second cycle will be quoted and displayed.

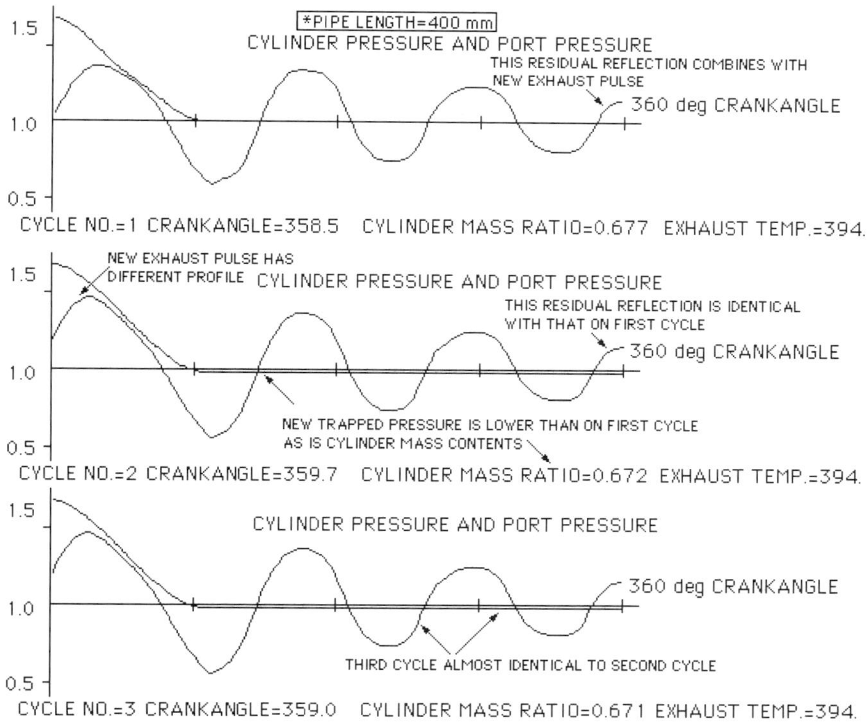

Fig. 2.19 Three individual cycles of a calculation showing that residual effects are important.

(f) The effect of changing pipe length at the same cyclic speed

Fig. 2.20 illustrates the cylinder and pipe pressures for various pipe lengths from 150 to 400 mm, all other data values being common with the original data set. It will be observed that this has a marked effect on the cylinder mass ratio and on the pressure diagrams. The lowest CMR value is for a pipe length of 200 mm. This effect will also be seen in the value of trapping pressure, the lowest pressure giving the lowest CMR value. The number and amplitude of the residual reflections is also greatly influenced by pipe length. It will also be observed that the reflections decay in amplitude with each succeeding reflection. This is caused more from the reflection of an expansion wave at an open end of a plain pipe, where a contraction coefficient of 0.6 applies to the particle inflow, than it does from any friction effects along the pipe. In fact, the friction factor, f, in the calculation is set at zero, as Blair and Goulburn(2.14) showed that a value between 0 and 0.004 would be applicable for smooth pipes. The reason for this is that the mathematical interpolation action within the program for the ∂ and ß values also gives an apparent diminution to the wave amplitude in a manner similar to a friction effect.

(g) Effect of cyclic speed on cylinder and pipe pressure-time histories

This is demonstrated in Fig. 2.21 for the original data set, but with the cyclic speed varied from 2500 to 6000 cycles per minute. The effect is not dissimilar to changing pipe length at a common cyclic speed. For the pipe length of 250 mm, it can be seen that the cylinder is most effectively emptied at 5000 cycles per minute and the least effective speed is at 2500 cycles per minute. All of the pressure diagrams in Fig. 2.21 are for the second cycle of the calculation. The attenuation of the multiple reflections between the open and closed ends of the straight pipe is most evident at the lowest cyclic speed of 2500 cpm.

Perhaps more importantly, the reason for the poor cylinder emptying at the lower speeds is because a compression wave reflection arrives at the exhaust port and blocks it, gas-dynamically speaking. At 2500 cpm, the suction wave arrives too early, pulls the cylinder pressure below atmospheric pressure at about the peak port opening period, and the CMR value at that point is about 0.6. However, the ensuing reflection of that suction pulse as a compression wave at the open end is able to return and repack some of the cylinder mass contents. This is the first indication of an exhaust tuning action which would have a double benefit for a two-stroke engine. At the higher speeds, the action of the exhaust pipe assists with cylinder emptying and the scavenge process, if one had taken place in this simulation. In this case, the continued extraction of any fresh charge of air and fuel would impair the value of mass trapped and the resulting BMEP attained, as discussed in Sect. 1.0. However, in the 2500 cpm simulation, the suction pulse would arrive early, assist with cylinder emptying and the scavenge process, whereupon the subsequent blocking of the exhaust port by the later compression wave reflection would greatly assist the retention of any fresh charge of air and fuel within the cylinder. This action is known as "plugging" the exhaust port. Early racing two-stroke engines had short, straight, parallel exhaust pipes about 300 mm long for this very purpose.

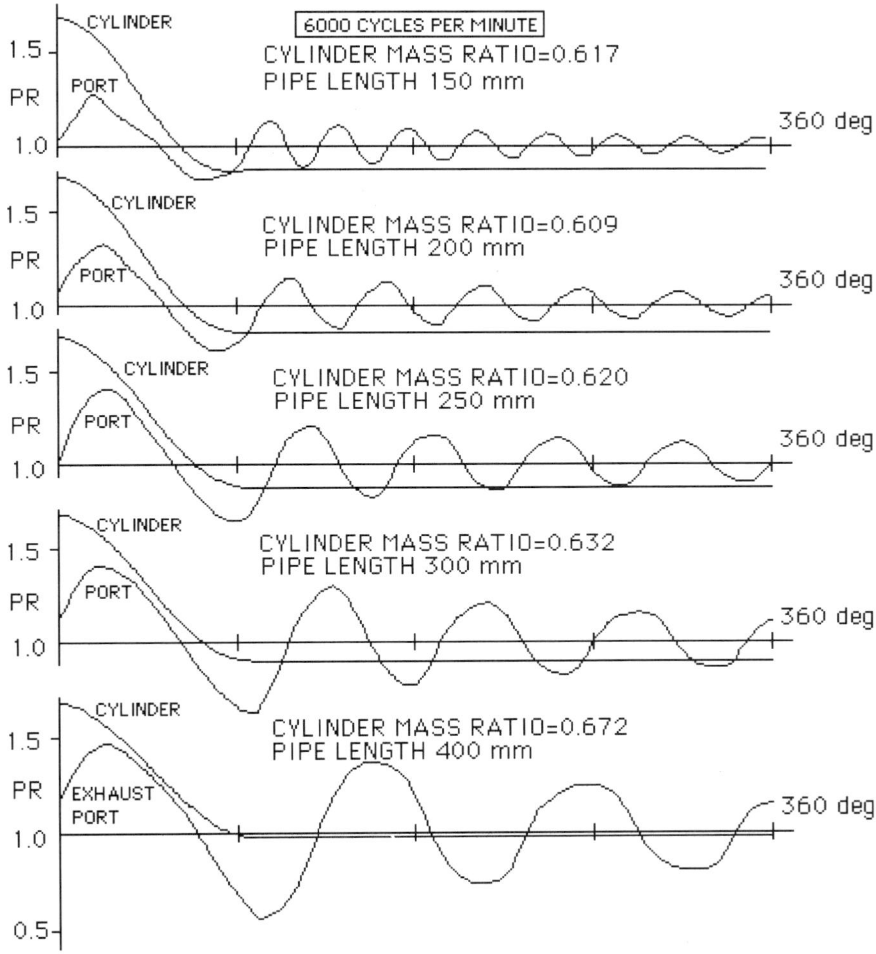

Fig. 2.20 Effect of pipe length on unsteady gas flow at the same cyclic speed.

In the 1930's the effects just discussed were advocated by Kadenacy, but without the benefit of knowledge of the behavior of pressure waves in the ducts of engines. Kadenacy, who held patents(2.21) on tuned exhaust pipes and diffusers attached to engines, attributed the enhanced cylinder pressure drop to the kinetic energy of the "outrushing gases extracting cylinder gas." It was Giffen(2.22), an Ulsterman, as were Kelvin and Reynolds, who showed in a classic paper in 1940 that the Kadenacy phenomenon was due to pressure wave action. In later years, the tuned expansion chamber was invented to enhance this effect, which will be discussed in detail in Chapters 5 and 6.

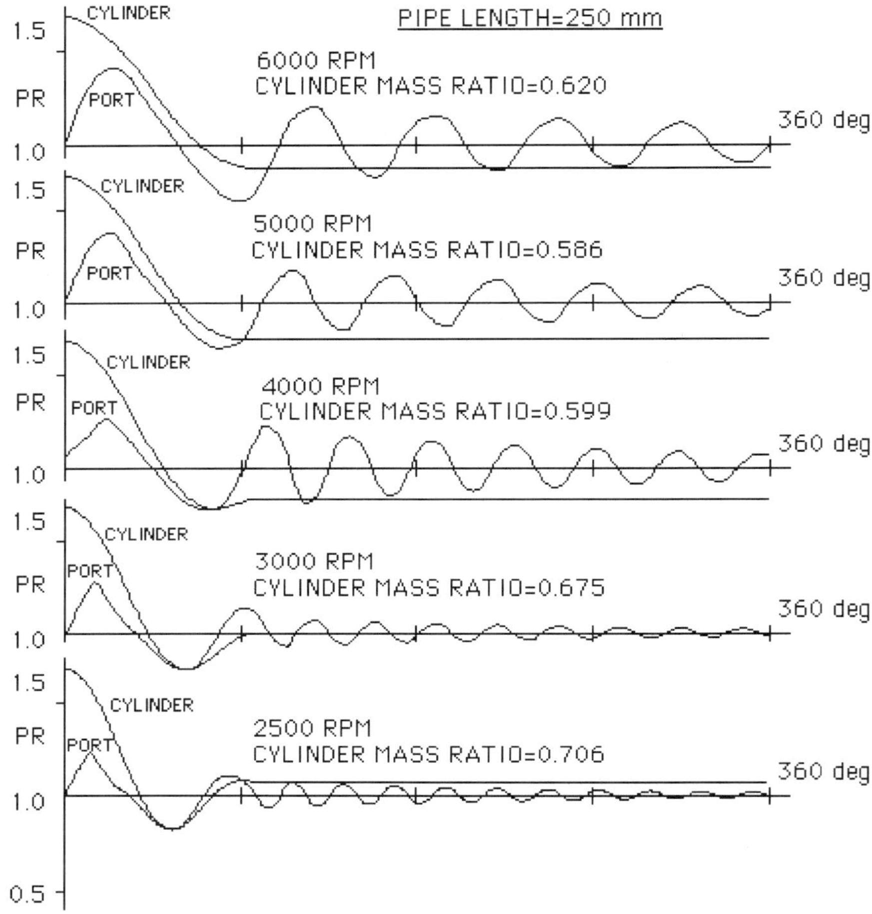

Fig. 2.21 Effect of cyclic speed on cylinder and exhaust pressure records.

(h) Effectiveness of cylinder emptying without an exhaust pipe

Yet another important question can be answered regarding the effectiveness of cylinder emptying: If the pipe were so long that there were no wave reflections, or so short that the discharge took place directly to the atmosphere, would the cylinder be more, or less, effectively emptied? With the computer program that question is easily answered by mathematically instructing the meshes that all ß values are unity. The result of this for the original data set is shown in brief for two cyclic speeds in Fig. 2.22 and in graphic form in Fig. 2.23.

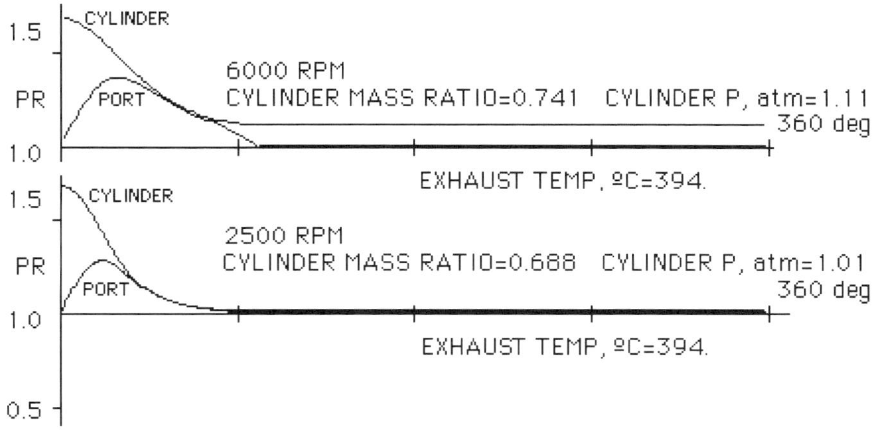

Fig. 2.22 Effect of cylinder blowdown into a pipe with no wave reflections.

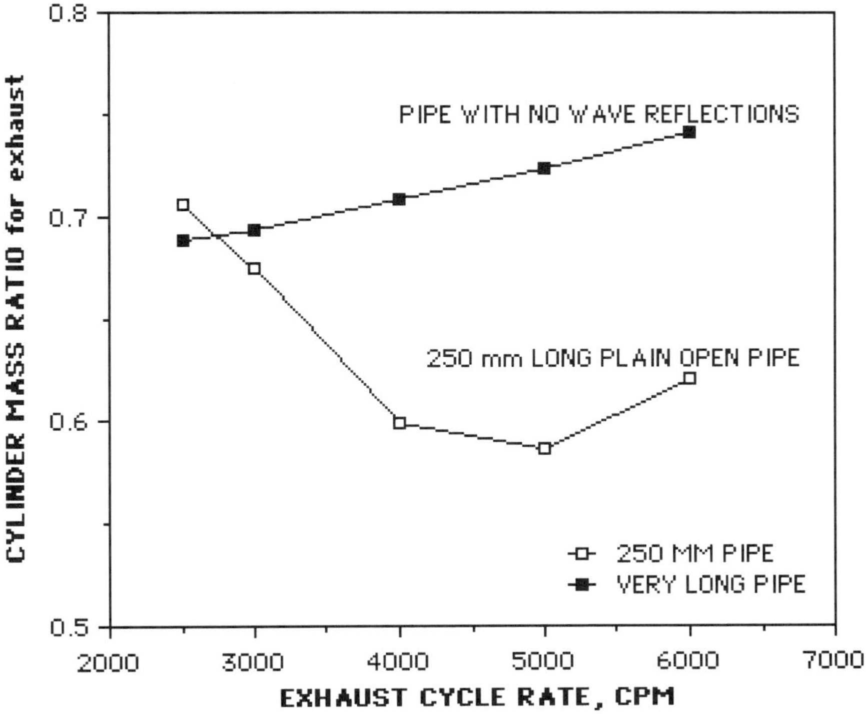

Fig. 2.23 Effect on cylinder mass of pressure waves in an exhaust pipe.

In Fig. 2.22, the second calculation cycle for both 6000 and 2500 cpm is shown. As might be expected, the cylinder pressure decays more closely to atmospheric pressure at the lowest possible cyclic speed when there is the longest time available for outflow in each cycle. At 6000 cpm, the cylinder pressure ratio has fallen to 1.11 atm by exhaust closing, whereas it has fallen to 1.01 atm at 2500 cpm. The result of carrying out this calculation for the range of cyclic speeds used (2500 to 6000 cpm) and comparing it with the equivalent situation where the pressure waves are propagating in the 250 mm pipe (as shown in Fig. 2.21), is shown in Fig. 2.23 in a graphic format. Apart from the 2500 cpm situation previously discussed, the 250 mm pipe is substantially more effective as a means of emptying the cylinder than either a "zero length" pipe or a "very long pipe" where reflections could not return in time to influence the exhaust process.

(i) Effect of throttling the exit of an exhaust pipe

Although it might seem to be a means of decreasing the effectiveness of emptying a cylinder, throttling an exhaust pipe is exactly what is done by attaching a turbocharger to it. A preliminary discussion of that has already taken place in Sect. 1.2.4. Although the simulation by Prog.2.1 does not include the scavenging of the cylinder by the compressor, the effect on the restriction of the exhaust flow can be simulated by assigning throttling diameters at the very end of the pipe and replacing the normal boundary condition of outflow to atmosphere by one of inflow through a restricted "pipe entrance" to another cylinder at atmospheric pressure. The Prog.2.1, EXHAUST, does not have this facility built in, but the interested reader may care to alter the software to carry out this calculation, thereby creating a new program as a tutorial exercise. In the software ProgLists.2.2 and 2.3, the appropriate lines have been indicated by REMinder statements.

In Fig. 2.24 the results of altering the pipe exit diameter in stages from the original data value of 25 mm to 12.5 mm are shown. As in the original data set, all of the calculations are conducted at 6000 cpm. The top graph is the original data set solution and is identical to that given previously as the top graph in Fig. 2.21. As the pipe exit is progressively restricted to 12.5 mm diameter, it can be observed that the suction reflections are progressively eliminated until the pipe end reflection criteria is almost one of neutrality! At the smallest pipe exit diameter of 12.5 mm, the main suction reflection is replaced by one of compression which opposes outflow during the closing period of the exhaust port. Any further restriction would result in increasingly larger compression wave reflections. The smallest value used here, where the exit diameter is half the pipe diameter, or a pipe exit area ratio of 0.25, corresponds to the normally acceptable restriction area ratio of the nozzle ring of a turbocharger(2.23). It will be observed that the cylinder pressure would only fall to 1.2 atm with this pipe exit area ratio of 0.25. Consequently, if one wished to scavenge this cylinder with a turbo-blower, one would need a compressor with an air supply pressure ratio of at least 1.5 atm to force a reasonable quantity of air through it during any scavenge period. Having done so, however, the high back-pressure provided by the restriction due to the turbine nozzle ring would ensure a high trapped mass characteristic within the cylinder; this would lead directly to a high IMEP and power output.

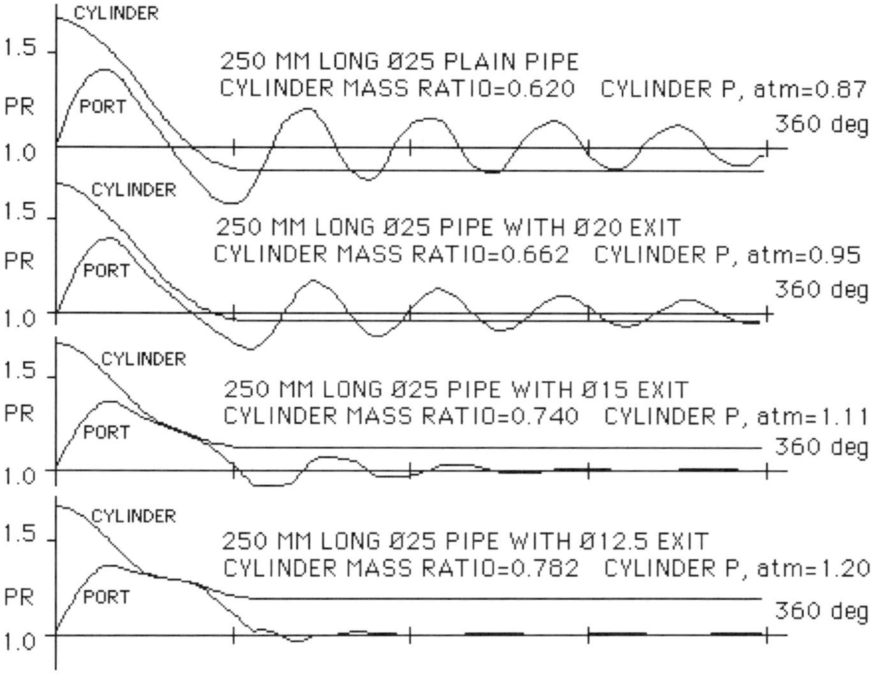

Fig. 2.24 Effect of restricting the outlet of an exhaust pipe to the atmosphere.

2.5.2 Simulation of "crankcase" inflow with Prog.2.2, INDUCTION

For the simulations which are described using Prog.2.2, INDUCTION, the following common data values are used:

cylinder pressure, 0.6 bar; cylinder temperature, 50ºC;
cylinder volume, 500 cc; cyclic speed, 6000 rev/min;
maximum port diameter, 27 mm; pipe diameter, 30 mm;
pipe length, 250 mm; total port opening period, 180º crank angle;
pipe reference temperature, 25ºC; pipe end is "bellmouth" type.

Any change of data value from any of the numbers above will be indicated in the following subsections of the text.

As with the inflow example given in Sect. 2.3.3, the pipe reference temperature, Tref, must be supplied as an input data value. This is normally the atmospheric temperature, To, or a value denoting some temperature drop into the pipe due to inflow from the atmosphere. It can also be a value representing some pipe wall heating effects, such as the case with a "hot-spot" in the intake manifold of a car engine; this latter effect is virtually incalculable(2.18) with any degree of confidence. In any event, the pipe reference pressure is clearly the atmospheric pressure, Po.

The several effects which can be observed from the INDUCTION program will be discussed below.

(a) Calculation stopped on cycle number 1 at 20.3º crank angle

In Fig. 2.25, the expansion wave created by the sub-atmospheric cylinder pressure is seen to be propagating towards the bellmouth end of the pipe to the atmosphere. The calculation has been stopped at an interesting point on the first cycle, for the front of the wave has not quite reached the open end. In the last two mesh points on the gas particle velocity diagram, the arrows indicating direction are still at the default undisturbed condition of zero velocity and pointing rightwards. In the remainder of the pipe the particle velocity is observed to be accelerating towards the port. In the lower diagram of pressure against crank angle, the cylinder pressure is already rising from the starting pressure of 0.6 bar, and the creation of a suction pressure wave as a function of time is observed. As the wave has not yet reached the open end, there are no reflections thus far, so the "inward pulse pressure ratio" diagram is seen to register an undisturbed set of values. Hence the "superposition pressure" diagram and the "outward pulse pressure ratio" diagram are identical.

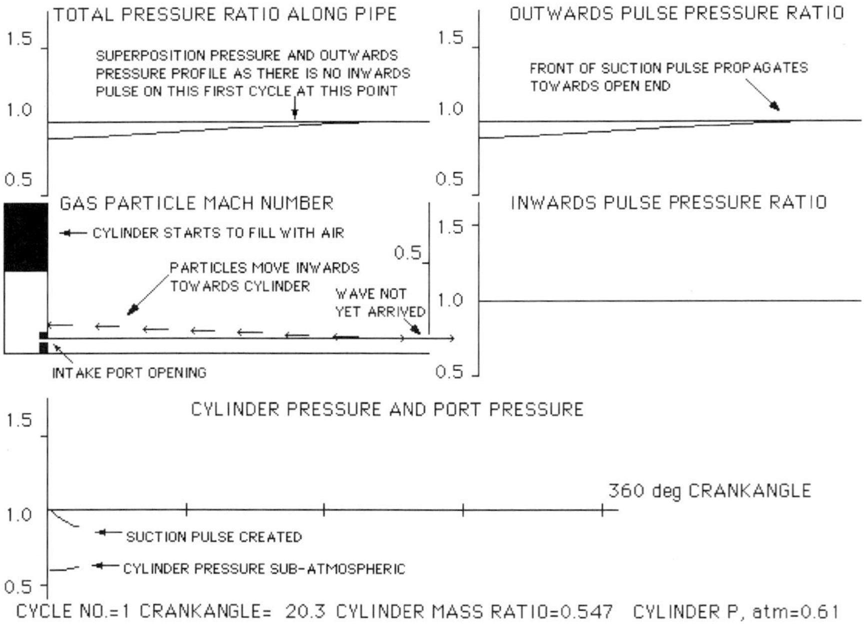

Fig. 2.25 Creation of an expansion wave in a pipe from a sub-atmospheric cylinder pressure.

(b) Calculation stopped on cycle number 1 at 40.1º crank angle

In Fig. 2.26, the calculation is stopped at a juncture where the front of the expansion wave reflection from the bellmouth open end is returning towards the cylinder port. It is a compression wave, as Sect. 2.3.2 would have predicted, and is colloquially referred to as a "ramming" wave. The effect is to increase the gas particle velocity in the pipe as a superposition of the inward compression and the outward expansion pressure waves. The result is the more rapid filling of the cylinder with air and this effect will be recorded by both cylinder pressure and the cylinder mass ratio, CMR.

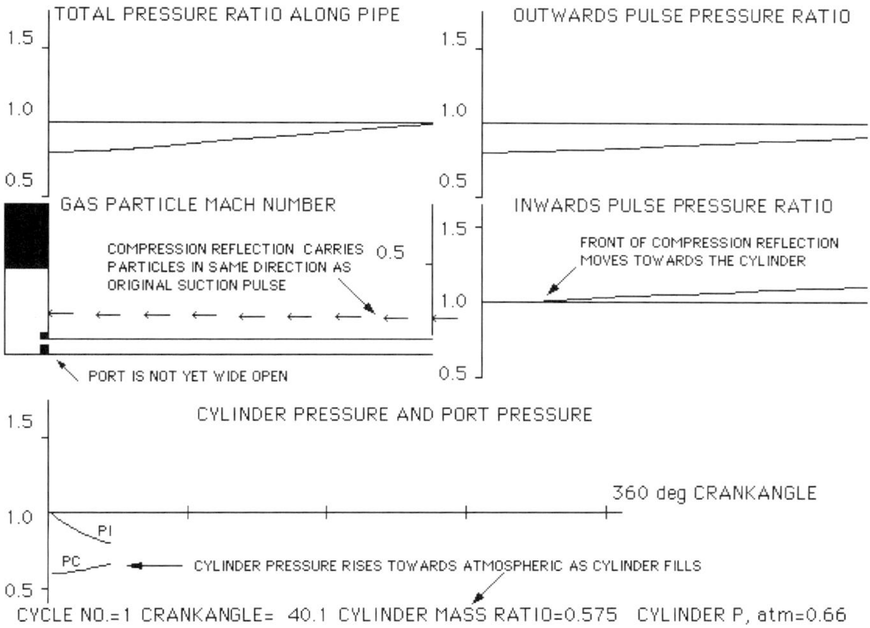

Fig. 2.26 Ramming reflection created at bellmouth open end enhances air flow to the cylinder.

(c) Calculation stopped on cycle number 1 at 90.5º crank angle

In Fig. 2.27, the calculation is stopped at a juncture where the port is virtually fully open and the ramming process is at its peak with this combination of pipe length at 250 mm and induction process cycling rate of 6000 cpm. The cylinder pressure has almost risen to the atmospheric value in the short space of 1/400 second, which is the highest it could ever rise in any steady flow process conducted over an "infinite" time period so that the cylinder and the atmospheric pressures are allowed to equalize. The suction pulse and the compression wave reflection fill the pipe so that the superposition diagram along the pipe looks as if it were close to atmospheric pressure at any point. As this is the diagram which would be recorded by a pressure transducer at a point in the pipe where that transducer is located, at this juncture the transducer would give little indication of the dramatic ramming action and rapid cylinder filling which is taking place. Hence, the accurate calculation of pressure diagrams is a more effective design tool than the information available from the experimentally determined result. The only experimental technique which will allow for proper correlation of experiment and theory in unsteady gas flow is to record the superposition pressure, PRs, diagram by a pressure transducer at some pipe location and to record the particle velocity, CGs, history with a fast response device, such as a laser velocimeter, at the same pipe location. From Eqs. 2.4.7 and 2.4.8 it will be observed that, in air for g=1.4:

$$PRs=Xs^7=[(\partial+\beta)/2]^7$$
$$CGs=2.5*Ao*(\partial-\beta)$$

Consequently, as shown by Blair(2.19), the simultaneous acquisition of PRs and CGs experimental data permits the direct correlation of measured and calculated ∂ and ß values. The presentation of measured superposition pressure diagrams from

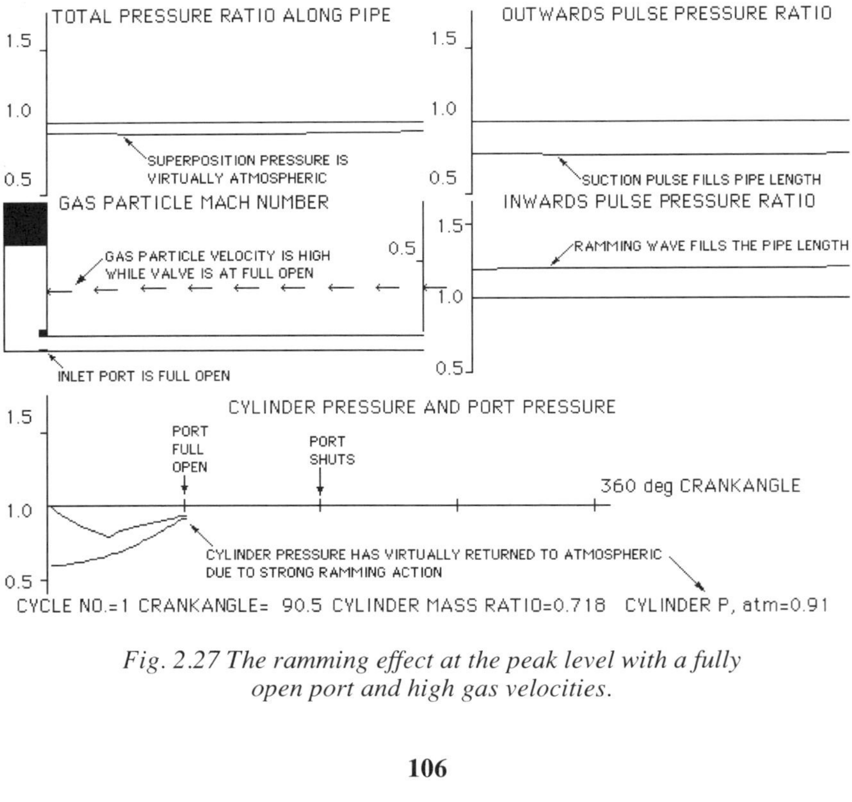

Fig. 2.27 The ramming effect at the peak level with a fully open port and high gas velocities.

a pressure transducer as the sole evidence of accuracy of a theoretical unsteady gas-dynamic calculation is insufficient evidence upon which to make a judgment of the accuracy of that theoretical approach. The literature is full of examples of this type of inadequate correlation of experiment and theory. In many ways this is understandable, for the measurement of unsteady particle velocity-time histories is not an easy experimental technique.

(d) The effect of different intake pipe lengths on cylinder filling

Fig. 2.28 shows the cylinder and port pressure ratio diagrams for three pipe lengths, the original data set value of 250 mm and two others at 200 and 300 mm. It can be seen that the 250 mm pipe is the most effective at the 6000 cpm calculation speed, for the cylinder mass ratio is the highest at 0.927. In all three cases the ramming is effective, for the cylinder pressures at the trapping point at intake port closure have been elevated above the atmospheric pressure value quite considerably. The highest trapped cylinder pressure corresponds to the greatest CMR value. The return of the peak of the ramming wave is the earliest in the case of the shortest pipe, 200 mm, but the optimum is given by the 250 mm pipe length case. After the port is shut, the subsequent reflections of the ramming wave between the open and closed ends can be clearly observed and it is the interaction of the latest residual reflection with the inflow process which will ultimately determine the CMR level. This effect, the combination of residual waves with a new process to the benefit of that process, is known as "resonance." For high specific performance engines for racing, the harnessing of this last "ounce" of pressure wave energy is the difference between winning and losing.

(e) Effect of a tuned intake pipe by comparison with a non-tuned pipe

Fig. 2.29 presents a series of calculations for the original intake data set conducted over the speed range of 1500 to 7500 cpm. The result is shown for both the cylinder mass ratio for induction, CMR, and the cylinder trapping pressure ratio, PCT. The same diagram gives the result of the same calculation for the same data set where all ß values are unity, i.e., the pipe is either so long that no reflections can ever return to affect the inflow process, or it is so short that reflections are immediately superposed at the port. The relationship between CMR and PCT is evident in both cases, illustrating that, as the state equation would dictate, cylinder pressure is the predominant controller of system mass in most situations.

The influence of the tuned pipe, over the untuned pipe, is beneficial at the higher speeds, but at the lower speeds the pressure wave action deteriorates the filling of the cylinder. Naturally, it would be possible to use a longer pipe and fill the cylinder in a better fashion at either 1500 or 3000 cpm, but that would almost certainly not enhance the cylinder filling at the higher speeds.

In the absence of any pressure wave reflections, it can be seen that the cylinder filling process is controlled solely by the time available for that filling process when the port is open; the higher the speed, the lower the CMR level.

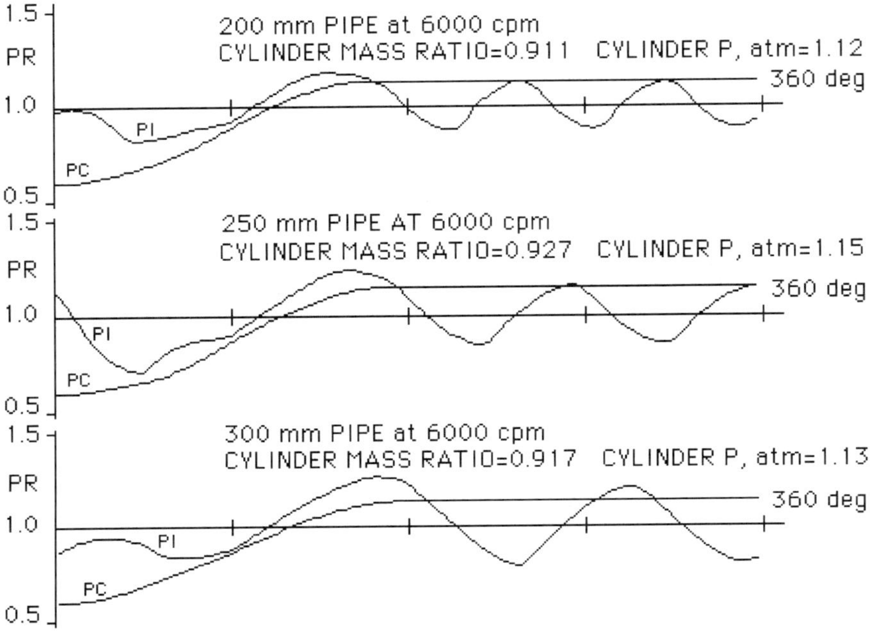

Fig. 2.28 Effect of intake pipe length on cylinder mass and pressure diagrams.

(f) The effect of throttling the intake process

In a real engine, except for a compression ignition or diesel power unit, the control over cylinder filling and power output at any speed is carried out by throttling the intake pipe. The effect of this is simulated in this calculation by presenting various area ratios at the pipe exit to atmosphere in the same manner as discussed in Sect. 2.5.1(i); otherwise the data used is the original induction data set. The results of such calculations are presented in Fig. 2.30 for full throttle, i.e., the original data set values in their entirety, and for throttle area ratios of 0.4, 0.2, and 0.1. As the throttle is progressively closed, it can be seen that the ramming reflections are ultimately eliminated and the trapped cylinder pressure becomes increasingly sub-atmospheric. As is intended in a real engine, this gives lower mass values of air filling the cylinder. In the conventional two-stroke engine case, it is the crankcase which is being filled initially at various levels of scavenge ratio, SR, or delivery ratio, DR, before the onward transmission of that air charge into the cylinder.

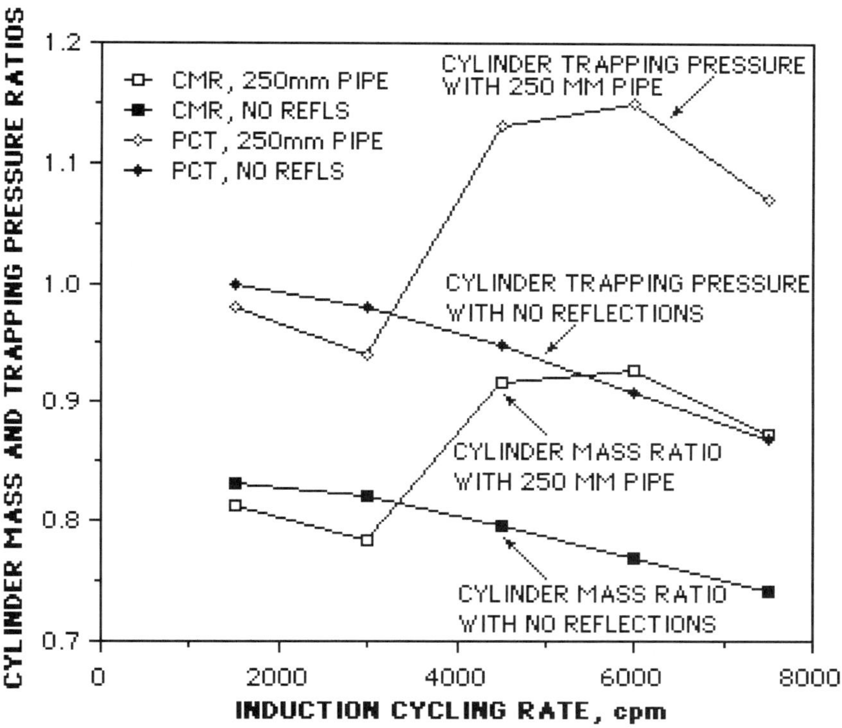

Fig. 2.29 Effect of intake pipe on ramming over a speed range.

(g) Effect of using a plain or a bellmouth end for an intake pipe

Fig. 2.31 presents the calculation result of replacing the bellmouth end criteria in the original data set with a plain end to the pipe. This reduces the effectiveness of the ramming process to an extent which might not be too important for an industrial engine yet would be highly significant for a racing engine. It can be seen that it is the ramming and residual reflections which are principally affected, as their decay after port closure is clearly visible. This arises because any expansion wave reflection process at the open end provides particle inflow, and that particle inflow process produces a vena-contracta at some level which is less than the full pipe area. Thus, the ramming reflection at a plain pipe end is reduced in amplitude by comparison with the bellmouth end. It will be recalled from the theoretical discussion in Sect. 2.3.2 and from Fig. 2.5 that the bellmouth pipe end case has superposition occurring over the full pipe area.

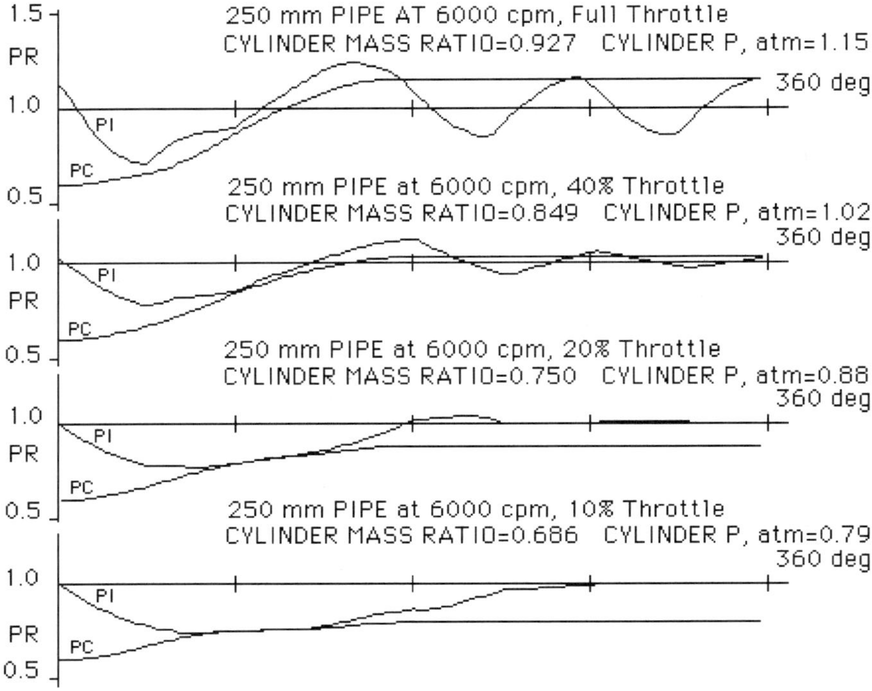

Fig. 2.30 Effect of throttling an intake pipe at the open end to atmosphere.

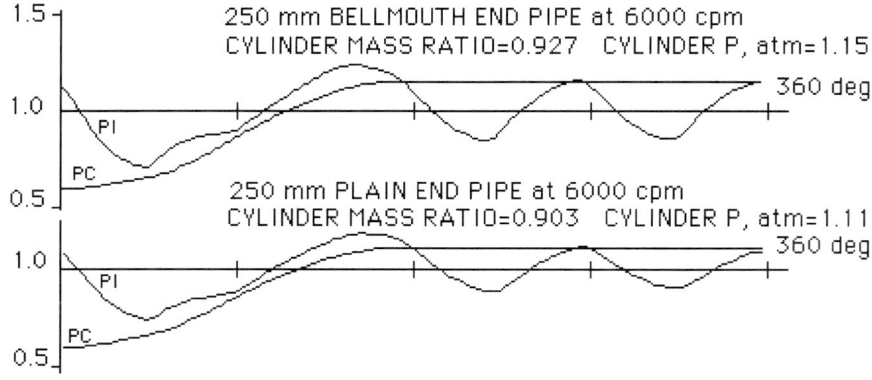

Fig. 2.31 Comparison of plain and bellmouth ended pipes for intake flow.

2.6 Unsteady gas flow and the two-stroke engine

By now the reader should have formed the opinion that the behavior of pressure wave action is sufficiently different from steady flow characteristics and that it is essential to have a thorough understanding of the mechanism in order to be able to design any reciprocating ic engine. This is particularly true of two-stroke engines, where cylinder emptying and filling are occurring within the same time frame of the open cycle period. This means that the resulting performance characteristics are heavily dependent upon the pressure-time characteristics in the plane of the ports, and particularly at the exhaust port. The theoretical solutions of unsteady gas flow derived in this chapter will be referred to frequently in the remainder of this book, because this is actually the manner in which an engine is emptied of exhaust gas and filled with fresh charge.

In the opening paragraph, the dramatic example was cited of removing the exhaust pipe from a racing two-stroke engine with the power output being halved. From the subsequent discussion, the reader should comprehend that this action halved the cylinder trapping pressure by reducing the trapped cylinder pressure by the same amount. How this is possible will be illustrated using a complete engine simulation in Chapter 5.

Before that point, it will be necessary to investigate in Chapter 3 the filling of the cylinder with a fresh charge of air through the scavenging process introduced in Sect. 1.2, and to comprehend the nature of the combustion process which will be discussed in Chapter 4.

NOTATION FOR CHAPTER 2

NAME	SYMBOL	SI UNIT	OTHER UNITS
Charging Efficiency	CE		
Crankcase Compression Ratio	CCR		
Dimensionless acoustic velocity	a		
Dimensionless particle velocity	u		
Dimensionless position	S		
Dimensionless time	Z		
Geometric Compression Ratio	GCR		
Mach Number	MN		
Pressure Amplitude Ratio	X		
Pressure Ratio	PR		
Ratio of Specific Heats	g		
Riemann variables	∂,ß		
Scavenging Efficiency	SE		
Scavenge Ratio	SR		
Trapped Compression Ratio	TCR		
Trapping Efficiency	TE		

NAME	SYMBOL	SI UNIT	OTHER UNITS
Acoustic Velocity	A	m/s	
Brake Mean Effective Pressure	BMEP	Pa	bar, atm, psi
Density	D	kg/m^3	
Engine Rotation Rate	RPS	rev/s	rev/min, rpm
Enthalpy	H	J	
Gas Constant	R	J/kgK	
Gas Constant for air	Ra	J/kgK	
Indicated Mean Effective Pressure	IMEP	Pa	bar, atm, psi
Internal Energy	U	J	
Length	s	m	
Mass Flow Rate	MG	kg/s	
Mesh length	L	m	
Particle Velocity	CG	m/s	
Pressure	P	Pa	bar, atm, psi
Propagation Velocity	CP	m/s	
Specific Enthalpy	h	J/kg	
Specific Heat at Constant Volume	cV	J/kgK	
Specific Heat at Constant Pressure	cP	J/kgK	
Temperature	T	K	ºC, ºF

REFERENCES FOR CHAPTER 2

2.1 S. Earnshaw, "On the Mathematical Theory of Sound," Phil.Trans.Roy.Soc., Vol.150, p. 133, 1860.

2.2 F.K. Bannister, "Pressure Waves in Gases in Pipes," *Ackroyd Stuart Memorial Lectures*, University of Nottingham, 1958.

2.3 G. Rudinger, Wave Diagrams for Non-Steady Flow in Ducts, Van Nostrand, New York, 1955.

2.4 R.S. Benson, The Thermodynamics and Gas Dynamics of Internal Combustion Engines, Volumes 1 and 2, Clarendon Press, Oxford, 1982.

2.5 F.J. Wallace, M.H. Nassif, "Air Flow in a Naturally Aspirated Two-Stroke Engine," *Proc.I. Mech.E.*, Vol.1B, p. 343, 1953.

2.6 G.P. Blair, W.L. Cahoon,"A More Complete Analysis of Unsteady Gas Flow Through a High Specific Output Two-Cycle Engine," SAE Automotive Congress, Detroit, January, 1972, SAE Paper No.720156.

2.7 J.H. McConnell, "Unsteady Gas Flow Through Naturally Aspirated Four-Stroke Cycle Internal Combustion Engines," Doctoral Thesis, The Queen's University of Belfast, March, 1974.

2.8 P. De Haller, "The Application of a Graphic Method to some Dynamic Problems in Gases," *Sulzer Tech. Review*, Vol.1, p. 6, 1945.

2.9 A. Jones, Doctoral Thesis, University of Adelaide, Australia, 1979.

2.10 R.S. Benson, R.D. Garg, D. Woollatt, "A Numerical Solution of Unsteady Flow Problems," *Int.J.Mech.Sci.*, Vol.1, p. 253, 1960.

2.11 D.R. Hartree, "Some Practical Methods of Using Characteristics in the Calculation of Non-Steady Compressible Flow," US Atomic Energy Commission Report, AECU-2713, 1953.

2.12 M. Chapman, J.M. Novak, R.A. Stein, "A Non-Linear Acoustic Model of Inlet and Exhaust Flow in Multi-Cylinder Internal Combustion Engines," Winter Annual Meeting, ASME, Boston, Mass., 83-WA/DSC-14, November, 1983.

2.12 P.J. Roache, Computational Fluid Dynamics, Hermosa Publishers, USA, 1982.

2.13 R. Courant, K. Friedrichs, H. Lewy, Translation Report NYO-7689, *Math.Ann*, Vol.100, p. 32, 1928.

2.14 G.P. Blair, J.R. Goulburn, "The Pressure-Time History in the Exhaust System of a High-Speed Reciprocating Internal Combustion Engine," SAE Mid-Year Meeting, Chicago, Illinois, May 15-19, 1967, SAE Paper No. 670477.

2.15 G.P. Blair, M.B. Johnston, "Unsteady Flow Effects in the Exhaust Systems of Naturally Aspirated, Crankcase Compression, Two-Stroke Cycle Engines," SAE Farm, Construction and Industrial Machinery Meeting, Milwaukee, Wisconsin, September 11-14, 1968, SAE Paper No. 680594.

2.16 G.P. Blair, J.R. Goulburn, "An Unsteady Flow Analysis of Exhaust Systems for Multi-Cylinder Automobile Engines," SAE Mid-Year Meeting, Chicago, Illinois, May 19-23, 1969, SAE Paper No. 690469.

2.17 G.P. Blair, M.B. Johnston, "Simplified Design Criteria for Expansion Chambers for Two-Cycle Gasoline Engines," SAE Automotive Engineering Congress, Detroit, Michigan, January 12-16, 1970, SAE Paper No. 700123.

2.18 J.F. Bingham, G.P. Blair, "An Improved Branched Pipe Model for Multi-Cylinder Automotive Engine Calculations," *Proc.I.Mech.E.*, Vol. 199, No. D1, 1985.

2.19 A.J. Blair, G.P. Blair, "Gas Flow Modelling of Valves and Manifolds in Car Engines," *Proc.I.Mech.E.*, International Conference on Computers in Engine Technology, University of Cambridge, April, 1987, C11/87.

2.20 J.S. Richardson, "Investigation of the Pulsejet Engine Cycle," Doctoral Thesis, The Queen's University of Belfast, March, 1981.

2.21 M. Kadenacy, British Patents 431856, 431857, ..., 484465, 511366.

2.22 E. Giffen, "Rapid Discharge of Gas from a Vessel into the Atmosphere," *Engineering*, August 16, 1940, p. 134.

2.23 G.P. Blair, "Unsteady Flow Characteristics of Inward Radial Flow Turbines," Doctoral Thesis, The Queen's University of Belfast, May, 1962.

Chapter 3

Scavenging the Two-Stroke Engine

3.0 Introduction

In Chapter 1, particularly in Sect. 1, there is a preliminary description of scavenging for the various types of two-stroke engines. This chapter continues that discussion in greater depth and provides the reader with practical advice and computer programs to aid the design process for particular types of engine.

3.1 Fundamental theory

The original paper on scavenging flow was written by Hopkinson(3.1) in 1914. It is a classic paper, written in quite magnificent English, and no serious student of the two-stroke engine can claim to be such if that paper has not been read and absorbed. It was Hopkinson who conceived the notion of "perfect displacement" and "perfect mixing" scavenge processes. Benson(1.4) also gives a good account of the theory and expands it to include the work of himself and Brandham. The theory used is quite fundamental to our conception of scavenging, so it is repeated here in abbreviated form. Later it will be shown that there is a problem in correlation of these simple theories with measurements.

The simple theories of scavenging all postulate the ideal case of scavenging a cylinder which has a constant volume, SV, as shown in Fig. 3.1, with a fresh air charge in an isothermal, isobaric process. It is obvious that the real situation is very different from this idealized postulation, for the reality contains gas flows occurring at neither constant cylinder volume, constant cylinder temperature, nor constant cylinder pressure. Nevertheless, it is always important, theoretically speaking, to determine the ideal behavior of any system or process as a marker of its relationship with reality.

In Fig. 3.1 the basic elements of flow are represented. The entering air, A, can enter either a space called the "perfect scavenge volume" where it will be quite undiluted with exhaust gas, or a "mixing volume" where it mixes with the exhaust gas, or it can be directly short-circuited to the exhaust pipe providing the worst of all scavenging situations.

In this isothermal and isobaric process, the incoming air density, Dair, the cylinder gas density, Dc, and the exhaust gas density, Dex, are identical. Therefore, from the theory previously postulated in Sect. 1.5, the values of scavenge ratio, scavenging efficiency, trapping efficiency, and charging efficiency become func-

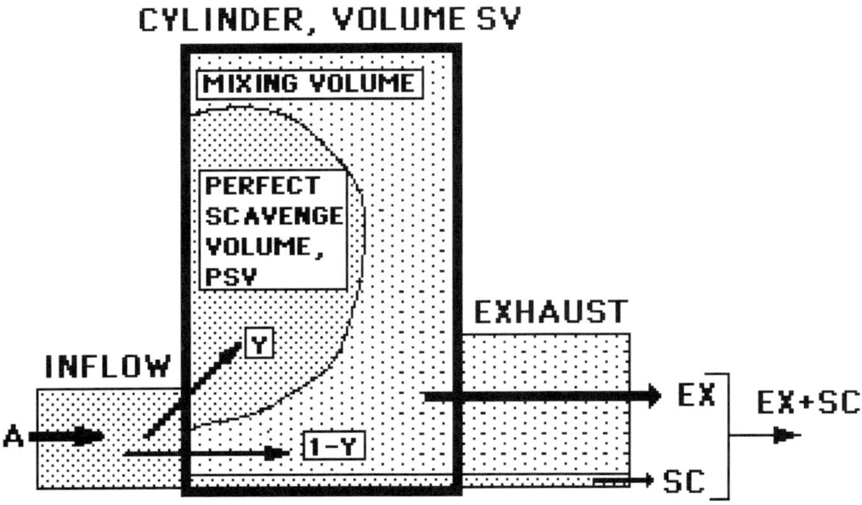

Fig. 3.1 Physical representation of isothermal scavenge model.

tions of the volume of the several components, rather than the mass values as seen in Eqs. 1.5.1-1.5.9. The following equations illustrate this point.

The first is for scavenge ratio, SR, derived from Eq. 1.5.7:

$$SR=(Vas*Dair)/(SV*Dair)=Vas/SV \tag{3.1.1}$$

The cylinder volume does not have to be the swept volume, SV, but it is clearly the first logical option. The volume of fresh air charge to be supplied is Vas.

The second is for scavenging efficiency, SE, derived from Eq. 1.5.9., where the volume of air trapped is Vta and the volume of exhaust gas trapped is Vex:

$$SE=(Vta*Dair)/(Vta*Dair+Vex*Dair)$$
$$=(Vta)/(Vta+Vex) \tag{3.1.2}$$

Hence as, Vta+Vex=SV, this equation is further simplified to:

$$SE=Vta/SV \tag{3.1.3}$$

The relationship for trapping efficiency, TE, follows from Eq. 1.5.13 as:

$$TE=(Vta*Dair)/(Vas*Dair)=Vta/Vas \tag{3.1.4}$$

The expression for charging efficiency, CE, follows from Eq. 1.5.15 as:

CE=(Vta*Dair)/(SV*Dair)=Vta/SV (3.1.5)

In this ideal scavenge process, it is clear that from manipulation of these above equations:

CE=SE (3.1.6)
TE=SE/SR (3.1.7)

3.1.1 Perfect displacement scavenging

In the perfect displacement process from Hopkinson(3.1), all fresh charge entering the cylinder is retained and "perfectly displaces" the exhaust gas. This air enters the perfect scavenge volume, PSV, shown in Fig. 3.1. Corresponding to the present theoretical presumption of perfect scavenging, Y=1 and the mixing volume illustrated contains only exhaust gas. Therefore, if all entering volumes of fresh charge, Vas, are less than the cylinder volume, SV (i.e., Vas<SV):

Vta=Vas (3.1.8)
SE=SR (3.1.9)
TE=1 (3.1.10)

If the scavenge ratio exceeds unity (i.e., Vas>SV), then:

Vta=SV (3.1.11)
SE=1 (3.1.12)
TE=SE/SR=1/SR (3.1.13)

3.1.2 Perfect mixing scavenging

This was Hopkinson's second concept. In this scenario the entering air had no perfect displacement characteristic, but upon arrival in the cylinder proceeded to mix "perfectly" with the exhaust gas. The resulting exhaust gas effluent, EX, was composed of the mixed cylinder charge. To analyze this process, consider the situation at some point in time where the instantaneous values of air supplied have been Vas, and the scavenge ratio and scavenging efficiency values to date are SR and SE, respectively. Upon the supply of a further increment of air, dVas, this will induce an exhaust flow of equal volume, dVex. The volume of air retained in the cylinder, dVta, due to this flow increment is given by:

dVta=dVas-SE*dVex (3.1.14)

Differentiation of Eq. 3.1.3 for scavenging efficiency gives:

dSE=dVta/SV (3.1.15)

Substituting dVta from Eq. 3.1.15 into Eq. 3.1.14, and taking into account that dVex is equal to dVas, produces:

$$SV*dSE=dVas-SE*dVas \qquad (3.1.16)$$

Rearranging this differential equation so that it may be integrated, and using a differential form of Eq. 3.1.1 for SR yields:

$$dSE/(1-SE)=dVas/SV=dSR \qquad (3.1.17)$$

This equation may be integrated on the left-hand side from the integration limits of 0 to SE, and on the right-hand side from 0 to SR, to give:

$$loge(1-SE)=-SR \qquad (3.1.18)$$

The further manipulation of this equation produces the so-called "perfect mixing" equation:

$$SE=1-e^{(-SR)} \qquad (3.1.19)$$

From Eq. 3.1.7, the trapping efficiency during this process is:

$$TE=[1-e^{(-SR)}]/SR \qquad (3.1.20)$$

3.1.3 Combinations of perfect mixing and perfect displacement scavenging

Benson and Brandham(3.2) suggested a two-part model for the scavenging process. The first part is to be perfect displacement scavenging until the air flow has reached a scavenge ratio value of SR=PS. At that point, when the theoretical situation in the cylinder is considered to be much like that in Fig. 3.1, the perfect scavenge volume and the exhaust gas are "instantaneously" mixed together. From that point on, until the fresh air flow process is concluded, a perfect mixing process takes place. This is described in great detail by Benson in his book(1.4). He also presents the theoretical analysis of such a flow, conducted in a very similar manner to that above. The result of that analysis reveals:

$$SE=SR \quad \text{for} \quad 0<SR<PS \qquad (3.1.21)$$

and

$$SE=1-(1-PS)e^{(PS-SR)} \quad \text{for} \quad SR>PS \qquad (3.1.22)$$

In other words, in terms of the symbolism shown in Fig. 3.1, in the first part of the Benson two-part model, Y=1, and in the second part, Y=0.

3.1.4 Inclusion of short-circuiting of scavenge air flow in theoretical models

In the book by Benson(1.4), the theory for the Benson-Brandham two-part model described in the preceding section is extended to include short-circuiting of the flow directly to the exhaust, as illustrated in Fig. 3.1. A portion of the incoming scavenge flow, SC, is diverted into the exhaust duct without scavenging exhaust gas or mixing with it in the cylinder. This results in modifications to Eqs. 3.1.21 and 22 to account for the fact that any cylinder scavenging is being conducted by an airflow of reduced proportions, viz (1-SC)*SR. After such modifications, Eqs. 3.1.21 and 22 become:

$$SE=(1-SC)*SR \quad \text{for} \quad 0<SR<(1-SC)*PS \qquad (3.1.23)$$

and $$SE=1-(1-PS)e^{(PS-(1-SC)*SR)} \quad \text{for} \quad (1-SC)*SR>PS \qquad (3.1.24)$$

3.1.5 The application of simple theoretical scavenging models

In the first computer program presented in this chapter, Eqs. 3.1.1-24 are in effect combined within a simple package. This is given in the Computer Program Appendix as Prog.3.1 and ProgList.3.1 called the BENSON-BRANDHAM MODEL. It is self-explanatory in use, with the program prompting the user for all of the data values required. As an example, the output plotted in Figs. 3.2 and 3.3 is derived using the program. Apart from producing the data to plot the perfect displacement scavenging and perfect mixing scavenging lines, two further examples are shown for a perfect scavenging period, PS, of 0.5, but one is with zero short-circuiting and the second is with SC equal to 10%. These are plotted in Figs. 3.2 and 3.3 as SE-SR and TE-SR characteristics. To calculate a perfect mixing characteristic, all that needs to be specified is that PS and SC are both zero, because that makes Eq. 3.1.24 identical to Eq. 3.1.19 within the program.

From an examination of the two figures, it is clear that a TE-SR characteristic provides a better visual picture for the comparison of scavenging behavior than does a SE-SR graph. This is particularly evident for the situation at light load, or low SR levels, where it is nearly impossible to tell from Fig. 3.2 that the 10% short-circuit line has fallen below the perfect mixing line. It is very clear, however, from Fig. 3.3 that such is the case. Further, it is easier to come to quantitative judgments from TE-SR characteristics. Should the two lines in Fig. 3.3 be the TE-SR behavior for two real, or theoretical, engine cylinders, then it would be possible to say positively that the BMEP, or power potential of one cylinder would be at least 10% better than the other. This could be gauged visually as the order of trapped mass improvement, for that is ultimately what trapping efficiency implies. It would not be possible to come to this judgment so readily from the SE-SR graph.

On the other hand, the SE-SR characteristic also indicates the purity of the trapped charge, as that is the alternative definition of scavenging efficiency, and in this case one which is totally accurate. As flammability limits of any charge are connected with the level of trapped exhaust gas, the SE value indicates the potential for the cylinder to fire that charge upon ignition by the spark plug. Should that not happen, then the engine will "four-stroke," or "eight-stroke" until there has been sufficient scavenging processes for the SE value to rise to the appropriate level to fire. In very broad terms, it is unlikely that a cylinder homogeneously filled with air-fuel mixture and exhaust gas to a SE value of less than 0.4 will be successfully fired by a spark plug. This is bound to be the case at light loads or at low SR levels, and is a common characteristic of carburetted two-stroke si engines. A more extended discussion of this topic takes place in Sect. 4.1.3. In such a situation, where the engine commences to fire unevenly, there will be a considerable loss of air-fuel

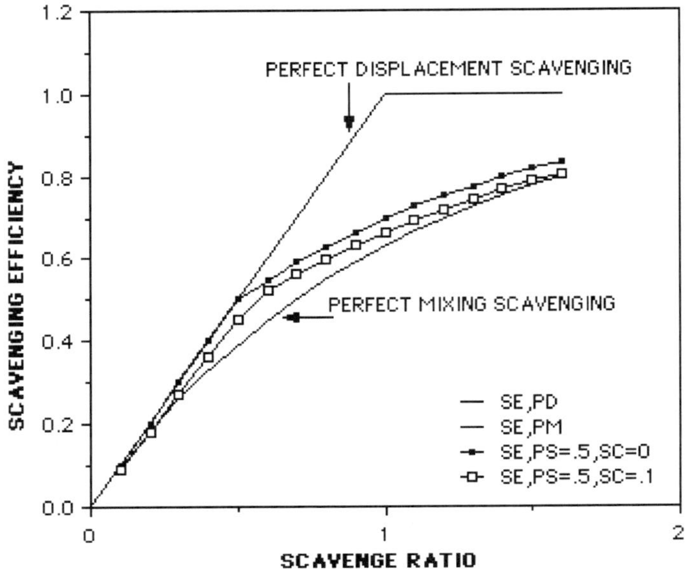

Fig. 3.2 Benson-Brandham model prediction of SE-SR characteristics.

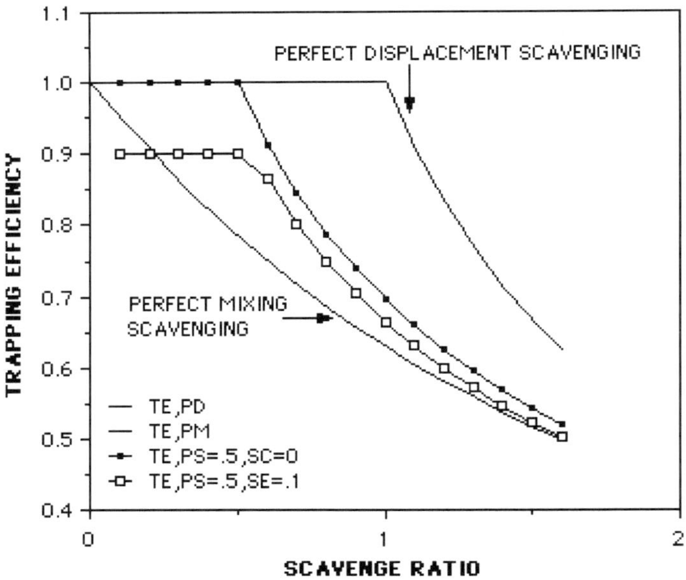

Fig. 3.3 Benson-Brandham model prediction of TE-SR characteristics.

mixture through the exhaust port, to the considerable detriment of specific fuel consumption and the level of exhaust emissions of unburned hydrocarbons. To further emphasize this important point, in the case of the two-cylinder characteristics cited in Fig. 3.3, there would be a small loss of fresh charge with one of them, and at least 10% loss with the other, should the SR value for each scavenge process be less than 0.5 while the engine was in this "four-stroking" mode. This makes it very important to be able to design or develop cylinders with high trapping efficiencies at light loads and low SR values. It should be remembered that the SR axis is, in effect, the same axis as throttle opening.

The literature is full of alternative theoretical models to that suggested by Benson and Brandham. Should the reader wish to study this matter further, the technical publications of Baudequin and Rochelle(3.3) and Changyou and Wallace(3.4) should be perused, as well as the references contained within those papers.

We will return to this subject later in this chapter, as it is proposed to demonstrate that these simple two-part models of scavenging provide poor correlation with experimentally derived results.

The reader is probably wondering where, in all of the models postulated already in this chapter, does loop or cross or uniflow scavenging figure as an influence on the ensuing SE-SR or TE-SR characteristics. The answer is that they do in terms of PS and SC values, but none of the literature cited thus far provides any numerical values of use to the engine designer or developer. The reason is that this requires correlation of the theory with relevant experiments conducted on real engine cylinders; until recent times this had not occurred. The word "relevant" in the previous sentence is used precisely; it means an experiment conducted as the theory prescribes, by an isobaric, isothermal and isovolumic process. Irrelevant experiments, in that context, are quite common, as the experimental measurement of SE-SR behavior by the author(3.7) and (3.13), Asanuma(3.8) and Booy(3.9) in actual firing engines will demonstrate; such experiments are useful and informative, but they do not assist with the assignment of theoretical PS and SC values to the engine cylinders described in those publications.

3.2 Experimentation in scavenging flow

Since the turn of the twentieth century, the engineers involved with the improvement of the two-stroke engine took to devolving experimental tests aimed at improving the scavenging efficiency characteristics of the engines in their charge. This seemed to many to be the only logical methodology because the theoretical route, of which Sect. 3.1 could well be described as the knowledge base pre-1980, did not provide specific answers to specific questions. Many of these test methods were of the visualization kind, employing colored liquids as tracers in "wet" tests and smoke or other visible particles in "dry" tests. This author experimented with both methods, but always felt the results to be more subjective than conclusive.

Some of the work was impressive in its rigor, such as that of Dedeoglu(3.10) or Rizk(3.11) as an example of liquid simulation techniques, or by Ohigashi and Kashiwada(3.12) and Phatak(3.19) as an example of gas visualization technology.

The first really useful technique for the improvement of the scavenging process in a particular engine cylinder, be it a loop or cross or uniflow scavenged design, was proposed by Jante(3.5). Although the measurement of scavenging efficiency in the firing engine situation (3.7)(3.8)(3.9) (Plate 3.3) is also an effective development and research tool, it comes too late in terms of the time scale for the development of a particular engine. The cylinder with its porting has been designed and constructed. Money has been spent on casting patterns and on the machining and construction of a finished product or prototype. It is somewhat late in the day to find that the SE-SR characteristic is, possibly, less than desirable! Further, the testing process itself is slow, laborious and prone to be influenced by extraneous factors such as the effect of dissimilarly tuned exhaust pipes or minor shifts in engine air-fuel ratio. What Jante(3.5) proposed was a model test on the actual engine cylinder, or a model cylinder and piston capable of being motored, which did not have the added complexity of confusing the scavenging process with either combustion behavior or the unsteady gas dynamics associated with the tuning process.

3.2.1 The Jante experimental method of scavenge flow assessment

The experimental approach described by Jante(3.5) is sketched in Fig. 3.4. A photograph of an experimental apparatus for this test employed at QUB some years ago is shown in Plate 3.1. It shows an engine, with the cylinder head removed, which is being motored at some constant speed. The crankcase provides the normal pumping action and a scavenging flow exits the transfer ports and flows towards, and out of, the open cylinder bore. At the head face is a comb of pitot tubes, which is indexed across the cylinder bore to provide a measured value of vertical velocity at various points covering the entire bore area. Whether the pitot tube comb is indexed radially or across the bore to give a rectangular grid pattern for the recording of the pitot tube pressures is immaterial. The use of pitot tubes for the recording of gas velocities is well known, and the theory or practice can be found in any textbook on fluid mechanics(3.14). The comb of pitot tubes shown in Fig. 3.4 can be connected to a variety of recording devices, ranging from the simplest, which would be an inclined water manometer bank, to some automatic data logging system through a switching valve from a pressure transducer. Irrespective of the recording method, the velocity of gas, V, in line with the pitot tube is given by:

$$V=4*\sqrt{Hw} \quad m/s \qquad (3.2.1)$$

where Hw is the pitot-static pressure difference in units of mm of water.

The engine can be motored at a variety of speeds and the throttle set at various opening levels to vary the value of scavenge ratio, SR. The most important aspect of the resulting test is the shape of the velocity contour map recorded at the open cylinder bore. Jante(3.5) describes various significant types of patterns, and these are shown in Fig. 3.5. There are four velocity contour patterns, (a)-(d), shown as if they were recorded from various loop scavenged engines with two side transfer ports and one exhaust port. The numerical values on the velocity contours are in m/s units.

Jante describes pattern (a) as the only one to produce good scavenging, where the flow is ordered symmetrically in the cylinder about what is often referred to as the "plane of symmetry." The velocity contours increase in strength from zero at the

Plate 3.1 A comb of pitot tubes for the Jante test on a motorcycle engine.

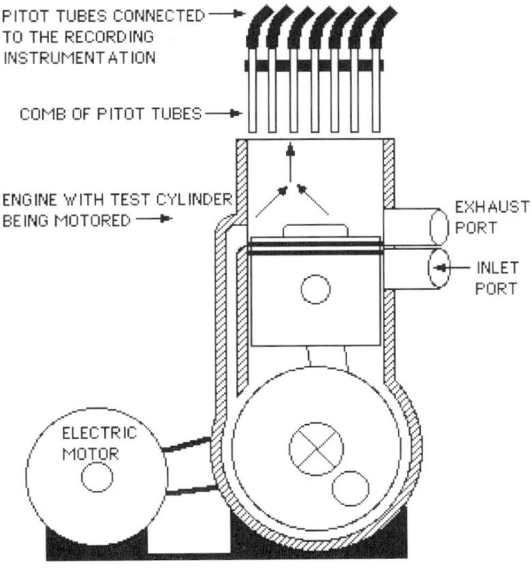

Fig. 3.4 Experimental configuration for the Jante test.

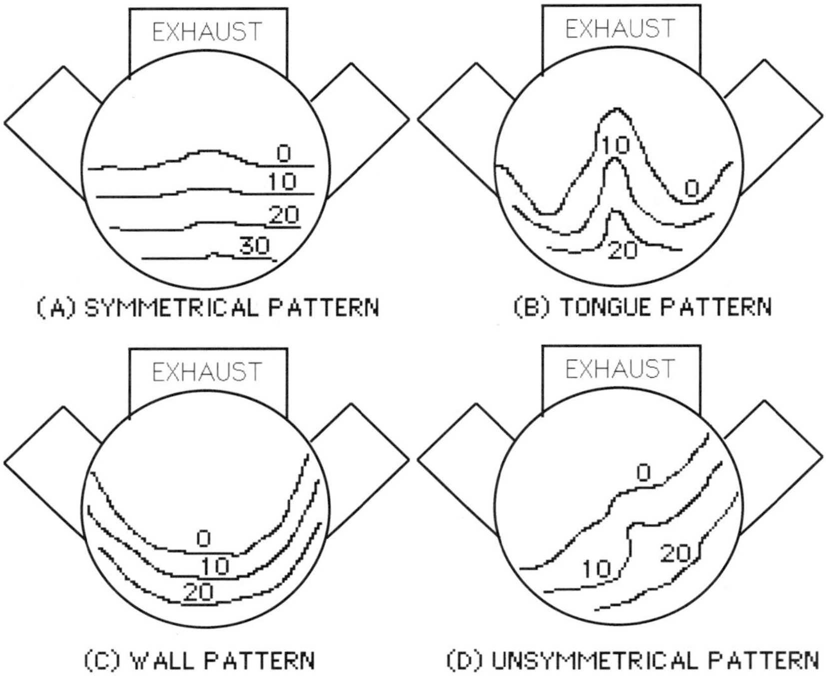

Fig. 3.5 Typical velocity contours observed in the Jante test.

center of the cylinder to a maximum at that side of the cylinder opposite to the exhaust port; in the "trade" this side is usually referred to as the "back of the cylinder." (As with most professional specializations, the use of jargon is universal!) The other patterns, (b)-(d), all produce bad scavenging characteristics. The so-called "tongue" pattern, (b), would give a rapid and high-speed flow over the cylinder head face and proceed directly to, and presumably out of, the exhaust port. The pattern in (c), dubbed the "wall" pattern, would ultimately enfold large quantities of exhaust gas and become a classic mixing process. So, too, would the asymmetrical flow shown in (d).

The author, at one time, took a considerable interest in the use of the Jante test method as a practical tool for the improvement of the scavenging characteristics of engines. At QUB there is a considerable experimental and theoretical effort in the development of actual engines for industry. Consequently, up to 1980, this method was employed as an everyday development tool. At the same time, a research program was instituted at QUB to determine the level of its effectiveness, and the results of those investigations are published in technical papers(1.23)(3.13). In an earlier paper(3.6) the author had published the methodology adopted experimentally at QUB for the recording of these velocity contours, and had introduced several analytical parameters to quantify the order of improvement observed from test to test and from engine to engine. It was shown in references (1.23) and (3.13) that the criterion of "mean velocity," as measured across the open cylinder bore, did determine the ranking order of scavenging efficiency in a series of cylinders for a 250cc Yamaha motorcycle engine. In that work the researchers at QUB also correlated the results of the Jante testing technique with scavenging efficiency measurements acquired under firing conditions with the apparatus shown in Plate 3.3. The work of Nuti and Martorano(3.33) confirmed that cylinders tested using the Jante simulation method correlated well with scavenging efficiency measurements acquired under firing conditions by cylinder gas sampling. In particular, they agreed that the "mean velocity" criterion proposed by the author(3.6) was an accurate indicator of the scavenging behavior under firing conditions.

Plate 3.2 The single cycle gas scavenging experimental apparatus at QUB.

The main advantages of the Jante test method are that (a) it is a test conducted under dynamic conditions, (b) it is a test which satisfies the more important, but not all, of the laws of fluid mechanics regarding similarity (see Sect. 3.2.2), (c) it is a test on the actual hardware relevant to the development program for that engine, (d) it does not require expensive instrumentation or hardware for the conduct of the test, (e) it has been demonstrated to be an effective development tool.

The main disadvantages of the Jante test method are that (f) it is a test of scavenging flow behavior conducted without the presence of the cylinder head, which clearly influences the looping action of the gas flow in that area, (g) it is difficult to relate the results of tests on one engine to another engine even with an equal swept volume, or with an unequal swept volume, or with a differing bore-stroke ratio, (h) it is even more difficult to relate "good" and "bad" velocity contour patterns from loop to cross scavenged engines, (i) it has no relevance as a test method for uniflow scavenged engines, where the flow is deliberately swirled by the scavenge ports at right angles to the measuring pitot tubes, (j) because it is not an absolute test method producing SE-SR or TE-SR characteristics, it provides no information for the direct comparison of the several types of scavenging flow conducted in the multifarious geometry of engines in existence, (k) because it is not an absolute test method, no information is provided to assist the theoretical modeller of scavenging flow during that phase of a mathematical prediction of engine performance.

3.2.2 Principles for successful experimental simulation of scavenging flow

Some of the points raised in the last paragraph of Sect. 3.2.1 require amplification in more general terms.

Model simulation of fluid flow is a science(3.14) developed to deal with, for example, the realistic testing of model aircraft in wind tunnels. Most of the rules of similarity which compare the model to be tested to reality are expressed in terms of dimensionless quantities such as Reynolds, Froude, Mach, Euler, Nusselt, Strouhal, and Weber numbers, to name but a few. Clearly, the most important dimensionless quantity requiring similarity is the Reynolds number, for that determines whether the test is being conducted in either laminar or turbulent flow conditions, and, as the real scavenging flow is demonstrably turbulent, at the correct level of turbulence. To be pedantic, all of the dimensionless quantities should be equated exactly for precision of simulation. In reality, in any simulation procedure, some will have less relevance than others. Dedeoglu(3.10), in examining the applicability of these several dimensionless factors, maintained that the Strouhal number was an important similarity parameter.

Another vitally important parameter which requires similarity in any effective simulation of scavenging flow is the motion of the piston as it opens and closes the transfer ports. The reason for this is that the gas flow commences as laminar flow at the port opening, and becomes fully developed turbulent flow shortly thereafter. That process, already discussed in Chapters 1 and 2, is one of unsteady gas-dynamic flow where the particle velocity varies in time from zero to a maximum value and then returns to zero at transfer port closure. If one attempts to simulate the scavenging flow by holding the piston stationary, and subjects the cylinder to steady flow of either gas or liquid, then that flow will form attachments to either the piston crown or the cylinder walls in a manner which could not take place in the real velocity-time situation. This effect has been investigated by many researchers, Rizk(3.11), Percival(3.16), Sammons(3.15), Kenny(3.18) and Smyth(3.17), and they have concluded that, unless the test is being conducted in steady flow for some specific reason, this is not a realistic simulation of the scavenging flow.

3.2.3 Absolute test methods for the determination of scavenging efficiency

To overcome the disadvantages posed by the Jante test method, it is preferable to use a method of assessment of scavenging flow which will provide measurements of scavenging, trapping and charging efficiency as a function of scavenge ratio. It would also be preferable to conduct this test isolated from the confusing effects of combustion and exhaust pipe tuning characteristics. This implies some form of model test, but this immediately raises all of the issues regarding similarity discussed in Sect. 3.2.2. A test method which does not satisfy these criteria, or at least all of the vitally important dimensionless criteria, is unacceptable.

It was Sammons(3.15) who first postulated the use of a single-cycle apparatus for this purpose. Because of the experimental difficulty of accurately measuring oxygen concentrations by gas analysis in the late 1940's (a vital element of his test method), his proposal tended to be forgotten. It was revived in the 1970's by Sanborn(3.13), who, together with researchers at QUB, investigated the use of a single cycle apparatus using liquids to simulate the flow of fresh charge and exhaust gas. Sanborn continued this work(3.21) and other researchers investigated scavenging flow with this experimental methodology(3.22).

At QUB there was a growing realization that the occasional confusing result from the liquid-filled apparatus was due to the low Reynolds numbers found during the experiment, and that a considerable period of the simulated flow occurred during laminar or laminar-turbulent transition conditions. Consequently, the QUB researchers reverted to the idea postulated by Sammons(3.15), for in the intervening years the invention of the paramagnetic analyzer, developed for the accurate determination of oxygen proportions in gas analysis, had realized the experimental potential inherent in Sammons' original technique. (As a historical note, for it illustrates the frailty of the human memory, the author had forgotten about, or had relegated to the subconscious, the Sammons paper and so reinvented his experimental method.)

However, as will become more evident later, the QUB apparatus incorporates a highly significant difference from the Sammons experimental method, i.e., the use of a constant volume cylinder during the scavenge process.

The QUB apparatus sketched in Fig. 3.6, and appearing in Plate 3.2, is very thoroughly described and discussed by Sweeney(3.20). The salient features of its operation are a constant volume crankcase and a constant volume cylinder during the single cycle of operation from tdc to tdc at a known turning rate. The equivalent rotational rate is 700 rev/min. The cylinder is filled with air to represent exhaust gas, and the crankcase is filled with carbon dioxide to represent the fresh air charge. Here, one similarity law is being well satisfied, in that the typical density ratio of crankcase air charge to cylinder exhaust gas is about 1.6, which corresponds almost exactly with the density ratio of 1.6 for the carbon dioxide to air used in the experiment. The Reynolds numbers are well into the turbulent zone at the mid-flow position, so yet another similarity law is satisfied.

The gases in the cylinder and the crankcase are at atmospheric temperature at the commencement of the process. The crankcase pressure is set at particular levels to

Plate 3.3 QUB apparatus for the measurement of scavenging efficiency under firing conditions.

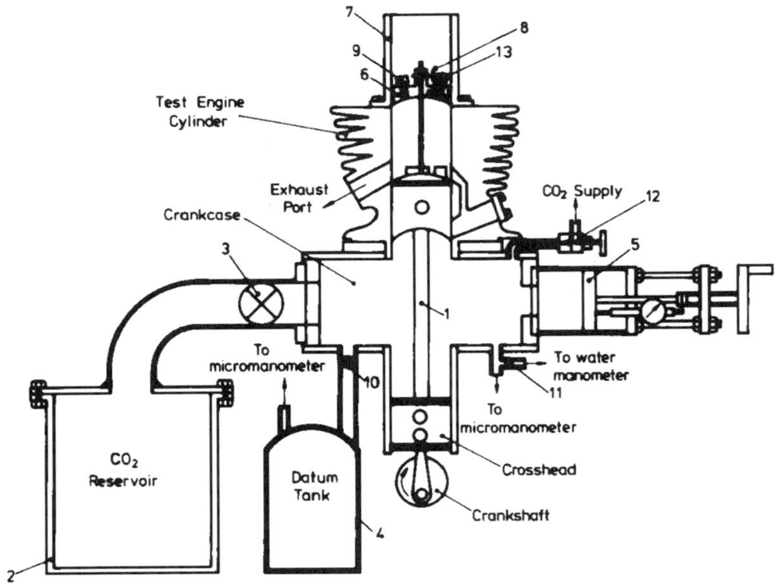

Fig. 3.6 QUB single cycle gas scavenging apparatus.

produce differing values of scavenge ratio, SR, during each experiment. At the conclusion of a single cycle, the crankshaft is abruptly stopped at tdc by a wrap-spring clutch brake, retaining under the movable cylinder head the trapped charge from the scavenging flow. The movable cylinder head is then released from the top piston rod and depressed so that more than 75% of the trapped gas contents are forced through a paramagnetic oxygen analyzer, giving an accurate measurement from a representative gas sample.

If MO_2 is the measured oxygen concentration in the trapped charge, RNO is the molecular ratio of nitrogen to oxygen in air, and KC is a correction coefficient for the slight paramagnetism exhibited by carbon dioxide, then Sweeney(3.20) shows that the scavenging efficiency, SE, is given by:

$$SE=[1-0.01*(1+RNO)*MO_2]/[1-KC*(1+RNO)] \tag{3.2.2}$$

The correction coefficient for carbon dioxide in the presence of oxygen, KC, is a negative number, -0.00265. The value of RNO is traditionally taken as 79/21, or 3.762.

The scavenge ratio, SR, is found by moving the piston, Fig. 3.5(5), inward at the conclusion of the single cycle experiment until the original crankcase state conditions of pressure and temperature are restored. A dial gauge accurately records the piston movement. As the piston area is known, the volume of charge which left the crankcase, Vcc, is readily determined. The volume of charge, Vcy, which scavenged the cylinder, is then calculated by the state equation:

$$Pcc*Vcc/Tcc=Pcy*Vcy/Tcy \tag{3.2.3}$$

The temperatures, Tcy and Tcc, are identical and equal to the atmospheric temperature, Tat. The cylinder pressure, Pcy, is equal to the atmospheric pressure, Pat.

The cylinder volume can be set to any level by adjusting the position of the cylinder head on its piston rod. The obvious value at which to set that volume is the cylinder swept volume, SV.

The scavenge ratio, SR, is then:

$$SR=Vcy/SV \tag{3.2.4}$$

In the final paragraph of Sect. 3.1, the desirability was emphasized of conducting a relevant scavenging experiment in an isothermal, isovolumic and isobaric fashion. This experiment satisfies these criteria as closely as any experiment ever will. It also satisfies all of the relevant criteria for dynamic similarity. It is a dynamic experiment, for the piston moves and provides the realistic attachment behavior of scavenge flow as it would occur in the actual process in the firing engine. Sweeney(3.20) (3.23) demonstrates the repeatability of the test method and of its excellent correlation with experiments conducted under firing conditions. Some of those results are worth repeating here, for that point cannot be emphasized too strongly.

The experimental performance characteristics, conducted at full throttle for a series of modified Yamaha DT250 engine cylinders, are shown in Fig. 3.7. Each cylinder has identical engine geometry so that the only modifications made were to the directioning of the transfer ports and the shape of the transfer duct. Neither port timings nor port areas were affected so that each cylinder had almost identical SR characteristics at any given rotational speed. Thus, the only factor influencing engine performance was the scavenge process. That this is significant is clearly evident from that figure, as the BMEP and BSFC behavior is affected by as much as 15%. When these same cylinders were tested on the single cycle gas scavenging rig, the SE-SR characteristics were found to be as shown in Fig. 3.8. The figure needs closer examination, so a magnified region centered on a SR value of 0.9 is shown in Fig. 3.9. Here, it can be observed that the ranking order of those same cylinders is exactly as in Fig. 3.7, and so too is their relative positions. In other words, Cylinders 14 and 15 are the best and almost as effective as each other, so too are cylinders 9 and 7 but at a lower level of scavenging efficiency and power. The worst cylinder of the group is cylinder 12 on all counts. The "double indemnity" nature of a loss of 8% SE at a SR level of 0.9, or a TE drop of 9%, is translated into the BMEP and BSFC shifts already detailed above at 15%.

A sustained research and development effort has taken place at QUB in the experimental and theoretical aspects of scavenging flow. For the serious student of the subject, the papers published from QUB form a series, each using information and thought processes from the preceding publication. That reading list, in consecutive order, is (3.6), (1.23), (3.13), (3.20), (1.10), (3.23), (1.11), and (3.17).

3.2.4 Comparison of loop, cross and uniflow scavenging

The QUB single cycle gas scavenge permits the accurate and relevant comparison of SE-SR and TE-SR characteristics of different types of scavenging. From some of those previous papers, and from other experimental work at QUB hitherto unpublished, test results for uniflow, loop and cross scavenged engine cylinders are presented to illustrate the several points being made. At this stage the most important issue is the use of the experimental apparatus to compare the various methods of scavenging, in order to derive some fundamental understanding of the effectiveness of the scavenging process conducted by these several methodologies. In Sect. 3.3 the information gained will be used to determine the theoretical relevance of this experimental data in the formulation of a model of scavenging to be incorporated within a complete theoretical model of the firing engine.

Figs. 3.10 and 3.11 give the scavenging and trapping characteristics for six engine cylinders, as measured on the single cycle gas scavenging rig. It will be observed that the test results fall between the perfect displacement line and perfect mixing line from the theories of Hopkinson(3.1). By contrast, as will be discussed in Sect. 3.3, some of the data presented by others(3.3)(3.4) lie below the perfect mixing line.

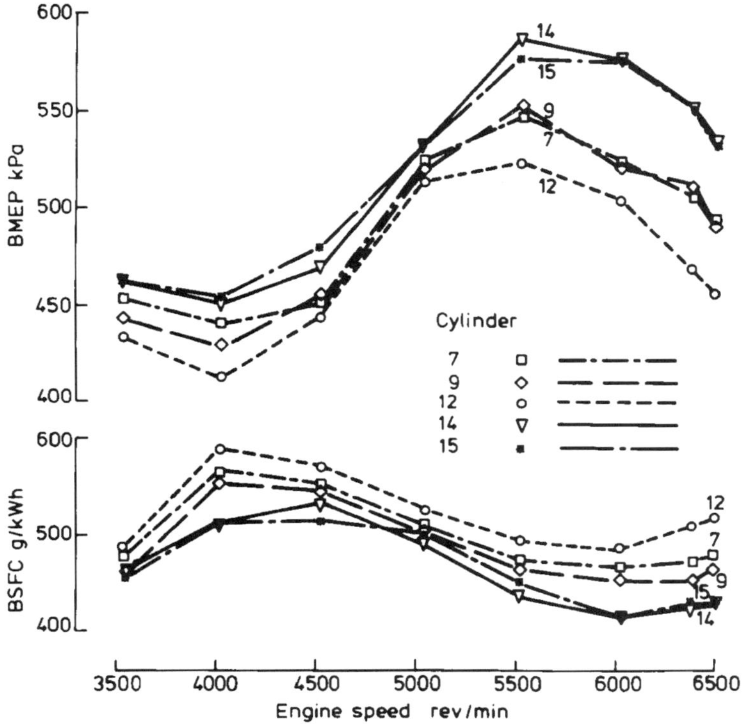

Fig. 3.7 Measured performance characteristics from firing tests.

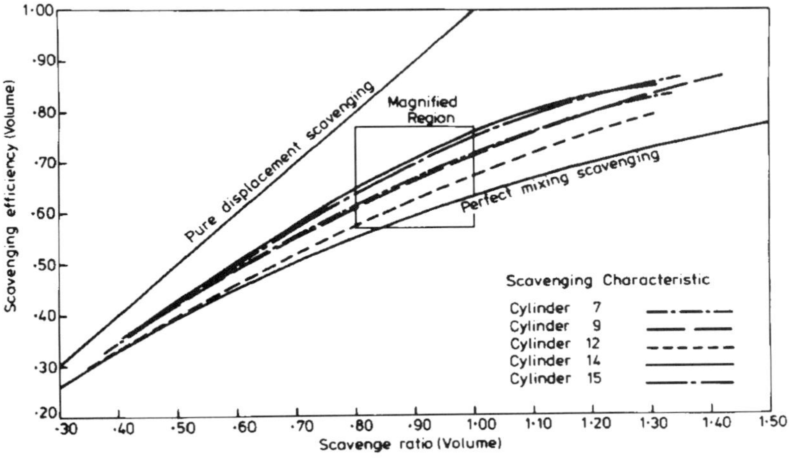

Fig. 3.8 Measured isothermal scavenging characteristics.

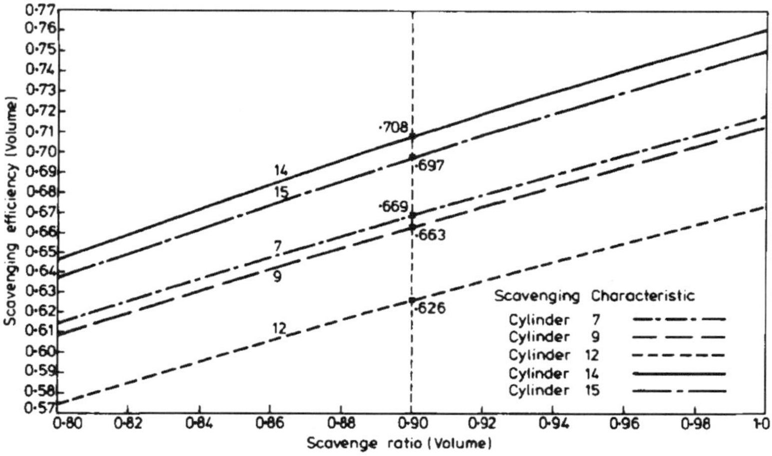

Fig. 3.9 Magnified region of SE-SR characteristics in Fig. 3.8.

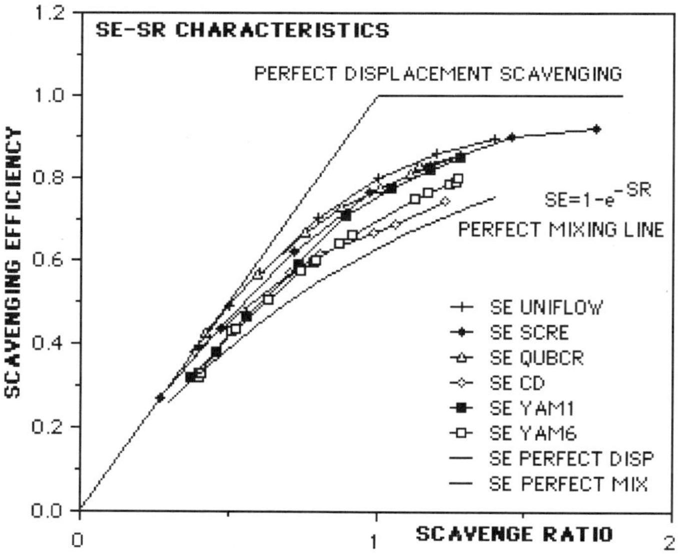

Fig. 3.10 SE-SR characteristics for the test cylinders.

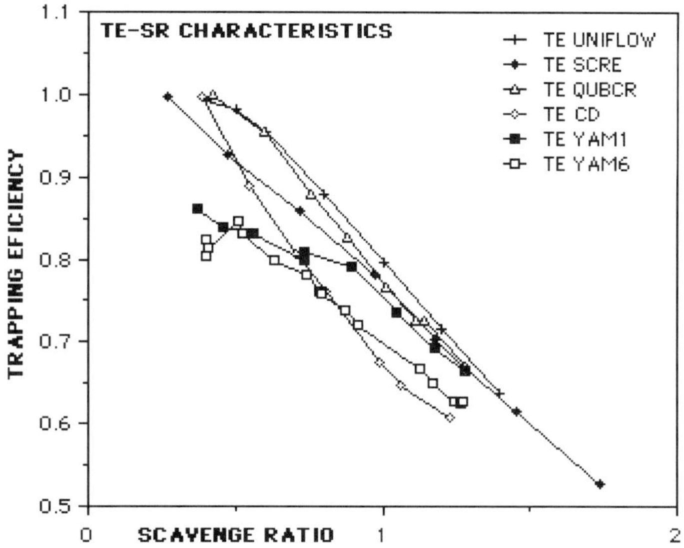

Fig. 3.11 TE-SR characteristics for the test cylinders.

The cylinders used in this study, in order of their listing on Figs. 3.10 and 3.11, are:

(a) A ported uniflow scavenged cylinder of 302 cc swept volume, called UNIFLOW. It has a bore stroke ratio of 0.6, and the porting configuration is not dissimilar to that found in the book by Benson(1.4, Vol.2, Fig. 7.7, p.213). The engine details are proprietary and further technical description is not possible.

(b) A loop scavenged cylinder of 375 cc swept volume, called SCRE. This engine has three transfer ports, after the fashion of Fig. 3.38.

(c) A 250 cc QUB cross scavenged cylinder, called QUBCR, and previously described in (1.10) in considerable detail. The detailed porting geometry is drawn in that paper. The general layout is almost exactly as illustrated in Fig. 1.4.

(d) A 250 cc classic cross scavenged cylinder, called CD, and previously described in (1.10) as the classic deflector. The detailed porting geometry is drawn in that paper. The general layout is almost exactly as illustrated in Fig. 1.3.

(e) A 250 cc loop scavenged cylinder, modified Yamaha DT 250 cylinder no.1, and called here YAM1, but previously discussed in (3.20). The detailed porting geometry is drawn in that paper.

(f) A 250 cc loop scavenged cylinder, modified Yamaha DT 250 cylinder no.6, and called here YAM6, but previously discussed in (3.20). The detailed porting geometry is drawn in that paper.

The disparate nature of the scavenging characteristics of these test cylinders is clearly evident. As might be expected, the uniflow scavenged cylinder is very good, indeed it is the best, but not by the margin suggested by Changyou(3.4). The behavior of the two types of cross scavenged cylinders is interesting, ranging from an equal-best situation at SR=0.4 to near "perfect mixing" mediocrity for the CD cylinder when SR=1.4. The QUBCR cylinder has a scavenging behavior which is quite similar to the best loop scavenged cylinder, SCRE. This SCRE cylinder is seen to have very superior scavenging to YAM1, even though the internal porting configuration is not too dissimilar.

When assessing the effectiveness of scavenging at light load, i.e., a low BMEP condition, it is worth remembering that the scavenge ratio axis is directly related to throttle opening. By definition, a small throttle opening will permit a reduced air flow to be induced into the engine; a low SE gives a low charging efficiency, CE, and hence a low torque output under firing conditions. Some of the points made in connection with the good low-speed and low-power characteristics for conventional cross scavenged engines in Sect. 1.2.2 can now be seen to be accurate. Only the CD, QUBCR and UNIFLOW engines have high trapping efficiencies at low scavenge ratios, whereas the best of loop scavenged engines, SCRE, is some 6% inferior in this SR region. In Sect. 3.1.5, this matter was discussed in terms of the ability of the engine to have the best possible power, fuel economy and emission characteristics at idle speed or at light load. There are measured performance characteristics which substantiate the statements above; a QUBCR engine and a loop scavenged engine with scavenging behavior at least equal to the SCRE cylinder, each of 400cc swept volume and both carburetted with identical exhaust systems, have idle fuel flow rates of 150 and 275 g/h, respectively.

The inferior nature of the scavenging of the cross scavenged engine, CD, at full throttle or a high SR value, is easily seen from the diagrams. It is for this reason that such power units have generally fallen from favor as outboard motors. However, in Sect. 3.5.2 it is shown that classical cross scavenging can be optimized in a superior manner to that already reported in the literature(1.10) and to a level which is rather higher than a mediocre loop scavenged design. It is probably true to say that it is easier to develop a satisfactory level of scavenging from a cross scavenged design, by the application of the relatively simple empirical design recommendations of Sect. 3.5, than it is for a loop scavenged engine.

The modified Yamaha cylinder, YAM6, has the worst scavenging overall. Yet, if one merely physically examined the cylinders YAM1, YAM6 and SCRE, it is doubtful if the opinion of any panel of "experts" would be any more unanimous on the subject of their scavenging ability than they would be on the quality of the wine being consumed with their dinner! This serves to underline the important function served by an absolute test for scavenging and trapping efficiencies.

3.3 Comparison of experiment and theory of scavenging flow

3.3.1 Analysis of experiments on the QUB single cycle gas scavenging rig

As has already been pointed out, the QUB single cycle gas scavenging rig is a classic experiment conducted in an isovolumic, isothermal and isobaric fashion. Therefore, one is entitled to compare the measurements from that apparatus with the theoretical models of Hopkinson(3.1), Benson and Brandham(3.2), and others(3.3), to determine how accurate they may be for the modelling of two-stroke engine scavenging.

Eq. 3.1.19 for perfect mixing scavenging is repeated here:

$$SE=1-e^{(-SR)} \quad (3.1.19)$$

Manipulation of this equation shows:

$$\log_e(1-SE)=-SR \quad (3.3.1)$$

Consideration of this equation for the analysis of any experimental SE and SR data should reveal a straight line of equation form y=mx+c, with a slope of value -1 and an intercept at y=0 and x=0.

The Benson-Brandham model contains a perfect scavenging period, PS, before total mixing occurs, and a short-circuited proportion, SC. This resulted in Eqs. 3.1.23 and 24, repeated here:

$$\text{If } (1-SC)*SR<PS \text{ then } SE=(1-SC)*SR \quad (3.1.23)$$

$$\text{If } (1-SC)*SR>PS \text{ then } SE=1-(1-PS)*e^{(PS-(1-SC)*SR)} \quad (3.1.24)$$

Manipulation of this latter equation reveals:

$$\log_e(1-SE)=(SC-1)*SR+\log_e(1-PS)+PS \quad (3.3.2)$$

Again, further consideration of this equation on the straight line theory that y=mx+c shows that test data of this type should give a slope of (SC-1) and an intercept of $\log_e(1-PS)+PS$. Any value of short-circuiting other than 0 and the maximum possible value of 1 would result in a line of slope m, where 0>m>-1. The slope of such a line could not be less than -1, as that would produce a negative value of the short-circuiting component, SC, which is clearly theoretically impossible.

Therefore it is interesting to examine the experimentally determined data presented in Sect. 3.2.4. The analysis is based on plotting $\log_e(1-SE)$ as a function of SR from the experimental data for two of the cylinders, SCRE and YAM6, as examples of "good" and "bad" loop scavenging, and this is shown in Figs. 3.12 and 3.13. It is gratifying that the experimental points fall on a straight line. In reference (1.11), the entire set of cylinders shown in Figs. 3.10 and 3.11 are analyzed in this manner and are shown to have a similar quality of fit to a straight line. What is less gratifying, in terms of an attempt at correlation with a Benson and Brandham type of theoretical model, is the value of the slope of any of the lines. All of the slopes lie in the region between -1 and -2. A summary of the findings is shown in Fig. 3.14 for selected cylinders, representing (a) the uniflow cylinder as an example of the best scavenging yet observed, (b) the SCRE cylinder as an example of very good loop scavenging, (c) the YAM1 cylinder as an example of quite good loop scavenging, (d) the YAM6 cylinder as an example of bad loop scavenging, (e) the CD cylinder as a somewhat mediocre example of conventional cross scavenging, and (e) the QUBCR cylinder as an example of QUB cross scavenging with radial ports. The slope which is nearest to correlation with a Benson-Brandham type of theoretical model is the CD cylinder with a slope of -1.0104, yet it is still on the wrong side of the -1 boundary value. All the rest have values closer to -2 and, clearly, the better the scavenging the closer the value is to -2. Therefore, there is no correlation possible with any of the "traditional" models of scavenging flow, as all of those models would seriously underestimate the quality of the scavenging in the experimental case.

It is possible to take the experimental results from the single cycle rig, plot them in the logarithmic manner illustrated and derive mathematical expressions for SE-SR and TE-SR characteristics representing the scavenging flow of all significant design types, most of which are noted in Fig. 3.14. This is particularly important for the theoretical modeller of scavenging flow who has previously been relying on models of the Benson-Brandham type.

The straight line equations, as seen in Figs. 3.12 and 3.13, and in reference (1.11), are of the form:

$$\log_e(1\text{-}SE)=M*SR+C \tag{3.3.3}$$

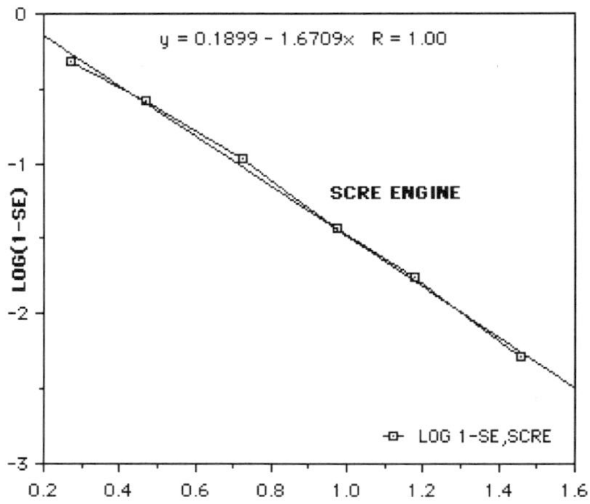

Fig. 3.12 Log$_e$(1-SE)-SR plot for the SCRE cylinder.

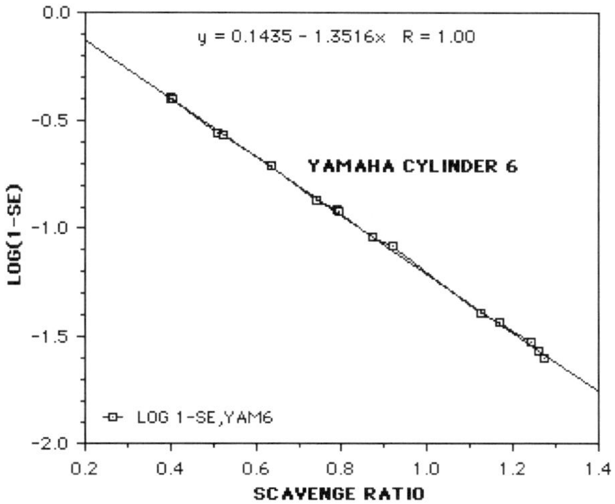

Fig. 3.13 Log_e(1-SE)-SR plot for the modified Yamaha cylinder No. 6.

NAME	TYPE	COMMENTS	SLOPE, M	INTERCEPT, C
UNIFLOW	PORTED UNIFLOW	VERY GOOD SCAVENGING	-1.7827	0.2094
SCRE	3 PORT LOOP	GOOD SCAVENGING	-1.6709	0.1899
YAM1	5 PORT LOOP	QUITE GOOD SCAVENGING	-1.6993	0.3053
YAM6	5 PORT LOOP	BAD LOOP SCAVENGING	-1.3516	0.1435
CD	CLASSIC CROSS	NORMAL CROSS TYPE	-1.0104	-0.117
QUBCR	QUB CROSS	QUB RADIAL PORTS	-1.6325	0.1397

Fig. 3.14 Experimental values of slope and intercept from log_e(1-SE)-SR plots.

Manipulating this equation shows:

$$SE=1-e^{(M*SR+C)} \tag{3.3.4}$$

Therefore, the modeller can take the appropriate M and C values for the type of cylinder which is being simulated from Fig. 3.14 and derive realistic values of scavenging efficiency, SE, and trapping efficiency, TE, at any scavenge ratio level, SR. What this analysis thus far has failed to do is to provide the modeller with information regarding the influence of "perfect displacement" scavenging, or "perfect mixing" scavenging, or "short circuiting" on the experimental or theoretical scavenging behavior of any of the cylinders tested.

3.3.2 A simple theoretical scavenging model which correlates with experiments

The problem with all of the simple theoretical models presented in previous sections of this chapter, is that the theoretician involved was under some pressure to produce a single mathematical expression or a series of such expressions. Much of this work took place in the pre-computer age, and readers who emanate from those slide rule days will appreciate that pressure. Consequently, even though Benson or Hopkinson knew perfectly well that there could never be an abrupt transition from perfect displacement scavenging to perfect mixing scavenging, this type of theoretical "fudge" was essential if Eqs. 3.1.8 to 3.1.24 were to ever be realized and be "readily soluble," arithmetically speaking, on a slide rule. Today's computer and calculator oriented engineering students will fail to understand the sarcasm inherent in the phrase, "readily soluble." It should also be said that there was no experimental evidence against which to judge the validity of the early theoretical models, for the experimental evidence contained in paper (3.20), and here as Figs. 3.10-14, has only been available since 1985.

Consequently, once the need was established, the production of a simple theoretical model of scavenging which could be solved on a microcomputer and did not contain a step change in flow behavior, was a relatively straightforward process. This is given as Prog.3.2 called BLAIR SCAVENGING MODEL and in the Computer Program Appendix as ProgList.3.2. This model of scavenging does not contain a step-function behavior but allows the scavenging process to proceed in a continuous fashion from start to finish, albeit containing the necessary elements of perfect displacement scavenging, mixing scavenging and short-circuiting. Such a theoretical postulation would be considered logical from the more advanced CFD calculations carried out by Sweeney(3.23) and discussed in Sect. 3.4.

The proposed model of scavenging flow is illustrated in Fig. 3.16, with reference to the geometry shown in Fig. 3.15. The variations of perfect scavenging and short-circuiting are functions of scavenge ratio. At any given level of scavenge ratio, SR, a proportion of the entering increment of scavenge air, dVas, is going to be short-circuited with value S, while the remainder of the entering air goes into the perfect scavenge volume, PSV. Of the air which enters the perfect scavenge volume, a proportion of value Y stays, and the remainder, proportion 1-Y, strips off the "surface" of the perfect scavenge volume and enters the mixing volume. The calculation method is very simple, as an examination of the computer program written in BASIC in ProgList.3.2 will reveal. The calculation proceeds to enter air increments of dVas equal to 1, up to any desired maximum value, with the cylinder volume set at V=1000. Hence the incremental change of scavenge ratio, dSR, is dVas/V, and the new scavenge ratio, SR, is increased by dSR. The entering increment of airflow, dVas, is distributed in various directions:

(a) the amount to be short-circuited, dSC, where dSC=S*dVas,

(b) the remainder to be distributed, dVrem, where dVrem=dVas-dSC,

(c) the amount perfectly scavenged, dVd, where dVd=Y*dVrem,

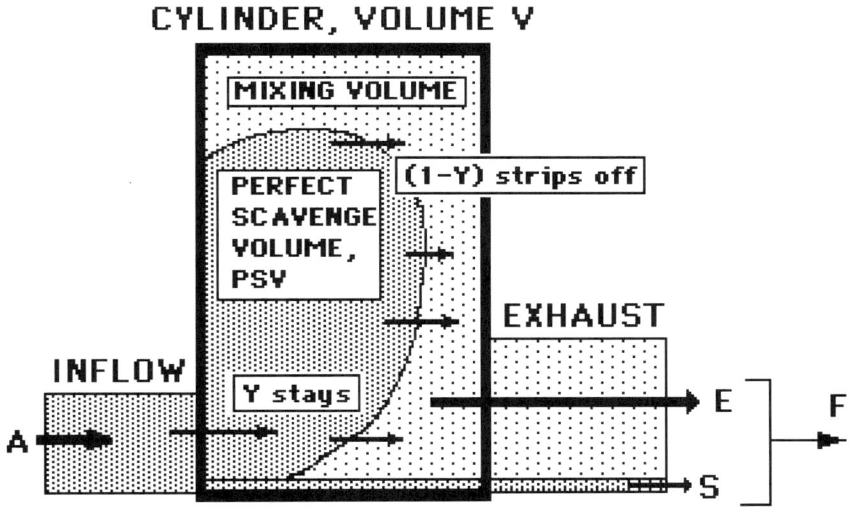

Fig. 3.15 Blair model of the scavenging flow.

(d) the amount to go to the mixing volume, dVam, where dVam=(1-Y)*dVrem. This results in new values of volume for the displacement and mixing volumes, Vd and Vmix, respectively. The actual values are given by:

$$Vmix_{new}=Vmix_{old}-dVd \quad (3.3.5)$$
$$Vd_{new}=Vd_{old}+dVd \quad (3.3.6)$$

The total amount of air in the cylinder is found by tracking the volume of air in both the displacement and the mixing volumes. In the displacement volume the scavenging efficiency is always unity, whereas in the mixing volume it is SEmix. This latter value is found by considering the following flow equations:

Flow from mixing region to exhaust, E, is given by E=dVrem. (3.3.7)

Air flow from mixing region to exhaust, Ea, is given by:

$$Ea=SEmix*dVrem \quad (3.3.8)$$

The new value of air retained in the mixing volume, Vam, is given by:

$$Vam_{new}=Vam_{old}+dVam-Ea \quad (3.3.9)$$

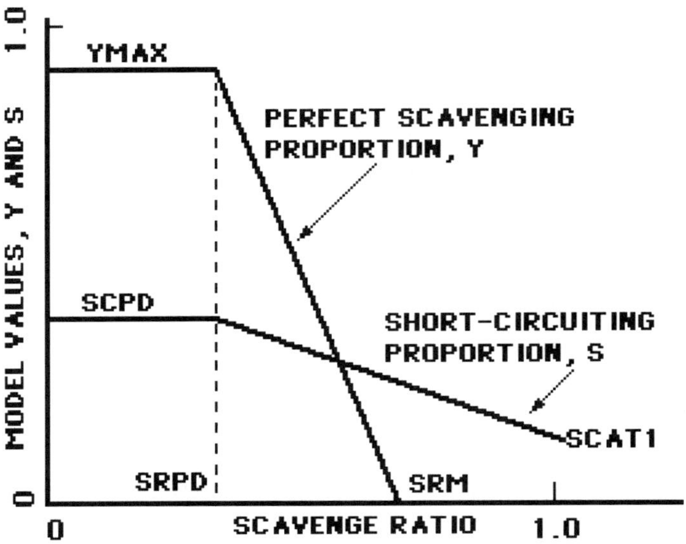

Fig. 3.16 Blair model of perfect scavenging and short-circuiting.

The next incremental value of the scavenging efficiency in that area, SEmix, is then:

SEmix=Vam/Vmix (3.3.10)

The overall scavenging and trapping efficiencies for the cylinder, SE and TE, are found by:

SE=(Vam+Vd)/V and TE=SE/SR (3.3.11)

In the computer example quoted in ProgList.3.2, the calculation proceeds in dVas increments of unity for 1400 steps for a cylinder volume of 1000; i.e., the dSR steps are of 0.001 value and are pursued until the scavenge ratio, SR, is 1.4.

The numerical values of short-circuiting and displacement scavenging, S and Y, for a particular level of scavenge ratio are determined from data representing a model of the type sketched in Fig. 3.16. The model is as follows:

For 0<SR<SRPD, Y=YMAX. (3.3.12)
For SRPD<SR<SRM, YMAX>Y>0 in a straight line relationship. (3.3.13)
For 0<SR<SRPD, S=SCPD. (3.3.14)
For SRPD<SR<1, SCPD>S>SCAT1 in a straight line relationship. (3.3.15)

It should be noted that it is possible to have the short-circuit proportion increase with scavenge ratio by this form of model, and in certain cases that was found to be the only logical method of fitting the theory to the experiment. For the majority of engines, correlation was found to be a very simple procedure, with SRPD=0, YMAX=1 and SCAT1=0, leaving only the variables SRM (where Y=0) and SCPD (at SR=0) to be manipulated in order to achieve good correlation with experiment.

Figs. 3.17 and 3.18 show the results of applying this model to the SCRE and the YAM6 cylinders, the analysis of their experimentally determined SE-SR characteristics having been shown in Figs. 3.12 and 3.13. The close correlation of theory and experiment is evident. On each diagram are the theoretical values of M and C, the slope and intercept of the best fit straight line on the loge(1-SE)-SR graph for the theoretical solution. These values can be compared with the experimental values from Figs. 3.12 and 3.13. First, they are in close agreement with experiment and, second, the values are in the band from -1 to -2. All of the cylinders shown in Fig. 3.10 are analyzed by the author and presented elsewhere(1.11). The correlation level is equally as good for the remainder of the cylinders.

There are some further important points to be made:

(a) This theoretical model correlates well with a relevant experiment, as decreed necessary in Sect. 3.1.5.

(b) The model is relatively simple to "tune" for a particular cylinder, e.g., the SCRE cylinder has the single parameter, SRM, set at 0.74 and the rest set at zero. In simpler words, this means that the scavenging process starts with perfect scavenging, Y=1 at SR=0, and decays linearly until Y=0 at SR=0.74, at which point an equal quantity of the incoming scavenge air strips off the "perfect scavenge volume" into the mixing volume. The case of the YAM6 cylinder is similar, except that there is short-circuiting taking place, decaying linearly from a maximum of 17% at the onset of scavenging to zero at SR=1.

(c) Some of the characteristics which emerge are in line with predictions of a more detailed nature, such as the computational gas-dynamic model suggested by Sher(3.24) and (3.25). In one of these papers, Sher shows that the shape of the charge purity characteristic at the exhaust port during the scavenge process should have a characteristic profile. The more linear the profile the worse is the scavenging, the more 'S' like the profile the better is the scavenging. Fig. 3.19 shows the purity of the exiting charge at the exhaust port as predicted by the author's model of scavenging, and the profiles are exactly as suggested by Sher(3.24). The worst scavenging, as will be recalled, is from the traditional cross scavenged cylinder CD and the YAM6 loop scavenged cylinder. The best scavenging is from the SCRE and UNIFLOW cylinders.

(d) The simple model of scavenging as suggested by the author has good correlation with experimentation, and the model interrelates the relative contributions of perfect displacement scavenging, perfect mixing scavenging and short-circuiting. The theoretical modeller has available the relevant information on the significant characteristics of each contribution for loop, cross and uniflow scavenging, including a range of scavenging effectiveness for each type of scavenging system.

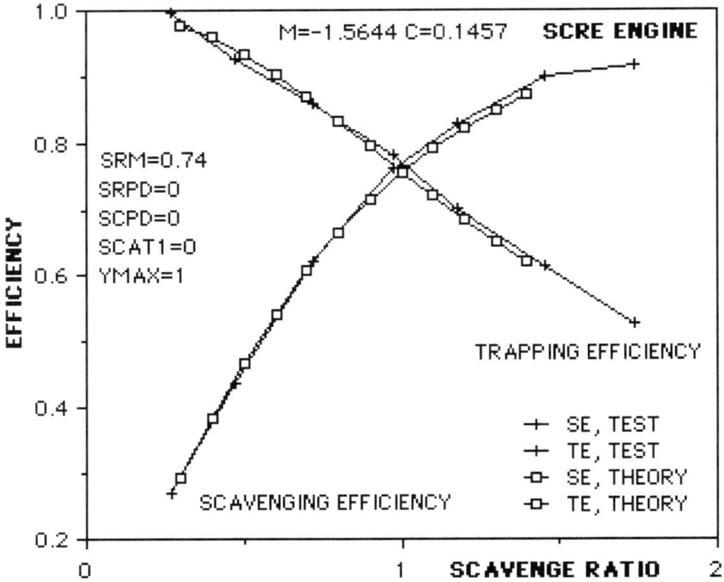

Fig. 3.17 Theory and experiment for the SCRE cylinder.

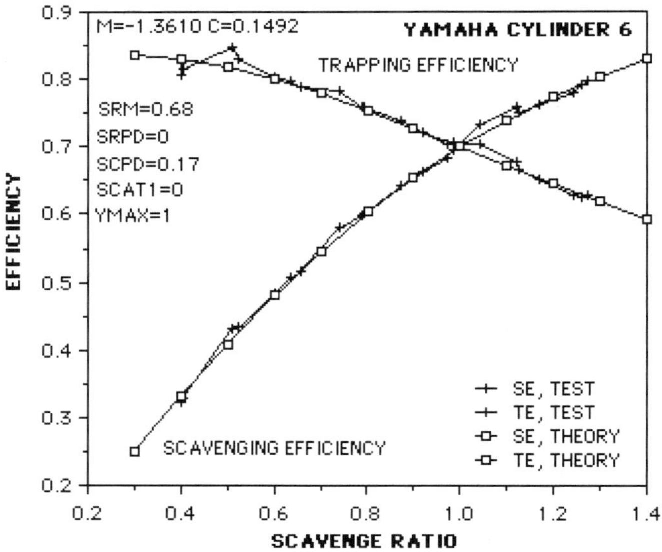

Fig. 3.18 Theory and experiment for the modified Yamaha cylinder No. 6.

3.4 Computational fluid dynamics

In recent years a considerable volume of work has been published on the use of Computational Fluid Dynamics, or CFD, for the prediction of in-cylinder and duct flows in (four-stroke cycle) ic engines. Typical of such publications are those by Brandstatter(3.26), Gosman(3.27) and Diwakar(3.28).

At The Queen's University of Belfast, much experience has been gained from the use of a general-purpose CFD code called PHOENICS. The structure and guiding philosophy behind this theoretical package has been described by its originator, Spalding(3.29). PHOENICS is developed for the simulation of a wide variety of fluid flow processes. It can analyze steady or unsteady flow, laminar or turbulent flow, flow in one, two, or three dimensions, and single- or two-phase flow. The turbulent flow is predicted by what is referred to as a k-epsilon model to estimate the effective viscosity of the fluid. The program divides the control volume of the calculated region into a large number of cells, and then applies the conservation laws of mass, momentum and energy over each of these regions. The mathematical intricacies of the calculation have no place in this book, so the interested reader is referred to the publications of Spalding(3.29) or Gosman (3.27), and the references those papers contain.

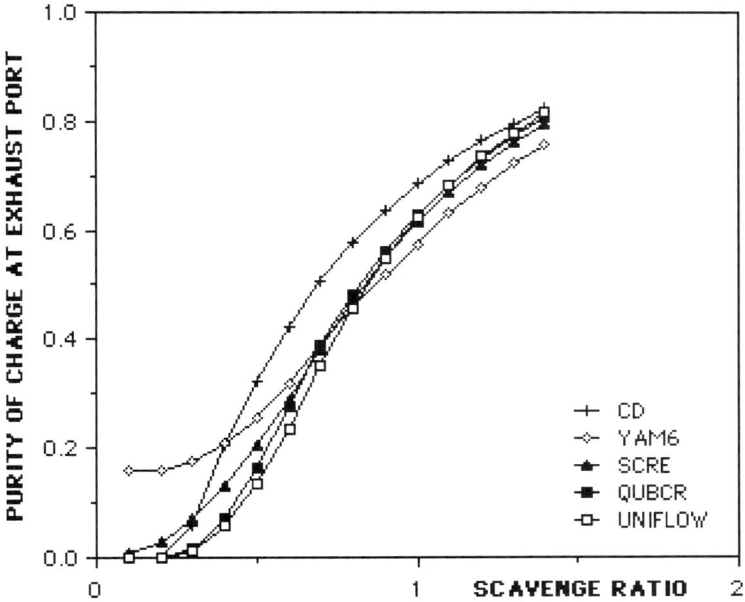

Fig. 3.19 Theoretically determined values of exhaust port charge purity.

A typical computational grid structure employed for an analysis of scavenging flow in a two-stroke engine is shown in Fig. 3.20. Sweeney(3.23) describes the operation of the program in some detail, so only those matters which are relevant to the current discussion will be dealt with here.

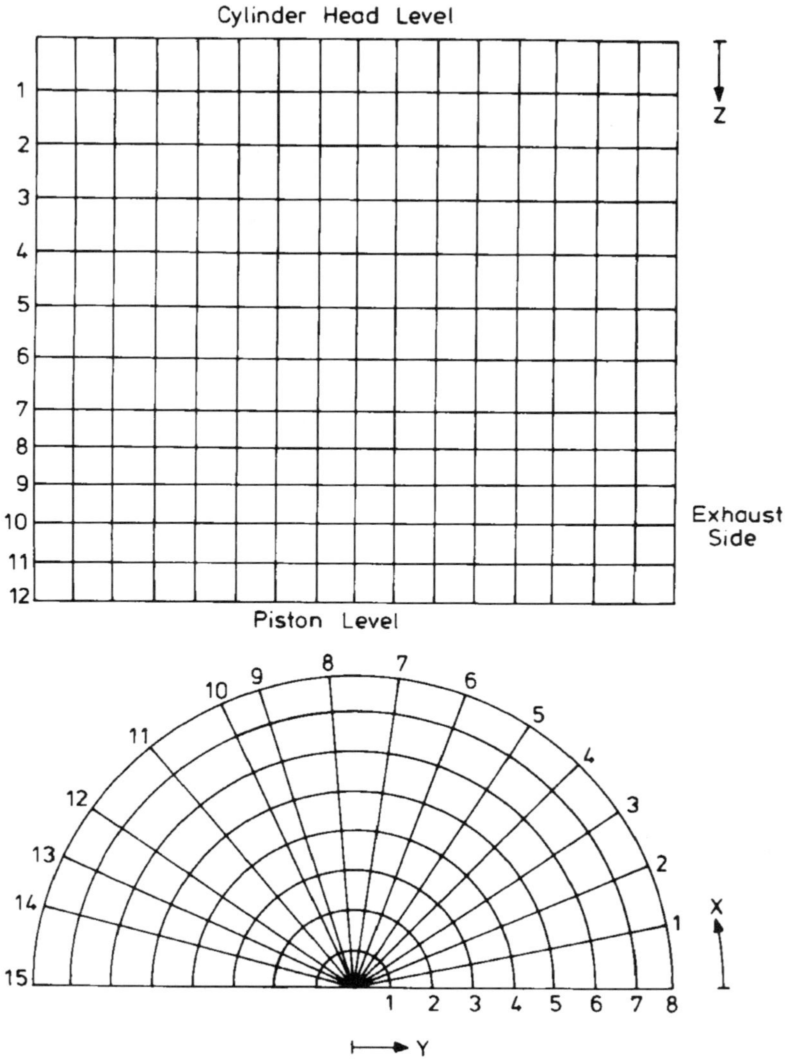

Fig. 3.20 Computational grid structure for scavenging calculation.

In the preceding sections, there has been a considerable volume of information presented regarding experimentally determined scavenging characteristics of engine cylinders on the single cycle gas scavenging rig. Consequently, at QUB it was considered important to use the PHOENICS CFD code to simulate those experiments and thereby determine the level of accuracy of such CFD calculations.

Further gas-dynamics software was written at QUB to inform the PHOENICS code as to the velocity and state conditions of the entering and exiting scavenged charge at all cylinder port boundaries. The flow entering the cylinder was assumed by Sweeney(3.23) to be "plug flow," i.e., the direction of flow of the scavenge air at any port through any calculation cell at the cylinder boundary was in the designed direction of the port. A later research paper by Smyth(3.17) shows this to be inaccurate, particularly for the main transfer ports in a loop scavenged design, and there is further discussion of that in Sect. 3.5.4.

CFD calculations use considerable computer time, and the larger the number of cells the longer is the calculation time. To be more precise, with the cell structure as shown in Fig. 3.20, and 60 time steps from transfer port opening to transfer port closing, the computer run time was 120 minutes on a VAX 11/785 mainframe computer. As one needs 7 individual calculations at particular values of scavenge ratio, SR, from about 0.3 to 1.5, to build up a total knowledge of the scavenging characteristic of a particular cylinder, that implies about 14 hours of computer run time.

A further technical presentation in this field, showing the use of CFD calculations to predict the scavenging and compression phases of in-cylinder flow in a firing engine, has been presented by Ahmadi-Befrui(3.30). In that paper, the authors also used the assumption of plug flow for the exiting scavenge flow into the cylinder and pre-calculated the state-time conditions at all cylinder-port boundaries using an unsteady gas-dynamic calculation of the type described in Chapter 2 and shown in Chapter 5. Nevertheless, their calculation showed the effect on the in-cylinder flow behavior of a varying cylinder volume during the entire compression process up to the point of ignition.

However, the insight which can be gained from this calculation technique is so extensive, and the accuracy level is also so very impressive, there is every likelihood that it will become the standard design method for the optimization of scavenging flow in two-stroke engines in the years ahead.

Examples of the use of the calculation are given by Sweeney(3.23) and a precis of the findings is shown here in Figs. 3.21-30. The examples selected as illustrations are modified Yamaha cylinders, nos. 14 and 12, the best and the worst of that group whose performance characteristics and scavenging behavior has already been discussed in Sect. 3.2.3.

A comparison of the measured and calculated SE-SR and TE-SR profiles are given in Figs. 3.29 and 3.30. It will be observed that the order of accuracy of the calculation, over the entire SR spectrum, is very high indeed. For those who may be familiar with the findings of Sweeney et al(3.23) it will be observed that the order of accuracy of correlation of the theoretical predictions with the experimental data is considerably better in Figs. 3.29 and 30 than it was in the original paper. The reason for this is that the measured data for the variance of the flow from the port

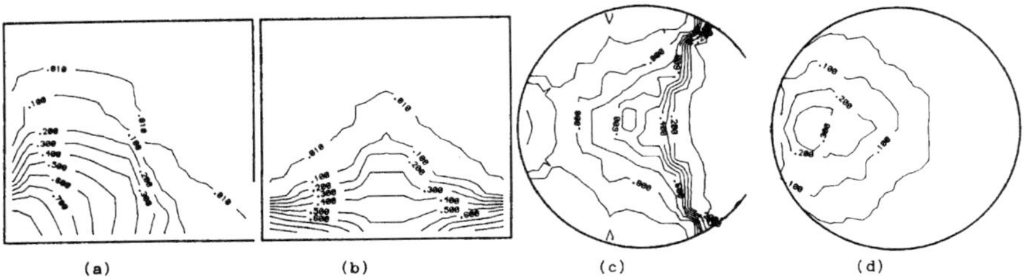

Fig. 3.21 Yamaha cylinder No. 14 charge purity plots at 39° bbdc.

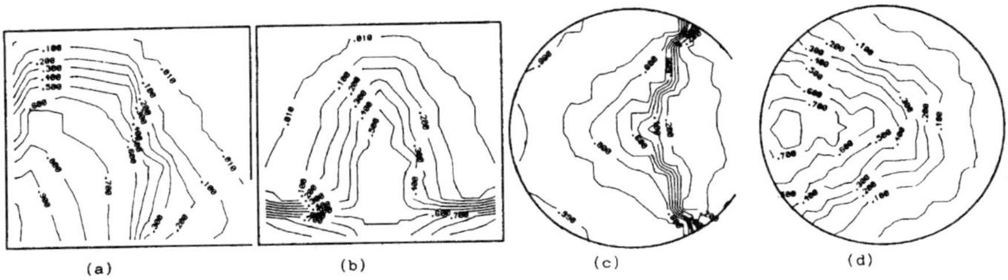

Fig. 3.22 Yamaha cylinder No. 14 charge purity plots at 29° bbdc.

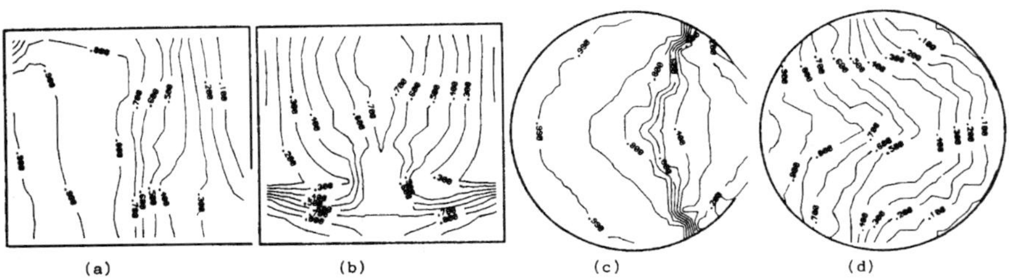

Fig. 3.23 Yamaha cylinder No. 14 charge purity plots at 9° bbdc.

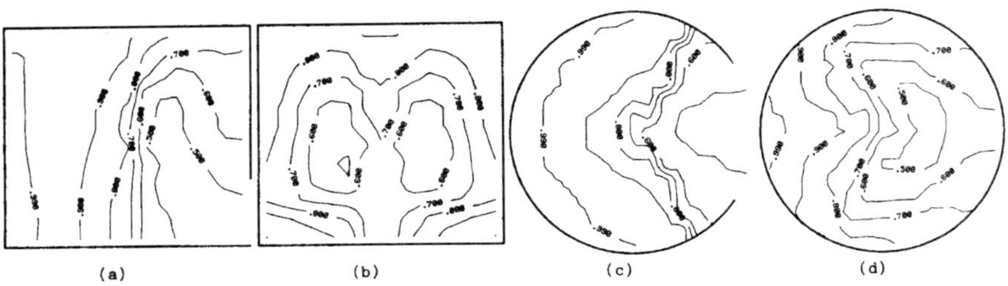

Fig. 3.24 Yamaha cylinder No. 14 charge purity plots at 29° abdc.

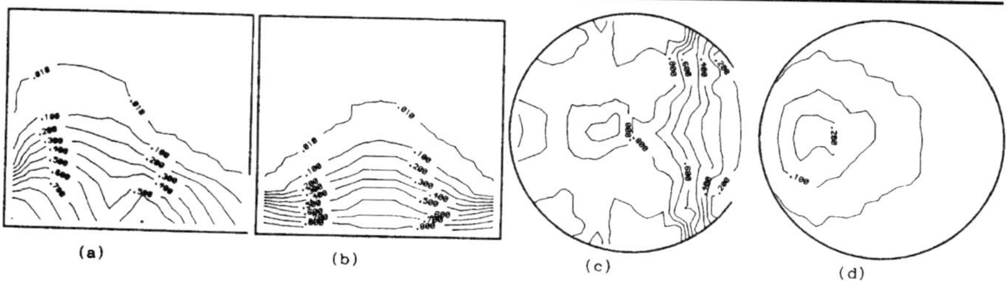

Fig. 3.25 Yamaha cylinder No. 12 charge purity plots at 39° bbdc.

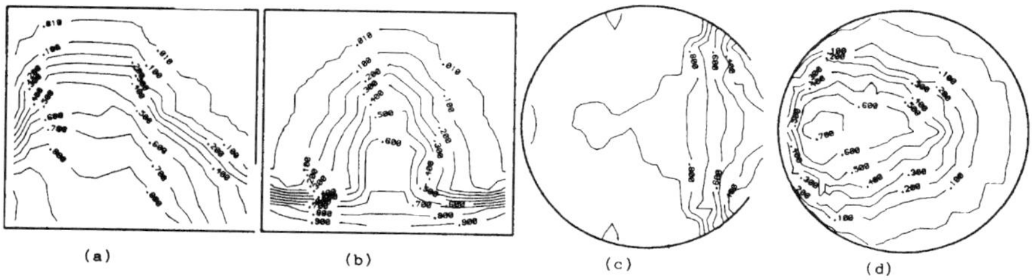

Fig. 3.26 Yamaha cylinder No. 12 charge purity plots at 29° bbdc.

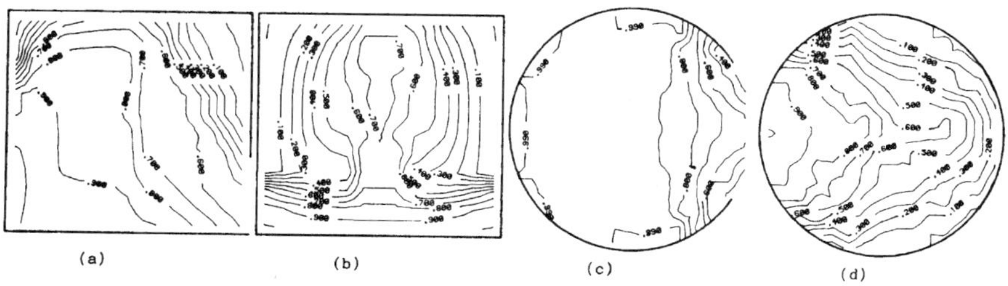

Fig. 3.27 Yamaha cylinder No. 12 charge purity plots at 9° bbdc.

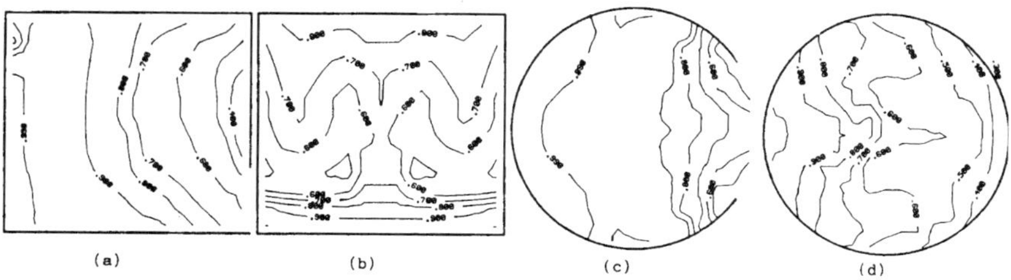

Fig. 3.28 Yamaha cylinder No. 12 charge purity plots at 29° abdc.

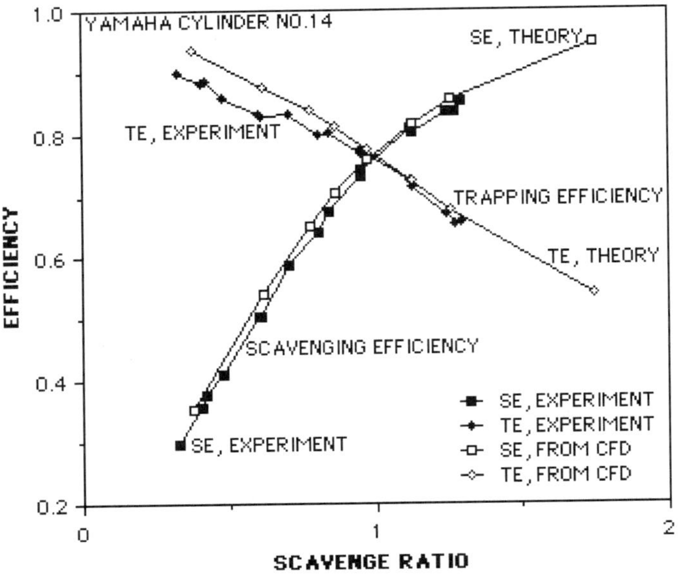

Fig. 3.29 Comparison of experiment and CFD theory for Yamaha cylinder No. 14.

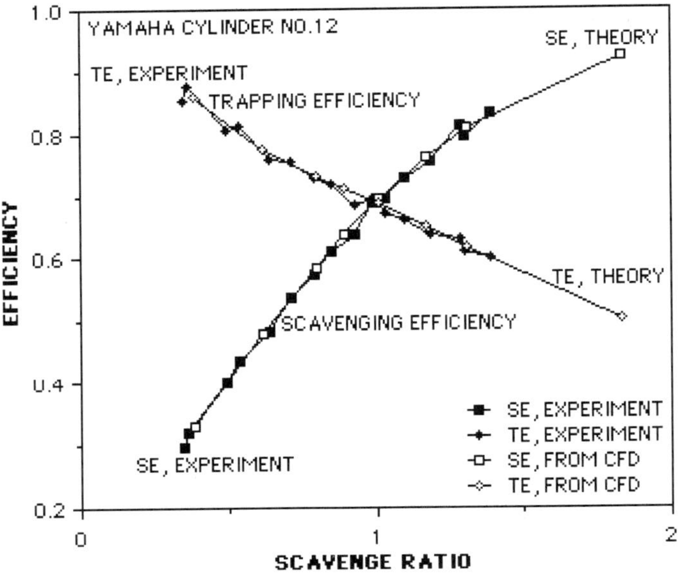

Fig. 3.30 Comparison of experiment and CFD theory for Yamaha cylinder No. 12.

design direction acquired by Smyth(3.17) and Kenny(3.31) was applied to the CFD calculations for those same modified Yamaha cylinders to replace the plug flow assumption originally used. It is highly significant that the accuracy is improved, for this moves the CFD calculation even closer to becoming a proven design technique.

In terms of insight, Figs. 3.21-3.24 and 3.25-3.28 show the in-cylinder charge purity contours for cylinders 14 and 12, respectively. On each figure are four separate cylinder sections culled from the cells in the calculation. The cylinder section (a) is one along the plane of symmetry with the exhaust port to the right. The cylinder section (b) is at right angles to section (a). The cylinder section (c) is on the surface of the piston with the exhaust port to the right. The cylinder section (d) is above section (c) and halfway between the piston crown and the cylinder head. In Fig. 3.21(a) and Fig. 3.25(a) it can be seen from the flow at 39 deg. bbdc that the short-circuiting flow is more fully developed in cylinder 12 than in cylinder 14. The SE level at the exhaust port for cylinder 14 is 0.01 whereas it is between 0.1 and 0.2 for cylinder 12. This situation gets worse by 29 deg. bbdc, when the SE value for cylinder 14 is no worse than 0.1, while for cylinder 12 it is between 0.1 and 0.4. This flow characteristic persists at 9 deg. bbdc, where the SE value for cylinder 14 is between 0.1 and 0.2 and the equivalent value for cylinder 12 is as high as 0.6. By 29 deg. abdc, the situation has stabilized with both cylinders having SE values at the exhaust port of about 0.6. For cylinder 12, the damage has already been done to its scavenging efficiency before the bdc piston position, with the higher rate of fresh charge flow to exhaust by short-circuiting. This ties in precisely with the views of Sher(3.24) and also with the author's simple scavenge model in Sect. 3.3.2, regarding the SE-SR or SE-crank angle profile at the exhaust port, as shown in Fig. 3.19.

The one aspect of a good scavenging cylinder such as cylinder 14 by comparison with a bad scavenging cylinder 12, is the sharp boundary between the "up" and the "down" parts of the looping flow. This is easily seen in Figs. 3.23 and 3.27. It is a recurring feature of all loop scavenged cylinders with good scavenging behavior, and this has been observed many times by the research team at QUB.

At the same time, examining those same figures but in cylinder section (d) closer to the cylinder head level, there are echos of Jante's advice on good and bad scavenging. Cylinder 12 would appear to have a "tongue," whereas cylinder 14 would conform more to the "good" Jante pattern as sketched in Fig. 3.5(a). Interested readers may wish to note that the measured Jante velocity profiles for cylinders 12 and 14 are given by Blair(1.23), and cylinder 12 does have a pronounced "tongue" pattern at 3000 rev/min.

3.5 Scavenge port design

By this time any reader involved in actual engine design will be formulating the view that however interesting the foregoing parts of this chapter may be, there has been little practical advice dispensed thus far. The author would contend that the best advice has been presented, which would be to construct a single cycle gas test rig, at best, or a Jante test apparatus if the funds will not run to the optimum test method. Actually, by comparison with many test procedures, neither are expensive forms of experimental apparatus. However, the remainder of this chapter is devoted to practical advice on scavenge port design for two-stroke engines, with the section on uniflow scavenging being the thinnest, as the author cannot claim extensive expertise in this area. The section on loop scavenging is the largest, with a major computer design program presented.

3.5.1 Uniflow scavenging

The only type which will be discussed is that for scavenging from the transfer ports which is deliberately aimed at inducing charge swirl. This form of engine design is most commonly seen in ci power units such as the Rootes TS3 or the GM diesel truck engines. The basic layout would be that of Fig. 1.5(a), where the scavenge air enters around the bdc position. One of the simplest port plan layouts is shown as a section through the scavenge ports in Fig. 3.31. The eight ports are configured at a mutual 45° with the port guiding edge being a tangent to a swirl circle

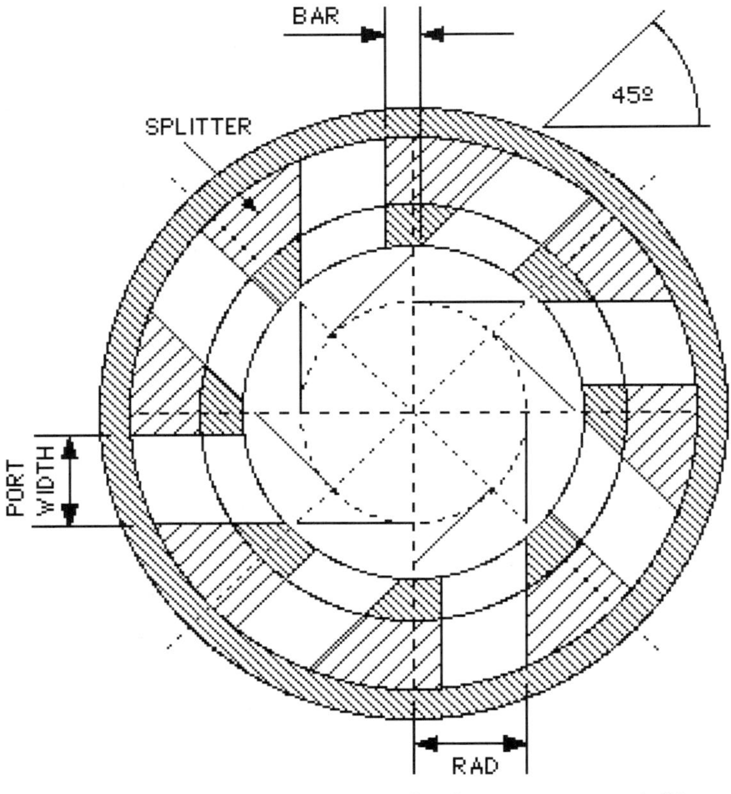

Fig. 3.31 Uniflow scavenge port plan layout at a mutual 45°.

of radius RAD, a typical value for which would be in the range 0.3*BO<RAD<0.4. The port width, PWID, is governed by the permissible value of the inter-port bar, BAR, and is determined by the mechanical strength of the liner, the stiffness of the piston rings in being able to withstand flexure into the ports, and the strong possibility of being able to dispense with pegging the piston rings. Piston rings are normally pegged in a two-stroke engine to ensure that the ends of the ring do not catch in any port during rotation. Because of the multiplicity of small ports in the uniflow case, elimination of pegging greatly reduces the tendency for the piston rings to stick, as will tend to be the case if they are held to a fixed position or if they are exposed to hot exhaust gas during passage over the exhaust ports. It is clear that uniflow scavenging utilizes the maximum possible area of the cylinder for scavenge porting by comparison with any variant of loop or cross scavenging, as shown in Figs. 1.2-1.4.

The actual design shown in Fig. 3.31, where the ports are at 45° and the RAD value is sketched at one third of the cylinder bore dimension, BO, is a very useful starting point for a new and unknown design. The use of a splitter in the "scavenge belt" may not always be practical, but it is a useful optimum to aim at, for the splitter helps to retain the gas motion closer to the design direction at the lower SR values, where the gas velocities are inherently reduced. The optimum value of the orientation angle, here shown as 45°, is very much affected by the bore-stroke ratio.

If the number of ports are PNO, then a useful guide to the designer is given by the ratio of the total effective port width to the bore dimension, WR, where:

$$WR=PNO*PWID/BO \tag{3.5.1}$$

The ratio WR is defined as the Port Width Ratio. As will be evident in the following sections, it is difficult to raise this ratio much in excess of 1.25 for loop and cross scavenged engines. The higher this ratio, the lower can be the port timings to flow a desired scavenge ratio, SR, at the design speed. It is clear from Fig. 3.31 that the value of WR is considerably greater than unity, as the uniflow engine does not share this cylinder plane with the exhaust port. If the engine is an opposed piston uniflow design, the port timings for both scavenge and exhaust ports can be modest indeed, even for high scavenge ratios. In actual fact, the WR value for Fig. 3.31 is exactly 2.0.

3.5.2 Conventional cross scavenging

The cross scavenged engine usually has a fairly simple port layout. The scavenge ports and the exhaust ports are often drilled from the exhaust port side in a single operation, thus simplifying the manufacturing process. Also, if this is carried out with some precision, it increases the accuracy of the port timing events. The basic layout is shown in Fig. 3.32 with further amplification in Fig. 1.3. Fig. 3.32 is a plan section through the scavenge and exhaust ports, with a view on the piston crown. Further amplification also is given by Plates 1.6 and 4.2. Fig. 3.32 shows splitters between the scavenge ports; this is the preferred geometry so that the gas flow follows the design directions, but it is rarely seen in practice. Although the ports are separate, the scavenge flow normally emanates from the crankcase and flows to the transfer ports in a common duct. The "ears" of the deflector are usually centered about a diameter or are placed slightly towards the scavenge side. The edge of the ears are normally chamfered at 45º and given clearance from the bore so that they may enter the combustion chamber above the tdc point without interference. Contrary to popular belief, the author(1.10) showed that while this apparently gave a direct short-circuit path to the exhaust port, the actual SE-SR characteristic was hardly affected. Consequently, any loss of engine performance caused by chamfering 50% off the ears of the deflector and giving them the necessary side clearance must be due to a further deterioration of the combustion efficiency by introducing even longer and more tortuous flame paths.

In Chapters 6 and 7 there will be discussions on the determination of port timings and port heights for all types of engines. In terms of Fig. 3.32, it will be assumed that the scavenge port height at bdc is PHT, and there are a number of scavenge ports,

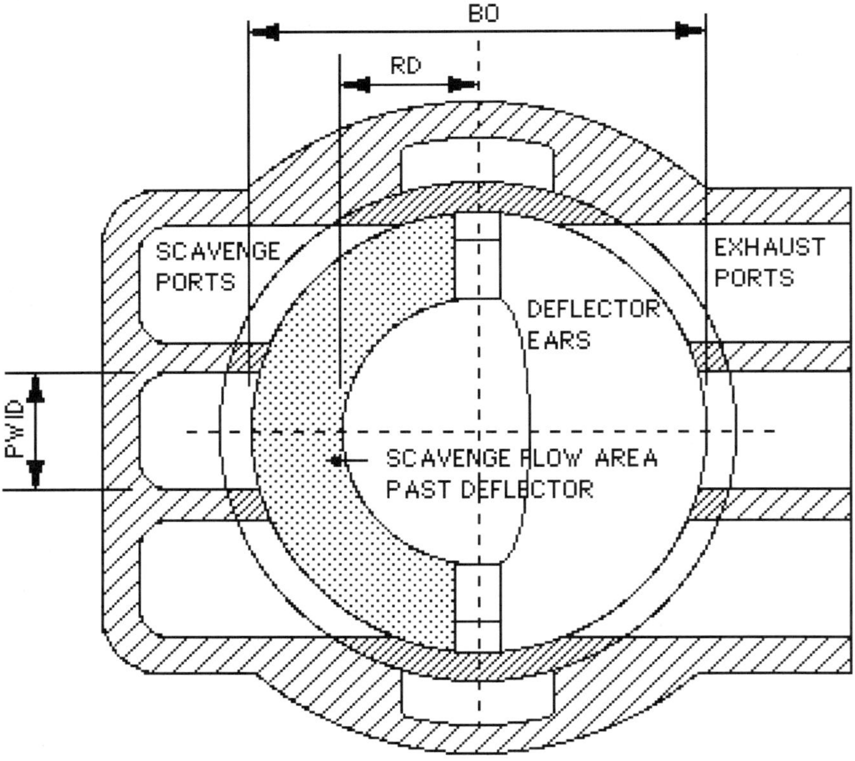

Fig. 3.32 Transfer port and deflector plan layout for conventional cross scavenging.

PNO. The almost universal value for PNO is 3. If the ports are rectangular in profile, then the maximum scavenge flow area, SFA, by ignoring corner radii, is approximately:

$$SFA=PNO*PWID*PHT \tag{3.5.2}$$

Where the maximum area of the ports at bdc are segments of drilled holes, the designer is forced to calculate that area for use in Eq. 3.5.2. The value of SFA determines one of the two important design criteria, the value for RD, the radius of the deflector. So that the flow exits the scavenge ports, makes as smooth a transition as possible through 90º and goes towards the cylinder head, the value of the deflector flow area should approximate the area of the scavenge ports. The shaded area in Fig. 3.32, showing the scavenge flow area past the deflector, is that which is under scrutiny and is labeled DFA. The best deflection process would be where a restriction took place, but that would reduce the scavenge ratio, SR. The rule of thumb for guidance in this matter can be expressed as Deflection Ratio, which is the ratio of the scavenge flow area to the deflector flow area, KD, where:

$$KD=DFA/SFA \tag{3.5.3}$$

The values of KD for the best possible answer for conventional cross scavenging is where $1.10<KD<1.20$. It should be remembered from Sect. 3.2.4 that the reported behavior of traditional cross scavenging is not very good. By optimizing such factors as deflection ratio it is possible to improve upon that level of scavenging effectiveness. Using a KD factor within the range indicated, it becomes possible to compute a value for the deflector radius, RD, assuming that the deflector ears are approximately on a diameter of the cylinder bore:

$$RD=\sqrt{[(BO^2/4)-(2*KD*SFA/\pi)]} \tag{3.5.4}$$

The vital importance of the deflection ratio in the design of the piston crown and the porting of a cross scavenged engine is illustrated by Fig. 3.33. This graph shows the results of two scavenge tests for a cross scavenged engine with a deflector of the traditional type, but where the deflection ratios are markedly different at 1.65 and 1.15. The engine with a deflection ratio of 1.65 is that previously presented in Figs. 3.10 and 3.11 as cylinder CD, whereas the one with the cylinder and piston deflector which has a deflection ratio of 1.15 is identical in every respect, except for the value of the radius RD. The experimental results in Fig. 3.33 are presented as TE-SR characteristics, and it is observed that the behavior of the classic type of cross scavenging is greatly improved by this optimization of the deflection ratio.

The second important design criterion is for the height of the deflector, DHT. The value for this parameter is determined through the ratio of port height to deflector height, KH, where:

$$KH=DHT/PHT \tag{3.5.5}$$

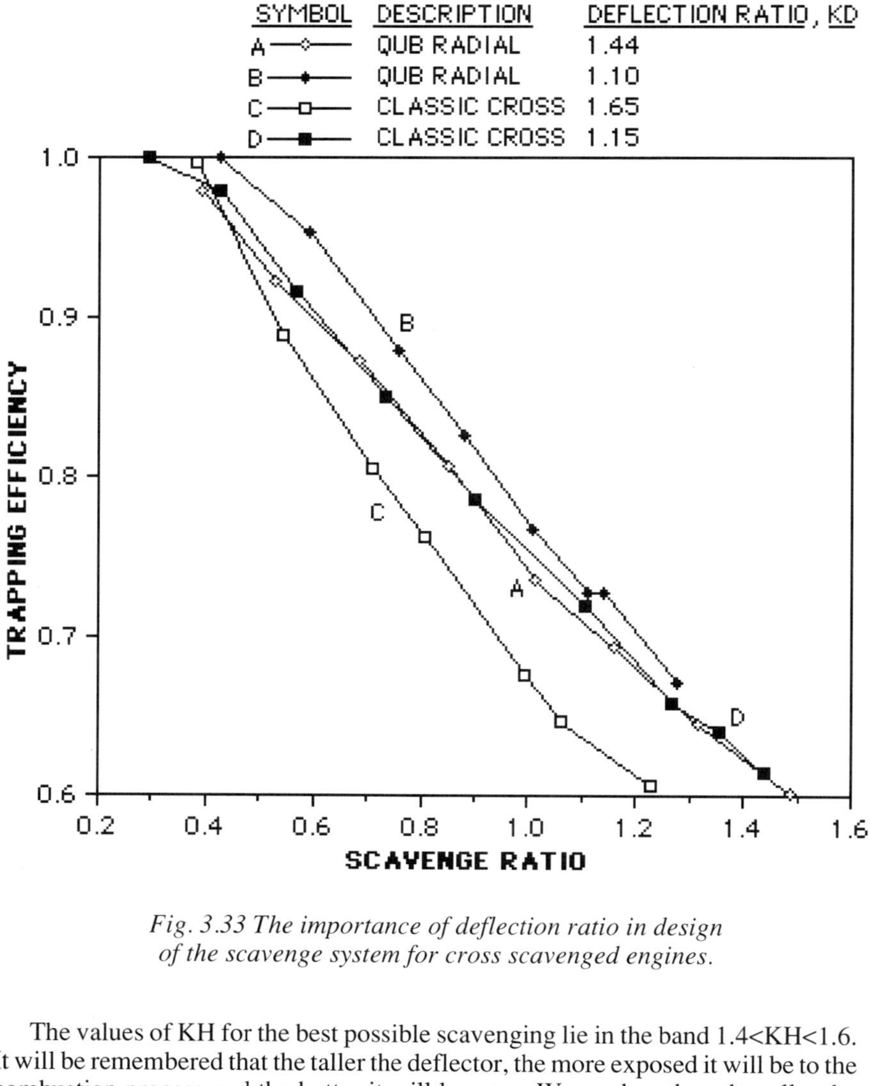

Fig. 3.33 The importance of deflection ratio in design of the scavenge system for cross scavenged engines.

The values of KH for the best possible scavenging lie in the band 1.4<KH<1.6. It will be remembered that the taller the deflector, the more exposed it will be to the combustion process and the hotter it will become. Worse than that, the taller the deflector the more complex the combustion space becomes which further deteriorates combustion efficiency. As there is no scavenging advantage in exceeding a KH value of 1.6, it is often wiser to keep the deflector as low as is practical.

The transition radius at the base of the deflector is not a particularly critical dimension as far as scavenging is concerned(1.10), and it should be made approximately one third of the deflector height.

The total scavenge port width ratio, WR, for this type of engine is usually somewhat less than unity, particularly in the straight drilled port design configuration. In Fig. 3.32 the numerical value is 0.80, which is typical of outboard motor practice.

3.5.3 QUB type cross scavenging

A preliminary discussion of this type of scavenging was made in Sect. 1.2.2 and shown in Fig. 1.4. It is clear that the radial disposition of the transfer ports will raise the port width ratio, WR, and the value for such designs is usually about 1.0. While one can use the straight-in port arrangement, it has been demonstrated that the scavenging of such a design is actually worse than conventional cross scavenging(1.10). There are further, unpublished engine tests from QUB which would substantiate that statement. The radial arrangement of scavenge ports has been shown(1.10) to have excellent scavenging characteristics by comparison with good quality loop scavenging and uniflow scavenging, as demonstrated in Figs. 3.10 and 3.11.

The criterion used by the designer to optimize the scavenging process, prior to conducting a confirmatory test on a single cycle test apparatus, is very similar to that followed for conventional cross scavenging. The deflection ratio, KD, calculated from Eq. 3.5.3 is still pertinent, and the determination of it, for this more complex geometry, is aided by a computer program, Prog.3.3. The design process is best illustrated by the example shown in Fig. 3.34 for a 70 mm bore and stroke "square" engine. The figure shows the screen display from the program, slightly accentuated to show the deflector flow area by shading that portion of the piston crown. By requiring input values for cylinder bore, stroke, con-rod length, transfer port timing and transfer port corner radii, the program accurately calculates the total scavenge port height and area, PHT and SFA, as 15.1 mm and 1062.8 mm^2, respectively. The deflector width and radius input data values of 65 and 120, respectively, give the deflector flow area, DFA, as 1231 mm^2. Consequently, the KD ratio is calculated as 1.16, which would be regarded as a satisfactory value. The value of deflector radius, selected here as 120 mm, and known to provide good scavenging for this particular bore dimension of 70 mm(1.10), can vary over a wide range from 30% to 80% greater than the cylinder bore dimension without unduly affecting the quality of scavenging.

It is difficult to overemphasize the importance of the deflection ratio in the design of cross scavenged engines, and further information with respect to the QUB type of cross scavenging is presented in Fig. 3.33. This shows the experimental results for trapping efficiency for two QUB type engines, one with a deflection ratio of 1.15 and the other with 1.44. The one with the KD value of 1.15 is the QUBCR cylinder discussed in Sect. 3.2.4 and Figs. 3.10 and 3.11. The higher KD ratio lowers the trapping efficiency of the QUB type of cross scavenged cylinder down to a level which places it equal to an optimized example of a classic cross scavenged design. As further evidence, the QUB type with the higher KD ratio of 1.44 was built as a firing engine, not just as a model cylinder for experimentation on the single cycle scavenge rig, and the resulting performance characteristics fell at least 10% short of the behavior reported for the QUB 400 loop scavenged research engine in

```
CURRENT INPUT DATA FOR DEFLECTOR PISTON CROWN AND PORTING
(B) BORE,mm= 70                (S) STROKE,mm= 70          (C) CON-ROD,mm= 140
(T) TRANSFER OPENS,deg atdc= 118     (SM) SIDE/MIDDLE PORT WIDTH RATIO = 1
(CR) PORT CORNER RADII,mm= 3          (BC) PORT BAR CHORD WIDTH,mm= 5
(WD) DEFLECTOR WIDTH,mm= 65           (DR) DEFLECTOR RADIUS,mm= 120

OUTPUT DATA
SQUISH AREA RATIO=0.680
PORT WIDTH/BORE RATIO=1.03
DEFLECTION RATIO=1.16
TRANSFER HEIGHT & AREA=     15.1  1062.8
SIDE, MIDDLE PORT WIDTHS,mm=23.9  23.9
SIDE PORT ANGLE,deg=40.0
MIDDLE PORT ANGLE,deg=40.0
DEFLECTOR ANGLE,deg=136.4
DEFLECTION RATIO,DY/BORE=0.378
DEFLECTOR POSITION,mm,DX & DY=43.5 26.5
PORT-DEFLECTOR ANGLE,deg= 6.1
SQUISH AREA,mm2,=2617
DEFLECTOR FLOW AREA,mm2,=1231
```

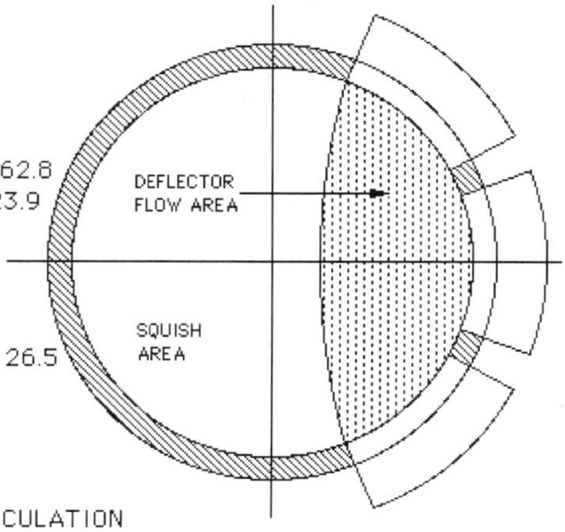


```
TYPE CODE FOR NEW DATA AND A RECALCULATION
OR TYPE CODE FOR PRINTING(P), FILING(F) OR QUITTING(Q) ??
```

Fig. 3.34 QUB type cross scavenging design through program Prog.3.3, QUB CROSS PORTS.

Chapters 5 and 7; in other words, they were somewhat disappointing. To emphasize the point, a QUB type deflector piston engine of the square dimensions of 70 mm bore and stroke and with a deflection ratio of 1.16, with a transfer port layout very similar to that shown in Fig. 3.34, was constructed and experimentally produced the identical levels of brake specific fuel consumption, power, trapping efficiency and exhaust emission characteristics to the QUB 400 loop scavenged engine.

As described previously for the classic cross scavenged cylinder, the deflection ratio criterion for the QUB type is the same:

$$1.1<KD<1.2$$

Because the design of this deflector piston influences both scavenge and combustion behavior, the value of Squish Area Ratio, SAR, is an important design parameter. This is defined in the same manner as for any engine, namely:

$$SAR=(\text{piston area inducing squish flow})/(\text{cylinder bore area}) \quad (3.5.6)$$

The squish area is indicated in Fig. 3.34, and the value is calculated as 0.680 in the output data table. For this type of piston, acceptable SAR values fall in the range $0.65<SAR<0.75$. Further discussion of this matter takes place when combustion chamber design is debated in Chapter 4.

The detail selection of the transfer port geometry occurs with the input data values of side-to-middle port width ratio, SM, and the inter-port bar width, BC. If at all possible, it is recommended that SM be equal to unity(1.10). The BC data value is determined from the mechanical considerations of the piston ring passing over the bar and the ensuing requirement for the rings not to trap in the port nor to excessively wear either the ring or the port bar during the required engine life span. To put an approximate number on the value of BC, a 3mm bar for a 40 mm bore and a 6mm bar for a 90 mm bore would be regarded as typical custom and practice. From all such considerations, the KD value for deflection ratio is produced with the individual port widths and angles subtended at the bore center. The port width ratio, WR, for this particular design is 1.03, and is usually 20% higher than a conventional cross scavenged engine. The position of the deflector center on the bore center-line, DY and DX, at 43.5 and 26.5 mm, respectively, completes all of the required geometry of the piston crown, in port plan dimensions at least. Like the criterion for the conventional deflector piston given in Eq. 3.5.5, the KH range for deflector height lies between 1.4 and 1.6, and if this was followed up for a combustion chamber design in Chapter 4, one would expect to see a deflector height between 21 and 24 mm.

The design process is aided by the computer program immediately drawing the sketch of the piston and the porting to scale on the computer screen. Most designers (and the author is no exception to this rule) make their final decisions and acceptances by eye, having satisfied themselves that the numerical values of the design criteria fall within acceptable boundaries. Such boundaries are dictated by custom, practice, past experience, new information and, most important of all, the instinct for innovation. In the case of the QUB cross scavenging engine, such a design type is relatively new, and further design information and experience is being gained almost daily. Therefore, the design criteria quoted numerically may well change with the passage of time and further experience, now that several such engines are in production with further units being designed and developed.

3.5.4 Loop scavenging

The design process for a loop scavenged two-stroke engine to ensure good scavenging characteristics is much more difficult than either the uniflow or the cross scavenged engine. In the first place, the number of potential porting layouts is much greater, as inspection of Figs. 1.2 and 3.35 will prove, and these are far from an exhaustive list of the geometrical combinations which have been tried in practice. In the second place, the difference between a successful transfer port geometry and an unsuccessful one is barely discernible by eye. If the reader cares to peruse the actual port plan layouts of Yamaha DT250 enduro motorcycle engine cylinders 1, 2, 3, 4, 5, 6, 7, 8, 9,10,12,14,15, presented in papers from QUB (1.23), (3.13) and (3.23), it is doubtful if the best scavenging cylinders could have been predicted in advance of conducting either the firing tests or the single cycle experimental assessment of their SE-SR behavior.

In a loop scavenge design, it is the main port which controls the scavenge process and it is necessary to orient that port correctly. The factors which influence that

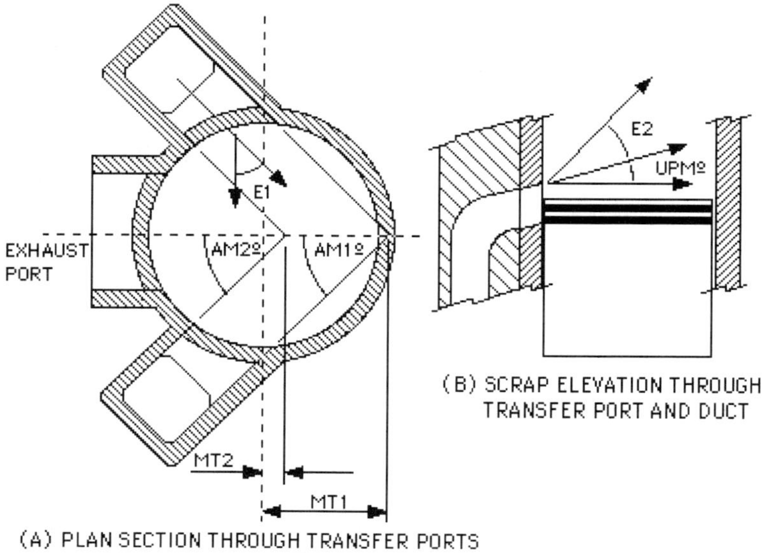

Fig. 3.35 Nomenclature for design for Prog.3.4, LOOP SCAVENGE DESIGN.

control are illustrated in Fig. 3.35. These geometrical values are the angles AM1 and AM2 and their "target" points on the plane of symmetry, MT1 and MT2, respectively. A further factor is the upsweep angle of the main port, UPM. Until recent times and the publication of the papers from QUB by Smyth et al(3.17) and Kenny et al(3.31), it was suspected that the exiting gas flow from the main ports did not follow the design direction of the port, but deviated from it in some manner. In particular, it was speculated that the plan angle of the exiting fresh charge would be contorted by an angle E1 from the design direction, due to the entrainment or short-circuiting of the flow with the exiting exhaust gas flow. That the flow might have a deviation of angle E2 in the vertical direction was regarded as more problematic. Smyth(3.17) showed that the value of E1 varied from 24.5º to 10.4º from the port opening point to full port opening at bdc, and that the value of E2 varied from 34º to 14º over the same range. As these measurements were taken in a particular engine geometry, the answer from Smyth's experiments cannot be regarded as universal. Nevertheless, as his engine had rather good scavenging for a simple two-port design, it is of the type sketched in Fig. 3.35, with AM1 and AM2 values of 50º, and as the values of gas flow deviation quoted are considerable in magnitude, it must be assumed that most engines have somewhat similar behavior. The actual experimental data presented by Kenny et al(3.31), and measured by laser doppler velocimetry in an optical access cylinder, are shown in Fig. 3.36.

As no designer can predict such gas flow deviation behavior, the recommendations for angles and targets for the main port, which have been demonstrated to

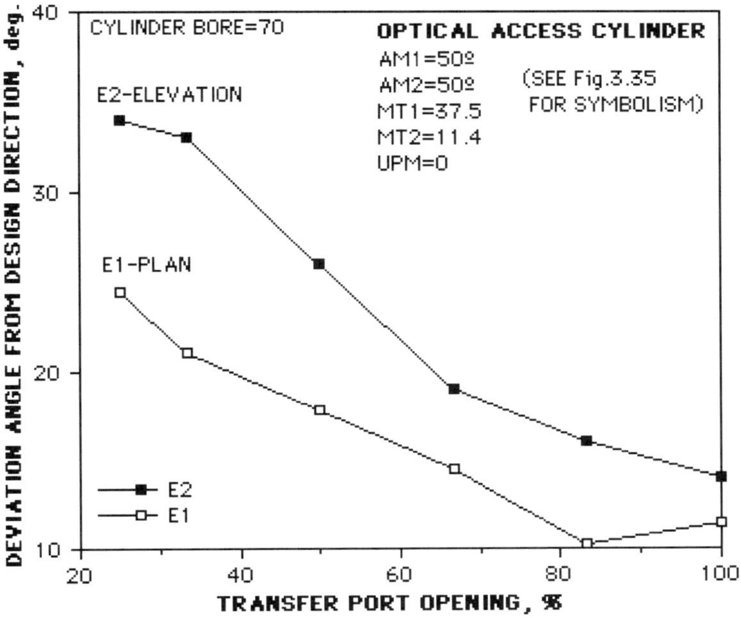

Fig. 3.36 Deviation angles from the design direction for loop scavenge flow.

provide successful scavenging characteristics, should be followed. In the not too distant future, accurate CFD calculations will replace much of the empirical advice in the succeeding sections. Nevertheless, as CFD calculations are expensive in terms of computer time, designers will always have a need for some better starting point which is based more on experimental evidence than intuition.

Finally, for those who wish to examine the full extent of porting design seen in two-stroke engines, albeit mostly loop scavenged two-stroke motorcycle engines, the comprehensive design compendia(3.32) emanating from the Technical University of Graz deserve study as a source of reference on what has been produced.

3.5.4.1 The main transfer port

The orientation of the main port, which is the scavenge port immediately beside the exhaust port, is already stated as the designer's first priority and this has some potential for empirical guidance. If one examines the port plan layouts of all of the cylinders known to provide successful scavenging, several factors stand out quite noticeably:

(a) The upsweep angle of the main port, UPM, is rarely larger than 10º.

(b) The value of AM2 is usually between 50º and 55º.

(c) The target point for MT2 is usually between 10% and 15% of the cylinder bore dimension, BO.

(d) The target point for MT1 is approximately on the edge of the cylinder bore.

(e) The port is tapered to provide an accelerating flow though the port, i.e., AM1>AM2, and AM1 is rarely larger than 70º.

(f) The larger the angle, AM1, the more the target point, MT1, is inside the cylinder bore. The narrower the angle, AM1, the farther outside the cylinder bore is the target point, MT1. The target length, MT1, is a function of the angle, AM1, and a useful empirical relationship is:

$$MT1=BO*\{[(50-AM1)/275]+0.55\} \quad (3.5.7)$$

To formalize some of the criteria stated above, the range of values for AM1 is usually in the band 50<AM1<70.

Eq. 3.5.7 shows a bore edge intersection for MT1 at a value for AM1 of 63.75º.

(g) In multiple transfer port layouts, i.e., greater than three, it is not uncommon to find the main port with parallel sides.

3.5.4.2 Rear ports and radial side ports

Such ports are sketched in Fig. 3.37, and the main design feature is that they should have upsweep angles between 50º and 60º to ensure attachment of the flow to the cylinder wall. Jante(3.5) gives a good discussion of such an attachment phenomenon, often known as the Coanda effect.

3.5.4.3 Side ports

Such ports are sketched in the bottom half of Figs. 1.2 and 3.37. They can have straight sides or, more usually, the side nearest to the main port has a similar slope to the main port, AM1. The objective is to have the opposing flow paths meet at the back wall, attach to it, and flow up to the cylinder head in a smooth manner. Consequently, an upsweep angle of between 15º and 25º helps that process to occur. Such ports, in conjunction with a rear port, help the rear port flow to attach to the back wall by providing a strong pressure differential across the face of the rear port.

3.5.4.4 The use of Prog.3.4, LOOP SCAVENGE DESIGN

The reader is provided with a useful design program for all of the porting layouts illustrated in Figs. 1.2 and 3.37. The program is self-explanatory and the typical outputs shown in Figs. 3.38 and 3.39 are exactly what the operator sees on the computer screen. The opening screen display asks the operator to decide on the number of transfer ports in the design, i.e., two, three, four or five, and from then on the design procedure will take the form of changing a particular data value until the operator is satisfied that the aims and objectives have been achieved. Such objectives would typically be the satisfaction of the criteria in Sects. 3.5.4.1 to 3.5.4.3 and to achieve the desired value of port width ratio, WR.

Upon completion of the design process, the printer will provide hard copy of the screen and, as can be seen from Fig. 3.39, give some useful manufacturing data for

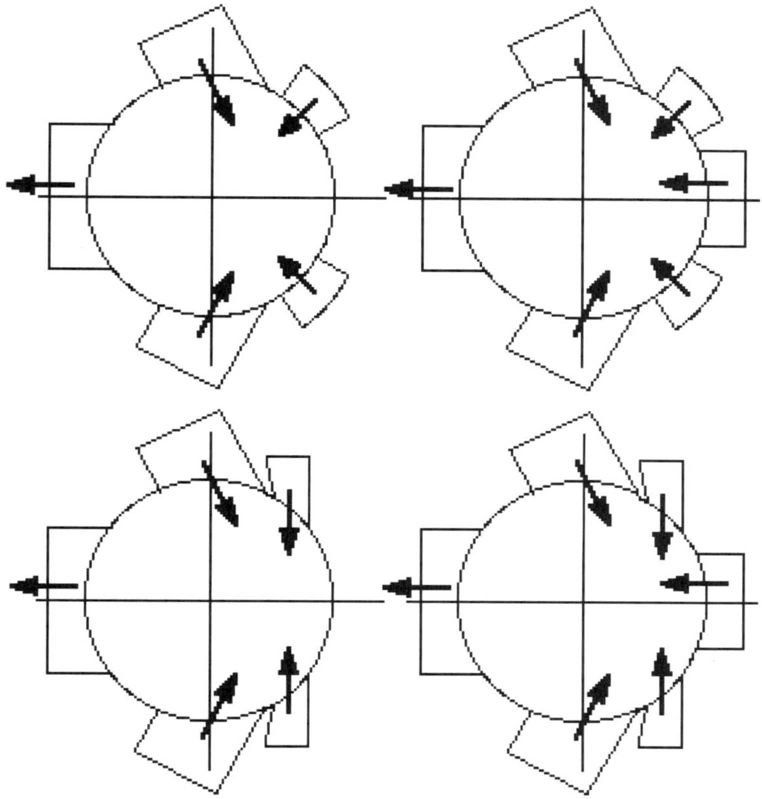

Fig. 3.37 Together with Fig. 1.2 this shows the port layouts for design by Prog.3.4, LOOP SCAVENGE DESIGN.

the cylinder liner. This becomes even more pertinent for four and five port layouts, where the structural strength of some of the port bars can become a major factor.

It is interesting to note that in the actual design shown for a 70 mm bore cylinder in this three transfer port layout, the WR value of 1.06 is about the same as for the QUB deflector piston engine; it was 1.03. It is clear that it would be possible to design a four or five port layout to increase the WR value to about 1.3. It will be remembered that the uniflow engine had an equivalent value of 2, although if that uniflow engine is of the long stroke type, or is an opposed piston engine, it will require a WR value at that level to achieve equality of flow area with the other more conventional bore-stroke ratio power units.

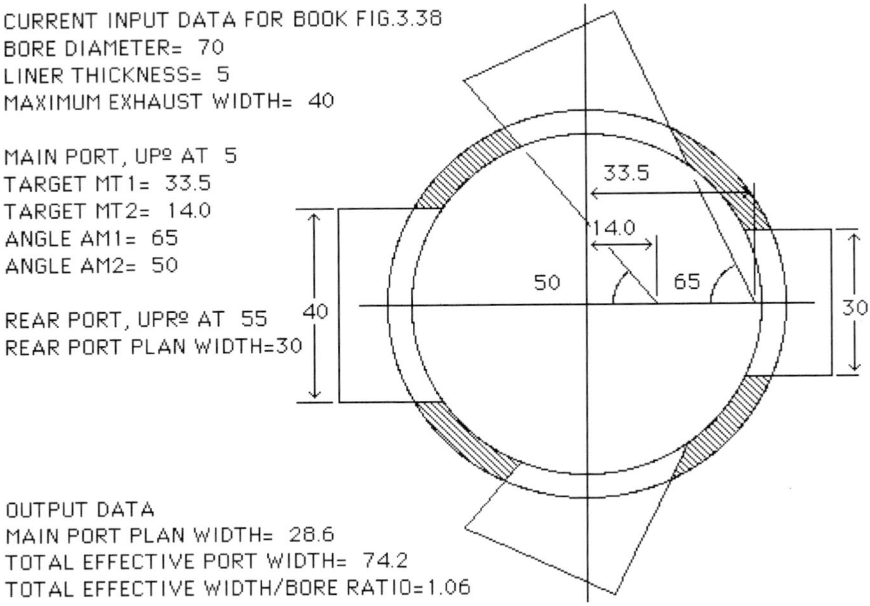

Fig. 3.38 First page of output from Prog.3.4, giving the scavenge geometry of the ports.

NOTATION FOR CHAPTER 3

NAME	SYMBOL	SI UNIT	OTHER UNITS
Charging Efficiency	CE		
Crankcase Compression Ratio	CCR		
Deflection Ratio	KD		
Geometric Compression Ratio	GCR		
Port Width Ratio	WR		
Squish Area Ratio	SAR		
Scavenging Efficiency	SE		
Scavenge Ratio	SR		
Trapped Compression Ratio	TCR		
Trapping Efficiency	TE		
Brake Mean Effective Pressure	BMEP	Pa	bar, atm, psi

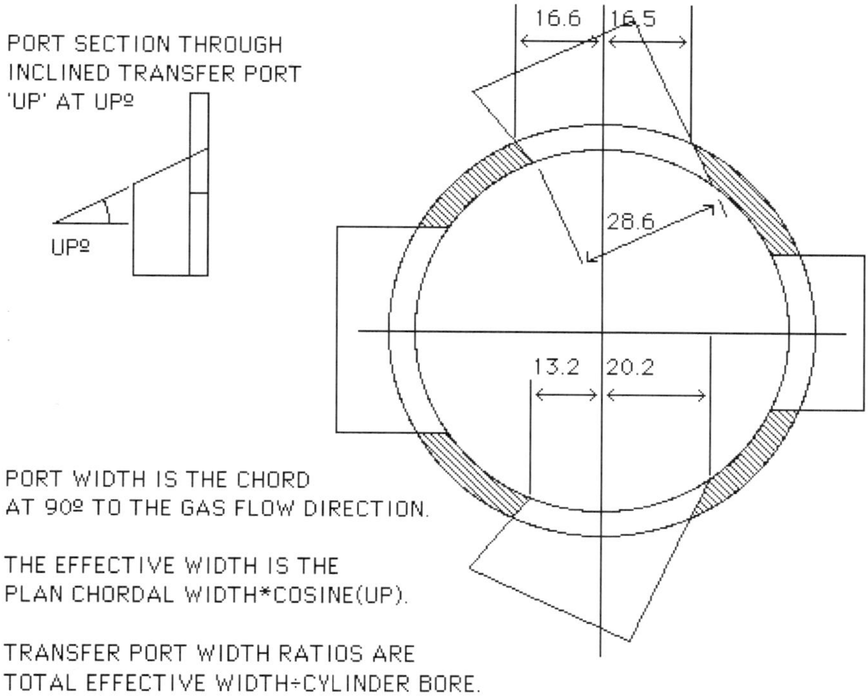

Fig. 3.39 Second page of output from Prog.3.4, giving port edge machining locations.

NAME	SYMBOL	SI UNIT	OTHER UNITS
Brake Specific Fuel Consumption	BSFC	kg/Ws	kg/kWh,lb/hp.hr
Clearance Volume	CV	m	mm, in
Cylinder Bore	BO	m	mm, in
Cylinder Stroke	ST	m	mm, in
Density	D	kg/m^3	
Engine Rotation Rate	RPS	rev/s	rev/min
Pressure	P	Pa	bar, atm, psi
Swept Volume	SV	m^3	cm^3, cc, in^3
Temperature	T	K	ºC, ºF
Trapped Swept Volume	TSV	m^3	cm^3, cc, in^3
Volume	V	m^3	cm^3, cc, in^3

REFERENCES FOR CHAPTER 3

3.1 B. Hopkinson, "The Charging of Two-Cycle Internal Combustion Engines," *Trans. NE Coast Instn. Engrs. Shipbuilders*, Vol. 30, 1914, p. 433.

3.2 R.S. Benson, P.J. Brandham, "A Method for Obtaining a Quantitative Assessment of the Influence of Charge Efficiency on Two-Stroke Engine Performance," *Int.J.Mech.Sci.*, Vol.11, p.303, 1969.

3.3 F. Baudequin, P. Rochelle, "Some Scavenging Models for Two-Stroke Engines," *Proc.I.Mech.E.*, Vol.194, 1980, pp.203-210.

3.4 C. Changyou, F.J.Wallace, "A Generalized Isobaric and Isochoric Thermodynamic Scavenging Model," SAE Intnl. Off-Highway Vehicle Meeting, Milwaukee, Wisconsin, September 1987, SAE Paper No.871657.

3.5 A. Jante, "Scavenging and Other Problems of Two-Stroke Spark-Ignition Engines," SAE Mid-Year Meeting, Detroit, Michigan, May, 1968, SAE Paper No. 680468.

3.6 G.P. Blair, "Studying Scavenge Flow in a Two-Stroke Cycle Engine," SAE Farm, Construction and Industrial Machinery Meeting, Milwaukee, Wisconsin, September, 1975, SAE Paper No.750752.

3.7 G.P. Blair, M.C. Ashe, "The Unsteady Gas Exchange Characteristics of a Two-Cycle Engine," SAE Farm, Construction and Industrial Machinery Meeting, Milwaukee, Wisconsin, September, 1976, SAE Paper No.760644.

3.8 T. Asanuma, S. Yanigahara, "Gas Sampling Valve for Measuring Scavenging Efficiency in High Speed Two-Stroke Engines," *SAE Transactions*, Vol.70, p.420, Paper T47, 1962.

3.9 R.R. Booy, "Evaluating Scavenging Efficiency of Two-Cycle Gasoline Engines," SAE Paper No. 670029, 1967.

3.10 N. Dedeoglu, "Model Investigations on Scavenging and Mixture Formation in the Dual-Fuel or Gas Engine," *Sulzer Tech. Review*, Vol.51, no.3, p.133, 1969.

3.11 W. Rizk, "Experimental Studies of the Mixing Processes and Flow Configurations in Two-Cycle Engine Scavenging," *Proc.I.Mech.E.*, Vol.172, p.417, 1958.

3.12 S. Ohigashi, Y. Kashiwada, "A Study on the Scavenging Air Flow through the Scavenging Ports," *Bull.J.S.M.E.*, Vol.9, No.36, 1966.

3.13 D.S. Sanborn, G.P. Blair, R.G. Kenny, A.H. Kingsbury, "Experimental Assessment of Scavenging Efficiency of Two-Stroke Cycle Engines," SAE Intnl. Off-Highway Vehicle Meeting, Milwaukee, Wisconsin, September, 1980, SAE Paper No.800975.

3.14 V.L. Streeter, Fluid Mechanics, McGraw-Hill, 3rd Edition, Tokyo, 1962.

3.15 H. Sammons, "A Single-Cycle Test Apparatus for Studying 'Loop-Scavenging' in a Two-Stroke Engine," *Proc.I.Mech.E.*, Vol.160, 1949, p.233.

3.16 W.H. Percival, "Method of Scavenging Analysis for Two-Stroke Cycle Diesel Cylinders," *Transactions SAE*, Vol.62, p.737, 1954.

3.17 J.G. Smyth, R.G. Kenny, G.P. Blair,"Steady Flow Analysis of the Scavenging Process in a Loop Scavenged Two-Stroke Cycle Engine–a Theoretical and Experimental Study," SAE International Off-Highway & Powerplant Congress, Milwaukee, Wisconsin, September 12-15, 1988, SAE Paper No.881267.

3.18 R.G. Kenny, "Scavenging Flow in Small Two-Stroke Cycle Engines," Doctoral Thesis, The Queen's University of Belfast, May 1980.

3.19 R.G. Phatak, "A New Method of Analyzing Two-Stroke Cycle Engine Gas Flow Patterns," SAE International Congress, Detroit, Michigan, February, 1979, SAE Paper No.790487.

3.20. M.E.G. Sweeney, R.G. Kenny, G.B. Swann, G.P. Blair, "Single Cycle Gas Testing Method for Two-Stroke Engine Scavenging," SAE International Congress, Detroit, Michigan, February, 1985, SAE Paper No.850178.

3.21 D.S. Sanborn, W.M. Roeder, "Single Cycle Simulation Simplifies Scavenging Study," SAE International Congress, Detroit, Michigan, February, 1985, SAE Paper No.850175.

3.22 C. Mirko, R. Pavletic, "A Model Method for the Evaluation of the Scavenging System in a Two-Stroke Engine," SAE International Congress, Detroit, Michigan, February, 1985, SAE Paper No.850176.

3.23 M.E.G. Sweeney, R.G. Kenny, G.B. Swann, G.P. Blair, "Computational Fluid Dynamics Applied to Two-Stroke Engine Scavenging," SAE Intnl. Off-Highway Vehicle Meeting, Milwaukee, Wisconsin, September, 1985, SAE Paper No.851519.

3.24 E. Sher, "An Improved Gas Dynamic Model Simulating the Scavenging Process in a Two-Stroke Cycle Engine," SAE International Congress, Detroit, Michigan, February, 1980, SAE Paper No.800037.

3.25 E. Sher, "Prediction of the Gas Exchange Performance in a Two-Stroke Cycle Engine," SAE International Congress, Detroit, Michigan, February, 1985, SAE Paper No.850086.

3.26 W. Brandstatter, R.J.R. Johns, G. Wrigley, "The Effect of Inlet Port Geometry on In-Cylinder Flow Structure," SAE International Congress, Detroit, Michigan, February, 1985, SAE Paper No.850499.

3.27 A.D. Gosman, Y.Y. Tsui, C. Vafidis, "Flow in a Model Engine with a Shrouded Valve—A Combined Experimental and Computational Study," SAE International Congress, Detroit, Michigan, February, 1985, SAE Paper No.850498.

3.28 R. Diwakar, "Three-Dimensional Modelling of the In-Cylinder Gas Exchange Processes in a Uniflow-Scavenged Two-Stroke Engine," SAE International Congress, Detroit, Michigan, February, 1987, SAE Paper No.850596.

3.29 D.B. Spalding, "A General Purpose Computer Program for Multi-Dimensional One and Two-Phase Flow," *Mathematics and Computers in Simulation*, Vol.23, 1981, p.267-276.

3.30 B. Ahmadi-Befrui, W. Brandstatter, H. Kratochwill, "Multidimensional Calculation of the Flow Processes in a Loop-Scavenged Two-Stroke Cycle Engine," SAE International Congress and Exposition, Detroit, Michigan, February, 1989, SAE Paper No. 890841.

3.31 R.G. Kenny. J.G. Smyth, R. Fleck, G.P. Blair, "The Scavenging Process in the Two-Stroke Engine Cylinder," Dritte Grazer Zweiradtagung, Technische Universitat, Graz, 13-14 April, 1989.

3.32 F. Laimbock, "Kenndaten und Konstruktion Neuzeitlicher Zweiradmotoren", Heft 31a, Heft 31b, Heft 31c, Institut for Thermodynamics and Internal Combustion Engines, Technische Universitat Graz, Austria, 1989.

3.33 M. Nuti, L. Martorano, "Short-Circuit Ratio Evaluation in the Scavenging of Two-Stroke S.I. Engine," SAE International Congress and Exposition, Detroit, Michigan, February, 1985, SAE Paper No. 850177.

Chapter 4

Combustion in Two-Stroke Engines

4.0 Introduction

This chapter will mainly deal with combustion in the spark ignition engine. The subject will be dealt with on a practical design basis, rather than as a fundamental treatise on the subject of combustion. It is not stretching the truth to say that combustion, and the heat transfer behavior which accompanies it, are the phenomena least understood by the average engine designer. The study of combustion has always been a specialized topic which has often been treated at a mathematical and theoretical level beyond that of all but the dedicated researcher. Very often the knowledge garnered by research has not been disseminated in a manner suitable for the designer to use. This situation is changing, for the advent of CFD design packages will ultimately allow the engine designer to predict combustion behavior without being required to become a mathematics specialist at the same time. For those who wish to study the subject at a fundamental level, and to be made aware of the current state of the science, the references (1.3), (1.4), and (4.1)-(4.5) will help to provide a starting point for that learning process. There seems little point in repeating the fundamental theory of combustion in this chapter of this book, because such theory is well covered in the literature. However, this chapter will cover certain aspects of combustion theory which are particularly applicable to the design and development of two-stroke engines and which are rarely found in the standard reference textbooks.

The first objective of this chapter is to make the reader aware of the difference between the combustion process in a real engine from either the theoretical ideal introduced in Sect. 1.5.8, or from an explosion, which is the commonly held view of engine combustion by the laity.

The second aim of the author is to introduce the reader to the analysis of cylinder pressure records for the heat release characteristics which they reveal, and how such information is the most realistic available for the theoretical modeller of engine behavior. In this section, other possible theoretical models of combustion are presented and discussed.

The third objective is to make the reader aware of the effect of various design variables such as squish action on detonation and combustion rates and the means at hand to design cylinder heads for engines taking these factors into account.

Penultimately, the chapter contains computer programs to assist with this design process and the reader is introduced to their use.

Finally, there is a section on the latest theoretical developments in the analysis of combustion processes, for it will not be many years before our understanding of engine combustion will be on a par with, say, unsteady gas dynamics or scavenging flow. The par level referred to is where the theory will design combustion chambers for two-stroke engines and the resulting engine will exhibit the designed combustion performance characteristics. It is the author's opinion that this level is approaching, but has not yet been achieved.

4.1 The spark-ignition process

4.1.1 Initiation of ignition

It is a well-known fact that a match thrown onto some spilled gasoline in the open atmosphere will ignite the gasoline and release a considerable quantity of heat with a significant rise in temperature. The gasoline is observed to burn as a vapor mixed with air above the remaining liquid, but rapidly vaporizing, gasoline. The procedure, for those who have witnessed it (although on safety grounds the author is not recommending that the experiment be conducted), commences when the lighted match arrives at the vapor cloud above the liquid, and the ignition takes place with a "whoosh," apparently a major or minor "explosion" depending on the mass fraction of the gasoline which has been allowed to evaporate before the arrival of the match. This tends to leave us with the impression that the gasoline-air mixture within the cylinder of an ic engine will "explode" upon the application of the spark at the sparking plug. That the flammability characteristics of a petrol-air mixture are a decidedly critical phenomena should be obvious to all who have had difficulty in starting either their lawnmower or their automobile!

What then are the requirements of an ignition process? Why does an engine fire up? When does it not fire up? The technical papers on combustion and the engineering textbooks tend to bypass such fundamental concepts, so that this author felt that a paragraph or two would not be amiss, especially as more difficult concepts will only be understood against a background of some fundamental understanding.

Fig. 4.1 depicts a two-stroke engine where a spark has ignited an air-fuel mixture and has produced a flame front burning its way through the mixture. For this to happen, like the match example before it, there had to be a gasoline vapor and air mixture, of the correct mass proportions, within the spark gap when that spark occurred. The energy in the spark provided a local rise in temperature of several thousand degrees Kelvin, which caused any gasoline vapor present to be raised above its Auto-Ignition Temperature. The auto-ignition temperature of any hydrocarbon fuel is that temperature where the fuel now has sufficient internal energy to break its carbon-hydrogen bond structure and be oxidized to carbon dioxide and steam. In the case of gasoline the auto-ignition temperature is about 220ºC. The compression process prior to the ignition point helps to vaporize the gasoline, whose maximum boiling point is about 200ºC. The mass of gasoline within the spark gap, having commenced to break down in an exothermic reaction, raises the local temperature and pressure. The reaction, if it was stoichiometric, would be as given previously in Eq. 1.5.16. The actual reaction process is much more complex than this, with the gasoline molecule breaking down in stages to methane and aldehydes. Immediately after the ignition point, the initial flame front close to the spark plug has been established and heats the unburned layers of gasoline vapor-air mixture surrounding it principally by radiation, but also by convection heat transfer and by the mixture motion propelling itself into the flame front. This induces further layers of mixture to reach the auto-ignition temperature and thus the flame front moves through the combustion chamber until it arrives at the physical extremities of the chamber(4.24)(4.25)(4.26). The velocity of this flame front has been recorded(4.3)(1.20) in two-stroke engines between 20 and 50 m/s. It will be observed that this is hardly an "explosive" process, although it is sufficiently rapid to allow the engine to burn its fuel reasonably efficiently even at the highest engine speeds. In the example quoted by Kee(1.20), the flame speed, FV, in an engine of similar physical geometry to Fig. 4.1, was measured at 24.5 m/s at 3000 rpm. The longest flame path, DE, from the spark plug to the extremity of the combustion chamber in an engine of 85 mm bore, was about 45 mm. The crankshaft rotation angle, AC, for this flame transmission to occur is given by:

$$AC = 6 * RPM * DE / FV \qquad (4.1.1)$$

In the example quoted by Kee(1.20), the flame travel time to the chamber extremity is 33º, as AC is 6*3000*0.045/24.5 from Eq. 4.1.1. That does not mean that the combustion process has been completed in 33º, but it does mean that initiation of combustion has taken place over the entire combustion space for a homogeneous charge. In Sect. 4.2.3, it will be shown that the travel time, in this case cited as 33º, coincides with the maximum rate of heat release from the fuel.

4.1.2 Air-fuel mixture limits for flammability

The flammability of the initial flame kernel has a rather narrow window for success(4.25)(4.26). The upper and lower values of the proportion by volume of gasoline vapor to air for a flame to survive are 0.08 and 0.06, respectively. As one is supplying a "cold" engine with liquid fuel, by whatever device ranging from a carburetor to a fuel injector, the <u>vaporization</u> rate of that gasoline due to the compression process is going to be highly dependent on the temperatures of the cylinder wall, the piston crown and the atmospheric air. Not surprisingly, in cold climatic conditions, it takes several compression processes to raise the local temperature sufficiently to provide the statistical probability of success. Equally, the provision of a high-energy spark to assist that procedure has become more conventional(4.6). At one time, ignition systems had spark characteristics of about 8 kV with a rise time of about 25 μs. Today, with "electronic" or capacitor discharge ignition systems those characteristics are more typically at 20 kV and 4μs, respectively. The higher voltage output and the faster spark rise time ensure that sparking will take place, even when the electrodes of the spark plug are covered in liquid gasoline.

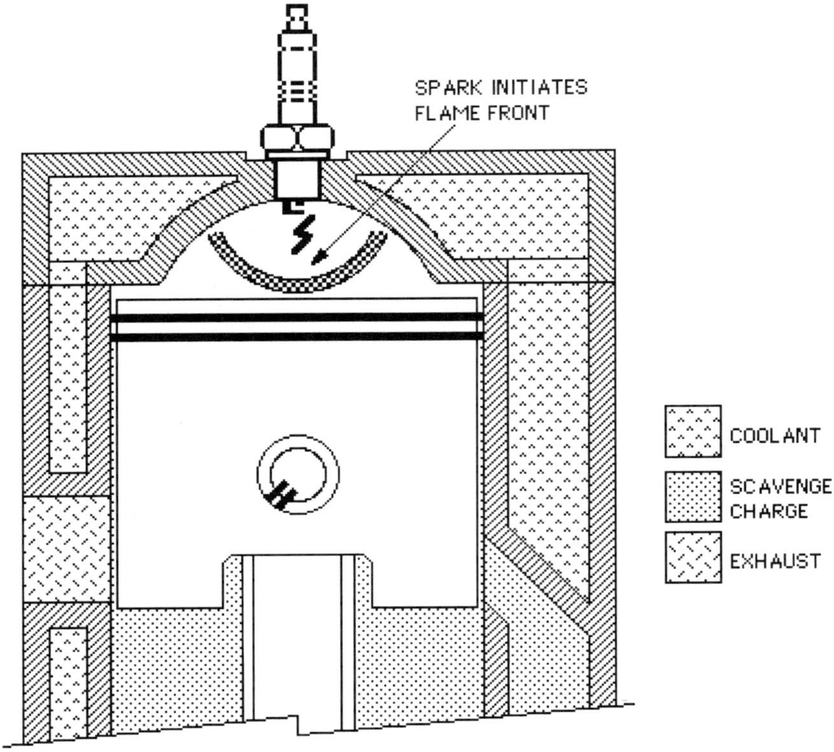

Fig. 4.1 Initiation of combustion in a spark-ignition two-stroke engine.

Under normal firing conditions, if the fuel vapor-air mixture becomes too lean, e.g., at the 0.06 volume ratio quoted above, then a flame is prevented from growing due to an inadequate initial release of heat. When the spark occurs in a lean mixture, the mass of fuel vapor and air which is ignited in the vicinity of the spark is too small to provide an adequate release of heat to raise the surrounding layer of unburned mixture to the auto-ignition temperature. Consequently, the flame does not develop and combustion does not take place. In this situation, intermittent misfire is the normal experience as unburned mixture forms the bulk of the cylinder contents during the succeeding scavenge process and will supplement the fuel supplied by it.

Under normal firing conditions, if the fuel vapor-air mixture becomes too rich, e.g., at the 0.08 volume ratio quoted above, then the flame is prevented from growing due to insufficient mass of air present at the onset of ignition. As with any flame propagation process, if an inadequate amount of heat is released at the critical inception point, the flame is snuffed out.

4.1.3 Effect of scavenging efficiency on flammability

In Sect. 3.1.5 there is mention of the scavenging efficiency variation with scavenge ratio. As the engine load, or BMEP, is varied by altering the throttle opening thereby producing changes in scavenge ratio, so too does the scavenging efficiency change. It will be observed from Fig. 3.10, even for the best design of two-stroke engines, that the scavenging efficiency varies from 0.3 to 0.95. Interpreting this reasonably accurately as being "charge purity" for the firing engine situation, it is clear that at light loads and low engine rotational speeds there will be a throttle position where the considerable mass of exhaust gas present will not permit ignition of the gasoline vapor-air mixture. When the spark occurs, the mass of vapor and air which is ignited in the vicinity of the spark is too small to provide an adequate release of heat to raise the surrounding layer of unburned mixture to the auto-ignition temperature. Consequently, the flame does not develop and combustion does not take place. The effect is somewhat similar to the lean burning misfire limit discussed above.

During the next scavenging process the scavenging efficiency is raised as the "exhaust residual" during that process is composed partly of the unburned mixture from the misfire stroke and partly of exhaust gas from the stroke preceding that one. Should the new SE value prove to be over the threshold condition for flammability, then combustion takes place. This skip firing behavior is called "four-stroking." Should it take a further scavenge process for ignition to occur, then that would be "six-stroking," and so on. Of course, the scavenging processes during this intermittent firing behavior eject considerable quantities of unburned fuel and air into the exhaust duct, to the detriment of the specific fuel consumption of the engine and its emission of unburned hydrocarbons.

4.1.4 Detonation or abnormal combustion

Detonation occurs in the combustion process when the advancing flame front, which is pressurizing and heating the unburned mixture ahead of it, does so at such a rate that the unburned fuel in that zone achieves its auto-ignition temperature before the arrival of the actual flame front. The result is that unburned mixture combusts "spontaneously" and over the zone where the auto-ignition temperature has been achieved. The apparent flame speed in this zone is many orders of magnitude faster, with the result that the local rise of pressure and temperature is significantly sharp. This produces the characteristic "knocking" or "pinking" sound and the local mechanical devastation which it can produce on piston crown or cylinder head can be considerable. Actually, "knocking" is the correct terminology for what is actually a detonation behavior over a small portion of the combustion charge. A true detonation process would be one occurring over the entire compressed charge. However, as detonation in the strictly defined sense does not take place in the spark-ignition ic engine, the words "knocking" and "detonation" are interchanged without loss of meaning in the literature to describe the same effect just discussed.

The effect is related to compression ratio, because the higher the compression ratio, the smaller the clearance volume, the higher the charge density, and at equal flame speeds, the higher the heat release rate as the flame travels through the mixture. Consequently, there will be a critical level of compression ratio where any unburned mixture in the extremities of the combustion chamber will attain the auto-ignition temperature. This effect can be alleviated by various methods, such as raising the octane rating of the gasoline used, promotion of charge turbulence, squish effects, improvement of scavenging efficiency, stratifying the combustion process and, naturally, lowering the trapped compression ratio. Some of these techniques will be discussed later.

It is clear that the combustion chamber shape has an influence on these matters. Plates 4.1 and 4.2 show the combustion chambers typical of loop scavenged and conventionally cross scavenged engines. The loop scavenged engine has a compact chamber with no hot nooks or crannies to trap and heat the unburned charge, nor a hot protuberant piston crown as is the case with the cross scavenged engine. In the cross scavenged engine there is a high potential for fresh charge to be excessively heated in the narrow confines next to the hot deflector, in advance of the oncoming flame front.

Compounding the problem of combustion design to avoid "knocking" in the two-stroke engine is the presence of hot exhaust gas residual within the combustible charge.

Plate 4.1 Compact combustion chamber for a loop scavenged two-stroke engine.

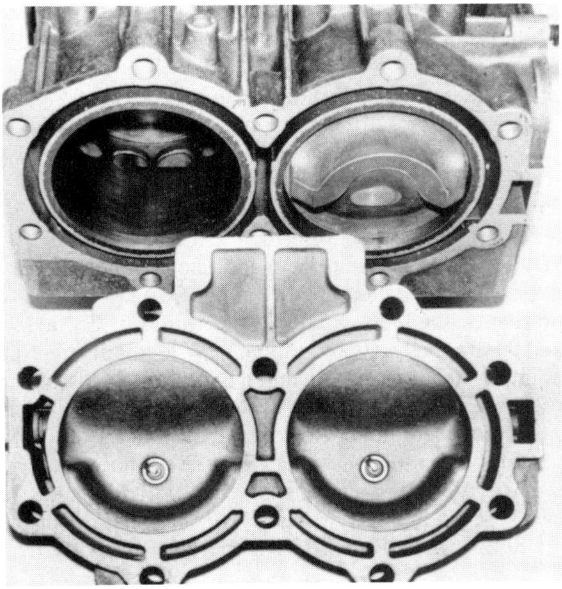

Plate 4.2 The complex combustion chamber of the conventional cross scavenged engine.

4.1.5 Homogeneous and stratified combustion

The conventional spark-ignition two-stroke engine burns a homogeneous charge. The air-fuel mixture is supplied to the cylinder via the transfer ports with much of the fuel already vaporized during its residence in the "hot" crankcase. The remainder of the liquid fuel vaporizes during the compression process so that by the time ignition takes place, the combustion chamber is filled with a vapor-air-exhaust gas residual mixture which is evenly distributed throughout the combustion space. This is known as a homogeneous combustion process.

Should the fuel be supplied to the combustion space by some other means, such as direct in-cylinder fuel injection, then, because all of the vaporization process will take place during the compression process, there is a strong possibility that at the onset of ignition there will be zones in the combustion space which are at differing air-fuel ratios. This is known as a stratified combustion process. This stratification may be deliberately induced, for example, to permit the local efficient burning of a small mass of air and fuel of the correct proportions to overcome the problems of "four-stroking."

It is also possible to utilize charge stratification to help alleviate detonation behavior. If the extremities of the combustion chamber contain air only, or a very lean mixture, then there exists the possibility of raising engine thermal efficiency with a higher compression ratio, while lowering the potential for detonation to occur.

Further detailed discussion on this topic can be found in Chapter 7, which covers the design of engines for good fuel economy and exhaust emissions.

4.2 Heat released by spark-ignition combustion

4.2.1 The combustion chamber

The combustion process is one described thermodynamically as a heat addition process in a closed system. It occurs in a chamber of varying volume proportions whose minimum value is the clearance volume, CV. This was discussed earlier in Sect. 1.4.2, and Eqs. 1.4.4 and 1.4.5 detail the values of CV in terms of the swept volume, SV, and the trapped swept volume, TSV, to attain the requisite parameters of geometric compression ratio, GCR, and trapped compression ratio, TCR. In Fig. 4.2, the piston is shown positioned at top dead center, TDC, so the clearance volume, CV, is seen to be composed of a bowl volume, VB, and a squish volume, VS. The piston has a minimum clearance distance from the cylinder head which is known as the squish clearance, SC. The areas of the piston which are covered by the squish band and the bowl are AS and AB, respectively, and this squish action gives rise to the concept of a Squish Area Ratio, SAR, where:

$$SAR=(\text{area squished})/(\text{cylinder bore area})=AS/(AS+AB) \qquad (4.2.1)$$

4.2.2 Heat release prediction from cylinder pressure diagram

In Sect. 1.5.8 there is a discussion of the theoretically ideal thermodynamic engine cycle, the Otto Cycle, and the combustion process is detailed as occurring at constant volume, i.e., the imaginary explosion. The reality of the situation, in Figs. 1.14 and 1.15, is that the engine pressure diagram as measured on the QUB LS400 research engine shows a time-dependent combustion process. This is pointed out in Sect. 4.1.1, where the flame speed is detailed as having been measured at 25 m/s at the same test conditions.

It is possible to analyze the cylinder pressure diagram and deduce from it the heat release rate for any desired period of time or crankshaft angular movement. Consider the thermodynamic system shown in Fig. 4.3, where the piston has been moved through a small crankshaft angle producing known volume changes V1 to V2 in the combustion chamber. These volumes are known from the crankshaft, connecting rod and cylinder geometry. The combustion process is in progress and a quantity of heat, dQr, has been released during this time step and a quantity of heat, dQl, has been lost through heat transfer to the cylinder walls and coolant at the same time. The cylinder pressure change has been measured as changing from P1 to P2 during this time step. The internal energy of the cylinder gas undergoes a change

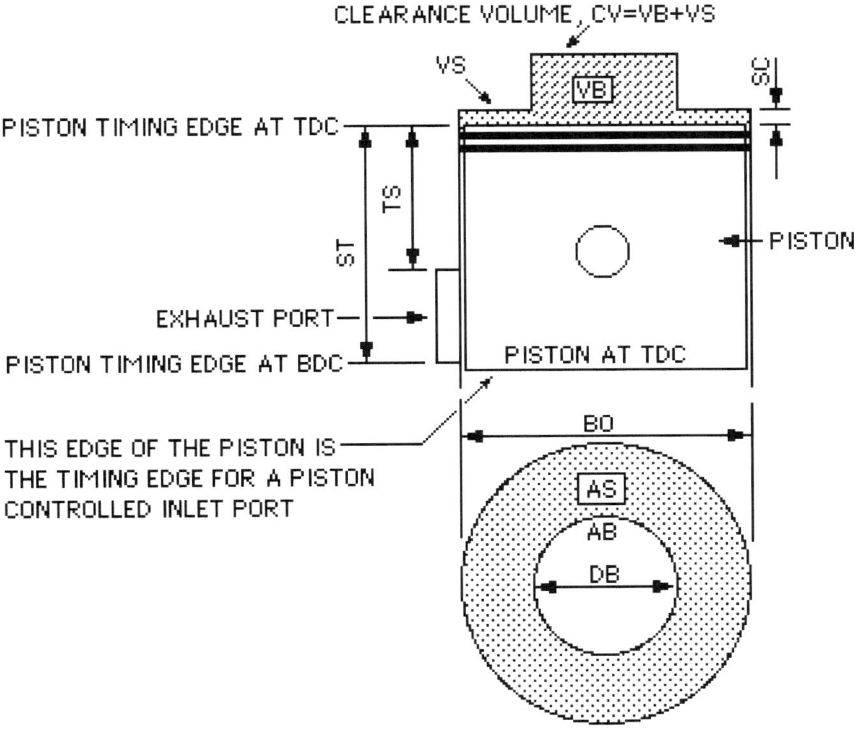

Fig. 4.2 Details of combustion chamber and cylinder geometry.

from U1 to U2, as does the temperature from T1 to T2. The work done on the piston is dW during the time interval. The first law of thermodynamics for this closed system states that, during this time step, such events are related by:

$$dQr-dQl=U2-U1+dW \quad (4.2.2)$$

Where m is the mass of gas in the cylinder and cV is the specific heat at constant volume, the internal energy change is:

$$U2-U1=m*cV*(T2-T1) \quad (4.2.3)$$

The pressure and temperature values at each point can be derived through the equation of state as:

$$P2*V2=m*R*T2 \quad (4.2.4)$$
$$P1*V1=m*R*T1$$

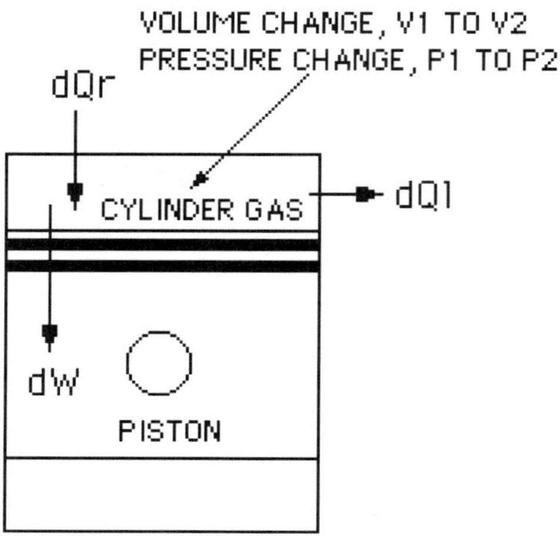

Fig. 4.3 Thermodynamic system during combustion.

where R is the gas constant.

Thence, as cV=R/(g-1), and g is the ratio of specific heats for the cylinder gas:

U2-U1=(P2*V2-P1*V1)/(g-1) (4.2.5)

The average work done on the piston during this interval is:

dW=(P1+P2)*(V2-V1)/2 (4.2.6)

Consequently, substituting Eqs. 4.2.5 and 4.2.6 into Eq. 4.2.2:

dQr-dQl= (P2*V2-P1*V1)/(g-1) + (P1+P2)*(V2-V1)/2 (4.2.7)

If the combustion process had not occurred, then the compression or expansion process would have continued in a normal fashion. The polytropic compression process, seen in Figs. 1.15 and 1.16, is taking place at PV^n=(constant). The index n is known from the measured pressure trace. Hence, P2 would not have been achieved, but the value P2a would have applied, where:

P2a=P1*VRN (4.2.8)

where the term VRN is defined as:

VRN=VR^n (4.2.9)

and the value VR is the incremental volume ratio:

$$VR=V1/V2 \qquad (4.2.10)$$

For such a non-heat addition process, where dQr is zero, the first law of thermodynamics shown in Eq. 4.2.7 could be rewritten to calculate the heat loss, dQl, by substituting P2a for P2:

$$-dQl=P1*\{[(V2*VRN-V1)/(g-1)]+[(VRN+1)*(V2-V1)/2]\} \qquad (4.2.11)$$

On the assumption that this heat loss behavior continues during the combustion process, substitution of this value for dQl into Eq. 4.2.7 yields an expression for the heat released, dQr:

$$dQr=(P2-P1*VRN)*\{[V2/(g-1)]+[(V2-V1)/2]\} \qquad (4.2.12)$$

The value of the polytropic exponent, n, is known from an analysis of the pressure trace. A value of the ratio of specific heats for the cylinder gas, g, could be taken as being between 1.35 and 1.38, but could be determined quite accurately from an exhaust gas analysis, and even more accurately from a chemical analysis of a sample of the cylinder gas at the end of combustion. It will be noted that the experimentally determined values of the polytropic exponents, n, for the compression and expansion processes for the QUB LS400 engine shown in Fig. 1.15, are 1.17 and 1.33, respectively.

Rassweiler and Withrow(4.7) presented this theory in a slightly simpler form as:

$$dQr=(P2-P1*VRN)*V2/(g-1) \qquad (4.2.13)$$

The error between Eqs. 4.2.12 and 4.2.13 is small, and Kee(1.20) shows it to be less than 1% for crank angle steps of 1º during the experimental pressure record analysis.

The cylinder pressure record is digitized, the values of dQr are determined, and the sum of all of the values of dQr at each crank angle step is evaluated to provide a value of the total heat released, QR. The ratio of dQr to QR at each step is the value of the mass fraction of the fuel which has been burned during that time step, MFB.

There has been a considerable volume of publications on heat release characteristics of engines, mostly four-stroke engines. To mention but four, one of the original papers related to diesel engines was presented by Lyn(4.8); others in more recent times were given by Hayes(4.11), Martorano(4.27), and Lancaster(4.12).

4.2.3 Heat release from a two-stroke loop scavenged engine

Sects. 1.5.8 and 9 present experimentally determined cylinder pressure diagrams in Figs. 1.14-1.16 from a QUB LS400 engine. This is a 400 cc loop scavenged, spark-ignition, carburetted, research engine, watercooled, of 85 mm bore and 70 mm stroke. The exhaust system is untuned; in short, it is a simple muffler or silencer.

Therefore, the trapping characteristics are those provided by the scavenging process and not through any dynamic assistance from a tuned exhaust pipe. The combustion chamber employed for these tests is of the "hemisphere" type shown in Fig. 4.14 with a trapped compression ratio of 6.62. It is, of course, not a true hemisphere, that is simply the jargon for all such cylinder heads which have sections which are segments of a sphere. For a particular test at full throttle, the cylinder pressure records were taken by Kee(1.20) at 3000 rev/min and analyzed for the values of indicated mean effective pressure, heat release rate and mass fraction burned. The flame speed is recorded during this test. It will be observed that at 24.5 m/s, or somewhat less than 55 mph, the combustion process could in no fashion be compared to an explosion!

For the record, the actual test values at this speed at wide open throttle (WOT) are tabulated below:

IMEP=6.80 bar, BMEP=6.28 bar, BSFC=435 g/kWh, air-fuel ratio=13.0,
delivery ratio=0.85, exhaust CO=2.1%, exhaust HC=4750 ppm,
exhaust O_2=7.7%, flame speed=24.5 m/s.

The analysis of the cylinder pressure trace for the heat release rate is shown in Fig. 4.4, and that for the mass fraction burned is shown in Fig. 4.5. It is quite clear that there is an ignition delay period of 10º before the heat release becomes manifest in the combustion chamber. The heat release rate peaks at 10º atdc, 20º after the heat release commences properly at 10º btdc. In Sect. 4.1.1, where this particular example was discussed in terms of flame travel time, it was adjudged that the flame had travelled throughout the chamber in 33º crank angle. It can be seen from Fig. 4.4 that this coincides with the time from ignition to the peak of the heat release rate, and that position also coincides with the point in Fig. 4.5 where 50% of the mass fraction was burned.

4.2.4 Combustion efficiency

It is possible to assess values of combustion efficiency from such test results, where this is defined as:

BE=(Heat released by fuel)/(Heat available in fuel) (4.2.14)

The symbol notation derives from the words "Burn Efficiency." This discussion is a continuation of that commenced in Sect. 1.5.7, and is best illustrated in terms of the experimental data presented already for the QUB LS400 engine. Here, the trapping efficiency was evaluated according to the procedures described in Sect. 1.6.3 and Eq. 1.6.22. The mixture was enriched at the same speed and throttle opening to ensure that the exhaust oxygen level recorded was due to scavenge losses only. A value of 7.7% oxygen was noted at an air-fuel ratio of 10.91 and this translates to a trapping efficiency of 60.2%. As the engine has a BSFC of 435 g/kWh

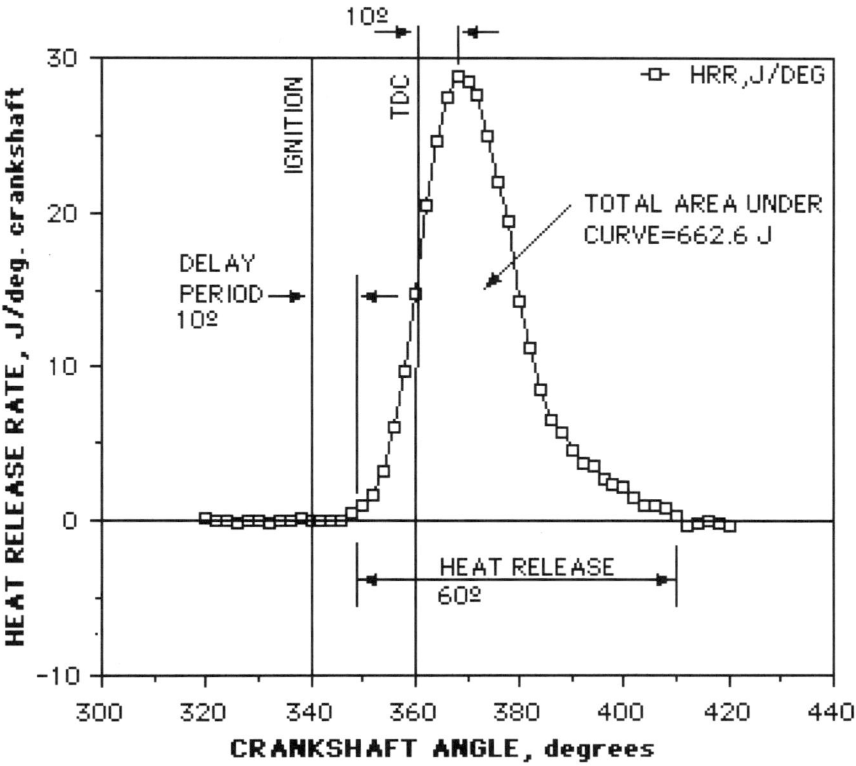

Fig. 4.4 Heat release rate for QUB LS400 at WOT at 3000 rpm.

at a BMEP of 6.28 bar from an engine of 397.2 cc swept volume running at 50 rev/s, the cyclic consumption of fuel is given from the following sequence of equations, all of which were introduced in Sect. 1.5.7 and Sect. 1.6.1:

Fuel supplied, kg/s=BSFC*BPR
Fuel supplied, kg/s=BSFC*BMEP*SV*RPS
Fuel supplied/cycle, kg=(fuel supplied/s)/RPS
Heat energy supplied/cycle, J=(fuel supplied/cycle)*(lower calorific value)
Heat energy trapped/cycle, J=TE* (heat energy supplied/cycle)
Heat available in fuel, J= heat energy trapped/cycle
=TE*CAL*BSFC*BMEP*SV (4.2.15)

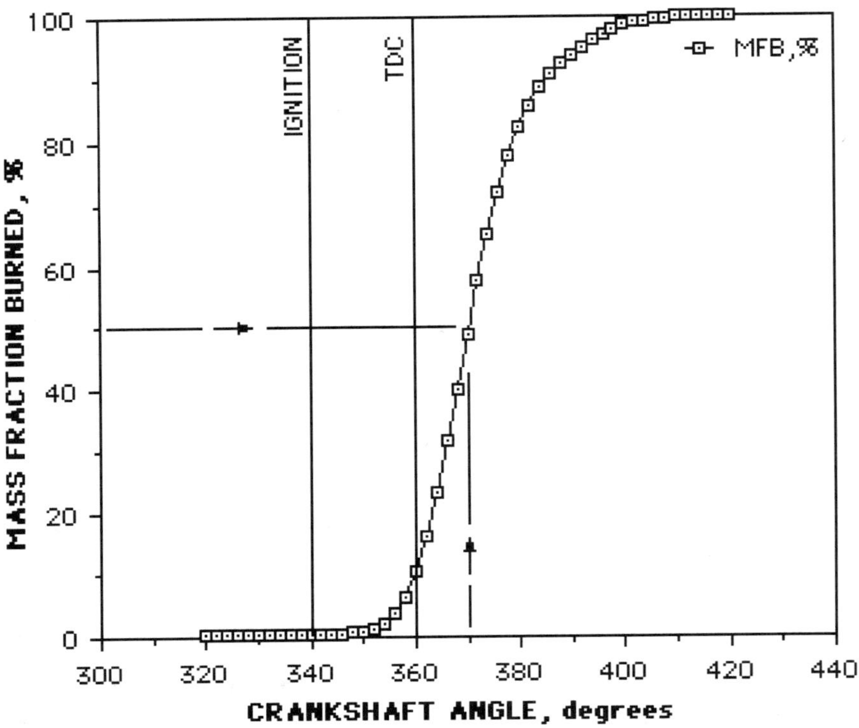

Fig. 4.5 Mass fraction burned in QUB LS400 engine at WOT and 3000 rpm.

All of the symbols and notation are exactly as used in Chapter 1. The gasoline fuel used has a lower calorific value of 43 MJ/kg. For the actual example quoted for the QUB LS400 unit, insertion of the experimental data into Eq. 4.2.15 reveals:

Heat available in fuel, J

$=0.602*43*10^6*(0.435/1000/3600)*(6.28*10^5)*(397.2*10^{-6})$

$=780.23$ J

The heat released from the fuel as recorded on the heat release diagram for that engine, Fig. 4.4, is the total area under the heat release curve. This is calculated from Fig. 4.4 as 662.6 J. Consequently, the combustion efficiency of this engine at that particular test point is:

$BE=662.6/780.23$

$=0.849$

4.3 Modelling the combustion process theoretically

For the engine designer who wishes to predict the performance characteristics of the engine, it is clear that the theory already given in Chapter 2 on unsteady gas flow and in Chapter 3 on scavenging permits the preparation of a computer model which will predict the thermodynamic and gas-dynamic behavior of an engine at any throttle opening and any rotational speed. All of this theory pertains to the open cycle, from exhaust port opening to exhaust port closing. The remaining part of the computer analysis which is required must cover the closed cycle period, when the compression, combustion and expansion phases take place. The discussion on modelling the compression and expansion phases is in Chapter 5, as is the implementation of a model for the combustion process. However, it is pertinent in this chapter on combustion to point out the various possibilities which exist for formulating such a model. There are three main methods which are possible and each of these will be discussed in turn:

(a) a heat release model
(b) a one-dimensional flame propagation model
(c) a three-dimensional combustion model

4.3.1 A heat release model of engine combustion

One of the simplest models of engine combustion is to use the heat release analysis presented in Sect. 4.2.2, in reverse. In particular, Eq. 4.2.12 would become a vehicle for the prediction of the pressure, P2, at the conclusion of any incremental step in crankshaft angle, if all of the other parameters were known as input data. This relies upon the assumption of data values for such factors as the combustion efficiency and the polytropic exponents during compression and expansion. At the same time, a profile has to be assumed for the heat release during the combustion phase as well as a delay period before that heat release commences.

At The Queen's University of Belfast, the use of heat release models has been used for many years, since their original publication on the subject(4.9) to the latest at the time of writing(4.10), some ten years later. That it is still found to be effective is evident from the accuracy of correlation between measurement and calculation of engine performance characteristics described in those publications.

The heat release characteristics of homogeneously charged spark-ignition two-stroke engines have been found to be remarkably similar, which is not too surprising as virtually all of the engines have rather similar combustion chambers. Such would not be the case for four-stroke cycle engines, where the combustion chamber shape is dictated by the poppet valve mechanisms involved. A uniflow two-stroke engine of the type illustrated in Fig. 1.5 would conform more to a four-stroke combustion model than one for a two-stroke engine.

Furthermore, the profile of the heat release rate as a function of crank angle is known not to be heavily dependent on load, or IMEP. Naturally, the peak value of heat release rate is dependent on load, but the profile would not appear to be so.

Therefore, the experimentally determined heat release rate curve shown in Fig. 4.4, typical of many which have been observed, can be postulated as a potential theoretical model of combustion. Such a model is extracted from the experimental data and displayed in Fig. 4.6. The heat release period is P°, with a rise time of $P^{\circ}/3$ equally distributed about tdc. The delay period is approximately $P^{\circ}/6$, but it must be remembered that the optimum ignition timing for any engine is one which will start the process so that the ensuing cylinder pressure rise is distributed about tdc to give the highest IMEP. Therefore, in modelling terms, the ignition timing and the delay period fall into the category of second-order importance, as it is the phasing and duration of the heat release profile about tdc which will ensure the highest IMEP by a theoretical calculation. While this method makes it unnecessary, the delay period can be predicted theoretically (4.13). The selection of the period P° is by far the most important criterion and it is the author's experience that this is rarely less than 45°, nor greater than 65°, irrespective of speed and load (IMEP). It will be noted that the heat release profile has a "tail" of magnitude $P^{\circ}/3$, falling to zero from one sixth of the maximum value of heat release.

The total area of the profile in Fig. 4.6 is $(14/36)*HM*P^{\circ}$, where HM is the maximum rate of heat release. The actual value of HM in Fig. 4.4 is 28.8 J/deg and the period, P°, is 60°. The area under the model profile in Fig. 4.6, if that same peak

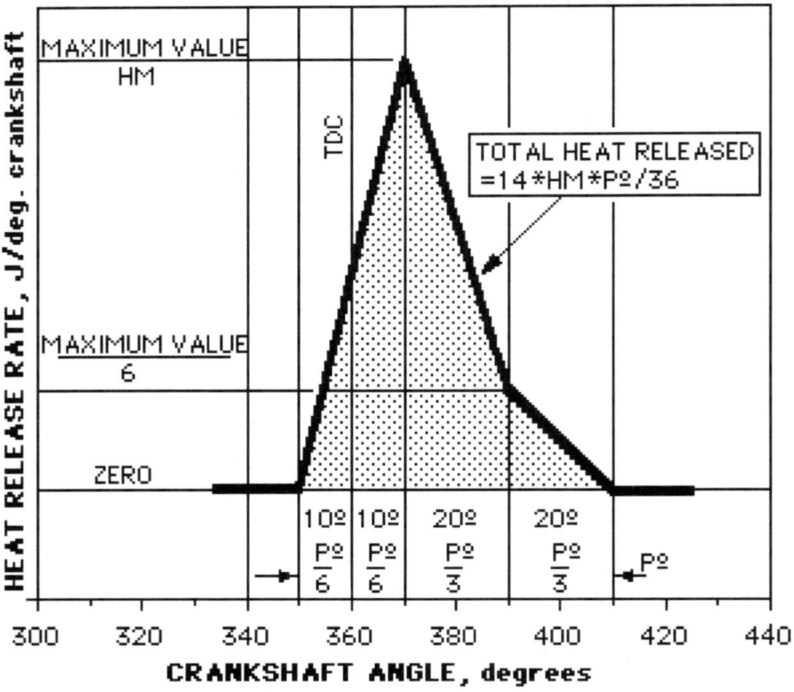

Fig. 4.6 Possible model of heat release rate for theoretical calculations.

value is maintained, is 672 J, which corresponds well with the measured value of 662.6 J. For a theoretical total heat release of 662.6 J, one would predict from the model a HM value of 28.4 J/deg.

One of the simplifying assumptions in the use of a heat release model for combustion is that the prediction of the heat loss to the cylinder walls and coolant is encapsulated in the selection of the polytropic exponents for the compression and expansion processes. This may be regarded by some as an unnecessary simplification, perhaps even an inaccurate one. However, as all models of heat transfer are more or less based on empirical forms, the use of experimentally determined polytropic exponents could actually be regarded as a more realistic assumption. This is particularly true for the two-stroke engine, where so many engine types are similar in construction. The modeller, in the absence of better experimental evidence from a particular engine, could have some confidence in using polytropic indices of 1.25 and 1.35 for the compression and expansion phases, respectively, and 1.38 for the numerical value of g in Eq. 4.2.12, together with a value between 0.85-0.9 for the combustion efficiency, BE.

The difficult nature of heat transfer calculations in ic engines is best illustrated by the proliferation of "empirical equations" in the literature, all bearing prestigious names, such as those attributed to Eichelberg, Woschni, or Annand, to mention but three. This will be more evident in the ensuing discussion on flame propagation models of combustion.

4.3.2 A one-dimensional model of flame propagation

One of the simplest models of this type was proposed by Blizard(4.2) and is of the eddy entrainment type. The model was used by Douglas(4.13) at QUB and so some experience was gained of the use of such models(5.13), by comparison with the simpler heat release model discussed in the previous section. The model is based on the propagation of a flame as shown in Fig. 4.1, and as already discussed in Sect. 4.1.1. The model assumes that the flame front entrains the cylinder mass at a velocity which is controlled by the in-cylinder turbulence. The mass is burned at a rate which is a function of both the laminar flame speed, LFV, and the turbulence velocity, TV.

The assumptions made in this model are:

(a) The flame velocity, FV, is:

$$FV=LFV+TV \tag{4.3.1}$$

(b) The flame forms a portion of a sphere centered on the spark plug.

(c) The thermodynamic state of the unburned mass which has been entrained is identical to that fresh charge which is not yet entrained.

(d) The heat loss from the combustion chamber is to be predicted by convection and radiation heat transfer equations based on the relative surface areas and thermodynamic states of burned and unburned gases.

(e) The mass fraction of entrained gas which is burned, MEB, at any given time, t, after its entrainment is to be estimated by an exponential relationship of the form:

$$MEB=e^{(-t/F)} \qquad (4.3.2)$$

where F=L/LFV and L is the turbulence micro-scale.

Clearly, a principal contributor to the turbulence present is squish velocity, of which there will be further discussion in Sect. 4.4. The theoretical procedure progresses by the use of complex empirical equations for the various values of laminar and turbulent flame speed, all of which are determined from fundamental experiments in engines or combustion bombs (4.1) (4.4) (4.5).

For spark ignition engines the heat transfer from the chamber during this combustion process is based upon Annand's work(4.15) (4.16) (4.17). For diesel engines, it should be added that the heat transfer equation by Woschni(4.18) is usually regarded as being more effective for such theoretical calculations. Typical of the heat transfer theory proposed by Annand is his expression for the Nusselt number, leading to a conventional derivation for the convection heat transfer coefficient, HC:

$$Nu=a*Re^{0.7} \qquad (4.3.3)$$

where, a=0.26 for a two-stroke engine, or, a=0.49 for a four-stroke engine.

Annand also considers the radiation heat transfer coefficient, HR, to be given by:

$$HR=4.25*10^{-9}*(T^4-TW^4)/(T-TW) \qquad (4.3.4)$$

However, the value of HR is many orders of magnitude less than HC, to the point where it may be neglected for two-stroke cycle engine calculations.

It is clear from this brief description of a turbulent flame propagation model that it is much more complex than the heat release model posed in Sect. 4.3.1. As the physical geometry of the clearance volume must be specified precisely, and all of the chemistry of the reaction process followed, the calculation will require much more computer time. By using this theoretical approach, the use of empirically determined coefficients and correctors has increased greatly over the earlier proposal of a simple heat release model to simulate the combustion process. It is very questionable if the overall accuracy of the calculation, for example, in terms of IMEP, has been much improved over the simpler model. There is no doubt that valuable understanding will be gained, in that the user obtains data from the computer calculation on such important factors as the proportions of the ensuing exhaust gas emissions and the flame duration. This type of calculation is much more logical when applied in three-dimensional form and allied to a CFD calculation for the gas behavior throughout the cylinder. This is discussed briefly in the next section.

4.3.3 Three-dimensional combustion model

From the previous comments it is clearly necessary that reliance on empirically determined factors for heat transfer and turbulence behavior, which refer to the combustion chamber as a whole, will have to be exchanged for a more microscopic examination of the entire system if calculation accuracy is to be enhanced. This is possible by the use of a combustion model in conjunction with a computational fluid dynamics model of the gas flow behavior within the chamber. Computational fluid dynamics, or CFD, was introduced to the reader in Chapter 3, where it was shown to be a powerful tool to illuminate the understanding of scavenge flow within the cylinder.

That the technology is moving towards providing the microscopic in-cylinder gas-dynamic and thermodynamic information is seen in the paper by Ahmadi-Befrui(4.21). Fig. 4.7 is taken directly from that paper and it shows the in-cylinder velocities, but the calculation holds all of the thermodynamic properties of the charge as well, at a point just before ignition. This means that the prediction of heat transfer effects at each time step in the calculation will take place at the individual calculation mesh level, rather than by empiricism for the chamber as a whole, as was the case in the preceding sections. For example, should any one surface or side of the combustion bowl be hotter than another, the calculation will predict the heat transfer in this microscopic manner giving new values and directions for the motion of the cylinder charge; this will affect the resulting combustion behavior.

This calculation can be extended to include the chemistry of the subsequent combustion process. Examples of this have been published by Amsden et al(4.20) and Fig. 4.8 is an example of their theoretical predictions for a direct injection,

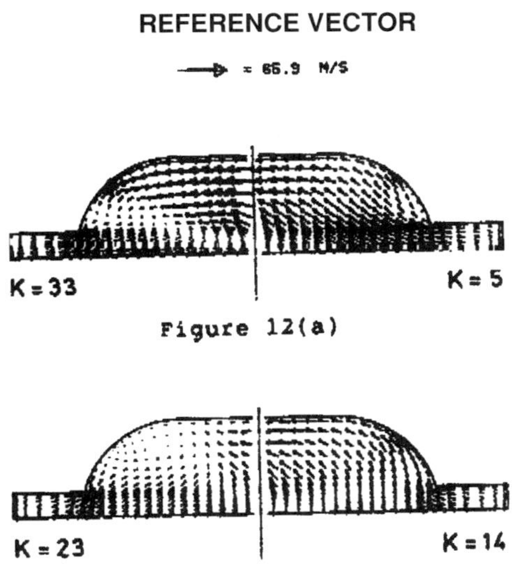

Figure 4.7 CFD calculations of velocities in the combustion chamber prior to ignition.

stratified charge, spark-ignition engine. Fig. 4.8 illustrates a section through the combustion bowl, the flat-topped piston and cylinder head. Reading across from top to bottom, at 28º btdc, the figure shows the spray droplets, gas particle velocity vectors, isotherms, turbulent kinetic energy contours, equivalence ratio contours, and the octane mass fraction contours. The paper(4.20) goes on to show the ensuing combustion of the charge. This form of combustion calculation is preferred over any of those models previously discussed, as the combustion process is now being theoretically examined at the correct level. However much computer time such calculations require, they will become the normal design practice in the future, for computers are increasingly becoming more powerful yet more compact and relatively inexpensive.

4.4 Squish behavior in two-stroke engines

It has already been stated in Sect. 4.1.4 that detonation is an undesirable feature of ic engine combustion, particularly when the designer attempts to operate the engine at a compression ratio which is too high. As described in Sect. 1.5.8, the highest thermal efficiency is to be obtained at the highest possible compression ratio. Any technique which will assist the engine to run reliably at a high compression ratio on a given fuel must be studied thoroughly and, if possible, designed into the engine and experimentally tested. Squish action, easily obtained in the uncluttered cylinder head area of a conventional two-stroke engine, is one such technique. A photograph of squish action, visually enhanced by smoke in a motored QUB type deflector piston engine, is shown in Plate 4.3.

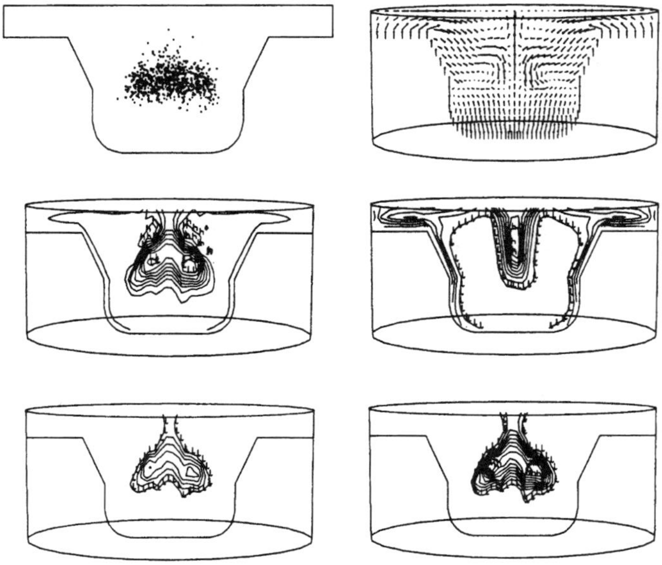

Fig. 4.8 Detailed calculation of gasoline combustion in a fuel-injected SI engine.

Plate 4.3 The vigorous squish action in a QUB type cross scavenged engine at the end of the compression stroke.

4.4.1 A simple theoretical analysis of squish velocity

From technical papers such as that given by Ahmadi-Befrui et al(4.21), it is clear that the use of a CFD model permits the accurate calculation of squish velocity characteristics within the cylinder. An example of this is seen in Fig. 4.7, a CFD calculation result by Ahmadi-Befrui et al(4.21) for in-cylinder velocities at the point of ignition at 20º btdc. This insight into the squish action is excellent, but it is a calculation method which is unlikely to be in common use by the majority of designers, at least for some time to come. Because a CFD calculation uses much time on even the fastest of super-computers, the designer needs a more basic program for everyday use. Indeed, to save both super-computer and designer time, the CFD user will always need some simpler guidance tool to narrow down to a final design before employing the fewest possible runs of the CFD package. An analytical approach is presented below which is intended to fulfill that need. This is particularly helpful, principally because the conventional two-stroke engine has a simple cylinder head, so the designer can conceive of an almost infinite variety of combustion chamber shapes. This is obvious from an examination of some of the more basic shapes shown in Fig. 4.10.

There have been several attempts to produce a simple analysis of squish velocity, often with theory which is more empirically based than fundamentally thermodynamic. One of the useful papers which has been widely quoted in this area is that by Fitzgeorge(4.22). Experimental measurements of such phenomena are becoming more authoritative with the advent of instrumentation such as laser doppler anemometry, and the paper by Fansler(4.23) is an excellent example of what is possible by this accurate and non-intrusive measurement technique. However, the following theoretical procedure is one which is quite justifiable in thermodynamic terms, yet is remarkably simple.

Figs. 4.2 and 4.9 represent a compression process inducing a squished flow between the piston and the cylinder head. The process commences at trapping, i.e., exhaust port closure. From Sect. 1.5.6, the value of trapped mass, Mtr, is known and is based upon an assumed value for the trapped charge pressure and temperature, Ptr and Ttr. At this juncture, the mass will be evenly distributed between the volume subtended by the squish band, VStr, and the volume subtended by the bowl, VBtr. The actual values of VS1 and VB1 at any particular piston position, as shown in Fig. 4.9, are a matter of geometry based on the parameters illustrated in Fig. 4.2. From such input parameters as squish area ratio, SAR, the values of squish area, AS, bowl area, AB, and bowl diameter, DB, are calculated from Eq. 4.2.1.

For a particular piston position, as shown in Fig. 4.9, the values of thermodynamic state and volumes are known before an incremental piston movement takes place.

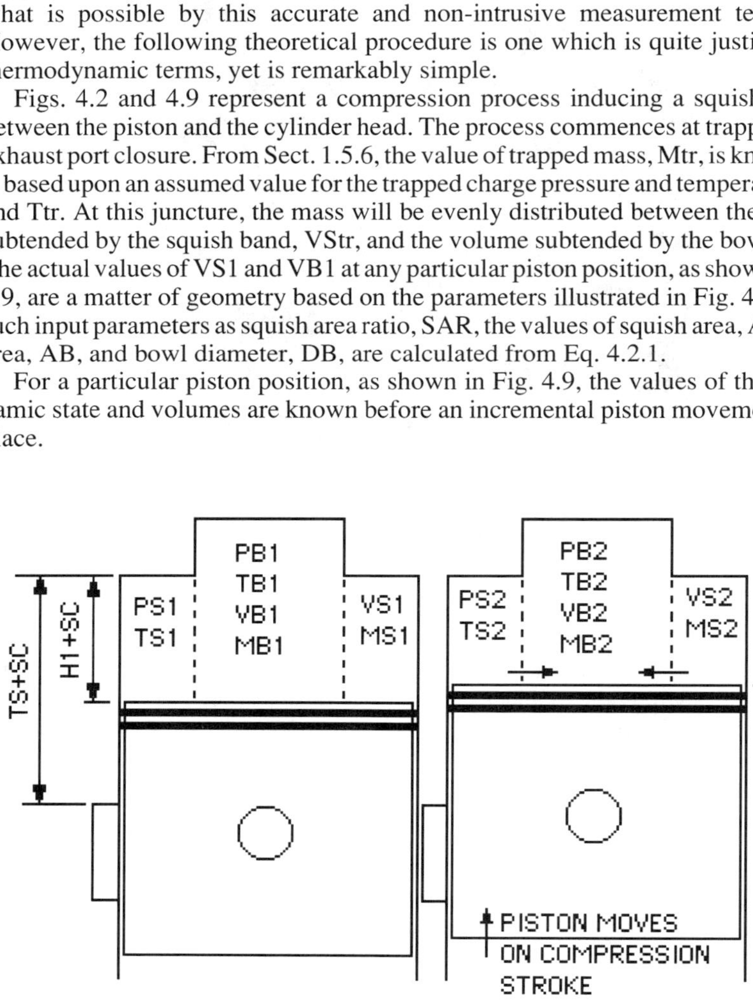

Fig. 4.9 Simple theoretical model of squish behavior.

The fundamental thesis behind the calculation is that all state conditions equalize during any incremental piston movement. Thus, the values of cylinder pressure and temperature are PC1 and TC1, respectively, at the commencement of the time step, dt, which will give an incremental piston movement, dH, an incremental crank angle movement, dC°, and a change of cylinder volume, dV.

However, if equalization has taken place from the previous time step, then:

PC1=PS1=PB1
TC1=TS1=TB1
VC1=VS1+VB1
DC1=DS1=DS2
Mtr=MS1+MB1

For the next time step the compression process will occur in a polytropic manner with an exponent, n, as discussed in Sect. 4.2.2. If the process is considered ideal, or isentropic, that exponent is g, the ratio of specific heats for the cylinder gas. Because of the multiplicity of trapping conditions and polytropic compression behavior, it is more logical for any calculation to consider engine cylinders to be analyzed on a basis of equality. Therefore, the compression process is assumed to be isentropic, with air as the working fluid, and the trapping conditions are assumed to be at the reference pressure and temperature of 1 atm and 20°C. In this case the value of g is 1.4. The value of the individual compression behavior in the squish and bowl volumes, as well as the overall macroscopic values, follows:

$PS2=PS1*(VS1/VS2)^g$
$PB2=PB1*(VB1/VB2)^g$
$PC2=PC1*(VC1/VC2)^g$
$DS2=DS1*(VS2/VS1)^g$
$DB2=DB1*(VB2/VB1)^g$
$DC2=DC1*(VC2/VC1)^g$

The values DS2, DB2 and DC2 are the new densities in the relevant volumes.

If a squish action takes place, then the value of squish pressure, PS2, is greater than either PC2 or PB2, for:

PS2>PC2>PB2

The squish pressure ratio, SPR, causing gas flow to take place, is found from:

$$SPR=PS2/PB2 \tag{4.4.1}$$

At this point, consider that a gas flow process takes place within the time step so that the pressures equalize in the squish band and the bowl, equal to the average cylinder pressure. This implies movement of mass from the squish band to the bowl, so that the mass distributions at the end of the time step are proportional to the volumes, as follows:

MS2=Mtr*VS2/VC2 (4.4.2)

During the course of the compression analysis, the mass in the squish band was considered to be the original value, MS1, so the incremental mass squished, dMS, is given by:

dMS=MS1-MS2 (4.4.3)

This incremental movement of mass, occurring in time, dt, is further evaluated to determine the squish velocity, SQV, by:

dMS=(mass rate of flow)*dt
=(DC2*AS*SQV)*dt (4.4.4)
=(DC2*AS*SQV)*dCº/(360*RPS) (4.4.5)

The remaining factor to be evaluated is the value of AS, the area through which the incremental mass, dMS, will pass during the time step, dt. This will have various values depending on the type of cylinder head employed, irrespective of the parameters previously mentioned. For example, consider the cases of the basic cylinder head types (a)-(d), as shown in Fig. 4.10.

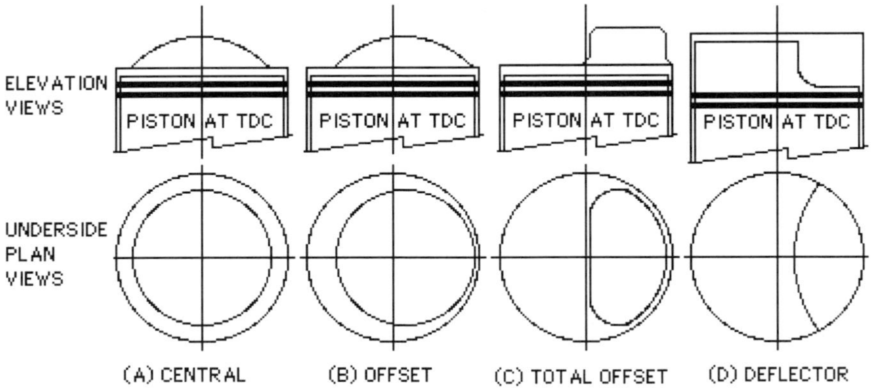

Fig. 4.10 Combustion chamber types analyzed in Prog.4.1, SQUISH VELOCITY.

For cylinder head type (a), the central chamber, the following analysis for the squish flow area, AS, will apply. The flow will be radial through a segment of the cylinder, and if the flow is considered to take place midway through the incremental step, then:

$$AS=\pi*DB*[SC+(H1+H2)/2] \quad (4.4.6)$$

For cylinder head type (b), the offset chamber, the following analysis for the squish flow area, AS, will apply. If the chamber edge approximately coincides with the cylinder bore, then about one quarter of the potential flow path is blocked by comparison with a central chamber. The flow will still be radial through a segment of the cylinder, and if the flow is considered to take place midway through the incremental step, then:

$$AS=0.75*\pi*DB*[SC+(H1+H2)/2] \quad (4.4.7)$$

For cylinder head type (c), the total offset chamber, the following analysis for the squish flow area, AS, will apply. The flow will be approximately across a cylinder diameter, and if the flow is considered to take place midway through the incremental step, then:

$$AS=BO*[SC+(H1+H2)/2] \quad (4.4.8)$$

For cylinder head type (d), the deflector chamber, the following analysis for the squish flow area, AS, will apply. The flow will be along the axis of the cylinder and across the width of the deflector, and if the flow is considered to take place midway through the incremental step, then:

$$AS=DFW*[SC+(H1+H2)/2] \quad (4.4.9)$$

where DFW is the deflector width.

It is clear that even for equal values of SAR, higher squish velocities will be given for those chambers with more offset.

It is also possible to determine the turbulence energy induced by this flow, on the assumption that it is related to the kinetic energy created. The incremental kinetic energy value at each time step is dKE, where:

$$dKE=0.5*dMS*SQV^2 \quad (4.4.10)$$

The total value of turbulence kinetic energy squished, SQKE, is then summed for all of the calculation increments over the compression process from trapping to tdc.

4.4.2 Evaluation of squish velocity by computer

The equations in Sect. 4.4.1 are programmed in Prog.4.1 and listed in the Computer Program Appendix as ProgList.4.1 entitled "SQUISH VELOCITY." All of the combustion chamber types shown in Fig. 4.10 can be handled by this program. The operator is prompted to type in the chamber type by name, either central, offset, total offset, or deflector. Of course, there will appear a design for a combustion chamber which is not central, yet is not sufficiently offset as to be described by category (b) in Fig. 4.10. Such a chamber would be one where there is still a significant radial clearance from the bore edge so that the squish flow can proceed in a radial fashion around the periphery of the bowl. In that case, and it is one requiring user judgment, the "chamber type?" prompt from the computer program should be answered with "central."

A typical output from the program is shown in Fig. 4.11, where all of the relevant data input values are observed, with the output values for maximum squish velocity and maximum squish pressure ratio at the crank angle position determined for these maxima, and the total kinetic energy which has been squished. The screen picture is dynamic, i.e., the piston moves from trapping to tdc, and the squish velocity graph is also dynamically created. By such means the operator obtains an enhanced design feel for the effect of squish action, for it is upon such insight that real design experience is built.

CYLINDER HEAD TYPE IS CENTRAL
BORE,mm= 70
STROKE,mm= 70
CON-ROD,mm= 125
ENGINE SPEED, rev/min= 5000
EXHAUST OPENS,ºatdc= 104

CYLINDER HEAD GEOMETRY
TRAPPED COMPRESSION RATIO= 7
SQUISH AREA RATIO= .5
SQUISH CLEARANCE,mm= 1.2

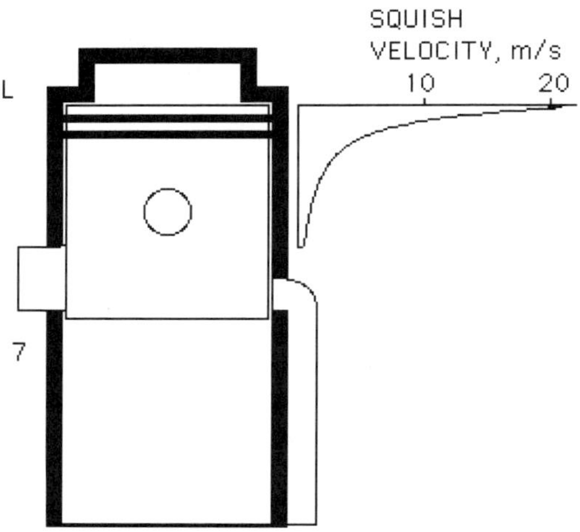

OUTPUT DATA
MAXIMUM SQUISH VELOCITY, m/s= 20.9
MAXIMUM SQUISH PRESSURE RATIO=1.094
BOTH OCCURRED AT, deg.btdc= 10
SQUISHED KINETIC ENERGY, mJ= 6.94

Fig. 4.11 Typical screen and printer output from Prog.4.1, SQUISH VELOCITY.

The sketch of the engine, cylinder, and cylinder head which appears on the screen is drawn to scale, although the head is graphically assumed to be of the "bathtub" type in each case. Thus, the physical changes which are incurred by altering bore-stroke ratio, squish area ratio, squish clearance, exhaust port timing and trapped compression ratio, are immediately obvious to the designer at the same time as the squish velocity characteristics.

4.4.3 Design of combustion chambers to include squish effects

To emphasize the points already made regarding squish action, the computer program was used to predict the squish effects in various combustion chamber configurations when mounted on a common engine block. The engine, selected at random, has bore, stroke and rod dimensions of 70, 70 and 125 mm, respectively. It runs at 5000 rev/min and has an exhaust closing point at 104º btdc. The chamber types (a), (b) and (c) have SAR values of 0.5, while the QUB deflector type (d) has a SAR value of 0.7 and a deflector width of 67 mm. Various squish clearances are now inserted as data in the computer program, ranging from a mechanically possible narrow clearance of 1.2 mm up to 3 mm. The results are plotted in Figs. 4.12 and 4.13 and the common data are printed on the figures.

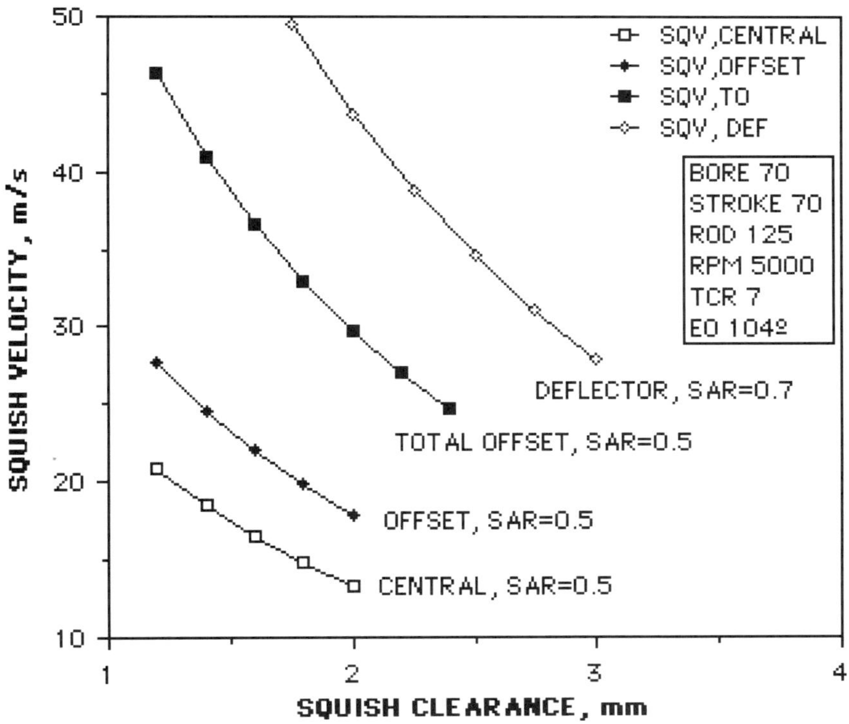

Fig. 4.12 Squish velocity in various combustion chamber types.

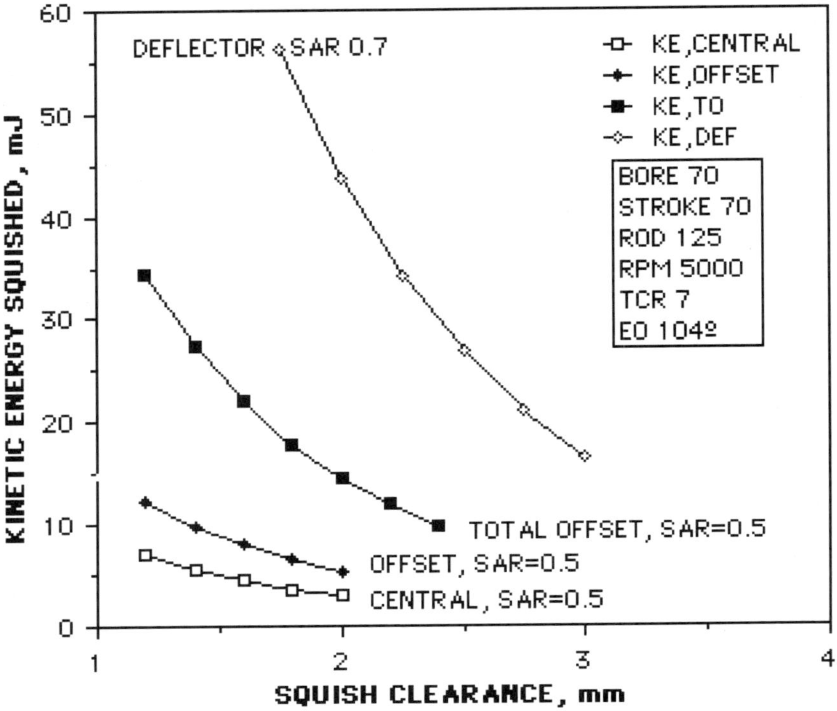

Fig. 4.13 Squished kinetic energy in various combustion chambers.

There is considerable disparity between the values of squish velocity for the various types of chamber, even for those of equal SAR and squish clearance. The total offset and deflector chambers can have squish velocity values which are at least double that of a central chamber. Naturally, as the kinetic energy values are based on the square of the velocity, the SQKE graph enhances the level of disparity among the several chambers. It is interesting to compare the order of magnitude of the highest value of SQKE, shown in Fig. 4.13 as 56 mJ, with that given by Amsden(4.20) and shown here as Fig. 4.8. The highest contour value of turbulent kinetic energy given in Fig. 4.8 is 263 mJ and the lowest value is 32 mJ. However, that engine also had considerable swirl motion to the in-cylinder gas flow.

In Eq. 4.3.1, it is reported that the total flame velocity is the sum of the laminar flame velocity and the turbulence velocity. If that statement has any authenticity then the flame velocities observed in various combustion chambers must bear some correspondence with squish velocity. Kee(1.20) reports flame speeds measured in a loop scavenged engine under the same test conditions as already quoted in Sect. 4.2.3. The chamber was central and the measured flame speed was 24.5 m/s. With the data inserted for the combustion chamber involved, Prog.4.1 predicted maximum squish velocity and squished kinetic energy values of 11.6 m/s and 2.61 mJ, respectively.

In further tests using a QUB type deflector piston engine of the same bore and stroke as the loop scavenged engine above, Kee(1.20) reports that flame speeds were measured at the same speed and load conditions at 47 m/s within the chamber and 53 m/s in the squish band. The average flame velocity was 50 m/s. With the data inserted for the deflector combustion chamber involved, Prog.4.1 predicted maximum squish velocity and squished kinetic energy values of 38.0 m/s and 50.2 mJ, respectively. In other words, the disparity in squish velocity as calculated by the theoretical solution is seen to correspond to the measured differences in flame speed from the two experimental engines. That the squish action in a QUB type cross scavenged engine is quite vigorous can be observed in Plate 4.3.

If the calculated squish velocity, SQV, is equated to the turbulence velocity, TV, of Eq. 4.3.1 and subtracted from the measured flame velocity, FV, in each of Kee's experimental examples, the laminar flame velocity, LFV, is predicted as 12.9 and 12.0 m/s for the loop scavenged and QUB deflector piston cases, respectively. That the correspondence for LFV is so close, as it should be for similar test conditions, reinforces the view that the squish velocity has a very pronounced effect on the rate of burning and heat release in two-stroke engines. It should be added, however, that any calculation of laminar flame velocity (4.2) (4.5) for the QUB 400 type engine would reveal values somewhat less than 3 m/s. Nevertheless, flame speed values measured at 12 m/s would not be unusual in engines with quiescent combustion chambers, where the only turbulence present is that from the past history of the scavenge flow and from the motion of the piston crown in the compression stroke.

The design message from this information is that high squish velocities lead to rapid burning characteristics and that rapid burning approaches the thermodynamic ideal of constant volume combustion. There is a price to be paid for this, evidenced by more rapid rates of pressure rise which can lead to an engine with more vibration and noise emanating from the combustion process. Nevertheless, the designer has available a theoretical tool, in the form of Prog.4.1, to tailor this effect to the best possible advantage for any particular design of two-stroke engine.

One of the beneficial side effects of squish action is the possible reduction of detonation effects. The squish effect gives high turbulence characteristics in the end zones and, by inducing locally high squish velocities in the squish band, increases the convection coefficients for heat transfer. Should the cylinder walls be colder than the squished charge, the end zone gas temperature can be reduced to the point where detonation is avoided, even under high BMEP and compression ratio conditions. For high performance engines, such as those used for racing, the design of squish action must be carried out by a judicious combination of theory and experimentation. A useful design starting point for loop scavenged engines with central combustion chambers is to keep the maximum squish velocity between 15 and 20 m/s at the peak power engine speed. If the value is higher than that, the mass trapped in the end zones of the squish band may be sufficiently large and, with the faster flame front velocities engendered by a too-rapid squish action, may still induce detonation. As with most design procedures, a compromise is required, and that compromise is different depending on the performance requirements of the engine over the entire speed and load range.

4.5 Design of combustion chambers with the required clearance volume

Perhaps the most important factor in geometrical design, as far as a combustion chamber is concerned, is to ensure that the subsequent manufactured component has precisely the correct value of clearance volume. Further, it is important to know what effect machining tolerances will have on the volume variations of the manufactured combustion space. One of the significant causes of failure of engines in production is the variation of compression ratio from cylinder to cylinder. This is a particular problem in multi-cylinder engines, particularly if they are high specific performance units, in that one of the cylinders may have a too-high compression ratio which is close to the detonation limit but is driven on by the remaining cylinders which have the designed, or perhaps lower than the designed, value of compression ratio. In a single- or twin-cylinder engine the distress caused by knocking is audibly evident to the user, but this is much less likely to be the case for a six- or eight-cylinder outboard engine.

So that the designer may assess all of these factors, as well as be able to produce the relevant data for the prediction of squish velocity by Prog.4.1, there are six design programs included in the Computer Program Appendix to cover the main possibilities usually presented to the designer as combustion chamber options. The physical geometry of these six combustion chambers is shown in Fig. 4.14, and all but one are intended primarily for use in loop scavenged engines. Clearly Prog.4.7 is specifically for use for QUB type deflector piston engines. The names which are appended to each type of combustion chamber, in most cases being the jargon applied to that particular shape, are also the names of the computer programs involved. To be pedantic, the name of computer program, Prog.4.4, is "BATHTUB CHAMBER" and it appears in the listings as ProgList.4.4. Actually, the "bowl in piston" combustion system would also be found in uniflow scavenged engines, as well as loop scavenged engines, particularly if they are diesel power units.

An example of the calculation is given in Fig. 4.15 for the hemisphere combustion chamber, a type which is probably the most common for loop scavenged two-stroke engines. A photograph of this type of chamber appears in Plate 4.1, where the bowl and the squish band are evident. A useful feature of this calculation is the data insertion of separate spherical radii for the piston crown and the squish band. A blending radius between the bowl and the squish band is also included as data. The output reflects the tapered nature of the squish band, but it is within the designer's control to taper the squish band or have it as a parallel passage. In the calculation for squish velocity, a tapered squish clearance would be represented by the mean of the two data output values; in the case shown it would be (1.6+2.1)/2, or 1.85 mm. Similarly, the designer can tailor the geometry of any of the central types of chamber to have no squish band, i.e., a zero value for the squish area ratio, should that form be desirable in some particular circumstance.

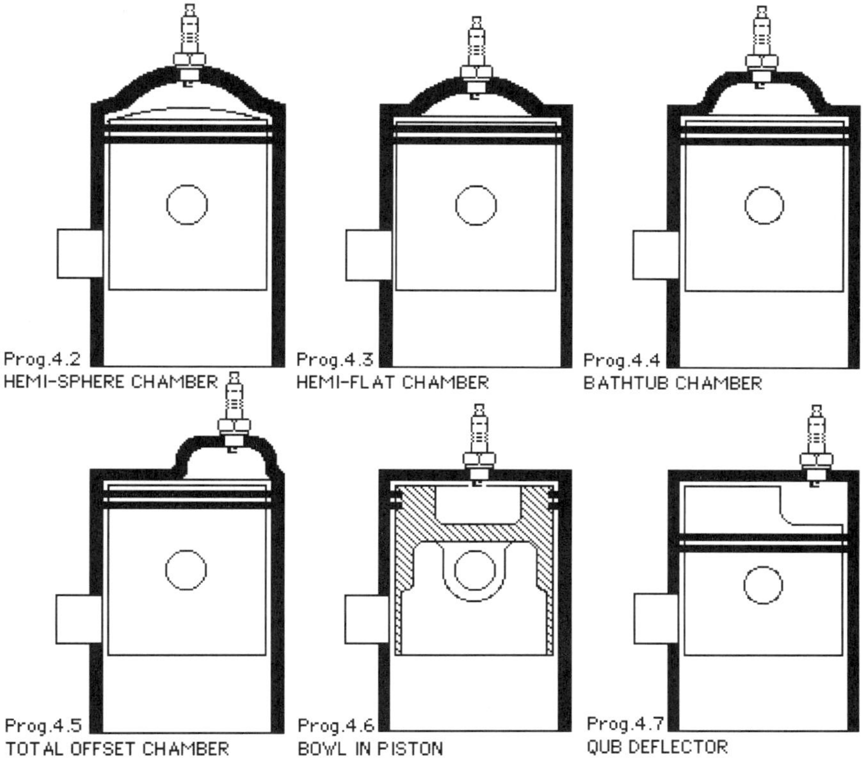

Fig. 4.14 Combustion chamber types which can be designed using the programs.

As with most of the programs associated with this book, the sketch on the computer screen is drawn to scale so that the designer may estimate by eye, as well as by number, the progress to a final solution. In the operation of any of these six programs, the computer screen will produce an actual example upon start-up, just like that in Fig. 4.15, and invite the designer to change any data value until the required geometry is ultimately produced. Finally, the printer output on the Macintosh Imagewriter (or the particular printer employed) is precisely what the operator sees on the screen, apart from the title typed in by the user.

4.6 Some general views on combustion chambers for particular applications

The various figures in this chapter, such as Figs. 4.10, 4.14 and 4.16 and Plates 4.1-4.3, show most of the combustion chambers employed in spark-ignition two-stroke engines. This represents a plethora of choices for the designer and not all are

```
CURRENT INPUT DATA FOR 2-STROKE 'SPHERICAL' BOWL
(B) BORE,mm= 72
(S) STROKE,mm= 60
(C) CON-ROD,mm= 120
(E) EXHAUST CLOSES,deg btdc= 103

BOWL DETAILS,mm
(SH) PISTON TO HEAD CLEARANCE= 1.6
(RP) PISTON CROWN RADIUS= 180
(RS) SQUISH BAND RADIUS= 140
(RC) CHAMBER RADIUS= 26
(RB) BLENDING RADIUS= 4
(SR) SQUISH AREA RATIO= .45

OUTPUT DATA
SWEPT VOLUME,cc=244.3
TRAPPED SWEPT VOLUME,cc=164.3
TRAPPED STROKE,mm=40.4
CLEARANCE VOLUME,cc=26.1
GEOMETRIC COMPRESSION RATIO=10.4
TRAPPED COMPRESSION RATIO= 7.3
MIN & MAX SQUISH CLEARANCE,mm= 1.6     2.1
TYPE CODE FOR NEW DATA AND A RECALCULATION
OR TYPE 'P' FOR PRINT-OUT, 'F' FOR FILING DATA, OR 'Q' FOR QUITTING? Q
```

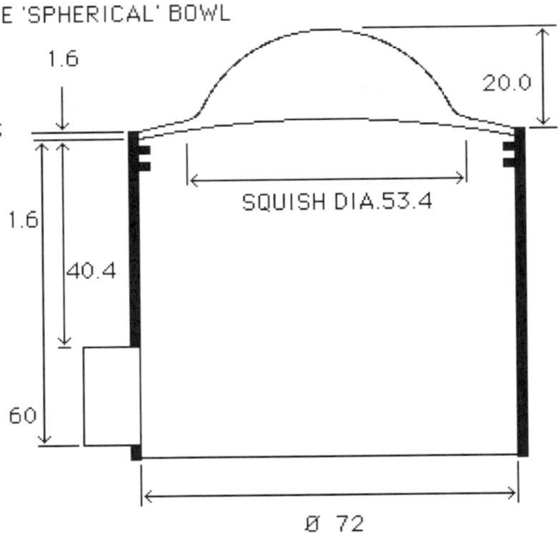

Fig. 4.15 Screen and printer output from computer design program, Prog.4.2.

universally applicable for both homogeneous and stratified charge combustion. The following views should help to narrow that choice for the designer in a particular design situation.

4.6.1 Stratified charge combustion

There will be further discussion in Chapter 7 on stratified charging of the cylinder, as distinct from the combustion of a stratified charge. A stratified charge is one where the air-fuel ratio is not common throughout the chamber at the point of combustion. This can be achieved by, for example, direct injection of fuel into the cylinder at a point sufficiently close to ignition that the vaporization process takes place as combustion commences or proceeds. By definition, diesel combustion is stratified combustion. Also by definition, the combustion of a directly injected charge of gasoline in a spark-ignition engine is potentially a stratified process. All such processes require a vigorous in-cylinder air motion to assist the mixing and vaporization of air with the fuel. Therefore, for loop and cross scavenged engines, the use of a chamber with a high squish velocity capability is essential, which means that all of the offset, total offset or deflector designs in Fig. 4.10, and the bowl in piston design in Fig. 4.14, should be examined as design possibilities. In the case of the uniflow scavenged engine, the bowl in piston design is particularly effective because the swirling scavenge flow can be made to spin even faster in the bowl towards the end of compression. This characteristic makes it as attractive for two-stroke diesel power units as for direct injection four-stroke diesel engines. It is also a strong candidate for consideration for the two-stroke spark-ignition engine with the direct injection of gasoline, irrespective of the method of scavenging.

4.6.2 Homogeneous charge combustion

The combustion of a homogeneous charge is in many ways an easier process to control than stratified charge combustion. On the other hand, the homogeneous charge burn is always fraught with the potential of detonation, causing both damage and loss of efficiency. If the stratified charge burning is one where the corners of the chamber contain air only, or a very lean mixture, then the engine can be run in "lean-burn, high compression" mode without real concern for detonation problems. The homogeneous charge engine will always, under equality of test conditions, produce the highest specific power output because virtually all of the air in the cylinder can be burned with the fuel. This is not the case with any of the stratified charge burning mechanisms, and the aim of the designer is to raise the air utilization value to as high a level as possible, with 90% regarded as a good target value for that criterion.

For homogeneous charge combustion of gasoline, where there is not the need for excessive air motion due to squish, the use of the central, bathtub, or offset chambers is the designer's normal choice. This means that the designer of the QUB deflector chamber has to take special care to control the squish action in this very active combustion system. Nevertheless, for high specific output engines, particularly those operating at high rotational speeds for racing, the designer can take advantage of higher flame speeds by using the dual ignition illustrated in Fig. 4.16(a).

For many conventional applications, the central squish system shown in Fig. 4.16(b) is one which is rarely sufficiently explored by designers. The designer is not

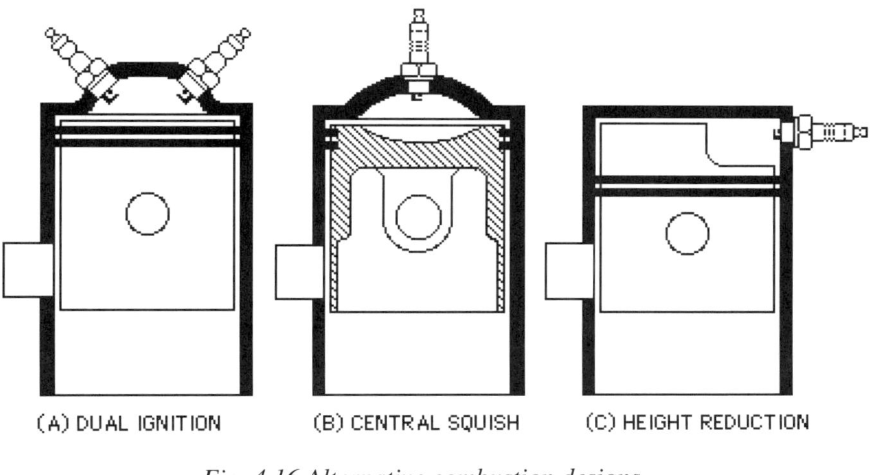

Fig. 4.16 Alternative combustion designs.

being urged to raise the squish velocity value, but to direct the squished charge into the middle of the chamber. As shown in the sketch, the designer should dish the piston inside the squish band and proportion the remainder of the clearance volume at roughly 20% in the piston and 80% in the head. By this means, the squished flow does not attach to the piston crown, but can raise the combustion efficiency by directly entering the majority of the clearance volume. Should such a chamber be tried experimentally and fail to provide an instant improvement to the engine performance characteristics, the designer should always remember that the scavenging behavior of the engine is also being altered by this modification.

For combustion of other fuels, such as kerosene or natural gas, which are not noted for having naturally high flame speed capabilities, the creation of turbulence by squish action will speed up the combustion process. In this context, earlier remarks in the discussion on stratified charge combustion become applicable in this homogeneous charge context.

4.6.2.1 Combustion chambers for conventional cross scavenged engines

It is regretted that little direct advice can be given on this topic, for the simple truth is that the complex shapes possible from the deflector design are such that universal recommendations are almost impossible. Perhaps the only common thread of information which has appeared experimentally over the years is that combustion chambers appear to perform most efficiently when placed over the center of the cylinder, with very little squish action designed to come from either the scavenge side or the exhaust side of the piston. Indeed, some of the best designs have been those which are almost quiescent in this regard.

NOTATION FOR CHAPTER 4

NAME	SYMBOL	SI UNIT	OTHER UNITS
Air-Fuel Ratio	AF		
Crankcase Compression Ratio	CCR		
Combustion Efficiency	BE		
Geometric Compression Ratio	GCR		
Mass Fraction Burned	MFB		
Nusselt Number	Nu		
Ratio of Specific Heats	g		
Reynolds Number	Re		
Scavenge Ratio	SR		
Scavenging Efficiency	SE		
Squish Area Ratio	SAR		
Squish Pressure Ratio	SPR		
Trapped Compression Ratio	TCR		
Trapping Efficiency	TE		
Trapped Air-Fuel Ratio	TAF		
Brake Mean Effective Pressure	BMEP	Pa	bar, atm, psi

NAME	SYMBOL	SI UNIT	OTHER UNITS
Brake Power Output	BPR	W	kW, hp
Brake Specific Fuel Consumption	BSFC	kg/Ws	kg/kWh, lb/hp.hr
Clearance Volume	CV	m^3	cm^3
Convection Heat Transfer Coefficient	HC	W/m^2K	
Cylinder Bore	BO	m	mm, in
Cylinder Stroke	ST	m	mm, in
Density	D	kg/m^3	
Engine Rotation Rate	RPS	rev/s	rev/min
Flame Velocity	FV	m/s	
Fuel Calorific Value (lower)	CAL	J/kg	kJ/kg
Fuel Consumption Rate	FC	kg/s	kg/h
Gas Constant	R	J/kgK	
Heat Release Period	P°	deg	
Indicated Mean Effective Pressure	IMEP	Pa	bar, atm, psi
Laminar Flame Velocity	LFV	m/s	
Maximum Heat Release Rate	HM	J/deg	
Piston Area Squished	AS	m^2	mm^2
Pressure	P	Pa	bar, atm, psi
Radiation Heat Transfer Coefficient	HR	W/m^2K	
Specific Heat at constant volume	cV	J/kgK	kJ/kgK
Squish Kinetic Energy	SQKE	J	mJ
Squish Velocity	SQV	m/s	
Swept Volume	SV	m^3	cm^3, cc, in^3
Temperature	T	K	°C, °F
Trapped Swept Volume	TSV	m^3	cm^3
Turbulence Velocity	TV	m/s	

REFERENCES FOR CHAPTER 4

4.1 B. Lewis, G. Von Elbe, Combustion, Flames and Explosions of Gases, Academic Press, 1961.

4.2 N.C. Blizard, J.C. Keck, "Experimental and Theoretical Investigation of Turbulent Burning Model for Internal Combustion Engines," *SAE Transactions*, Paper No. 740191

4.3 T. Obokata, N. Hanada, T. Kurabayashi, "Velocity and Turbulence Measurements in a Combustion Chamber of SI Engine under Motored and Firing Conditions by LDA with Fibre-Optic Pick-up," Society of Automotive Engineers, International Congress and Exposition, Detroit, Michigan, February 23-27, 1987, SAE Paper No. 870166.

4.4 W.G. Agnew, "Fifty Years of Combustion Research at General Motors," *Prog.Energy Combust.Sci.*, Vol. 4, pp. 115-155, 1978.

4.5 R.J. Tabaczynski, "Turbulence and Turbulent Combustion in Spark-Ignition Engines," *Prog.Energy Combust.Sci.*, Vol. 2, p. 143, 1977.

4.6 M.S. Hancock, D.J. Buckingham, M.R. Belmont, "The Influence of Arc Parameters on Combustion in a Spark-Ignition Engine," Society of Automotive Engineers, International Congress and Exposition, Detroit, Michigan, February 24-28, 1986, SAE Paper No. 860321.

4.7 G.M. Rassweiler, L. Withrow, "Motion Pictures of Engine Flames Correlated with Pressure Cards," SAE Paper No. 800131, 1980, reissued at that time but originally presented in SAE Jl., 42,185,1938!

4.8 W.T. Lyn, "Calculations of the Effect of Rate of Heat Release on the Shape of Cylinder Pressure Diagram and Cycle Efficiency," *Proc.I. Mech.E*, No. 1, 1960-61, pp. 34-37.

4.9 G.P. Blair, "Prediction of Two-Cycle Engine Performance Characteristics," SAE Farm, Construction and Industrial Machinery Meeting, Milwaukee, Wisconsin, September, 1976, SAE Paper No.760645

4.10 R. Fleck, R.A.R. Houston, G.P. Blair, "Predicting the Performance Characteristics of Twin Cylinder Two-Stroke Engines for Outboard Motor Applications," Society of Automotive Engineers International Off-Highway & Powerplant Congress, Milwaukee, Wisconsin, September 12-15, 1988, SAE Paper No. 881266.

4.11 T.K. Hayes, L.D. Savage, S.C. Sorensen, "Cylinder Pressure Data Acquisition and Heat Release Analysis on a Personal Computer," Society of Automotive Engineers, International Congress and Exposition, Detroit, Michigan, February 24-28, 1986, SAE Paper No. 860029.

4.12 D.R. Lancaster, R.B. Krieger, J.H. Lienisch, "Measurement and Analysis of Engine Pressure Data," *Trans. SAE*, Vol. 84, p. 155, 1975, SAE Paper No. 750026.

4.13 R. Douglas, "Closed Cycle Studies of a Two-Stroke Cycle Engine," Doctoral Thesis, The Queen's University of Belfast, May, 1981.

4.14 R.J. Tabaczynski, S.D. Hires, J.M. Novak, "The Prediction of Ignition Delay and Combustion Intervals for a Homogeneous Charge, Spark Ignition Engine," Society of Automotive Engineers, International Congress and Exposition, Detroit, Michigan, February, 1978, SAE Paper No. 780232.

4.15 W.J.D. Annand, "Heat Transfer in the Cylinders of Reciprocating Internal Combustion Engines," *Proc.I. Mech.E.*, Vol. 177, p. 973, 1963.

4.16 W.J.D. Annand, T.H. Ma, "Instantaneous Heat Transfer Rates to the Cylinder Heat Surface of a Small Compression Ignition Engine," *Proc.I. Mech.E.*, Vol. 185, p. 976, 1970-71.

4.17 W.J.D. Annand, D. Pinfold, "Heat Transfer in the Cylinder of a Motored Reciprocating Engine," Society of Automotive Engineers, International Congress and Exposition, Detroit, Michigan, February, 1980, SAE Paper No. 800457.

4.18 G. Woschni, "A Universally Applicable Equation for the Instantaneous Heat Transfer Coefficient in the Internal Combustion Engine," SAE Paper No. 670931, November, 1967.

4.19 A.A. Amsden, J.D. Ramshaw, P.J. O'Rourke, J.K. Dukowicz, "KIVA: A Computer Program for Two- and Three-Dimensional Fluid Flows with Chemical Reactions and Fuel Sprays," Los Alamos National Laboratory Report, LA-102045-MS, 1985.

4.20 A.A. Amsden, T.D. Butler, P.J. O'Rourke, J.D. Ramshaw, "KIVA–A Comprehensive Model for 2-D and 3-D Engine Simulations," Society of Automotive Engineers, International Congress and Exposition, Detroit, Michigan, February, 1985, SAE Paper No. 850554.

4.21 B. Ahmadi-Befrui, W. Brandstatter, H. Kratochwill, "Multidimensional Calculation of the Flow Processes in a Loop-Scavenged Two-Stroke Cycle Engine," Society of Automotive Engineers, International Congress and Exposition, Detroit, Michigan, February, 1989, SAE Paper No. 890841.

4.22 D. Fitzgeorge, J.L. Allison, "Air Swirl in a Road-Vehicle Diesel Engine," *Proc.I. Mech.E.*, Vol. 4, p. 151, 1962-63.

4.23 T.D. Fansler, "Laser Velocimetry Measurements of Swirl and Squish Flows in an Engine with a Cylindrical Piston Bowl," Society of Automotive Engineers, International Congress and Exposition, Detroit, Michigan, February, 1985, SAE Paper No. 850124.

4.24 G.F.W. Zeigler, A. Zettlitz, P. Meinhardt, R. Herweg, R. Maly, W. Pfister, "Cycle Resolved Two-Dimensional Flame Visualization in a Spark-Ignition Engine," Society of Automotive Engineers, International Fuels and Lubricants Meeting and Exposition, Portland, Oregon, October 10-13, 1988, SAE Paper No. 881634.

4.25 P.O. Witze, M.J. Hall, J.S. Wallace, "Fiber-Optic Instrumented Spark Plug for Measuring Early Flame Development in Spark Ignition Engines," Society of Automotive Engineers, International Fuels and Lubricants Meeting and Exposition, Portland, Oregon, October 10-13, 1988, SAE Paper No. 881638.

4.26 R. Herweg, Ph. Begleris, A. Zettlitz, G.F.W. Zeigler, "Flow Field Effects on Flame Kernel Formation in a Spark-Ignition Engine," Society of Automotive Engineers, International Fuels and Lubricants Meeting and Exposition, Portland, Oregon, October 10-13, 1988, SAE Paper No. 881639.

4.27 L. Martorano, G. Chiantini, P. Nesti, "Heat Release Analysis for a Two-Spark Ignition Engine," International Conference on the Small Internal Combustion Engine, Paper C372/026, Institution of Mechanical Engineers, London, 4-5 April, 1989.

Chapter 5

Computer Modelling of Engines

5.0 Introduction

In the first four chapters of this book, the reader is introduced to the two-stroke engine, to the unsteady nature of the gas flow in, through and out of it, to the nature of the scavenging process and its potential for retention of the fresh charge supplied each cycle, and to the combustion process where the characteristics of the burning process in a two-stroke engine are explained. In each case, be it cylinder geometry, unsteady gas flow, scavenging or combustion, theoretical models are described which could be solved on a digital computer. The purpose of this chapter is to bring together those separate models to form a complete model of the two-stroke engine. This simulation will be able to predict all of the pressure, temperature and volume variations with time within an engine of specified geometry, and calculate the resulting performance characteristics of power, torque, fuel consumption and air flow.

Because of the (almost) infinite number of conceivable combinations of intake and exhaust geometry which could be allied to various selections of engine geometry, it will not be possible to provide the reader with a computer solution to cope with every eventuality. Therefore, there are two major computer programs presented in this chapter which can deal with the majority of engine simulations. They are both for single-cylinder, two-stroke engines with piston port control of the exhaust, transfer and inlet ports. The first program, Prog.5.1, has a simple box silencer of the type found on any industrial engine, chainsaw or small outboard motor. The second program, Prog.5.2, has a tuned exhaust pipe of the type found on the high-performance motorcycle, go-kart or hydroplane engine used for competition.

Within the chapter, it is proposed to use these programs to illustrate many of the points made in earlier chapters regarding the design of various types of engines and to highlight those further principles not previously raised regarding the design of two-stroke engines.

Perhaps one of the most useful aspects of engine modelling is that the simulation allows the designer to imagine the unimaginable, by being able to see on a computer screen the temporal variations of pressure, volume and gas flow rate which take place during the engine cycle. Many of the parametric changes observed in this visual manner, together with their net effect on power output and fuel consumption, provide the designer with much food for design thought. Computer output is traditionally in the form of numbers and graphs of the analysis which has been conducted. In the era when such analysis was conducted on mainframe computers, the designer acquired the output as a package of printout and pictorial information, and then pondered over it for its significance. This post-analytical approach, regarded at the time as a miracle of modern technology, was much less effective as a design aid than the 1980's micro-computer presenting the same information in the same time span, but allowing the designer to watch the engine "run" on the computer screen in a slow-motion mode. It is from such physical insights that a designer conceives of future improvements in, say, power or fuel consumption or exhaust emissions. Let us not forget that it is the human brain which synthesizes while the computer merely analyzes. The pace of development of stand-alone computer systems, in terms of performance for cost, is such that the comments made above are virtually out of date as they are being typed into just such a computer!

5.1 Structure of a computer model

The key elements of any computer program to model a two-stroke engine are:

(a) The physical geometry of the engine so that all of the port areas, the cylinder volume and the crankcase volume are known at any crankshaft angle during the rotation of the engine for several revolutions at a desired engine speed.

(b) The composition of a model of the unsteady gas flow in the inlet, transfer and exhaust ducts of the engine.

(c) The composition of a model of the thermodynamic and gas-dynamic behavior within the cylinder and in the crankcase of the engine while the ports are open, i.e., the open cycle period.

(d) The composition of a model of the thermodynamic behavior within the cylinder of the engine while the ports are closed, i.e., the closed cycle period.

(e) The composition of a model of the scavenge process so that the proportion of fresh charge which is retained within the cylinder can be predicted.

Almost all of the material related to the above five topics has already been described and discussed in the earlier chapters, and all that remains to be done is to reorient the reader to each subject in turn, bringing them together for the preparation of a major program to describe a complete engine.

5.1.1 Physical geometry required for an engine model

It will be recalled that a detailed description of this is carried out in Sects. 1.4.3-6, which formed the subject of three programs, Progs.1.1-1.3. The geometry required for this new program is even more detailed, requiring all that was needed for those earlier programs as well as the information presented in Figs. 5.1 and 5.2. The complete list of geometrical data required for the program is given in Figs. 5.3 and 5.4 for the four engines which will be used as worked examples in this chapter. Although they will be discussed more fully later in the chapter, the four engines

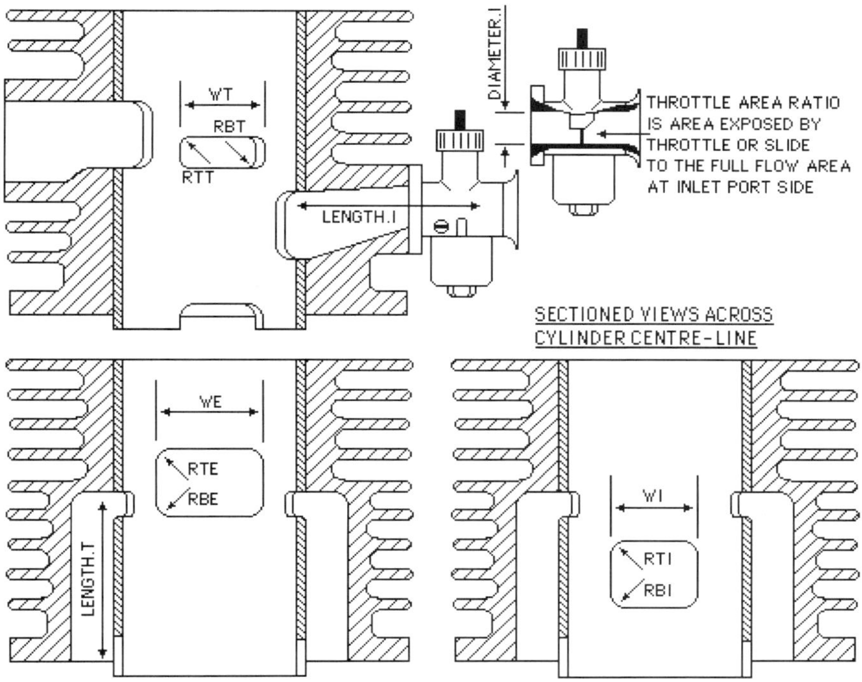

Fig. 5.1 Nomenclature for cylinder porting dimensions in computer engine models.

labelled in Fig. 5.3 as QUB400, SAW, HUSQ4, and 125GP, are a QUB 400 single-cylinder research engine, a chainsaw engine, the Husqvarna 250 MK4 motocross engine described even more completely in reference (3.7), and an archetypical 125 cc Grand Prix racing motorcycle engine, respectively.

Turning first to Fig. 5.3 and the first block of data listed, the values required there are the macroscopic data which can describe the cylinder, the position of the piston at any crank angle, and the clearance volumes in the cylinder and crankcase. As described in Sect. 1.4, the values required are for the cylinder bore, stroke, connecting rod length between bearing centers, the trapped compression ratio in the cylinder and the compression ratio in the crankcase. The exhaust pipe temperature is the reference temperature as detailed in Sect. 2.5.1.

The next four blocks of data are best discussed in conjunction with Figs. 5.1 and 5.2. Fig. 5.1 shows various sectioned views of a two-stroke engine cylinder where the areas and timings of the exhaust, transfer and inlet ports are controlled by the relevant upper and lower timing edges of the piston. The heights of the ports, either total or exposed at any instant, are a function of the engine geometry in the first block

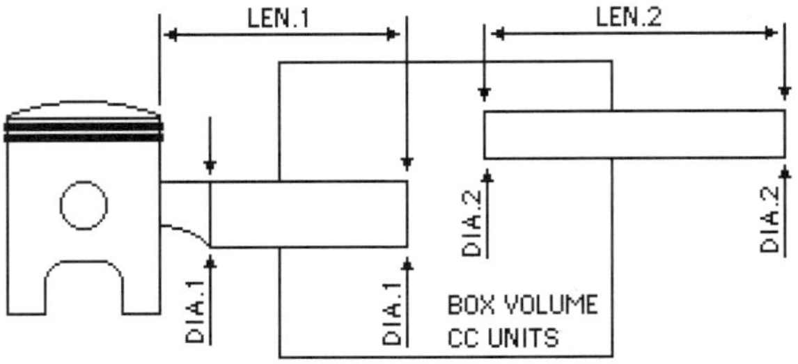

(A) NOMENCLATURE FOR EXHAUST PIPES AND BOX FOR Prog.5.1

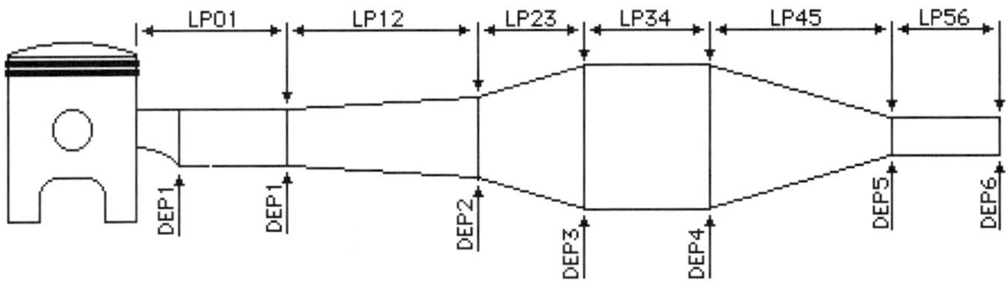

(B) NOMENCLATURE FOR THE DIMENSIONS OF A TUNED PIPE FOR Prog.5.2

Fig. 5.2 Nomenclature and dimensions of exhaust systems for the computer programs in Chapter 5.

of data and of the crankshaft angles at opening and fully open (see Sect. 1.4). To determine the port areas at any instant, the port width is required at any piston position during the cycle, and the request is made for data on "number of ports," "effective width of each port in the gas flow direction," and the "top" and "bottom" corner radii of each port. The instantaneous area of any open port at any crankshaft position is required within the computer model to calculate the AK area ratio value from a port to a pipe (see Eq. 2.3.17, Fig. 2.6 and Fig. 6.1).

The effective port width of each port, WE, WT and WI in Fig. 5.1 for exhaust, transfer or inlet ports, requires some further explanation. For a cylinder which has more than one exhaust or inlet port, it is quite common to have those ports of equal effective width, but this is rarely the case for transfer ports. Fig. 3.38 provides a perfect example of this situation. In that illustration, the engine has three scavenge ports, the two side ports with an effective width of 28.6 mm and a rear port of 30 mm effective width; however, neither of these numbers took into account the upsweep angles, which were 5 and 55 degrees, respectively. The effective port area, as explained in that figure, is the exposed port height at any point multiplied by the cosine of the upsweep angle and the effective port width at that point. In this

DATA NAME	QUB400	SAW	HUSQ4	125GP
BORE, mm	85.0	56.0	69.5	54.0
STROKE,mm	70.0	41.0	64.5	54.0
CON-ROD, mm	125.0	73.0	120.0	105.0
ENGINE SPEED, rpm	3000	7500	VARIOUS	11500
EXHAUST TEMP, ºC	400	400	550	650
TRAPPED CR	6.7	7.5	7.0	7.2
CRANKCASE CR	1.45	1.4 & 1.5	1.5	1.5
EXHAUST PORT OPENS, ºATDC	96	100	90	81
FULLY OPEN, ºATDC	180	180	180	180
NUMBER OF PORTS	1	1	1	1
WIDTH EACH PORT, WE	54.8	31.5	42.0	41.0
TOP RADIUS, RTE	12	6	8	12
BOTTOM RADIUS, RBE	12	6	8	12
TRANSFER PORT OPENS, ºATDC	118	119	120	114
FULLY OPEN, ºATDC	180	180	180	180
NUMBER OF PORTS	6	2	5	4
WIDTH EACH PORT, WT	15.3	20.0	15.4	15.0
TOP RADIUS, RTT	3	3	3	3
BOTTOM RADIUS, RBT	3	3	3	3
SCAVENGING TYPE	VARIOUS	YAM1	YAM6	YAM1
INLET PORT OPENS, ºBTDC	60	74	80	98
FULLY OPEN, ºBTDC	0	26	0	0
NUMBER OF PORTS	1	1	2	1
WIDTH EACH PORT, WI	45.0	30.5	21.0	35.0
TOP RADIUS, RTI	6	3	3	10
BOTTOM RADIUS, RBI	6	3	3	10
TRANSFER LENGTH.T	90.0	60.0	90.0	60.0
TRANSFER AREA RATIO	1.35	1.20	1.40	1.40
INLET LENGTH.I	200.0	75.0	185.0	150.0
THROTTLE AREA RATIO	1 & 0.25	0.75	1.0	1.0
CARBURETTER DIAMETER.I	34.0	24.0	36.0	36.0
COMBUSTION EFFICIENCY	0.85	0.85	0.80	0.80
COMBUSTION PERIOD, deg	60.0	50.0	55.0	50.0
AIR-TO-FUEL RATIO	13	13	AS EXPT.	12
COMPRESSION INDEX, NC	1.25	1.25	1.25	1.25
EXPANSION INDEX, NE	1.35	1.35	1.35	1.35

Fig. 5.3 Cylinder and engine geometry used in the computer programs in Chapter 5.

NAME OF DATA	LEN.1	DIA.1	LEN.2	DIA.2	BOX VOLUME, cc
QUB400 ENGINE	200	38	200	38	10000
CHAINSAW ENGINE	90	28	90	18	500

(A) EXHAUST BOX DIMENSIONS FOR ENGINES USED IN COMPUTER MODELS

NAME OF DATA	LP01	LP12	LP23	LP34	LP45	LP56
HUSQVARNA MK4	75	385	275	195	280	230
125 GRAND PRIX	64	341	116	93	200	200

NAME OF DATA	DEP1	DEP2	DEP3	DEP4	DEP5	DEP6
HUSQVARNA MK4	49	66	110	100	28	28
125 GRAND PRIX	38	78	112	112	20	20

(B) TUNED EXHAUST PIPE DIMENSIONS FOR ENGINES USED IN COMPUTER MODELS

Fig. 5.4 Dimensions of exhaust systems for the computer programs in Chapter 5.

example, the programmer using Prog.5.1 or 5.2 would insert the data value for number of transfer ports as 3, and the effective port width as [2*28.6*cos(5º)+30*cos(55º)]/3, or 74.2/3, i.e., 24.7. By this somewhat convoluted logic, the programmer is saved from a multiplicity of data entries for each port, its width and upsweep angle; yet within the computer calculation at the appropriate crank angle the correct transfer port area is being calculated by the inclusion of the precise number of top and bottom corner radii. The logician will point out that each of the transfer ports mentioned might have differing top or bottom corner radii also! The author's defense is that it would be a very rare engine in which this is the case for transfer ports, and it is unheard of for multiple exhaust or inlet ports. Further, as has already been mentioned, it is possible to specify a program to cater to every geometrical whim of the designer, but it would be at the expense of writing a rather pedantic book about that and little else!

5.1.2 Geometry relating to unsteady gas flow within the engine model

As outlined in Sect. 5.1(b), it is necessary to specify the lengths and areas of any pipe or duct, and the volumes of any "boxes" which are allied to the exhaust, transfer or inlet system. The exhaust systems to be used with the two engine models in Progs.5.1 and 5.2 are illustrated in Fig. 5.2. The transfer and inlet duct are shown in Fig. 5.1.

The exhaust system for Prog.5.1 is sketched in Fig. 5.2(a). It is a simple box type silencer, with parallel pipes of differing length and area at entry and exit to the box. The engines which will be used in the illustration of this type of system are the QUB 400 unit and a chainsaw engine; the power unit data values appropriate to Fig. 5.2(a) are listed in Fig. 5.4(a). The silencer for the QUB 400 engine is a generously proportioned device with the box volume twenty-five times larger than the cylinder swept volume, with an outlet pipe equal in area to the inlet pipe from the engine. On the other hand, the chainsaw engine, which has a swept volume of 101 cc, has a silencer volume which is only 500 cc, i.e., five times larger than the cylinder swept volume, and an exit pipe which is but 41% of the pipe area at the inlet to that exhaust box. This is typical of such power units where bulk and weight are at a marketing premium.

The exhaust system for the high specific output engine types within Prog.5.2 is illustrated in Fig. 5.2(b), and the data values appropriate to the Husqvarna MK4 and 125 Grand Prix engines are listed in Fig. 5.4(b).

The data value needed for the transfer duct length is sketched in Fig. 5.1 as LENGTH.T and that for the inlet tract as LENGTH.I. It is assumed that the transfer ducts are not tapered and that the area along the duct is decreed as a function of the total transfer port area at full opening at bdc. Hence, as will be seen in block five of the data in Fig. 5.3, the programmer is requested to decide on a data value for the area ratio of the transfer ports to the transfer duct, called TRANSFER AREA RATIO. Typical values for this factor range from 1.2 to 1.5.

For the inlet duct, the length, LENGTH.I, is that from the piston face to the throttle body or venturi of the carburetor, and the duct is assumed to be linearly tapered from the full inlet port area to the carburetor flow diameter, the value of which is also observed in the same block of data in Fig. 5.3. Not all inlet ducts are designed in this manner, but the methodology represented by the data format given is the one which most readily fits the majority of cases. In the opinion of the author it is the logical manner in which they should be designed, i.e., that the area change with length is a constant.

Within an actual program, e.g., in ProgList.5.1, the unsteady gas flow portion is to be observed:

(a) In Subroutine NINT, where each pipe is divided into distance meshes. The exhaust pipe mesh is set at 30 mm and the others are proportioned at a somewhat lower dimensional value which is related to the reference speed of sound in those cooler ducts or pipes, in order that the actual time interval of calculation is approximately equal at each time step in all gas flow ducts during the calculation cycle (see Sect. 2.4.1(b)).

(b) In Subroutine AREA, where the areas and change of area between meshes is determined for all pipes (see Sect. 2.4.1(d)).

(c) In Subroutine STAB, where the minimum time step, dZ, is determined (see Sect. 2.4.1(d)).

(d) In Subroutine PIPE, where the interpolation for the ∂ and ß values is carried out for each mesh point (see Sect. 2.4.1(d)).

(e) In Subroutine PORT, where the determination of the reflection behavior at cylinder port boundaries and at pipe ends is carried out using the UFLO data bank (see Sect. 2.4.1(c)).

5.1.3 The open cycle model within the computer programs

The structure of the model for the open cycle behavior in the cylinder and crankcase is almost exactly as detailed in Sect. 2.4.2. There are some further considerations required for implementation within an engine context, by comparison with the simulation experiment conducted in Chapter 2. In that chapter, examples of intake and exhaust flow were used where the temperature changes, or levels, between the cylinder and a pipe were not unduly significant. However, in a firing engine at exhaust opening, and for the subsequent scavenging process, there are very significant changes in gas temperature within the cylinder and in the exhaust pipe. As the analysis of filling and emptying of the cylinder and crankcase shows, as seen in Sect. 2.4.2, the procedure is dominated by the enthalpy of inflow and outflow and of the internal energy in the system (see Eqs. 2.4.18-37). As enthalpy and internal energy are functions of the specific heats at constant pressure and volume and their ratio, cP, cV and g, respectively, it is important to ensure that their relationship with temperature is included. Any textbook on thermodynamics will state that the values of cP, cV and g are independent of temperature for "perfect gases." However, gases such as air and exhaust gas are real gases, as will be observed in Figs. 5.5-5.7. In those graphs, it is evident that there are considerable variations of cV, cP, and g with temperature. For example, when an exhaust port opens the cylinder temperature will be about 2000K and this will be followed later by the flow of scavenge air at about 350K. If cP, cV and g were to be treated as constants at the numerical values of 1005 and 718 J/kgK and 1.4 often quoted for air (indeed used in that fashion in Sect. 2.4.2), this would induce a considerable error for thermodynamic calculations in the simulation of real engines. The reader must not then assume that the analysis in Sect. 2.4.2 is incorrect, for that is not the case. What is necessary is that the correct numerical values of cP, cV and g will be used at any stage during the use of the equations created in that section. In that context, the information contained in Figs. 5.5-5.7 is invaluable. The third-order polynomial mathematical fit to the information plotted in those diagrams can be programmed into any computer solution of such thermodynamic systems. The information as expressed on the diagrams for the cP, cV, and g values for both air and exhaust gas, as functions of absolute temperature, T, is as follows:

for air,

$$cP_{air}=900.39+0.31904*T-8.9293E-5*T^2+8.8328E-9*T^3 \text{ J/kgK} \quad (5.1.1)$$

for exhaust gas,

$$cP_{ex}= 950.18+0.4254*T-1.1019E-4*T^2+9.6896E-9*T^3 \text{ J/kgK}$$

for air,

$$cV_{air}=613.29+0.31904*T-8.9293E-5*T^2+8.8328E-9*T^3 \text{ J/kgK} \quad (5.1.2)$$

for exhaust gas,

cV_{ex}= 659.58+0.4254*T-1.1019E-4*T^2+9.6896E-9*T^3 J/kgK

for air,

g_{air}=1.4506-1.6384E-4*T+5.8378E-8*T^2-7.3897E-12*T^3 (5.1.3)

for exhaust gas,

g_{ex}=1.4221-1.8752E-4*T+6.9668E-8*T^2-9.099E-12*T^3

The values plotted for exhaust gas are those which pertain to the stoichiometric combustion of octane, as given in the combustion equation, Eq. 1.5.16. The molecular proportions of the individual constituents of the exhaust gas are known from the right-hand side of that equation, and the individual properties of each constituent gas are to be found in many thermodynamic texts(5.11). The average

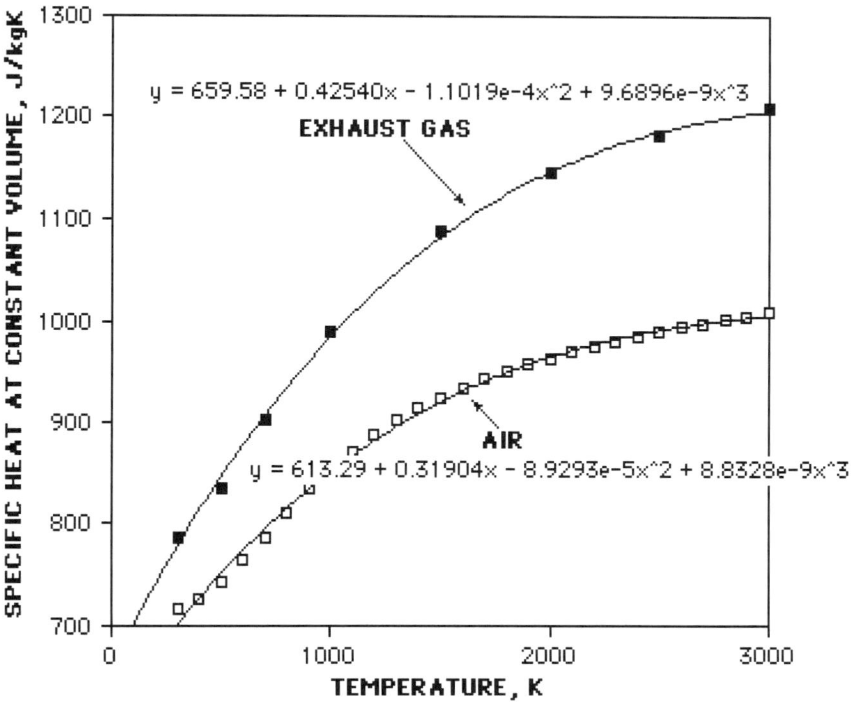

Fig. 5.5 Variation of specific heat at constant volume with temperature.

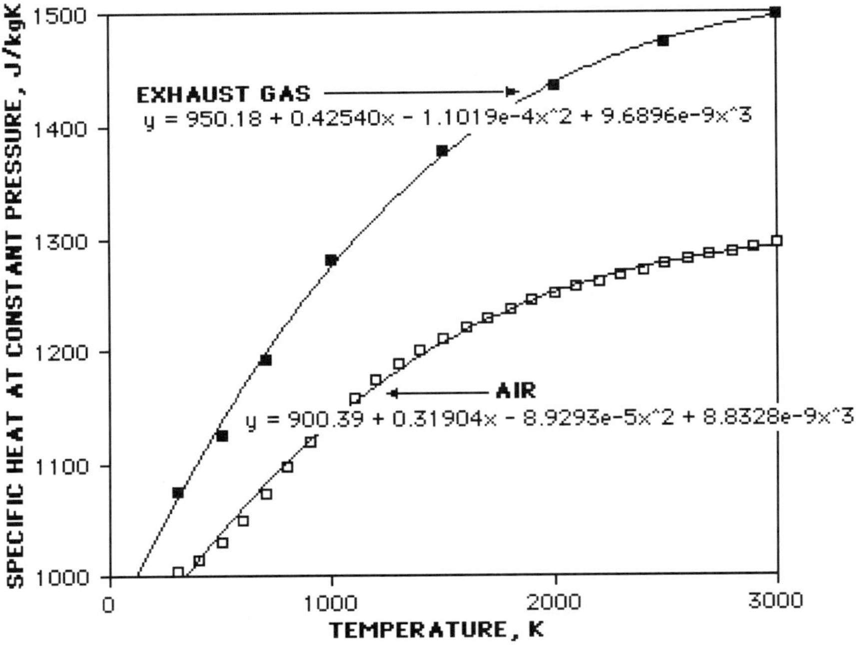

Fig. 5.6 Variation of specific heat at constant pressure with temperature.

properties of the mixture making up the exhaust gas are calculated from the mass proportions and properties of each constituent; the results are the plots shown in Figs. 5.5-5.7. Actually, the properties of air are determined in precisely the same manner, for it is also a gas mixture composed principally of oxygen and nitrogen.

Students of thermodynamics will note that the difference between the cP and cV values, which is the gas constant for either air or exhaust gas, R, is not a function of temperature, and is found as follows:

In general, for any gas,

$$R_{gas}=cP_{gas}-cV_{gas}.$$

Therefore, for air and exhaust gas,

$$R_{air}=cP_{air}-cV_{air}=900.39-613.29=287.1 \quad J/kgK,$$
$$R_{ex}=cP_{ex}-cV_{ex}=950.18-659.58=290.6 \quad J/kgK.$$

In actual fact, the numerical value of the gas constant, R, for a gas of a particular chemical composition has a variation with respect to compressibility effects but, in

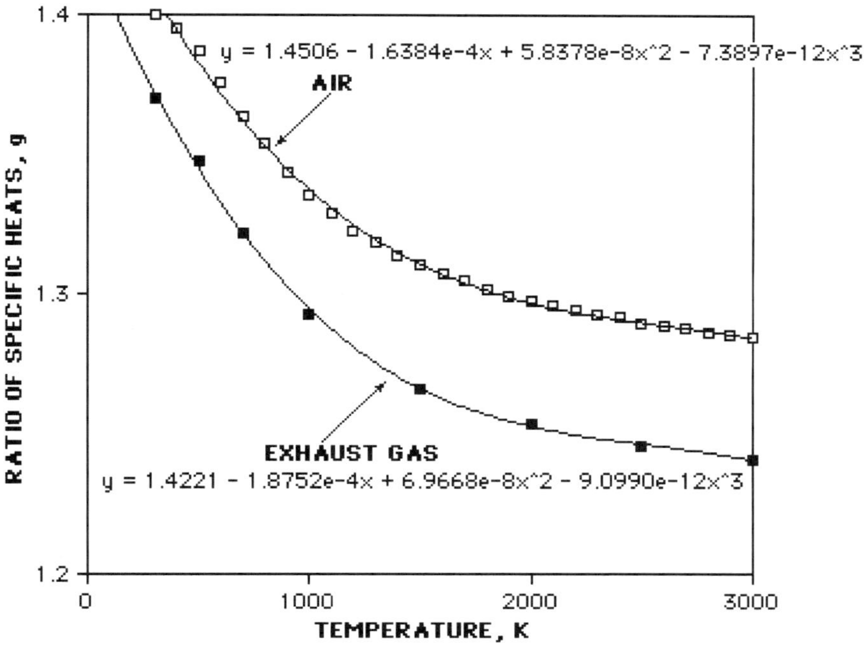

Fig. 5.7 Variation of ratio of specific heats with temperature.

terms of its application in thermodynamics theory to gas flow and combustion in ic engines, the effect can be neglected(5.11). The third-order polynomial fit to the known values for air is more than sufficiently accurate for calculation purposes, considering that the gas in the cylinder is never "pure air." For those who would wish to program an engine model in the most accurate terms, the proportionate mix of air and exhaust gas within the cylinder and flowing out of the exhaust port is known at every step in the calculation and it is possible to precisely describe the thermodynamic properties of the gases. Within the relatively simple engine models associated with this book, all gases are treated as having the properties of air, although the thermodynamically correct values of cP, cV and g are calculated as functions of temperature and used appropriately at each time step.

Apart from this significant alteration to the method of calculation of the gas properties, the thermodynamic solution for open systems, as postulated in Sect. 2.4.2, is exactly that which is used in the engine models in Progs.5.1 and 5.2 and can be found contained within a somewhat more sophisticated version of the original Subroutine BOX.

The same subroutine is used for the analysis of the behavior in the crankcase, the cylinder during the open cycle, and for the exhaust silencer box in the case of Prog.5.1.

5.1.4 The closed cycle model within the computer programs

The theoretical basis for the closed cycle model is described in Sect. 4.3.1 and is used in precisely that form in Progs.5.1 and 5.2. The model for the heat release characteristic employed is as shown in Fig. 4.6. It will be seen from the data set in the final block of Fig. 5.3 that the values of the heat release period, P°, and combustion efficiency, BE, which were determined from the experimental work with the QUB 400 engine in Sect. 4.2.3, have been used for a wide variety of engine types, as represented by the four engines employed as illustrative examples. The numerical values of the polytropic indices for compression and expansion, nc and ne, which are recommended in Sect. 4.3.1 for general use, are 1.25 and 1.35, respectively. These are used for all four engine examples in Fig. 5.3. A trapped air-fuel ratio, TAF, of 13.0, which happens to be the experimentally determined value for the QUB 400, is used for two of the four engine examples. The experimental values of air-fuel ratio for the Husqvarna MK4, which are stated in the paper(3.7), are used here in the theoretical analysis; they range from 11.1 to 11.5. For the 125 racing engine data, a conventional value of 12 is used to give realism to the results, as that would be the air-fuel ratio normally associated with peak power output in high-performance competition engines.

During the discussion in Chapter 2 on combustion models, it is pointed out that the engine modeller needs a relatively simple, almost universal but reasonably accurate model of the burning process(5.13). It is the author's contention that the form of model used here is of that type. Considering that it is to be used for four different engines whose specific power outputs range from 20 kW/liter to 200 kW/liter, with almost the same values employed for P°, BE, TAF, nc and ne, such a combustion model can be deemed to be satisfactorily universal should the result of such a calculation be reasonably accurate.

Another useful feature of the proposed heat release model of combustion is the automatic phasing of the combustion process about tdc, without the need to insert an ignition timing to accomplish the phasing of that burning process. As observed in Fig. 4.6, the first slope of the heat release rate profile is equally distributed about the tdc position. This accomplishes the phasing effect required, as the ignition timing in a real engine is experimentally determined so that it optimizes the cylinder pressure distribution from combustion about tdc. As the actual value of that ignition timing is heavily influenced by factors beyond the scope of this present modelling calculation, such as squish velocity, combustion chamber shape, surface-to-volume ratio, piston-created turbulence, scavenge-created turbulence, etc., it is more logical to simulate the end effect of optimizing the ignition timing rather than inserting the ignition timing as a misleading data input. At some time in the future, when CFD calculations become the norm for combustion calculations, more sophisticated combustion modelling will be possible with a consequential improvement in accuracy. Needless to add, when that day comes, the programmer must also be prepared to face the further task of inserting into the computer data bank a considerably detailed specification of the cylinder head, combustion chamber and coolant passage geometry.

The use of a simple heat release model of combustion can also be used for calculations for two-stroke diesel engines with some simple modifications to the phasing of the heat release about tdc, the total period and the trapped air-fuel ratio. The following recommendations are made without further proof being presented of their validity. The combustion period, P°, should be set at 40°. The combustion efficiency, BE, can be left at 0.85. The polytropic indices, nc and ne, can remain at 1.25 and 1.35. The diagram should be phased later so that the first slope of the heat release rate is proportioned at 0.33 and 0.66 around tdc rather than the 50/50 distribution employed for petrol burning. The trapped air-fuel ratio, TAF, should be set at the conventional diesel engine value for best thermal efficiency and exhaust smoke levels, in other words at about 20. At lighter loads, the TAF value will approach 50. It should be remembered also that the compression ignition engine is unthrottled and the throttle area ratio will always be unity.

Within the computer programs presented in the ProgLists.5.1 and 5.2, the combustion model is found in Subroutine CYLINDER.

5.1.5 The simulation of the scavenge process within the engine model

In Chapter 3 on scavenging flow, the fundamental impression which the reader should have gained is that scavenging is a gas exchange process conducted in a volumetric fashion. In other words, throughout Chapter 3, the reader is introduced to theory and experimentation which is based on the replacement of volumes of exhaust gas by volumes of fresh charge. Experimental evidence is presented from the single cycle gas scavenging rig on the differing volumetric scavenging characteristics of uniflow scavenged, good and bad loop scavenged, and cross scavenged power units. These are summarized in Fig. 3.14 as numerical values of slope and intercept from $\log_e$(1-SE)-SR plots of that experimental evidence. For any computer model of a firing two-stroke engine to be realistic, it is essential that such, or similar, information be programmed into the model to allow the designer to determine the degree of relevance to any design of attaining good or bad, or using different types of, scavenging in the final product.

To accomplish the preparation of a model of scavenging for the firing engine model, consider the situation during a scavenge process as sketched in Fig.5.8. The fresh charge is entering the cylinder at the same time as a mixture composed of exhaust gas and some fresh charge is leaving. In a perfect scavenge process, only exhaust gas would be exiting the cylinder. As shown on the sketch, the information available from the computer calculation at any crank angle for a time step, Δt, is extensive and relates to all of the relevant properties and conditions in the cylinder, transfer duct, and the exhaust pipe. If the swept volume is SV, the clearance volume, CV, and the atmospheric density is Da, then the reference masses to compute delivery ratio, DR, and scavenge ratio, SR, from the nomenclature established in Sect. 1.5, are:

Reference masses for DR and SR are,

Mdr=SV*Da (5.1.4)

Msr=(SV+CV)*Da. (5.1.5)

During the time step the following can be calculated:
Volume increment of charge exiting transfer duct,

$$\Delta VF=\Delta t*CGt1*Ft1. \quad (5.1.6)$$

Mass increment of fresh charge exiting transfer duct,

$$\Delta MF=Dt1*\Delta VF. \quad (5.1.7)$$

The value of CGt1 is the superposition particle velocity at the first mesh point in the transfer duct, Ft1 is the duct area at that point, and Dt1 is the superposition density at the same location.

At the conclusion of the time step under consideration, the SR value at that point in the open cycle, SRt, is given by the summation of all of the previous and present mass flow increments:

$$SRt=\Sigma[\Delta MF*\Delta t/Msr] \quad (5.1.8)$$

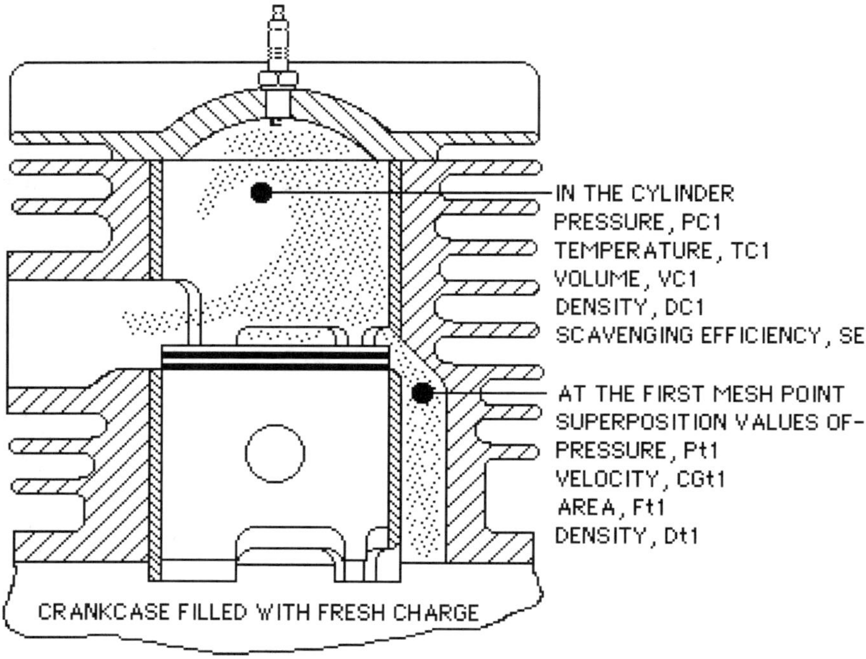

Fig. 5.8 Modelling the scavenge process in the firing engine.

A similar situation would exist for the calculation of the instantaneous value of delivery ratio, DRt, although this is normally computed at the inlet port boundary to the crankcase:

$$DRt= \Sigma[\Delta MF*\Delta t/Mdr] \tag{5.1.9}$$

Naturally, in any accurate calculation by a computer engine model, these mass flow rates should be identical when summed over a complete cycle at any of the cylinder or crankcase boundaries. Consequently, the value of SRt and DRt at the conclusion of a computed engine cycle are the net levels of scavenge ratio and delivery ratio for the engine on that cycle.

However, both of these computations of fresh charge flow are for mass flow, while Chapter 3 teaches that the scavenging behavior will be a function of the volume flow at any instant by comparison with a reference volume. In the case of the single cycle rig described in Sect. 3.2.3, this is indisputably the swept volume, for the experiment is conducted at constant volume, with that volume set equal to the swept volume of the test cylinder in question. Therefore, the first logical approach to estimate scavenge ratio by volume, SRv, would be to sum the increments of ΔVF as a function of the instantaneous cylinder volume at that crank angle:

$$SRv=\Sigma[\Delta VF/VC1] \tag{5.1.10}$$

However, the volume which exits the transfer duct is that calculated at properties of the first mesh position in the transfer duct, and this fresh charge will clearly occupy a larger volume upon entering the cylinder which is at a lower pressure and lower density. Therefore, Eq. 5.1.10 must be modified to take this into account, which is rather difficult mathematically. In reality, as the fresh charge enters in a stratified manner, it will not immediately assume the density of the cylinder charge at that time step. Therefore, there is little point in following slavishly the dictates of the information for cylinder properties which would be provided by Subroutine BOX at that juncture. The only property which would be instantaneously common to both the "cold" fresh charge which entered and the "hot" cylinder charge in situ is the cylinder pressure. Thus, it would be logical to estimate that the fresh charge volume entering the cylinder would expand by the ratio of the transfer duct pressure (superposition, at that first mesh) to the cylinder pressure. Hence, the first logical assumption for an expansion ratio, EXPAND, for this entering fresh charge would be:

$$EXPAND=Pt1/PC1 \tag{5.1.11}$$

The ensuing modification to Eq. 5.1.10 is then given by:

$$SRv=\Sigma[EXPAND*\Delta VF/VC1] \tag{5.1.12}$$

There is some evidence from correlation with known experimental data at QUB(1.22)(3.13) that this gives values of SRv, and consequently SE information to the combustion model, which are somewhat low. It has been found by some experience that a better value for the correction coefficient EXPAND is given by a density ratio proportioned as follows between the densities in the cylinder and the transfer duct:

EXPAND=Dt1/(0.6*Dt1+0.4*DC1) (5.1.13)

Using Eq. 5.1.12, the computer calculation has available the SRv data as a function of both time and crank angle, so it is quite logical to use the previously determined experimental data for any of the cylinders tested in Chapter 3 to determine the scavenging efficiency, SE. This can be applied either at the conclusion of the entire process at trapping, or during its progress from transfer port opening to transfer port closing. The value of scavenging efficiency at any instant during the scavenge process, SE, based on the SRv value at the same juncture, would be, from Eq. 3.3.4:

$$SE=1-e^{(M*SRv+C)} \qquad (5.1.14)$$

In Prog.5.1 and 5.2, the reader will find this set of equations in the subroutine TRPORTOPEN. Both of the models for EXPAND are programmed there, so the user can select which of the two versions of EXPAND are most suitable for the design problems undertaken. All of the program examples presented within this chapter are computed using the version of EXPAND presented in Eq. 5.1.11. In a Utopia, the proper course of action would be to halt the unsteady gas-dynamic program at each calculation increment and side-step to a full CFD program of the type discussed in Sect. 3.4. At this stage of development of the microcomputer, the designer does not have the time to wait during the three or four hour intermission while this accurate scavenging calculation takes place. Perhaps during the next half decade, when computers will get faster exponentially, this will become a viable calculation option. In the interim, the simplistic model of scavenging discussed above will have to suffice for unsteady gas-dynamic engine models.

At the data input stage, the user is requested to name which of the scavenging systems described in Fig. 3.14 is to be employed in the subsequent calculation. The name in question is that used in the left-hand column of Fig. 3.14 and the character string is recognized by the program which translates it internally into the numerical values of the appropriate slope, M, and intercept, C, for that type of scavenging system.

To cope with any backflow which may occur into the transfer ducts towards the end of the blowdown phase, the computer calculation is instructed to sum all transfer duct flow and no computation of DRt, SRt or SRv will take place until the net transfer duct outflow has returned to zero. This would indicate that all exhaust gas backflow has been returned to the cylinder so that true scavenging of fresh charge can then commence.

At the trapping point at exhaust port closure, the mass of fuel trapped is determined from the scavenging efficiency and the cylinder mass at that juncture, Mtr. As discussed in Sect. 1.5.2, there is a fundamental difference between the purity of a trapped charge and the scavenging efficiency. In this form of engine modelling, the only option open to the programmer is to consider that they are identical, i.e., all of the oxygen in the cylinder is consumed by the previous combustion process. Thus, from Eq. 1.5.11, the mass of air and fuel trapped, Mtas and Mtf, is computed by:

Mtas=Mtr*SE (5.1.15)
Mtf=Mtas/TAF (5.1.16)

The values of trapping and charging efficiency are determined by applying Eqs. 1.5.13 and 1.5.14 directly.

From the last two equations (referring back to Sect. 1.5), it is assumed there is no air remaining in the cylinder exhaust gas from combustion during a previous cycle, i.e., the Mar value of Eq. 1.5.8. In other words, the value of scavenging efficiency being employed theoretically is also the charge purity.

5.1.6 Deducing the overall performance characteristics

The procedures used in the engine models presented in this chapter are as described in a paper by Blair(4.9) and in Sects. 1.5 and 1.6 of this text. The indicated work in the cylinder and the crankcase is assessed by the cyclic summation of the PdV term at each time step in the calculation. This permits the direct calculation of indicated mean effective pressure, IMEP, and the pumping mean effective pressure in the crankcase, PMEP. As discussed in Sect. 1.6.1, it implies that a separate term must be produced for the friction loss, FMEP, in order to predict the resulting brake mean effective pressure, BMEP (see Eq. 1.6.9). The expression for friction work employed in Progs.5.1 and 5.2 are for friction losses in an engine whose power output is being measured directly at the crankshaft. In the equation below, it is seen as being linearly related to piston speed(1.3)(5.12):

FMEP=100*ST*RPM (5.1.17)

The cyclic deduction of specific fuel consumption characteristics follows directly from the knowledge of the fuel mass trapped and the power which is calculated as being produced. The theory used to obtain this information is presented in Sects. 1.5.10 and 1.6.1.

5.2 Using Prog.5.1, "ENGINE MODEL No.1"

This engine model has the simple box type of exhaust silencer illustrated in Fig. 5.2(a). The basic data for the engine and its porting is listed in Fig. 5.3 and illustrated in Fig. 5.1. It should be noted that the program runs much faster in the compiled mode. Upon activating the compiled program, it will request the user to insert the data in blocks, exactly as in Fig. 5.3, or as an entire file of data previously stored.

Before the program will run, the user is asked to check the data on the computer screen to ensure its accuracy in thought or insertion. The result of running such a program for various sets of data, i.e., the QUB400 engine or the SAW engine, is demonstrated below, for many basic design principles are to be found in the ensuing output data from the computer calculation.

Apart from the normal output in the form of a printout of the required performance characteristics of the engine represented by the data input, the computer screen depicts the events as they occur during the calculation, as shown in Fig. 5.9. The screen will show the dynamic variations of cylinder, crankcase and exhaust port pressure ratio during the open cycle from exhaust port opening to closure. To prevent programming problems, the maximum cylinder pressure ratio plotted during the open cycle is 2.0 atm. The screen will then be cleared and a second picture will appear showing the events for the remainder of the cycle for the crankcase and inlet port pressure ratio. On this same diagram is presented the cylinder pressure ratio and the delivery ratio as it progresses. The cylinder pressure ratio is plotted at 1/40 of the scale of the crankcase and inlet pressures and its 1.0 atm level is at the base of the diagram; in other words, each tick on the left-hand scale represents 4 atm for the cylinder pressure diagram. The pressure ratio scale for crankcase and inlet port pressure is at the left and delivery ratio at the right. The top and bottom halves of Fig. 5.9 show the encapsulation of the computer screen during the fifth engine cycle. The print underneath each picture appears at bdc or tdc on each cycle as the calculation proceeds, so the designer has progressive information on the net effect of the dynamic events on the computer screen upon the engine performance characteristics.

5.2.1 The analysis of the data for the QUB 400 engine at full throttle

The exhaust system for the QUB 400 single-cylinder research engine has a short pipe of 200 mm to a generously proportioned silencer box which poses little restriction to the exhaust flow. This is evident in Fig. 5.9, where a suction reflection from the open pipe end into the exhaust box arrives back at the exhaust port and assists with the scavenging of the cylinder. A further compound reflection of that appears around bdc as a compression wave and this opposes the scavenging flow to a certain extent.

This can be examined further in Figs. 5.10 and 5.11 where the calculation has been indexed for the instantaneous values of scavenge ratio, by either mass or volume, and of scavenging efficiency, which results from the use of Eq. 5.1.14 within the program. In Fig. 5.10, scavenge ratio by volume, SRv, and scavenging efficiency, SE, are plotted as a function of crank angle from transfer port opening to closing. It can be seen that there is backflow from the cylinder during the first 10 degrees of opening, as both SRv and SE are zero at that juncture. That this would occur could also have been deduced from Fig. 5.9, for the cylinder pressure is higher than the crankcase pressure in this period. The effectiveness of the suction reflection in assisting scavenge flow is observed in Fig. 5.10, for the scavenge ratio by volume rises rapidly before bdc and has reached unity by bdc. As the final value is 1.15, with some backflow occurring towards the end of the scavenge period, it is seen that the

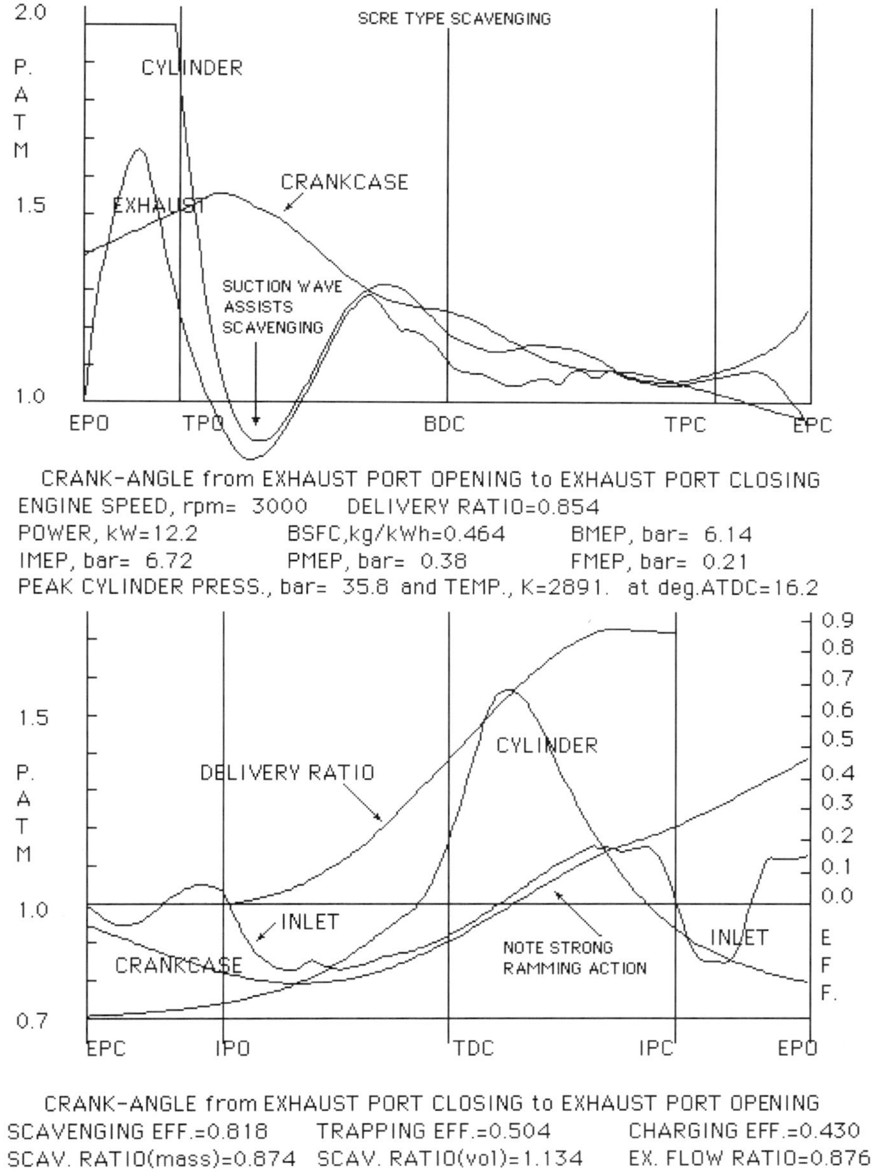

Fig. 5.9 Predicted performance characteristics of the QUB 400 engine at full throttle.

majority of the scavenge flow enters the cylinder before bdc. From the viewpoint of retaining fresh charge within the cylinder, this early supply of air and fuel into the cylinder is not an optimum procedure, for it gives the longest possible time for the fresh charge to mix with the cylinder contents and escape through the exhaust port. Fig. 5.10 shows a plot of the individual characteristics for three of the different scavenge types calculated, namely SCRE, YAM6 and CD. It will be recalled that these refer to "very good" and "poor" loop scavenging and to deflector piston cross scavenging, respectively. Although the SRv characteristic is virtually common for all three scavenging types, the ensuing SE behavior is dissimilar, for this is controlled by Eq. 5.1.14 . The employment of the scavenge model provides the closed cycle model with the appropriate information on the purity of the cylinder charge. Within the program, as described in Sect. 5.1, this will translate into varying levels of trapped air and fuel mass, with this latter value leading to a prediction of the heat to be released for combustion. The end effect of the differing scavenge behavior produces the ensuing performance characteristics for the various types of scavenging.

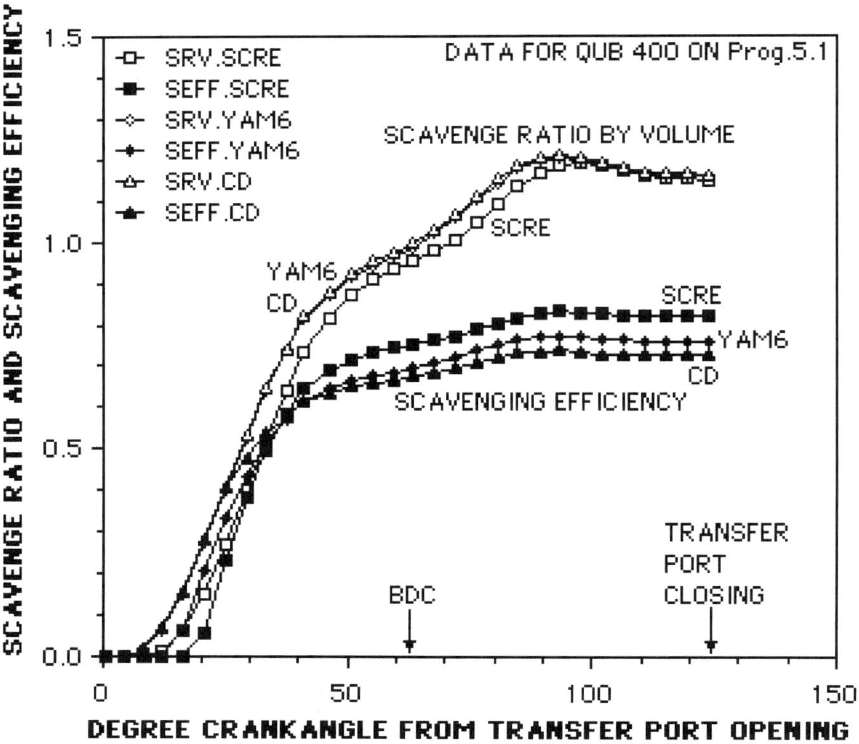

Fig. 5.10 Scavenging efficiency and scavenge ratio during the open cycle.

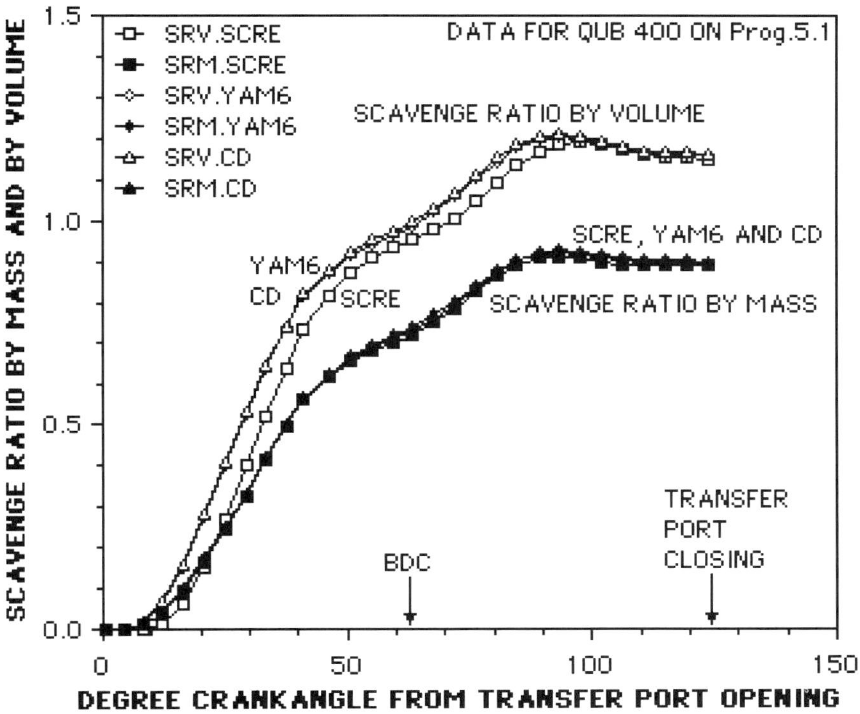

Fig. 5.11 Scavenge ratio by mass and volume during the open cycle.

These performance characteristics are summarized from the computer output in Fig. 5.12. As might be expected, the best scavenging produces the best performance characteristics. It is the order of magnitude of that variation which the designer should note. From the best to the worst of loop scavenging, i.e., from an SCRE to a YAM6 type scavenging system, the variation in power is from 12.1 to 11.2 kW, i.e., an 8% drop. The consequential change in specific fuel consumption is from 463 to 506 g/kWh, i.e., a 9% deterioration. These variations would not be as great as those predicted by experimental work at QUB(1.10) (3.13) (3.18) (3.20), where 15% would have been observed for power and fuel consumption changes for those same scavenging types, albeit in a quite different engine with a tuned exhaust pipe running at a higher engine speed. Nevertheless, the designer is informed, again, of the importance of attaining good scavenging in ensuring the best possible performance characteristics for a given engine.

The calculation system can also differentiate between different types of scavenging systems. The best scavenging system is seen to be the uniflow process. However, as is forecast in Chapter 3, its superiority over the best possible loop scavenged

DATA FOR CALCULATION FROM QUB 400 SINGLE CYLINDER RESEARCH ENGINE AT 3000 rpm

TYPE	POWER kW	IMEP bar	BMEP bar	BSFC g/kWh	DR	SEFF	TEFF	CEFF	PMAX bar
EXPT.	12.4	6.80	6.24	437	0.85	-	0.60	-	36.9
UNIFLOW	12.37	6.81	6.22	452	0.84	0.84	0.52	0.44	36.2
SCRE	12.11	6.68	6.10	463	0.85	0.82	0.51	0.43	35.7
YAM1	11.87	6.56	5.98	473	0.85	0.80	0.50	0.42	35.3
YAM6	11.21	6.23	5.65	506	0.86	0.75	0.47	0.40	34.1
QUBCR	12.11	6.69	6.10	463	0.85	0.82	0.51	0.43	35.7
CD	10.76	6.00	5.42	531	0.86	0.72	0.45	0.39	33.3

Fig. 5.12 Effect of differing scavenge models on the performance characteristics.

process is not as great as might have been imagined from a general reading of the literature(1.6). In this context, it is very important to bear in mind that the engines involved in both this calculation and in the firing tests, are all over-square engines; should the uniflow scavenged engine have an undersquare bore-stroke ratio, it is very possible that the experimental answer could be somewhat different. The indifferent scavenging available from a conventional deflector piston engine, type CD, is seen in Fig. 5.12 to lose 12% in power and 13% in BSFC to the SCRE type of good loop scavenging, and it is significantly worse than the YAM6 type scavenging system. To compound the potential problem, the conventional deflector piston calculation is conducted as if it had the same combustion characteristics as a loop scavenged engine, which may not be the case in practice. On the other hand, the calculations for the QUB type of cross scavenging, QUBCR, demonstrate that it is scavenging as well as the SCRE type of loop scavenging.

In Fig. 5.12, the top line of information coded EXPT is that for the experimentally determined performance parameters of the engine. The correlation between the SCRE calculated data and the experimental facts is very good for virtually every parameter, with the exception of the trapping efficiency, TE. This term, like SE, is heavily dependent upon the application of either Eq. 5.1.11 or Eq. 5.1.13.

That the combustion model is performing as postulated can be assessed by comparing the level of peak cylinder pressure, PMAX, measured at 36.9 bar and calculated at 35.7 bar. In Sect. 4.2, the experimental data from this engine is used to suggest the model of combustion currently employed in Progs.5.1 and 5.2. It would signal a poor future for combustion modelling if it could not compute accurately the combustion behavior in the very engine from which the model is concocted! The combustion diagram calculated for the QUB 400 engine with SCRE type scavenging can be seen in Fig. 5.9. Further information presented for the fifth engine cycle shows that the peak cycle pressure occurs at 16.2° atdc. This corresponds precisely with the experimental data from the same engine.

Further information on scavenging is presented in Fig. 5.11. Here is plotted the scavenge ratios by volume, SRv, and by mass, SR, during the scavenge period from transfer port opening to closure. It will be seen that the SR data is virtually identical, irrespective of the type of scavenge system. The SRv data is larger numerically, as would be expected for the volumetric process and the differences during calculation, for either the SCRE, YAM6, or CD system is controlled by the application of Eqs. 5.1.11-14. The engine with the SCRE scavenging runs at a lower level of SRv until the concluding period before port closure, even though the SR behavior is identical to the others. This is because there is less volume expansion possible for the scavenge flow between the transfer duct and the cylinder, as the more powerful SCRE engine has a higher cylinder pressure during the blowdown phase.

It is useful to examine the crankcase, inlet and airflow curves for an engine, as exemplified by those of the QUB 400 engine in Fig. 5.9. These curves would be regarded as very acceptable in design terms for the reasons which will be detailed below. The delivery ratio level is high, at 0.86, for a large capacity single-cylinder engine which is operating without a tuned exhaust system to aid the scavenging flow. Although the exhaust system does provide a suction pulse during blowdown, it is rather small in scale by comparison with those available from the tuned systems to be discussed in Sect. 5.3. The good delivery ratio behavior comes from a well optimized inlet system with the correct port timings, port area, and carburetor flow diameter allied to a strong ramming characteristic. The net effect is a delivery ratio characteristic which rises smoothly to a peak level, already stated as somewhat high in general, and maintained at that level until port closure due to a ramming reflection which is above the crankcase pressure almost until port closure.

The user of this program is always advised to examine the calculation output for balance between the mass flows into, through and out of the engine, namely the printed values of delivery ratio, scavenge ratio, and exhaust flow ratio. In Fig. 5.9 it can be seen that there is a good mathematical balance during the fifth cycle of calculation, as the DR, SR and exhaust flow ratio values are 0.851, 0.876 and 0.876, respectively.

5.2.2 The QUB 400 engine at quarter throttle

The output for the fifth calculation cycle at this throttled condition is shown in Fig. 5.13 for the same data as in Fig. 5.9, with the single exception of setting the throttle area ratio at 0.25 instead of 1.0.

As would be expected, the airflow rate into the engine is reduced significantly from 0.85 to 0.49. Like the calculation conducted in Sect. 2.5.2(f), the suction pressure in the crankcase and the inlet duct at the port is longer and deeper with a virtual absence of a ramming reflection from the throttled inlet pipe end. In that situation, the subsidiary reflections between the open and closed ends in the port closed period are also virtually non-existent. The reduced air mass flow rate, as seen from the lower profile in Fig. 5.13 by comparison with Fig. 5.9, is the net effect. This translates into reduced values of SRv, down from 1.13 to 0.73, a drop in scavenging efficiency, down from 0.82 to 0.64, but a higher value of trapping efficiency, up from 0.51 to 0.57. This latter effect could have been forecasted from the measured TE-SR and SE-SR curves for the SCRE scavenging system in Figs. 3.10 and 3.11.

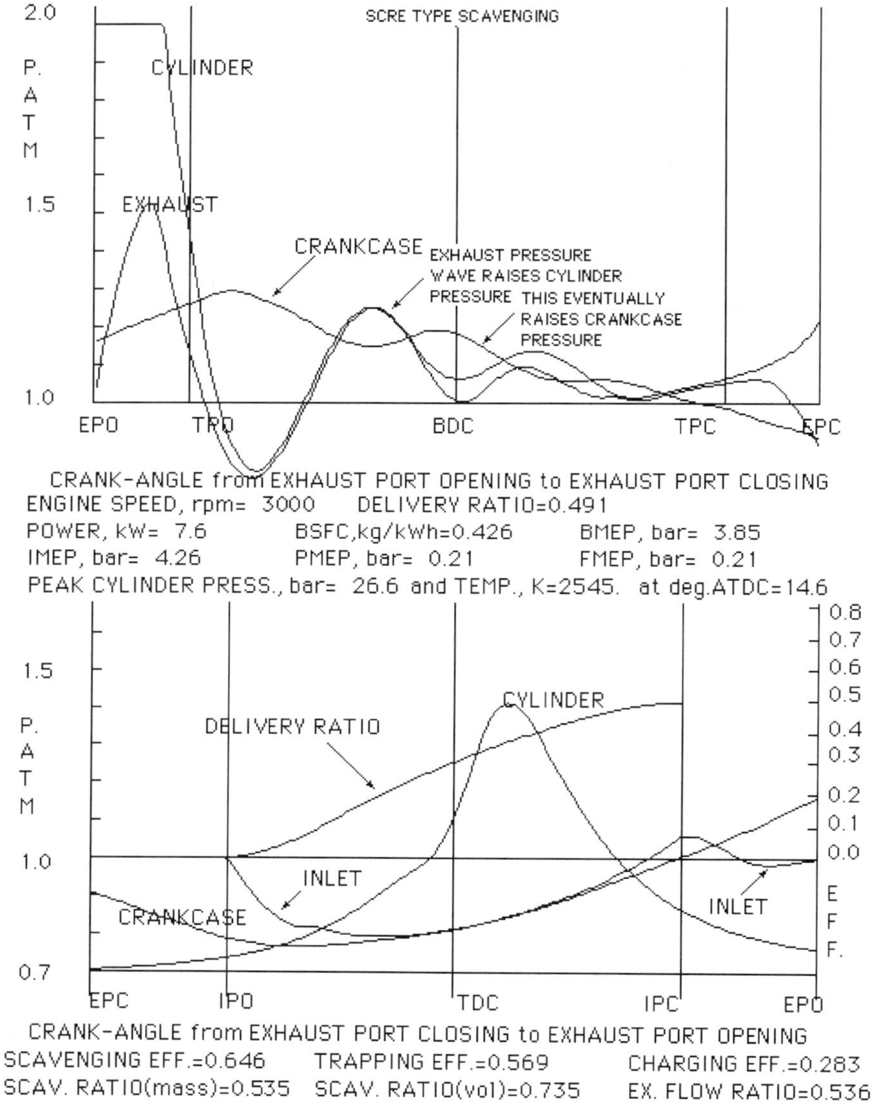

Fig. 5.13 Predicted performance characteristics of the QUB 400 engine at quarter throttle.

The friction loss remains constant at 0.21 bar MEP, while the pumping loss is reduced from 0.38 bar to 0.20 bar MEP. This latter prediction, the reduction of pumping work in a throttled crankcase compression two-stroke engine, is in direct contrast to a four-stroke engine, where throttling the inlet system increases the pumping work. In the two-stroke engine, although the crankcase suction pumping loss is increased during throttling, it is alleviated by taking place at a compression ratio of, in this typical example, 1.4. The pressure-volume envelope on the suction phase is not grossly different for the throttled and unthrottled cases. If it had been a four-stroke engine, that suction would be occurring at a, say, 8:1 compression ratio with a consequent considerable increase in the work expended. On the crankcase compression phase for the two-stroke case, the much reduced pressure-volume envelope during the compression process is clearly evident in the throttled example, and it is this which ultimately lowers the overall PMEP.

As expected, the engine produces less IMEP and power due to the lower mass of trapped air charge. The power output is down from 12.2 kW to 7.6 kW, and BMEP is reduced from 6.1 bar to 3.8 bar. Thanks to the reduced pumping loss and to the increase in trapping efficiency, the brake specific fuel consumption is improved by 15% from 463 to 426 g/kWh.

Of interest is the influence of exhaust system pressure dynamics upon the crankcase pressure. The compression pressure reflection at bdc is seen to raise the cylinder pressure above the crankcase pressure and this, via transmission with time through the transfer duct, also gives a pressure bulge within the crankcase. This has a knock-on effect, as that keeps the crankcase pressure higher during suction when the piston motion is upward which decreases air inflow at the same throttle opening. It also raises the value of crankcase pumping work during the compression phase. The same effect, to a lesser degree, is seen to occur at full throttle in Fig. 5.9.

5.2.3 The chainsaw engine at full throttle

The data for the model to be analyzed by Prog.5.1 is given in Figs. 5.3 and 5.4(a). In Fig. 6.1 there is a sketch of port areas related to crankshaft angle, as part of a general discussion on the topic, however the actual geometry used to create Fig. 6.1 is that for the chainsaw engine under discussion. As a further aid to physical visualization, Fig. 6.2 shows a computer generated sketch of this same chainsaw engine cylinder. The data could be regarded as being virtually a universal chainsaw engine design, for there are many such engines in production with very similar porting and exhaust systems. The hallmark of these small high-speed engines, used for such tools as chainsaws, brush cutters and snow blowers, is a small, compact silencer with a limited volume so that the bulk of the entire device does not make it unwieldy. The porting layout of many of these engines is remarkably common, with a piston controlled inlet port, usually two transfer ports and a single exhaust port. The main aim of the designer is to attain the requisite power output at as low a sound and vibration level as possible, for most of these tools are hand-held. Fuel consumption is also an important criterion, for although the engines are small in terms of both cylinder capacity and total power output, they are often used professionally in excess of 1000 hours per annum. In other words, the total fuel consumed can be considerable and a reduction in the specific fuel consumption rate is noticeable. This is particularly true as most of these hand-held tools have quite small capacity fuel tanks, so the operator tends to observe, and mentally record, the time interval between refueling.

There is one curiosity in the data bank in Fig. 5.3 which may raise a query in the minds of some readers. The engine data apparently specifies that the engine is running at 3/4 throttle, rather than at full throttle. The reason for this is that such engines use diaphragm carburetors with a very pronounced venturi giving a restriction ratio of 25% to the carburetor flow diameter, even when the butterfly throttle valve is wide open! Such are the mathematical subterfuges to which the programmer must resort when it is necessary to accurately describe some particular engine geometry.

It will be seen from the data table that, although the port timings are not dissimilar to the QUB 400 engine, the exhaust box silencer is small and restrictive. As further acquaintance with the technical jargon, silencers are often referred to in automotive literature as "mufflers." The volume of this box is merely five times larger than the cylinder capacity, and the outlet pipe area from the box is just 40% of that of the exhaust downpipe connecting the box silencer to the exhaust port. The pipes at either side of the muffler are also very short, each being 90 mm.

The output of the computer calculation using Prog.5.1 is presented in Fig. 5.14 for a crankcase compression ratio of 1.4 . It will be seen that the restrictive exhaust box has a profound influence on the cylinder and exhaust pressure diagrams. The build-up of exhaust back-pressure during the scavenge phase opposes the crankcase blowdown with the result that the crankcase pressure is still above the atmospheric pressure by exhaust port closure. This reduces the strength of the suction to half of the crankcase pumping action which translates into a low delivery ratio of 0.487. Weaker suction pulses provide lesser intake ramming behavior which compounds the problem. As the cylinder pressure remains high during the scavenge period, the expansion of the scavenging volume flow is also limited and the SRv value is only marginally higher than the scavenge ratio level, i.e., SRv is 0.614 for a SR level of 0.537. The resulting scavenging efficiency is correspondingly low at 0.52. Nevertheless, the engine produces 3.66 bar BMEP at 7500 rpm and a power output of 4.6 kW, with a commendable BSFC of 444 g/kWh. This would be regarded as a normal performance characteristic for such an engine.

With the restrictive exhaust system having been highlighted as the limiter on engine performance, the logical design option is to examine an engine design with one which is less restrictive. If that is not a viable design option, due to both engine bulk and noise legislation limitations, then an alternative is to attempt to incorporate a stronger air pumping action. Thus a data change is made to give a crankcase with a compression ratio of 1.5, and the calculation is repeated with the result shown in Fig. 5.15. The delivery ratio is considerably higher at 0.562, the scavenging efficiency is increased to 0.564, the power rises to 5.4 kW, and the BMEP to 4.3 bar. Although the pumping MEP has risen to 0.29 bar from 0.21 bar, the increase of IMEP from 4.18 to 4.90 bar more than compensates for the greater pumping work.

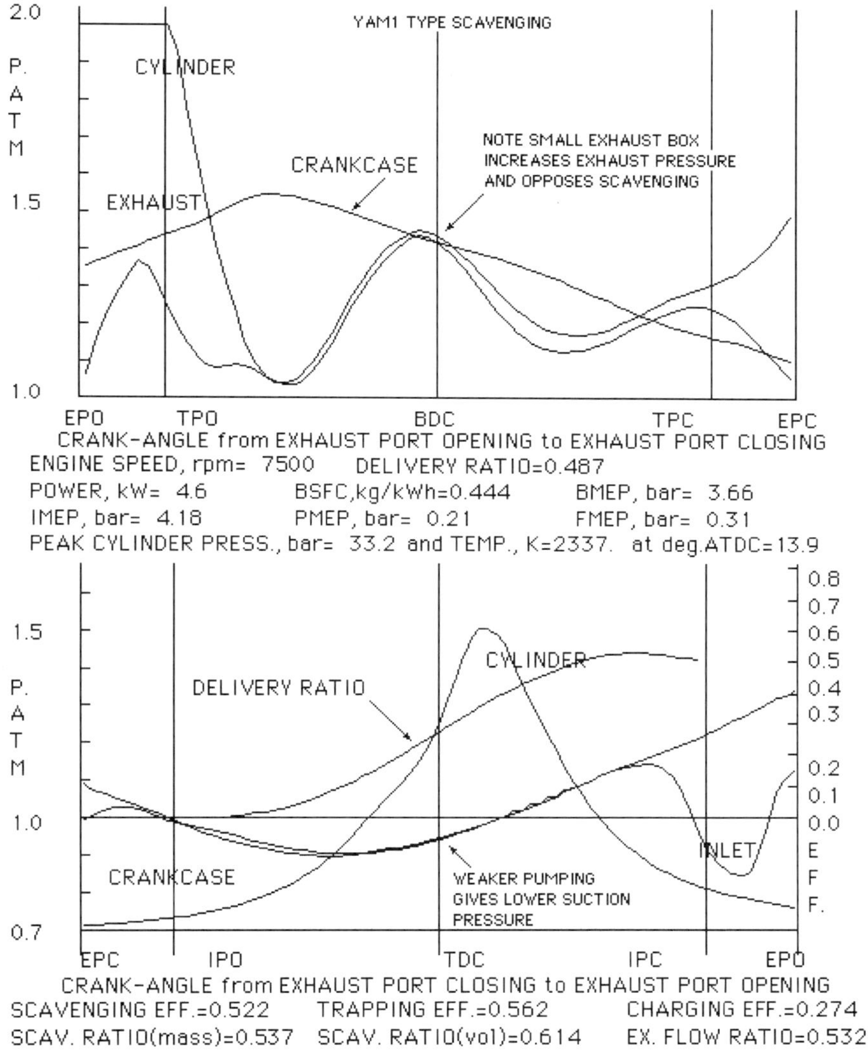

Fig. 5.14 Predicted performance characteristics of the chainsaw engine with a CCR of 1.4.

The result is a reduction of brake specific fuel consumption from the original design value at 444 g/kWh to 435 g/kWh. The stronger pumping action on both the compression and suction phases of the crankcase operation due to the higher CCR is clearly evident by comparing Fig. 5.14 with Fig. 5.15.

The overall performance characteristics of the chainsaw engine at full throttle over a speed range of 5000 to 9000 rev/min is calculated by altering the data input for engine speed in the opening data block in Fig. 5.3. This has been done at 1000 rpm intervals and the result for both crankcase compression ratio values of 1.4 and 1.5 is presented in Fig. 5.16. The output data for power, BMEP, BSFC, and delivery ratio at each engine speed is presented in graphical form. The actual values plotted at each calculation speed are the average of the performance characteristics determined for the third, fourth and fifth engine cycles. The variation of performance characteristics with each succeeding cycle is small for an engine of the industrial type with an untuned exhaust system; in Sect. 5.4 a quite different form of cyclic behavior will be demonstrated. In the reality of a firing engine, as in a computer calculation, there are cyclic variations of all the performance characteristics but this is particularly true for the combustion diagram.

In Fig. 5.16, as found at the particular speed of 7500 rpm plotted in Figs. 5.14 and 5.15, the higher crankcase compression ratio of 1.5 performs better across the entire engine speed range. The peak power output at 8000 rpm has been raised by 18% with a marginal improvement of 4% in specific fuel consumption. It reinforces the opinion that the crankcase pumping action is one of the most significant design parameters for a high-speed industrial engine of this type(5.1).

The almost linear fall of DR with the lower crankcase compression ratio of 1.4 results in a 20% drop in power at 9000 rpm. At 5000 rpm there is very little difference in performance between the two variations of the engine, emphasizing the point that it is for the design of a high-speed engine that attention must be paid to the optimization of the crankcase pumping action. Equally, it could be that the designer may have a particular reason to have the power level fall off sharply at high speed, as there may be some legal safety requirement to be met regarding the upper speed which can be attained by an operator using the device. In this case also, the designer can tailor the CCR level to meet such a need, for the sharp fall in power from 8000 to 9000 rpm in the lower CCR example is evidence that this is technically possible.

5.2.4 Concluding discussion on Prog.5.1, "ENGINE MODEL NO.1"

It is clear that this model allows the designer to produce predicted engine performance characteristics for a parameter change which could be as major as crankcase compression ratio, or as minor as a radius variation in the top corners of a port. By such means, the designer can tailor the performance characteristics of the engine to meet the market needs of the device and thus eliminate many of the experimental "cut and try" techniques so common in industry. This does not mean that the unsteady gas-dynamic program is so accurate as to replace experimentation, such is definitely not the case, but it does permit a more rapid understanding of the

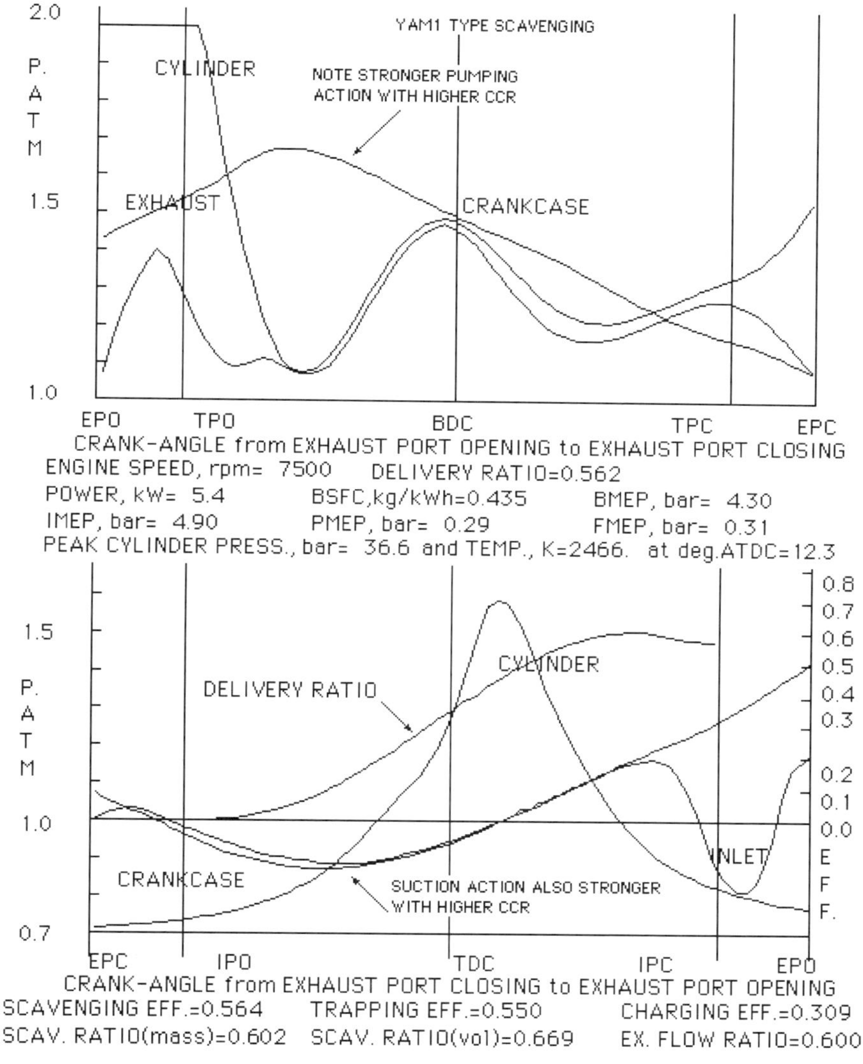

Fig. 5.15 Predicted performance characteristics of the chainsaw engine with a CCR of 1.5.

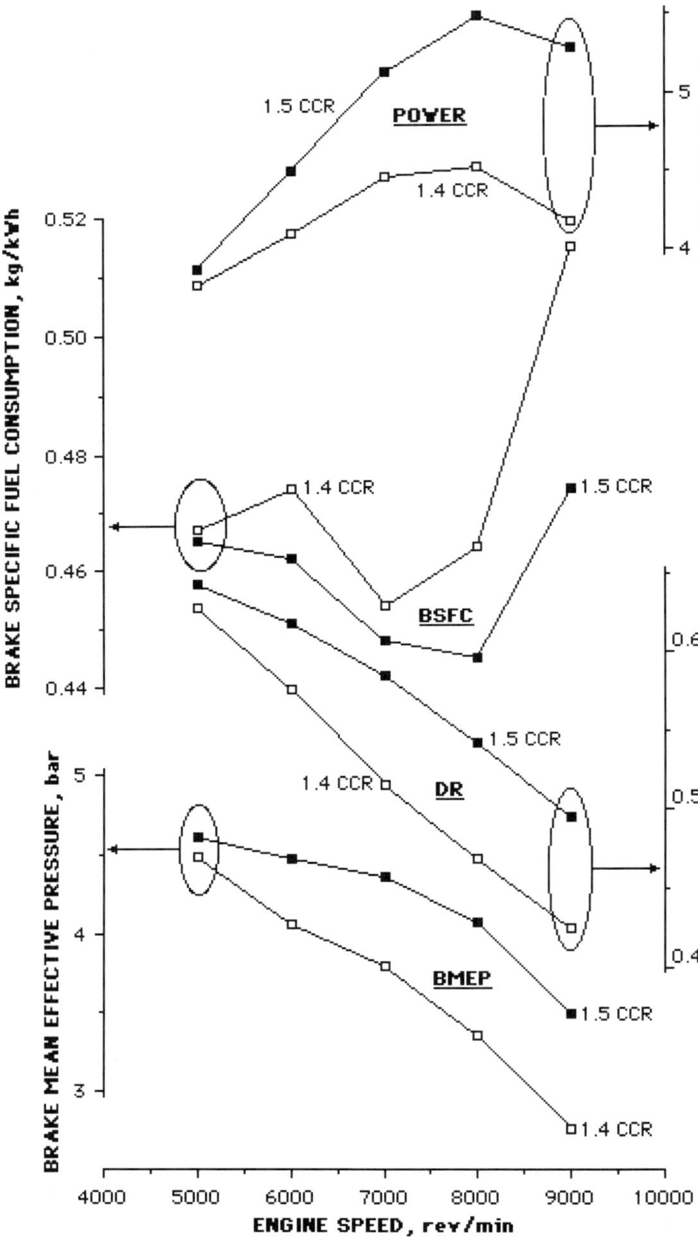

Fig. 5.16 Calculated chainsaw performance with different CCR values.

potential behavior of any engine undergoing design or development. It is from such understanding that design and development problems are solved. In the design case, the problem can often be eliminated before the first drawings are made, casting dies purchased, or a prototype machined and constructed. In the experimental situation, the simulation often provides the answer to a baffling problem; in the chainsaw example just discussed, the opening gambit of the experimenter might be to test a series of engines with longer inlet timings in a fruitless attempt to raise the delivery ratio. In either situation, the potential for savings of time, effort and money is considerable.

One word of caution is required regarding scavenging behavior. Within the calculation packages, the designer can allow the "paper" engine to be scavenged as well as a uniflow engine or as badly as a classic deflector piston engine. The only means of ensuring that the actual engine has a scavenging system as good as can be attained, within the design limitations of the engine in question, is to conduct tests using the theoretical or experimental techniques discussed in Chapter 3.

A general conclusion can be made regarding the design of engines with untuned exhaust systems, that in the absence of any effective engine tuning by pressure waves, apart from the ramming effect in the intake system, the BMEP or torque characteristic with speed is effectively controlled by the shape of the delivery ratio curve and the quality of the scavenging process.

5.3 Using Prog.5.2, "ENGINE MODEL No.2"

From Fig. 5.2(b), it can be seen that this computer model contains a tuned exhaust pipe of the type commonly seen on high-performance racing motorcycle engines; it is also fitted to high-performance road-going motorcycles with silencers of varying degrees of effectiveness attached to the ends of the tuned pipes. Further discussion on this particular topic can be found in Chapter 8. As remarked in Sect. 1.3.2, this form of tuned exhaust pipe was successful in the late 1950's on the MZ racing motorcycles from Zschopau in East Germany. The author does not know who invented this type of exhaust system, colloquially referred to as an "expansion chamber," for racing two-stroke engines, but it is known that Walter Kaaden, the Chief Engineer of MZ, played a major role in its early development and testing(1.1). At the same time, the author has clear memories of seeing a 125 cc racing Montesa from Spain which had a very similar type of exhaust system, and this was several years before the advent of the MZ machines. The application of the expansion chamber on a QUB 500 cc single-cylinder racing machine(5.4)(5.5) is shown in Plate 5.1.

The result of the use of this type of tuned exhaust system, allied to a disc valve inlet system with its asymmetrical inlet port timing, was that the two-stroke engine employed for motorcycle racing produced more specific power output than the four-stroke engine. This situation persists to this day in all forms of motorcycle racing and competition.

At QUB there has been, and still exists, considerable research in the understanding, design and development of tuned exhaust systems for all kinds of two-stroke engines(1.13)(2.6)(2.15)(3.7)(4.9)(4.10)(5.2-5.5). Therefore, the computer model

Plate 5.1 The QUB 500 single-cylinder 68 bhp engine with the expansion chamber exhaust slung underneath the motorcycle (photo by Rowland White).

represented by Prog.5.2 is the result of many years of theoretical research allied to experimentation with high specific output engines. Although Prog.5.2 does not contain induction system variations to handle disc valve or reed valve induction systems, the program permits the designer to model all of the engine porting and exhaust geometry of a high-performance engine. There will be discussion in a later chapter on the design of racing engines and the interpretation of calculated data from a power unit with piston controlled induction for the design of reed valves and disc valves. It has been amply demonstrated by Blair et al(5.9) and Fleck et al(4.10)(1.13) that engine modelling can incorporate reed valve induction, and predict both the reed valve motion and the delivery ratio. The incorporation of disc valve induction into a computer program like Prog.5.2 is quite straightforward, as can be observed from the relevant design discussions in Chapter 6.

It is rather interesting to consider that in Sect. 5.2.3, dealing with the chainsaw type of engine, adverse comment is passed on the debilitating effect on engine performance posed by a restrictive exhaust system. Here, if one interprets the data in a simple fashion for the expansion chamber proposed in Fig. 5.4(b) for the 125 Grand Prix engine, the exhaust pipe at the exhaust port has a diameter of 38 mm and an outlet diameter to the atmosphere of just 20 mm. There are those who would logically consider that to be "restrictive"! The following discussion will reveal that this area restriction at outlet is vital to the attainment of a high specific power output.

Before commencing to discuss the behavior of expansion chamber exhaust systems, it is important to describe the format of the computer screen output from

Prog.5.2. The program will produce line-print output of all input data and calculation output, including all of the screen graphs. The latter is the principal output of the analysis and is the basis of Figs. 5.17-5.20 regarding the Husqvarna MK4 engine.

Taking Fig. 5.17 as an example of the screen output from Prog.5.2 while it is running, the x-axis is the crankshaft angle commencing at exhaust port opening, EO, for one complete cycle; exhaust port closure is marked as EC. The other timing events are for the transfer port, i.e., opening and closing is TO and TC, respectively, and for the inlet port, i.e., opening and closing is IO and IC, respectively. The markers for top and bottom dead center are self-explanatory as TDC and BDC. The diagram on the computer screen is refreshed at the conclusion of each cycle. There are two sets of pressure ratio-time histories being written simultaneously. The top one with its scale at the left is for cylinder pressure, exhaust pipe pressure (that value next to the exhaust port), and crankcase pressure. To prevent a programming problem, the maximum cylinder pressure plotted is at 2.0 atm. The bottom set, with its scale of pressure ratio at the right, is for the crankcase pressure and for the inlet pipe pressure in the pipe next to the intake port. Unlike Prog.5.1, the cylinder pressure diagram during combustion does not appear, because the major interest is in the tuning effect of the exhaust system and its control over engine power and torque. This latter information is presented at the base of the diagram at bdc on each cycle, as seen in Fig. 5.17 for the fifth engine cycle in that example. There, the information includes the peak cycle maximum pressure, temperature and crank angle location and includes that for SE, TE, and CE on each cycle as for Prog.5.1.

5.3.1 Analysis of data for a Husqvarna motorcycle engine using Prog.5.2

In a previous paper(3.7), Blair and Ashe presented experimental and theoretical data for a series of exhaust systems attached to a Husqvarna 250 cc motocross racing engine. That paper, which is also to be found in (1.7), contains much more information on the actual engine than is proposed to be given here, including data on scavenging efficiency measured under firing conditions with the type of apparatus shown in Plate 3.3, and the measured performance characteristics of all of the engine and exhaust variations tested. To illustrate the potential for correlation of measured and calculated scavenging efficiency by an unsteady gas-dynamic engine model, the model for the EXPAND function as expressed in Eq. 5.1.13 is used in this particular set of calculations epitomized by Figs. 5.17-5.20. The purpose of using the data from the Husqvarna engine, in particular the so-called MK4 variant, is (a) to demonstrate that the relatively simple Prog.5.2 will accurately calculate the behavior of an engine whose performance is known, and (b) to illustrate the pressure wave motion within an expansion chamber which yields the high specific power characteristics. The output in Figs. 5.17-5.20 are for the Husqvarna MK4 engine data displayed in Figs. 5.3 and 5.4(b) at the engine speeds of 4000, 5500, 6500 and 8000 rpm.

(a) The pressure wave action in the expansion chamber at 6500 rpm

Examining first Fig. 5.19, where this 250 cc swept volume engine is calculated to produce 7.76 bar BMEP and 20.6 kW of power, this is manifestly a high specific power output of some 82 kW/liter or 110 bhp/liter. The pressure wave action in the

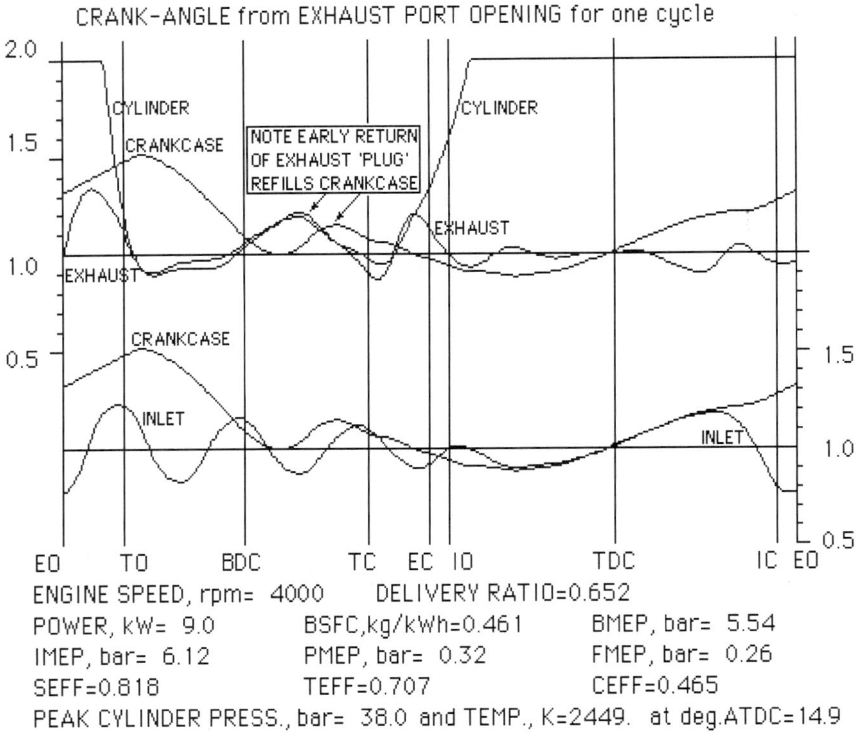

Fig. 5.17 Predicted performance characteristics of the Husqvarna MK4 engine at 4000 rpm.

exhaust system provides a strong suction action throughout the scavenge period, with the crankcase pressure pulled down to the atmospheric line by the end of the scavenge period at transfer port closure. The next action of the exhaust system is to give a positive compression wave at the exhaust port in the period before the piston closes that port physically. Thus all of the fresh charge supplied to the cylinder, and retained in it up to a point at or about transfer port closure, is now gas-dynamically kept inside the cylinder by this "plugging" exhaust pressure pulse. Indeed, as the exhaust port pressure is actually briefly above the cylinder pressure in this period, some of the fresh charge which had been lost to the exhaust pipe during scavenging is "stuffed" back into the cylinder. From whence comes these useful pressure reflections?

The suction reflection at the exhaust port is caused by the exhaust pulse travelling through the diffuser section in Fig. 5.2(b), indicated by lengths LP12 and LP23. As Eq. 2.4.19 would predict from the positive DF/F terms within it, the continual area increase with length sends back a series of ß reflections as expansion waves to the

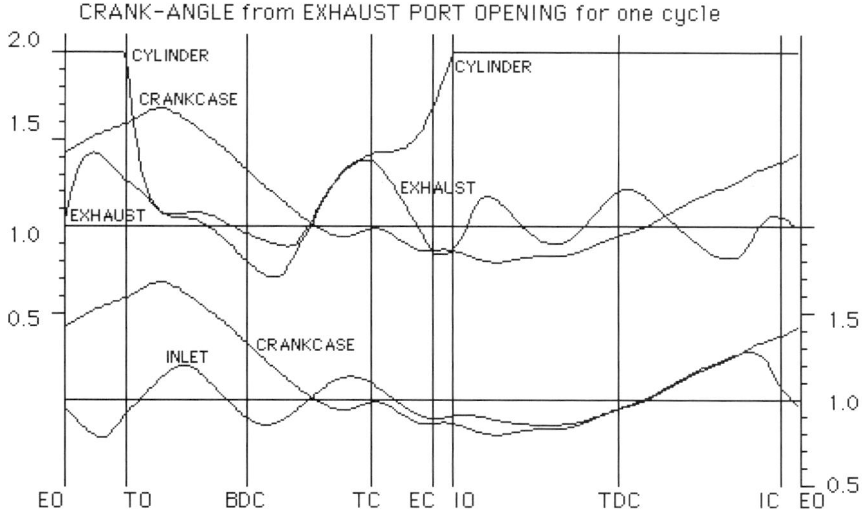

ENGINE SPEED, rpm= 5500 DELIVERY RATIO=1.067
POWER, kW=19.7 BSFC,kg/kWh=0.458 BMEP, bar= 8.78
IMEP, bar= 9.64 PMEP, bar= 0.50 FMEP, bar= 0.35
SEFF=0.918 TEFF=0.602 CEFF=0.632
PEAK CYLINDER PRESS., bar= 55.3 and TEMP., K=2975. at deg.ATDC=14.4

Fig. 5.18 Predicted performance characteristics of the Husqvarna MK4 engine at 5500 rpm.

engine, i.e., at a sub-atmospheric pressure. The reflection process in the diffuser is rather like having a continuous series of small sudden area expansions like those in Sect. 2.3.4(a). The use of a diffuser is a more efficient method of reflecting pressure wave energy than that obtained by conducting it completely at one physical location, such as at a sudden enlargement in a pipe or at an open end to the atmosphere.

The "plugging" pulse is a compression wave reflection which emanates from the onward progression of the original exhaust pulse into the nozzle portion of the pipe indicated by length LP45 in Fig. 5.2(b). Here, Eq. 2.4.19 would dictate that the negative values of DF/F for each mesh in that section would produce ß reflections which are above unity, i.e., compression waves which are greater than the atmospheric pressure. The process, like that in the diffuser, is efficient because it is spread over the length of the nozzle section. The pipe outlet diameter can then be quite restrictive so as to attain the strongest plugging pressure characteristic. The compression pressure wave reflection will reflect to and fro between the exhaust port and the nozzle during the closed cycle period, however, the expansion chamber must empty in good time before the advent of yet another exhaust pulse. Thus, the design criterion for tail pipe diameter is that which will give the strongest plugging

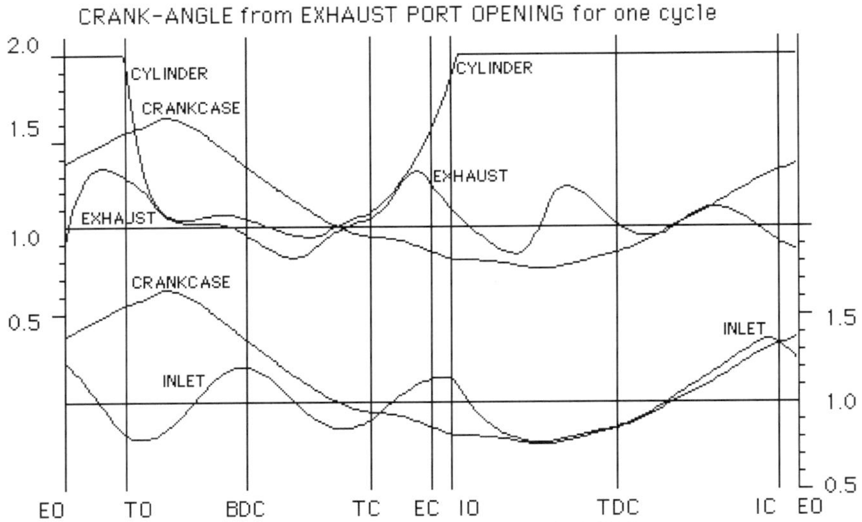

ENGINE SPEED, rpm= 6500 DELIVERY RATIO=1.036
POWER, kW=20.6 BSFC,kg/kWh=0.519 BMEP, bar= 7.76
IMEP, bar= 8.64 PMEP, bar= 0.46 FMEP, bar= 0.42
SEFF=0.902 TEFF=0.555 CEFF=0.566
PEAK CYLINDER PRESS., bar= 50.2 and TEMP., K=2951. at deg.ATDC=15.1

Fig. 5.19 Predicted performance characteristics of the Husqvarna MK4 engine at 6500 rpm.

reflection while not allowing the chamber pressure to rise to a level which would seriously deteriorate the strength of the suction reflection behavior in the diffuser in succeeding cycles(2.17).

(b) The effect of the exhaust dynamics on charge trapping

The trapping pressure ratios in the cylinder in Figs. 5.17-5.20 are 1.3, 1.7, 1.6 and 1.35 atm. These trapping pressures are closely related to the BMEP which is produced. As discussed in Sect. 1.5.6, the principal control of trapped mass is by trapping pressure. By contrast, the QUB 400 engine in Fig. 5.9 is seen to trap at 1.25 atm without the benefit of a tuned exhaust system and by piston action only. The quality of the scavenging process also plays a part in this behavior, as it affects the fresh charge retention characteristics up to the point of any exhaust port plugging by the pressure waves in an expansion chamber type of exhaust system.

Of general interest is the level of specific fuel consumption from this engine with a tuned exhaust pipe. The trapping efficiency is 0.555 and the BSFC is 0.519 kg/kWh even though the air-fuel ratio is quite rich at 11.17(3.7). The QUB 400 engine with an untuned exhaust, but a better scavenging system, has a trapping efficiency of 0.511 and a BSFC of 0.462 kg/kWh with an air-fuel ratio set quite lean at 13. This contrast reveals the obvious potential to produce good power and efficiency

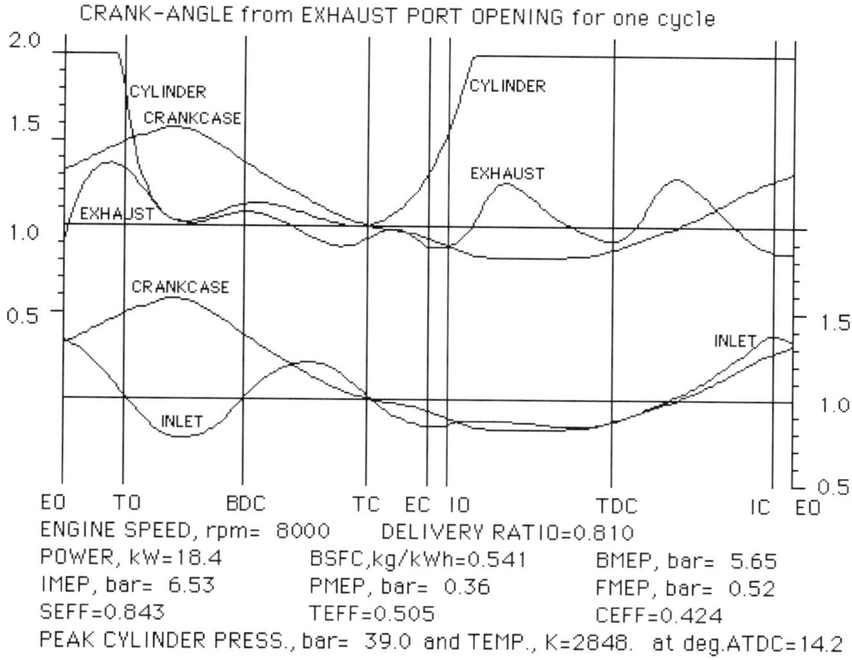

Fig. 5.20 Predicted performance characteristics of the Husqvarna MK4 engine at 8000 rpm.

characteristics for the expansion chamber type of exhaust system mounted on an engine not designed specifically as a racing unit but with a more modest delivery ratio and an optimized scavenge system.

(c) The effect of engine speed on the behavior of the expansion chamber

Figs. 5.17-5.20 show the prediction of pressure wave action at 4000, 5500, 6500, and 8000 rpm and these can be compared with the measured pressure-time graphs in the published paper(3.7). That original paper contains experimental and theoretical data at other engine speeds. The following table shows the net effect on predicted engine performance when this particular expansion chamber exhaust system is being used.

Predicted Performance Characteristics of the Husqvarna MK4 Engine

Speed, rpm	power, kW	BMEP, bar	BSFC, kg/kWh	DR	SE
4000	9.0	5.54	0.461	0.652	0.818
5500	19.7	8.78	0.458	1.067	0.918
6500	20.6	7.76	0.519	1.036	0.902
8000	18.4	5.65	0.541	0.810	0.843

From this table it is seen that the peak effect of the exhaust system is to enhance the flow of fresh charge as evidenced by the delivery ratio, and the cylinder charging as evidenced by BMEP, at 5500 and 6500 rpm. It is clearly less effective on both counts at 4000 and 8000 rpm. The exhaust system dynamics produce a peak effect over a fairly narrow range of engine speeds. Further examination of the more complete data in the paper(3.7) would show that this effective speed range is about 2500 rpm wide. It is possible to widen this speed range by suitable control over the exhaust port timing duration, and this topic is discussed in Sect. 7.3.3.2.

The reasons for the relatively narrow speed band of high torque becomes quite clear when Figs. 5.17-5.20 are examined in a more detailed fashion. In Figs. 5.18 and 5.19, where the engine is running at 5500 and 6500 rpm, the excellent tuning effects already discussed in (a) and (b) are observed, indeed the behavior is most pronounced at 5500 rpm where the torque peak occurs at 8.78 bar BMEP.

At 4000 rpm, in Fig. 5.17, the exhaust plugging reflection arrives at the exhaust port much too early to be effective, indeed the exhaust port plugging action which does take place is due to a second reflection of the original exhaust pulse travelling to and fro between the exhaust port and the tail pipe nozzle. The situation is actually worse than that, for the early return of the plugging reflection around the bdc period opposes the scavenge flow. This refills the crankcase in exactly the same manner which is described for the chainsaw engine in Sect. 5.2.3. The knock-on effect, or deterioration of the induction of fresh air and fuel into the crankcase, is also experienced in this case, as the DR is considerably less at 4000 than it is at 5500 rpm.

At 8000 rpm, the plugging reflection arrives too late to assist with fresh charge retention within the cylinder and the cylinder is exposed to a continual extraction of fresh charge from it. This is due to the presence of the suction reflection from the diffuser being at the exhaust port for the duration of its opening. This is the reason for the delivery ratio at 8000 rpm being higher than at 4000 rpm, as a normal crankcase pumping action unaffected by exhaust system dynamics produces a falling DR characteristic with increasing speed, viz Fig. 5.16. In general, one can expect a tuned exhaust system of the expansion chamber type to provide a usable speed band of high torque output of between 2000 and 3000 rpm.

(d) The accuracy of computer models of high-performance engines

To open the discussion, it is useful to compare the measured and calculated performance behavior of the Husqvarna MK4 engine. The original experimental data in (3.7) was presented in Imperial units, so they have been translated into SI units for the table below:

Measured Performance Characteristics of the Husqvarna MK4 Engine

Speed, rpm	power, kW	BMEP, bar	BSFC, kg/kWh	DR	SE
4000	7.3	4.44	0.555	0.621	0.81
5500	18.9	8.43	0.496	1.059	0.91
6500	19.8	7.45	0.527	0.993	0.92
8000	17.5	5.36	0.554	0.772	0.90

A comparison of the measured and calculated performance characteristics shows that Prog.5.2 will model a high-performance two-stroke engine in a manner sufficiently accurate for design purposes. The correspondence is particularly good in the region where the exhaust tuning is at its most effective, between 5500 and 6500 rpm. It is not so good for the prediction of performance at the low-speed untuned condition of 4000 rpm, although it does illustrate the extent to which the exhaust plugging effect deteriorates the delivery ratio. The theoretical figures show computed levels of scavenging efficiency which correspond in a very satisfactory manner with those which were measured and reported by Blair and Ashe(3.7). The calculated level is obtained by adopting the second model of the volumetric expansion of the scavenge flow expressed in Eq. 5.1.13; both versions of the EXPAND function are programmed as such in ProgList.5.2 as alternatives for use by the reader. This particular topic is the subject of much research(3.31) at this point in time, and technical papers will appear in the future to provide more detailed advice.

Further comparison of the calculated pressure-time histories from this program can be made with those determined experimentally which illustrate the original technical paper(3.7). It will be seen that the correspondence is sufficiently close to give the reader confidence in using Prog.5.2 for design purposes.

5.4 Single-cylinder high specific output two-stroke engines

This section will be used as an introduction to the specialized topic of racing engine design in Chapter 6. It is being opened here because Prog.5.2 is being employed to illustrate the phenomenon of resonance in the expansion chamber exhaust systems of engines for track or road racing.

From the previous discussion on the Husqvarna MK4 engine, it is seen that an expansion chamber exhaust system attached to each cylinder of a two-stroke engine provides a high BMEP and power characteristic from the engine. This presumes that all of the other parts of the engine have been designed to take advantage of the action of the exhaust system. In the case of the Husqvarna 250 motocross engine, the engine was designed for use in cross-country motorcycle racing, or motocross. The design requirement is for an engine with a high specific output over as broad a speed range as possible, for the rider needs to be able to accelerate from very tight turns and from race situations where indexing the correct gear may prove to be physically difficult at high speed over rough terrain! Consequently, the engine speed at which peak power is developed is normally not as high as that which might conceivably be employed. For the design of an engine for road or track racing, the designer must attempt to achieve the highest possible power output from an engine of a given maximum engine swept volume, and often within rules which dictate the maximum number of cylinders which can be deployed to reach that total displacement. In either circumstance, irrespective of the number of cylinders which are used, the designer of the road racing engine attempts to produce an engine which will achieve the maximum possible BMEP at the highest engine rotational speed mechanically feasible.

The maximum rotational speed is determined by the loads which can be tolerated by the mechanical components of the engine, and can usually be expressed as having some simple relationship with the mean piston speed. This is discussed initially in Sect. 1.7.1. As the designers of the mechanical components take advantage of the latest advances in materials technology, and the oil companies provide improved lubricants, that relationship with piston speed tends to change with time. Some years ago, 11000 rpm was regarded as the highest safe rotational speed for a 125 cc cylinder with a 54 mm stroke; today, that figure would be 12000 rpm for the same stroke and cylinder capacity. The mean piston speeds in each case are 19.8 and 21.6 m/s.

For such high rotational speeds, long port timings must be employed to pass the requisite gas flow through the engine. If one examines the data set for the 125 Grand Prix engine in Fig. 5.3, it will be seen that the exhaust and inlet ports are open for a longer period than they are shut, and are considerably greater than for the Husqvarna MK4 power unit. Fig. 6.7 shows a computer generated sketch of the cylinder of this same Grand Prix engine, revealing the physical implications of such large ports and long port timings. In Chapter 6 this subject will be explored in more depth, so at this juncture it will be simply stated that the data blocks described as the 125 Grand Prix engine (in Fig. 5.3 for the engine and Fig. 5.4(b) for the exhaust system) are typical of any one of a dozen designs raced in the 1970's and 80's. Although this discussion seems to be biased towards racing motorcycles, the debate is equally applicable to engines for go-karts, hydroplanes, or snowmobiles.

Upon entering this data into Prog.5.2, the user will observe that as the pressure-time histories unfold on the computer screen, engine cycle by engine cycle, the early promise of high power output would appear to be unlikely. For example, Fig. 5.21 shows the output for the third calculation cycle from start-up. In almost all calculations for Prog.5.1, and also for the Husqvarna MK4 data in Prog.5.2, the program could be stopped at this point as virtually every cycle from that point onwards to infinity would be identical with that determined by the third or fourth cycle; in short, the calculation had reached equilibrium. In Fig. 5.21, the result is disappointing, as the BMEP is a mere 7.10 bar and the delivery ratio is just 0.83. This is a lesser performance level than the Husqvarna MK4 engine so it is certainly no Grand Prix race winner.

Little of note happens for the next two cycles, but during the sixth calculation cycle the picture changes markedly; the result of that sixth engine cycle is shown in Fig. 5.22. In that cycle, the suction pressure wave around bdc has become deeper and a stronger plugging pulse has returned a little earlier. The effect of the strong suction caused by the exhaust pipe should be emphasized, for the pressure ratio from crankcase through to the cylinder is approaching 3:1, which is well above that required to produce sonic particle velocity and choked flow. On this cycle, the engine BMEP is up to 8.87 bar and the power has risen by 25%, with the delivery ratio up from 0.83 to 0.91. The large secondary exhaust pressure wave reflection around the tdc period seems to have reduced in amplitude from the third cycle.

On the next two calculation cycles, the seventh and eighth, the picture strengthens to a level which is destined to be about equilibrium, and to a level of BMEP and

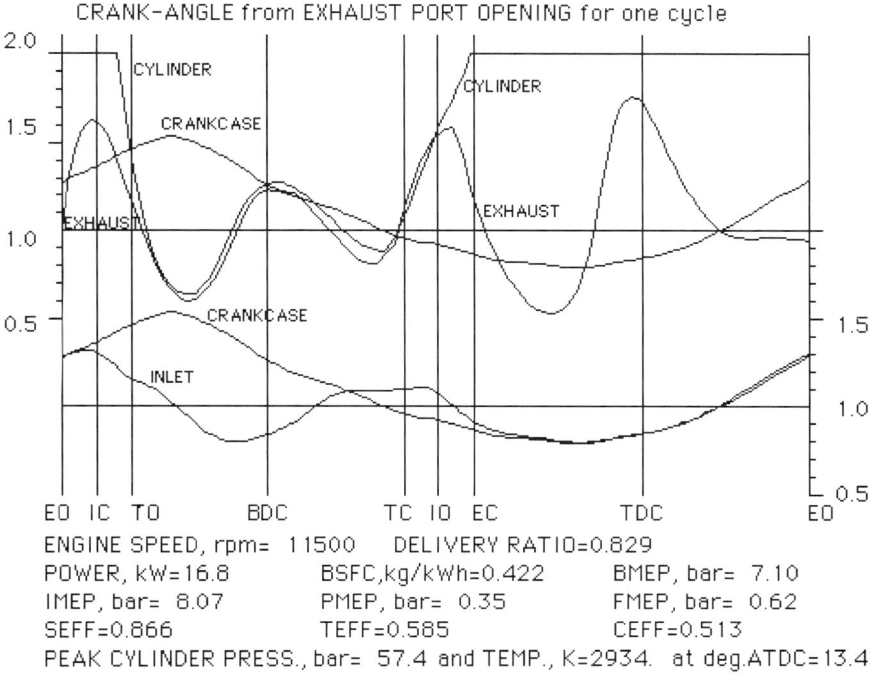

Fig. 5.21 The 125 cc Grand Prix engine on the third calculation cycle.

power which which will place it in a race-competitive category. Fig. 5.23 illustrates the eighth cycle where the power maximizes. The BMEP is at 11.29 bar, the power at 26.8 kW (35.9 bhp) and the delivery ratio is up to 1.1. This is a specific power output of 215 kW/liter (287 bhp/liter), more than double that of the Husqvarna MK4 power unit. It will be noticed that the secondary exhaust reflection around the tdc period has been reduced and broadened.

On subsequent calculation cycles, the program shows performance oscillations about this power output level, rising and falling between 26 and 28 kW. Of general interest are the very good (for a racing engine, that is) specific fuel consumption characteristics.

The internal pressure wave motion within the pipe, due to the disposition and shape of the diffusers and nozzles, is such that secondary pressure waves are being combined with the next outgoing exhaust pulse, making it more extensive in both amplitude and time. This produces suction reflections from diffuser and nozzle which are also larger in both amplitude and time. The combination of lesser reflections with major pressure waves is referred to as "resonance." The end effect is a theoretical prediction of high specific power output by Prog.5.2.

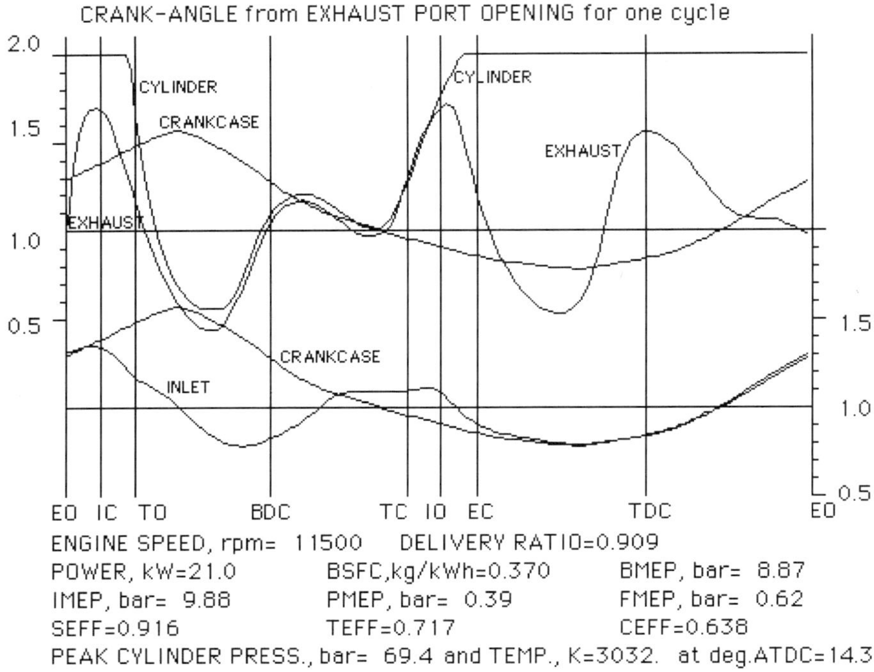

Fig. 5.22 The 125 cc Grand Prix engine on the sixth calculation cycle.

Experimentally, a similar pressure wave resonance behavior is observable in racing engines at the highest engine speed test points. The engine will be held by the dynamometer at an engine speed just beyond the point of normal maximum power, and the power recorded will be as disappointing as that in Fig. 5.21, until the dynamometer load record will suddenly swing up to a level such as that seen in Fig. 5.23, and remain there. This experimental phenomenon, which is slightly different from the theoretical example in that it is associated with a corresponding elevation in exhaust gas temperature as the BMEP rises on succeeding cycles, is referred to colloquially as "uptune." The similarity to the theoretical example is that the pressure waves recombine as the exhaust gas temperature rises with each succeeding cycle to reform a pressure wave resonance pattern in the manner of the progression of Figs. 5.21-5.23. It is also thought, but not known for certain, that some of the short-circuited air-fuel mixture, pulled into the exhaust pipe by the exceptionally strong suction reflections from the diffuser, actually burns in the exhaust pipe, and that this initiates the temperature rise which ultimately results in the "uptune effect." This effect is particularly important in racing engines, as it can extend the usable engine speed power band of the machine by as much as 10%. To the reader, this may not appear to be an impressive margin, but it will be so to the competitor in second place.

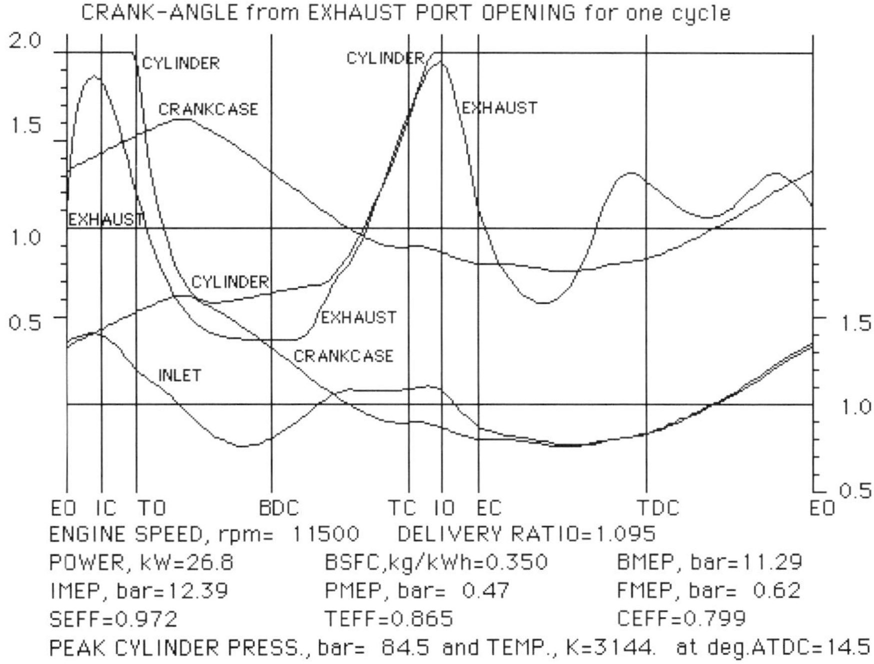

Fig. 5.23 The 125 cc Grand Prix engine on the eighth calculation cycle.

Perhaps the most important aspect of the foregoing demonstration is that Prog.5.2, "ENGINE MODEL No. 2," is shown to be capable of analyzing engines of the highest specific power output, of providing understanding of the complex pressure wave motion in the exhaust systems of such engines, and of being a theoretical tool which can be used for the design and development of every aspect of the geometry of these power units.

5.5 Computer modelling of multi-cylinder engines

Up to this point in the discussion, all of the engines which have been modelled have been single-cylinder engines, or those multi-cylinder engine units which have independent intake and exhaust systems. For example, all of the discussion in the previous section on the racing 125 Grand Prix engine is equally applicable to 250 cc twin-cylinder and 500 cc four-cylinder power units, for such motorcycle engines have completely separate intake and exhaust systems. The same separation of exhaust and intake ducting is also to be observed in engines for racing snowmobiles, hydroplanes, go-karts or remotely piloted vehicles (RPV). There is room in such vehicles for the bulk which is inherent in the expansion chamber design, although later discussion in Chapter 6 will demonstrate that some contraction of that bulk is possible. For certain applications, this exhaust system bulk on a multi-cylinder engine is either unacceptable, or the narrow speed band providing high power output is unusable in the particular application. A particular example would be a multi-cylinder outboard motor, either in two, three, in-line four, V4, V6 or V8 cylinder configurations, all of which are to be found in the marketplace(1.12)(4.10). A further example would be of an automobile, where past history has provided some case histories(3.5)(5.6) and the future may provide some more.

At some expense in terms of programming complexity and of computer run-times, there is nothing inherently difficult in programming any multi-cylinder configuration, for the fundamental theory of unsteady gas flow at a three-way branch has already been presented in Sect. 2.3.5. The extension of that theory to cover any number of branched pipes at a common junction is straightforward and is covered by Bingham(2.18) or Benson(1.4).

The main reason for discussing it further at this point in the text, is to illustrate that the main design information which has been disseminated thus far for single-cylinder engines, namely that exhaust pressure action which assists good trapping behavior without deteriorating the scavenge and intake flow will provide high power output and good specific fuel consumption, is also applicable and achievable for compact exhaust systems in certain multi-cylinder engine configurations. More importantly, such characteristics are predictable by a computer model, which permits the analysis of all possible combinations of engine cylinders and exhaust and intake manifolding.

Fig. 5.24 illustrates the result of a computer model of a three-cylinder in-line two-stroke engine(5.14). The engine data is exactly as that for the QUB 400 engine in Fig. 5.3. The exhaust manifold and pipe system is sketched in Fig. 5.24 and shows a pipe system which has parallel pipes, all of which are of a uniform diameter of 44 mm. The exhaust pipe length from the piston face at the exhaust port to the center of the first junction is 120 mm for each cylinder and the length of the common exhaust from that junction to the open pipe end in the silencer box is 240 mm. The outlet tail pipe is 90 mm long. The box volume is 10 liters, which is 8.3 times larger than the total swept volume of the engine at 1200 cc. The engine which is being modelled is a three-cylinder in-line outboard motor with a common compact manifold and downpipe to a volume above the gearcase followed by a short underwater outlet exhaust pipe(1.12).

The first part of the computer output from the model is in the same format as for Prog.5.1, and shows the open cycle in Fig. 5.24 from exhaust port opening to closure on the fifth calculation cycle. The presence of the plugging pulse is clearly seen, arriving along the manifold from another cylinder firing 120º later. The cylinder trapping pressure is elevated to 1.75 atm. Even though the box volume is only 8.3 times larger than the total swept volume, the presence of a suction reflection around bdc is evident and is contributing to the extraction of exhaust gas and giving assistance to the scavenge flow. The only difference between this pressure diagram and that from a good expansion chamber design is the magnitude of the suction pressure available around bdc, for the plugging pulse at trapping is equally as strong.

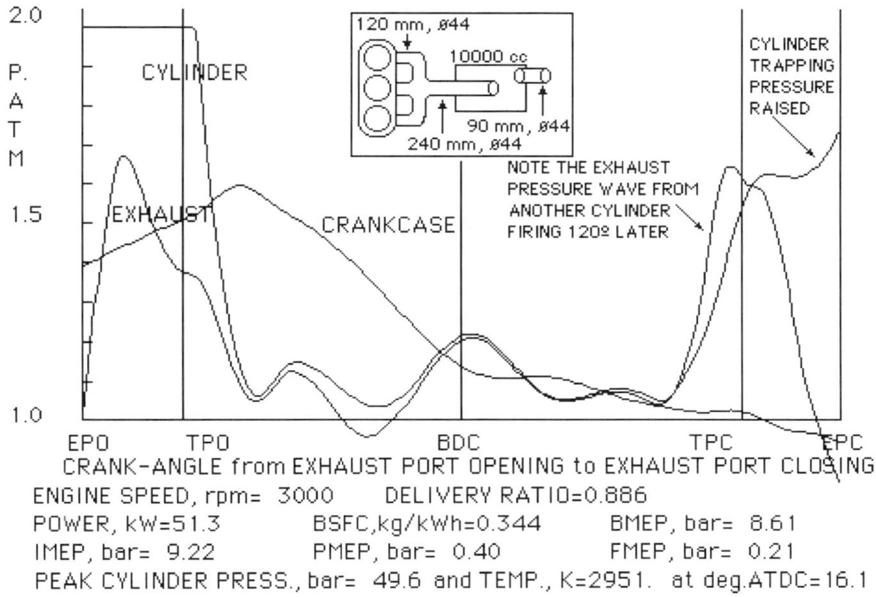

Fig. 5.24 The QUB 400 data used in a three-cylinder engine modelling program.

This form of exhaust port plugging, by utilizing an exhaust pulse from another cylinder blowing down at a convenient phasing in a compact manifold of a multi-cylinder engine, is referred to as "cross-charging" or "cross-stuffing."

As the engine speed is identical with that calculated for the single-cylinder QUB 400 engine data, it is informative to compare both the pressure diagrams and the performance characteristics from Fig. 5.9. The same engine data as a single-cylinder engine programmed in Prog.5.1 had power and fuel consumption characteristics of 6.22 bar and 0.462 kg/kWh. For a single-cylinder engine scavenged with air and fuel and an untuned exhaust system, that is a good full-throttle characteristic for both power and fuel economy. In three-cylinder form, the BMEP is increased to 8.61 bar and the specific fuel consumption is reduced to 0.344 kg/kWh. In other words, the power is up by 38% and BSFC is reduced by 26% from the same design of engine cylinder but incorporated into a three-cylinder formation. The air flow through the engine is also increased slightly, for the delivery ratio has gone up by 3%.

The three-cylinder in-line engine is ideal for this form of cross charging, for the cylinders fire at 120º intervals and exhaust port durations are usually between 160º and 180º in such multi-cylinder outboard motor or automobile engine designs. Thus, the cross-stuffing pulse arrives at the exhaust port perfectly phased to trap the cylinder charge. Furthermore, this tuning action is almost independent of speed so that the range of speed over which good torque and fuel economy are available is much wider than would be the case if the expansion chamber exhaust was attached to a single-cylinder engine, or to each cylinder of a multi-cylinder power unit.

As Flaig and Broughton(1.12) indicate for the OMC V8 300 hp outboard motor, it is possible to have good cross-stuffing characteristics within a four-cylinder engine or a V8 layout, where each cylinder fires at 90° intervals. In this case, as shown in Plate 5.2 for this very engine, it is necessary to have some path length available for the exhaust pulse between the relevant cylinders so that the phasing of the cross-charging process is optimized. This implies that the usable speed range would not be as wide as that of a three-cylinder, or V6, two-stroke engine. The short extension exhaust pipe, leading from the branched junction to the four cylinders of each bank of the Vee into the generous volume of exhaust box above the gearcase of the outboard, is clearly visible in the photograph.

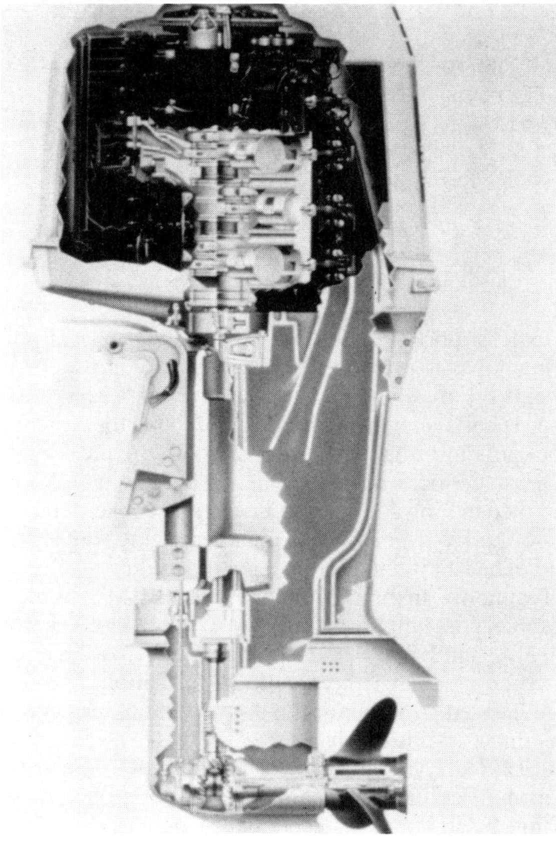

Plate 5.2 A cut-away view of a 300 hp V8 outboard motor (courtesy of Outboard Marine Corporation).

NOTATION FOR CHAPTER 5

NAME	SYMBOL	SI UNITS	OTHER UNITS
Air-Fuel Ratio	AF		
Charging Efficiency	CE		
Crankcase Compression Ratio	CCR		
Delivery Ratio	DR		
Geometric Compression Ratio	GCR		
Index of compression	nc		
Index of expansion	ne		
Pressure Ratio	PR		
Ratio of Specific Heats	g		
Scavenge Ratio	SR		
Scavenge Ratio by volume	SRv		
Scavenging Efficiency	SE		
Trapped Air-Fuel Ratio	TAF		
Trapped Compression Ratio	TCR		
Trapping Efficiency	TE		
Area	F	m^2	
Brake Mean Effective Pressure	BMEP	Pa	bar, atm, psi
Brake Specific Fuel Consumption	BSFC	kg/Ws	kg/kWh,lb/hp.hr
Clearance Volume	CV	m^3	cm^3
Cylinder Bore	BO	m	mm, in
Cylinder Stroke	ST	m	mm, in
Density	D	kg/m^3	g/cc, lb/in^3
Engine Rotation Rate	RPS	rev/s	rev/min, rpm
Friction Mean Effective Pressure	FMEP	Pa	bar, atm, psi
Gas Constant	R	J/kgK	
Indicated Mean Effective Pressure	IMEP	Pa	bar, atm, psi
Particle Velocity	CG	m/s	
Power	BPR	W	kW, bhp
Pressure	P	Pa	bar, atm, psi
Pumping Mean Effective Pressure	PMEP	Pa	bar, atm, psi
Swept Volume	SV	m^3	cm^3, in^3
Temperature	T	K	ºC, ºF
Torque	BTRQ	Nm	Nm, ft.lbf
Trapped Swept Volume	TSV	m^3	cm^3
Trapped Charge Mass	Mtr	kg	
Trapped Air Mass	Mtas	kg	
Trapped Fuel Mass	Mtf	kg	
Volume	V	m^3	

REFERENCES FOR CHAPTER 5

5.1 F. Nagao, Y. Shimamoto, "Effect of the Crankcase Volume and the Inlet System on the Delivery Ratio of Two-Stroke Cycle Engines," SAE Automotive Congress, Detroit, Michigan, January, 1967, SAE Paper No. 670030.

5.2 G.P. Blair, "Gas Flow Techniques Applied to a Sleeve-Valve Engine Design," *Automotive Design Engineering*, Vol. 9, April, 1970.

5.3 G.P. Blair, M.B. Johnston, "The Development of a High Performance Motor Cycle Engine," *Proc.I. Mech.E.*, 1970-1971, Vol. 185, 20/71, pp. 273-283.

5.4 G.P. Blair, W.L. Cahoon, "The Design and Initial Development of a High Specific Output, 500cc Single Cylinder, Two-Stroke, Racing Motorcycle Engine," SAE Automotive Congress, Detroit, Michigan, January, 1971, SAE Paper No. 710082.

5.5 G.P. Blair, "Further Development of a 500cc Single Cylinder Two-Cycle Engine for Motorcycle Racing and Motocross Applications," SAE Farm, Construction and Industrial Machinery Meeting, Milwaukee, Wisconsin, September, 1974, SAE Paper No. 740745.

5.6 R. Mellde, J. Eklund, K. Knutsson, "Use of the Two-Stroke Cycle Engine as a Power Unit for a Passenger Car," *SAE Transactions*, Vol. 74, SAE Paper No. 650008.

5.7 V. Sathe, P.S. Myers, O.A. Uyehara, "Parametric Studies using a Two-Stroke Engine Cycle Simulation," SAE Automotive Congress, Detroit, Michigan, January, 1970, SAE Paper No. 700124.

5.8 R.B. Krieger, R.R. Booy, P.S. Myers, O.A. Uyehara, "Simulation of a Crankcase Scavenged Two-Stroke, SI Engine and Comparison with Experimental Data," SAE Automotive Congress, Detroit, Michigan, January, 1969, SAE Paper No. 690135.

5.9 G.P. Blair, E.T. Hinds, R. Fleck, "Predicting the Performance Characteristics of Two-Cycle Engines Fitted with Reed Induction Valves," SAE International Off-Highway Vehicle Meeting, Milwaukee, Wisconsin, September, 1979, SAE Paper No. 790842.

5.10 D.S. Sanborn, "Paper Powerplants Promote Performance Progress," SAE International Automotive Congress, Detroit, Michigan, February, 1975, SAE Paper No. 750016.

5.11 G.J. Van Wylen, R.E. Sonntag, Fundamentals of Classical Thermodynamics, Wiley, 2nd Edition, SI Version, New York, 1976.

5.12 K.J. Patton, R.G. Nitschke, J.B. Heywood, "Develpoment and Evaluation of a Friction Model for Spark-Ignition Engines," SAE International Automotive Congress, Detroit, Michigan, February, 1989, SAE Paper No. 890836.

5.13 H. Naji, R. Said, R.P. Borghi, "Towards a Turbulent Combustion Model for Spark-Ignition Engines," SAE International Automotive Congress, Detroit, Michigan, February, 1989, SAE Paper No. 890672.

5.14 K. Landfahrer, D. Plohberger, H. Alten, L.A. Mikulic, "Thermodynamic Analysis and Optimization of Two-Stroke Gasoline Engines," SAE International Automotive Congress, Detroit, Michigan, February, 1989, SAE Paper No. 890415.

Chapter 6

Empirical Assistance for the Designer

6.0 Introduction

In the first five chapters of this book, the reader is introduced to the two-stroke engine, to scavenging, unsteady gas flow, combustion, and then to the combination of all of the preceding topics into computer modelling of a complete engine. Having just finished reading Chapter 5, the reader can be forgiven for feeling somewhat bemused. In that chapter, the reader is assured that modelling reveals the "inside story" on engine behavior, and hopefully that is how it is presented. At the same time the reader will be conscious that, for the engine models presented in that chapter to be of practical use in a design mode, the designer will to have to produce real input data in the form of the banks of numbers shown in Figs. 5.3 and 5.4.

Some readers may feel that it will be as simple as opening the program for ENGINE MODEL NO.1, inserting the numbers for the engine and geometry as the mood takes them, and letting the computer inform them of the outcome regarding the pressure wave action and the performance characteristics. Of course that is an option open to the designer, but the author, and possibly also the designer's employer, would not consider it as the most rapid approach to optimize a design to meet a particular need. There are 48 different data numbers which can be changed in those engine models for a single calculation, and the statistical probability of achieving an optimized engine design by an inspired selection of those numbers is about on a par with successfully betting on racehorses. Other readers may feel so overwhelmed at the range of choices available for the data value of any one parameter, that they could spend more time trying to estimate a sensible data value than using the engine model for the predictive task.

What is required is, hopefully, what is to be presented in this chapter, namely some empirical guidance for the data values to be used in the engine models in particular and engine design in general. There are the academic purists who will react adversely to the revelation that this textbook, written by a university Professor and which can be read by undergraduate students, will contain empirical guidance and advice. The contrary view is that the design and creation of engineering artifacts involves the making of decisions which are based just as much on past experience as they are in theoretical analyses. For those whose past experience is limited, such as undergraduate engineering students, they require that advice and guidance, both in lectures and from textbooks, from those with that know-how.

The empirical guidance to be given will range from the simplistic, such as the maximum port width that the designer can select without encountering piston ring breakage, to the complex, such as the design of the entire geometry of an expansion chamber to suit a given engine. Actually, what will be regarded as simplistic by some will be considered by others as inconceivable. To quote the apparently simple example above, the experienced engineer in industry will have little problem in deciding that each exhaust port maximum width could be as large as 70% of the cylinder bore dimension, provided that suitably large corner and top edge radii are used to compress the piston ring inside the timing edges. On the other hand, the archetypical undergraduate university student will not have that past experience to fall back on as empirical guidance and has been known to spend inordinate amounts of time peering at the computer screen showing Prog.5.1, unable to make a decision.

There exists the danger that the empirical guidance to be given will be regarded as the final design, the "quick-fix formula" so beloved by a human race ever desirous of cutting corners to get to a solution of a problem. Let a word of caution be sounded in that case, for no simplistic approach will ever totally provide an exact answer to a design question. A good example of that is the use of the heat release model of combustion within the engine modelling programs presented in Chapter 5. While the quality of the answer is acceptable in design terms today, the thinking engineer will realize the limitations imposed by such an approach and strive to incorporate ever more sophisticated models of combustion into those engine design programs, particularly as CFD theory becomes more practical and desk-top computers calculate ever faster.

The empirical advice will incorporate matters relating to the engine modelling programs of Chapter 5 and to designs for which those programs do not cater, such as reed valves and disc valves for the control of the induction process. The following is a list of the topics to be covered:

(a) The estimation of the data for porting characteristics to meet a design requirement for a given engine cylinder at a given engine speed to produce a particular power output.

(b) The translation of the data in (a) into port timings, widths, heights and areas for a particular engine geometry.

(c) The estimation of the data for the geometry of an expansion chamber exhaust system for a high specific output engine so that it is tuned at a desired engine speed and matched to the flow requirements of the engine.

(d) The preliminary design of reed valve and disc valve induction systems for two-stroke engines.

6.1 Design of engine porting to meet a given performance characteristic

The opening part of this section will discuss the relationship between port areas and the ensuing gas flow, leading to the concept of Specific Time Area as a means of empirically assessing the potential for that porting to produce specific performance characteristics from an engine. The second part will deal with the specific time area analysis of ports in engines of a specified geometry and the predictive information to be gleaned from that study. The third part will examine the results of that study by its analysis through an engine modelling program.

6.1.1 Specific time areas of ports in two-stroke engines

Fig. 6.1 is a graphical representation of the port areas as a function of crank angle in a two-stroke engine. The actual data presented is for the chainsaw engine design discussed in Chapter 5 and the geometrical data given in Fig. 5.3. On the grounds that engineers visualize geometry, that same cylinder is presented in Fig. 6.2. Therefore, it is the movement of the piston past those ports, be they exhaust, transfer or inlet, as the crankshaft turns which produces the area opening and closing relationships presented in Fig. 6.1.

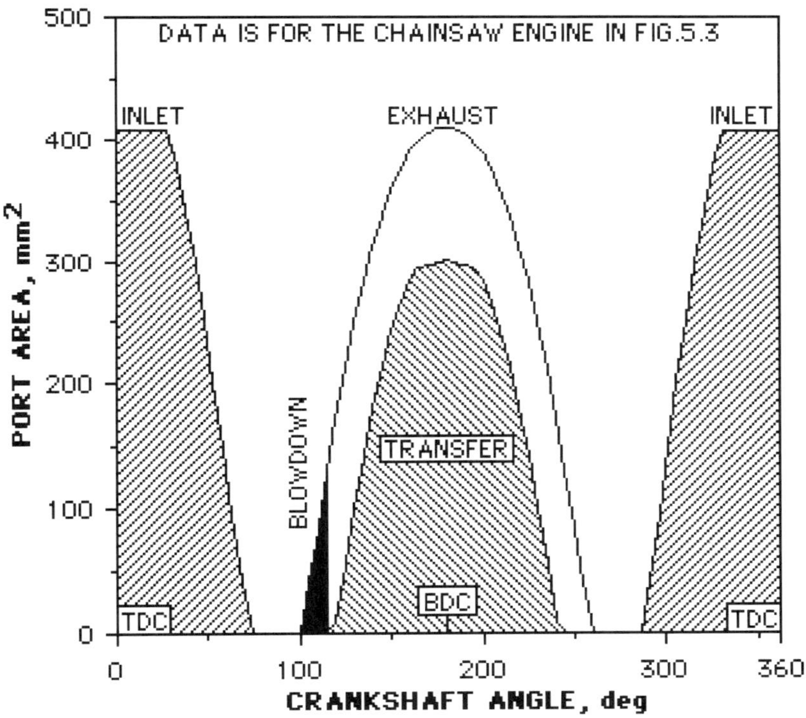

Fig. 6.1 Port areas by piston control in a two-stroke engine.

Some of the numbers generated are of general interest. Note that the maximum area of the inlet port and the exhaust port are about equal. The transfer area is clearly less than the inlet and exhaust areas. The area reached by the exhaust port at the point of transfer port opening, i.e., at the official end of the blowdown period, is about one quarter of the maximum area attained by the exhaust port. The flat top to the inlet area profile is due to that particular design being fully uncovered for a period of 52º around tdc. The symmetry of the port area diagrams around the tdc or bdc positions

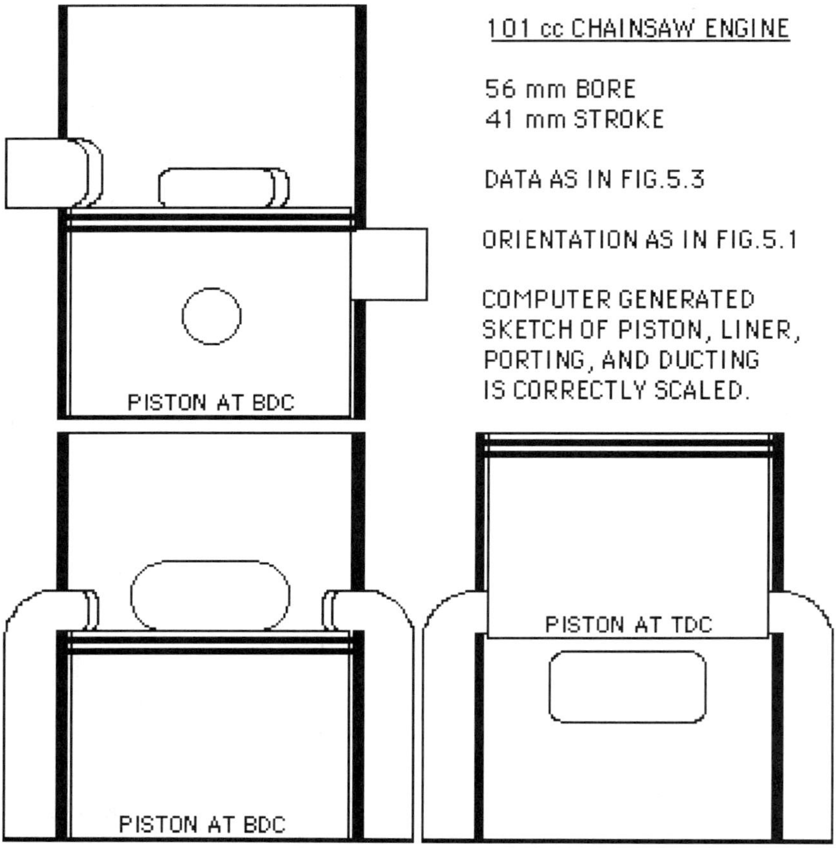

Fig. 6.2 Computer generated picture of chainsaw engine cylinder.

is created by the piston control of all gas flow access to the cylinder; this would not be the case for ports controlled by a disc valve, reed valve or poppet valve mechanism (see Sect. 1.3).

It has already been demonstrated that the performance characteristics of an engine are related to the mass flow rates of gas travelling through the ports of the engine. For example, in Sect. 5.2.3, the chainsaw engine is shown to produce more power because the delivery ratio is increased by improving the pumping action of the crankcase. The following theory for gas flow through a port or valve is applicable to that process. Let A be the instantaneous area of an open port at any crank angle, let CG be the particle velocity through it, and D be the density of the gas at that instant. The continuity equation (see Sect. 2.1.4) dictates that the mass rate of flow, M, is:

$$M=D*A*CG \tag{6.1.1}$$

If this situation persists for a small time interval, dt, the increment of flow to pass through would be:

$$\Delta M=M*dt$$

The total gas flow, MG, to pass through the port is the summation of all flow increments for the entire duration of the port opening, either t in time units or ø in crankshaft angle units. This is written mathematically as:

$$\begin{aligned} MG &= \Sigma(\Delta M) \\ &= \Sigma[M*dt] \\ &= \Sigma[D*A*CG*dt] \\ &= \Sigma[D*A*d\phi*CG*(dt/d\phi)] \end{aligned} \tag{6.1.2}$$

This equation applies to all ports and to all gas flow through the engine. Within the engine and gas-dynamic models of Chapters 2 and 5, this summation exercise is carried out at each calculation step so that the values of delivery ratio, scavenge ratio, etc., are ultimately determined.

As most two-stroke engines have the symmetrical porting geometry shown in Fig. 6.1, and if the variations of density and gas velocity with time are somewhat similar from engine to engine, then there might exist some mathematical function linking the remaining terms to the gas flow transmitted. Perhaps more important, as the gas flow which is transmitted is related to that which is trapped and burned with fuel to produce power, the relationship could extend to a more direct linkage to engine performance.

For example, the application of Eq. 6.1.2 to the inlet port within an engine model predicts the total air mass ingested into the engine on each cycle. This must have a direct relationship to the work output from that cycle when that air is burned with fuel in the correct proportions. This can be seen in Fig. 5.16 from the close correspondence between the delivery ratio and BMEP with engine speed for the two versions of the chainsaw engine.

The net work output per cycle, from Fig. 1.16, is given by:

$$\text{Work/cycle}=BMEP*SV$$

If the relationship is linear, then the division of air mass ingested by work output should be "a constant":

$$\{\Sigma[D*CG*(A*dt)]\}/\{BMEP*SV\}=\text{“a constant”} \tag{6.1.3}$$

On the assumption that the variations of D and CG are somewhat similar, even for different engines, incorporating these so-called constants into the right-hand side of Eq. 6.1.3 reduces it to:

$$\Sigma[(A*dt)/SV]/(BMEP)=\text{"a constant"} \qquad (6.1.4)$$

This can also be written as:

$$\Sigma[(A*d\phi)*(dt/d\phi)/SV]/(BMEP)=\text{"a constant"} \qquad (6.1.5)$$

The top line of the left-hand side of Eq. 6.1.4 is known as the "time-integral of area per unit swept volume" for the port of a two-stroke engine. This expression is often shortened colloquially to Specific Time Area. The units of "time-integral of area per unit swept volume," or specific time area, can be seen to be (m^2*s/m^3), or s/m.

The evaluation of this expression is best seen from the variation in Eq. 6.1.5. The summation of the term (A*dø) is the port area shown in Fig. 6.1 and is the shaded area for the inlet port or for any of the ports on that diagram. The term, dt/dø, is the fixed relationship between crankshaft angle and time given by:

$$\begin{aligned} dt/d\phi &= \text{time per crankshaft angle} \\ &= (\text{time per cycle})/(\text{crankshaft degrees per cycle}) \\ &= 1/(RPS*360) \end{aligned} \qquad (6.1.6)$$

On the assumption that the mass of gas which flows into an engine on each cycle must also flow through it and ultimately out of it, the evaluation of "time-integral of area per unit swept volume" for inlet ports, transfer ports and exhaust ports can be carried out and an attempt made to relate that data to the BMEP produced by an engine of a known swept volume, SV, at a known engine speed, RPS. That a relationship exists for mass flow equality at all ports has already been stated at the end of Sect. 5.2.1, where it is shown to be a calculation requirement for precise equality between the delivery ratio at the inlet port, the scavenge ratio at the transfer ports and the exhaust flow ratio at the exhaust port.

Another important flow parameter can be evaluted by calculating the "time-integral of area per unit swept volume" for the blowdown period between the exhaust port opening and the transfer port opening. For the cylinder pressure to fall from the release point to a value approaching that in the scavenge ducts in order to avoid excessive backflow into the transfer ports, a certain proportion of the cylinder mass must pass out of the exhaust port during that period. The higher the BMEP in a given engine, the greater will be the trapped charge mass and pressure, and therefore, so will be the need for a larger specific time area to let the cylinder pressure fall appropriately during the blowdown phase of the exhaust process. The blowdown value of (A*dø) is indicated in Fig. 6.1.

At QUB over the years a considerable volume of data has been collected from engine design and testing, not only from in-house designs but from production engines as part of R and D programs. A small fraction of this "expert system" data is illustrated in Figs. 6.3-6.6 for the specific time areas related to brake mean effective pressure. On those diagrams, the engine cylinder sizes range from 50 cc to 500 cc, the speeds at which the performance data was recorded range from 3000 to 12000 rpm, and the BMEP spread is from 3.5 bar to 11 bar. The graphical scatter is, not unexpectedly, considerable. The reasons are fairly self-evident, ranging from engines which have over- or under-designed ports, to engines with exhaust systems attached in varying degrees of tuning level from the Grand Prix racing type to that for a road-going motorcycle. Further complications arise, such as in one of the illustrations in Chapter 5 where the chainsaw engine produced two different BMEP levels with a common porting and exhaust system but a different crankcase pumping action; this factor would not emerge in any analysis of the specific time area of that chainsaw engine. Despite all of that, in each of these figures, trends for a BMEP-specific time area relationship can be seen to exist and a straight line opinion as to the best fit is drawn on each diagram; in the case of the transfer ports a band of a higher and a lower level is shown as the most informed view of that relationship.

The relationships from those diagrams are rewritten below.

For inlet ports, where the specific time area is labelled as INTA:

BMEP, bar=774*INTA-1.528 s/m (6.1.7)

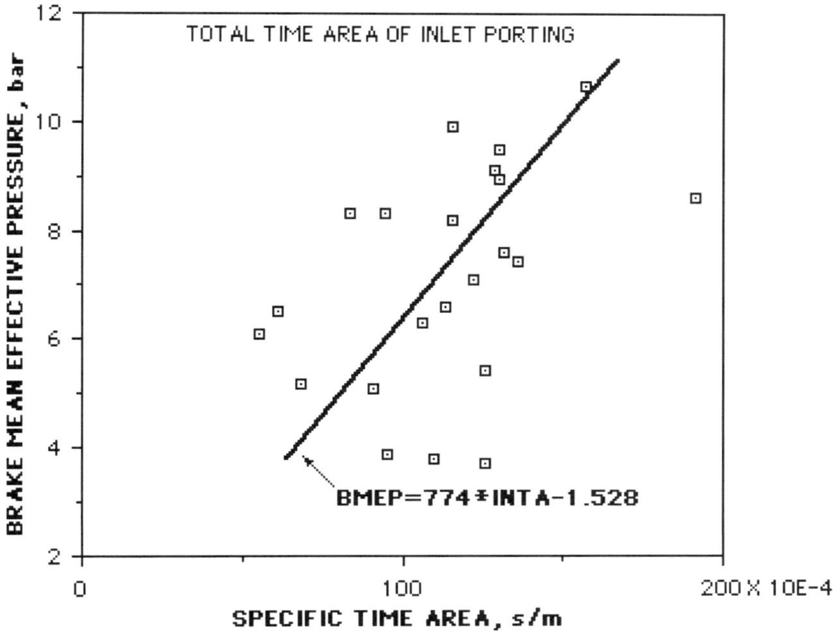

Fig. 6.3 Specific time areas for inlet ports from measured data.

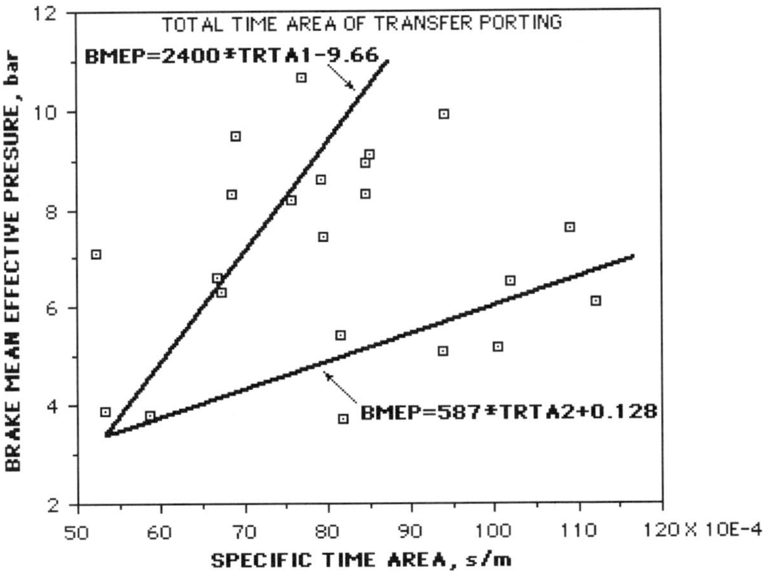

Fig. 6.4 Specific time areas for transfer ports from measured data.

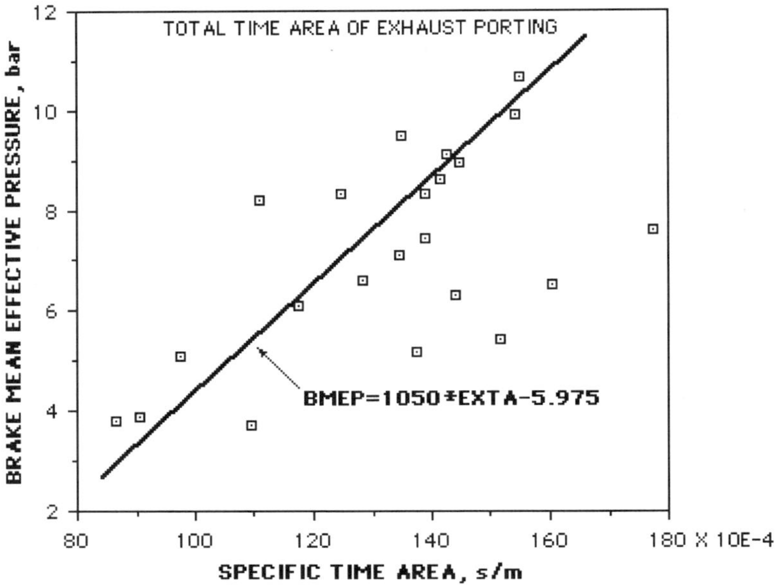

Fig. 6.5 Specific time areas for exhaust ports from measured data.

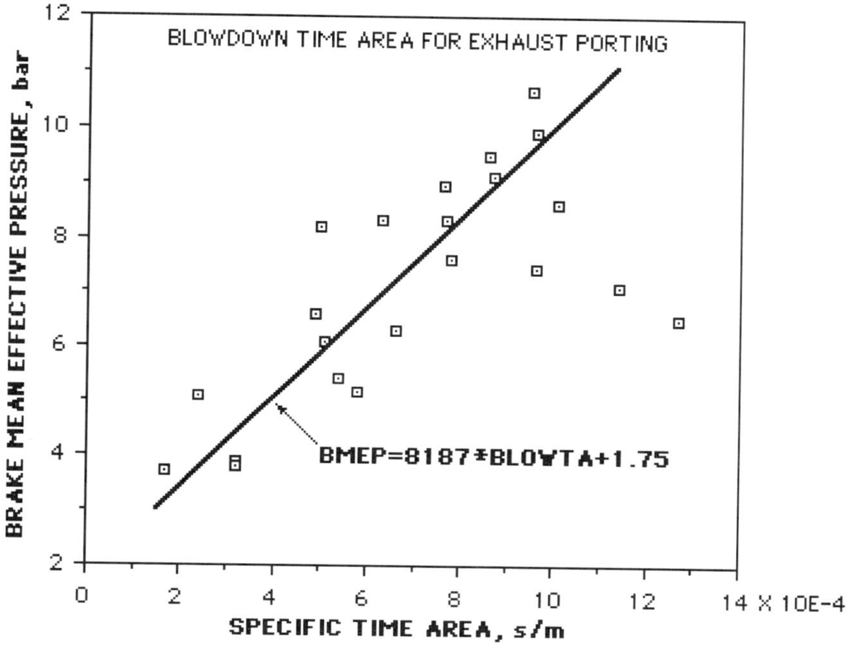

Fig. 6.6 Specific time areas for blowdown from measured data.

For transfer ports, where the higher level of specific time area is labelled as TRTA1:

BMEP, bar=2400*TRTA1-9.66 s/m (6.1.8)

For transfer ports, where the lower level of specific time area is labelled as TRTA2:

BMEP, bar=587*TRTA2+0.128 s/m (6.1.9)

For exhaust ports, where the specific time area is labelled as EXTA:

BMEP, bar=1050*EXTA-5.975 s/m (6.1.10)

For exhaust blowdown, where the specific time area is labelled as BLOWTA:

BMEP, bar=8187*BLOWTA+1.75 s/m (6.1.11)

However, while these functions are easily solved using a pocket calculator, a simple computer program is given in the Computer Program Appendix as Prog.6.1, "TIMEAREA TARGETS." For those not very familiar with programming in Basic this straightforward computer program will serve as another useful example of data insertion, simple calculation, and data presentation on both the computer screen and the printer.

As an example of the use of this program and of the analysis represented by Eqs. 6.1.7-6.1.11, consider the case of the two engines studied in Chapter 5, the chainsaw engine and the Grand Prix engine. These engines are at opposite ends of the BMEP spectrum. Imagine that a design is to be formulated for two new engines to satisfy the very criteria which are known to be attainable by virtue of the engine modelling analysis carried out in Chapter 5 on these very same engines. The design brief might read as follows:

(a) A chainsaw engine is needed to produce 5.4 kW at 8000 rpm, with a BMEP of 4 bar as a potentially obtainable target. That this would necessitate an engine of 101 cc swept volume is found from Eq. 1.6.6 in Sect. 1.6.1.

(b) A Grand Prix engine is needed to compete in the 125 cc World Championships, where 26.5 kW is required to be competitive (as of 1988!). For mechanical reasons, it is decided to try to produce this power at 11500 rpm. Using Eq. 1.6.6, this translates into the production of 11 bar BMEP.

Upon inserting the information on the BMEP level for each engine into Eqs. 6.1.7-6.1.11, or using Prog.6.1, the following information on specific time area is obtained for each engine. All units are s/m.

Specific time area requirements predicted by Prog.6.1

Parameter	EXTA	BLOWTA	TRTA1	TRTA2	INTA
Chainsaw engine, 8000 rpm	0.0095	0.00027	0.0057	0.0066	0.0071
Grand Prix engine, 11500 rpm	0.0162	0.00113	0.0086	0.0185	0.0162

It will be noted that the specific time area requirements for the porting of the Grand Prix engine are much larger than for the chainsaw engine. Although the cylinder sizes are only 25% different, the larger cylinder of the Grand Prix engine is expected to produce 2.75 times more BMEP at an engine speed that has only 70% of the time available for filling it. To assist with the visualization of what that may imply in terms of porting characteristics, a computer generated sketch of the actual cylinder of the 125 Grand Prix engine is produced from the data bank in Fig. 5.3; this is shown in Fig. 6.7. A comparison with the similar sketch for the chainsaw engine in Fig. 6.2 reveals the considerable physical differences in both port timings and area. It is clear that the large ports needed for a Grand Prix engine are open much longer than they are closed.

The next step is to analyze the porting characteristics of these two engines given in the data bank in Fig. 5.3, and to determine if the criteria which are noted above have any relevance to those which are known to exist.

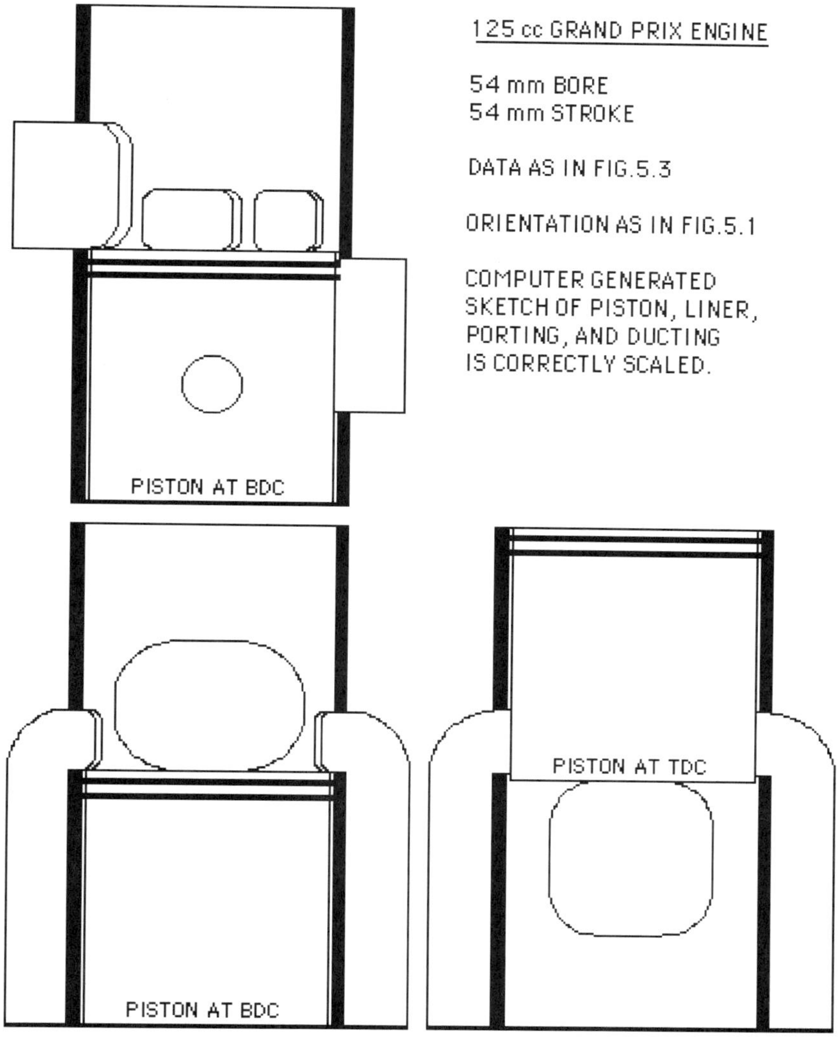

Fig. 6.7 Computer generated picture of 125 Grand Prix engine cylinder.

6.1.2 The determination of specific time area of engine porting

Discussion of the calculation of port areas in a two-stroke engine has been a recurring theme throughout this book, most recently in Sect. 5.1.1 for the engine modelling computer programs. The calculation procedure for specific time area is encapsulated within any program which will compute the area of any port in an engine as a function of crankshaft angle or time. The value required is for the

∫(A*dø), i.e., the shaded areas in Fig. 6.1 for inlet, transfer, exhaust or blowdown. As it is unlikely that the ∫(A*dø) will be determined by a direct mathematical solution due to the complexity of the relationship between the instantaneous value of A as a function of ø, the computer solution by summation at crankshaft intervals of one or two degrees, i.e., Σ(A*dø), will provide adequate numerical accuracy. When the Σ(A*dø) is determined, that value is inserted into the appropriately combined Eqs. 6.1.5 and 6.1.6 as follows:

Specific time area, s/m= Σ[A*dø]/(RPS*360*SV) (6.1.12)

Those who wish to use this equation should note that the units are strictly SI, and that the area A is in m^2 and SV in m^3. By sheer coincidence, one can enter the value of A in mm^2 and SV in cc and the units are self-correcting to s/m.

The data required to perform such a calculation is the physical geometry of the piston, connecting rod and crankshaft and the porting for the engine. This data is typically provided by the first four blocks of numbers in Fig. 5.3. That data is sufficient to produce Fig. 6.1, Fig. 6.2 and Fig. 6.7 by a computer program, which is demonstrably in command of all of the necessary information. To assist with the deduction of specific time area values for proposed or existing engines, a computer program is given in the Computer Program Appendix as Prog.6.3, "TIME-AREAS." The data insertion format for the various blocks of data is identical with that for either of the engine models, Progs.5.1 or 5.2, so it is relatively easy to use in practice. The program output is in the form of screen or line-printer output without graphics, so the following is a precis of the use of Prog.6.3 for the analysis of the chainsaw and the Grand Prix engine geometry presented in Fig. 5.3.

Specific time area of the engine data in Fig.5.3

Specific time area, s/m	Exhaust	Blowdown	Transfer	Inlet
Chainsaw engine at 8000 rpm	0.00907	0.00032	0.00534	0.00965
Grand Prix motor at 11500 rpm	0.01549	0.00095	0.00776	0.0157

It can be seen that the specific time area values from the data bank of Fig. 5.3 for the actual engines are very similar to those from the table of data predicted as being necessary to produce the required performance characteristics. In other words, the specific time areas predicted as requirements and as examples are sufficiently close to give some confidence in the predictive tool as an initial design step. There are some differences which will be debated below with the designer's options and methodologies when tackling the initial steps in the design process.

The topic of specific time area is not mentioned very frequently in technical papers. However, a useful reference in this context, which would agree with much of the above discussion, is that by Naitoh and Nomura(6.5)

6.2 The use of the empirical approach in the design process

Each relevant topic will be discussed in turn, and examples of the approach adopted for the solutions will be oriented towards the practical design problems raised by the Grand Prix and chainsaw power units.

6.2.1 The use of specific time area information

Consider first the Grand Prix power unit. The actual exhaust specific time area value is almost identical with the required value. Should it be decided to raise the exhaust specific time area from the actual 0.01549 to the "required" 0.0162, it would be unwise to do it by widening the exhaust port or by reducing the radii in the corners. The exhaust port is already dangerously wide from the point of view of the mechanical reliability of the piston ring passing by the timing edge of the exhaust port. Should the ring strike it, catastrophic damage to the engine would result. Therefore, one could raise the exhaust timing by 1.5 degrees to 79.5 opening after tdc, which would produce the necessary increase of exhaust time area. This would also increase the blowdown specific time area from the actual 0.00095 to the required 0.00113. The end result on engine performance could then be assessed by running Prog.5.2, or by modifying a real engine cylinder and experimentally determining the outcome.

A similar argument can be mounted for the transfer ports in the Grand Prix motor, where the requirement as predicted by the empirical Prog.6.1 is for a specific time area of 0.0086 and the actual value for the Fig. 5.3 data is 0.00776. This could be "improved" by retaining the transfer timing at a 114 deg. atdc opening timing and increasing the total effective transfer port width by 8 mm, or 12.5%, from 60 to 68 mm. This would have the dual advantage of increasing the transfer time area without affecting the blowdown time area. If a designer elects to change the transfer timing, it is vital to remember that this always produces a consequential alteration to the blowdown behavior.

The inlet system of the Grand Prix engine is quite closely matched at a required time area of 0.0162 s/m to an actual value of 0.0157. Precise matching could be accomplished by increasing the inlet timing by the relatively minor change of 1º. Perhaps the most important aspect of this discussion is that the reader will realize how closely the majority of the engine geometry shown in Fig. 5.3 could have been created by the combined use of Progs.6.1 and 6.3, allied to the judicious use of some experience.

6.2.2 The acquisition of the basic engine dimensions

Imagine the experienced designer faced with the task of creating the data shown in Fig. 5.3. The design criteria for swept volume, power and speed have already been declared for both engines.

For various mechanical reasons, the "square" engine dimensions of 54 mm bore and stroke have been decided for the Grand Prix engine, although a bore-stroke ratio of up to 1.2 could also have been adopted without an excessive penalty on port areas at equal port timings, and the engine would probably operate more easily with the lower piston speed. The decision on the connecting rod length is fairly straightforward: it should be as short as possible to assist with the creation of a high crankcase compression ratio. In practice, the mechanical limitations of flywheel diameter and crankpin diameter necessary to withstand the mechanical loadings, and the necessity to have the piston clear the flywheel at bdc, will dictate a connecting rod center distance between 1.85 and 2.05 times longer than the stroke length. Let it be assumed that the 105 mm length currently selected is eminently practical. In any event, the minimum value of 100 mm, or the maximum of 110 mm, would have very little effect on the ensuing time area values.

For the chainsaw engine, the oversquare layout of 56 mm bore and 41 mm stroke is selected because this is a hand-held power tool and the minimum mechanical vibration level is not just desirable, it is a legislative requirement in many countries. Mechanical vibration is a function of the out-of-balance forces in an engine and are related to the inertial forces. Where Mb is the out-of-balance mass of the piston, connecting rod, etc., the maximum value of such forces, F, are given by:

$$F=Mb*ST*RPS^2$$

It can be seen that this is apparently a linear function of the engine stroke, ST, and so the lowest possible stroke length is used to minimize the vibration felt by the operator. Thus bore-stroke ratios of up to 1.4 are quite common in the design of hand-held power tools, even though the potential BMEP attainment of the engine is compromised by the reduced port areas available for any given set of port timings. As an alleviating factor, the highest crankcase compression ratio will always be produced by the largest bore-stroke ratio, and that has already been shown in Chapter 5 to be a vital factor in the design of this type of motor.

Where there are no overriding considerations of the type discussed above, the two-stroke engine designer should opt for a bore-stroke ratio of between 1.0 and 1.2. However, for diesel engines, where combustion is totally important, bore-stroke ratios are conventionally at 0.9 for high-speed engines and can be at 0.5 for low-speed marine engines. In the latter special case, the values of bore and stroke are gross by comparison with the chainsaw engine, as typically the bore is 900 mm and the stroke is 1800 mm!

6.2.3 The width criteria for the porting

The designers of high-performance two-stroke engines tend to favor the use of single exhaust and inlet ports, if at all possible. Although a divided port for exhaust or inlet is seen in many designs, it is always a highly stressed feature and the lubrication of the hot, narrow exhaust bridge is never an easy proposition. Furthermore, as the need for large ports with long timings is self-evident for a high specific output engine, from Fig. 6.7 and the required specific time area criterion, every mm of bore circumference around the bdc position is needed for ports and not port bridges, so that the port opening periods can be kept as short as possible.

The designer of the Grand Prix motor will opt for a single exhaust port whose width is 75% of the bore dimension, and insert large corner radii to compress the piston ring into its groove before it passes the timing edge on the compression stroke. The net physical effect of the data chosen is seen in Fig. 6.7. The top edge itself can also be radiused to enhance this effect.

In the case of the chainsaw engine of Fig. 6.2, and other such motors of a more modest BMEP and power requirement, where the designer is not under the same pressure for effective port width at the bdc position, the typical criterion for the effective exhaust port width is to employ 55% of the bore dimension. For many such engines, a long and reliable engine life is an important feature of the design, so the ports are made to ensure a smooth passage of the rings over the port timing edges. See Chapter 8 for further comments on the design of this port for noise considerations.

The inlet port width of the racing engine is deduced in much the same manner, with the port width to bore criterion for a single port being about 65% of the bore dimension. Often this inlet port width conveniently coincides with the carburetor flow diameter, so the designer allows that dimension to be used, permitting a straight gas flow path for two sides of the intake duct. For the racing engine design with a wide single inlet port, the corner radii of the port need to be very generous around the bdc period by the upper timing edge of that port if a piston ring is not to be trapped. As for the exhaust port, Chapter 8 includes further comment on noise generation from the induction system with respect to this port design.

The transfer port width for calculation purposes can be decided relatively easily. The geometrical layout can be effected by using Prog.3.4, although the experienced designer will know that it is mechanically possible to pack in a total effective transfer port width of between 1.2 and 1.35 times the bore dimension in most loop scavenged designs. The original design of the Grand Prix engine in the data bank of Fig. 5.3 is for a 1.2 ratio, i.e., four transfer ports giving a total effective width of 60 mm. There is discussion in Sect. 5.1.1 regarding the mathematical insertion of that data as four ports each of 15 mm width even though the ports are actually not of equal width. The designer, experienced or not, will wish to determine the port width criteria in some detail, and for this tedious exercise Prog.3.4 is invaluable. Fig. 6.8 shows the result of a few moments of time spent at the computer screen, confirming that the total port width selected at 60 mm is mechanically acceptable and that a potential width of 68 mm is available with a little attention to design detail. It will also be observed that the rear transfer ducts must curve around the inlet port, for the back bar is only 25 mm wide and the inlet port is 35 mm across at the widest point. Because the width of the ports are proportioned in the ratio of 5:3 in favor of the main transfer port, the actual width of the main port is relatively narrow and cannot trap a piston ring; the area can be maximized by inserting small corner radii, hence the selection of the 3 mm corner radii in the data bank in Fig. 5.3. During the operation of Prog.3.4, the user should always bear in mind the advice given in Sect. 3.5.4.1 regarding the orientation of the main transfer ports.

6.2.4 The port timing criteria for the engine

Having determined the port widths which are available for the various ports, the designer will then index Prog.6.3 to calculate the specific time areas of those ports. Naturally, the reader has the option of writing a personal numerical solution to the theory presented above for the derivation of specific time area. The values for port timings are estimated and simply inserted as data into Prog.6.3, then subsequently modified until the values for the deduced specific time area match those predicted as being required by Prog.6.1 to provide the requisite performance characteristics. The program also automatically produces the port height and area information for all ports. This is also helpful, for the inlet port maximum area is a very useful guide to carburetor size, i.e., its flow diameter. As an approximate guide, if one matches

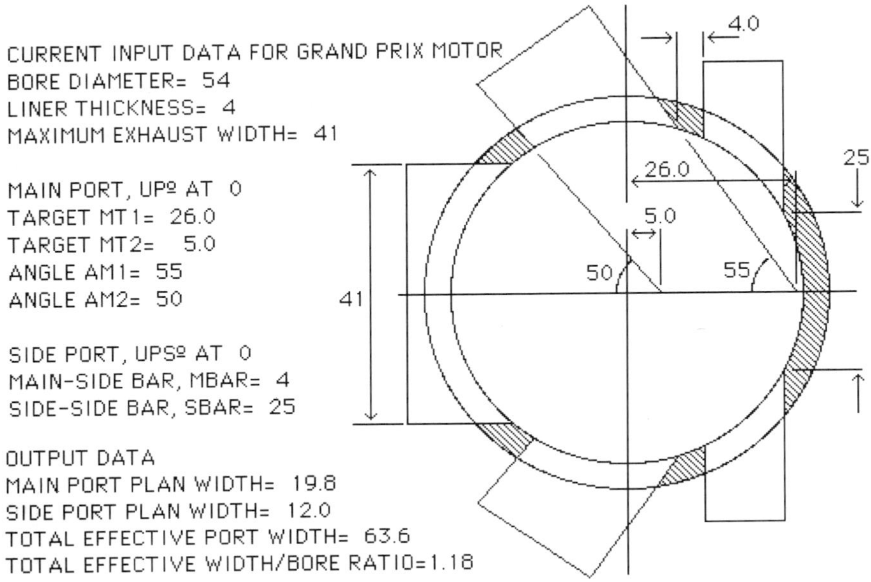

Fig. 6.8 Scavenge port geometry for a 125 Grand Prix engine used as a design example.

ports in this manner, the carburetor flow area, i.e., that shown as DIAMETER.I of Fig. 5.1, can be set to 85% of the inlet port flow area.

In the case of the Grand Prix engine, it has already been pointed out that the values originally selected for Fig. 5.3 are very close to those which would have to be used to match the criteria precisely. The engine modelling program, Prog.5.2, is employed in Chapter 5 with that original data and the creation of the necessary power output is witnessed in Fig. 5.23. In that sense, the specific time area approach proves a satisfactory method of creating a matched design for the porting, but the production of the exhaust pipe data set has still to be discussed below.

In the case of the chainsaw engine, it will be recalled from the discussion in Chapter 5 that the main problem for that type of engine is the creation of an adequate pumping action from the crankcase to provide the necessary flow of fresh air charge. The crankcase compression ratio is raised and found to have a beneficial effect on the delivery ratio level (see Sect. 5.2.3 and Fig. 5.16). That makes the prediction from Prog.6.1 for the specific time area targets for the chainsaw engine rather interesting. In the tables above, the required exhaust and blowdown values are sufficiently well matched so as not to cause comment, however, the values of actual transfer and inlet specific time area are at some variance with that required. The transfer port actual value is 0.00534, whereas the minimum requirement is for TRTA1 to be 0.0057. Considering that good pumping action is needed, that value could be elevated much closer to the larger TRTA2 criterion by increasing the total effective transfer port width to 48 mm from 40 mm. This would raise the specific time area to 0.00645, which is still slightly less than the TRTA2 value for that design.

It is the inlet port criterion for the chainsaw design which is more difficult to believe at first sight. The actual value from the data in Fig. 5.3 is 0.00965 and the requirement is for INTA to be 0.0071, as observed in the two tables above. This recommendation is for a reduction in the inlet port timing if the present port width is retained. As the engine needs all of the fresh charge possible, reducing the inlet port timing seems an unlikely design prospect. Yet, if Fig. 5.15 is reexamined, it can be seen that the delivery ratio curve in the lower diagram does rise to a maximum and then decay to a lower level at the crankcase trapping point. Could a narrower port timing trap the charge better? The inlet port specific time area requirement can be met by a port opening at 65º btdc which becomes fully open at tdc with no dwell, while retaining the original width and corner radii; this would give an inlet time area value of 0.00736 s/m. The answer to the question regarding the superiority or otherwise of these new designs for the inlet and transfer port will not be solved by the empirical programs presented in this chapter, so the new data is inserted into Prog.5.1 and the performance characteristics computed again. As this new calculation is run with a crankcase compression ratio, CCR, of 1.5, with the new widths and timings for the transfer and inlet port as deduced above while all other data remains as in Fig. 5.3, the predicted performance data is compared appropriately in Fig. 6.9. This latest design for the chainsaw engine, with the modified transfer port width and inlet port timing, is designated "SAW MK2" in Fig. 6.9. The original design of Chapter 5, but with a crankcase compression ratio of 1.5, is referenced in Fig. 6.9 as "1.5 CCR." The result of that comparison illustrates that there is a gain of delivery ratio with the modified inlet and transfer porting and this translates into better overall engine performance characteristics of BMEP, power, and fuel economy. The specific time area approach provided better matching of the porting throughout the engine in a manner which would have proved difficult to arrive at instinctively or by inspired guesswork.

It should always be remembered that empirical design criteria are based on past history and experience. By definition, the answers emanating from empiricism are not always correct and the result of any design decision should be checked by an overall engine modelling program and, ultimately, the acid test of dynamometer testing of a firing engine. At the same time, it is an unwise move to totally ignore recommendations which are based on the accumulation of past history and experience. In that regard, examination of that which already exists in production always provides a guide to the possible, and a study of the design data compendia from T.U. Graz(3.32) can assist in the design process.

6.2.5 The selection of the exhaust system dimensions

The subject of exhaust systems for two-stroke engines is one of considerable complexity, not simply because the unsteady gas dynamics are quite difficult to trace within the human mind, but also because the optimum answer in any given application is arrived at from a combination of the desirable gas dynamics and the

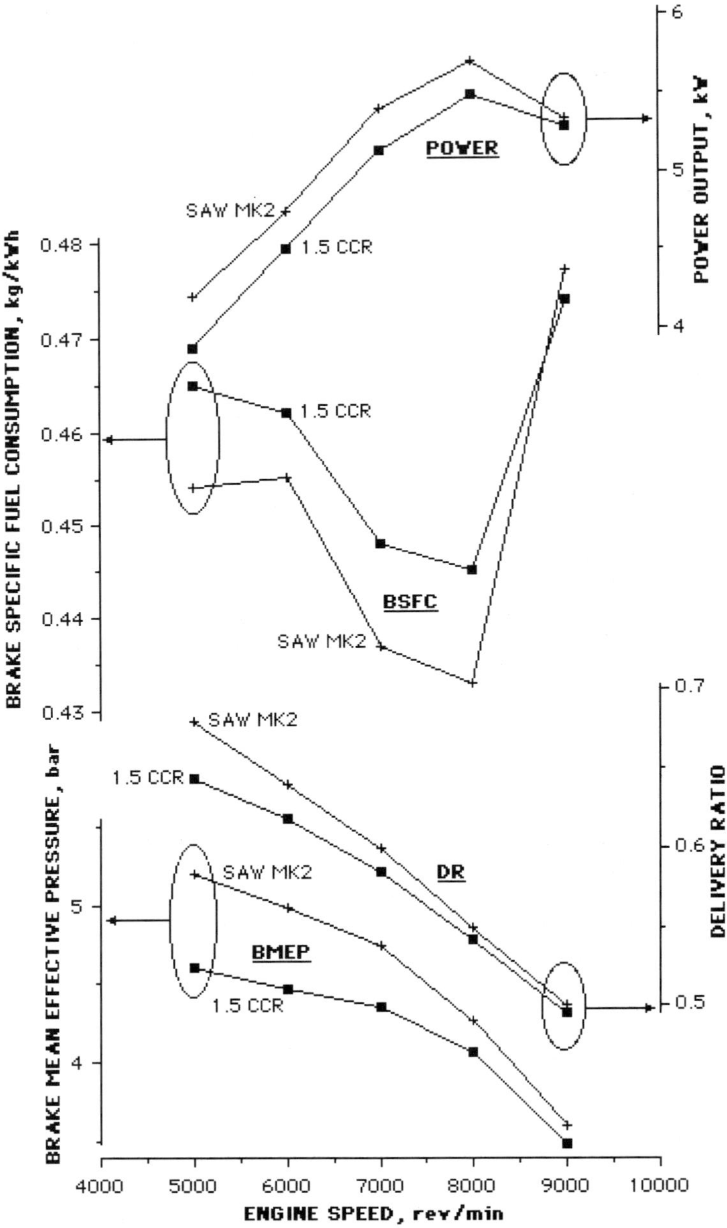

Fig. 6.9 Improvement of chainsaw performance with alternative porting.

necessity of physical constraints caused by the market requirements. A good example is the chainsaw engine discussed as a design example within this book. The conclusion in Chapter 5 is that the compact exhaust system necessary for a hand-held tool is a serious limitation on the engine power output. The designer of that unit is well aware of the problem, but is unable to silence the engine adequately within a given package volume constraint without impairing performance.

As is evident from the above statements and from the discussion in Chapter 5, the complexity of the pressure wave reflections in a tuned system are such that it is difficult to keep track of all of the possible combinations of the pulsations within the mind. However, as Chapter 5 demonstrates, the computer will carry out that task quite well, but only after the numbers for the lengths and diameters have been inserted into it for subsequent analysis. That puts the problem of selecting those numbers right back into the human domain, which implies some form of basic understanding of the process. Incidentally, this applies equally to the design of the compact chainsaw system as it does to the expansion chamber for the racing engine.

6.2.5.1 The exhaust system for an untuned engine, as in Prog.5.1

The first objective is to allow the creation of the exhaust pulse without restriction, thus the downpipe diameter should be at least as large in area as the total exhaust port effective area. If the system is untuned then there is almost certainly a silencer to be incorporated within it. As exhaust noise is a function of exhaust pulse amplitude and also of the rate of rise of that exhaust pulse, there is little point in having a downpipe from the engine which is the minimum criteria diameter. Thus the downpipe can be comfortably sized between 1.2 and 1.4 times the port area. The silencer box, if it has a volume at least 10 times the volume of the cylinder swept volume and does not have an outlet pipe which is unduly restrictive, can be situated at an appropriate distance from the engine so that a suction reflection, of whatever sub-atmospheric amplitude, can assist the engine to dispose of the exhaust gas and promote a stronger flow of fresh charge through the motor. The length of the downpipe from the piston face to the entrance to the silencer box is labelled LP, in mm units for convenience in this empirical calculation. This pipe length can be easily assessed from an empirical knowledge of the following: the total exhaust period in degrees, EP; the probable exhaust pipe temperature in deg.C, TexC; the engine speed of rotation in revolutions per minute, RPM.

The front of all pressure waves travel at the local speed of sound, Ao, which from Eq. 2.1.1 is:

$$Ao=\sqrt{[g*R*(TexC+273)]} \quad m/s$$

For engines below 5 bar BMEP the value of TexC will be in the region of 350º to 450ºC, and R and g can be taken to be 287 J/kgK and 1.4, respectively.

Although the exhaust pulse peak will propagate at supersonic velocity, the suction reflection will travel at subsonic speed and the average can be approximated to the local speed of sound (see Sect. 2.1.3). If one assumes that the exhaust pulse peak occurs at 15º after exhaust port opening, one would wish the peak of the suction reflection to return at or about the bdc piston position. This gives a reflection period in degrees crankshaft of RP:

RP=(EP/2-15) deg

The time, t, taken for this double length travel along the pipe LP must equal the reflection period, where:

t=(RP/360)*(60/RPM)=2*(LP/1000)/Ao

Rearranging for the pipe length, LP, which is in mm units:

LP=41.7*(EP-30)*Ao/RPM (6.2.1)

If the appropriate numbers are inserted into this equation from the chainsaw data, i.e., RPM=8000, TexC=400, and EP=160, then LP is derived as 352 mm. There being no possibility of installing a 352 mm long downpipe on a chainsaw, the designer is left to dimension a silencer box which will have the least possible restriction effect on gas flow from the engine while still silencing it to the legislative requirement. Fortunately the two aims are mutually compatible, for the largest possible box which can be ingeniously designed into the available space will give the best power output and the best silencing effect; see Chapter 8 for an expanded discussion on noise and silencing.

For engines which are not so constrained by package volume, such as a generator set, or even a brush cutter if one uses the handle in an innovative fashion, the designer can obtain some power advantage by using Eq. 6.2.1.

6.2.5.2 The exhaust system of the high-performance engine as in Prog.5.2

Having been through the fundamentals of unsteady gas flow in Chapter 2 and the results of computer engine modelling of high-performance engines in Chapter 5, the reader will find this empirical study of flow in an expansion chamber a lighter mathematical affair. A useful reference for further perusal is the technical papers by Naitoh et al(6.4)(6.5) and by the author(2.17). As stated in the introduction, it is the designer who has to provide the numbers for all of the data slots in Fig. 5.4(b) and so the fundamental understanding gained in Chapters 2 and 5 must be translated into the ability to deduce appropriate data for the calculation. To aid with the empirical design of an expansion chamber, a computer program, Prog.6.2, is included in the Computer Program Appendix.

Fig. 6.10 shows the basic action which is taking place in an optimized system, such as in the Husqvarna MK4 engine at 5500 rpm in Fig. 5.18. The sketch is drawn with time on the vertical axis and distance on the horizontal axis. The port timings appropriate to that time scale for that particular speed are marked on the time axis as EPO, TPO, BDC, TPC and EPC, for the porting events from exhaust and transfer opening through bdc to transfer and exhaust closing, respectively. The important factor in this discussion is that the scale on the vertical axis is time, not crankshaft angle. The exhaust pulse is seen to leave the exhaust port and to propagate rightwards towards the atmosphere, primarily sending suction reflections back towards the engine from the diffuser, and compression wave reflections from the tail nozzle. The exhaust pulse will propagate in a parallel pipe with no reflections, but on arrival at a change in section area, be it gradual or abrupt, reflections will occur in the manner described in Chapter 2. In the optimized design this primary reflection behavior ensures that (a) the peak of the suction reflection returns to the exhaust port around the bdc period to aid the scavenging process and (b) all of the plugging pulse returns to the exhaust port in time to hold in the fresh cylinder charge before the exhaust port actually closes. The exhaust pressure-time history of the racing engine in Fig. 5.23 can be regarded as a classic example of that statement.

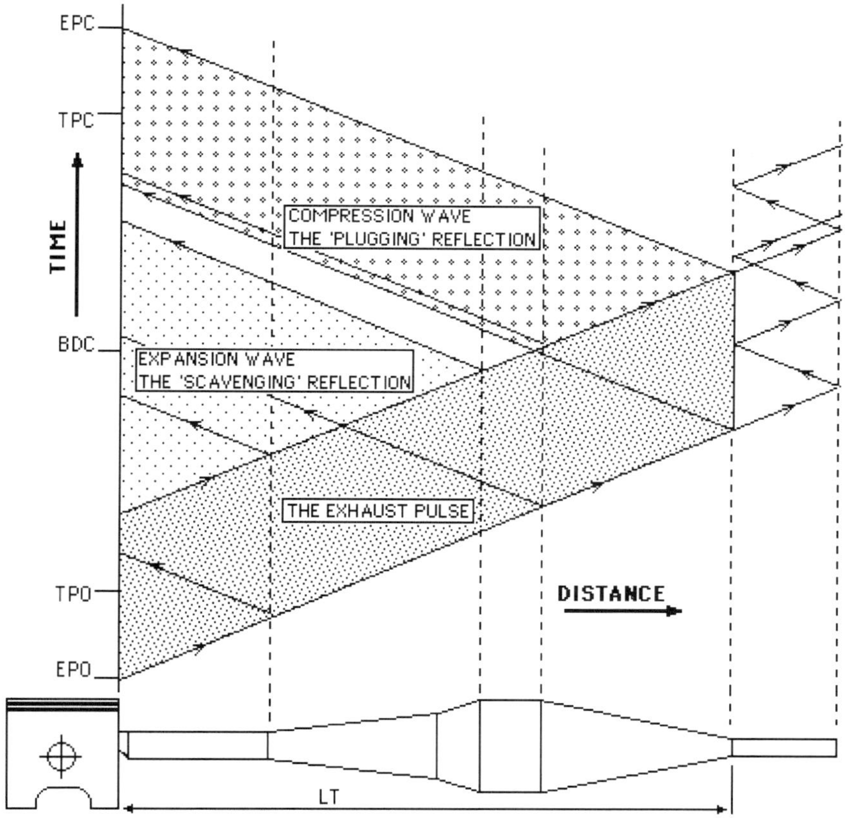

Fig. 6.10 Optimum pressure wave phasing in an expansion chamber exhaust pipe.

This picture of time and distance is one which will remain ostensibly similar at all engine speeds, therefore it is the crankshaft timing points which move on the vertical axis as a linear function of the engine rotation rate. At low engine speeds, such as the Husqvarna MK4 at 4000 rpm in Fig. 5.17, the time span from EPO to EPC will increase, moving the marker for EPC and the others "farther up" the time axis allowing the plugging pulse to interfere with the scavenging process and contribute nothing to the trapping of charge in the cylinder as the port is being closed by the piston. At high engine speeds beyond that for an optimized behavior, such as the Husqvarna MK4 at 8000 rpm in Fig. 5.20, the time span from EPO to EPC will be shortened so that the marker for EPC will be one half of the distance up the time axis that it would be at 4000 rpm. The result at a high engine speed beyond the optimum, as already seen in Fig. 5.20, is that the plugging pulse appears at the exhaust port after it is already closed so that the trapping process receives no benefit from its arrival.

One of the methods available to broaden that band of speed, over which the plugging pulse would provide a significant gas trapping action, is for the pipe to remain the same length dimension physically but for the exhaust opening period, EP, to be reduced at low engine speeds by a valving mechanism. Such a mechanism to control the exhaust port timing and duration is described in Sect. 7.3.3.2 and shown in Fig. 7.17. An alternative design approach is to be able to vary the tuned length, LT, as a function of engine speed and this has been done in some go-kart racing machines with trombone-like sliding sections within a tuned exhaust pipe. Both of these approaches to broadening the effective power band of the engine rely on some linear relationship between the exhaust period, EP, and the tuned length, LT. That this is the case, empirically speaking, is seen from Eq. 6.2.2 and Fig. 6.10.

The principal aim of the empirical design process is to phase the plugging reflection correctly for the engine speed desired for peak power, which means calculating the tuning length from piston face to tail pipe entry, LT. The next important criterion is to proportion the tail pipe exit area as a function of the exhaust port area so that the pipe will empty in a satisfactory manner before the arrival of the next exhaust pulse. The third factor is to locate the end of the diffuser and the beginning of the tail nozzle so that the spread of the suction and plugging effects are correctly phased. The fourth function is to locate the second half of the diffuser so that the reverse reflection of the plugging pulse from it, and the primary reflection of it from the exhaust port, can recombine with the next outgoing exhaust pulse, thereby producing the resonance effect demonstrated in Figs. 5.21-5.23. This will only happen in racing engines with very long exhaust port periods, typically in excess of 190º.

Nevertheless, a two-stage diffuser is still an efficient method of reflecting the exhaust pulse and the shallower first stage reduces the energy losses from shock formation at the entrance to that diffuser. The shocks form as high-amplitude compression exhaust waves meet strong suction waves which are progressing back to the engine from the diffuser. As seen in Sect. 2.2.1, this raises the local gas particle velocity, and in the diffuser entrance area this can attempt to exceed a Mach number of unity. The result is a shock which effectively "destroys" the wave energy, and although the energy reappears as heat, this does nothing for the all-important pressure wave action. The slower rate of area expansion of the first stage of the diffuser, i.e., lower dF/F terms in Eq. 2.4.17, smears the reflection process more widely and efficiently over the entire length of the diffuser.

The entire empirical design process is founded in very simple equations which have been tried and tested for many years in the theoretical and experimental development of expansion chambers for racing engines at QUB. As remarked earlier, it should be remembered that this is an empirical approach, not a precise calculation, and is the starting point for a theoretical design process using Prog.5.2, preferably in combination with the experimental testing of the firing engine.

As this is an empirical calculation for lengths conventionally discussed in mm units, the data is inserted and produced by the calculation in those units. The data which is required for the calculation are in blocks, the first of which is: engine bore, stroke, connecting rod length, exhaust port total opening period EP in deg. crankshaft units, number of exhaust ports, maximum width of each exhaust port, and radii in the corners of each exhaust port.

The above set of data permits the calculation of the exact exhaust port area. Incidentally, it is presumed that the designer has followed some design process to this point and the data being used is matched to the required performance of the engine. With this knowledge of the exhaust port flow area, equivalent to a diameter, EXD, the empirical calculation describes this as the ideal exhaust pipe initial diameter. In a racing pipe, where maximizing the plugging pressure gives the highest trapping efficiency, there seems little point in having the first downpipe diameter much larger than 1.05*EXD, for in unsteady gas flow, the larger the pipe, the lower the pressure wave amplitude for the same mass rate of gas flow (see Sect. 2.1.4). Knowing the ideal value of EXD, the calculation ascribes values for the tailpipe diameter so that the pipe will provide the requisite plugging reflection, yet still empty the system during each cycle. Using the geometrical nomenclature of Fig. 5.2(b):

DEP5=0.6*EXD
if this is a road-racing type of engine, or about 11 bar BMEP peak.

DEP5=0.65*EXD
if this is a motocross type of engine, or about 9 bar BMEP peak.

DEP5=0.7*EXD
if this is an enduro type of engine, or about 8 bar BMEP peak.

The latter will have a comprehensive silencer following the tuned section. Motocross and road racing engines have rudimentary silencers, and the design philosophy being presented is not at all affected by the attachment of a simple absorption silencer section in conjunction with the tailpipe. That, as with all such debate on silencers, is to be discussed in Chapter 8.

The next set of data is required to calculate lengths, all of which will be proportioned from the tuned length, LT, to be calculated: engine speed for peak power, RPM, in rev/min units; exhaust gas temperature, TexC, as estimated in the pipe mid-section, in ºC units.

For those who may not have immediate access to temperature data for expansion chambers, the potential values have a loose relationship with BMEP. Therefore, if the reader expects to have a design approach a Grand Prix racing engine power level, then an exhaust temperature of 650ºC can be anticipated. If the engine is a lesser rated device, such as an enduro motorcycle, then a temperature of 500ºC could be safely used in the calculation. As the speed of sound is the real criterion, then as it is proportional to the square root of the Kelvin temperature, the selection of exhaust pipe temperature is not quite so critical as it might appear to be. The speed of sound, Ao, is given by Eq. 2.1.1:

$$Ao=\sqrt{[g*R*(TexC+273)]}$$

The calculation of LT is based on the return of the plugging pulse within the exhaust period. The most effective calculation criterion, based on many years of empirical observation, is the simplest. The reflection period is decreed to be the exhaust period and the speed of wave propagation is the local speed of sound in the pipe mid-section. Much more complex solutions have been used from time to time for this empirical calculation, but this approach gives the most reliable answers in terms of experimental verification from engine testing.

As with Eq. 6.2.1, the time, t, taken for this double length travel along the tuned length, LT, must equal the reflection period, where:

$$t=(EP/360)*(60/RPM)=2*(LT/1000)/Ao$$

Rearranging for the pipe length, LT, which is in mm units:

$$LT=83.3*EP*Ao/RPM \text{ mm} \qquad (6.2.2)$$

The remainder of the solution is straightforward, following the nomenclature of Fig. 5.2(b). The end of the diffuser is positioned as in Fig. 6.10, about two-thirds of the distance between the exhaust port and the tail-pipe entry.

LP01=0.10*LT
LP12=0.41*LT
LP23=0.14*LT
LP34=0.11*LT
LP45=0.24*LT
LP56=LP45

The preferred value of DEP1=K1*EXD. The value for K1 ranges from 1.125 for broadly tuned enduro type engines to 1.05 for road-racing engines of the highest specific power output.

The preferred value of DEP3=K2*EXD. The normal design criterion is for the value of DEP4 to be the same as DEP3. The value for K2 ranges from 2.25 for broadly tuned enduro type engines to 3.25 for road-racing engines of the highest specific power output.

The remaining diameter to be determined is that for DEP2 and it is estimated by the following expression:

$$DEP2=DEP3*[LP12/(LP12+LP23)]^{1.33} \tag{6.2.3}$$

It is self-evident that all diameters in the previous discussion are the internal diameters of the exhaust pipe, and that all lengths are distances along the center-line of a pipe which may be curved in any of its segments.

Some years ago, the author presented a paper concerning the empirical design of expansion chambers(2.17). It is instructive to compare the above discussion with that paper and to note that the empiricism has actually become simpler with the passage of both time and experience.

For those who wish to have these calculations performed automatically, the Computer Program Appendix lists Prog.6.2, "EXPANSION CHAMBER." This carries out the useful dual function of deducing all of the above dimensions and data and the computer screen and line printer will show a dimensioned sketch of the exhaust system. An example of that is shown for a prediction of an expansion chamber for the 125 Grand Prix engine used as one of the working examples in these chapters. The screen/line-print output is shown in Fig. 6.11. If the data predicted by the empirical design program is compared with those listed in Fig. 5.4(b), a considerable level of close correspondence will be observed. It will be recalled that the engine modelling program predicted a high specific power performance for this engine and exhaust system in Fig. 5.23. This should give the reader some confidence that this simple program, Prog.6.2, albeit overlaid with some considerable practical experience, enables the rapid and efficient preparation of the geometry of an expansion chamber for further complete analysis by the engine modelling program, Prog.5.2.

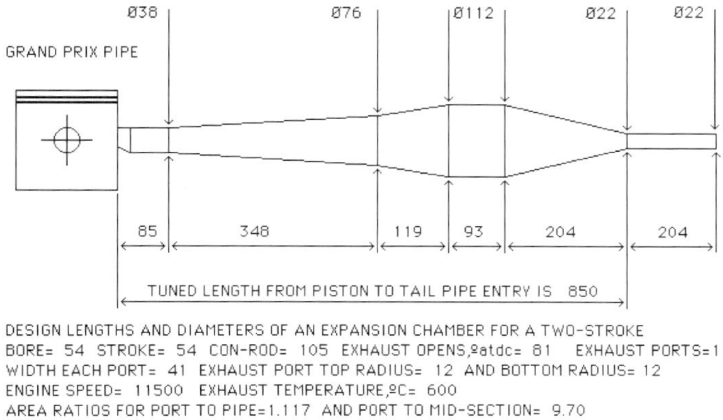

Fig. 6.11 Example of design of expansion chamber for a 125 Grand Prix engine from Prog.6.2.

Another useful feature of this empirical program, Prog.6.2, is that there is available a second page of line-print output. This gives the manufacturing data for the various cones of the expansion chamber so that they may be marked out and rolled from flat sheet metal, then subsequently seam welded into the several cones. The cones and parallel sections are butt welded to form a pipe for either competition or dynamometer testing. An example of the actual line printer output for the Grand Prix engine design of Fig. 6.11 is shown in Fig. 6.12. In that figure, RX and RY are the radii from the center of a circle marked on the metal sheet, and Ø is the angle of the flat sheet segment. This information alone is quite sufficient to permit the complete marking of the sheet metal. The extra information for the chords, CHX and CHY, allows an alternative method of marking out the flat sheet metal segment, or provides a cross-check on the accuracy of setting out the angle Ø.

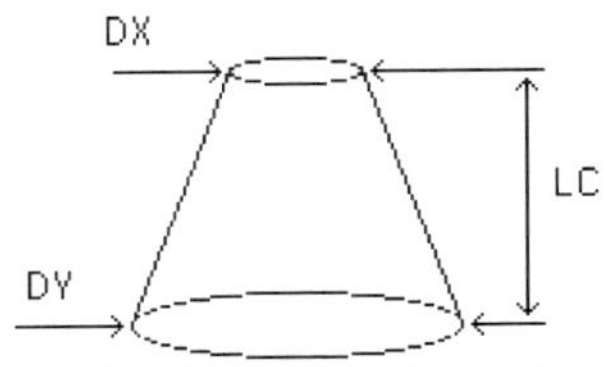

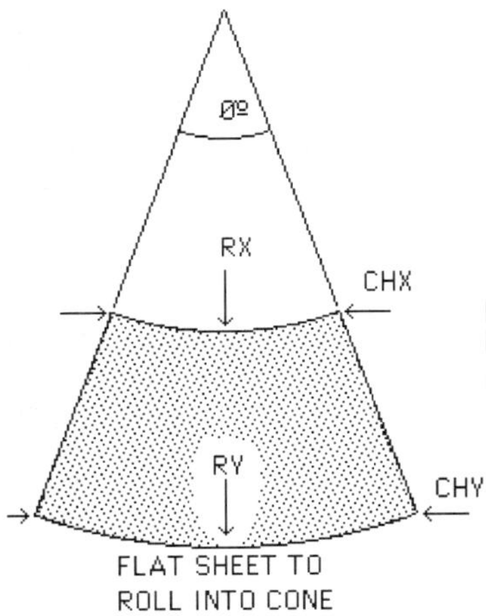

	DIFFUSER CONES		TAIL PIPE CONE	
	CONE1	CONE2		
DX	38	76	22	
DY	76	112	112	CONE
LC	348	119	204	DIMENSIONS
RX	350	249	49	
RY	699	368	253	
CHX	119	229	62	FLAT PLATE
CHY	237	339	324	DIMENSIONS
Øº	19.5	55	80	

Fig. 6.12 Example of manufacturing data available as computer output from Prog.6.2.

6.2.6 Concluding remarks on data selection

By the methods discussed above the information is selected for the data bank for the main engine geometry and porting, as typified by the set in Figs. 5.3 and 5.4. Those who pursue this empirical approach by using the several computer packages for the creation of the data bank necessary for the operation of an engine modelling program will discover that with a little practice the data bank can be assembled in a remarkably short time. This approach reduces the number of times which the engine modelling programs need to be run to create an optimized design to meet the perceived requirement for the performance characteristics. As the engine modelling programs take the longest time to run on the computer, indexing in the first instance the virtually instantaneous response of the empirical design programs is an effective use of the designer's time. Perhaps more important, it tends to produce porting and exhaust systems which are well matched. An example of that is the design for the chainsaw, where the transfer and inlet ports would have been better matched by an initial application of the empirical approach.

There will be the natural tendency, about which the reader is cautioned earlier, to regard the empirical deductions as a final design for either the porting or the exhaust system. That warning is repeated here, for the author has made that mistake sufficiently often to realize its validity!

6.3 The design of alternative induction systems for the two-stroke engine

In Sect. 1.3, there is an introduction to the use of mechanical devices which permit asymmetrical timing of the exhaust or induction process. The three types discussed are poppet valves, disc valves and reed valves. Needless to add, the inventors of this world have produced other ingenious devices for the same purpose, but the three listed have withstood the test of time and application. It is proposed to discuss the design of disc valves and reed valves in this section, but not poppet valves for they are covered in the literature of four-stroke cycle engines to an extent which would make repetition here just that. As the design of reed valves and disc valves is not covered extensively in the literature, it is more logical that the appropriate space within this book should be devoted to the "unknown."

The use of reed valves for the induction system has always been common for outboard motors and increasingly so for motorcycle and other types of two-stroke engines. The incorporation of a reed valve into an engine is shown in Fig. 1.7(b) and the details of a reed block assembly are sketched in Fig. 6.13, with a photograph giving further illustration in Plate 6.1. The technology of the design has improved greatly in recent times with the use of new materials and theoretical design procedures. The new materials, such as plastics reinforced by either glass-fiber or carbon-fiber, are effective replacements for the conventional use of spring steel for the reed petal. In particular, any failure of a plastic reed petal in service does not damage the engine internally, whereas it would in the case of a steel reed.

The use of disc valves for the induction system has been confined mostly to high-performance racing engines although there are some production examples of lesser specific power output. The initial discussion in Chapter 1 gives a sketch of the installation in Fig. 1.7(a) and Plate 1.8. A more detailed drawing of the disc and the inlet port which it uncovers is shown in Fig. 6.14. There is not a large body of technical information available on the design and development of disc valves, but the papers by Naitoh et al(6.4)(6.5) and the book by Bossaglia(6.3) contain valuable insights to the design and development process.

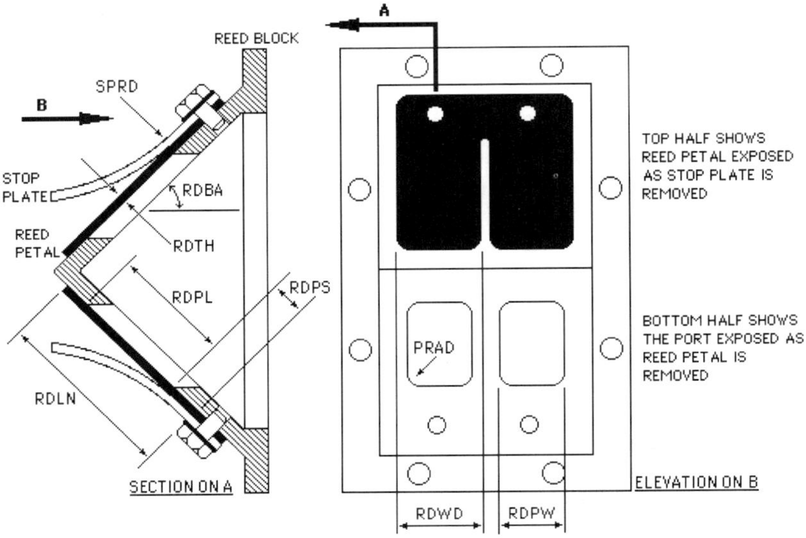

Fig. 6.13 Significant dimensions of reed valves and block required for design purposes.

Plate 6.1 A reed valve block with six steel reeds and stop-plate. The block is rubber coated to reduce both noise and damage to bouncing reeds.

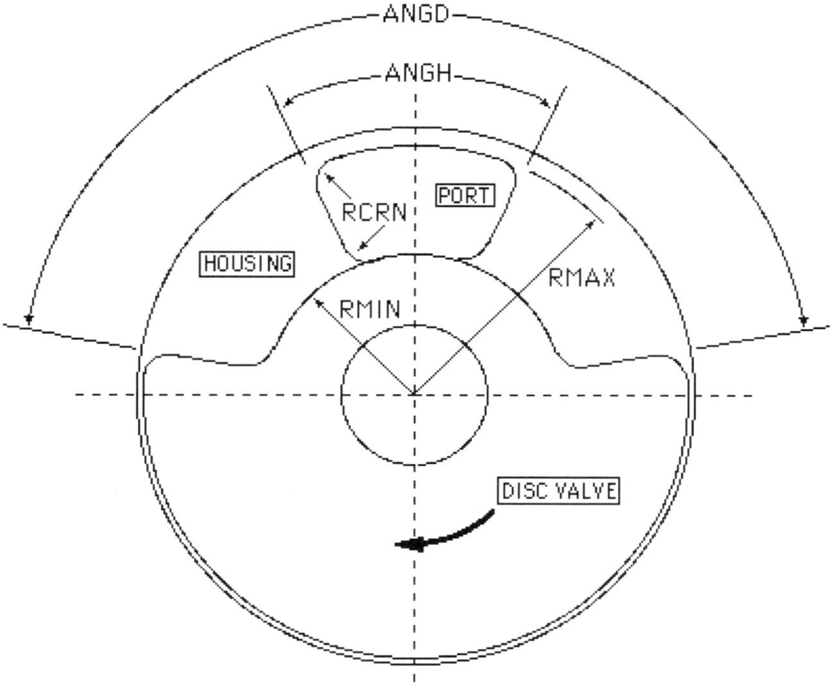

Fig. 6.14 Controlling dimensions of disc valve timings and inlet port area.

The incorporation of these induction systems into an engine modelling program is straightforward in the case of the disc valve and complex for a reed valve design. That it has been successfully achieved for the reed valve case is evident from the technical papers presented(1.13)(4.10)(5.9). Actually there is a considerable body of literature on reed valves when used in air and refrigeration compressors as automatic valves; the technical paper by MacClaren(6.6) will open the door to further references from that source. The modelling of the disc valve case within an engine design program, such as Prog.5.1 or 5.2, is regarded as requiring a simple extension to the program to accommodate the inlet port area geometry as uncovered by the disc. The discussion in Sect. 6.4 clarifies this statement.

However, just as the parameters for the engine in Figs. 5.3 and 5.4 have to be assembled in some logical manner, it is necessary to be able to arrive at an initial decision on the dimensions of the reed or disc valve in advance of either the use of a computer modelling package or the experimental testing of prototype devices. The following discussion should be of practical assistance within that context.

6.3.1 The empirical design of reed valve induction systems

The reed valve induction system is installed in the engine between the atmosphere and the crankcase of the engine, as shown in Figs. 1.7(b) and 6.13. If the carburetor is the fueling device, it is placed between the reed valve and the atmosphere. If low-pressure fuel injection is used, the injector nozzle can be situated in the same position as a carburetor, but it is also possible to inject the fuel directly into the crankcase. A photograph of a cylinder of a motorcycle engine with a reed induction system appears in Plates 1.9 and 4.1 and a closer view of the block, petals and stop plate is shown in Plate 6.1. The typical opening and closing characteristics of this automatic valve are illustrated in Fig. 1.8(c). To expand this information, and to illustrate the point made above regarding the incorporation of a reed valve model into an engine simulation program, a few of the results from the experimental and theoretical paper by Fleck et al(1.13) are presented here as Figs. 6.15-6.17. The engine used as the research tool in this paper is the high-performance YPVS RD350LC twin-cylinder Yamaha motorcycle engine, with a peak BMEP of nearly 8 bar at 9000 rpm. Each cylinder has a block holding four reed petals and ports. In this case, designated as RV1, they are steel reed petals of 0.20 mm thickness, i.e., from Fig. 6.13 the dimension RDTH=0.20.

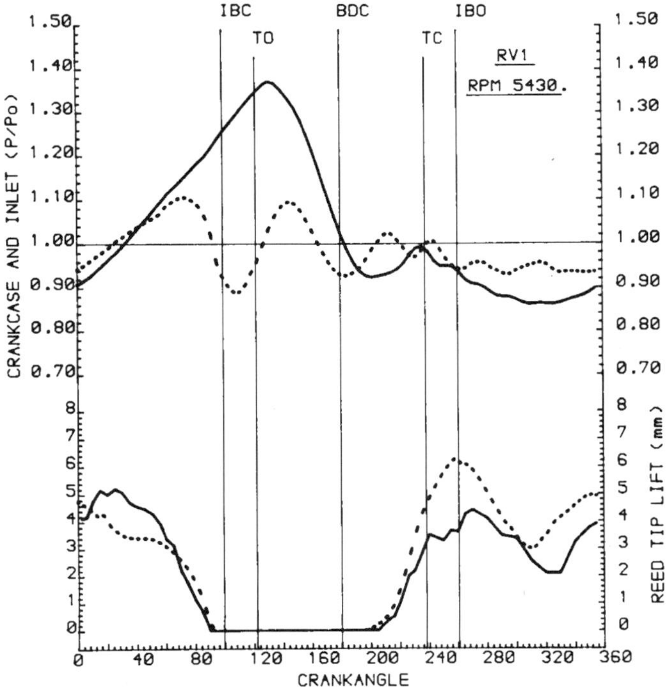

Fig. 6.15 The crankcase and inlet port pressure and reed lift behavior at a low engine speed.

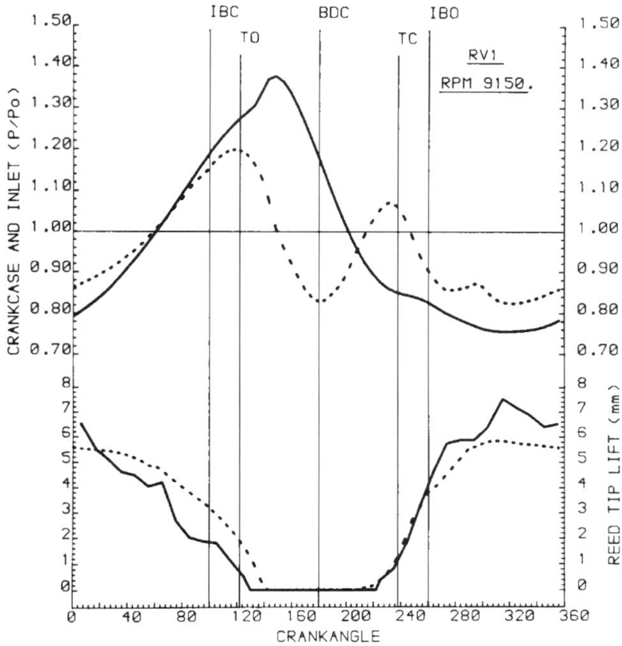

Fig. 6.16 The crankcase and inlet port pressure and reed lift behavior at a high engine speed.

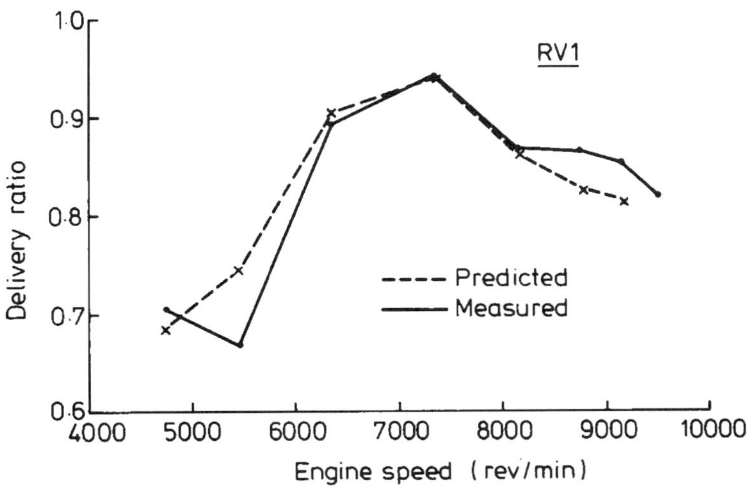

Fig. 6.17 The comparison between measured and calculated delivery ratio from an engine model incorporating a reed valve simulation.

The upper portion of Fig. 6.15 shows the crankcase and inlet port pressure at a "low" engine speed of 5430 rev/min. Those pressure diagrams are predicted by the engine simulation program and one can see that they are very similar in profile to those observed for piston ported engines in Chapter 5. In short, there is nothing unusual about the pressure difference across the reed valve by comparison with that observed for a piston controlled induction system. The solid line in the lower half of that figure shows the measured reed tip lift in mm, and that predicted by the reed valve simulation within the computer program is the dashed line. The close correspondence between the calculation and measurement of reed lift is evident, as is the resulting calculation and measurement of delivery ratio over the entire speed range of the engine in Fig. 6.17. The timing of opening and closing of the reed petal at the low speed of 5430 rpm shows the reed opening at 160º btdc and closing at 86º atdc, which is an asymmetrical characteristic about tdc.

At the "high" engine speed of 9150 rpm, the reed petal opens at 140º btdc and closes at 122º atdc. This confirms the initial view, expressed in Sect. 1.3.4, that the reed petal times the inlet flow behavior like a disc valve at low engine speeds and as a piston controlled intake port at high engine speeds. In other words, it is asymmetrically timed at low speeds and symmetrically timed at high speeds.

Within this same paper by Fleck et al (1.13), there is given a considerable body of experimental and theoretical evidence on reed valve characteristics for steel, glass- and carbon-fiber reinforced composites when used as reed petal materials.

To help with the correlation of the discussion in this book and that paper, the nomenclature for the reed block and petals shown in Fig. 6.13 is identical in both instances. The paper(1.13) also gives the dimensions of the engine with its tuned expansion chamber exhaust system and associated measured power data, which the reader will find instructive as another working example for the comparison of Prog.6.2 with experiment.

6.3.2 The use of specific time area information in reed valve design

From the experimental and theoretical work at QUB on the behavior of reed valves, it has been found possible to model in a satisfactory fashion the reed valve in conjunction with an engine modelling program. This implies that all of the data listed as parameters for the reed valve block and petal in Fig. 6.13 has to be assembled as input data to run that engine model. Naturally, this data would replace the block of inlet port data for a piston controlled port in Fig. 5.3. The data set for a reed valve is even more extensive than that for a piston controlled intake port, adding to the complexity of the data selection task by the designer before running the engine model. This places further emphasis on the use of some empirical design approach to obtain a first estimate of the design parameters for the reed block and petals, before the insertion of that data set into an engine modelling calculation. It is proposed to pass on to the reader the data selection experience of the author, in the form of an empirical design for reed valves as a pre-modelling exercise.

In Sect. 6.1 there is a discussion on specific time area and its relevance for exhaust, transfer and intake systems. The flow through a reed valve has to conform to the same logical approach. In particular, the value of specific time area for the reed petal and reed port during its period of opening must provide that same numerical value if the flow of air through that aperture is to be sufficient to provide equality of delivery ratio with a piston controlled intake port. The aperture through the reed valve assembly is seen in Fig. 6.13 to be composed of two segments, the effective reed port area in the flow direction as if the reed petal is not present, and the effective flow area past the reed petal when it has lifted to its maximum, caused either by the gas flow or as permitted by the stop plate. Clearly, there is little point in not having these two areas matched. For example, if the design incorporates a large reed port area but a stiff reed which will barely lift under the pressure differential from the intake side to the crankcase, very little fresh charge will enter the engine. Equally, the design could have a large flexible reed which lifted easily but exposed only a small reed port, in which case that too would produce an inadequate delivery ratio characteristic. In a matched design, the effective reed port area will be larger than the effective flow area past the petal, but not by a gross margin. Therefore, the empirical design process is made up of the following elements:

(a) Ensuring that the effective reed port area has the requisite specific time area, on the assumption that the reed petal will lift at an estimated rate for an estimated period.

(b) Ensuring that the reed will lift to an appropriate level based on its stiffness characteristics and the forcing pressure ratio from the crankcase.

(c) Ensuring that the natural frequency of vibration of the reed petal is not within the operating speed range of the engine, thereby causing interference with criterion (b) or mechanical damage to the reed petal by the inevitable fatigue failure.

The data required for such a calculation is composed of the data sketched in Fig. 6.13 and the parameters listed in Fig. 6.18. Of general interest, Fleck et al(1.13) report the physical properties of both composites and steel when employed as reed petal materials, as recorded in a three-point bending test. They show that a glass-fiber reinforced composite has a Young's Modulus, YMOD, of 21.5 GN/m^2 and a density of 1850 kg/m^3. The equivalent data for carbon-fiber reinforced plastic and steel is measured in the same manner and by the same apparatus The value of Young's Modulus for steel is 207 GN/m^2 and its density is 7800 kg/m^3. The value of Young's Modulus for a carbon-fiber reinforced plastic is 20.8 GN/m^2 and its density is 1380 kg/m^3. Within the paper there are more extensive descriptions of the specifications of the GRP and carbon-fiber composite materials actually used as the reed petals.

The opening assumption in the calculation is that the specific time area required is the same as that targeted in Eq. 6.1.7 as INTA related to BMEP.

INTA=(BMEP+1.528)/774

The theoretical relationship for specific time area, Eq. 6.1.12, is repeated below:

Specific time area, s/m= INTA=Σ[A*dø]/(RPS*360*SV)

INPUT DATA NAME	SYMBOL	CHAINSAW	125 GRAND PRIX	125 GRAND PRIX
SWEPT VOLUME, cc	SV	101	125	125
ENGINE SPEED, rpm	RPM	8000	11500	11500
BMEP, bar	BMEP	4.5	11	11
CRANKCASE COMPRESSION RATIO	CCR	1.45	1.5	1.5
PETAL MATERIAL		STEEL	CARBON FIBRE	STEEL
PETAL NUMBER	RDNO	2	6	6
PETAL THICKNESS, mm	RDTH	0.20	0.60	0.25
PETAL WIDTH, mm	RDWD	22	23	23
PETAL FREE LENGTH, mm	RDLN	30	38	38
PORTS NUMBER	RDPNO	2	6	6
PORT WIDTH, mm	RDPW	19	20	20
PORT LENGTH, mm	RDPL	23	28	28
CORNER RADII, mm	PRAD	3	3	3
LENGTH FROM CLAMP, mm	RDPS	5	8	8
REED BLOCK ANGLE, deg	RDBA	30	26	26
YOUNG'S MODULUS GN/m2	YMOD	207.0	20.8	207.0
REED PETAL DENSITY, kg/m3	RDRH	7850	1380	7850
OUTPUT DATA				
REQUIRED FLOW AREA, mm2	RFA	378	1396	1396
DESIGNED FLOW AREA, mm2	DFA	435	1468	1468
DESIGN CARBURETTER DIAMETER, mm	CARB	19.5	36	36
NATURAL FREQUENCY, Hz, 1ST ORDER,	FREQ1	185	261	144
2ND ORDER	FREQ2	1159	1633	903
TIP LIFT RATIO	TIPR	0.20	0.17	0.23
STOP PLATE RADIUS	SPRD	45	58	58

Fig. 6.18 Tabular result of using the reed valve design program, Prog.6.4.

A standard opening period, Ω, of 200° crankshaft is assumed with a lift-to-crank angle relationship which is of isosceles triangle form. The time for each degree of crank rotation is given by dt/dø. From Eq. 6.1.12, the A*dø term can be evaluated with swept volume inserted as cm^3, SVcc, and if the area A is in mm^2 units, the units for the time-area INTA remain as s/m.

$$\int A*d\phi = 6*INTA*SVcc*RPM \qquad mm^2.deg \qquad (6.3.1)$$

The required reed port area, RPAreq, is given by determining the area of an isosceles triangle of opening which spans an assumed Ω=200°:

$$RPAreq=2*(\int A*d\phi)/\Omega=12*INTA*SVcc*RPM/\Omega \qquad mm^2 \qquad (6.3.2)$$

At this point the designer has to estimate values for the reed port dimensions and determine a flow area which will match that required from the time-area analysis. That step is to assign numbers to the reed port dimensions shown in Fig. 6.13, not forgetting that the effective flow area of a sloping reed port is the trigonometrical sine of the projected area of the port. The notation for the data is also given in Fig. 6.18. The effective reed port area is declared as RPAdes:

$$RPAdes=[RDPL*RDPW-PRAD^2*(1-0.25*\pi)]*RDPNO*\sin(RDBA) \quad (6.3.3)$$

The required area, RPAreq, and that estimated, RPAdes, are compared and the port dimensions adjusted until the two values match from Eqs. 6.3.2 and 6.3.3. If anything, one should always err slightly on the generous side in apportioning reed port area. Without carrying out this form of design calculation, one tends to err on the restrictive side because the eye, viewing it on a drawing board or a CAD screen, tends to see the projected plan area and not the all-important effective area in the flow direction.

The carburetor flow area can be estimated from this required area in the same manner as is effected for a piston controlled intake port, although with a larger restriction factor of 0.7 from the reed port area to the carburetor flow area. This is to create a greater pressure differential across the reed, assisting it to respond more rapidly to the fluctuating crankcase pressure. The carburetor flow diameter, notated as CARB in Fig. 6.18, is deduced by:

$$CARB=\sqrt{(0.7*4*RPAreq/\pi)} \quad (6.3.4)$$

The next requirement is for calculations to satisfy the vibration and amplitude criteria (b) and (c) above. These are connected to theory found in texts on vibrations or strength of materials such as that by Morrison and Crossland(6.1) or Tse et al(6.2). The second moment of area of a reed petal, MI, is given in units of m^4:

$$MI= RDWD*RDTH^3/12 \quad (6.3.5)$$

The mass per unit length of the reed petal, Mo, is given in kg/m units:

$$Mo= RDWD*RDTH*RDRH \quad (6.3.6)$$

The natural frequencies of vibration of a uniform profile reed petal are quoted by Fleck et al(1.13) as FREQ1 and FREQ2 for the first two modes of vibration, in Hz:

$$FREQ1=(1.758/\pi)\sqrt{\{YMOD*MI/(Mo*RDLN^4)\}} \quad (6.3.7)$$
$$FREQ2=(11.015/\pi)\sqrt{\{YMOD*MI/(Mo*RDLN^4)\}} \quad (6.3.8)$$

The natural frequency of the forcing function on the reed is created by the pressure loading across the reed from intake pipe to crankcase, and is equal to the engine speed in cycles/second, i.e., RPS. The first-order natural frequency is the one to which some attention should be paid, for it is the author's experience that it tends to be quite close to the maximum speed of engine operation of a well-matched design. As long as the reed natural frequency is 20% higher than engine natural frequency, there is little danger of accelerated reed petal damage through fatigue.

The final step is to ensure that the reed petal will lift sufficiently during the pumping action of the crankcase. This is precisely what the engine modelling program and reed simulation of Fleck et al(1.13) is formulated to predict, and provides in Figs. 6.15-6.17. Therefore, the empirical approach is aimed at getting a first estimate of that lift behavior. Fleck et al(1.13) show reed tip lift to length ratios, TIPR, in the region of 0.15-0.25. That will be used as the lift criterion for the reed petal undergoing design scrutiny when it is treated as a uniformly loaded beam by a proportion of the pressure differential estimated to be created by the crankcase compression ratio, CCR.

The mean pressure differential, Pload, assumed to act across the reed is found by estimating that this is a linear function of CCR in atmosphere units, where 18% is regarded as the mean effective value of the maximum:

$$\text{Pload}=0.18*(\text{CCR}-1)*101325 \quad \text{N/m}^2 \qquad (6.3.9)$$

From Morrison and Crossland(6.1), the deflection, DEFL, at the end of a uniformly loaded beam is given by:

$$\text{DEFL}=\text{Pload}*\text{RDWD}*\text{RDLN}^4/(8*\text{YMOD}*\text{MI}) \qquad (6.3.10)$$

The tip lift ratio, TIPR, is then:

$$\text{TIPR}=\text{DEFL}/\text{RDLN} \qquad (6.3.11)$$

An earlier caution is repeated, that any use of the above equations referring to the dynamics of reed lift, i.e., Eqs. 6.3.5-6.3.11, should be strictly carried out in SI units.

The last geometrical dimension to be calculated is the stop plate radius which should not permit the tip of the reed to move past a TIPR value of 0.3, but should allow the reed petal to be tangential to it should that lift actually occur. This relationship is represented by the following trigonometrical analysis, where the limit criterion is TIPQ=0.3:

$$\text{SPRD}=(1-\text{TIPQ}^2)*\text{RDLN}/(2*\text{TIPQ}) \qquad (6.3.12)$$

6.3.3 The design process programmed into a package, Prog.6.4

This calculation is not as convenient to carry out with an electronic calculator as that for the expansion chamber, as a considerable number of cycles of estimation and recalculation are required before a matched design emerges. Therefore, a computer program has been added to those already presented, Prog.6.4, "REED VALVE DESIGN." This interactive program has a screen output which shows a plan and elevation view to scale of the reed block and reed petal under design consideration. A typical example is that illustrated in Fig. 6.19, which is also a design of a reed block and petals for the 125 Grand Prix engine already discussed and used throughout Chapters 5 and 6 as a design example. The top part of the sketch on the computer screen shows an elevation section through one half of the reed block, although it could just as well represent a complete block if that were the design goal. The lower half of the sketch shows a projected view looking normally on to the reed port and the petal as if it were transparent, with the darkened area showing the clamping of the reed by the stop plate. The scaled dimension to note here is the left-hand edge of the dark area, for that is positioned accurately; some artistic licence has been used to portray the extent of the clamped area rightwards of that position.

On the left are all of the input and output data values of the calculation, any one of which can be readily changed, whereupon the computer screen refreshes itself, virtually instantaneously, with the numerical answers and the reed block image. It will be noted that the values of YMOD and RDRH are missing from those columns. The data values for Young's Modulus and material density are held within the program as permanent data and all that is required is to inform the program, when it asks, if the petal material is steel, glass-fiber, or carbon-fiber. The information is recognized as a character string within the program and the appropriate properties of the reed material are indexed from the program memory.

The reed valve design program, Prog.6.4, is used to empirically design reed blocks and petals in advance of using the data in engine modelling programs. This program is useful to engineers, as it is presumed that they are like the author, designers both by "eye" and from the numerical facts. The effectiveness of such modelling programs(1.13) has already been demonstrated in Figs. 6.15-6.17, so it is instructive to examine three empirical designs for the engine examples used in these chapters, i.e., two for the 125 Grand Prix engine and one for the chainsaw engine to replace the piston controlled intake port debated hitherto.

The result of using the program for this purpose is found in Figs. 6.18 and 6.19. The considerable difference in bulk of the reed block for the two engines is apparent, even though the engines are only 25% different in swept volume but quite different in terms of power and BMEP attainment. The chainsaw engine will function satisfactorily with two reed petals and ports and a 20 mm carburetor. The Grand Prix engine needs six larger reed petals and ports and a 36 mm carburetor. The steel reed petal on the chainsaw is 30 mm free length and 0.20 mm thick, which gives a first natural frequency of 185 Hz, against the engine maximum forcing frequency of 9000/60, or 150 Hz; this reed petal will safely function at the maximum engine speed seen in Fig. 6.9.

On the other hand, one of the two designs for the Grand Prix engine has a steel reed of 38 mm free length and, although thicker at 0.25 mm, has a natural frequency of 144 Hz. As the engine has a natural forcing frequency of 192 Hz, i.e., 11500/60, this reed is in serious danger of early fracture from resonance induced fatigue as it passes through a resonant frequency each time the machine is revved to 11500 rpm in each gear when employed as a racing motorcycle engine. Of some interest is the photograph in Plate 4.1, which is of a 250 cc twin-cylinder racing motorcycle engine, i.e., each cylinder is 125 cc like the design being discussed. On this engine is a reed valve induction system which can be observed in the lower right hand corner of the photograph.

The alternative design for the 125 Grand Prix engine has a carbon-fiber reinforced plastic reed which is the same length as the steel reed but is 0.60 mm

CYLINDER SV, cc= 125   SPEED, rpm= 11500   BMEP, bar= 11   CCR= 1.5
PETAL MATERIAL IS CARBON-FIBRE
PETAL THICKNESS 'RDTH', mm= .6
REED BLOCK ANGLE 'RDBA', deg= 26
PETAL NUMBER 'RDNO'= 6
PORTS NUMBER 'RDPNO'= 6
PETAL WIDTH 'RDWD', mm= 23
PORT WIDTH 'RDPW', mm= 20
CORNER RADIUS 'PRAD', mm= 3
PETAL LENGTH 'RDLN', mm= 38
PORT LENGTH 'RDPL', mm= 28
LENGTH FROM CLAMP 'RDPS', mm= 8
OUTPUT DATA
REQUIRED FLOW AREA, mm2=1396.
DESIGNED FLOW AREA, mm2=1468.
CARBURETTER DIA., mm= 36.
REED NATURAL FREQUENCY, Hz= 261.
ENGINE NATURAL FREQUENCY, Hz= 192.
TIP LIFT RATIO (0.15-0.25)=0.17
STOP PLATE RADIUS 'SPRD', mm= 58.

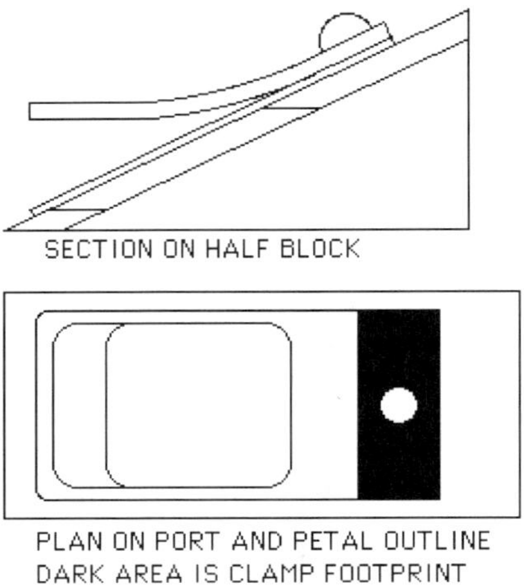

Fig. 6.19 Computer screen output from Prog.6.4, "REED VALVE DESIGN," for the 125 Grand Prix engine.

thick. It does not lift so far as the steel reed, but its natural frequency is 261 Hz; hence, it is obviously a safer design with which to proceed to either the modelling or experimental stage.

6.3.4 Concluding remarks on reed valve design

The second-order natural frequency is shown in Fig. 6.18, but not on the screen printout. It will be seen that this is at a frequency level well beyond any normal engine forcing frequency. It is held within the program as information and the user can get it printed should that be desired. Actually, Fleck et al(1.13) report that the high-speed video pictures taken of reed valve motion show considerable evidence of second-order mode vibration but not of resonance at that frequency. The theoretical simulation by computer modelling also records second-order mode amplitude levels of some significance.

It is not uncommon to find flat plate reed valves being used in small outboard motors and in other low specific power output engines. A flat plate reed block is just that, a single plate holding the reed petals at right angles to the gas flow direction. The design procedure within Prog.6.4 uses a reed block angle, RDBA, of 40º and proceeds with the design as usual. Ultimately the plan view of the working drawing will appear with the reed plate, per petal and port that is, exactly as the lower half of the sketch in Fig. 6.19.

With this inside knowledge of the reed design and behavior in advance of engine modelling or experimentation, the engineer is in a sound position to tackle the next stage of theoretical or experimental design and development. Should the reader assemble a computer program to simulate reed valve motion similar to that of Fleck et al(1.13) by using the theoretical approach described in the technical papers from QUB, then this section will permit a better initial selection of the data for use within that model.

6.4 The empirical design of disc valves for two-stroke engines

This subject is one which the author rarely discusses without a certain feeling of nostalgia, because it was the motorcycle racing engines from MZ in East Germany, with their disc valve inlet systems and expansion chamber exhaust pipes, which appeared on the world's racing circuits during his undergraduate student days. They set new standards of specific power performance. It was, literally, a triumph of intellect and professional engineering over lack of finance, resources, materials and facilities. For many years thereafter, the disc valve intake system was acknowledged to be the superior method of induction control for high specific output engines. However, it was not long before it was learned how to produce equal, if not superior performance, from piston control of the intake system. The current technical position is that the reed valve has supplanted the piston controlled intake port for racing engines and is probably the most popular intake control method for all engines from the cheapest 2 hp brushcutter to the most expensive 300 hp V8 outboard motor. Nevertheless, the disc valve engine, either in the flat form made popular by MZ and identical to the Rotax design shown in Plate 1.8, or the rotary sleeve version conceived by QUB(5.2)(5.3) (Plate 1.1), may become more popular again on the racing circuits. Racing engine design has a fashion element, as well as engineering logic, at work. Therefore, it is important that a design procedure for disc valve induction be among the designer's options and that engine modelling programs are available to predict the behavior of the disc valved engine. This enables direct comparisons to be made of engine performance when fitted with all of the possible alternatives for the induction process.

As with other data required for engine modelling programs, the designer needs to be able to establish empirical factors so that a minimum of guesswork is required when faced with the too-numerous data bank of a major computer program or when rapid optimization is required for the design. The following discussion sets out a logic procedure which by now should be familiar, for it repeats the same basic empirical methodology used for the piston controlled intake and the reed valve induction systems.

6.4.1 Specific time area analysis of disc valve systems

Fig. 1.7(a), Fig. 6.13, and Fig. 6.20 show the mechanical positioning, porting control and timing of the disc valve intake device. The earlier explanations in this chapter regarding the influence of the specific time area on the BMEP attainable are just as relevant here, as the disc valve merely gives an asymmetrical element to the port area diagrams of Fig. 6.1 for piston controlled ports. Consequently, Fig. 6.20 shows that it is possible to (a) open the intake port early, thereby taking full advantage of induction over the period when the crankcase experiences a sub-atmospheric pressure and (b) shut it early before the piston forces out any trapped air charge on the crankcase compression stroke. Actually, this is an important feature, for the intake pressure wave is weaker by virtue of taking place over a longer time period, and so the ramming wave is not so strong as to provide the vigorous ramming effect seen for the piston controlled induction system. In Fig. 6.13, the angle required to open the port into the engine is represented by ANGH in degrees of crankshaft rotation. The total cutaway angle on the disc is shown as ANGD. If the disc valve opens at DVO deg. btdc and closes at DVC deg. atdc, then the total opening period is DVP, where:

$$DVP=DVO+DVC \tag{6.4.1}$$

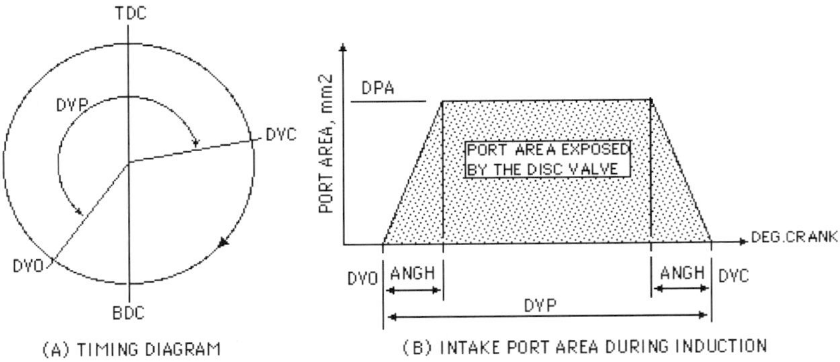

Fig. 6.20 Disc valve system timing diagram and areas during the intake process.

It is also straightforward geometry to show that the relationship between DVP, ANGD and ANGH is given by:

$$DVP=ANGD+ANGH \tag{6.4.2}$$

The port opens and closes in ANGH degrees as shown in Fig. 6.20(b) and, either precisely for a port as sketched in Fig. 6.13, or approximately for any other shape such as a round hole, that port area rises linearly to a maximum value of DPA, the total inlet port area. Such geometry is easily programmed into an overall engine model typified by Prog.5.1. Using the nomenclature of Fig. 6.13, the total value of the maximum port area, DPA, is calculated from its occupancy of a segment of a circular annulus, less the corner radii residues:

$$DPA=[ANGH*\pi*(RMAX^2-RMIN^2)/360]-[RCRN^2*(1-\pi/4)] \qquad (6.4.3)$$

The $\int A*d\varnothing$ is directly calculable from the trapezoid in Fig. 6.20(b) as:

$$\int A*d\varnothing=DPA*(DVP-ANGH) \qquad (6.4.4)$$

From the previous discussion regarding the relevance of specific time area, the numerical level for the intake in question as a function of BMEP is given by Eq. 6.1.7 and repeated here as:

$$INTA=(BMEP+1.528)/774 \qquad s/m \qquad (6.4.5)$$

When the target requirement of BMEP for a given swept volume, SV, at an engine speed of RPM is declared, there is available a direct solution for the design of the entire device. From Eq. 6.1.12:

$$\text{Specific time area, s/m}= INTA=\Sigma[A*d\varnothing]/(RPS*360*SV) \qquad (6.4.6)$$

If one glances ahead at Fig. 6.21, and notes the data listed in the upper left-hand side of that figure as being all of the relevant input data for the calculation, it is seen that the unknown data value which emerges from the calculation is the value of the outer port edge radius, RMAX. This is the most convenient method of solution. Eqs. 6.4.1-6.4.6 are combined, estimates are made of all data but RMAX, and the result is a direct solution for RMAX. This is not as difficult as it might appear, for disc valve timings are not subject to great variations from the lowest to the highest specific power output level. For example, DVO and DVC would be at 150 and 65 for a small scooter engine and at 140 and 80, respectively, for a racing engine. For completeness, the actual solution route for the determination of RMAX and the carbureter flow diameter, CARB, is given below:

$$\int(A*d\varnothing)=6*INTA*SV*RPM \qquad (6.4.7)$$

$$DPA=\int(A*d\varnothing)/(DVP-ANGH) \qquad (6.4.8)$$

$$RMAX=\sqrt{\{[360*(Kexp*DPA+(RCRN^2*(1-\pi/4)))/(\pi*ANGH)]+RMIN^2\}} \qquad (6.4.9)$$

$$CARB=\sqrt{[4*DPA*\sqrt{Kexp}/\pi]} \qquad (6.4.10)$$

The value of the expansion coefficient, Kexp, is very similar to that used for the reed valve case and is in the range of 1.35-1.45. Actually, in the reed valve design theory it is employed as the reciprocal, at 0.70. In the computer solution presented in Prog.6.5, the actual value used is 1.44.

It should be noted, as with the solution of similar equations, if the data for swept volume in Eq. 6.4.7 is inserted in the conventional units of cm^3, then the next line will evaluate the port area DPA as mm^2. If strict SI units are used, then SV is in m^3 in Eq. 6.4.7 and Eq. 6.4.8 will produce the output for DPA in the strict SI units of m.

6.4.2 A computer solution for disc valve design, Prog.6.5

So that the designer may concentrate more on the design process and less on the arithmetic tedium of solution, the above equations are programmed into a computer program, Prog.6.5, "DISC VALVE DESIGN," listed in the Computer Program Appendix as ProgList.6.5. As with the majority of the programs presented in this book, maximum advantage is taken of the screen graphics capability of the desktop computer. An example of the use of the calculation is given in Fig. 6.21, which shows the computer screen image of the program being used to design a disc valve option for induction to the 125 Grand Prix engine, one of the power units used throughout these chapters as a working example of a high specific output engine. The screen displays the correctly scaled drawing of the disc valve and the intake port, based on both the input data dimensions inserted by the designer and the output data as calculated by the program. Indeed, the computer screen picture is almost identical to the view given of the left-hand cylinder in Plate 1.8.

The input data is listed in the upper half of the printed portion of the diagram. As discussed above, the performance target requirements based on cylinder swept volume and the BMEP to be attained at a peak power engine speed are both needed, and are followed by an estimate of the disc valve timings for the opening and closing positions. The latter are not difficult to estimate for the reasons already stated. The principal parameters requiring design experience are those for the minimum radius, RMIN, and the port opening angle, ANGH.

```
SWEPT VOLUME, cc= 125    ENGINE BMEP, bar= 11    SPEED, rpm= 11500
DISC VALVE OPENS, deg.btdc= 140
DISC VALVE CLOSES, deg.atdc= 80
DISC MINIMUM RADIUS, 'RMIN', mm= 28
PORT CORNER RADII, 'RCRN', mm= 8
PORT ANGLE, 'ANGH', deg.= 56

OUTPUT DATA
MAXIMUM RADIUS, 'RMAX', mm= 58.
DISC ANGLE, 'ANGD', deg=164.
CARBURETTER DIA., mm= 36.
PORT HEIGHT, mm= 30.
PORT WIDTH at mid-height, mm= 35.
ASPECT RATIO, width/height=1.20
```

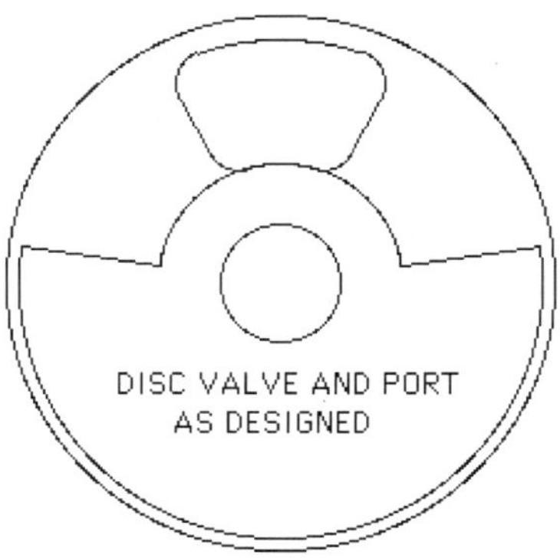

Fig. 6.21 Computer screen display from Prog.6.5, "DISC VALVE DESIGN," for a 125 Grand Prix engine.

The minimum radius, RMIN, is determined by the geometry of the engine crankshaft. The reader should refer to Fig. 1.7(a), where it can be seen that the value of this minimum radius is decided by the path of the gas flow past the outer diameter of the engine's main bearing on that side of the crankshaft. Therefore, this element of the disc valve design must be carried out in conjunction with the design of the crankshaft and crankcase. It will be an iterative process, going to and fro between the two design procedures, until a satisfactory compromise has been achieved for the disc valve and the crankshaft with its bearings and oil seals.

The output data listed on the screen, or subsequently line-printed, gives the ensuing values for RMAX, the carbureter flow diameter, and the disc valve cutaway angle, ANGD. It also gives the port aspect ratio at the mid-port height. The aspect ratio is defined as width to height, and the designer should attempt to retain it within the band of 1.0 to 1.25 by suitably adjusting the input parameter of ANGH. The gas flow from the (usually) round carburetor exit area to the port area needs a minimum expansion path to follow, so that the coefficient of discharge of the junction is as high as possible. A simple guide to follow is to arrange the carburetor flow diameter to be approximately equal to the width of the intake port at mid-height.

It should be noted that the dimension RMAX is not the outer diameter of the disc but the outer radius of the port. The actual disc needs an outer diameter which will seal the face of the intake port during the crankcase compression stroke. This will depend on several factors, such as the disc material, its thickness and stiffness, but a value between 3 and 5 mm of overlap in the radial direction would be regarded as conventional.

6.5 Concluding remarks

The reader, having reached this juncture in the book, will realize that the end of a design cycle has been achieved. The first chapter introduced the subjects in general, the second elaborated on unsteady gas flow, the third on scavenging, the fourth on combustion, the fifth on engine modelling, and this, the sixth, on the assembly of information so that the engine modelling can be performed in a logical manner. It is now up to the reader to make the next move.

Some will want to use the various computer programs as soon as possible, depending on their level of experience or need. Some will want to use the programs from Chapter 2, so that they can become more familiar with unsteady gas flow as a means of understanding engine behavior. Some will apply directly the engine modelling programs of Chapter 5 to the design process for actual engines. The book is liberally sprinkled with the data for all programs so that the initial runs can be effected by checking that their computer listing has been inserted accurately and that the results tally with those which illustrate this book.

The most important global view expressed here by the author for the reader to consider is that the actual engine, like the engine model, uses the "separate" technologies of Chapters 1 to 4 to behave like the paper engines of Chapter 5. The best engine designer is one who can think in terms of a paper engine turning in his mind, each turn bringing images of the effect of changing the unsteady gas flow regime, the scavenging regime, and the combustion regime together with the mutual interaction of each of those effects. This implies a thorough understanding of all of those topics, and the more thorough that understanding, the more complex can be the interaction which the mind will handle. Should one of those topics be missing or poorly understood, then the quality of the mental debate will suffer. It is hoped that the visual imagery provided by the computer programs presented in this book will accelerate the understanding of all of these topics.

As pointed out frequently thus far, our understanding of many phenomena is in its infancy and it is only twenty-five years on from the black art of much of the 1950's technology. It behooves the thinking designer to follow and to study the progress of technology through the technical papers which are published in this and related fields of endeavor.

The remainder of this book will concentrate on selected specialized topics, and particularly on the future of the design and development of two-stroke engines.

NOTATION FOR CHAPTER 6

NAME	SYMBOL	SI UNITS	OTHER
Crankcase Compression Ratio	CCR		
Delivery Ratio	DR		
Ratio of Specific Heats	g		
Scavenging Efficiency	SE		
Scavenge Ratio	SR		
Trapped Compression Ratio	TCR		
Trapping Efficiency	TE		
Area	A	m^2	
Brake Mean Effective Pressure	BMEP	Pa	bar, atm, psi
Brake Specific Fuel Consumption	BSFC	kg/Ws	kg/kWh, lb/hp.hr
Clearance Volume	CV	m^3	cm^3
Crankshaft angle	ø	degree	
Cylinder Bore	BO	m	mm, in
Cylinder Stroke	ST	m	mm, in
Density	D	kg/m^3	g/cc, lb/in^3
Engine Rotation Rate	RPS	rev/s	rev/min, rpm, rps
Friction Mean Effective Pressure	FMEP	Pa	bar, atm, psi
Gas Constant	R	J/kgK	
Power	BPR	W	kW, bhp
Particle velocity	CG	m/s	
Pressure	P	Pa	bar, atm, psi
Pumping Mean Effective Pressure	PMEP	Pa	bar, atm, psi
Swept Volume	SV	m^3	cm^3, in^3
Speed of sound	Ao	m/s	
Specific time area, blowdown	BLOWTA	s/m	
Specific time area, exhaust	EXTA	s/m	
Specific time area, transfer	TRTA	s/m	
Specific time area, inlet	INTA	s/m	

NAME	SYMBOL	SI UNITS	OTHER
Temperature	T	K	ºC, ºF
Time	t	s	
Trapped Swept Volume	TSV	m^3	cm^3
Volume	V	m^3	
Young's Modulus	YMOD	N/m^2	GN/m^2

REFERENCES FOR CHAPTER 6

6.1 J.L.M. Morrison, B. Crossland, An Introduction to the Mechanics of Machines, Longman, 2nd Edition, London, 1971.

6.2 F.S. Tse, I.E. Morse, R.T. Hinkle, Mechanical Vibrations, Theory and Applications, 2nd Edition, Allyn and Bacon, 1978.

6.3 C. Bossaglia, Two-Stroke High Performance Engine Design and Tuning, Chislehurst and Lodgemark Press, London, 1972.

6.4 H. Naitoh, M. Taguchi, "Some Development Aspects of Two-Stroke Cycle Motorcycle Engines," SAE Mid-Year Meeting, Detroit, Michigan, June, 1966, SAE Paper No.660394

6.5 H. Naitoh, K. Nomura, "Some New Development Aspects of Two-Stroke Cycle Motorcycle Engines," SAE Automotive Congress and Exposition, Detroit, Michigan, January, 1971, SAE Paper No.710084.

6.6 J.F.T. MacClaren, S.V. Kerr, "Automatic Reed Valves in Hermetic Compressors," IIR Conference, Commission III, September, 1969.

Chapter 7

Reduction of Fuel Consumption and Exhaust Emissions

7.0 Introduction

Throughout the development of the internal combustion engine, there have been phases of concentration on particular aspects of the development process. In the first major era, from the beginning of the twentieth century until the 1950's, attention was focused on the production of ever greater specific power output from the engines, be they two- or four-stroke cycle power units. To accomplish this, better quality fuels with superior octane ratings were prepared by the oil companies so that engines could run at higher compression ratios without risk of detonation. Further enhancements were made to the fund of knowledge on materials for engine components, ranging from aluminum alloys for pistons to steels for needle roller bearings, so that high piston speeds could be sustained for longer periods of engine life. The text of this book thus far has concentrated on the vast expansion of the knowledge base on gas dynamics, thermodynamics and fluid mechanics which has permitted the design of engines to take advantage of the improvements in materials and tribology. Each of these developments has proceeded at an equable pace. For example, if a 1980's racing engine had been capable of being designed in 1920, it would have been a case of self-destruction within ten seconds of start-up due to the inadequacies of the fuel, lubricant, and materials from which it would have been assembled at that time. Should it have lasted for any length of time, at that period in the 1920's, the world would have cared little that its fuel consumption rate was excessively high, or that its emission of unburned hydrocarbons or oxides of nitrogen was potentially harmful to the environment!

The current era is one where design, research and development is increasingly being focused on the fuel economy and exhaust emissions of the internal combustion engine. The reasons for this are many and varied, but all of them are significant and important.

The world has a limited supply of fossil fuel of the traditional kind, i.e., that which emanates from prehistorical time and is available in the form of crude oil capable of being refined into the familiar gasoline or petrol, kerosene or paraffin, diesel oil and lubricants. These are the traditional fuels of the internal combustion engine and it behooves the designer, and the industry which employs him, to develop more efficient engines to conserve that dwindling fossil fuel reserve. Apart from ethical considerations, many governments have enacted legislation setting limits on fuel consumption for various engine applications.

The population of the world has increased alarmingly, due in no small way to a more efficient agriculture which will feed these billions of humans. That agricultural system, and the transportation systems which back it up, are largely efficient due to the use of internal combustion engine driven machinery of every conceivable type. This widespread use of ic engines has drawn attention to the exhaust emissions from its employment, and in particular to those emissions which are harmful to the environment and the human species. For example, carbon monoxide is toxic to humans and animals. The combination of unburned hydrocarbons and nitrogen oxides, particularly in sunlight, produces a visible smog which is harmful to the lungs and the eyes. The nitrogen oxides are blamed for the increased proportion of the rainfall containing acids which have a debilitating effect on trees and plant growth in rivers and lakes. Unburned hydrocarbons from marine engines are thought to concentrate on the beds of deep lakes, affecting in a negative way the natural development of marine life. The nitrogen oxides are said to contribute to the depletion of the ozone layer in the upper atmosphere, which potentially alters the absorption characteristics of ultraviolet light in the stratosphere and increases the radiation hazard on the earth's surface. There are legitimate concerns that the accumulation of carbon dioxide and hydrocarbon gases in the atmosphere increases the threat of a "greenhouse effect" changing the climate of the Earth.

One is tempted to ask why it is the important topic of today and not yesterday. The answer is that the engine population is increasing faster than people, and so too is the volume of their exhaust products. All power units are included in this critique, not just those employing reciprocating ic engines, and must also encompass gas turbine engines in aircraft and fossil fuel burning electricity generating stations. Actually, the latter are the largest single source of exhaust gases into the atmosphere.

The discussion in this chapter will be in two main segments. The first concentrates on the reduction of fuel consumption and emissions from the simple, or conventional, two-stroke engine which is found in so many applications requiring an inexpensive but high specific output powerplant such as motorcycles, outboard motors and chainsaws. There will always be a need for such an engine and it behooves the designer to understand the methodology of acquiring the requisite performance without an excessive fuel consumption rate and pollutant exhaust emissions. The second part of this chapter will focus on the design of engines with fuel consumption and exhaust pollutant levels greatly improved over that available from the "simple" engine. Needless to add, this involves some further mechanical complexity or the use of expensive components, otherwise it would be employed on the "simple" engine. As remarked in Chapter 1, the two-stroke engine has an inherently low level of exhaust emission of nitrogen oxides, and this makes it an attractive proposition for future automobile engines, provided that the extra complexity and expense involved does not make the two-stroke powerplant non-competitive with its four-stroke engine competitor.

Before embarking on the discussion regarding engine design, it is necessary to expand on the information presented in Chapter 4 on combustion, particularly relating to the fundamental effects of air-fuel ratio on pollutant levels and to the basic differences inherent in homogeneous and stratified charging, and homogeneous and stratified combustion.

7.1.1 Some fundamentals regarding combustion and emissions

Although much of the fundamental material regarding combustion is covered in Chapter 4, there remains some discussion which is specific to this chapter and the topics therein.

The first is to explain the origins of exhaust emission of carbon monoxide, unburned hydrocarbons and nitrogen oxides from the combustion process. The reader will recall the simple chemical relationship posed in Eq. 1.5.16 for the stoichiometric combustion of air and gasoline. Also, the reader should remember the discussion in Chapter 4, wherein it is stressed that the combustion of fuel and air occurs with vaporized fuel and air, but not liquid fuel and air. Gasoline is defined as octane, the eighth member of the family of paraffins (alkanes) whose general formula is C_xH_{2x+2}. The stoichiometric combustion equation is repeated here.

$$2*C_8H_{18} + 25*O_2 + (25*79/21)*N_2$$
$$=16*CO_2 +18*H_2O+ (25*79/21)*N_2 \qquad (7.1.1)$$

The air-fuel ratio, AF, emanating from this balanced equation is calculated as:

$$AF=(25*32+25*79*28/21)/(16*12+36*1)=15.06 \qquad (7.1.2)$$

However, should the combustion process not be conducted at the correct air-fuel ratio then a different set of exhaust gas components would appear on the right-hand side of this equation. For example, consider the situation where the air-fuel ratio is 20% lean of the stoichiometric value, i.e., there is excess air present during the combustion process. The Eq. 7.1.1 is modified as follows:

$$2*C_8H_{18} +1.2*\{25*O_2 +(25*79/21)*N_2\}$$
$$=16*CO_2 +18*H_2O+ 1.2*(25*79/21)*N_2 + 5*O_2 \qquad (7.1.3)$$

It is observed that the exhaust gas now contains oxygen which is due to the excess of air over that required to consume the fuel.

Consider the rich air-fuel ratio, for example at 20% rich of the stoichiometric value. The following "ideal" equation would ensue, based on the premise that the more active hydrogen consumes all of the oxygen before the carbon can so do.

$$2*C_8H_{18} + 0.8*\{25*O_2 + (25*79/21)*N_2\}$$
$$= 10*CO + 6*CO_2 +18*H_2O + 0.8*(25*79/21)*N_2 \qquad (7.1.4)$$

It can be seen that the exhaust gas now contains a significant fraction of carbon monoxide in this theoretical combustion of a rich mixture.

In summary, from these simple examples of applied chemistry, one would expect to see a zero level of carbon monoxide and oxygen in the exhaust gas after a "perfect" combustion process, but increasing quantities of carbon monoxide if the air-fuel mixture becomes richer, or oxygen if it is leaner. As with all real-life processes, no combustion process, even at the stoichiometric mixture, is ever quite "perfect" for the molecule of octane. A more realistic combustion analysis would reveal that some of the hydrocarbon molecule never breaks down completely, leaving unburned hydrocarbons, and that some of the carbon monoxide never achieves complete oxidation to carbon dioxide, even in the presence of excess air. Many of these effects are due to a phenomenon described as "dissociation"(4.1). A further experimental fact is the association of nitrogen with oxygen to form the nitrogen oxide pollutant, NO_x, and this undesirable result becomes more pronounced as the combustion temperature is increased. Thus, the stoichiometric equation, Eq. 7.1.1, is more realistically:

$$\begin{aligned} &2*C_8H_{18} + 25*O_2 + (25*79/21)*N_2 \\ &=Z1*CO_2 + Z2*H_2O + Z3*N_2 + Z4*CO \\ &\quad + Z5*C_xH_y + Z6*NO_x + Z7*O_2 \end{aligned} \tag{7.1.5}$$

It should be remembered that the molecular quantities of the pollutants, Z4-Z7, are quite small for a stoichiometric combustion situation. Exhaust gas analysis is usually conducted with a dry exhaust gas sample, i.e., the steam is normally removed. To put an approximate number on the values of Z4-Z7 for a "dried" exhaust gas sample of exhaust emanating from the combustion process alone, i.e., not confusing the hydrocarbon analysis with poorly scavenged fresh charge in the two-stroke engine, one would record values such as: 0.15% CO by volume, 1% O_2 by volume, unburned hydrocarbons as 600 ppm CH_4 (methane) equivalent, and 500 ppm NO equivalent. As a first approximation, ignoring the values of Z4-Z7 as being non-significant arithmetically, the total moles in the dried exhaust gas sample are principally derived from the carbon dioxide and the nitrogen, i.e., (16+25*79/21), or 110.05. The volumetric proportions of the pollutants are also (from Avogadro) molecular proportions. Therefore:

for CO	Z4=0.15*110.05/100=0.165
for O_2	Z5=1*110.05/100=1.10
for CH_4	Z6=600*110.05/1E6=0.066
for NO	Z7=500*110.05/1E6=0.055

It will be observed that the original coefficients for carbon dioxide, steam and nitrogen would be barely affected by this modified exhaust gas analysis. The proportion by volume of carbon dioxide in this dried exhaust gas sample would be given by:

$$\%vol\ CO_2=100*16/110.05=14.5$$

It is reasonably clear from the foregoing that, should the air-fuel ratio be set correctly for the combustion process to the stoichiometric value, even an efficient combustion system will still have unburned hydrocarbons, carbon monoxide and nitrogen oxides in the exhaust gas from the engine. Should the air-fuel ratio be set incorrectly, either rich or lean of the stoichiometric value, then the exhaust pollutant levels will increase. If the air-fuel mixture is very lean so that the flammability limit is reached and misfire takes place, then the unburned hydrocarbon and the carbon monoxide levels will be considerably raised. It is also clear that the worst case, in general, is at a richer air-fuel setting, because both the carbon monoxide and the unburned hydrocarbons are inherently present on theoretical grounds.

It is also known, and the literature is full of technical publications on the subject, that the recirculation of exhaust gas into the combustion process will lower the peak cycle temperature and act as a damper on the production of nitrogen oxides. This is a standard technique at this time for production four-stroke automobile engines to allow them to meet legislative requirements for nitrogen oxide emissions. In this regard, the two-stroke engine is ideally suited for this application, for the retention of exhaust gas is inherent from the scavenging process. This natural scavenging effect, together with the lower peak cycle temperature due to a firing stroke on each cycle, allows the two-stroke engine to produce very reduced nitrogen oxide exhaust emissions at equal specific power output levels.

Any discussion on exhaust emissions usually includes a technical debate on catalytic after-treatment of the exhaust gases for their added purification. In this chapter, there is a greater concentration on the design methods to attain the lowest exhaust emission characteristics before any form of exhaust after-treatment takes place.

As a postscript to this section, there may be readers who will look at the relatively tiny proportions of the exhaust pollutants in Eq. 7.1.5 and wonder what all the environmental, ecological or legislative fuss is about in the automotive world at large. Let such readers work that equation into yearly mass emission terms for each of the pollutants in question for the annual consumption of many millions of tons of fuel per annum. The environmental problem then becomes quite self-evident!

7.1.2 Homogeneous and stratified combustion and charging

The combustion process can be conducted in either a homogeneous or stratified manner, and an introduction to this subject is given in Sect. 4.1. The words "homogeneous" and "stratified" in this context define the nature of the mixing of the air and fuel in the combustion chamber at the period of the flame propagation through the chamber. A compression ignition or diesel engine is a classic example of a stratified combustion process, for the flame commences to burn in the rich environment of the vaporizing fuel surrounding the droplets of liquid fuel sprayed into the combustion chamber. A carburetted four-stroke cycle si engine is the classic example of a homogeneous combustion process, as the air and fuel at the onset of ignition are thoroughly mixed together, with the gasoline in a gaseous form.

Both of the above examples give rise to discussion regarding the charging of the cylinder. In the diesel case, the charging of the cylinder is conducted in a stratified manner, i.e., the air and the fuel enter the combustion chamber separately and any mixing of the fuel and air takes place in the combustion space. As the liquid fuel is sprayed in 35º before tdc it cannot achieve homogeneity before the onset of combustion. In the carburetted four-stroke cycle si engine example, the charging of the engine is conducted in a homogeneous fashion, i.e., all of the required air and fuel enter together through the same inlet valve and are considered to be homogeneous, even though much of the fuel is still in the liquid phase at that stage of the charging process

It would be possible in the case of the carburetted four-stroke cycle si engine to have the fuel and air enter the cylinder of the engine in two separate streams, one rich and the other a lean air-fuel mixture, yet, by the onset of combustion, be thoroughly mixed together and burn as a homogeneous combustion process. In short, the charging process could be considered as stratified and the combustion process as homogeneous. On the other hand, that same engine could be designed, viz the Honda CVCC type, so that the rich and lean air-fuel streams are retained as separate entities up to the point of ignition and the combustion process is also carried out in a stratified manner. The main point behind this discussion is to emphasize the following points:

(a) If a spark-ignition engine is charged with air and fuel in a homogeneous manner, the ensuing combustion process is almost inevitably a homogeneous combustion process.

(b) If an engine is charged with air and fuel in a stratified manner, the ensuing combustion process is possibly, but not necessarily, a stratified combustion process.

In the analysis conducted above for the combustion of gasoline (see Eqs. 7.1.1-7.1.5), the air-fuel ratio is noted as the marker of the relationship of that combustion process to the stoichiometric, or ideal. The reader will interpret that as being the ratio of the air and fuel supply rates to the engine. This will be perfectly accurate for a homogeneous combustion process, but can be quite misleading for a design where stratified charging is taking place.

Much of the above discussion is best explained by the use of a simple example illustrated in Fig. 7.1. The "engine" in the example is one where the combustion space can contain, or be charged with, 15 kg of air. Consider the "engine" to be a spark-ignition type and the discussion is pertinent for both two-stroke and four-stroke cycle engines.

At a stoichiometric air-fuel ratio for gasoline, this means that a homogeneously charged engine, followed by a homogeneous combustion process, would ingest 1 kg of octane with the air. This situation is illustrated in Fig. 7.1(a1) and (b1). The supplied air-fuel ratio and that in the combustion space are identical at 15.

If the engine had stratified charging, but the ensuing mixing process is complete followed by homogeneous combustion, then the situation is as illustrated in Fig. 7.1(a2) and (b2). Although one of the entering air-fuel streams has a rich air-fuel ratio of 10 and the other is lean at 30, the overall air-fuel ratio is 15, as is the air-fuel ratio in the combustion space during burning. The supplied air-fuel ratio and that in the combustion space are identical at 15. In effect, the overall behavior is the same as for homogeneous charging and combustion.

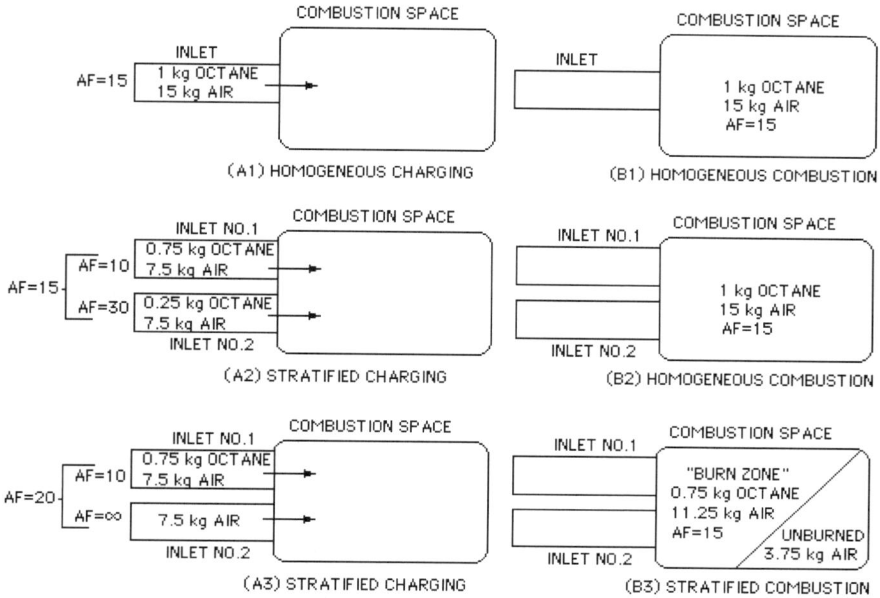

Fig. 7.1 Homogeneous and stratified charging and combustion.

If the engine has both stratified charging and combustion, then the situation portrayed in Fig. 7.1 (a3) and (b3) becomes a real possibility. At an equal "delivery ratio" to the previous examples, the combustion space will hold 15 kg of air. This enters in a stratified form with one stream rich at an air-fuel ratio of 10 and the second containing no fuel at all. Upon entering the combustion space, not all of the entering air in the second stream mixes with the rich air-fuel stream, but a sufficient amount does to create a "burn zone" with a stoichiometric mixture at an air-fuel ratio of 15. This leaves 3.75 kg of air unburned which exits with the exhaust gas. The implications of this are:

(a) The overall or supplied air-fuel ratio is 20, i.e., it gives no indication of the air-fuel ratio during the actual combustion process and is no longer an experimental measurement which can be used to optimize the combustion process. For example, many current production automobile engines have "engine management systems" which rely on the measurement of exhaust oxygen as a means of electronically controlling the overall air-fuel ratio to the stoichiometric value.

(b) The combustion process would release 75% of the heat available in the homogeneous combustion example, and it could be expected that the BMEP and power output would be similarly reduced. In the technical phrase used to describe this behavior, the "air-utilization" characteristics of stratified combustion are not as efficient as homogeneous combustion. The diesel engine is a classic example of this phenomenon, where the overall air-fuel ratio for maximum thermal efficiency is usually 25% greater than the stoichiometric value.

(c) The exhaust gas will contain a significant proportion of oxygen. Depending on the exhaust after-treatment methodology, this may or may not be welcome.

(d) The brake specific fuel consumption will be reduced, all other parameters being equal. The IMEP attainable is lower with the lesser fuel mass burned and, as the parasitic losses of friction and pumping are unaffected, the BSFC deteriorates.

(e) An undesirable combustion effect can appear at the interface between the burned and unburned zones. Tiny quantities of aldehydes and ketones are produced in the end zones, and although they would barely register as pollutants on any instrumentation, the hypersensitive human nose records them as unpleasant odors(4.4). Diesel engine combustion suffers from this complaint.

The above discussion may appear as one of praise for homogeneous combustion and derision of stratified combustion. Such is not the case, as the reality for the two-stroke engine is that stratified charging, and possibly also stratified combustion, will be postulated in this chapter as a viable design option for the reduction of fuel consumption and exhaust emissions. In that case, the goal of the designer becomes the maximization of air-utilization and the minimization of the potentially undesirable side-effects of stratified combustion. On the bonus side, the two-stroke engine has that which is lacking in the four-stroke cycle powerplant: an uncluttered cylinder head zone for the design and creation of an optimum combustion space.

7.2 The simple two-stroke engine

This engine has homogeneous charging and combustion and is spark-ignited, burning a volatile fuel such as gasoline, natural gas or kerosene. It is commonly found in a motorcycle, outboard motor or industrial engine and the fuel metering is conventionally via a carburetor. In general, the engine has fresh charge supplied via the crankcase pump. Indeed, the engine would be easily recognized by its inventor, Sir Dugald Clerk, as still embodying the modus operandum he envisaged; he would, it is suspected, be somewhat astonished at the level of specific power output which has been achieved from it at this juncture in the twentieth century!

The operation of this engine has been thoroughly analyzed in earlier chapters, and repetition here would be just that. However, to achieve the optimum in terms of fuel consumption or exhaust emissions, it is necessary to re-examine some of those operating characteristics. This is aided by Fig. 7.2.

The greatest single problem for the simple engine is the homogeneous charging, i.e., scavenging, of the cylinder with fuel and air. By definition, the scavenging process can never be "perfect" in such an engine because the exhaust port is open as fresh charge is entering the cylinder. At best, the designer is involved in a damage limitation exercise. This has been discussed at length in Chapter 3 and further elaborated on in Chapter 5 by the use of a computer model of the engine which incorporates a simulation of the scavenging process. It is proposed to debate this matter further so that the designer is familiar with all of the available options for the improvement of those engine performance characteristics relating to fuel economy and exhaust emissions.

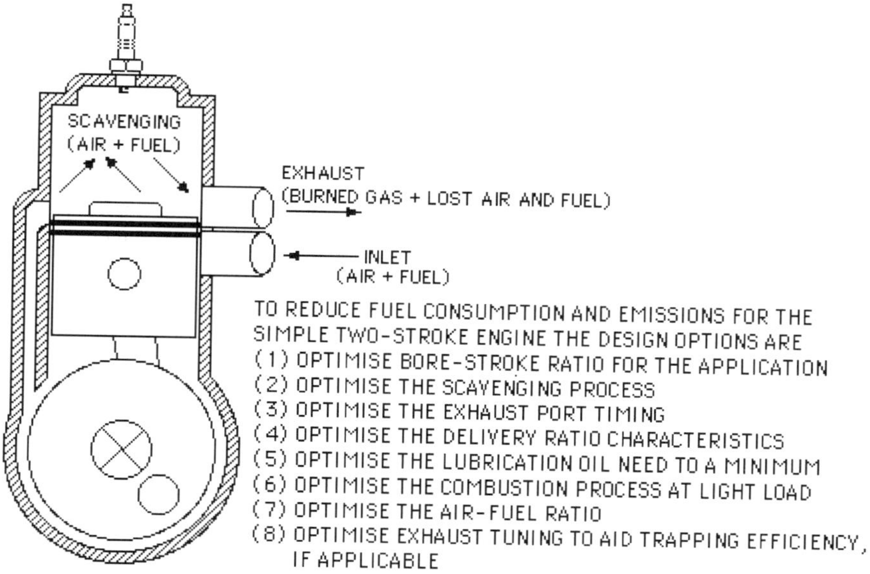

Fig. 7.2 The several design options to improve the simple two-stroke engine.

Linked to the scavenging problem is the necessity to tailor the delivery ratio curve of the engine to suit the application. Any excess delivery ratio over that required results in merely pumping air and fuel into the exhaust system.

One of the factors within design control is exhaust port timing and/or exhaust port area. Many engines evolve or are developed with the peak power performance requirement at the forefront of the process. Very often the result is an exhaust port timing which is excessively long even for that need. The end product is an engine with poor trapping characteristics at light load and low speed, which implies poor fuel economy and exhaust emissions. This subject will be debated further in this chapter, particularly as many of the legislative tests for engines are based on light load running as if the device were used as an automotive powerplant in an urban environment. Equally, the designer should never overlook the possibility of improving the trapping efficiency of any engine by suitable exhaust pressure wave tuning; this matter is fully covered in Chapters 2 and 5. Every 10% gain in trapping pressure is at least a 10% reduction in BSFC and an even larger proportionate improvement in hydrocarbon exhaust emissions.

Together with scavenging and delivery ratio there is the obvious necessity of tailoring the air-fuel ratio of the supplied charge to be as close to the stoichiometric as possible. As a 10% rich mixture supplies more power at some minor expense in BSFC, the production engine is often marketed with a carburetor set at a rich mixture level, more for customer satisfaction than for necessity. Legislation on exhaust emissions will change that manufacturing attitude in the years ahead, but the designer is often presented by a cost-conscious management with the simplest and most uncontrollable of carburetors as part of a production package beyond designer influence. The air-fuel ratio control of some of these mass-produced cheap carburetors is very poor indeed. That this can be rectified, and not necessarily in an expensive fashion, is evident from the manufacturing experience of the automobile industry since the so-called "oil crisis" of 1973.

Many of the simplest engines use lubricant mixed with the gasoline as the means of engine component oiling. In Great Britain this is often referred to as "petroil" lubrication. For many years, the traditional volumetric ratio of gasoline to oil was 25 or 30. Due to legislative pressure, particularly in the motorcycle and outboard field, this ratio is much leaner today, between 50 and 100. This is due to improvements in both lubricants and engine materials. For many applications, separate, albeit still total-loss lubrication methods, are employed with oil pumps supplying lubricant to selected parts of the engine. This allows gasoline-oil ratios to be varied from 200 at light loads to 100 at full load. This level of oiling would be closely aligned with that from equivalent sized four-stroke cycle engines. It behooves the designer to continually search for new materials, lubricants and methods to further reduce the level of lubricant consumption, for it is this factor which influences the exhaust smoke output from a two-stroke engine at cold start-up and at light load. In this context, the papers by Fog et al(7.22) and by Sugiura and Kagaya(7.25) should be studied as they contain much practical information.

One of the least understood design options is the bore-stroke ratio. Designers, like the rest of the human species, are prone to fads and fashions. The in-fashion of today is for over-square engines for any application, and this approach is probably based on the success of over-square engines in the racing field. Logically speaking, it does not automatically follow that it is the correct cylinder layout for an engine-driven, portable, electricity generating set which will never exceed 3000 or 3600 rpm.

If the application of the engine calls for its extensive use at light loads and speeds, such as a motorcycle in urban traffic or trolling for fish with an outboard motor, then a vitally important factor is the maintenance of the engine in a "two-stroke" firing mode, as distinct from a "four-stroke" firing mode. This matter is introduced in Sect. 4.1.3. Should the engine skip-fire in the manner described, there is a very large increase in exhaust hydrocarbon emissions. This situation can be greatly improved by careful attention during the development phase to combustion chamber design, spark plug location and spark timing. Even further gains can be made by exhaust port timing and area control, and this matter will be discussed later in this chapter.

To satisfy many of the design needs outlined above, the use of a computer based simulation of the engine is ideal. This is particularly true of items 1, 2, 3, 4, 6, and 8 listed in Fig. 7.2. The engine model presented in Chapter 5, Prog.5.1, will be used in succeeding sections to illustrate many of the points made above and to provide an example for the designer that such models are not primarily, or solely, aimed at design for peak specific power performance.

7.2.1 Typical performance characteristics of simple engines

Before embarking on the improvement of the exhaust emission and fuel economy characteristics of the simple two-stroke engine, it is important to present and discuss some typical measured data for such engines.

7.2.1.1 Measured performance data from a QUB 400 research engine

The first set of data to be presented is from the QUB 400 single-cylinder research engine(1.20). This is the same engine whose complete geometrical data is given in Figs. 5.3 and 5.4 and whose performance is analyzed in Sect. 5.2. The engine speed selected for discussion in Chapter 5 is 3000 rpm and it is appropriate that a complete set of measured performance characteristics at that same engine speed be given here, as Figs. 7.3-7.8. Figs. 7.3-7.5 are at full throttle and Figs. 7.6-7.8 are at 10% throttle opening area ratio. In each set are data, as a function of air-fuel ratio, for BMEP, BSFC, unburned hydrocarbon emissions as both ppm and bsHC values, and carbon monoxide and oxygen exhaust emission levels. It is worth remembering from the earlier discussion in Chapter 5 that this engine has no exhaust pressure wave tuning to aid the trapping efficiency characteristic, therefore, the performance characteristics attained are due solely to the design of the porting, scavenging, and combustion chamber. The engine does not have a high trapped compression ratio; the TCR value is somewhat low at 6.7. Even without exhaust pressure wave tuning, that this is not a low specific power output engine is evident from the peak BMEP level of 6.2 bar at 3000 rpm. Consider the measured data around the stoichiometric air-fuel ratio of 15.

That this simple two-stroke engine must have good scavenging characteristics is seen from the hydrocarbon emission levels, which are at 80 g/kWh at full throttle and 17 g/kWh at a light load of 2.65 bar BMEP. The raw HC emission data is 4200 ppm and 1250 ppm, respectively. The BSFC is at 0.40 kg/kWh at 6.2 bar BMEP and 0.30 kg/kWh at 2.65 bar BMEP. The carbon monoxide level is as low as 0.2% at full throttle and 0.1% at one-tenth throttle. The oxygen emission is 7.5% at full throttle and 3% at one-tenth throttle; it will be remembered that the majority of the oxygen emission derives from the air lost during the scavenge process with about 1% coming from the inefficiency of the combustion process. These are probably the best fuel consumption and emission values recorded at QUB, particularly for a simple engine capable of 6.2 bar BMEP without exhaust tuning. By comparing notes with colleagues in industrial circles, it is probable that these figures represent a "state of the art" position for a simple, single-cylinder, two-stroke engine at this point in history. The levels of fuel economy and exhaust emissions at the lighter load point of 2.65 bar BMEP would be regarded as particularly impressive, and there would be four-stroke cycle engines which could not improve on these numbers. It is noticeable that the carbon monoxide levels are at least as low as those from four-stroke cycle engines, but this is the one data value which truly emanates from the combustion process alone, and is not confused by intervening scavenging losses.

Nevertheless, the values of hydrocarbon emission are, by automotive standards, very high. The raw value of HC emission caused by combustion inefficiency alone should not exceed 400 ppm, yet it is 4200 ppm at full load. This is a measure of the

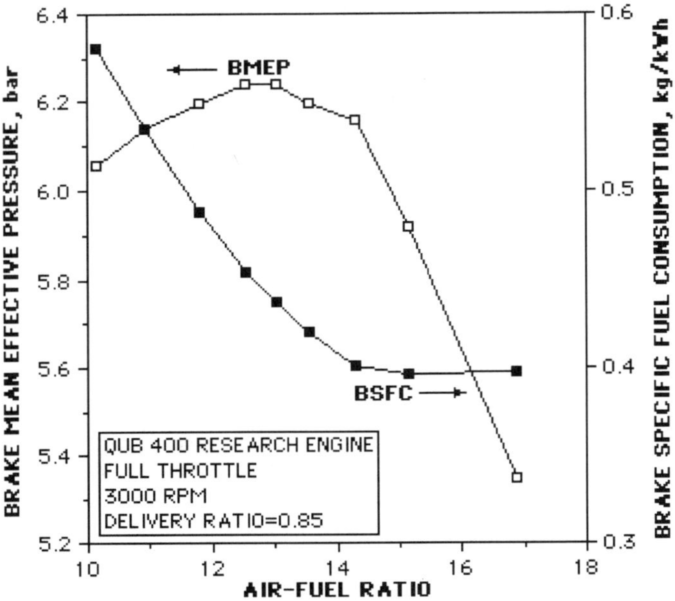

Fig. 7.3 Influence of air-fuel ratio on BMEP and BSFC at full throttle.

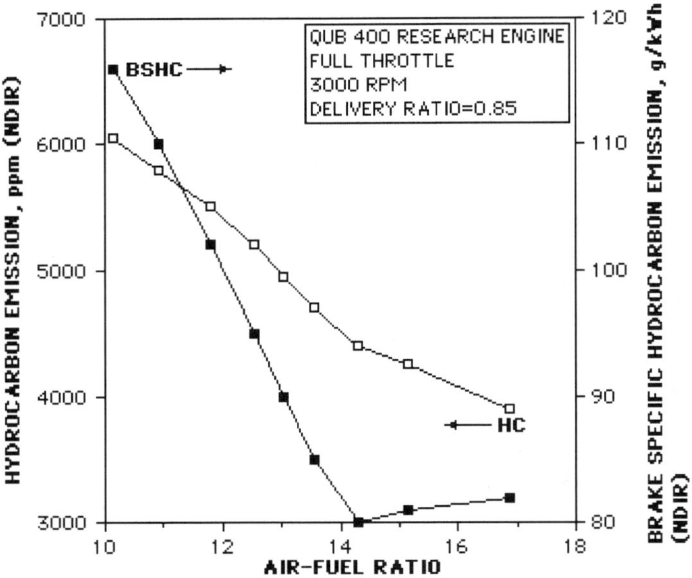

Fig. 7.4 Influence of air-fuel ratio on HC emission at full throttle.

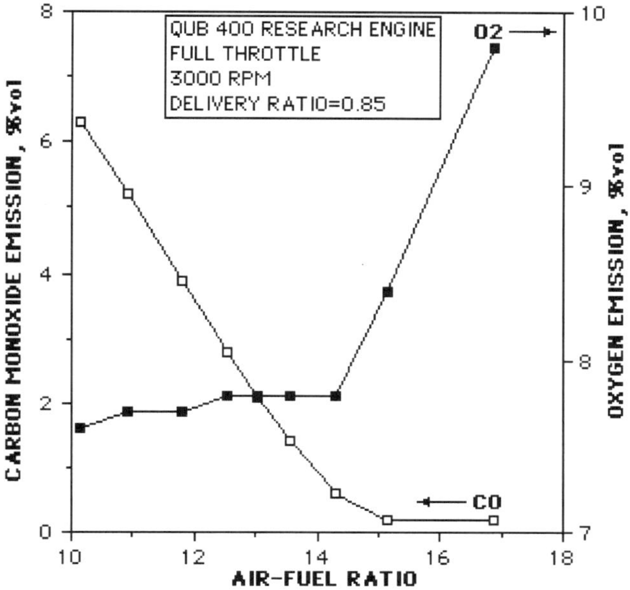

Fig. 7.5 Influence of air-fuel ratio on CO and O2 emission at full throttle.

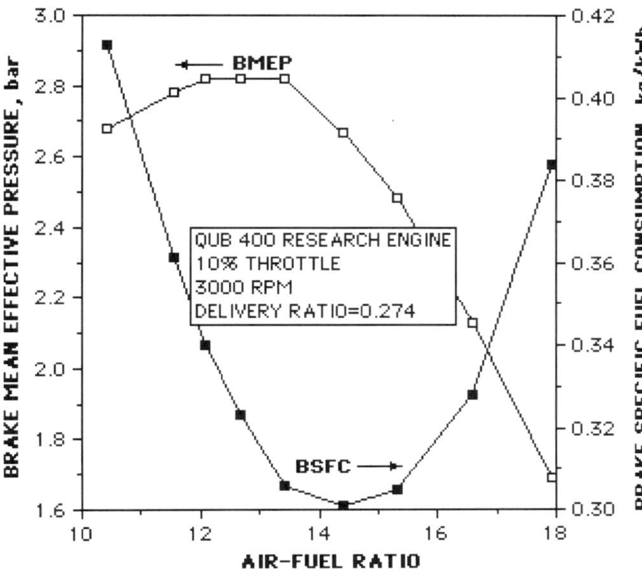

Fig. 7.6 Influence of air-fuel ratio on BMEP and BSFC at 10% throttle.

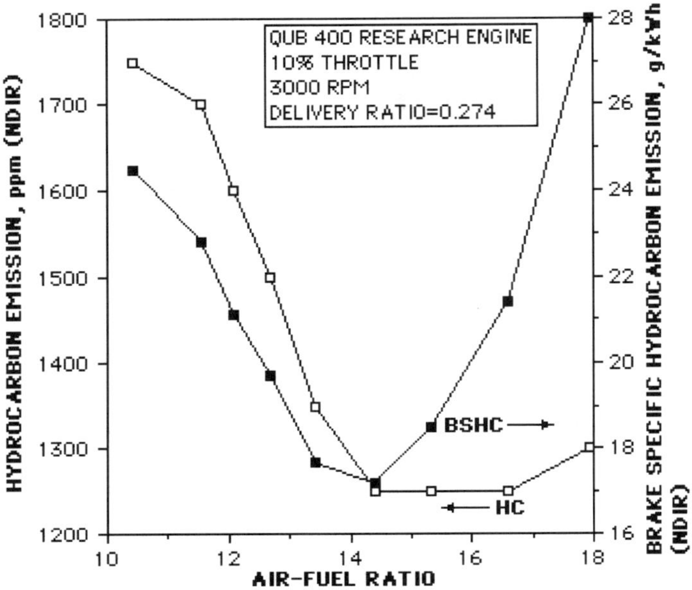

Fig. 7.7 Influence of air-fuel ratio on HC emission at 10% throttle.

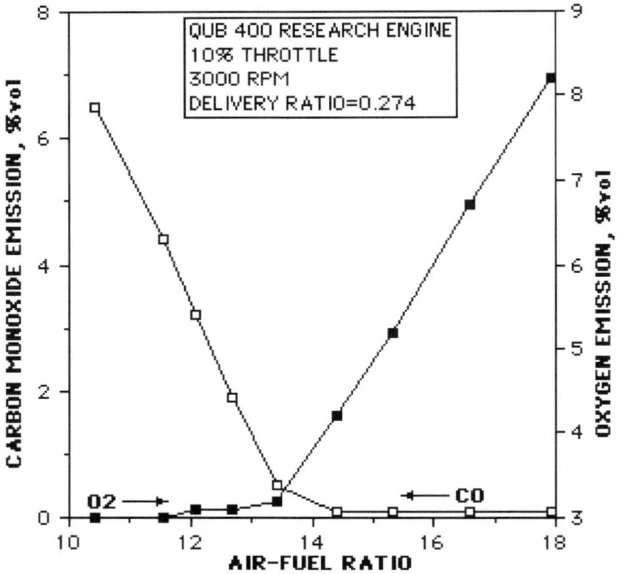

Fig. 7.8 Influence of air-fuel ratio on CO and O2 emission at 10% throttle.

ineffectiveness of homogeneous charging, i.e., scavenging, of a simple two-stroke engine. It forever rules out the use of simple two-stroke engines in automotive applications against a background of legislated emissions levels.

To reinforce these comments, it will be recalled from Sect. 1.6.3 that exhaust oxygen concentration can be used to compute trapping efficiency, TE. The data sets shown in Fig. 7.5 and 7.8 would yield a TE value of 0.6 at full throttle and 0.86 at one-tenth throttle. This latter is a remarkably high figure and attests to the excellence of the scavenge design. What these data imply, however, is that if the fuel were not short-circuited with the air, and all other engine behavioral factors remained as they were, the exhaust HC level could possibly be about 350 ppm, but the full throttle BSFC would actually become 240 g/kWh and the one-tenth throttle BSFC would be at 260 g/kWh. These would be exceptional BSFC values by any standards and they illustrate the potential attractiveness of an optimized two-stroke engine to the automotive industry.

To return to the discussion regarding simple two-stroke engines, Figs. 7.3-7.8 should be examined carefully in light of the discussion in Sect. 7.1.1 regarding the influence of air-fuel ratio on exhaust pollutant levels. As carbon monoxide is the one exhaust gas emission which is not distorted in level by the scavenging process, it is interesting to note that the theoretical predictions provided by Eqs. 7.1.1-7.1.4 for stoichiometric, rich, and lean air-fuel ratios, are quite precise. In Fig. 7.5, the CO level falls linearly with increasing air-fuel ratio and it levels out at the stoichiometric value. At one-tenth throttle in Fig. 7.8, exactly the same trend occurs.

The theoretical postulations in terms of the shape of the oxygen curve are also accurate. In Fig. 7.5 the oxygen profile is flat until the stoichiometric air-fuel ratio, and increases linearly after that point. The same trend occurs at one-tenth throttle in Fig. 7.8, although the flat portion of the curve ends at 14:1 air-fuel ratio rather than at the theoretical stoichiometric level of 15.

The brake specific fuel consumption and the brake specific hydrocarbon emission are both minimized at, or very close to, the stoichiometric air-fuel ratio.

All of the theoretical predictions from the relatively simple chemistry used in Eqs. 7.1.1-7.1.4 are shown to be relevant. In short, for the optimization of virtually any performance characteristic, the simple two-stroke engine should be operated at the stoichiometric air-fuel ratio with a tolerance of +0%, -6%. The only exception is maximum power or torque, where the optimum air-fuel ratio is observed to be at 13, i.e., 13% rich of the stoichiometric level.

7.2.1.2 Typical performance maps for simple two-stroke engines

It is necessary to study the more complete performance characteristics for simple two-stroke engines so that the designer is aware of the typical characteristics of such engines over the complete load and speed range. Such performance maps are presented in Figs. 7.9-7.11 from the publication by Batoni(7.1) and in Fig. 7.12 from the paper by Sato and Nakayama(7.2).

(a) The experimental data from Batoni(7.1)

In Figs. 7.9-7.11 the data is taken from a 200 cc motor scooter engine which has very little exhaust tuning to assist with its charge trapping behavior. The engine is carburetted and spark-ignited, and is that used in the familiar Vespa motor scooter.

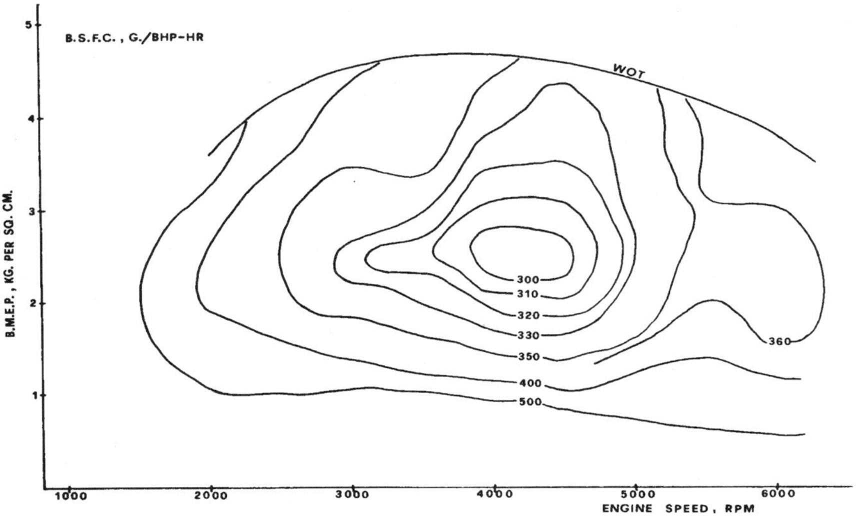

Fig. 7.9 Fuel consumption map for a 200 cc two-stroke engine.

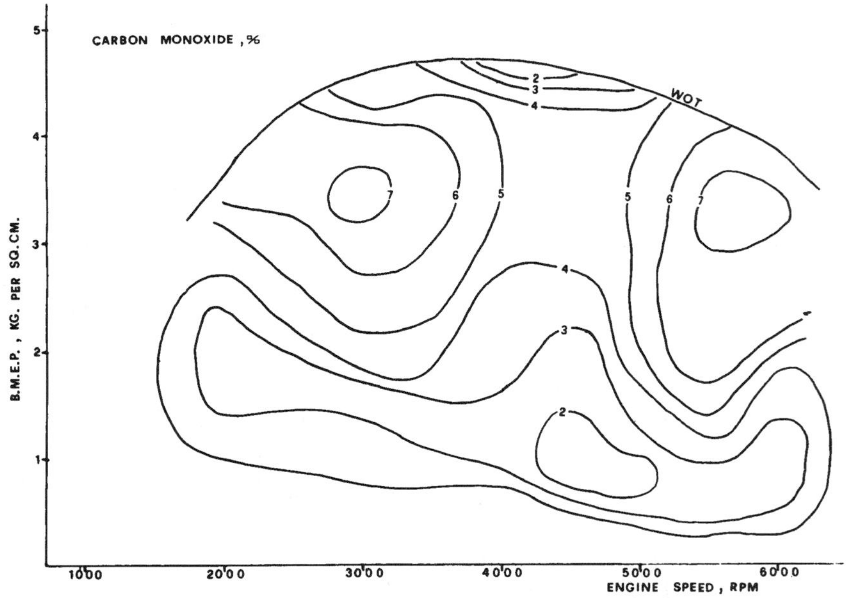

Fig. 2 - CO emission - production 2-stroke

Fig. 7.10 Carbon monoxide emission map for a 200 cc two-stroke engine.

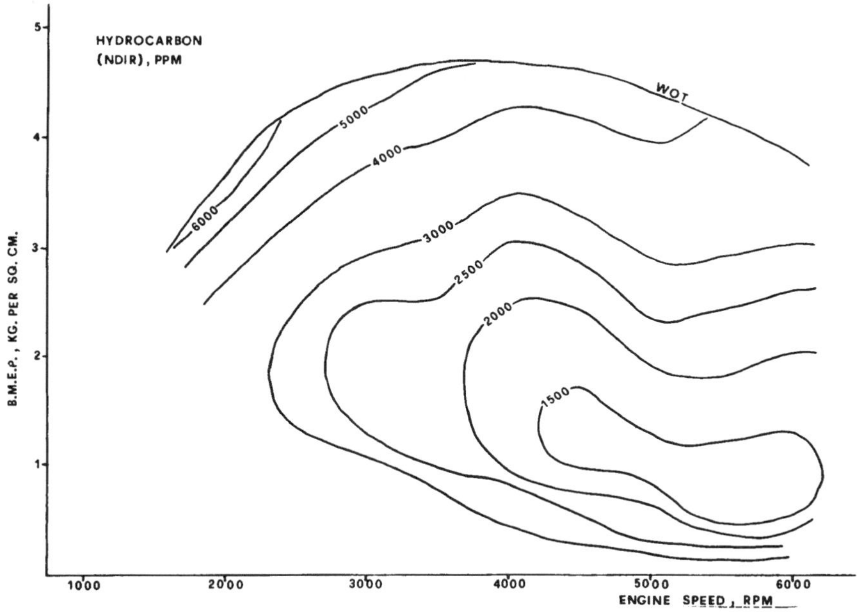

Fig. 7.11 Hydrocarbon map for a 200 cc two-stroke engine.

The units for BMEP are presented as kg/cm^2, and 1 kg/cm^2 is equivalent to 0.981 bar; the units of BSFC are presented as g/hp.hr, and 1 g/hp.hr is equivalent to 0.746 g/kWh.

The BMEP from this engine has a peak of 4.6 bar at 3500 rpm. It is observed that the best BSFC occurs at 4000 rpm at about 50% of the peak torque and is a quite respectable 0.402 kg/kWh. Below the 1 bar BMEP level the BSFC deteriorates to 0.67 kg/kWh. The map has that general profile which causes it to be referred to in the jargon as an "oyster" map.

The carbon monoxide emission map has a general level between 2 and 6%, which would lead one to the conclusion, based on the evidence in Figs. 7.5 and 7.8, that the air-fuel ratio used in these experimental tests was in the range of 12 to 13. By the standards of equivalent four-stroke cycle engines, this level of CO emission would be normal or even slightly superior for the two-stroke engine.

The hydrocarbon emission map, which has units in ppm from a NDIR measurement system, is directly comparable with Figs. 7.4 and 7.7, and exhibits values which are not dissimilar from those recorded for the QUB 400 engine. To be more specific, it would appear that the hydrocarbon emission levels from a simple two-stroke engine will vary from 5000 ppm at full load to 1500 ppm at light load; note that the levels quoted are those recorded by NDIR instrumentation. The recording of unburned hydrocarbons and other exhaust emission levels is discussed earlier in Sect. 1.6.2. As the combustion process is responsible for 300-400 ppm of those hydrocarbon emissions, it is clear that the scavenging process is responsible for the creation of between 3 and 10 times that emission level which would be produced by an equivalent four-stroke cycle power unit.

In general, the best fuel economy and emissions occur at load levels which are considerably less than the peak value. This is directly attributable to the trapping characteristic for fresh charge in a homogeneously scavenged engine. The higher the scavenge ratio, whether the particular scavenging process be a good one or a bad one, the greater the load and power, but the lower the trapping efficiency (see Fig. 7.13, the discussion in the next section, or any of the TE-SR characteristics presented in Chapter 3).

(b) The experimental data from Sato and Nakayama(7.2)

As the QUB 400 and the Batoni data does not contain information on nitrogen oxide emission, it is important that measurements are presented to provide the reader with the position occupied by the simple two-stroke engine in this regard. Such data has been provided by quite a few researchers, all of them indicating very low nitrogen oxide emission by a two-stroke powerplant. The data of Sato and Nakayama(7.2) is selected because it is very representative, is in the form of a performance engine speed map, and it refers to a simple carburetted two-stroke engine with a cylinder capacity of 178 cc. The actual engine has two cylinders, each of that capacity, and is employed in a snowmobile.

The measured data is given in Fig. 7.12. As would be expected, the higher the load or BMEP, the greater the peak cycle temperature and the level of the oxides of nitrogen. The values are shown as NO equivalent and measured as ppm on NDIR instrumentation. The highest value shown is at 820 ppm, the lowest is at 60 ppm, and the majority of the performance map is in the range from 100 to 200 ppm. This is much lower than that produced by the equivalent four-stroke engine, perhaps by as much as a factor of between 4 and 8. It is this inherent characteristic, introduced earlier in Sect. 7.1.1, that has attracted the automobile manufacturers to indulge in research and development of two-stroke engines; this will be discussed further in later sections of this chapter, as it will not be a "simple" two-stroke engine which is developed for such a market requirement.

7.3 Optimizing the emissions and fuel economy of the simple two-stroke engine

In Sect. 7.2, the problems inherent in the design of the simple two-stroke engine are introduced and typical performance characteristics are presented. Thus, the designer is now aware of the difficulty of the task which is faced, for even with the best technology the engine is not going to be competitive with a four-stroke engine in terms of hydrocarbon emission. In all other respects, be it specific power, specific bulk, specific weight, maneuverability, manufacturing cost, ease of maintenance, durability, fuel consumption, or CO and NO emissions, the simple two-stroke engine is equal, and in some respects superior, to its four-stroke competitor. There may be those who will be surprised to see fuel consumption in that list, but investigation shows that small capacity four-stroke engines are not particularly thermally efficient. The reason is that the friction loss of the valve gear begins to

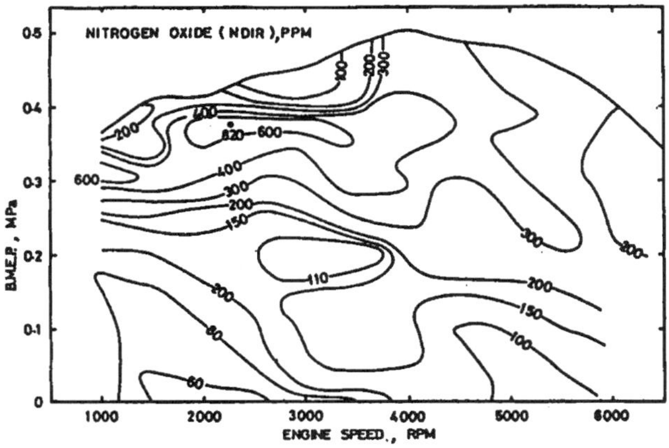

Fig. 7.12 Nitrogen oxides emission map for a 178 cc two-stroke engine.

assume considerable proportions as the cylinder size is reduced, and this deteriorates the mechanical efficiency of the engine.

Fig. 7.2 lists options which are open to the designer, and the remainder of this section will be devoted to their closer examination. In particular, the engine computer model will be used to illustrate the relevance of some of those assertions. This will reinforce much of the earlier discussion in Chapter 5.

7.3.1 The effect of scavenging behavior

In Chapter 5, and in this chapter, the QUB 400 single-cylinder research engine is used to illustrate much of the discussion. It is appropriate that it is used again to demonstrate the effects of design changes on engine performance, particularly as they relate to fuel economy and emissions. It is appropriate because the more complete the theoretical and experimental data given about a particular engine, it is hoped that the greater is the understanding gained by the reader of the design and development of any two-stroke engine.

In this section, the data for the physical geometry of the QUB 400 engine given in Figs. 5.3 and 5.4 are inserted into ENGINE MODEL NO.1, Prog.5.1, and are run over a range of throttle openings at 3000 rpm for two differing types of loop scavenging. The variation of throttle opening area ratio used in the calculations is from 0.15 to 1.0, and the individual values inserted for the data are 0.15, 0.2, 0.3, 0.4, 0.5, and 1.0. The scavenge systems tested are those listed as SCRE and YAM6, first introduced in Sect. 3.2.4. The SCRE scavenge type is a very good loop scavenged design, whereas the YAM6 type is shown to have rather indifferent scavenging qualities. The reader may be interested to note that the second model of the EXPAND function is used to describe the scavenging behavior within the computer program, this being introduced in Eq. 5.1.13; in the earlier exposition in Chapter 5, Eq. 5.1.11 is used for the analysis of behavior of the QUB 400 engine. The results of the calculations are given in Figs. 7.13 and 7.14.

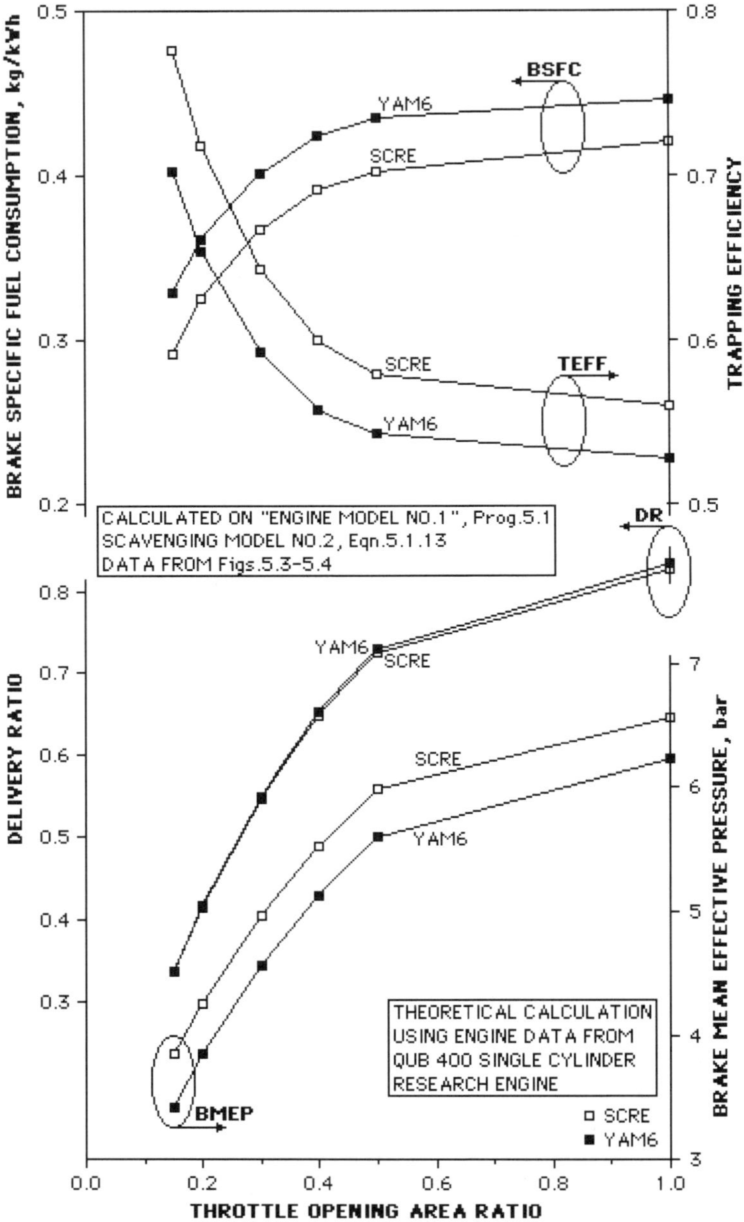

Fig. 7.13 Predicted performance characteristics of QUB 400 engine as a function of throttle opening area ratio.

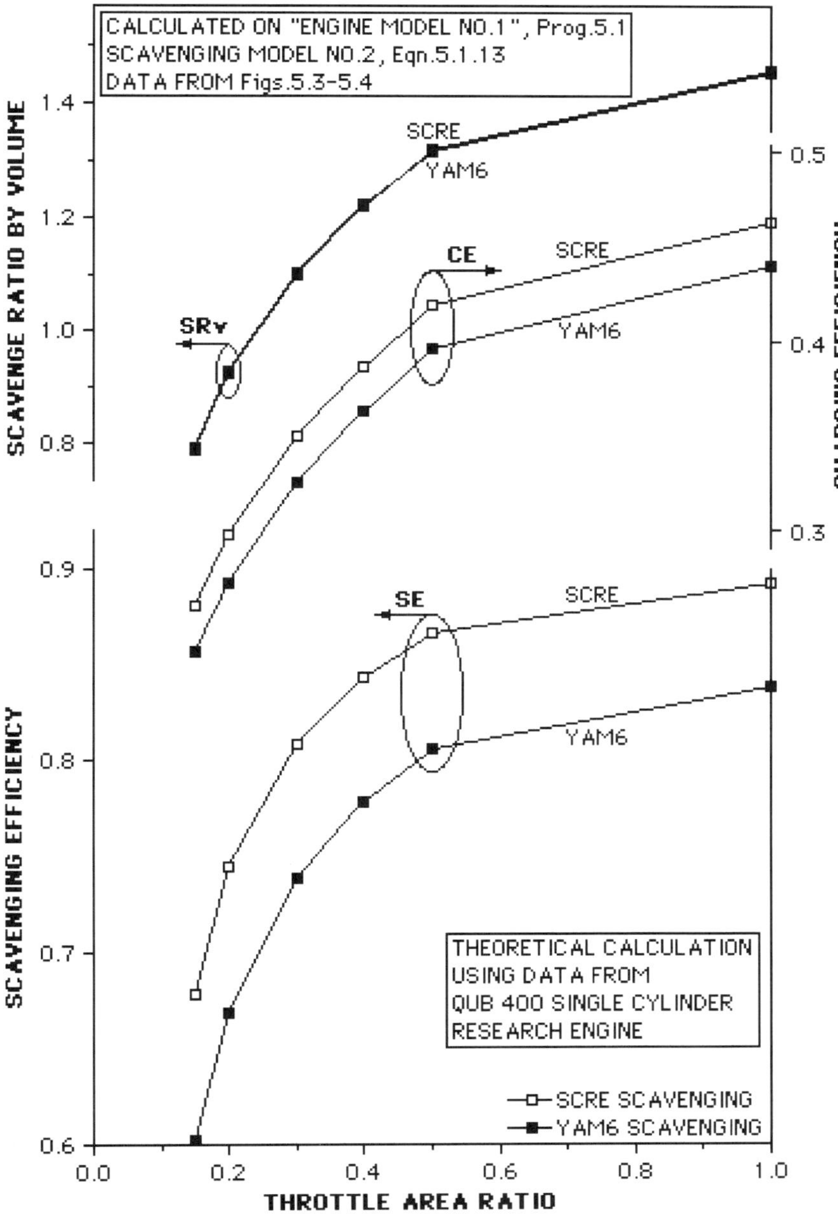

Fig. 7.14 Predicted scavenging behavior of QUB 400 engine as a function of throttle opening area ratio.

The principal variables being investigated are throttle opening, i.e., load variation controlled by delivery ratio, and the quality of the scavenging system employed. In Fig. 7.13, the close relationship between delivery ratio and BMEP is evident, this point having been discussed before as being the typical effect one observes for an engine which does not have any exhaust pressure wave tuning (see Sect. 5.2.3). The delivery ratio for the two scavenging types is identical, but the superior retention of fresh charge by the SCRE system is very evident. This translates into superior trapping efficiency and BSFC over the entire load range. The trapping efficiency profile with respect to delivery ratio, but more importantly and theoretically with respect to scavenge ratio (by volume) if the discussion in Chapter 3 is recalled, shows a decrease with increasing load, BMEP, and scavenge ratio. Due to this effect, the best specific fuel consumption occurs at the highest trapping efficiency. It is interesting to note that the scavenge ratio (by volume) values for the two scavenge systems are identical, but their disposition into charging efficiency and scavenging efficiency describes clearly the effectiveness of the SCRE, and the ineffectiveness of the YAM6, scavenging design. The improvements in BMEP and BSFC are proportionately greater for the SCRE than for the YAM6 design, as the load is reduced by throttle opening.

The fundamental messages to the designer regarding the options in this area are:

(a) Optimize the scavenging system to the highest level possible using the best theoretical and experimental tools available. It is recommended, as in Chapter 3, that a combination of experiment, using a single cycle gas scavenging rig, and theory, using CFD techniques, is employed in this regard.

(b) At the design stage, consider seriously the option of using an engine with a large swept volume and designing the entire porting and inlet system to operate with a low delivery ratio to attain a more modest BMEP at the design speed, while achieving the target power by dint of the larger engine swept volume. In this manner, with an optimized scavenging and air-flow characteristic, the lowest BSFC and exhaust emissions will be attained at the design point. The engine durability will also be improved by this methodology as the thermal loading on the piston will be reduced. There is a limit to the extent to which this design approach may be taken, as the larger engine will be operating closer to the misfire limit from a scavenging efficiency standpoint.

7.3.2 The effect of air-fuel ratio

Fig. 7.15 shows the result of employing the engine model, Prog.5.1, to predict the behavior of an engine, in this case the QUB 400 engine, at a throttle opening area ratio of 0.15 and an engine speed of 3000 rpm, over the range of air-fuel ratios from 10.5 to 17.5. The stoichiometric value is at 15. On the same figure, as a means of showing the relevance of the calculation, is plotted the same experimental data for a throttle opening of 0.1 which was previously presented in Fig. 7.6 for that same engine. This relatively simple theoretical computer model is seen to predict the variation with respect to air-fuel ratio quite well, and should give the designer confidence in its employment in this regard. The most important message to the

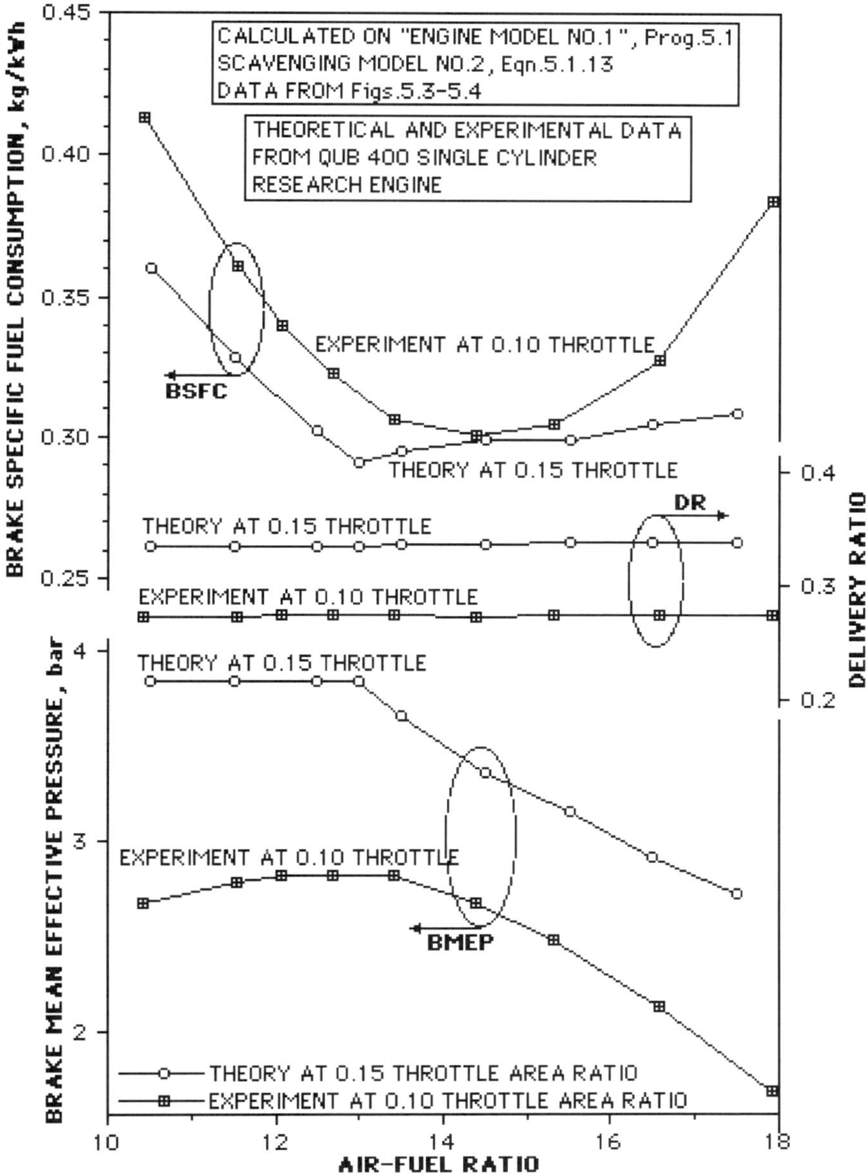

Fig. 7.15 QUB 400 engine at light load as a function of air-fuel ratio.

designer is the vital importance of having the fuel metered to the engine in the correct proportions with the air at every speed and load. There is at least as large variations of BSFC and BMEP with inaccurate fuel metering as there is in allowing the engine to be designed and manufactured with bad scavenging.

There is a tendency in the industry for management to insist that a cheap carburetor be installed on a simple two-stroke engine, simply because it is a cheap engine to manufacture. It is ironic that the same management will often take an opposite view for a four-stroke model within their product range, and for the reverse reason!

7.3.3 The effect of exhaust port timing and area

One of the more obvious methods of increasing the trapping efficiency of an engine is to lower the exhaust port timing, i.e., reduce both the period for scavenging flow and/or the area through which it passes, thereby lowering the delivery ratio, the BMEP attained, and the BSFC and exhaust emissions. Perhaps this obvious solution would accompany the suggestion in Sect. 7.3.1(b) of using a larger, more lightly loaded engine. Needless to add, if one applied this approach to the QUB 400 engine, it would no longer produce the desirable BMEP level of 6.24 bar from a high delivery ratio of 0.86 which is well trapped by a good scavenging system. However, it is informative to use the engine simulation model to examine the effect of a change to the exhaust port timing of the QUB 400 engine when the air throttle opening area ratio is at 0.15 and the engine speed is 3000 rpm. In other words, the data being used is exactly the same as in Sect. 7.3.2, but the exhaust port timing will be varied successively from the original standard value of 96º atdc opening, in 4º steps, until the exhaust port opening is at 116º atdc. At this point the blowdown period would be just 2º, for the transfer ports open at 118º atdc. The result of this calculation is shown in Fig. 7.16 for the variations of BMEP, BSFC, delivery ratio and trapping efficiency.

Rather surprisingly, the model predicts that the delivery ratio, at this low level of 0.33, would not be reduced further by lowering the exhaust port timing edge and, by inference, reducing its area. The trapping efficiency rises sharply, as does the BMEP produced, and the fuel economy is improved dramatically. From this calculation it is evident that the designer has to ensure that the lowest possible exhaust port timing is employed on any particular engine, consistent with attaining the peak power and speed required from the powerplant. This is a particularly subtle area for optimization, and one where it is vital to remember that neither one's experience, nor a computer simulation, nor any form of theoretical assistance will completely supplant a well organized test program conducted under the most realistic of experimental conditions. The reason for this is that the simple engine is always on the verge of four-stroking at light load and speed, and the first priority to attain good fuel economy and exhaust emissions is to ensure that the engine charges itself and fires evenly on each cycle. This can only be done effectively under experimental conditions.

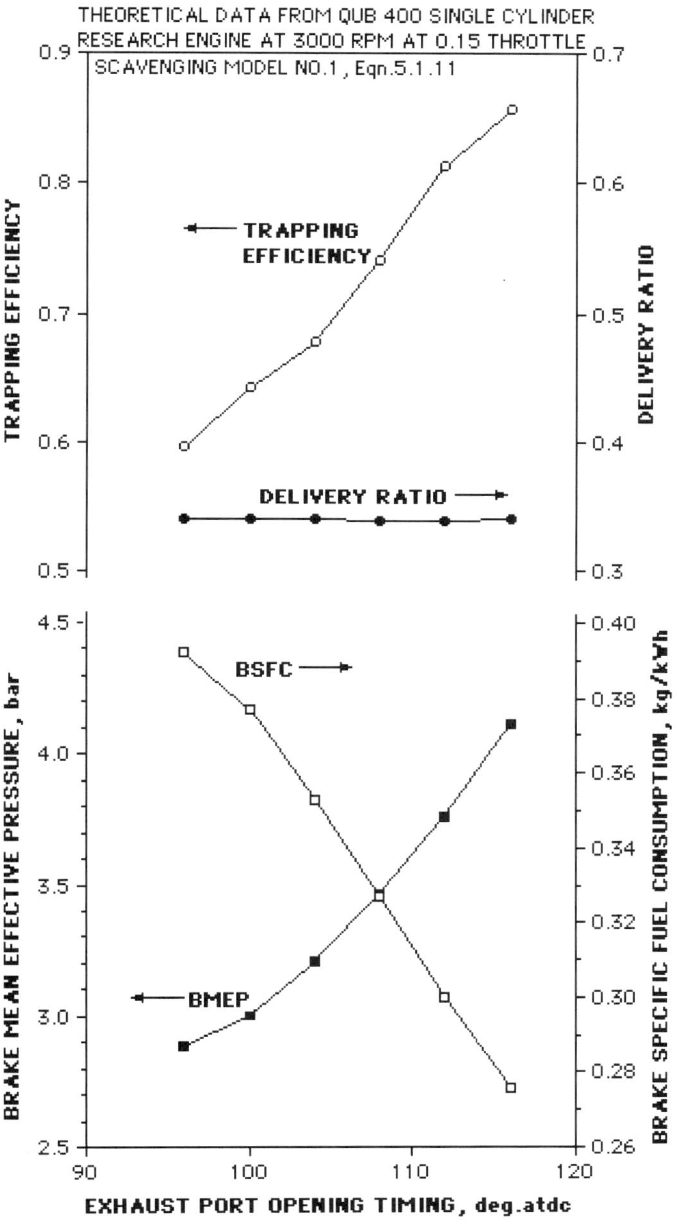

Fig. 7.16 Effect of changing exhaust port timing and area at light load on fuel economy and power output.

Nevertheless, the fundamental message to the designer is that control over the exhaust port timing and area has a dramatic influence on power output, fuel economy and exhaust emissions at light load and low engine speeds. This approach has been employed in several production engines, and the publication by Tsuchiya et al(7.3) discusses the subject in some detail.

There are two basic mechanical techniques to accomplish the design requirement for an exhaust port restriction to improve the light load behavior of the engine. The two methodologies are illustrated in Fig. 7.17 and discussed below.

7.3.3.1 The butterfly exhaust valve

The first of these, shown in Fig. 7.17(a) and (b) on the left of each diagram, is a butterfly valve and the concept is much like that described by Tsuchiya et al(7.3). This is a relatively simple device to manufacture and install, and has a good record of reliability in service. The ability of such a device to reduce exhaust emissions of unburned hydrocarbons is presented by Tsuchiya(7.3), and Fig. 7.18 is from that paper. Fig. 7.18 shows the reduction of hydrocarbon emissions, either as mass emissions in the top half of the figure or as a volumetric concentration in the bottom half, from a Yamaha 400 cc twin-cylinder road motorcycle at 2000 rpm at light load. The notation on the figure is for CR which is the exhaust port area restriction posed by the exhaust butterfly valve situated close to the exhaust port. The CR values range from 1, i.e., open as in Fig. 7.17(b), to 0.075, i.e., virtually closed as in Fig. 7.17(a). It is seen that the hydrocarbons are reduced by as much as 40% over a wide load variation at this low engine speed, emphasizing the theoretical indications discussed above.

Tsuchiya(7.3) reports that the engine behaved in a much more stable manner when the exhaust valve was employed at light load driving conditions in an urban situation.

7.3.3.2 The exhaust timing edge control valve

It is clear from Fig. 7.17 that the butterfly valve controls only the area at the exhaust port rather than the port opening and closing timing edges as well. On the right side of Fig. 7.17 is sketched a valve which fits closely around the exhaust port and can simultaneously change both the port timing and the port area, exactly as in the mathematical simulation discussed in Sect. 7.3.3. There are many innovative designs of exhaust timing edge control valve ranging from the oscillating barrel type to the oscillating shutter shown in Fig. 7.17. The word "oscillating" may be somewhat confusing, so it needs to be explained that it is stationary at any one load or speed condition but it can be changed to another setting to optimize an alternative engine load or speed condition. While the net effect on engine performance of the butterfly valve and the timing edge control valve is somewhat similar, the timing edge control valve carries out the function more accurately and effectively. Of course, the butterfly valve is a device which is cheaper to manufacture and install than the timing edge control device.

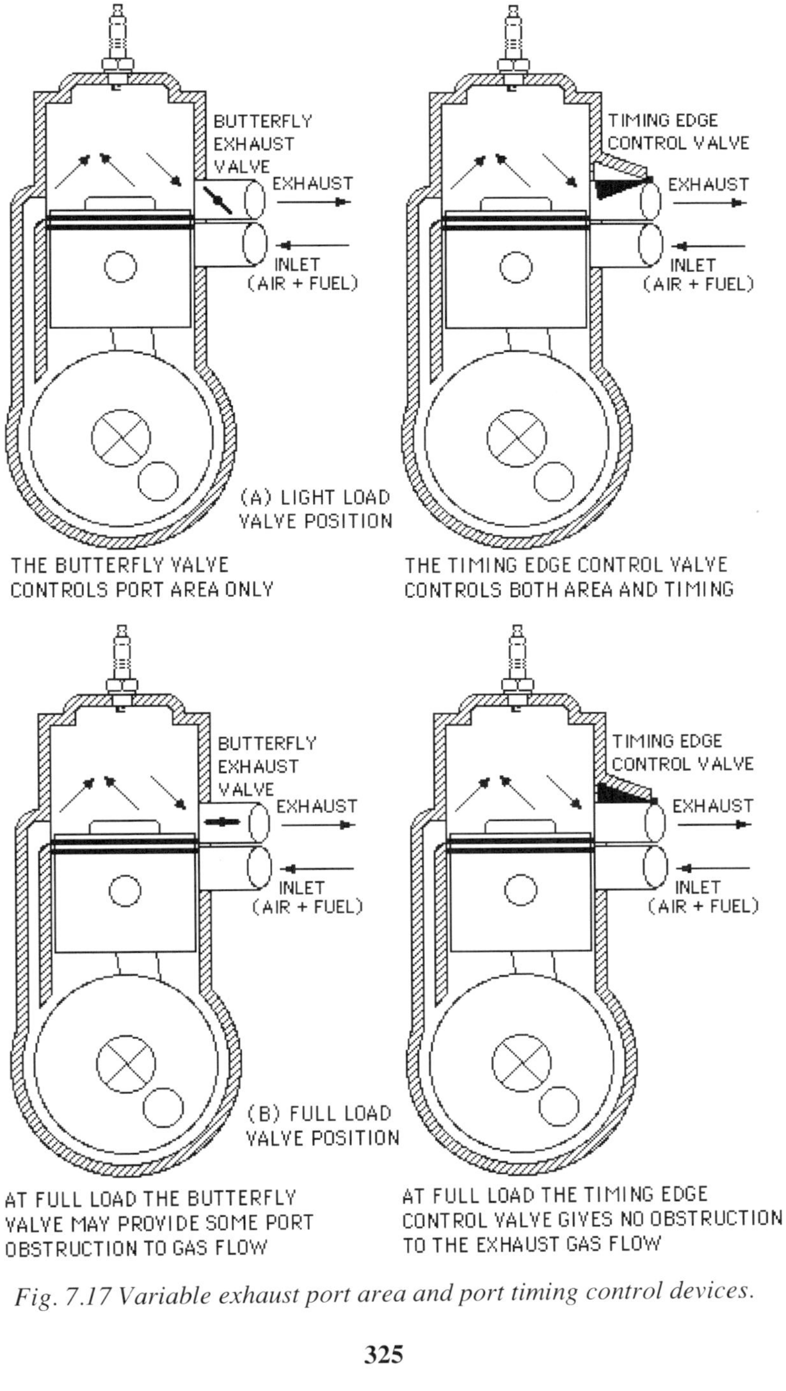

Fig. 7.17 Variable exhaust port area and port timing control devices.

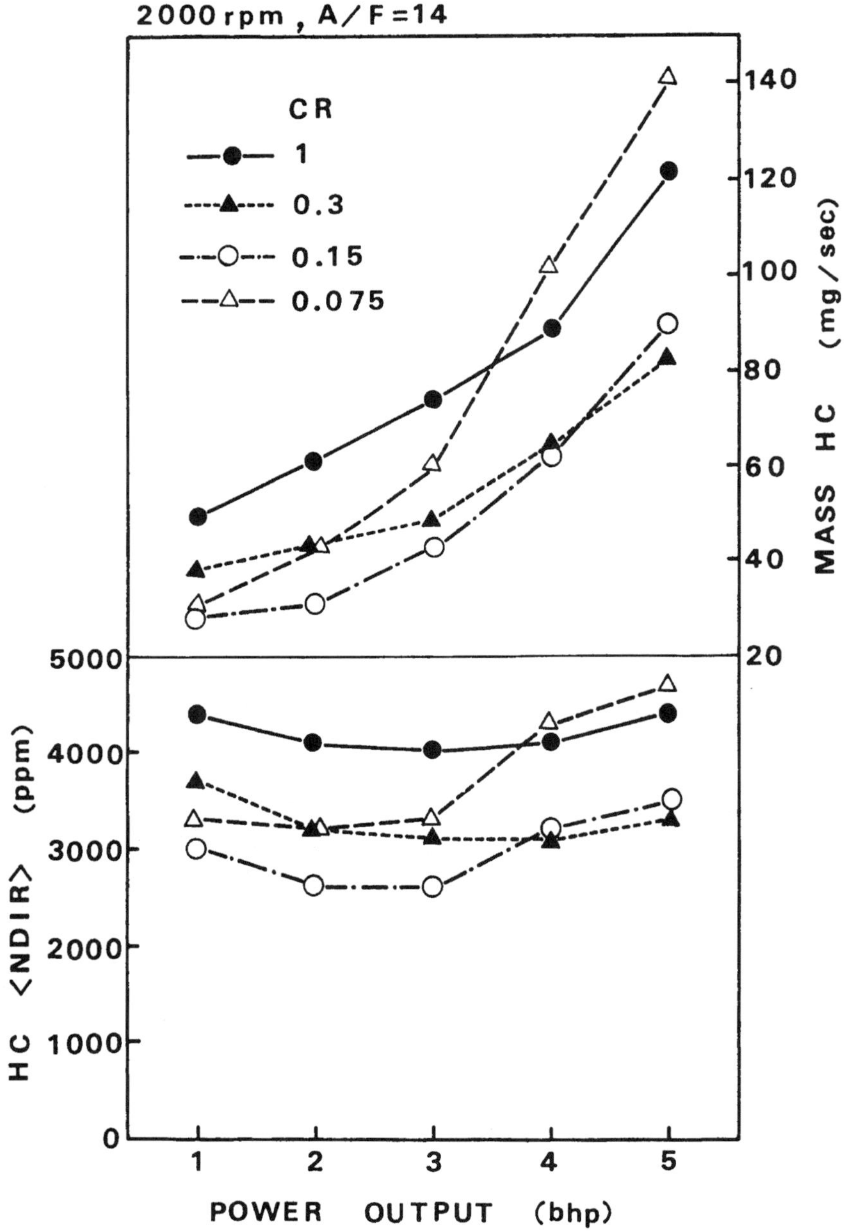

Fig. 7.18 Emissions reduction using a restriction at the exhaust port by a butterfly valve.

Hata and Iio(7.4) describe the use of such an exhaust timing control valve, although it is clear from the paper that the emphasis is more on the improvement of the power performance of a sports motorcycle than on reduction of fuel consumption and exhaust emissions. Indeed, informed readers will have observed these exhaust control valves on racing motorcycle engines, where their employment is intended to retune the expansion chamber exhaust system over a wider speed range. If Eq. 6.2.2 is re-examined, it is clear that if the exhaust period, EP, is reduced by a timing edge control valve, then for a fixed tuned length, LT, the engine speed for peak tuning effect is reset to a new and lower level. Reproducing Eq. 6.2.2:

$$LT=83.3*EP*Ao/RPM \qquad (6.2.2)$$

Rearranging this to find the tuned speed, RPM, if the other terms are variables or constants:

$$RPM=83.3*EP*Ao/LT \qquad (7.3.1)$$

Thus, for any fixed pipe length, if the exhaust control valve is in the position shown in Fig. 7.17(a) and the exhaust period, EP, is less than at the peak tuning speed, the pipe will provide well-phased plugging reflections at a lower speed. This raises the BMEP and power at that lower engine rate of rotation.

Nevertheless, in their paper Hata(7.3) shows a reduction of as much as 40% in BSFC at the lower end of the speed and load range when the exhaust control valve is employed.

7.3.4 Conclusions regarding the simple two-stroke engine

The main emphasis in the discussion above is that the simple two-stroke engine is capable of a considerable level of optimization by design attention to scavenging, carburetion, lubricants and lubrication, and exhaust timing and area control. However, the very best design will still have an unacceptably high emission of unburned hydrocarbons even though the carbon monoxide and nitrogen oxide levels are acceptably low, perhaps even very low. Indeed, it is possible that the contribution of internal combustion engines to the atmospheric pollution by nitrous oxide, N_2O, may be a very important factor in the future(7.24). There is every indication that any two-stroke engine produces this particular nitrogen oxide component in very small quantities by comparison with its four-stroke engine counterpart.

The simple two-stroke engine, optimized at best, has a low CO and NO_x exhaust pollutant level, but a high HC and O_2 exhaust emission output. This leaves the engine with the possibility of utilizing an oxidation catalyst in the exhaust to remove the hydrocarbons and further lower the carbon monoxide levels.

An early paper on this subject by Uchiyama et al(7.11) showed that a small Suzuki car engine could have the hydrocarbon emission reduced quite significantly by exhaust gas after-treatment in this manner. They reported an 80% reduction of the hydrocarbon exhaust emission by an oxidizing catalyst.

More recently, and aimed specifically at the simple two-stroke engine used in mopeds, chainsaws and small motorcycles, Laimbock(7.21) presents experimental data on the effect of using the advice given in this chapter. He shows the results for a 125 cc high-performance motorcycle engine when the scavenging and carburetion have been optimized and an exhaust timing edge control valve is used. For such small motorcycles there are emission control laws pending in Switzerland, Austria and Taiwan. The most severe of these is in Switzerland, where the machine must execute a driving cycle and emit no more than 8 g/km of CO, 3 g/km of HC and 0.1 g/km of NO_x. Laimbock(7.21) shows that a production 125 cc motorcycle engine, which has a peak BMEP of 8 bar at 9000 rpm and is clearly a high specific output power unit, has emissions on this cycle of 21.7 g/km of CO, 16.9 g/km of HC and 0.01 g/km of NO_x. Clearly this motorcycle is unsuitable for sale within such regulations. By optimizing the scavenging and carburetion, the same machine will have emission characteristics on the same cycle of 1.7 g/km of CO, 10.4 g/km of HC and 0.03 g/km of NO_x. Thus, the optimization procedures already discussed in the chapter lowered the CO and HC significantly, but raised the NO_x levels. The HC level is still unacceptable from a Swiss legal standpoint. By introducing an oxidation catalyst into the tuned exhaust pipe of this engine in the manner shown in Fig. 7.19, Laimbock(7.21) provides experimental evidence that the peak power performance of the motorcycle is barely affected, but the emissions are dramatically reduced. In this case, where the catalyst is of the oxidizing type, the test results on the Swiss driving cycle gave emission levels of 0.8 g/km of CO, 1.9 g/km of HC and 0.02 g/km of NO_x; such a machine is now well within the limits pending or proposed by many legislative bodies worldwide.

As far as fuel consumption is concerned, Laimbock(7.21) shows that the original 125 cc production motorcycle on the test driving cycle had a fuel consumption level of 20.8 km/liter, the model with improved scavenging and carburetion did 29.5 km/liter, while the final version with the exhaust catalysts fitted travelled 31.2 km/liter of gasoline.

There is no logical reason why a similar approach could not be successful for any type of simple two-stroke cycle engine.

7.4 The more complex two-stroke engine

Although the results of the most recent research and development of the simple two-stroke engine are impressive, such as that of Laimbock(7.21) or Kee(1.20), if the two-stroke engine is to have relevance in the wider automotive application, the raw level of unburned hydrocarbons in the exhaust system before catalytic after-treatment will have to be reduced to lower levels. Equally, levels of specific fuel consumption in the range of 250 to 300 g/kWh will be required over much of the speed and load range. In this case, it is essential that no fuel is ever lost to the exhaust system during scavenging, thereby deteriorating the thermal efficiency of the engine. Even the most miniscule quantity lost in this manner gives unacceptable levels of HC emission. Clearly, an engine designed to accomplish that set of criteria is going to have to be more complex in some manner.

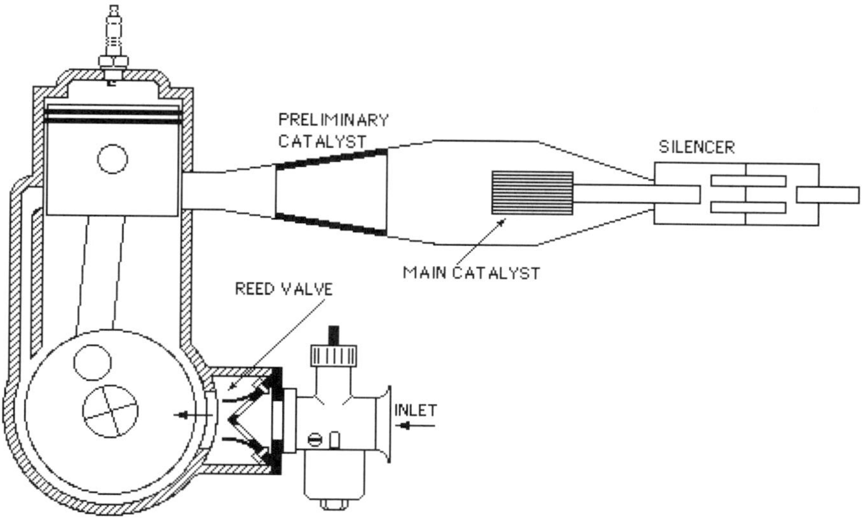

Fig. 7.19 Positioning of catalysts in expansion chamber suggested by Laimbock(7.21).

The Achilles' heel of the simple two-stroke engine is the loss of fuel when it is supplied in conjunction with the scavenge air. Remove this problem, albeit with added complexity, and the fuel economy and hydrocarbon emissions of the engine are significantly improved, as is theoretically pointed out in Sect. 7.2. The fundamental requirement in design terms is shown in Fig. 7.20. Somewhere in the cylinder head or cylinder wall is placed a "device" which will supply fuel, or a mixture of fuel and air, into the cylinder in such a manner that none of the fuel is lost into the exhaust duct during the open cycle period. Although the sketch shows a two-stroke engine with crankcase scavenging, this is purely pictorial. The fundamental principle would apply equally well to an engine with a more conventional automotive type of crankshaft with pressure oil-fed plain bearings and the scavenge air supplied by a pump or a blower; in short, an engine as sketched in Fig. 1.6 and described initially in Sect. 1.2.4.

The simplest idea which immediately comes to mind as a design solution is to use a diesel engine type of injection system and spray the fuel into the cylinder after the exhaust port is closed. This straightforward approach is sketched in Fig. 7.21. Naturally, the fuel injection system is not limited to that generally employed for diesel engines and several other types have been designed and tested, such as that proposed by Beck(7.16)

However, it is possible that this elegantly simple solution to the problem is not as effective as it might seem. In Chapter 4, the combustion process by spark ignition

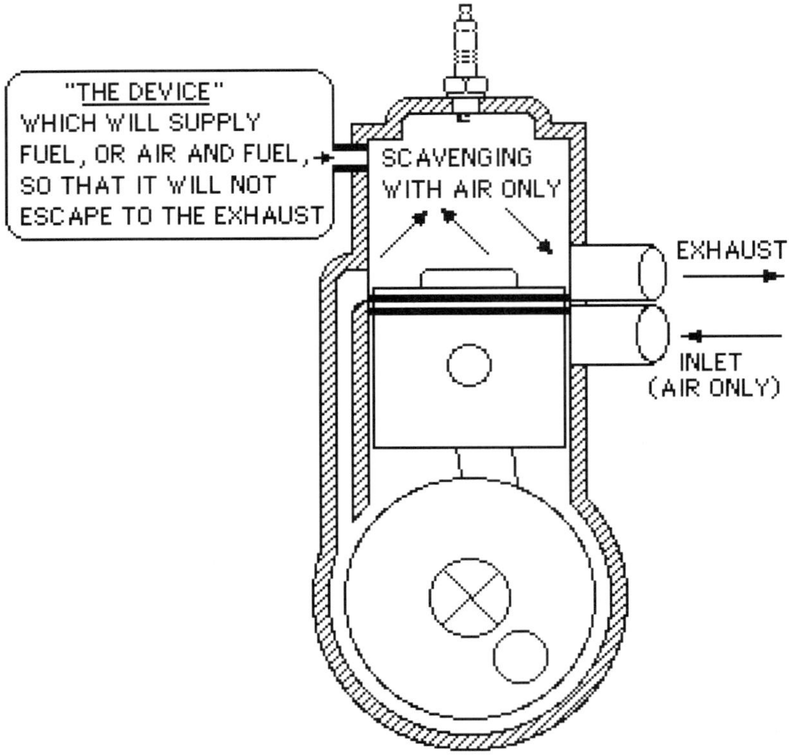

Fig. 7.20 The fundamental design principle for the reduction of hydrocarbon emissions and fuel consumption.

of a fuel and air mixture is detailed as being between a homogeneous mixture of fuel vapor and air. It is conceivable that a liquid fuel injected in even the smallest droplets, such as between 10 and 15 μm, may still take too long to mix thoroughly with the air and evaporate completely in the relatively short time period from exhaust port closure until the ignition point before the tdc position. Put in the simplest terms, if the end of the fuel injection of the gasoline is at 90º before the ignition point, and the engine is running at 6000 rpm, this implies a successful evaporation and mixing process taking place in 0.25/100 seconds, or 2.5 ms. That is indeed a short time span for such an operation when one considers that the carburetted four-stroke cycle engine barely accomplishes that effect in a cycle composed of 180º of induction period, followed by 180º of compression process, i.e., four times as long. In the simple two-stroke cycle engine, the induction process into a "warm" crankcase for 180º of engine rotation helps considerably to evaporate the fuel before the commencement of the scavenge process, but even then does not

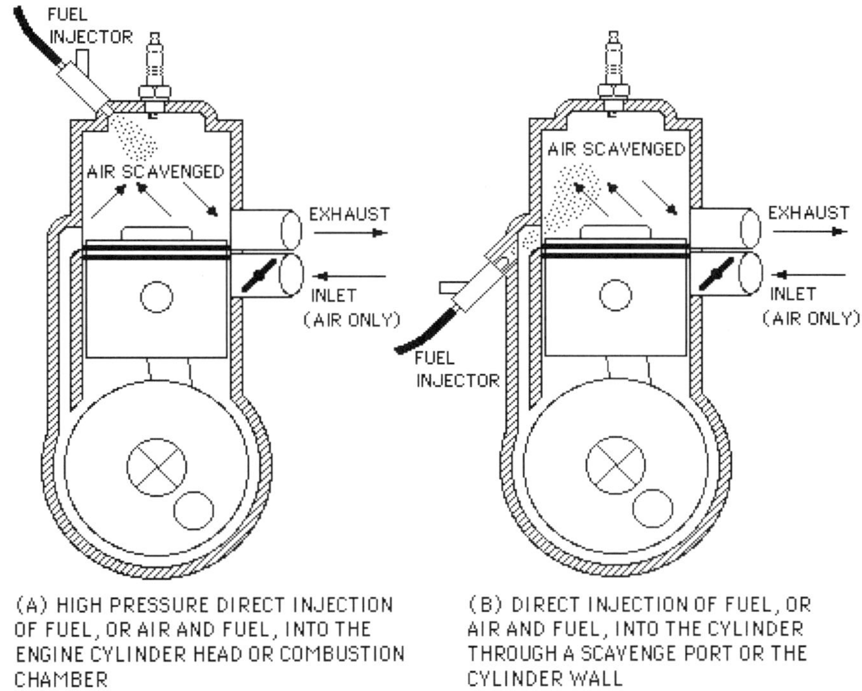

Fig. 7.21 The potential positioning of fuel injectors for direct in-cylinder injection of gasoline in a two-stroke engine.

complete the vaporization process, as is pointed out in the excellent presentation by Onishi et al(7.8). Further discussion on this will be found in Sect. 7.4.3 dealing with direct in-cylinder fuel injection.

This potential difficulty regarding the adequate preparation of the fuel and air mixture prior to the combustion process opens up several other solutions to this design problem. There are two basic approaches which are possible:

(a) The stratified charging of the cylinder by the "device" of Fig. 7.20 with a premixed charge of vaporized fuel and air followed by mixing with the trapped cylinder charge and a homogeneous combustion process.

(b) The stratified charging of the cylinder by the "device" of Fig. 7.20 with a premixed charge of vaporized fuel and air followed by mixing with the trapped cylinder charge and a stratified combustion process.

The fundamental advantages and disadvantages of stratified charging and stratified combustion processes have already been discussed in Sect. 7.1.2.

In the following section of this chapter, the various design approaches outlined above will be discussed in greater detail, presented together with the known experimental facts as they exist at this point in history. These are very early days in the development of this form of the two-stroke cycle engine.

7.4.1 The stratified charging and homogeneous combustion engine

This type of engine has been tested in various forms and by several research groups and organizations. The most significant are discussed below in terms of their applicability for future use in production as power units which will rival the four-stroke engine in hydrocarbon emissions and fuel consumption levels, but which must retain the conventional advantage in carbon monoxide and nitrogen oxide emissions.

7.4.1.1 The QUB stratified charging engine

This early work is presented in technical papers published by Blair and Hill(7.9) and by Hill and Blair(7.10). The fundamental principle of operation of the engine is illustrated by Fig. 7.22. The overriding requirement is to introduce a rich mixture of air and fuel into the cylinder during the scavenge process at a position which is as remote as possible from the exhaust port. Ideally, the remaining transfer ports would supply air only into the cylinder. The engine has two entry ports for air, a main entry for 80% of the required air into the crankcase, and a subsidiary one for the remaining air and for all of the necessary fuel into a long storage transfer port. That port and transfer duct would pump the stored contents of air and fuel into the cylinder during the succeeding scavenge process so that no fuel migrated to the crankcase. In the meantime, during the induction and pumping period, the fuel would have some residence time within the air and on the walls of the long rear transfer port so that some evaporation of the fuel would take place. In this manner the cylinder could be supplied with a pre-mixed and partially evaporated fuel and air mixture in a stratified process. The resultant mixing with the trapped charge of cylinder air and retained exhaust gas would permit a homogeneous combustion process.

The test results for the engine, shown in Figs. 7.23 and 7.24 for fuel consumption and BMEP levels at several throttle openings, reveal significantly low levels of fuel consumption. Most of the BMEP range from 2 bar to 5.4 bar over a speed range of 1500 to 5500 rpm, but the BSFC levels are in the band from 0.36 to 0.26 kg/kWh. These are particularly good fuel consumption characteristics, at least as good if not superior to an equivalent four-stroke cycle engine, and although the hydrocarbon emission levels are not recorded, they must be significantly low with such good trapping of the fuel within the cylinder. The power performance characteristics are unaffected by this stratified charging process, for the peak BMEP of this engine at 5.4 bar is quite conventional for a single-cylinder engine operating without a tuned exhaust system.

The mechanical nature of the engine design is relatively straightforward, and it is one eminently suitable for the conversion of a simple two-stroke cycle engine. The disadvantages are the extra complication caused by the twin throttle linkages

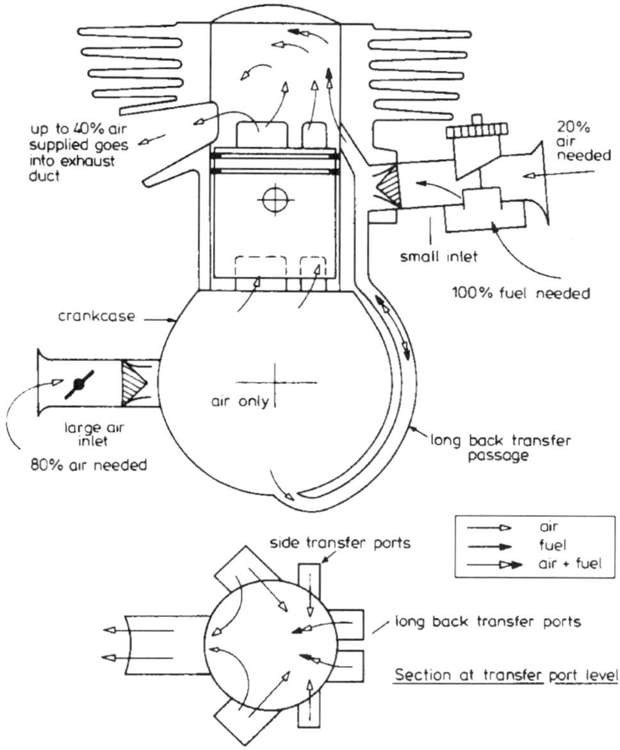

Fig. 7.22 Air and fuel flow paths in a QUB type stratified charging system.

and the accurate carburetion of a very rich mixture by a carburetor. The use of a low-pressure fuel injection system to replace the carburetor would simplify that element of the design at the further disadvantage of increasing the manufacturing costs.

7.4.1.2 The Piaggio stratified charging engine

The fundamental principle of operation of this power unit is shown in Fig. 7.25 and is described in much greater detail in the paper by Batoni(7.1). This engine takes the stratified charging approach to a logical conclusion by attaching two engines at the cylinder head level. The crankshafts of the two engines are coupled together in the Piaggio example by a toothed rubber belt. In the paper presented by Piaggio, one of the engines, the "upper" engine of the sketch in Fig. 7.25, has 50 cc swept volume, and the "lower" engine has 200 cc swept volume. The crankcase of both engines ingest air and the upper one inhales all of the required fuel for combustion of an appropriate air-fuel mixture in a homogeneous process. The crankcase of the upper engine supplies a rich mixture in a rotating, swirling scavenge process giving the fuel as little forward momentum as possible towards the exhaust port. The lower

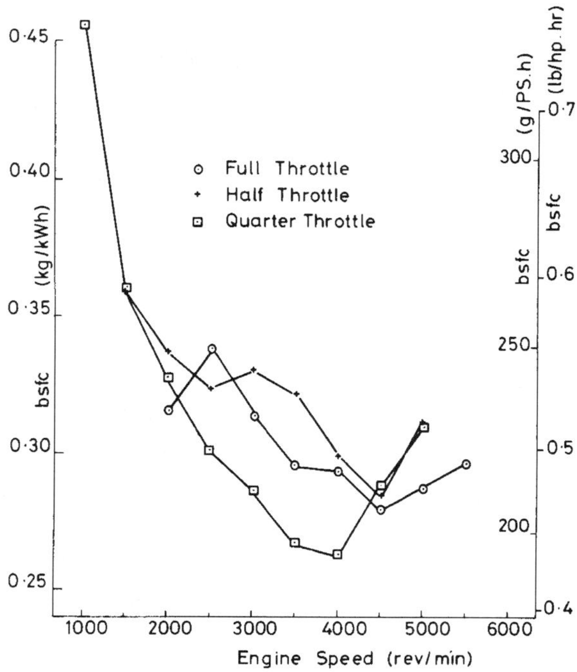

Fig. 7.23 Optimized fuel consumption levels for the QUB stratified charging engine.

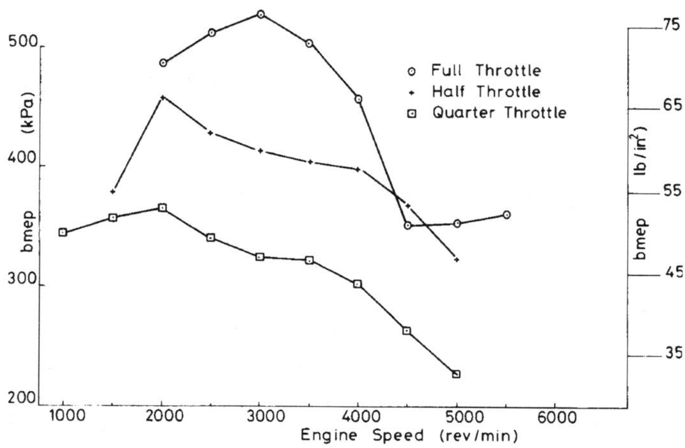

Fig. 7.24 BMEP levels at the optimized fuel consumption levels for the QUB stratified charging engine.

cylinder conducts a conventional loop scavenge process with air only. Towards the end of compression the mixing of the rich air-fuel mixture and the remaining trapped cylinder charge takes place, leading to a homogeneous combustion process.

The results of the experimental testing of this 250 cc Piaggio engine are to be found in the paper by Batoni(7.1), but are reproduced here as Figs. 7.26-7.28. A direct comparison can be made between this stratified charging engine and the performance characteristics of the 200 cc engine which forms the base of this new power unit. Figs. 7.9-7.11, already discussed fully in Sect. 7.2.1.2, are for the 200 cc base engine. Fig. 7.9 gives the fuel consumption behavior of the 200 cc base engine, Fig. 7.10 the CO emission levels and Fig. 7.11 the HC emission characteristics.

Fig. 7.26 shows the fuel consumption levels of the experimental engine (note that 1 g/kWh=0.746 g/bhp.hr). The lowest contour in the center of the "oyster" map is 240 g/hp.hr or 322 g/kWh. The units of BMEP on this graph are in kg/cm^2, which is almost exactly equal to a bar (1kg/cm^2=0.981 bar). These are quite good fuel consumption figures, especially when one considers that this engine is one of the first examples of stratified charging presented; the paper was published in 1978. The reduction of fuel consumption due to stratified charging is very clear when one compares Figs. 7.26 and 7.9. The minimum contour is lowered from 300 g/bhp.hr to 240 g/bhp.hr, a reduction of 33%. At light load, around 1 bar BMEP and 1500 rpm, the fuel consumption is reduced from 500 to 400 g/bhp.hr, or 20%. This condition is particularly important for power units destined for automotive applications as so many of the test cycles for automobiles or motorcycles are formulated to simulate urban driving conditions where the machine is accelerated and driven in the 15-50 km/h zone. The proposed European ECE-R40 cycle is such a driving cycle(7.21).

The reduction of hydrocarbon emissions is particularly impressive, as can be seen from a direct comparison of the original engine in Fig. 7.11 with the stratified charging engine in Fig.7.28. The standard engine, already discussed in Sect. 7.2.1.2, showed a minimum contour of 1500 ppm HC (C6, NDIR), at a light load but high speed point. In the center of the load-speed map in Fig. 7.11 the figures are in the 2500 ppm region, and at the light load point of 1 bar and 1500 rpm, the figure is somewhat problematic but 5000 ppm would be typical. For the stratified charging engine the minimum contour is reduced to 200 ppm HC, the center of the load-speed picture is about 500 ppm, and the all-important light load and speed level is somewhat in excess of 1000 ppm. This is a very significant reduction and is the level of diminution required for a successful automotive engine before the application of catalytic after-treatment.

A comparison of the carbon monoxide emission levels of the standard engine in Fig. 7.10 and of the stratified charging engine in Fig. 7.27 shows significant improvements in the two areas where it really matters, i.e., at light loads and speeds and at high loads and speeds. In both cases the CO emission is reduced from 2-3% to 0.2-0.3%, i.e., a factor of 10. The absolute value of the best CO emission at 0.2% is quite good, remembering that this experimental data was acquired in 1978.

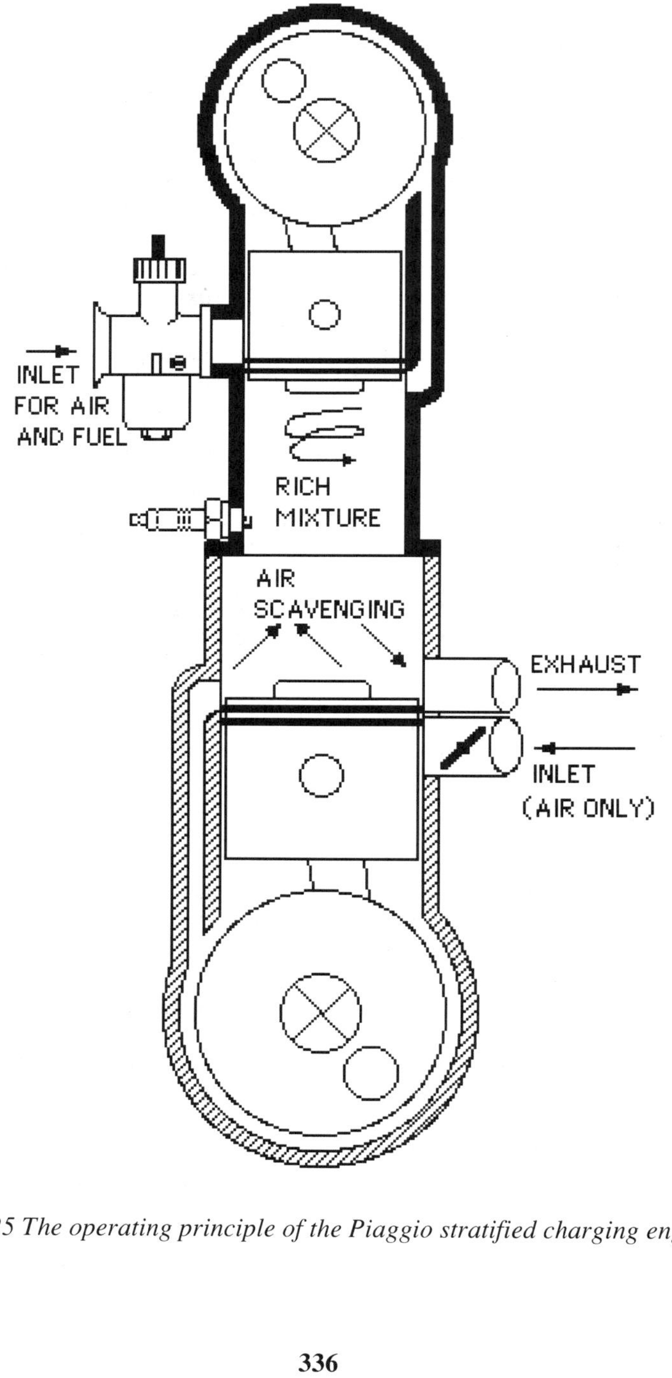

Fig. 7.25 The operating principle of the Piaggio stratified charging engine.

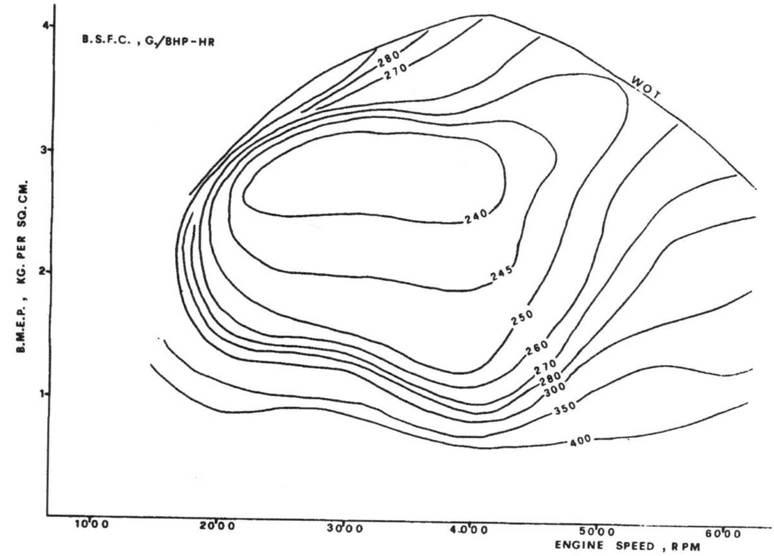

Fig. 7.26 Fuel consumption levels from the Piaggio stratified charging engine.

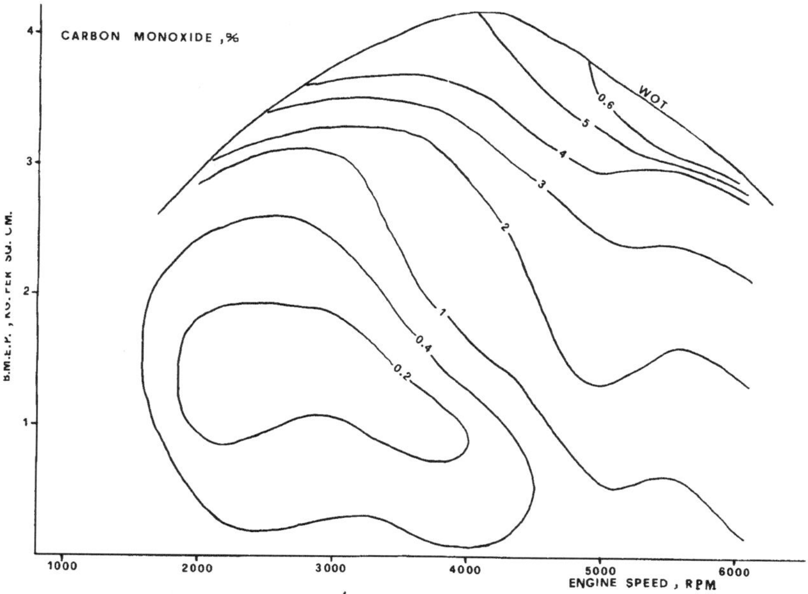

Fig. 7.27 Carbon monoxide emission from the Piaggio stratified charging engine.

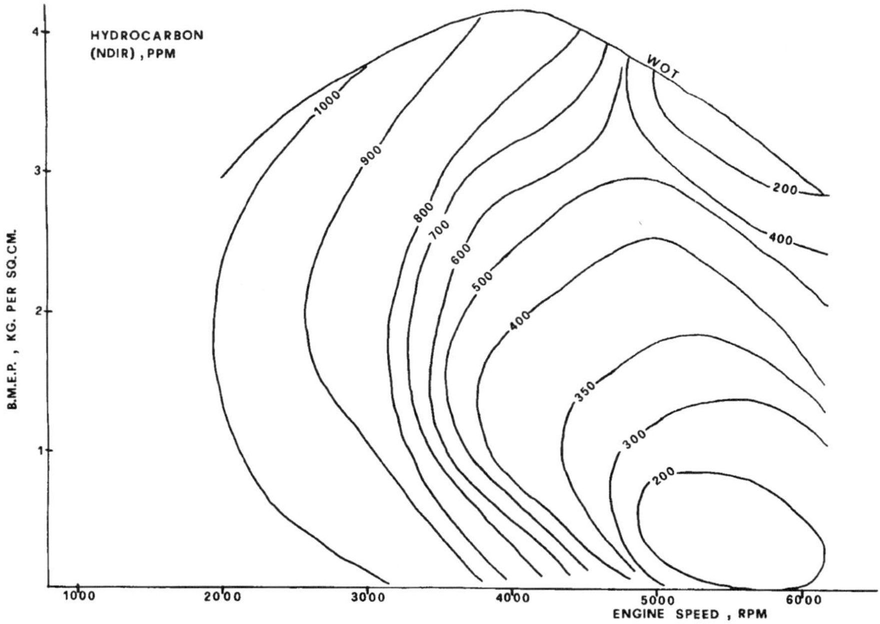

Fig. 7.28 Hydrocarbon emission levels from the Piaggio stratified charging engine.

It should also be noted that the peak BMEP of the engine is slightly reduced from 4.8 bar to 4.1 bar due to the stratified charging process, and there is some evidence that there may be some diminution in the air utilization rate of the engine. This is supplied by the high oxygen emission levels at full load published by Batoni(7.1, Fig. 8) where the value at 4 bar and 3000 rpm is shown as 7%. In other words, at that point it is almost certain that some stratified combustion is occurring.

This engine provides an excellent example of the benefits of stratified charging. It also provides a good example of the mechanical disadvantages which may accrue from its implementation. This design, shown in Fig. 7.25, is obviously somewhat bulky, indeed it would be bulkier than the equivalent four-stroke cycle engine. One of the profound advantages of the two-stroke engine is lost by this particular mechanical layout. An advantage of this mechanical configuration, particularly in a single-cylinder format, is the improved primary vibration balancing of the engine due to the opposed piston layout.

Nevertheless, a fundamental thermodynamic and gas-dynamic postulation is verified from this experimental data: stratified charging of a two-stroke engine is a viable and sound approach to the elimination of much of the excessive fuel consumption and raw hydrocarbon emission from a two-stroke engine.

7.4.1.3 An alternative mechanical option for stratified charging

The fundamental principle of stratified charging has been detailed above, but other researchers have striven to emulate the process with either less physical bulk or less mechanical complication than that exhibited by the Piaggio device.

One such engine is the double piston device, an extension of the original split-single Puch engine of the 1950's. Such an engine has been investigated by Ishihara(7.7). Most of these engines are designed in the same fashion as shown in Fig. 7.29. Instead of the cylinders being placed in opposition as in the Piaggio design, they are configured in parallel. This has the advantage of having the same bulk as a conventional twin-cylinder engine, but the disadvantage of having the same (or worse!) vibration characteristics as a single-cylinder engine of the same total swept volume. The stratified charging is at least as effective as in the Piaggio design, but the combustion chamber being split over two cylinder bores lends itself more to stratified burning than homogeneous burning. This is not necessarily a criticism.

However, it is clear that it is essential to have the cylinders as close together as possible, and this introduces the weak point of all similar designs or devices. The thermal loading between the cylinder bores is somewhat excessive if a reasonably high specific power output is to be attained.

Another design worthy of mention and study, which has considerable applicability for such designs where the cost and complexity increase cannot be excessive due to marketing and packaging requirements, is that published in the technical paper by Kuntscher(7.23). This design for a stratified charging system has the ability to reduce the raw hydrocarbon emission and fuel consumption from such engines as those fitted in chainsaws, mopeds, and small motorcycles.

7.4.1.4 The stratified charging engine proposed by Institut Français du Pétrole

This suggestion for stratified charging which emanates from IFP is probably the most significant yet proposed. The performance results are superior in most regards to four-stroke cycle engines, as is evident from the technical paper presented by Duret et al(7.18). The fundamental principle of operation is described in detail in that publication, a sketch of the engine operating principle is given in Fig. 7.30, and a photograph of their engine is shown in Plate 7.1. The engine in the photograph is a multi-cylinder unit and, in a small light car operating on the EEC fuel consumption cycle at 90 and 120 km/h, had an average fuel consumption of 30.8 km/liter (86.8 miles/Imperial gallon or 73.2 miles/US gallon).

The crankcase of the engine fills a storage tank with compressed air through a reed valve. This stored air is blown into the cylinder through a poppet valve in the cylinder head. At an appropriate point in the cycle, a low-pressure fuel injector sprays gasoline onto the back of the poppet valve and the fuel has some residence time in that vicinity for evaporation before the poppet valve is opened. The quality of the air-fuel spray past the poppet valve is further enhanced by a venturi

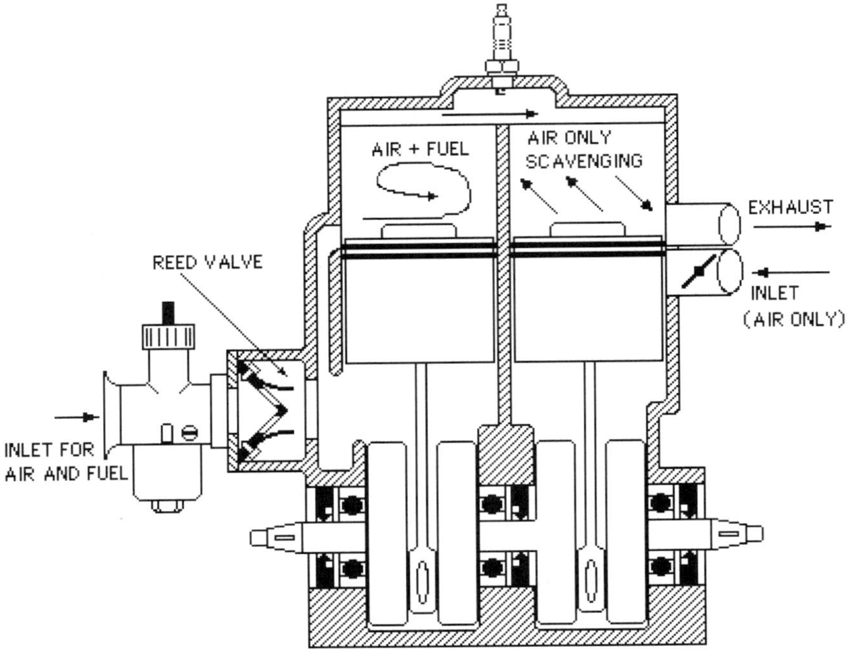

Fig. 7.29 Alternative stratified charging system of the double piston genre.

surrounding the valve seat. It is claimed that any remaining fuel droplets have sufficient time to evaporate and mix with the trapped charge before the onset of a homogeneous combustion process.

The performance characteristics for the single-cylinder test engine are of considerable significance, and are presented here as Figs. 7.31-7.33 for fuel consumption, hydrocarbons, and nitrogen oxides. The test engine is of 250 cc swept volume and produces a peak power of 11 kW at 4500 rpm, which realizes a BMEP of 5.9 bar. Thus, the engine has a reasonably high specific power output for automotive application, i.e., 44 kW/liter. In Fig. 7.31, the best BSFC contour is at 0.26 kg/kWh, which is an excellent result and superior to most four-stroke cycle engines. More important, the BSFC value at 1.5 bar BMEP at 1500 rpm, a light load and speed point, is at 0.4 kg/kWh and this too is a significantly low value.

The unburned hydrocarbon emission levels are shown in Fig. 7.32, and they are also impressively low. Much of the important legislated driving cycle would be below 20 g/kWh. When an oxidation catalyst is applied to the exhaust system, considerable further reductions are recorded, and this data is presented in Fig. 7.34.

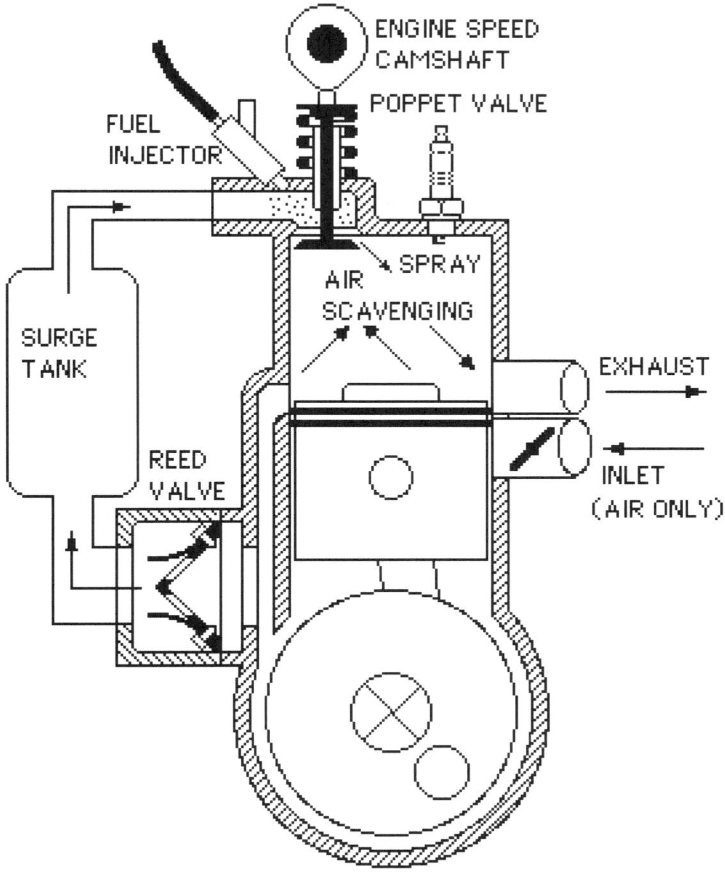

Fig. 7.30 Stratified charging system proposed by the Institut Français du Pétrole.

The conversion rate exceeds 91% over the entire range of BMEP at 2000 rpm, leaving the unburned hydrocarbon emission levels below 1.5 g/kWh in the worst situation.

Of the greatest importance is the nitrogen oxides emissions, and they remain conventionally low in this stratified charging engine. The test results are shown in Fig. 7.33. The highest level recorded is at 15 g/kWh, but they are less than 2 g/kWh in the legislated driving cycle zone.

The conclusions drawn by IFP are that an automobile engine designed and developed in this manner would satisfy the most stringent exhaust emissions

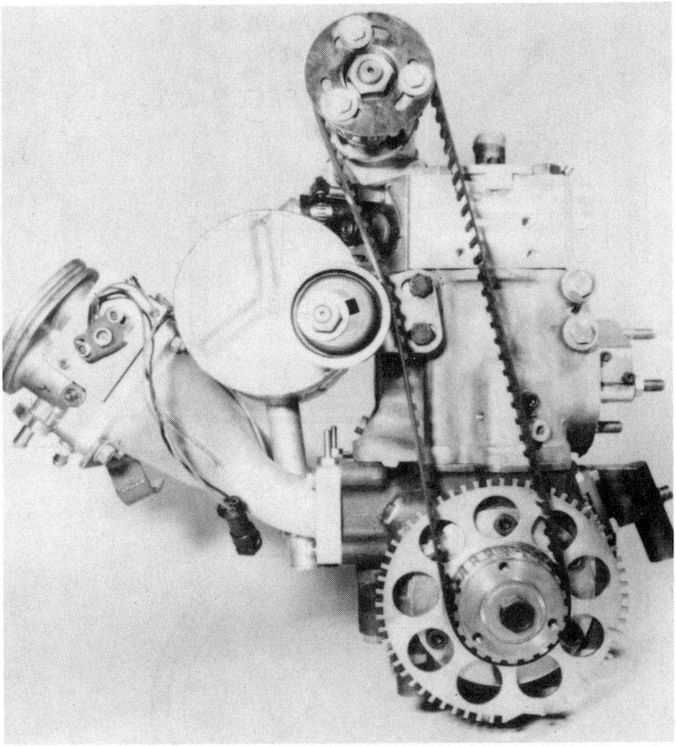

Plate 7.1 Stratified charging engine from the Institut Français du Pétrole (courtesy of Institut Français du Pétrole).

legislation for cars. More important, the overall fuel economy of the vehicle would be enhanced considerably over an equivalent automobile fitted with the most sophisticated four-stroke cycle spark-ignition engine.

The bulk of the engine is increased somewhat over that of a conventional two-stroke engine, particularly in terms of engine height. The complexity and manufacturing cost is also greater, but no more so than that of today's four-stroke engine equipped cars, or even some of the larger capacity motorcycles or outboard motors.

7.4.2 The stratified charging and stratified combustion engine

At QUB there is an investigation into an engine which has both stratified charging and stratified combustion. The basic principle of operation is shown in Fig. 7.35, and a photograph of the test engine is illustrated in Plate 7.2. The engine is of the two-piston type with the cylinder axes at 90º, and the geometrical shape has led

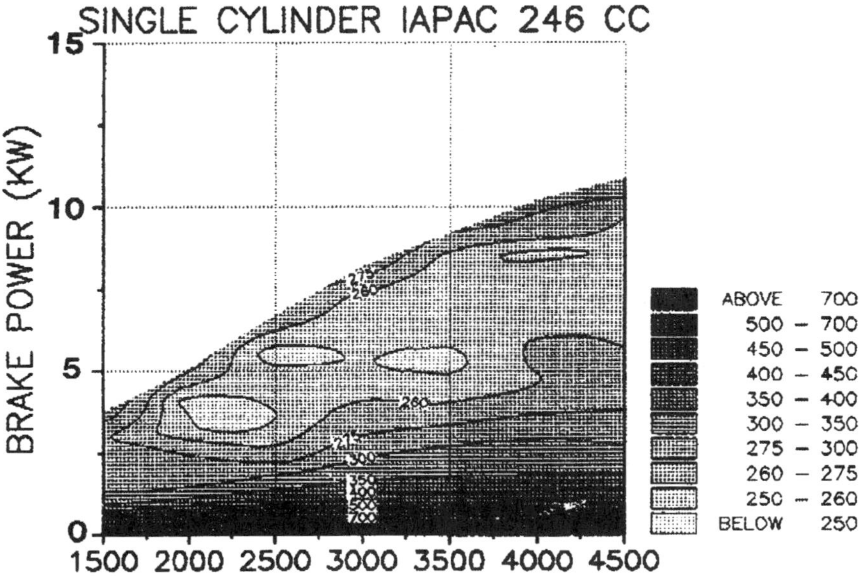

Fig. 7.31 Fuel consumption contours (g/kWh) of IFP stratified charging engine.

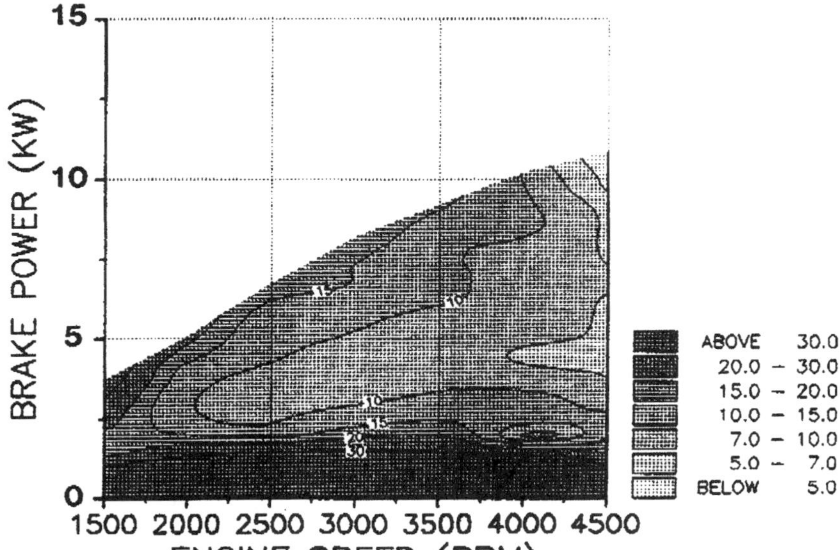

Fig. 7.32 Hydrocarbon emission contours (g/kWh) of IFP stratified charging engine.

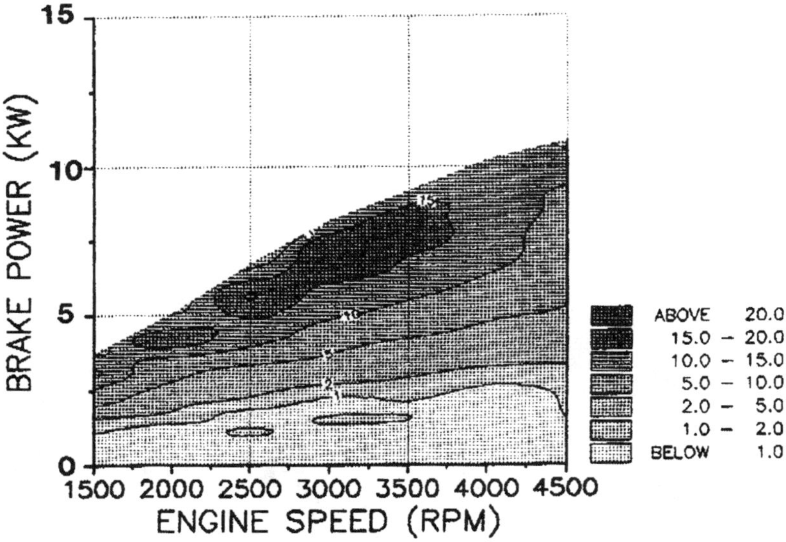

Fig. 7.33 Nitrogen oxide emission contours (g/kWh) of IFP stratified charging engine.

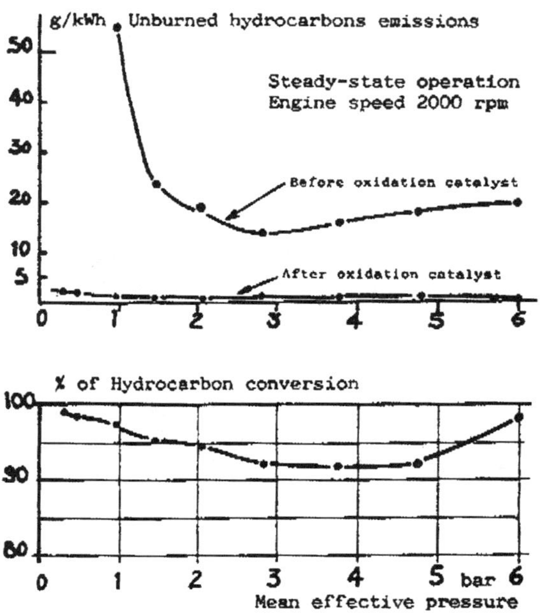

Fig. 7.34 Effect of oxidation catalyst on exhaust hydrocarbons in IFP engine.

Plate 7.2 Stratified charging and stratified combustion engine from QUB.

to the unit being described as an L-Head engine. The main, large capacity cylinder is scavenged with fresh air from the crankcase, or possibly also from a blower if the engine has a conventional automotive crankshaft. The top, smaller capacity cylinder scavenges the combustion chamber with air and fuel from its crankcase pump which is unthrottled. The fuel is supplied directly to that crankcase with a low-pressure fuel injector. The fuel and air entering the cylinder is swirled to enhance the mixing and evaporation of those fuel droplets not evaporated in the top crankcase. There is an orifice between the main cylinder and the combustion chamber, and the compression ratio in the main cylinder is much higher than that for the smaller capacity top cylinder. Thus, from the bdc position onwards, with the pistons in phase arriving together at tdc, all of the gas motion in the main cylinder is into the combustion chamber zone. This effectively stops any short-circuiting of fuel from the combustion space to the exhaust port during the trapping period. The orifice can be arranged to provide further swirling motion to the air-fuel mixture during the compression process.

Therefore, for lighter loads the main air throttle to the crankcase is progressively closed and an ever increasing proportion of the required charge is provided by the crankcase pump of the top cylinder. For higher loads at any engine speed, the throttle to the main crankcase pump is progressively opened and the bulk of the required air is supplied via the scavenge process in the main cylinder.

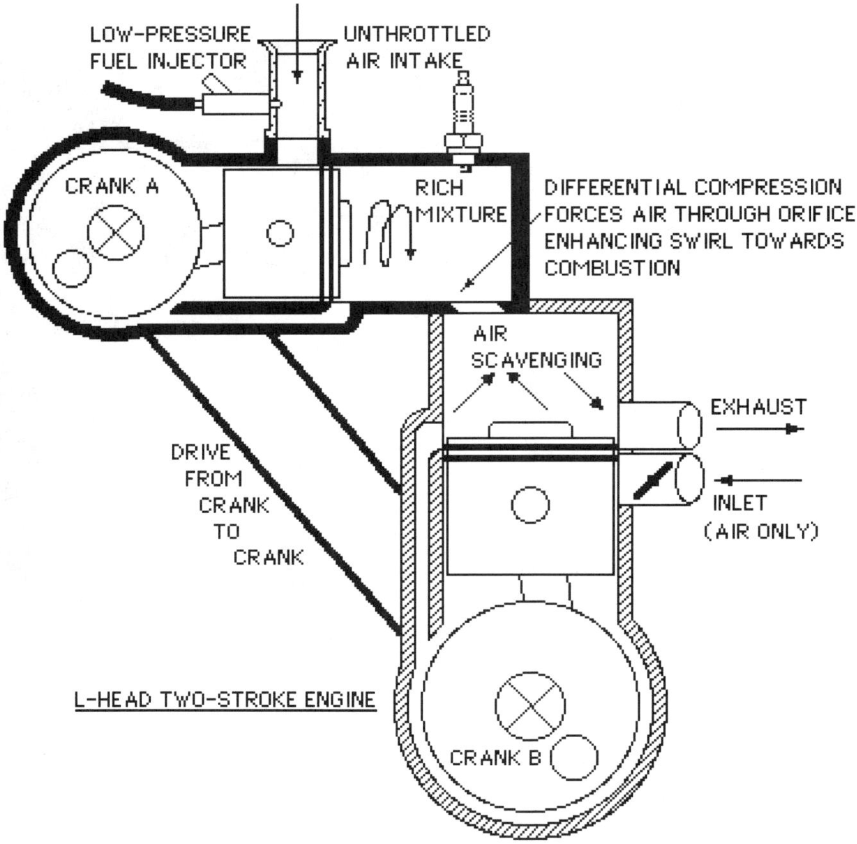

Fig. 7.35 The stratified charging and stratified combustion engine from QUB.

It is possible that this type of engine could be configured as in Fig. 1.6 with an automotive type crankshaft, yet be self-starting with a turbocharger attached to the exhaust port. In that situation the blower would be replaced by the compressor of the turbocharger and a considerable increase in thermal efficiency would ensue. The reason, as in turbocharged diesel engines vis-a-vis supercharged diesel engines, is that the power to supply the compressed air for scavenging comes from the energy of the exhaust gas and not from the engine crankshaft.

The main problem in all such engines employing stratified combustion, as has been fully described in Sect. 7.1.2, is to raise the air utilization rate to acceptable levels at full load. The flammability situation at light load for this type of two-stroke engine is, or should be, considerably superior to that provided by homogeneous combustion, as the charge purity in the combustion zone will be greater. This is a particularly relevant statement, so the serious student should consider the implications of it carefully and, if necessary, reread Sects. 7.1.2 and 4.1.

As far as mechanical complexity is concerned, the engine is more compact in height terms than either the Piaggio or the IFP design. Whether it is more economical to manufacture than either of those, or direct in-cylinder injection power units, has yet to be determined.

That the engine exhibits the theoretical characteristics postulated for stratified charging and combustion is seen from a sample of some of the preliminary test results acquired in the period from October 1988 to February 1989. The test engine is of 450 cc swept volume, made up of a 400cc main cylinder and a 50 cc top cylinder. The main engine is very similar to the QUB 400 research engine described elsewhere in this book. The exhaust system attached to the engine is untuned and the exhaust port does not have recourse to any of the valving described in Sect. 7.3.3. Naturally, the BMEP for the engine is assessed on the total swept volume capacity of the engine. It is instructive to compare the test results for the QUB 400 engine, presented in Figs. 7.4-7.8, with those for this power unit as an example of the level of exhaust emissions and fuel economy which were predicted in Sect. 7.2.1.1 for the QUB 400 engine when the short-circuited fuel loss is eliminated.

Fig. 7.36 is a composite picture of the full load behavior at 3000 rpm and light load at 1500 rpm. The full load is set by having the main air throttle at wide-open and the fueling is varied from a maximum to a minimum; beyond either of those limits, rich or lean, misfire occurs and the data is not recorded. That the combustion process is fairly rapid at 3000 rpm is evidenced by an ignition timing at 10º btdc. The light load at 1500 rpm is set by having the main throttle at 15% area ratio; the ignition timing is at 22º btdc. That the air utilization is not acceptable in the 3000 rpm full load tests is seen from the overall air-fuel ratio ranging from 25 to 42. Yet the highest BMEP at 3000 rpm is quite good at 4.3 bar and the best BSFC is also good at 0.295 g/kWh. The nitrogen oxides are also conventionally low. At the light load point at 1500 rpm the air-fuel ratio is what one would expect from a stratified burning process and varies from 13 to 20, where the minimum BMEP is 1.2 bar (100 kPa=1 bar). At these light load points the NOx is very low, about 80 ppm. It is interesting to note that the NOx levels are identical at equal BMEP at either 1500 or 3000 rpm.

The results of further testing at light load at 1500 rpm are presented in greater detail in Fig. 7.37, where the unburned hydrocarbon and carbon monoxide emissions are also included as test data. The HC emission is about 850 ppm NDIR(C6) which illustrates the almost total absence of lost fuel to the exhaust pipe caused by this stratified charging system. That the engine is firing evenly on every cycle is obvious and audible to the experimenter, and is clearly seen by the low levels of CO emission at 0.15%, by the low HC emission, and by the BSFC at 0.4 kg/kWh. Equally important from an engine management standpoint, the control of such an engine is quite straightforward due to its reduced sensitivity to air-fuel ratio.

At full load at 3000 rpm, a more complete set of experimental test data is given in Fig. 7.38. The peak BMEP is but 4.1 bar in this test, when the CO emission level

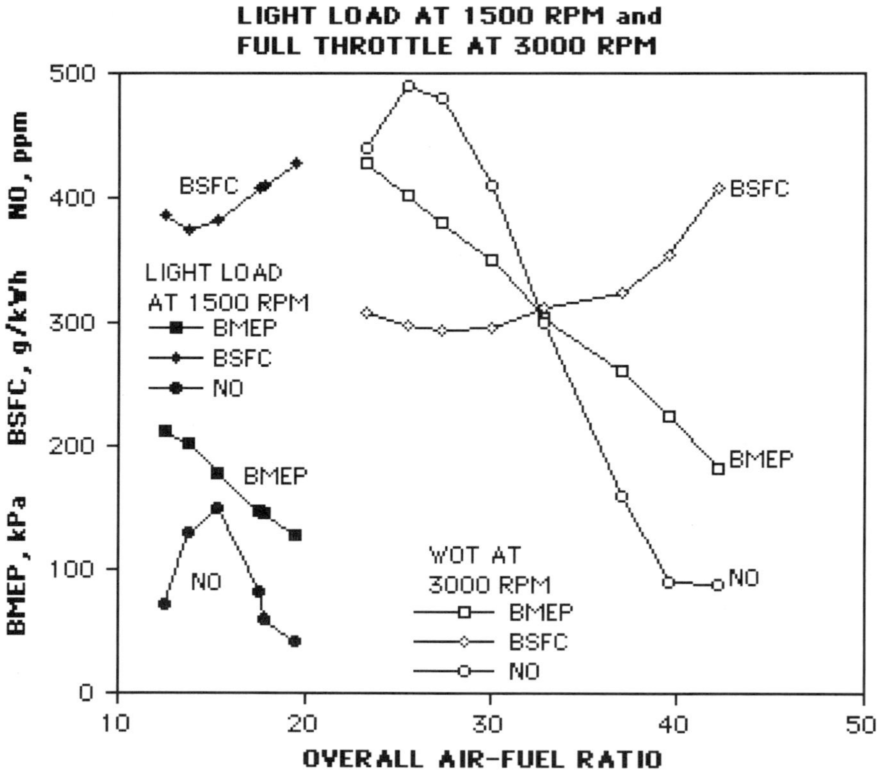

Fig. 7.36 L-Head behavior as a stratified charging and combustion engine.

is just 0.4%. This is at an air-fuel ratio of 25, indicating an air utilization rate which is unacceptably low in this configuration. The base engine when homogeneously charged is capable of 5.5 bar BMEP at this engine speed and when connected to the same (untuned) exhaust system. However, the HC emission is very low at 360-400 ppm and indicates the total absence of fuel short-circuiting to the exhaust port. The BSFC is very good at a best point of 0.295 kg/kWh and the CO emission falls to an excellent minimum of 0.1%. It is clear that the control strategy for such an engine is quite simple, for the only variable in this set of test data is the fueling rate with all other test parameters remaining constant throughout.

It is considered that such an engine type deserves further intensive investigation, for this design incorporating stratified combustion solves one of the basic problems inherent in homogeneous combustion, namely the ability of the engine to fire evenly and efficiently on every cycle at the lightest of loads at low engine speeds. This is

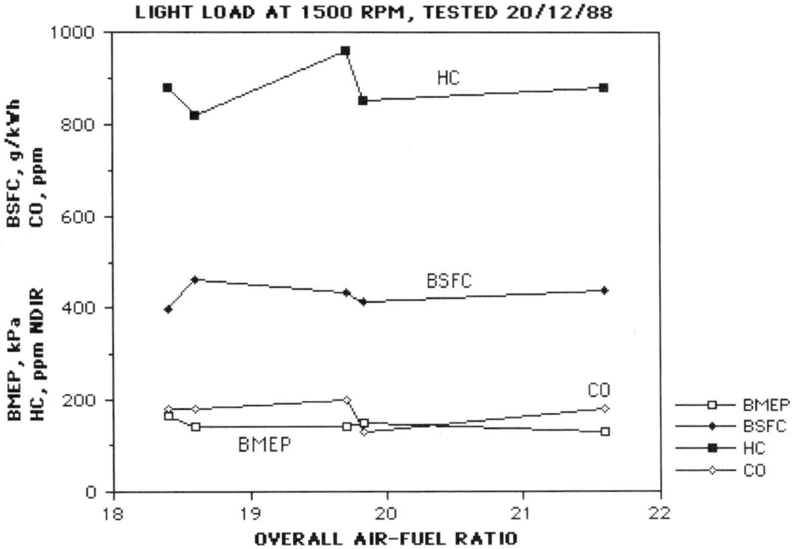

Fig. 7.37 QUB L-Head engine performance characteristics at light load and speed.

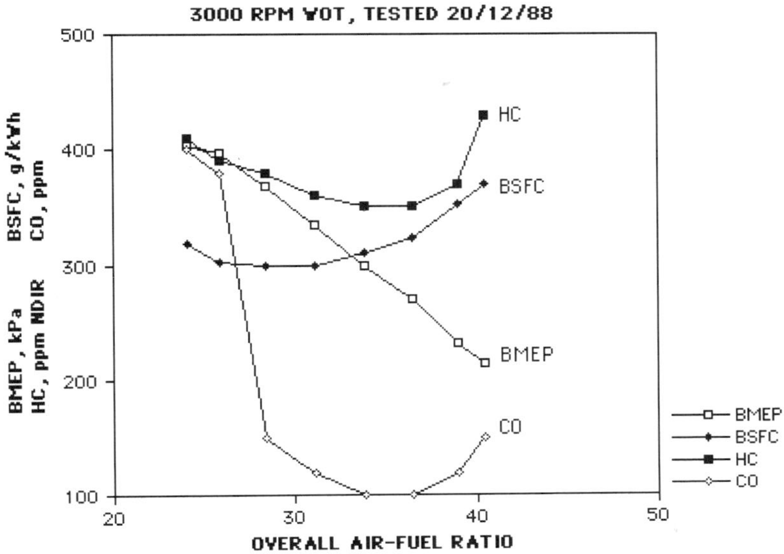

Fig. 7.38 QUB L-Head engine performance characteristics at full throttle at 3000 rpm.

a particularly important criterion for engines destined for an automotive application, as not only are the legislated driving cycles performed in this area but the actual vehicle engine must perform smoothly in an urban environment to be acceptable to the motorist.

7.4.3 Direct in-cylinder fuel injection

As was pointed out earlier, this is one of the obvious methods of reducing, or even eliminating, the loss of fuel to the exhaust port during the scavenge process. Fig. 7.21 illustrates the basic options regarding the positioning of such a fuel injector. The potential difficulties of evaporating that fuel spray in time for a homogeneous combustion process to occur have also been debated. The even more fundamental problem of attaining good flammability characteristics at light load and speed in a homogeneous combustion process has also been addressed. What, then, are the known experimental facts about in-cylinder fuel injection? Is the combustion process homogeneous, or does the desirable possibility of stratified burning at light load and speed exist? Does the fuel vaporize in time to burn in an efficient manner?

The answers to these questions are contained, in part, in the technical papers published by several authors. Possibly the most representative of the state of the art, at this early juncture in the experimental work in this field, are the papers presented by Fuji Heavy Industries from Japan. The first of these was published as long ago as 1972(7.17) and more recently by Sato and Nakayama(7.2) in 1987. Nuti(7.12) from Piaggio, and Plohberger et al(7.19) from AVL, have also published experimental data on this subject and they too should be studied. The papers just mentioned all use high-pressure liquid injection systems, not unlike that employed for diesel engines, but modified to attain smaller droplet sizes when using gasoline fuel. The droplet sizes required are of 10-15 μm mean diameter, usually measured by laser-based experimental techniques to acquire the Sauter Mean Diameter of the fuel droplets(7.16).

The results measured by Sato(7.2) are reproduced here as Fig. 7.39. The peak power performance characteristics of the engine are unaffected by the use of fuel injection, as is their reported levels of NO_x emission. However, considerable reductions in the BSFC values are seen in the center of the "oyster" map, with the best contour being lowered to 0.3 kg/kWh with fuel injection, from 0.38 kg/kWh when the engine was carburetted. The equivalent picture is repeated for the hydrocarbon emission levels with the carburetted engine showing some 3000 ppm HC/NDIR values, whereas the fuel injection engine is reduced to 400 ppm. This effect is also reported by Nuti(7.12) and the fuel consumption levels in his engine would be even lower at the best possible condition, these being BSFC values of 0.27 kg/kWh. These are significantly low levels of fuel consumption and hydrocarbon exhaust emissions.

However, Sato(7.2) also reports that the direct in-cylinder fuel injection did not improve the misfiring behavior (four-stroking) at light loads and speeds, and this can be seen in Fig. 7.39. In the lower left-hand corner of the "oyster" maps, at the light load and speed positions, the BSFC and hydrocarbon emission values are the same for the carburetted and fuel-injected engines. The values of BSFC are an

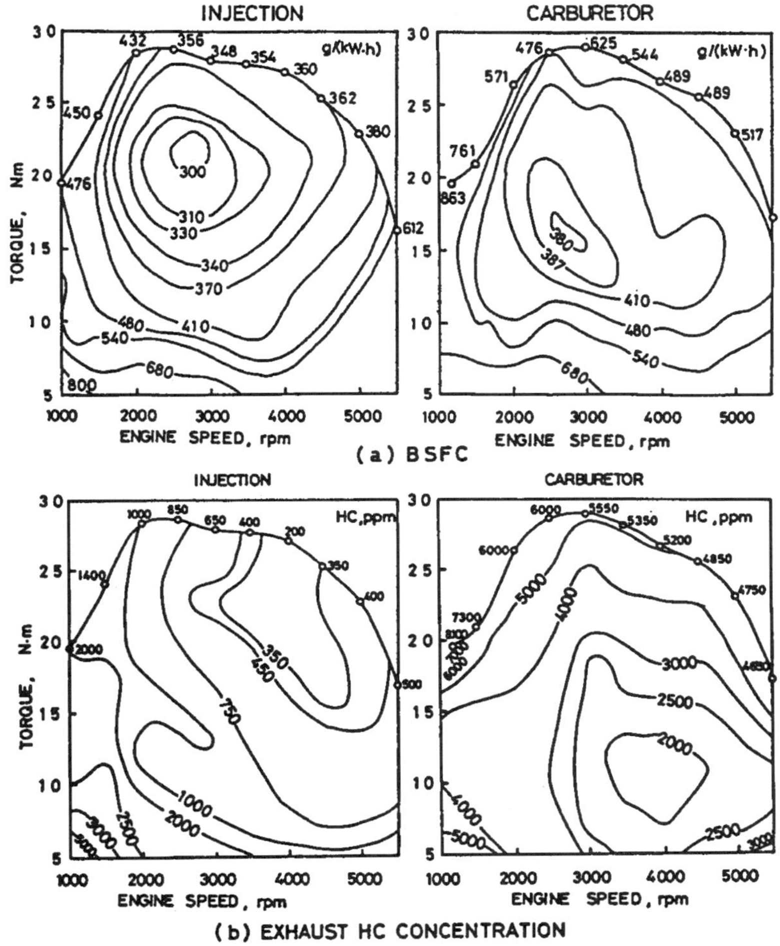

Fig. 7.39 Comparison of fuel consumption and hydrocarbon emissions between direct fuel injection and a carburetor.

unimpressive 0.5 to 0.6 kg/kWh and the unburned hydrocarbons are at 3000-5000 ppm. Plohberger(7.19) shows very complete test data at light load over the speed range, and below 3000 rpm his results confirm that reported by Sato(7.2): the direct in-cylinder injection of fuel has not solved this vital problem. Although Nuti(7.12) does not comment on this situation, his HC emission is reportedly only slightly better than that given by Sato(7.2) and his CO levels are rising rapidly in that zone. One must conclude from the experimental data presently available that the direct fuel injection of liquid gasoline does not provide stratified combustion at light loads and speeds, and thereby does not improve the emissions and fuel economy of the two-stroke engine at that important driving condition.

It would appear that there is sufficient time at higher speeds and loads to vaporize the gasoline and mix it with the air in time for a homogeneous combustion process to occur. The onset of injection required is remarkably early, and well before the trapping point at exhaust closure. Sato(7.2) describes the end of the dynamic injection process taking place at 10º abdc at 1000 rpm and 50º abdc at 5000 rpm. The exhaust port on this 356 cc twin-cylinder engine closed at 68º abdc.

7.4.3.1 Air-blast injection of fuel into the cylinder

This is actually the oldest technique for injecting fuel into diesel engines, employed by Dr. Diesel no less, and was replaced for diesel engines by the invention of the jerk pump giving liquid injection at the turn of this century. In a form suitable for the injection of gasoline into today's high-speed two-stroke engines, the method of operation could be by electromagnetic solenoid valves.

The method of operation of a modern air-blast fuel injector is sketched in Fig. 7.40. The injector is supplied with air at about 7 bar and liquid fuel at about 6 bar. The pressure difference between the air and the fuel would be carefully controlled so that any known physical movement of the fuel needle would deliver a controlled quantity of fuel into the sac before final injection. In Fig. 7.40(a) the injector is ready for operation. Both of the electromagnetic solenoids are independently activated by a control system as part of the overall engine and vehicle management system. The fuel solenoid is electronically activated for a known period of lift and time, as seen in Fig. 7.40(b), the result being that a precise quantity of fuel is metered into the sac behind the main needle. The fuel needle is then closed by the solenoid, most likely assisted by a spring, and the fuel awaiting final injection is heated within this space by conduction from the cylinder head; this period is shown in Fig. 7.40(c). At the appropriate juncture, the main needle is activated by its solenoid and the high-pressure air supply can now act upon the stored fuel and spray it into the combustion chamber; this is shown in Fig. 7.40(d). That needle too is closed by its solenoid by electronic triggering, and the needle is assisted by a spring to returns to its seat, as in Fig. 7.40(a). This completes the cycle of operation. The spray which is created by such an aerosol method is particularly fine, and droplet sizes in the range of 5-10 μm Sauter Mean Diameter have been reported for these devices. Indeed, the device emanating from IFP, and discussed in Sect. 7.4.1.4, is a mechanical means of accomplishing the same ends, albeit with a greater mass of air present during the heating phase, but the final delivery to the cylinder is at a lower velocity than that produced by the air-blast injector.

The advantage of such an injector is that it potentially eliminates one of the liquid injector's annoying deficiencies, that of "dribbling" or leaving a droplet on the nozzle exit at low flow rates. This "dribble" produces misfire and excessive HC emission. The disadvantage is that the long-term reliability and accuracy of fuel delivery of such a device is open to question. Solenoids exposed to high temperatures are not known for precise retention of their electromagnetic behavior, and this

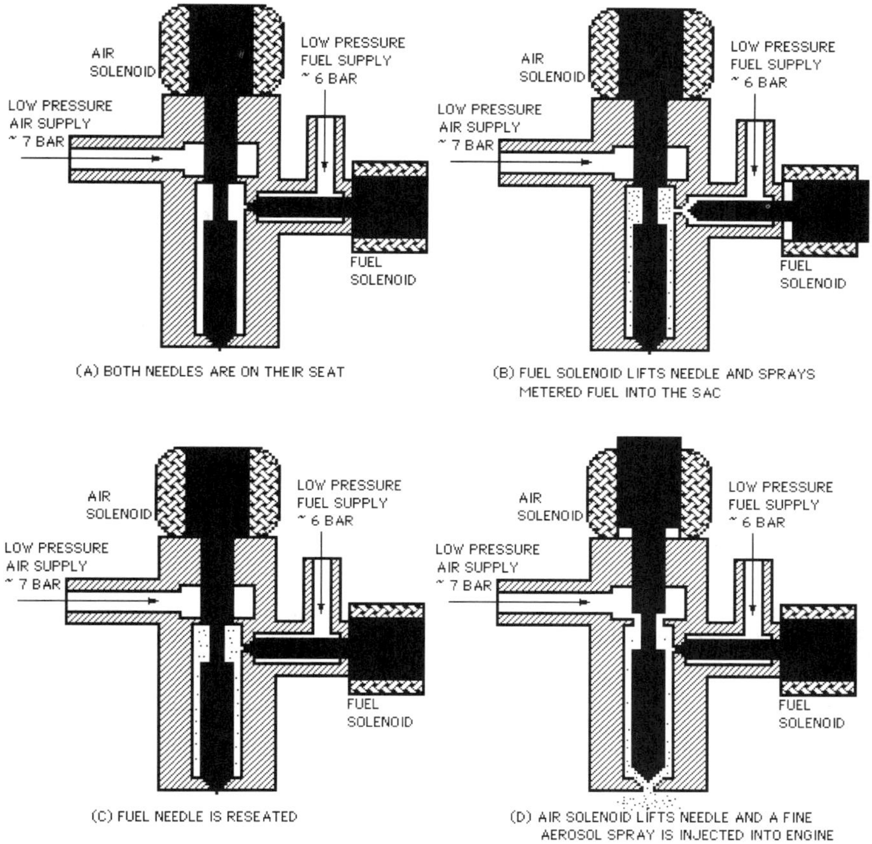

Fig. 7.40 Sequence of events in the operation of an air-blast fuel injector.

would be unacceptable in vehicle service. That is why the injector is placed in the cylinder wall in some designs, as in Fig. 7.21(b), away from the hot cylinder head. However this gives an undesirable subsidiary effect, as this leaves some residual hydrocarbons in the passage between the injector and the cylinder after the piston has passed by in the trapping process, and these are then carried into the cylinder on the next scavenge process or are exposed to the exhaust gas by the downward travel of the piston during the blowdown phase.

The Orbital Engine Company of Perth in Western Australia has been very active in the design and development of air-blast injectors and of vehicle engines fitted with such injectors. Although they have published(7.26) some technical information on this subject in a form which would answer the many questions posed in this chapter, the commercial information they provide would tend to suggest that they are aware of the need for such systems to provide stratified combustion at light loads and homogeneous burning at high BMEP and engine speed. Whether they have solved these difficult problems remains to be seen. In any event, this type of injector offers some hope, like the IFP engine, that the best of both combustion worlds is yet possible.

7.5 Concluding comments

It is too early in the development history of two-stroke engines for precise conclusions to be drawn regarding its future role; this process has really only been in progress for some thirty years. Nevertheless, the general level of improvement in performance characteristics of the simple engine has been greatly enhanced by new research and development techniques for cycle calculation, scavenging and combustion design, fueling methods and exhaust tuning. Even more remarkable are the performance characteristics of some of the advanced two-stroke engines aimed at future automotive applications, some with behavioral characteristics at the first experimental attempt which rival or exceed those available from current four-stroke engine practice. Often, some of these new two-stroke engines are produced on research budgets and by engineering teams whose size is a mere fraction of that being devoted to the onward development of an existing four-stroke production engine. The decade of the 1990's is going to be a very important period in the development of the two-stroke cycle engine and it is quite possible that an automobile fitted with such an engine will be in series production before the year 2000. In that context, the prophetic remark passed by Professor Dr. Alfred Jante, in an SAE paper(3.5) presented in May, 1968, is particularly significant: "The stricter that the exhaust emissions standards become, the more the two-stroke engine will regain importance."

REFERENCES FOR CHAPTER 7

7.1 G. Batoni, "An Investigation into the Future of Two-Stroke Motorcycle Engine," Society of Automotive Engineers, West Coast Meeting, San Diego, California, August, 1978, SAE Paper No. 780710.

7.2 T. Sato, M. Nakayama, "Gasoline Direct Injection for a Loop-Scavenged Two-Stroke Cycle Engine," Society of Automotive Engineers, International Off-Highway and Powerplant Congress and Exposition, Milwaukee, Wisconsin, September, 1987, SAE Paper No. 871690.

7.3 K. Tsuchiya, S. Hirano, M. Okamura, T. Gotoh, "Emission Control of Two-Stroke Motorcycle Engines by the Butterfly Exhaust Valve," Society of Automotive Engineers, International Off-Highway and Powerplant Meeting and Exposition, Milwaukee, Wisconsin, September, 1980, SAE Paper No. 800973.

7.4 N. Hata, T. Iio, "Improvement of the Two-Stroke Engine Performance with the Yamaha Power Valve System," Society of Automotive Engineers, International Off-Highway Meeting and Exposition, Milwaukee, Wisconsin, September, 1981, SAE Paper No. 810922.

7.5 S. Ohigashi, "Exhaust Pipe Contraction in Two-Stroke Engines," Japanese Patent No. Showa 38-19659 (1963).

7.6 H. Nishizaki, S. Koyama, "EK34 Type Engine and Idle Silence Valve for Subaru Rex," *Internal Combustion Engines*, Vol. 11, No. 128, Sept., 1972.

7.7 S. Ishihara, "An Experimental Development of a New U-Cylinder Uniflow Scavenged Engine," Society of Automotive Engineers, International Congress and Exposition, Detroit, Michigan, March, 1985, SAE Paper No. 850181.

7.8 S. Onishi, S.H. Jo, P.D. Jo, S. Kato, "Multi-Layer Stratified Scavenging (MULS)—A New Scavenging Method for Two-Stroke Engine," Society of Automotive Engineers, International Congress and Exposition, Detroit, Michigan, March, 1984, SAE Paper No. 840420.

7.9 G.P. Blair, B.W. Hill, A.J. Miller, S.P. Nickell, "Reduction of Fuel Consumption of a Spark-Ignition Two-Stroke Cycle Engine," Society of Automotive Engineers, International Congress and Exposition, Detroit, Michigan, March, 1983, SAE Paper No. 830093.

7.10 B.W. Hill, G.P. Blair, "Further Tests on Reducing Fuel Consumption with a Carburetted Two-Stroke Cycle Engine," Society of Automotive Engineers, International Off-Highway Meeting and Exposition, Milwaukee, Wisconsin, September, 1983, SAE Paper No. 831303.

7.11 H. Uchiyama, T. Chiku, S. Sayo, "Emission Control of Two-Stroke Automobile Engine," Society of Automotive Engineers, International Off-Highway Meeting and Exposition, Milwaukee, Wisconsin, September, 1977, SAE Paper No. 770766.

7.12 M. Nuti, "Direct Fuel Injection: An Opportunity for Two-Stroke SI Engines in Road Vehicle Use," Society of Automotive Engineers, International Congress and Exposition, Detroit, Michigan, February, 1986, SAE Paper No. 860170.

7.13 N.J. Beck, R.L. Barkhimer, M.A. Calkins, W.P. Johnson, W.E. Weseloh, "Direct Digital Control of Electronic Unit Injectors," Society of Automotive Engineers, International Congress and Exposition, Detroit, Michigan, March, 1984, SAE Paper No. 840273.

7.14 E. Vieilledent, "Low Pressure Electronic Fuel Injection System for Two-Stroke Engines," Society of Automotive Engineers, International Off-Highway Meeting and Exposition, Milwaukee, Wisconsin, September, 1978, SAE Paper No. 780767.

7.15 R. Douglas, G.P. Blair, "Fuel Injection of a Two-Stroke Cycle Spark Ignition Engine," Society of Automotive Engineers, International Off-Highway Meeting and Exposition, Milwaukee, Wisconsin, September, 1972, SAE Paper No. 820952.

7.16 N.J. Beck, W.P. Johnson, R.L. Barkhimer, S.H. Patterson, "Electronic Fuel Injection for Two-Stroke Cycle Gasoline Engines," Society of Automotive Engineers, International Off-Highway and Powerplant Congress and Exposition, Milwaukee, Wisconsin, September, 1986, SAE Paper No. 861242.

7.17 G. Yamagishi, T. Sato, H. Iwasa, "A Study of Two-Stroke Fuel Injection Engines for Exhaust Gas Purification," Society of Automotive Engineers, International Congress and Exposition, Detroit, Michigan, January, 1972, SAE Paper No. 720195.

7.18 P. Duret, A. Ecomard, M. Audinet, "A New Two-Stroke Engine with Compressor-Air Assisted Fuel Injection for High Efficiency Low Emissions Applications," Society of Automotive Engineers, International Congress and Exposition, Detroit, Michigan, March, 1988, SAE Paper No. 880176.

7.19 D. Plohberger, K. Landfahrer, L. Mikulic, "Development of Fuel Injected Two-Stroke Gasoline Engine," Society of Automotive Engineers, International Congress and Exposition, Detroit, Michigan, March, 1988, SAE Paper No. 880170.

7.20 S. Ishihara, "Experimental Development of Two New Types of Double Piston Engines," Society of Automotive Engineers, International Congress and Exposition, Detroit, Michigan, February, 1986, SAE Paper No. 860031.

7.21 F. Laimbock, "Der abgasame Hochleitungszweitaktmotor," Dritte Grazer Zweiradtagung, Technische Universitat, Graz, Austria, 13-14 April, 1989.

7.22 D. Fog, R.M. Brown, D.H. Garland, "Reduction of Smoke from Two-Stroke Engine Oils," International Conference on the Small Internal Combustion Engine, Paper C372/040, Institution of Mechanical Engineers, London, 4-5 April, 1989.

7.23 O.V. Kuntscher, A. Singer, "Mixture Injection Application for Avoiding Charge Exchange Losses in Two-Stroke Cycle Engines," International Conference on the Small Internal Combustion Engine, Paper C372/025, Institution of Mechanical Engineers, London, 4-5 April, 1989.

7.24 M. Prigent, G. De Soete, "Nitrous Oxide N_2O in Engine Exhaust Gases, A First Appraisal of Catalyst Impact," Society of Automotive Engineers, International Congress and Exposition, Detroit, Michigan, February, 1989, SAE Paper No. 890492.

7.25 K. Sugiura, M. Kagaya, "A Study of Visible Smoke Reduction from a Small Two-Stroke Engine using Various Engine Lubricants," Society of Automotive Engineers, Fuels and Lubricants Meeting, Tulsa, Oklahama, June, 1977, SAE Paper No. 770623.

7.26 K. Schlunke, "Der Orbital Verbrennungsprozess des Zweitaktmotors," Tenth International Motor Symposium, Vienna, April, 1989, pp.63-78.

Chapter 8

Reduction of Noise Emission from Two-Stroke Engines

8.0 Introduction

The subject of noise emission from an internal combustion engine, and its reduction, is a specialized topic. Many textbooks and technical papers have been written on the subject, so it is not possible to completely cover the matter in a single chapter in this book. Instead, it is intended to orient the reader to noise emission pertaining to two-stroke engines and to the problems of designing silencers for this particular type of power unit. Many texts discuss silencer design as if the only engine in existence is the spark-ignited, multi-cylinder, four-stroke automobile engine and merely describe the techniques applicable to that area. As that can create considerable misconceptions for the designer of silencers for two-stroke engines, this provides sufficient justification for the inclusion of this chapter within this book.

The opening section of this chapter gives some general background to the subject of noise, indeed it repeats in brief what may be found more completely in other books and papers which are suitably referenced for the wider education of the reader. This opening section is included so that the remainder of the chapter will be more immediately meaningful to the tyro.

The next section deals with the fundamental nature of pressure wave created noise emission and to the theoretical methods available for its prediction. The latter part of this section debates the future for the technology of silencer design.

The succeeding sections will cover the more empirical approaches to silencer design and incorporate pragmatic advice on the design of such devices for two-stroke engines.

8.1 Noise

In many texts, the reader will find "noise" described as "unwanted sound." This is a rather loose description, as that which is wanted by some may be unwanted by others. A 250 cc Yamaha V4 racing two-stroke, producing 72 unsilenced horsepower at 14000 rpm, and wailing its way up the Mountain in the Isle of Man TT race of 1967, produced a noise which was music to the ears of a thousand racing motorcycle fans. A nearby farmer, the owner of a thousand chickens vainly trying to lay eggs, viewed the same noise from an alternate standpoint. This simple example illustrates the quite subjective nature of noise assessment. Nevertheless, between the limits of the threshold of human hearing and the threshold of damage to the human ear, it is possible to physically measure the pressure level caused by sound and to assign an experimental number to that value. This number will not detail whether the sound is "wanted" or "unwanted." As already pointed out, to some it will be described as noise.

8.1.1 Transmission of sound

As initially discussed in Sect. 2.1, sound propagates in three dimensions from a source through the air or a gas as the medium of transmission. The fundamental theory for this propagation is to be found in Sect. 2.1.2. The speed of that propagation is given by A, where:

$$A=\sqrt{(g*R*T)}=\sqrt{(g*P/D)} \qquad (8.1.1)$$

The value for the ratio of specific heats, g, is 1.4 for air and 1.35 for exhaust gas when it is at a temperature around 350ºC. The value of the gas constant, R, is 287 J/kgK for air and 291 J/kgK for exhaust gas emanating from a stoichiometric combustion. Treating exhaust gas as air in calculations for sound wave attenuation in silencers produces little error of real significance. For example, at 400ºC:

$$A_{air}=\sqrt{(1.4*287*673)}=520 \text{ m/s}$$
$$A_{exh}=\sqrt{(1.35*291*673)}=514 \text{ m/s}$$

As exhaust gas in a two-stroke engine contains a significant proportion of air which is short-circuited during the scavenge process, this reduces the already negligible error even further.

8.1.2 Intensity and loudness of sound

The propagation of pressure waves is already covered thoroughly in Sect. 2.1, so it is not necessary to repeat it here. However, the propagation of these small pressure waves in air, following one after the other, varying in both spacing and amplitude, gives rise to the human perception of the pitch and of the amplitude of the sound. The frequency of the pressure pulsations produces the pitch and their amplitude denotes the loudness. The human ear can detect frequencies ranging from 20 Hz to 20 kHz, although the older one becomes that spectrum shortens to a maximum of about 12 kHz. For another introductory view of this topic, the reader should consult the books by Annand and Roe(1.8) and by Taylor(8.11)

More particularly, the intensity, E, is used to denote the physical energy of a sound, and loudness, Lp, is the subjective human perception of that intensity. The relationship between intensity and loudness is fixed for sounds which have a pure tone, or pitch, i.e., the sound is composed of sinusoidal pressure waves of a given frequency. For real sounds that relationship is more complex. The intensity of the sound, being an energy value, is denoted by units of W/m^2. Noise meters, being basically pressure transducers, record the "effective sound pressure level," which is the root-mean-square of the pressure fluctuation about the mean pressure caused by the sound pressure waves. This rms pressure fluctuation is denoted by dP, and in a medium where the density is D and the speed of sound is A, the intensity is related to the square of the rms sound pressure level by:

$$E=dP^2/(D*A) \qquad (8.1.2)$$

The pressure rise, dP, can be visually observed in Plates 2.1-2.3 as it propagates away from the end of an exhaust pipe.

The level of intensity which can be recorded by the human ear is considerable, ranging from 1 pW/m^2 to 1 W/m^2. The human eardrum, our personal pressure transducer, will oscillate from an imperceptible level at the minimum intensity level up to about 0.01 mm at the highest level when a sensation of pain is produced by the nervous system as a warning of impending damage. To simplify this wide variation in physical sensation, a logarithmic scale is used to denote loudness, and the scale is in the units called a Bel, the symbol for which is B. Even this unit is too large for general use, so it is divided into ten sub-divisions called decibels, the nomenclature for which is dB. The loudness of a sound is denoted by comparing its intensity level on this logarithmic scale to the "threshold of hearing," which is at an intensity, E0, of 1 pW/m^2 or a rms pressure fluctuation, dP0, of 0.00002 Pa, or 0.0002 μbar. Thus, intensity level of a sound, EL1, where the actual intensity is E1, is given by:

$$\begin{aligned} EL1 &= \log_{10}(E1/E0)\ B \\ &= 10*\log_{10}(E1/E0)\ dB \end{aligned} \qquad (8.1.3)$$

In a corresponding fashion, a sound pressure level, Lp1, where the actual rms pressure fluctuation is dP1, is given by:

$$\begin{aligned} Lp1 &= \log_{10}(E1/E0)\ B \\ &= 10*\log_{10}(E1/E0)\ dB \\ &= 10*\log_{10}[(dP1/dP0)^2]\ dB \\ &= 20*\log_{10}(dP1/dP0)\ dB \end{aligned} \qquad (8.1.4)$$

8.1.3 Loudness when there are several sources of sound

Let the reader imagine he is exposed to two sources of sound of intensities E1 and E2. These two sources would separately produce sound pressure levels of Lp1 and Lp2, in dB units. Consequently, from Eq. 8.1.4:

$$E1=E0*\text{antilog}_{10}(Lp1/10)$$
$$E2=E0*\text{antilog}_{10}(Lp2/10)$$

The absolute intensity experienced by the reader from both sources simultaneously is E3, where:

$$E3=E1+E2$$

Hence, the total sound pressure level experienced from both sources is Lp3, where:

$$\begin{aligned} Lp3 &= 10*\log_{10}(E3/E0) \\ &= 10*\log_{10}\{\text{antilog}_{10}(Lp1/10)+\text{antilog}_{10}(Lp2/10)\} \end{aligned} \qquad (8.1.5)$$

Consider two simple cases:

(a) The reader is exposed to two equal sources of sound which are at 100 dB.

(b) The reader is exposed to two sources of sound, one at 90 dB and the other at 100 dB.

First, consider case (a), using Eq. 8.1.5:

$$\begin{aligned} Lp3 &= 10*\log_{10}\{\text{antilog}_{10}(Lp1/10)+\text{antilog}_{10}(Lp2/10)\} \\ &= 10*\log_{10}\{\text{antilog}_{10}(100/10)+\text{antilog}_{10}(100/10)\} \\ &= 103.01 \text{ dB} \end{aligned}$$

Second, consider case (b):

$$\begin{aligned} Lp3 &= 10*\log_{10}\{\text{antilog}_{10}(Lp1/10)+\text{antilog}_{10}(Lp2/10)\} \\ &= 10*\log_{10}\{\text{antilog}_{10}(100/10)+\text{antilog}_{10}(90/10)\} \\ &= 100.41 \text{ dB} \end{aligned}$$

In case (a), it is clear that the addition of two sound sources, each equal to 100 dB, produces an overall sound pressure level of 103.01 dB, a rise of just 3.01 dB due to a logarithmic scale being used to attempt to simulate the response characteristics of the human ear. To physically support this mathematical contention, the reader will recall that the noise of an entire brass band does not appear to be so much in excess of one trumpet at full throttle.

In case (b), the addition of the second weaker source at 90 dB to the noisier one at 100 dB produces a negligible increase in loudness level, just 0.41 dB above the larger source. The addition of one more trumpet to the aforementioned brass band does not raise significantly the level of perception of total noise for the listener.

There is a fundamental message to the designer of engine silencers within these simple examples: if an engine has several different sources of noise, the loudest will swamp all others in the overall sound pressure level. The identification and muffling of that major noise source becomes the first priority on the part of the engineer. Expanded discussion on these topics are to be found in the books by Harris(8.12) and Beranek(8.13) and in this chapter.

8.1.4 Measurement of noise and the noise-frequency spectrum

An instrument for the measurement of noise, referred to as a noisemeter, is basically a microphone connected to an amplifier so that the system is calibrated to read in the units of dB. Usually the device is internally programmed to read either the total sound pressure level which is known as the linear value, i.e., dBlin, or on an A-weighted or B-weighted scale to represent the response of the human ear to loudness as a function of frequency. The A-weighted scale is more common and the units are recorded appropriately as dBA. To put some numbers on this weighting effect, the A-weighted scale reduces the recorded sound below the dBlin level by 30 dB at 50 Hz, 19 dB at 100 Hz, 3dB at 500 Hz, 0 at 1000 Hz, then increases it by about 1 dB between 2 and 4kHz, before tailing off to reduce it by 10 dB at 20 kHz. The implications behind this weighting effect are that high frequencies between 1000 and 4000 Hz are very irritating to the human ear, to such an extent that a noise recording 100 dBlin at around 100 Hz only sounds as loud as 81dB at 1000 Hz, hence it is recorded as 81dBA. To quote another example, the same overall sound pressure level at 3000 Hz appears to be as loud as 101 dB at 1000 Hz, and is noted as 101 dBA.

Equally common is for the noisemeter to be capable of a frequency analysis, i.e., to record the noise spectra over discrete bands of frequency. Usually these are carried out over one-octave bands or, more finely, over one-third octave bands. A typical one-octave filter set on a noisemeter would have switchable filters to record the noise about 31.25 Hz, 62.5 Hz, 125 Hz, 250 Hz, 500 Hz, 1000 Hz, 2000 Hz, 4000 Hz, 8000 Hz and 16000 Hz. A one-third octave filter set carries out this function in narrower steps of frequency change. The latest advances in electronics and computer-assisted data capture allow this process of frequency analysis to be carried out in even finer detail.

The ability of a measurement system to record the noise-frequency spectrum is very important for the researcher who is attempting to silence, say, the exhaust system of a particular engine. Just as described in Sect. 8.1.3 regarding the addition of noise levels from several sources, the noisiest frequency band in the measured spectrum is that band which must be silenced first and foremost as it is contributing in the major part to the overall sound pressure level. The identification of the frequency band of that major noise component will be shown later as the first step towards its eradication as a noise source.

The measurement of noise is a tedious experimental technique in that a set procedure is not just desirable, but essential. Seemingly innocent parameters, such as the height of the microphone from the ground during a test, or the reflectivity of the surface of the ground in the vicinity of the testing, e.g., grass or tarmac, can have a major influence on the numerical value of the dB recorded from the identical engine or machine. This has given rise to a plethora of apparently unrelated test procedures, such as those in the SAE Standards(8.16). In actual fact, the logic behind their formulation is quite impeccable and any reader embarking on a silencer design and development exercise will be wise to study them thoroughly and implement them during experimentation.

8.2 Noise sources in a simple two-stroke engine

The sources of noise emanating from a two-stroke engine are illustrated in Fig. 8.1. The obvious ones are the intake and the exhaust system, where the presence of gas pressure waves has been discussed at length in Chapter 2. As these propagate into the atmosphere they produce noise. The series of photographs in Chapter 2, Plates 2.1-2.4, illustrate the rapid nature of the pressure rise propagating into the atmosphere and towards the ear of the listener. The common belief is that the exhaust is the noisier of the two, and in general this is true. However, the most rudimentary of exhaust silencers will almost inevitably leave the intake system as the noisier of these two sources, so it also requires silencing to the same level and extent. If the intake system contains a reed valve, it produces its own unique high-frequency noise components, much like the vibrating reed in an oboe or a clarinet which have a special quality described occasionally as "honking." This adds to the design complexity of the intake silencer.

The reader will realize that any vibrating metal surface can act as a noise source, in the same manner as does the vibrating diaphragm of a loudspeaker, and thus the pressure signal emanating from the combustion pressure rise is transmitted through the cylinder walls and can be propagated away from the outer surface of the cylinder and the cylinder head. If the engine is air-cooled, the cylinder and head finning is ideally suited to become metal diaphragms for this very purpose. The need to control this form of noise transmission is obvious and the reader will observe that air-cooled motorcycle engines have rubber damping inserted between the cooling fins for this

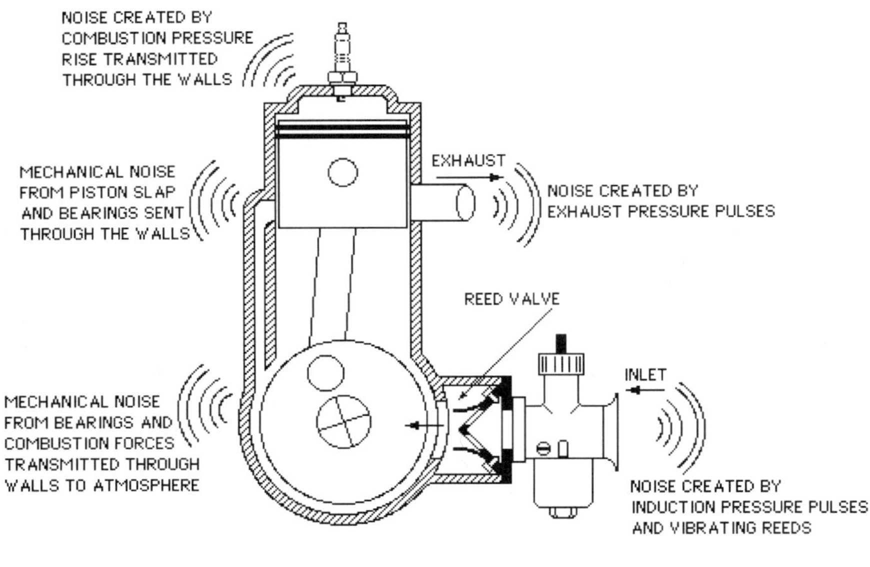

Fig. 8.1 Various sources of noise from a two-stroke engine.

very purpose. The problem is greatly eased by the use of liquid (water) cooling of a cylinder and cylinder head, as the intervening water layer acts as a damper on the noise transmission. That does not totally solve this problem, for the noise can be transmitted through the piston to the connecting rod, via the bearings to the crankcase walls, and ultimately to the atmosphere. The use of pressure-fed plain bearings on an engine crankshaft is superior to ball, roller, or needle roller bearings as a means of suppression of this type of noise transmission, because the hydrodynamic oil film in the plain bearing does not as readily transmit the combustion vibrations. However, the use of pressure-fed plain bearings does not lend itself to the design of a crankcase as the air pump of the simple two-stroke engine.

The rotating bearings of an engine produce noise characteristics of their own, emanating from the vibrations of the mechanical components. Another mechanical noise source is that from piston slap, as the piston rocks on its axis within the cylinder around the tdc and bdc positions.

The discussion regarding noise suppression in this chapter will concentrate on the gas pressure wave generated noise from the exhaust and the inlet systems.

8.3 Silencing the exhaust and inlet system of the two-stroke engine

In the matter of silencing, the two-stroke engine has some advantages and disadvantages by comparison with an equivalent four-stroke cycle engine. These are catalogued below in the first instance and discussed at greater length in succeeding sections as the need arises.

(a) Disadvantages

1) The engine operates at double the frequency of creation of gas pressure waves and humans dislike exposure to higher-frequency noise.

2) The ports in a two-stroke engine open faster than the poppet valves of the four-stroke engine and so the pressure wave fronts are steeper, thereby creating more high-frequency noise components within the sound spectrum.

3) Many two-stroke engines are used in applications calling for an engine with low bulk and weight, thereby further reducing the space available for muffling and potentially giving the engine type a reputation for being noisy.

4) Two-stroke engines use ball, roller, and needle roller element bearings and these tend to be noisy by comparison with pressure-fed hydrodynamic bearings.

(b) Advantages

1) The engine with the tuned exhaust pipe produces a high specific output, but this is achieved by choking the final outlet diameter, thereby simplifying the design of an effective exhaust silencer.

2) The crankcase pump induces air by pumping with a low compression ratio and, as this reduces the maximum values encountered in the air intake particle velocity-time history, it lowers the higher frequency content of the sound produced.

3) The peak combustion pressures are lower in the equivalent two-stroke cycle engine, so the noise spectrum induced by that lesser combustion pressure is reduced via all of the transmission components of the cylinder, cylinder head, piston and crankshaft.

8.4 Some fundamentals of silencer design

If the reader studies the textbooks or technical papers on acoustics and on silencer design, such as many of those referenced below, it will be to find that the subject is full of empirical design equations for the many basic types of silencers used in the field of internal combustion engines. However useful these may be, the reader will get the feeling that a fundamental understanding of the subject is not being acquired, particularly as the acoustic theory being applied is one oriented to the propagation of acoustic waves, i.e., waves of infinitesimal amplitude, rather than finite amplitude waves, i.e., waves of considerable amplitude to be found in the inlet and exhaust systems of the internal combustion engine. The subject matter to be found in the acoustic treatment of the theory is somewhat reminiscent of that for the topic of heat transfer, producing an almost infinite plethora of empirical equations for which the authors admit rather large error bands for their implementation in practice. This has always seemed to this author as being an unsatisfactory state of affairs. Consequently, a research program was instigated at QUB some years ago to determine if it was possible to predict the noise spectrum emanating from the exhaust systems of internal combustion engines using the approach of the calculation of the propagation of finite amplitude waves by the method of characteristics described in Chapter 2. This resulted in the technical publications by Blair, Spechko and Coates(8.1)-(8.3). A much more complete exposition of the work of Coates is to be found in his doctoral thesis(8.17).

8.4.1 The theoretical work of Coates(8.3)

In those publications, a theoretical solution is produced(8.3) which shows that the sound pressure level at any point in space beyond the termination of an exhaust system into the atmosphere is, not empirically, but directly capable of being calculated. The amplitude of the nth frequency component of the sound pressure, Pn, is shown to be primarily a complex function of: (a) the instantaneous mass flow rate leaving the end of the pipe system, q, and (b) the location of the measuring microphone in both distance and directivity from the pipe end, together with other parameters of some lesser significance.

For any sinusoidal variation, the mean square sound pressure level, Prms, of that nth frequency component is given by:

$$Prms_n^2 = Pn^2/2$$

From this the rms sound pressure level, Lpn, of the nth frequency component is found from:

$$Lpn = 10*\log_{10}[Prms_n^2/4*10^{-10}] \quad dB$$

The theoretical analysis hinges on being able to calculate the instantaneous mass flux-time history, q, at the termination of the pipe system to the atmosphere and to conduct a Fourier analysis(8.20) of it over a complete cycle. From this Fourier analysis it becomes possible to establish the component harmonics of the mass flow rate, q_n, and hence the corresponding pressure amplitudes, Pn.

It will be recalled from Chapter 2, that the unsteady gas flow calculation by the method of characteristics involves tracking Riemann variables by a mesh method of calculation along an entire pipe system, including in this case the presence of silencer components installed within it. Assuming that the mesh location at the termination of the pipe to atmosphere is denoted by z, then the instantaneous mass flow rate is given by q, where D is density, F is area, CG is superposition particle velocity and g is the ratio of specific heats:

$$q=Dz*CGz*Fz$$
$$=[Do*Xz^{(2/(g-1))}]*[(2/(g-1))*Ao*(Xz-1)]*Fz$$
$$=[Do*((\partial z+\beta z)/2)^{(2/(g-1))}]*[Ao*(\partial z-\beta z)/(g-1)]*Fz$$

This relationship for mass flow rate is the same as that deduced in Eq. 2.4.10. It will be observed that the accuracy of the solution depends on the ability of the theoretical calculation to trace the Riemann variables, ∂ and ß, to the very end of the most complex pipe and silencer system. How successful that can be may be judged from some of the results presented by Coates(8.3).

8.4.2 The experimental work of Coates(8.3)

The experimental rig used by Coates is described clearly in that technical paper(8.3), but a summary here will aid the discussion of the experimental results and their correlation with the theoretical calculations. The exhaust system is simulated by a rotary valve which allows realistic exhaust pressure pulses of cold air to be blown down into a pipe system at any desired cyclic speed for those exhaust pressure pulsations. The various pipe systems attached to the exhaust simulator are shown in Fig. 8.2, and are defined as SYSTEMS 1-4. Briefly, they are as follows:

SYSTEM 1 is a plain, straight pipe of 28.6 mm diameter, 1.83 m long and completely unsilenced.

SYSTEM 2 has a 1.83 m plain pipe of 28.6 mm diameter culminating in what is termed a diffusing silencer which is 305 mm long and 76 mm diameter. The tail pipe, of equal size to the entering pipe, is 152 mm long.

SYSTEM 3 is almost identical to SYSTEM 2 but has the entry and exit pipes reentering into the diffusing silencer so that they are 102 mm apart within the chamber.

SYSTEM 4 has what is defined as a side-resonant silencer placed in the middle of the 1.83 m pipe, and the 28.6 mm diameter through-pipe has 40 holes drilled into it of 3.18 mm diameter.

More formalized sketches of diffusing, side resonant and absorption silencers are to be found in Figs. 8.7-8.9. Further discussion of their silencing effect, based on an acoustic analysis, will be found in Sect. 8.5. It is sufficient to remark at this juncture that:

(a) The intent of a diffusing silencer is to absorb all noise at frequencies other than those at which the box will resonate. Those frequencies which are not absorbed are called the pass-bands.

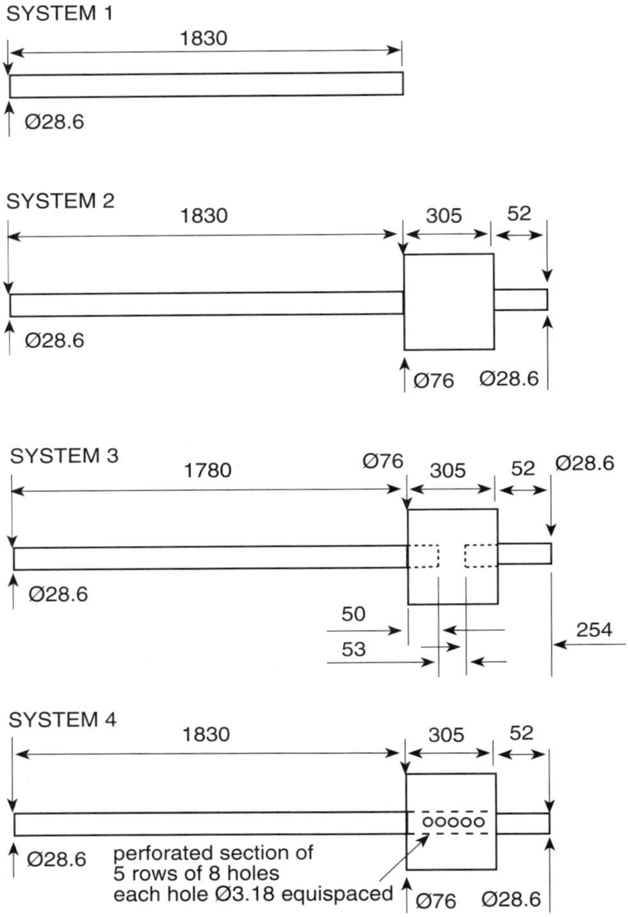

Fig. 8.2 Various exhaust systems and silencers used by Coates and Blair(8.3).

(b) The intent of a side-resonant silencer is to completely absorb noise of a specific frequency, such as the fundamental exhaust pulse frequency of an engine.

(c) The intent of an absorption silencer is to behave as a diffusing silencer, but to have the packing absorb the resonating noise at the pass-band frequencies.

The pressure-time histories within these various systems, and the one-third octave noise spectrograms emanating from these systems, were recorded. Of interest are the noise spectra and these are shown for SYSTEMS 1-4 in Figs. 8.3-8.6, respectively. The noise spectra are presented in the units of overall sound pressure level, dBlin, as a function of frequency. There are four spectrograms on any given figure: the one at the top is where the measuring microphone is placed directly in line with the pipe end and the directivity angle is declared as zero, and the others are at angles of 30, 60 and 90º. The solid line on any diagram is the measured noise spectra and the dashed line is that emanating from the theoretical solution outlined in Sect. 8.4.1. It can be seen that there is a very good degree of correlation between the calculated and measured noise spectra, particularly at frequency levels below 2000 Hz.

Of interest is the prediction of the mass flow spectrum at the termination of SYSTEM 1, which is shown in Fig. 8.10. It is predicted by the insertion of the appropriate data for the geometry and cylinder conditions of SYSTEM 1 into a slightly modified version of Prog.2.1, where the graphic output is altered to demonstrate the instantaneous mass flow rate at the termination to the atmosphere of a plain straight pipe. In other words, Eq. 2.4.10 is programmed as output data rather than the familiar pressure diagrams shown in Chapter 2. If one examines Fig. 8.10, the mass flow rate leaving the pipe end of SYSTEM 1 at a simulator rotor valve speed of 1000 rpm, i.e., a cycling rate of 2000 exhaust pulses created per minute as explained by Coates(8.3), one can see that there are four basic oscillations of the mass flow rate during one complete cycle. Thus, a Fourier analysis of this spectrum would reveal a fundamental frequency of 4*2000/60, or 133.3 Hz. If the theoretical postulations of Coates(8.3) have any validity, then this should show up clearly as the noisiest frequency in the measured noise spectra for that same rotor speed, as illustrated in Fig. 8.3. In that diagram, the fundamental, and noisiest frequency is visibly at 133 Hz. The theoretical solution of Coates, as would be expected from the above discussion because his calculation is also conducted by the method of

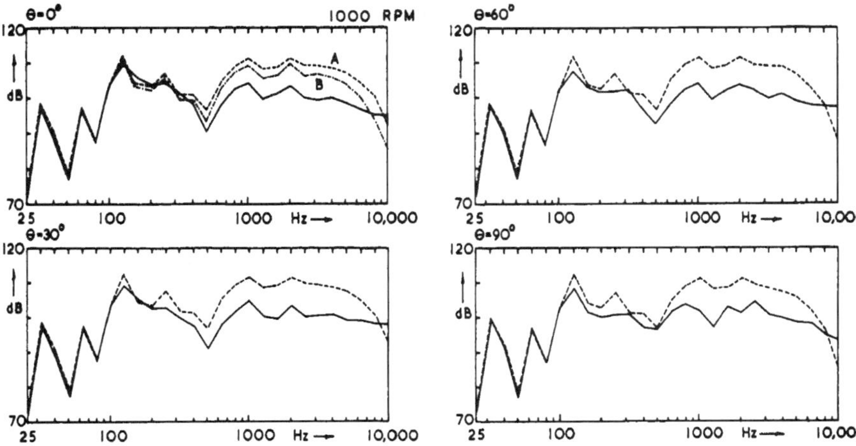

Fig. 8.3 One-third noise spectrogram from SYSTEM 1.

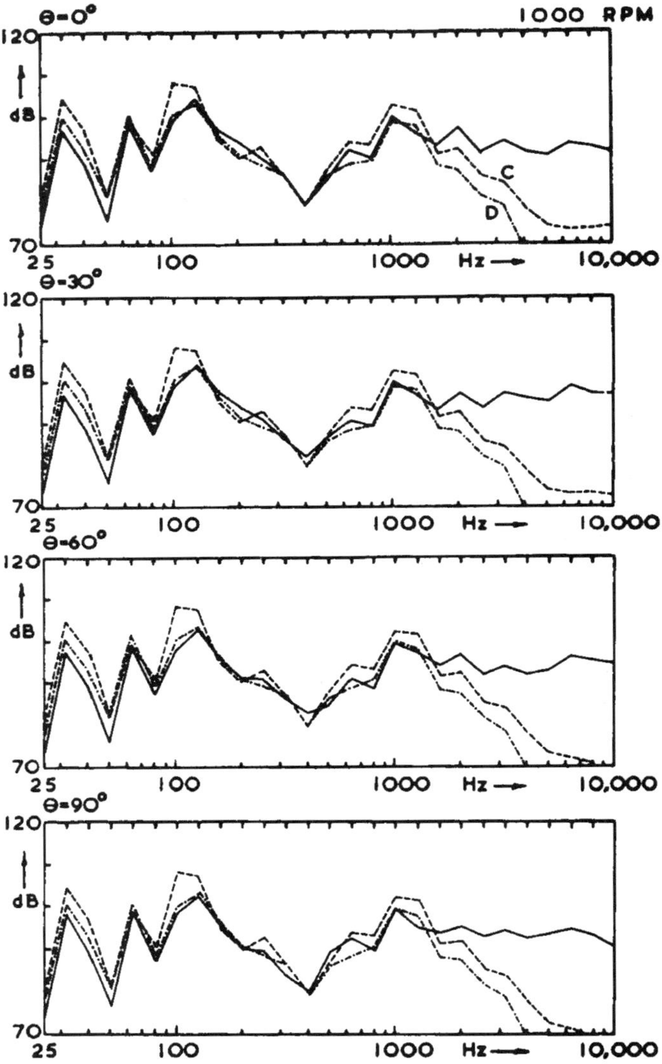

Fig. 8.4 One-third noise spectrogram from SYSTEM 2.

characteristics, also agrees with the prediction of 133 Hz as the noisiest frequency. Perhaps the most important conclusion to be drawn from this result is that the designer of silencing systems, having the muffling of the noisiest frequency as a first priority, can use the unsteady gas-dynamic modelling programs as a means of prediction of the mass flow rate at the pipe termination to atmosphere, be it the inlet

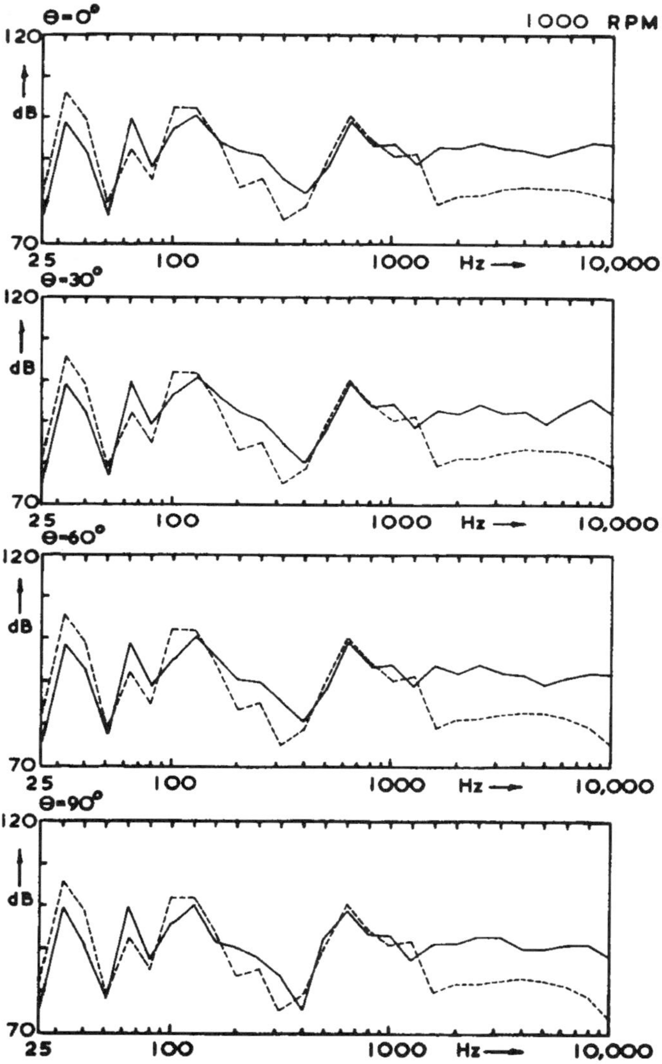

Fig. 8.5 One-third noise spectrogram from SYSTEM 3.

or the exhaust system, and from that mass flow rate calculation over a complete cycle determine the fundamental noisiest frequency to be tackled by the silencer.

Of direct interest is the silencing effect of the various silencer elements attached to the exhaust pipe by Coates, SYSTEMS 2-4, the noise spectra for which are shown

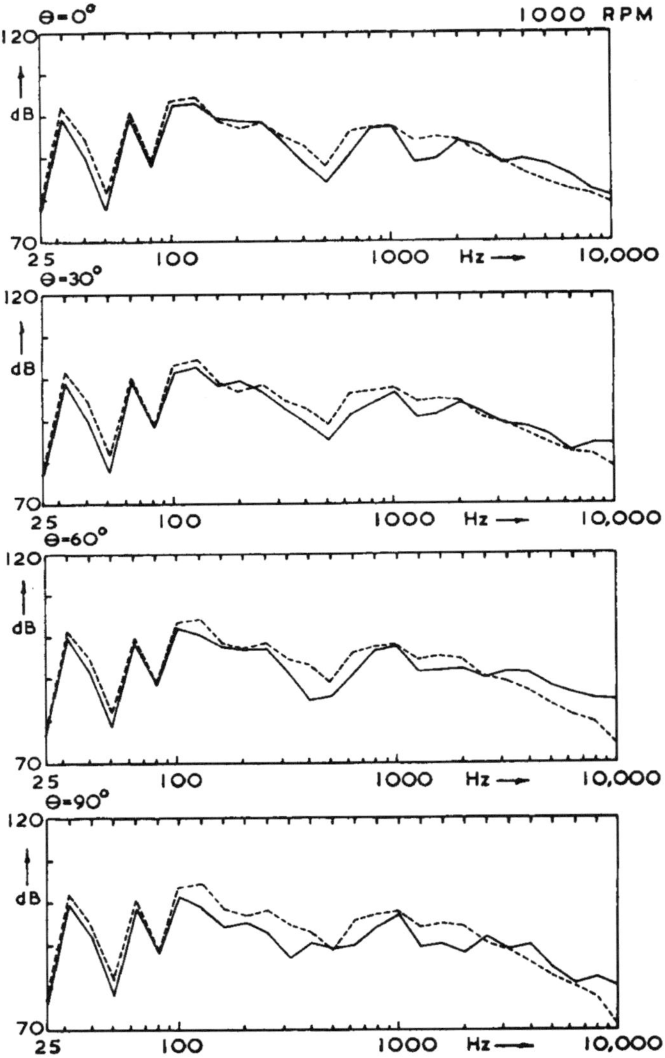

Fig. 8.6 One-third noise spectrogram from SYSTEM 4.

in Figs. 8.4-8.6. The SYSTEM 2, a simple diffusing silencer without reentrant pipes, can be seen to reduce the noise level of the fundamental frequency at 133 Hz from 116 dB to 104 dB, an attenuation of 12 dB. From the discussion in Sect. 8.1.2, an attenuation of 12 dB is a considerable level of noise reduction. It will also be observed that a large "hole," or strong attenuation, has been created in the noise

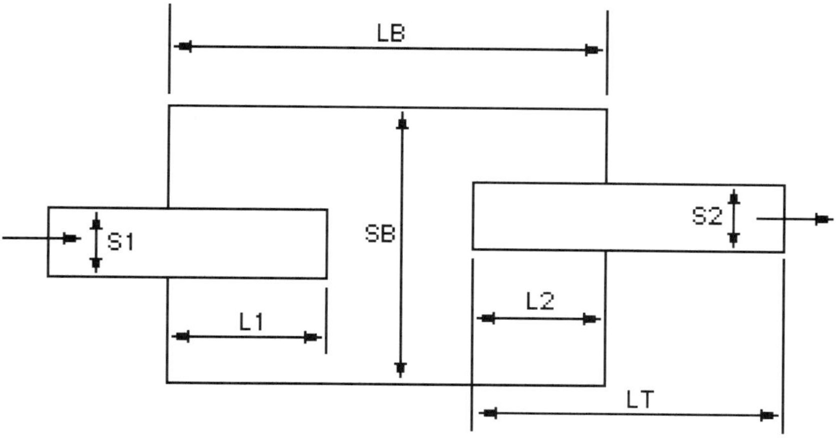

Fig. 8.7 Significant dimensions of a diffusing silencer element.

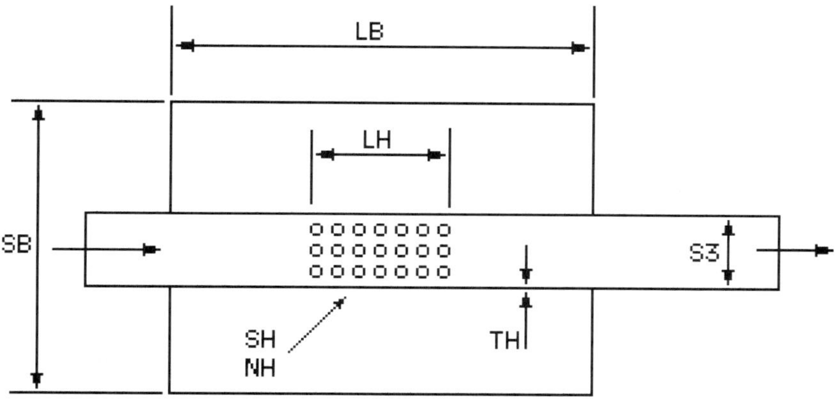

Fig. 8.8 Significant dimensions of a side-resonant silencer element.

spectra of SYSTEM 2 at a frequency of 400 Hz. In Sect. 8.5.1 this will be a source of further comment when an empirical frequency analysis is attempted for this particular design.

The noise spectra for SYSTEM 3, a diffusing silencer with reentrant pipes, is shown in Fig. 8.5. The noise level of the fundamental frequency of 133 Hz has been reduced further to 98 dB. The "hole" of high attenuation is now at 300 Hz and deeper than that recorded by SYSTEM 2.

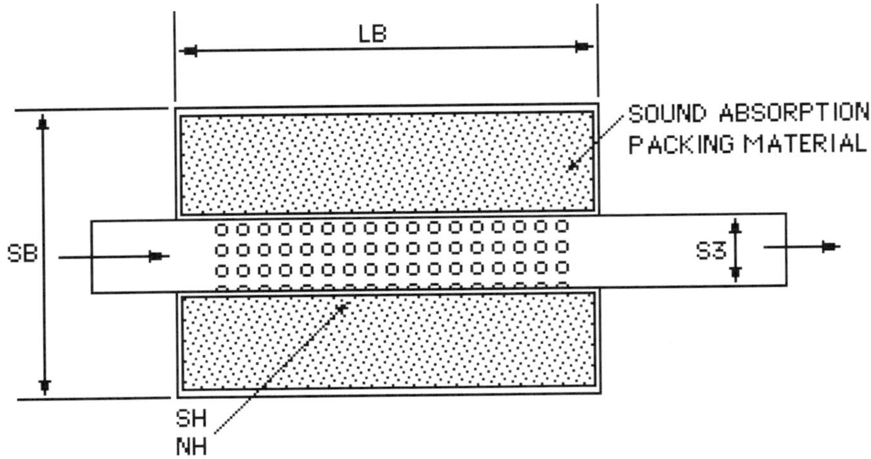

Fig. 8.9 Significant dimensions of an absorption silencer.

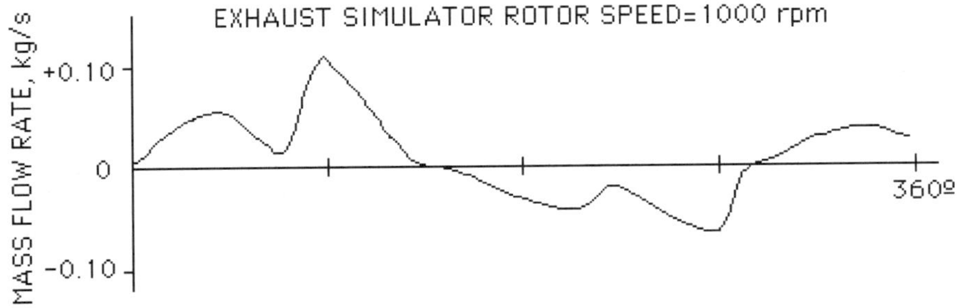

Fig. 8.10 Mass flow rate at the termination of pipe SYSTEM 1 to atmosphere.

The noise spectra for SYSTEM 4 is shown in Fig. 8.6. This is a side-resonant silencer. The noise level of the fundamental frequency of 133 Hz is nearly as quiet as SYSTEM 2, but a new attenuation hole has appeared at a higher frequency, about 500 Hz. This, too, will be commented on in Sect. 8.5.2 when an empirical acoustic analysis is presented for this type of silencer. It can also be seen that the noise level at higher frequencies, i.e., above 1000 Hz, is reduced considerably from the unsilenced SYSTEM 1.

The most important conclusion from this work by Coates is that the noise propagation into space from a pipe system, with or without silencing elements, could be predicted by a theoretical calculation based on the motion of finite amplitude waves propagating within the pipe system to the pipe termination to the atmosphere. In other words, there did not have to be a reliance by designers on empirically based acoustic equations for the design of silencers, be they for the intake or the exhaust system, for internal combustion engines.

8.4.3 Future work for the prediction of silencer behavior

It has always seemed to the author that this pioneering work of Coates(8.3)(8.17) has not received the recognition it deserves. Worse, it has tended to be ignored, due in part to those involved in silencer design not tackling their problems by the use of theory based on unsteady gas dynamics but persisting with theory based on acoustics, i.e., infinitesimal amplitude pressure waves. A subsidiary reason, and Coates comments on this, is that silencer elements tend to be somewhat complicated in geometrical terms and have gas particle flow characteristics which are more three-dimensional in nature. As unsteady gas-dynamic calculations conducted by the method of characteristics are one-dimensional, their potential for accurate prediction of the pressure wave reflection and transmission characteristics of real silencer elements is definitely reduced. Nevertheless, the fundamental accuracy of the theoretical premise by Coates is clearly demonstrated and should be the preferred route for design sophistication by the researchers of the subject to deal with the real problems posed by the complex geometry of silencers employed for internal combustion engines. The theoretical means for so doing is now at hand.

In Sect. 3.4 it is shown how Computational Fluid Dynamics (CFD) is employed to solve the three-dimensional flow behavior inside engine cylinders, with an unsteady gas-dynamic calculation controlling the flow at the entry and exit boundaries of the system. There is no logical reason why this same theoretical approach should not be employed for the design of the most complex of silencers. The engine, together with its inlet and exhaust systems, would be modelled by the unsteady gas-dynamics method given in Chapter 2, and described further in Chapter 5, up to the boundaries of the various silencers, and that information fed to a CFD program for three-dimensional pursuit through the entire silencer geometry. The result of the CFD calculation would then be the accurate prediction of the mass flow rate spectrum entering the atmosphere, from which the ensuing noise spectra can be correctly assessed by the methods of Coates(8.3) already presented briefly in Sect. 8.4.2. This would be a realistic design process for the engine, together with its ducting and silencers, rather than the empirically based acoustics still in vogue today(8.18). Acoustics theory could then be relegated to the useful role which befits all empiricism, namely providing more appropriate initial data for the geometry of silencer systems so that the CFD/unsteady gas-dynamics analysis would not be so wasteful of expensive human and computer time.

8.5 Theory based on acoustics for silencer attenuation characteristics

The behavior of silencers as treated by the science of acoustics is to be found in a plethora of texts and papers, of which the book by Annand and Roe(1.8) deserves the most attention by the reader. Further useful papers and books are to be found in the Reference Section at the end of this chapter. This section will concentrate on the characteristics of the most common type of exhaust silencers, the diffusing, the side-resonant and absorption silencers already shown in Figs. 8.7-8.9, and the low-pass intake silencer sketched in Fig. 8.11.

8.5.1 The diffusing type of exhaust silencer

The sketch of a diffusing silencer element is shown in Fig. 8.7. It has entry and exit pipes of areas S1 and S2, respectively, which have diameters DS1 and DS2 if they are of a circular cross-section. The pipes can be reentrant into the box with lengths L1 and L2, respectively. The final tail-pipe leaving the box has a length LT. The box has a length LB and a cross-sectional area SB, with that having a diameter DSB should the box be of a circular cross-section. It is quite common, for reasons of manufacturing simplicity, for silencers to have a circular cross-section for both the pipes and the box. If the pipes are of the reentrant type, then any theory calling for a computation of the volume of the box, VB, should take into account the box volume occupied by those pipes, thus:

$$VB=SB*LB-(S1*L1+S2*L2) \tag{8.5.1.1}$$

The transmission loss, or attenuation, of a diffusing silencer is basically a function of two parameters: the expansion ratio, Er, i.e., Er1=SB/S1 or Er2=SB/S2, and the relationship between the wavelength of the sound, Ω, to the length of the box, LB.

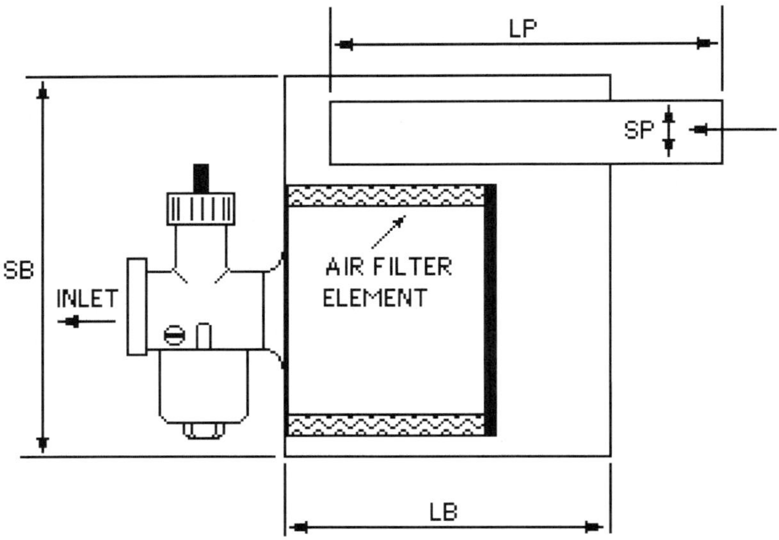

Fig. 8.11 Significant dimensions of an intake silencer element.

The wavelength of the sound is connected to the frequency of that sound, FR, and the acoustic velocity, A. The acoustic velocity is calculated by Eq. 8.1.1. The wavelength, Ω, is found from:

$$\Omega=A/FR \qquad (8.5.1.2)$$

As the frequency of the gas-borne noise arriving into the diffusing silencer varies, the box will resonate, much like an organ pipe, at various integer amounts of half the wavelength(1.8), in other words at 2*LB, 2*LB/2, 2*LB/3, 2*LB/4, etc. From Eq. 8.5.1.2, this means at frequencies of A/2*LB, 2*A/2*LB, 3*A/2*LB, 4*A/2*LB, and so on. At these frequencies the silencer will provide a transmission loss of zero, i.e., no silencing effect at all, and such frequencies are known as the "pass bands." It is also possible for the silencer to resonate in the transverse direction through the diametral dimension, DSB, and provide further pass-band frequencies at what would usually be a rather high frequency level.

There are many empirical equations in existence for the transmission loss of such a silencer, but Kato and Ishikawa(8.18) state that the theoretical solution of Fukuda(8.9) is found to be useful. The relationship of Fukuda(8.9) is as follows for the transmission loss of a diffusing silencer, TRd, in dB units:

$$TRd=10*\log_{10}[Er2*f(k,L)]^2 \quad dB \qquad (8.5.1.3)$$

$$\text{where,} \quad f(k,L)=\{\sin(k*LB)*\sin(k*LT)\}/\{\cos(k*L1)*\cos(k*L2)\} \qquad (8.5.1.4)$$

$$\text{and,} \quad k=2*\pi*FR/A \qquad (8.5.1.5)$$

Not surprisingly, in such empirical relationships there are many correcting factors offered for the modification of the basic relationship to cope with the effects of gas particle velocity, end correction effects for pipes which are reentrant or flush with the box walls, boxes which are lined with absorbent material but not sufficiently dense as to be called an absorption silencer, etc., etc. It will be left to the reader to pursue these myriad formulae which are to be found in the references.

To assist the reader with the use of the basic equations for design purposes, a simple computer program is included with this book in the Computer Program Appendix, Prog.8.1, "DIFFUSING SILENCER." This appears as ProgList.8.1. The attenuation equations which are programmed are those seen above from Fukuda, Eqs. 8.5.1.3-5.

To determine if such a design program is useful in a practical sense, an analysis of SYSTEM 2 and SYSTEM 3, emanating from Coates(8.3), is attempted and the results shown in Figs. 8.12 and 8.13. They show the computer screen output from Prog.8.1, which are plots of the attenuation in dB as a function of noise frequency up to a maximum of 4 kHz, beyond which frequency most experts agree the diffusing silencer will have little silencing effect. The theory would continue to predict some attenuation to the highest frequency levels, indeed beyond the upper threshold of hearing. Also displayed on the computer screen is the input data for the geometry of that diffusing silencer according to the symbolism presented in Fig. 8.7 and in this section. However, the input data is in the more conventional length dimensions in mm units and temperature in deg. C values; they are converted within the program to strict SI units before being entered into the programmed equations for the attenuation of a diffusing silencer according to Fukuda(8.9), Eqs. 8.5.1.3-5. The reader can check the input dimensions of SYSTEM 2 and SYSTEM 3 from Fig. 8.2.

The attenuation of SYSTEM 2, as predicted by the theory of Fukuda and shown in Fig. 8.12, has a first major attenuation of 14 dB at 320 Hz and the first two pass-band frequencies are at 550 and 1100 Hz. If one examines the measured noise frequency spectrum in Fig. 8.4, it is to find that an attenuation hole of 12 dB is created at a frequency of 400 Hz. Thus the correspondence with the theory of Fukuda, with regard to this primary criterion, is quite good and gives some confidence in its application for this particular function. There is also some evidence of the narrow pass band at 550 Hz and there is no doubt about the considerable pass-band frequency at 1100 Hz in the measured spectra. There is no sign of the predicted

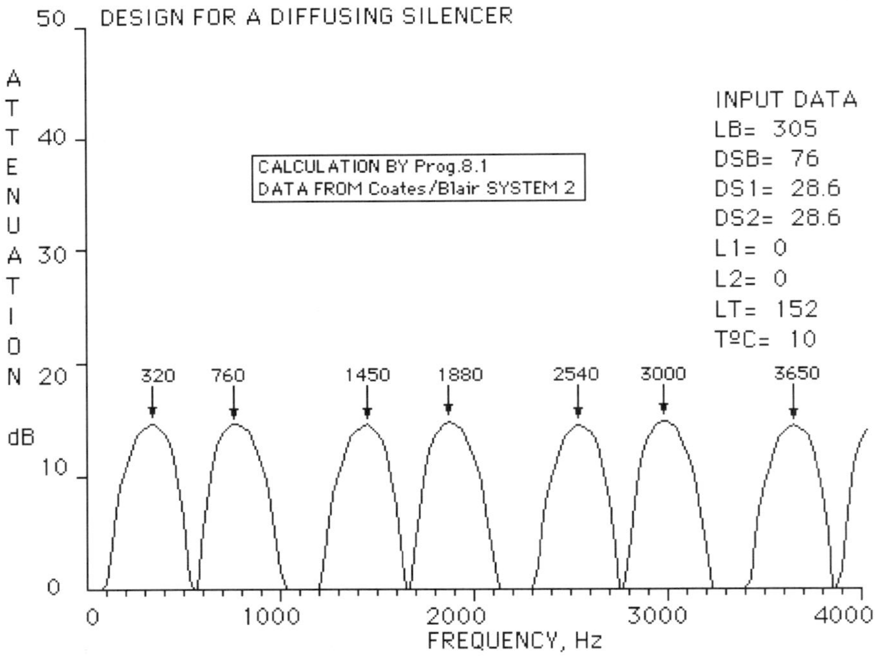

Fig. 8.12 Calculation by Prog.8.1 for the silencing characteristics of SYSTEM 2.

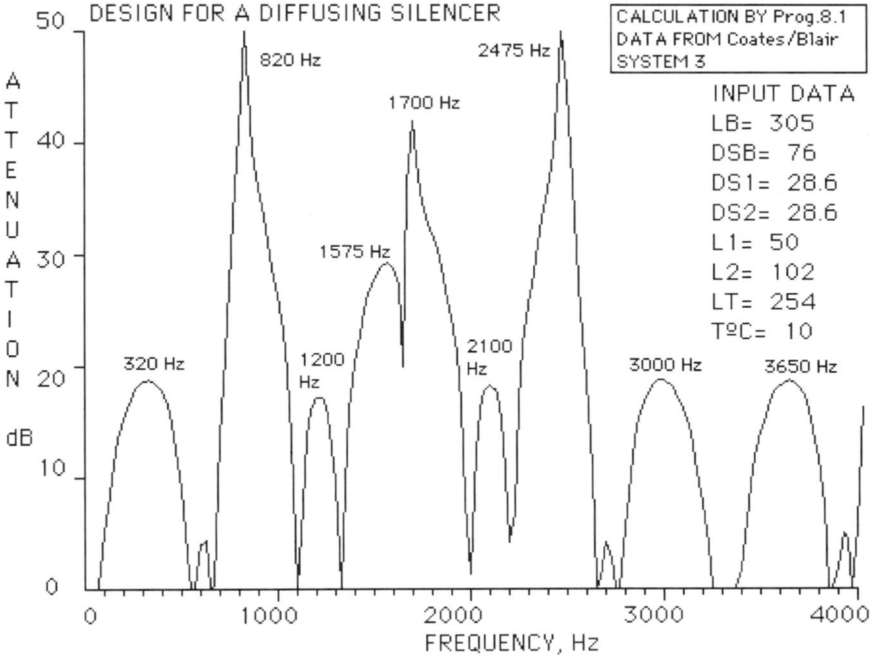

Fig. 8.13 Calculation by Prog.8.1 for the silencing characteristics of SYSTEM 3.

attenuation, nor the pass-band holes, at frequencies above 1.5 kHz in the measured spectrum. There is also no evidence from the theoretical solution of the reason for the attenuation in the measured spectrum of the fundamental pulsation frequency of 133 Hz, as commented on in Sect. 8.4.2. The general conclusion as far as SYSTEM 2 is concerned is that Prog.8.1 is a useful empirical design calculation method for a diffusing silencer up to a frequency of 1.5 kHz.

The attenuation of SYSTEM 3, as predicted by the theory of Fukuda and shown in Fig. 8.13, has a first major attenuation of 19 dB at 320 Hz, i.e., greater than that of SYSTEM 2. It is true that the measured silencing effect of SYSTEM 3 is greater than SYSTEM 2, but not so at the frequency of 400 Hz, where the measured attenuations are both at a maximum and are virtually identical. The theory would predict that the pass-band hole of SYSTEM 2 at 1100 Hz would not be so marked for SYSTEM 3, and that can be observed in the measured noise spectra. The theoretically predicted pass-band at 600 Hz is clearly seen in the measured noise diagram. Again, there is little useful correlation between theory and experiment after 1500 Hz.

The general conclusion to be drawn, admittedly from this somewhat limited quantity of evidence, is that Prog.8.1 is a useful empirical design program for diffusing silencers up to a frequency of about 1500 Hz.

8.5.2 The side-resonant type of exhaust silencer

The fundamental behavior of this type of silencer is to absorb a relatively narrow band of sound frequency by the resonance of the side cavity at its natural frequency. The sketch in Fig. 8.8 shows a silencing chamber of length LB and area SB, or diameter DSB if the cross-section is circular. The exhaust pipe is usually located centrally in the silencer, is of area S3, or diameter DS3 if it is a round pipe, and has a pipe wall thickness TH. The connection to the cavity, whose volume is VB, is via holes, or a slit, or by a short pipe in some designs. The usual practice is to employ a number of holes, NH, of area SH, or diameter DSH if they are round holes for reasons of manufacturing simplicity. The volume of the resonant cavity is VB where, if all cross-sections are circular:

$$VB=LB*SB-[\pi*LB*(DSB+2*TH)^2/4] \qquad (8.5.2.1)$$

According to Kato(8.18), the length, LH, that is occupied by the holes should not exceed the pipe diameter, DSB, otherwise the system should be theoretically treated as a diffusing silencer. The natural frequency of the side-resonant system, FRsr, is given by Davis(8.4) as:

$$FRsr=(A/2*\pi)*\sqrt{(K/VB)} \qquad (8.5.2.2)$$

where K, the conductivity of the opening, is calculated from:

$$K=NH*SH/(TH+0.8*SH) \qquad (8.5.2.3)$$

The attenuation or transmission loss in dB of this type of silencer, TRsr, is given by Davis(8.4) as:

$$TRsr=10*\log_{10}[1+Z^2] \qquad (8.5.2.4)$$

where the term Z is found from:

$$Z=\{(\sqrt{(K*VB)})/(2*S3)\}/\{(FR/FRsr)-(FRsr/FR)\} \qquad (8.5.2.5)$$

When the applied noise frequency is equal to the resonant frequency, FRsr, the value of the term Z becomes infinite as does the resultant noise attenuation in Eq. 8.5.2.4. Clearly this is an impractical result, but it does give credence to the view that such a silencer has a considerable attenuation level in the region of the natural frequency of the side-resonant cavity and connecting passage.

To help the reader to use these acoustic equations for design purposes, a simple computer program is included with this book in the Computer Program Appendix, Prog.8.2, "SIDE-RESONANT SILENCER." This is listed as ProgList.8.2. The attenuation equations which are programmed are those discussed above from Davis, Eqs. 8.5.2.2-5.

To demonstrate the use of this computer design program, and to determine if the predictions emanating from it are of practical use to the designer of side-resonant elements within an exhaust muffler for a two-stroke engine, the geometrical and experimental data pertaining to SYSTEM 4 of Coates(8.3) are inserted as data and the calculation result is illustrated in Fig. 8.14. The information in that figure is the computer screen picture as seen by the user of Prog.8.2. The information displayed is the input data for, in this instance, SYSTEM 4 and the reader can check the methodology of data input from Fig. 8.2, and the output data which is the attenuation in dB over a frequency range up to 4 kHz. The reader may wonder about 10°C being the declared data value for exhaust temperature, but the simulation work of Coates(8.3) was conducted by a rotor valve delivering very realistic exhaust pulses, but in cold air!

The calculated sound attenuation of SYSTEM 4, as seen in Fig. 8.14, shows a large peak of 50 dB transmission loss at 600 Hz, with further attenuation stretching to 2.5 kHz. If one examines the measured noise spectrum of SYSTEM 4, and compares it to the unsilenced SYSTEM 1, it can be seen that there is considerable

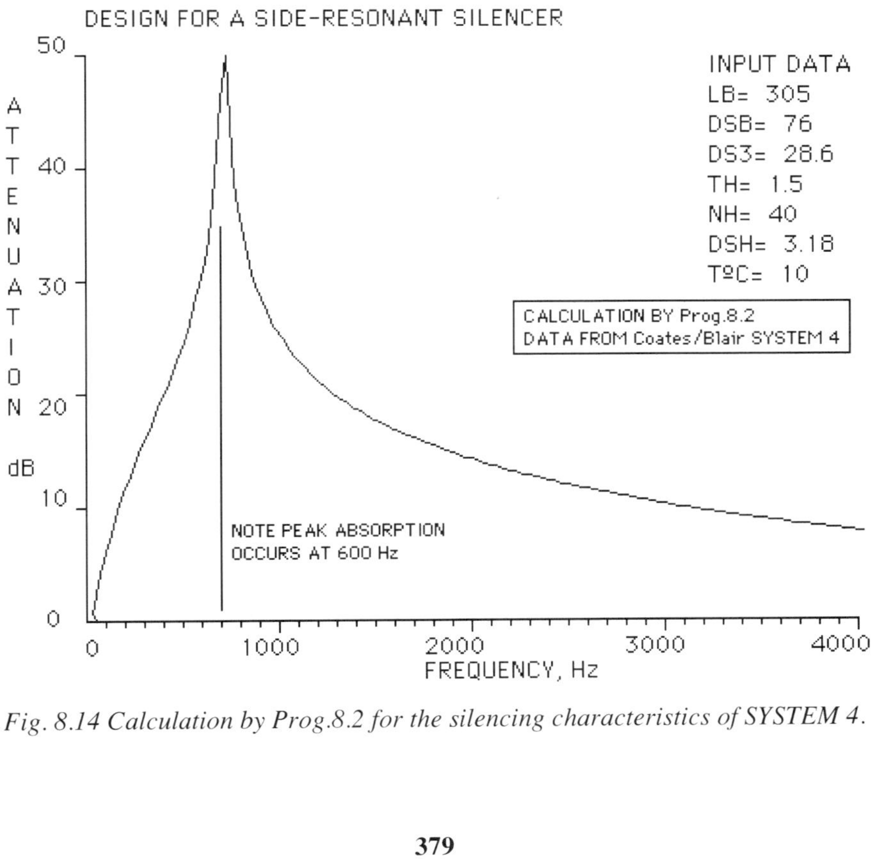

Fig. 8.14 Calculation by Prog.8.2 for the silencing characteristics of SYSTEM 4.

attenuation produced by this side-resonant silencer at 500 Hz, although it is not as profound as 50 dB. Of even greater interest to the designer is the visibly strong attenuation of SYSTEM 4 stretching up to 10 kHz, which is not present in either of the diffusing silencer designs discussed in Sect. 8.5.1. It will also be noted that the calculated attenuation stretches down to 100 Hz, i.e., below the fundamental frequency of 133 Hz determined for Coates' experiments in Sect. 8.4.2, and that the measured diagram does show attenuation of that frequency at least equal to the level observed for the diffusing silencers.

The evidence is that the theoretical work of Davis(8.4), as programmed in Prog.8.2, is of direct use in determining the fundamental muffling frequency of a side-resonant silencer and that such a device is very useful in practical terms for attenuating noise at frequencies above 1 kHz.

8.5.3 The absorption type of exhaust silencer

It is generally held that an absorption silencer, as illustrated in Fig. 8.9, acts as a diffusing silencer(8.18) and that the effect of the packing is to absorb noise in the pass bands appropriate to a diffusing silencer. The geometry of the system is for a pipe of area S3 passing through a box of length LB and area SB. If the cross-sections are circular, the diameters are DS3 and DSB, respectively. The holes through the pipe are normally round and are NH in number of area SH, or diameter DSH. The physical geometry is very similar to the side-resonant silencer, except the total cross-sectional area of all of the holes is probably in excess of five times the pipe area, S3. This is the reason for the theoretical acoustic analysis being more akin to a diffusing silencer than a side-resonant silencer, as the area ratio of NH*SH/S3 in a side-resonant silencer is normally less than unity. The experimental evidence is that the absorption silencer is very effective at attenuating high frequencies, i.e., above 2 kHz, and is relatively ineffective below about 400 Hz. Most authorities agree that the theoretical design of an absorption silencer, either by the methodology of Coates(8.3) or by an acoustic procedure, is somewhat difficult because it depends significantly on the absorption capability of the packing material. The packing materials are usually a glass-reinforced fiber material or mineral wool.

The reader should attempt the design using Prog.8.1, but arrange for as many holes of an appropriate diameter in the central pipe as is pragmatic. If the hole size is too large, say in excess of 3.5 mm diameter, the packing material within the cavity will be blown or shaken into the exhaust stream. If the hole size is too small, say less than 2.0 mm diameter, the particulates in the exhaust gas of a two-stroke engine will ultimately seal them over, thereby rendering the silencing system ineffective. The normal hole size to be found in such silencers is between 2.0 and 3.5 mm diameter.

One detailed aspect of design which needs special comment is the configuration of the holes in the perforated section of the silencer. The conventional manufacturing method is to roll, then seam weld, the central pipe from a flat sheet of perforated mild steel. While this produces an acceptable design for the perforated pipe, a superior methodology is to produce the pipe by the same production technique, but from a mild steel sheet in which the holes have been somewhat coarsely "stabbed," rather than cleanly excised. Indeed, the stabbed holes can be readily manufactured in pre-formed round pipe by an internal expanding tool. The result is shown in sketch form in Fig. 8.15 for the finished pipe section. This has the effect of reducing the turbulent eddies produced by the gas flowing in either direction past the sharp edges of the clean-cut holes of conventional perforations. This type of turbulence has an irritating high-frequency content (a whistle produces noise by this very edge effect).

8.5.3.1 Positioning an absorption silencer segment

The published literature(1.8) on the topic agrees on several facets of the design as commented on above, but the principal function is to remove the high-frequency end of the noise spectra. The high-frequency part of any noise spectrum emanates from two significant sources: the first is due to the sharp pressure rise at the front of the exhaust pulse, always faster in a two-stroke than a four-stroke engine (see Sect. 8.6.1), and the second is from the turbulence generated in the gas particle flow as it passes sharp edges, corners and protuberances into the gas stream. The considerable noise content inherent in turbulence is quite visible in Plate 2.4 where the whirling smoke-ring of particles is seen to follow the spherical pressure wave front into the atmosphere. Thus, an absorption silencer element should always be the final segment of a series of elements making up a complete muffler. The logic for that statement is that there is little point in muffling the high-frequency part of the noise spectrum by an absorption silencer element which is then followed by a reentrant pipe diffusing silencer, as this would produce further turbulence generated noise by the eddies in the wake from its sharp, protruding pipe edges.

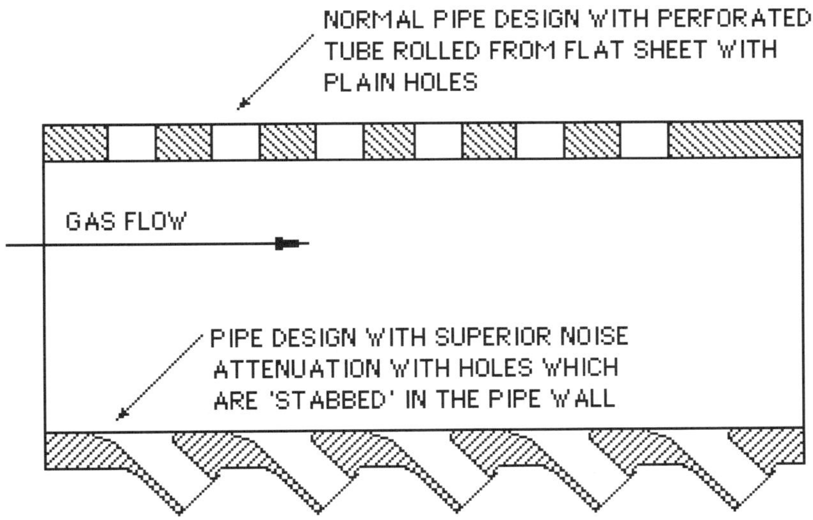

Fig. 8.15 Holes "stabbed" in the pipe wall provide better noise attenuation in absorption silencers.

8.5.3.2 A possible absorption silencer segment for a two-stroke engine

As will be evident to the most untrained of observers when listening to a four-stroke and two-stroke motorcycle passing by, the two-stroke machine delivers a sharper, i.e., higher frequency, quality of sound. The reason for this, apart from the natural doubling of the exhaust pulsing rate from the equivalent two-stroke engine, is the inherently faster rise time of the exhaust pulse of the two-stroke powerplant, as mentioned earlier; this is discussed in greater detail in Sect. 8.6.1. This puts an even greater emphasis on the suppression of high-frequency noise, yet the traditional design of two-stroke engine muffler does not appear to have a visibly larger absorption silencer segment. One design, which the author presents in Fig. 8.16, is for an annular type of absorption silencer which accomplishes several theoretical objectives within it. These objectives can be summarized as:

(a) The potential for the presentation of a greater number of holes to the exiting exhaust gas stream is improved by several orders of magnitude within the same space allocation of length LB, diameter DSB, and packing volume VB.

(b) The flow exits through an annulus of height XAN and the resulting potential for the lowering of the turbulence level is obvious, as the Reynolds Number of the flow for the same state conditions of particle velocity, density and viscosity will be reduced as a consequence in comparison to the same flow in a pipe of equal area but diameter DEP.

(c) The flow exits to the atmosphere as an annular sheet with a considerable surface area, sandwiched between layers of colder air which will be at a higher density and viscosity. This will provide further dissipation of the turbulent eddies and their high-frequency content, in the same fashion as a by-pass gas turbine jet engine encases the hot central jet efflux by a surrounding layer of colder air and is quieter as a consequence.

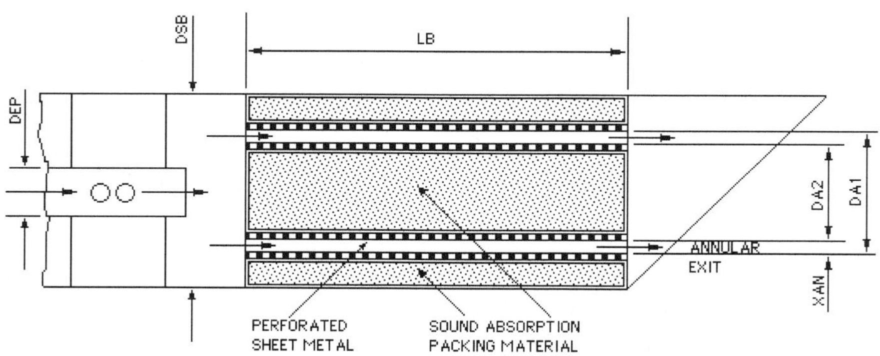

Fig. 8.16 Design of annular absorption silencer section of a two-stroke engine muffler.

To examine some of these contentions, particularly that for the lowering of the Reynolds Number, consider some of the listed dimensions in Fig. 8.16. Assuming that the correct flow area dimension has been selected up to the absorption silencer segment, the area of inflow to it is given by SEP, where:

$$SEP=\pi*DEP^2/4 \quad (8.5.3.2.1)$$

This should be equated to the flow area through the annulus, and perhaps the annulus area, SA, should be some factor, KA, greater in area to allow for the higher friction pressure drop due to the larger surface area which is presented to the flow. Thus:

$$SA=KA*SEP=\pi*(DA1^2-DA2^2)/4 \quad (8.5.3.2.2)$$
$$XAN=(DA1-DA2)/2 \quad (8.5.3.2.3)$$

The Reynolds Number, Re, is given by:

$$Re=D*CG*\o/\mu \quad (8.5.3.2.4)$$

where D, CG, and μ are density, particle velocity and viscosity, respectively.

The term ø is the characteristic length dimension, usually the "hydraulic diameter" of the particle flow path. In the case of the round entry pipe it is øp, simply the diameter DEP, and for the annulus it is øa, estimated from (3.14):

$$\o a=4*(\text{flow area})/(\text{wetted perimeter})$$
$$=4*SA/\{\pi*(DA1+DA2)\}$$
$$=2*XAN \quad (8.5.3.2.5)$$

Other authorities such as Hall(8.19) state that the calculation of Reynolds Number for annular ducts should be made using a characteristic length ø, composed of the shortest dimension plus one fourth of the hydraulic diameter. Thus the value of øa, as recommended by Hall, would be XAN+(2*XAN/4), or 1.5*XAN.

To put some order of magnitude on these contentions, suppose that the entry pipe, DEP, is 25 mm in diameter, the silencer segment body, DSB, is 90 mm in diameter, and it is proposed to site the diameter, DA1, at 60 mm diameter. If KA is set at 1.15, i.e., a 15% increase of flow area through the annulus by comparison with the original entry pipe, then from the above:

$$SA=1.15*\pi*(25^2)/4=565 \text{ mm}^2 =\pi*(60^2-DA2^2)/4$$

Hence, DA2=53.6 mm diameter and XAN=3.2 mm.

This annular dimension is on the borderline for potential carbon build-up and resultant clogging, so a value of XAN of 3.5 mm is probably a more pragmatic choice. The reduction of Reynolds Number from the entry pipe, Re_p, to the annulus flow, Re_a, can now be estimated as:

$$Re_p/Re_a=DEP/(2*XAN)=25/7=3.6$$

If the recommendation of Hall(8.19) is used for the characteristic length, øa, then this same reduction ratio becomes:

$$Re_p/Re_a=DEP/(1.5*XAN)=25/5.25=4.8$$

It is also somewhat instructive to predict the typical values of Reynolds Number which might exist in the entry pipe leading to the absorption silencer segment. The peak pressure ratio of the gas pressure pulses at this point should not be in excess of 1.1 atmospheres, otherwise the previous diffusing and side-resonant silencing segments are not performing their function very successfully. Recalling the discussion in Sect. 2.1.4, and assuming that the properties of the exhaust gas are similar to that of air where the reference temperature is considered to be 700 K (427 ºC), then the viscosity of air is known to be 0.000033 kg/ms at that temperature, and the particle velocity, CG, and density, D, are as follows:

Pressure amplitude ratio, $X=(P/Po)^{(1/7)}$
Acoustic velocity, $Ao=\sqrt{(g*R*To)}$
Reference density, $Do=Po/(R*To)$
Density, $D=Do*X^5$
Particle velocity, $CG=5*Ao*(X-1)$

Hence, for a peak pulse pressure ratio of 1.1, X=1.014, Ao=530 m/s, Do=0.5043 kg/m^3, D=0.54 kg/m^3, and CG=37 m/s.

From Eq. 8.5.3.2.4, the Reynolds Number in the entry pipe, Re_p, is found:

$$Re_p=D*CG*øp/\mu=0.54*37*(25/1000)/0.000033=15136$$

This indicates that the maximum Reynolds Number appearing at the absorption silencer segment is such that the flow is clearly turbulent, as the laminar/turbulent transition occurs at a Reynolds Number of about 2500. However, by the analysis conducted above, it would appear that the Reynolds Number in the annulus at the maximum flow rate is between 3150 and 4200, bringing it ever closer to the laminar region. To reinforce the point, by definition there is no turbulence possible in laminar flow.

By either method of calculation this is a significant level of reduction of Reynolds Number and, as this is the marker of turbulence level in fluid flow, one would expect to find such a device perform in the manner indicated in the discussion above and further attenuate the high-frequency noise content. The use of an annular, or laminar, silencing concept is not new, Annand and Roe(1.8) discuss the subject extensively, but the specific application to the absorption silencing segment of a production two-stroke engine is worthy of mention and design consideration.

8.5.4 Silencing the intake system

Probably the simplest and most effective form of intake silencer is of the type sketched in Fig. 8.11. The geometry illustrated is for a single-cylinder engine but the same arrangement can also apply to a multi-cylinder design, and the reader will find such a discussion in the paper by Flaig and Broughton(1.12). From that paper is reproduced the attenuation curve for a V8 two-stroke outboard motor, shown in Fig. 8.17. It can be seen that the peak attenuation required is mainly in the band from 400-600 Hz. It should be noted that this corresponds to a forcing frequency from the fundamental induction pulses of this eight-cylinder engine at 3000-4500 rpm.

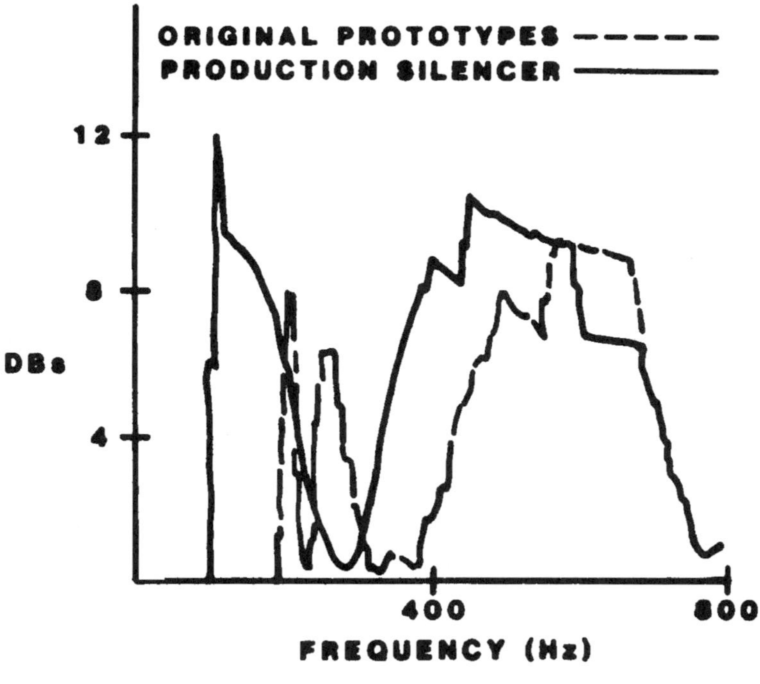

Fig. 8.17 Attenuation characteristics of an intake silencer for the OMC V8 outboard.

8.5.4.1 The acoustic design of the low-pass intake silencer

Fig. 8.11 supplies the basic mechanism of induction silencing, being a volume connected to the induction system, however many air intakes to the several cylinders there may be. On the atmospheric side of this box is a pipe of length LP and area SP. The air cleaner is placed within the box, helping to act as an absorption silencer of the various high-frequency components emanating from the edges of throttles or carburetor slides. This has no real effect on the silencing behavior of the box volume, VB.

This type of silencer is known as a low-pass device and has no silencing capability below its lowest resonating frequency. This sharp drop in noise attenuation behavior is clearly seen in Fig. 8.17 for the OMC V8 outboard motor.

The natural frequency of this type of silencer(1.8) is given by FRi, where:

$$FRi=(A/2\pi)*\sqrt{[SP/(LPi*VB)]} \qquad (8.5.4.1)$$

The effective length of the intake pipe, LPi, is related to the actual length, LP, and the diameter, DSP, by:

$$LPi=LP+\pi*DSP/4 \qquad (8.5.4.2)$$

The design of the silencer is accomplished by setting this natural frequency of the silencer, FRi, to correspond to the natural frequency of the induction pulses from the engine, FRe. That the natural frequency of the induction pulsations is indeed the rotational speed of a two-stroke engine can best be judged from an examination of Fig. 8.18. This diagram shows the mass flow rate characteristics for the same Husqvarna engine presented previously in Sect. 5.3.1. In particular, the data used for the calculation is identical to that for Fig. 5.19, except the calculation using Prog.5.2 is instructed to draw the mass flow rate at the carburetor end of the inlet tract, employing Eqs. 2.4.10 in the manner discussed in Sect. 8.4.1. The calculation is at 6500 rpm and it can be seen that the mass flow rate is predicted for one complete cycle. In Sect. 8.4.1 it was concluded that the relevancy of the work of Coates(8.3) is that a Fourier analysis of this mass flow rate is directly related to the noise spectrum in the atmosphere beyond the end of the inlet tract. Examining Fig. 8.18, it is obvious that there are two distinct frequencies of importance, the fundamental at engine speed because it is much larger than the subsidiary oscillations and another at six times the fundamental frequency to encompass the remaining five oscillations, all of which are of a somewhat similar, but lower, amplitude. Returning to the attenuation behavior of the intake silencer of the OMC V8 engine outboard motor in Fig. 8.17, it can be seen that the peak transmission losses occur in two distinct bands, i.e., at 100 Hz and 500 Hz, a ratio of five; the lower absorption frequency at 100 Hz fundamentally corresponds to an engine speed of 750 rpm.

From this discussion, it is clear that the engine forcing frequency of direct design interest, Fre, is that corresponding to the engine speed of rotation, i.e., rpm/60 and, as the intake silencer is of the low-pass type, the engine speed in question is at the lower end of the usable speed band. Annand and Roe(1.8) give some further advice on this particular matter.

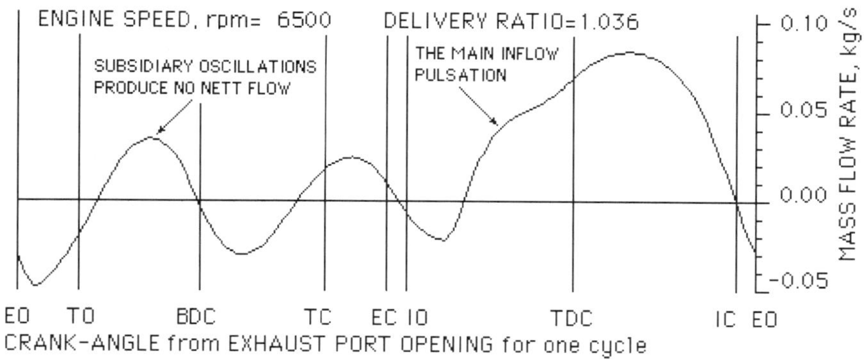

Fig. 8.18 Mass flow characteristics at the carburetor of the Husqvarna inlet tract.

There are several other criteria to be satisfied as well in this design process, such as ensuring that the total box volume and the intake pipe area are sufficiently large so as not to choke the engine induction process and reduce the delivery ratio. Such a calculation is normally accomplished using an unsteady gas-dynamic engine model, for example Prog.5.1, extended to include the intake and exhaust silencers(8.21). However, an approximate guide to such parameters is given by the following relationship, where Ncyl is the number of cylinders, SV is the swept volume of any one cylinder, and Dint is the flow diameter of the intake duct or the carburetor:

$$15*Ncyl*SV<VB<25*Ncyl*SV \quad (8.5.4.3)$$
$$0.6*Dint<DSP<0.8*Dint \quad (8.5.4.4)$$

To put this advice in numerical form, imagine a 175 cc single-cylinder motorcycle engine which is designed to run between 2400 and 6000 rpm, accelerating through the various gears in the gearbox. The carburetor on the engine has a 28 mm flow diameter. The lowest engine speed of direct interest is 2400 rpm, where the engine has a fundamental induction frequency, FRe, of 40 Hz. Therefore, for design purposes it will be assumed that this imaginary 175 cc motorcycle engine is unlikely to be operated at full throttle below 2400 rpm, consequently the minimum value for FRi is 40 Hz. The box volume is to be 20*SV*Ncyl from Eq. 8.5.4.3, or 3500 cc, giving VB at 0.0035 m^3. The pipe diameter, DSP, from Eq. 8.5.4.4, is to be 0.7*Dint, approximately 20 mm. This means that the area, SP, is π*(20/1000)/4, or 0.000314 m^2. The air temperature is at 20°C, so the acoustic velocity, A, from Eq. 8.1.1, is $\sqrt{(1.4*287*293)}$, or 343 m/s.

Using Eq. 8.5.4.1, where the minimum silencing frequency, FRi, is to be the required 40 Hz:

$$FRi=(A/2\pi)*\sqrt{[SP/(LPi*VB)]}$$

Inserting the data into this equation, the selection of which is debated above:

$$40=(343/2\pi)*\sqrt{[0.000314/(LPi*0.0035)]}$$

Thence, LPi=0.156 m, or 156 mm.

From Eq. 8.5.4.2, LP=156-π*20/4=140 mm.

The conclusion from this calculation is that a silencing pipe 140 mm in length and 20 mm internal diameter inserted into a box of 3.5 liters volume would provide an adequate intake silencing characteristic for this particular motorcycle engine. If the pipe is to be reentrant into the box, as shown in Fig. 8.11, the reader should remember that it is the effective internal box volume which is being employed as data.

It is important to emphasize that this is an empirical, not an absolute, design calculation and the experienced designer will know that this prediction is but a prelude to an intensive period of experimental development of the actual intake silencing system(1.12).

8.5.4.2 Shaping the intake port to reduce high-frequency noise

If the intake system of the engine is controlled by a port and piston skirt or is a disc valve intake engine as introduced initially in Sects. 1.0 and 1.3, the profile of the opening of that port is somewhat important. An expanded discussion on this matter is found in Sect. 8.6.1 for exhaust ports, where the topic is of even greater importance, so it will be sufficient at this stage to draw the reader's attention to Fig. 8.19. In that diagram, the designer has presented the option of shaping the opening or timing edge of the intake port so that the flow area-time characteristics of the port have a shallower profile. This provides an induction pulse pressure-time and mass flow-time behavior which is less steep, and this reduces the high-frequency content of the intake noise spectra(8.15). By definition this implies that, for equality of intake port time-areas, a shallow profile designed on the timing edge will require a longer port timing duration with a potential for reduction of delivery ratio. Nevertheless, this is a viable design option, particularly for such engine applications where there is little room available for a conventional low-pass intake silencer; a chainsaw is a typical example of such an application.

8.6 Silencing the exhaust system of a two-stroke engine

In Sect. 8.4.1 it was concluded that the relevancy of the work of Coates(8.3) is that a Fourier analysis of the instantaneous mass flow rate at the exhaust pipe termination to the atmosphere is directly related to the noise spectrum in the

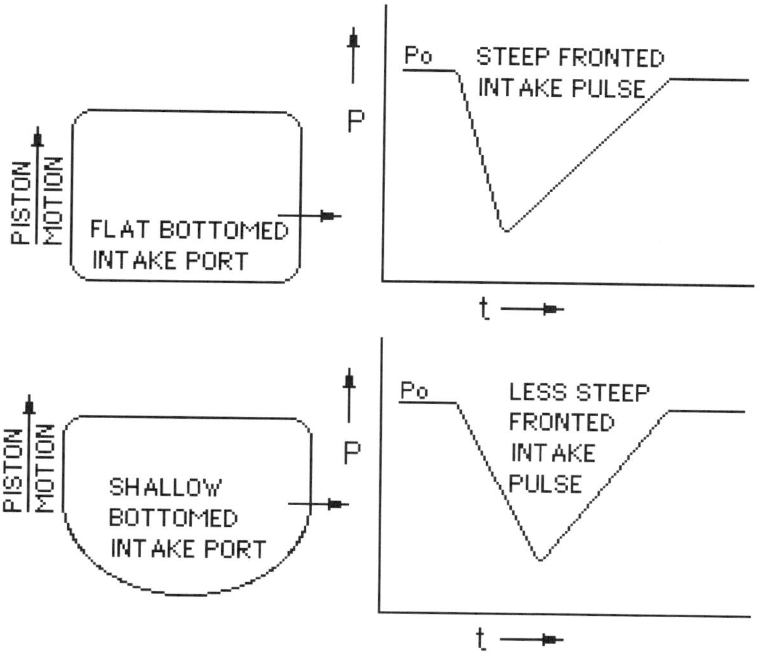

Fig. 8.19 Effect of intake port opening profile on induction pulse shape.

atmosphere. It is of interest to examine the mass flow-time characteristics for an engine already familiar to the reader from Chapter 5, the Husqvarna motocross engine with a tuned exhaust pipe. The input data and the computer program are precisely the same as that which predicted Fig. 5.19, except that Eqs. 2.4.10 are preprogrammed as output data for the mass flow rate of the exhaust gas at the outlet end of the tailpipe of the expansion chamber, and that result is plotted in Fig. 8.20 over one complete engine cycle at 6500 rpm. In short, the procedure which produced Fig. 8.18 for the discussion on induction noise is being repeated to illustrate the debate on exhaust noise suppression.

Fig. 8.20 shows the exhaust mass flow rate behavior at the pipe termination to atmosphere, and it is quite different from the equivalent induction graph in Fig. 8.18. Almost all of the flow oscillations are of outflow and are somewhat equal in magnitude, with virtually no oscillation into the inflow region. A Fourier analysis of this signal would record the primary, or fundamental, frequency as being five times that of the engine's exhaust pulsation rate, i.e., the engine speed. From this graph, the designer of a tuned exhaust pipe gleans the first important piece of information: the fundamental noisiest frequency needing to be suppressed is five times that of the engine firing frequency.

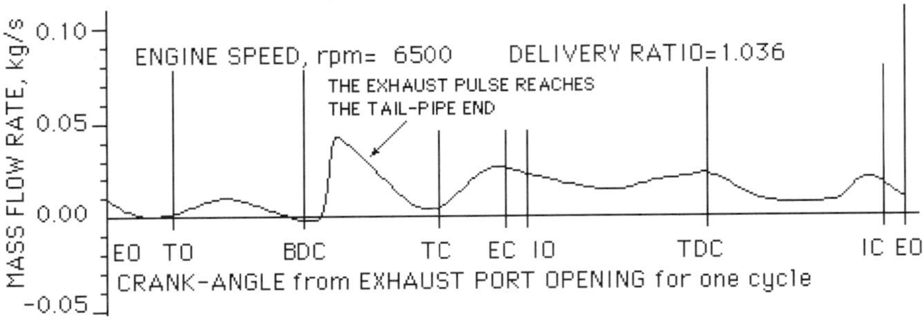

Fig. 8.20 Mass flow characteristics at the exhaust pipe end of the Husqvarna engine.

Also of note in Fig. 8.20 is the steep-front associated with the arrival of the main exhaust pulse, a situation further exacerbated by the natural tendency of compression waves to become steep-fronted (see Sect. 2.1.5). If a Fourier analysis were attempted, this steep-front would need a sinusoidal pulse of at least 20 times the engine frequency to approach a mathematical fit(8.20). This implies that there is considerable unpleasant noise, in terms of amplitude at a high frequency, due to the steep front of an exhaust pulse. In this case, the frequency level is at 20*6500/60, or 2167 Hz.

It can be seen by comparing the mass flow rate spectra for this unsilenced Husqvarna engine at 6500 rpm in Figs. 8.18 and 8.20 that the raw noise level emanating from the induction system is at least as high, if not higher, as that from the exhaust side. In point of fact, the peak mass flow rate is actually higher on the induction side. However, because of the more sinusoidal shape of the induction mass flow rate picture, there is little higher frequency content present than that of the oscillations one observes. As this is about 600 Hz, it will appear to be quieter than the exhaust system, particularly when recorded on a dBA scale.

8.6.1 The profile of the exhaust port timing edge

The first pragmatic step to be taken by the designer is to ensure that the front of the exhaust pulse rises as slowly as possible while still satisfying the performance criteria to be determined by the techniques described in Chapters 2-6. Fig. 8.21 shows a sketch of the potential to be gained by profiling the timing edge of the exhaust port. More germane to the issue is a calculation conducted using Prog.2.2, "EXHAUST," which is the program originally presented and discussed in Sect. 2.5 for the situation where the exhaust port opens for a total of either 45° or 60°, all other data being identical for the two calculations. The calculation from that program, as given in Sect. 2.5, normally illustrates the pressure diagrams but is reprogrammed for this particular example to provide the mass flow rate diagram emanating during one cycle from the end of a 500 mm straight pipe at 5000 rpm. The result is shown in Fig. 8.22, the mass flow rate diagrams at the pipe termination to atmosphere being

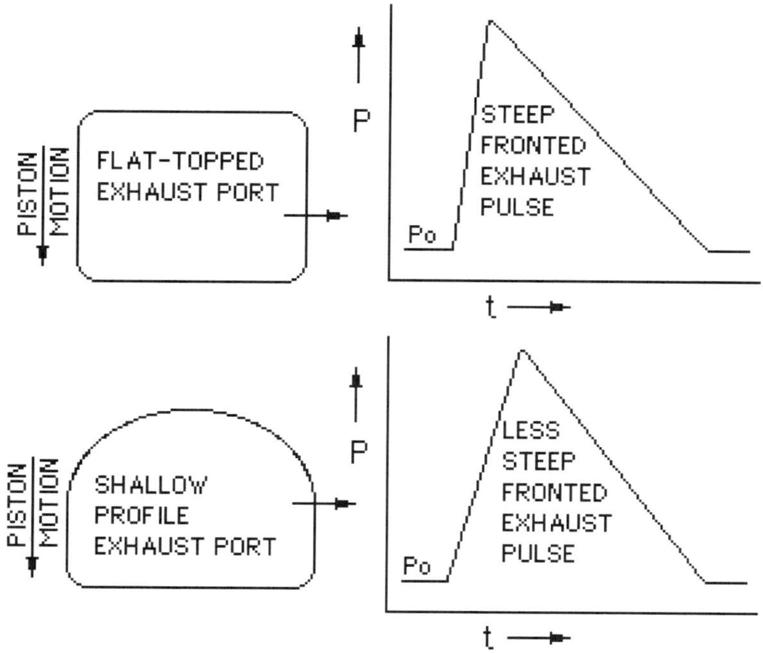

Fig. 8.21 Effect of exhaust port opening profile on exhaust pulse shape.

superimposed upon each other, to facilitate comparisons of potential noise spectra. The mass flow rate diagram for the more rapidly opening port is visibly steeper, thus considerable noise will occur at a higher frequency.

The effect of profiling the exhaust port in a chainsaw, and its ensuing behavior in both noise and power terms, is discussed in some detail in a paper by Johnston(8.14). It is important to emphasize that slowing down the exit rate of exhaust gas flow during the blowdown phase inevitably leads to longer port timing durations and some reduction of trapping efficiency. This can be alleviated to some extent by making the exhaust port wider without unduly increasing the total timing duration, as the shallower area-time profile necessary for noise reduction also permits a wider port without endangering the mechanical trapping of the piston ring by the timing edge (see Sect. 6.2.3).

Even with a suitably profiled exhaust port timing edge, the pressure-time rise characteristics of an exhaust pulse in a two-stroke power unit is much faster than that in a four-stroke cycle engine where a poppet valve is employed to release exhaust gas from the cylinder. As already remarked, the noise spectrum from a rapidly rising exhaust pressure wave is "rich" in high-frequency content, irritating to the human ear. A rifle shot is a classic example of that effect.

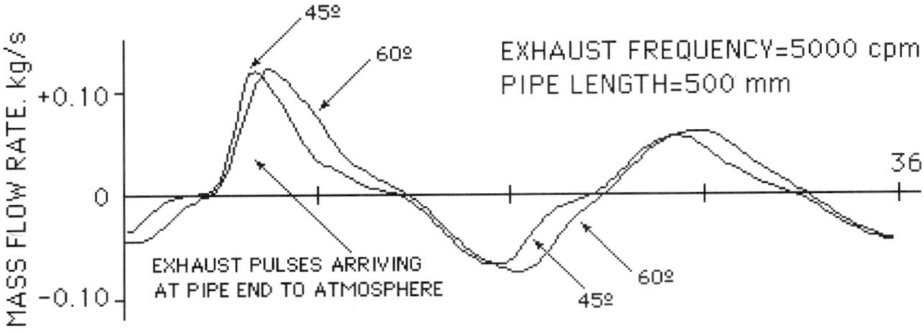

Fig. 8.22 Exhaust mass flow rate behavior for differing exhaust port opening rates.

8.6.2 Silencing the tuned exhaust system

The muffling of the tuned exhaust pipe is relatively straightforward and poses little difficulty in attaining the dual aims of good silencing while retaining the pressure wave tuning action necessary to attain a high BMEP and high specific power output. In Sect. 7.3.4 it is shown how even the introduction of a catalyst can be accommodated in a tuned pipe without deterioration of the many design goals. Fig. 7.19 shows how that is accomplished, including the presence of a silencer after the tuned section. This explanation is expanded here by Fig. 8.23. The design of the tuned section of the pipe is obtained following the design methods given in Chapters 5 and 6, and then the design of the silencing section is to be carried out by the recommendations to be found within this section. After the tuned section of the pipe, depending on the availability of space within the application for the power unit, there will be several chambers dedicated to the removal of specific acoustic frequencies or frequency bands. Typical of such chambers are those defined as diffusing silencers side-resonant silencers and absorption silencers. The detailed discussion of the noise suppression behavior of each of these types is to be found in Sects. 8.5.1-3.

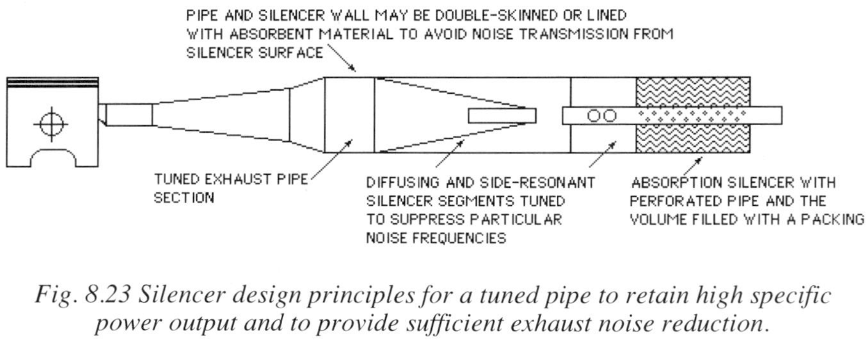

Fig. 8.23 Silencer design principles for a tuned pipe to retain high specific power output and to provide sufficient exhaust noise reduction.

Perhaps the most important point to be made at this introductory stage is that, having reduced the area of the exhaust pipe deliberately to attain a plugging reflection pulse for power performance reasons, the first silencing design action has already been taken to the benefit of the specific power output of the engine. To reinforce the point, the optimum tail pipe diameter, DEP5 in Fig. 5.2b, is always about half that of the exhaust pipe diameter, DEP1. In other words, the area of the exhaust pipe which left the engine cylinder has been reduced by a factor of about 4. This is quite unlike the situation for a four-stroke engine where any restriction of the exhaust pipe almost inevitably leads to a drop in delivery ratio and a consequential linear fall in power output. As a direct result, the silencing of a four-stroke engine is a more delicate design procedure. Equally, it implies that a high-performance street motorcycle with a two-stroke engine can be relatively easily designed to be legally quiet and the manufacturers of such machines ensure that this is the case. The somewhat noisy devices which cause irritation to the general public are the result of modifications by the owner, aimed more at machismo than machine efficiency!

The earlier design discussion on tuned expansion chambers did not dwell on the material to be used in its mechanical construction, but custom and practice shows that this is typically fabricated from mild steel sheet metal which is about 1.1-1.3 mm thick. This may be acceptable for a racing engine where the mass of the exhaust system as a ratio of the total machine mass must be as low as possible. For a silenced system, particularly as the exhaust pulses of a two-stroke engine are particularly steep-fronted, a thin metal outer skin of an expansion chamber and silencer would act as a metal loudspeaker diaphragm causing considerable noise transmission to the atmosphere. Therefore, the designer must inevitably introduce double-skinning of the most sensitive parts of the system, and often a layer of damping material is inserted between a double-skinned surface. Occasionally, it is found to be adequate to line the internal surface of the silencer with a high-temperature plastic damping compound, and this is clearly an economic alternative in some cases.

8.6.2.1 A design example for a silenced expansion chamber exhaust system

To reinforce the points made above, let the reader consider the design of a silenced and tuned expansion chamber system for an imaginary 125 cc road-going motorcycle engine. The machine will be designed to produce peak power at 7500 rpm and be capable of pulling top gear at 100 km/h on a level road at 6000 rpm. The speed of 6000 rpm also coincides with the engine speed attained during acceleration noise testing for legislation purposes(8.16). Not unnaturally, the manufacturer wishes to ensure that this is a well silenced condition, and clearly the engine will be operating at a reasonably high BMEP level at that point. The tuned pipe section of the pipe has been designed, it is assumed by the techniques already discussed in Chapters 5 and 6, and the optimum tail pipe diameter has already been settled at 16 mm diameter with a box mid-section of 80 mm diameter. Thus, the type of pipe and silencer which would be applicable to this design has been drawn in Fig. 8.24, showing a dimensioned sketch of the silencing section of the system and the ensuing transmission losses of the three boxes which make up the first portion of the silencer.

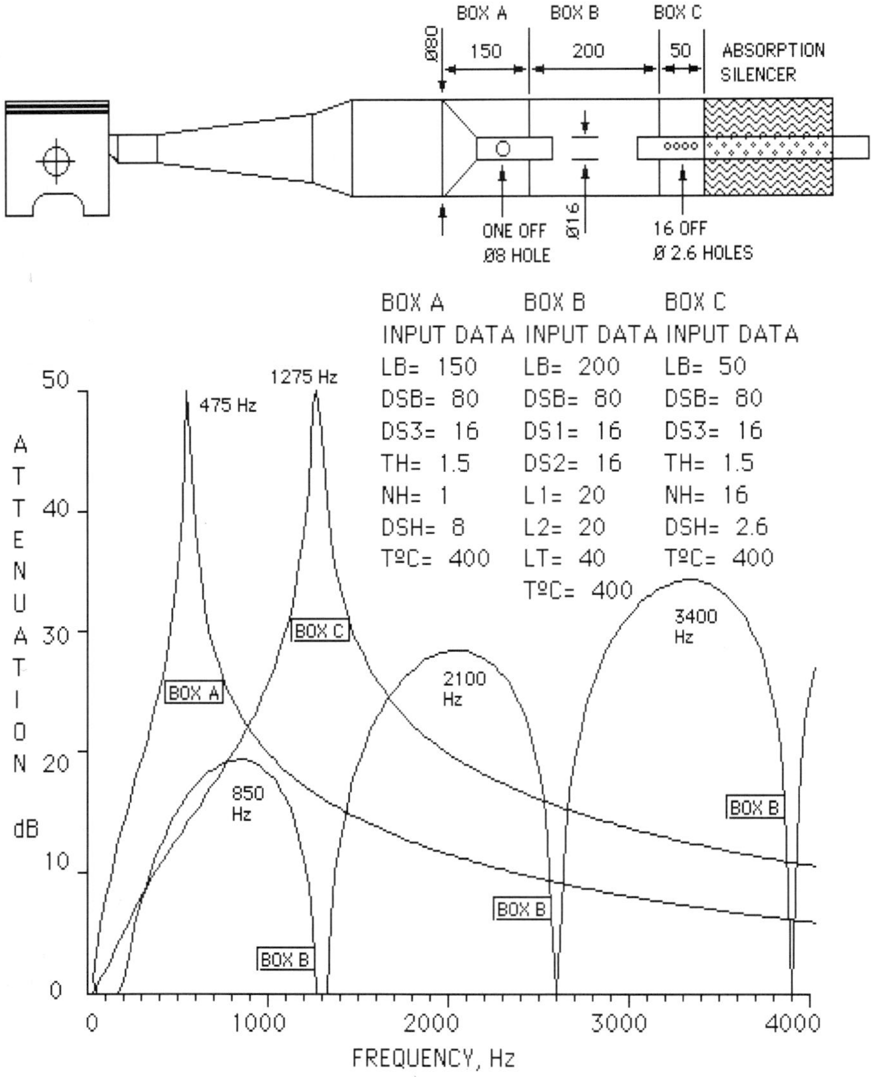

Fig. 8.24 Silencer design for a small motorcycle engine using Progs.8.1 and 8.2.

The first item on the design agenda is the decision regarding the fundamental noisiest frequency to be silenced. From the discussion in Sect. 8.6, it is clear that the frequency in question is five times that of the firing frequency, although the use of Prog.5.2 for the particular design of this imaginary 125 cc motorcycle engine, and with the program organized to predict the mass flow rate spectrum at the tail pipe outflow, would be a more accurate method of arriving at this judgment for this particular, or any given, powerplant. Assuming that the frequency multiplication factor is correct at five, the noisiest frequency will be at this fundamental value, i.e., 6000*5/60, or 500 Hz.

The second item on the design agenda is the disposition of the silencing segments. From Sect. 8.5.3, the absorption silencer will come last, before the exit to the atmosphere. The first segment should be that aimed at suppressing the fundamental noise, i.e., a side-resonant silencer, followed by a diffusing silencer which has more broad-band muffling effectiveness. There might be a need for a third box to cope with any pass bands which cannot be removed by the first two boxes.

The basic decisions having been made, the design of the first two boxes, A and B, is attempted. In Fig. 8.24 are the transmission losses of boxes A and B, together with the input data to Progs.8.1 and 8.2 for those boxes which correspond to the dimensioned sketch. The output from the individual calculations using Progs.8.1 and 8.2 are superimposed on top of each other.

The box A, a side-resonant silencer, has a single 8 mm diameter hole in the 16 mm pipe with a 1.5 mm thick wall. The exhaust temperature is assumed to be 400ºC. The reader should note that, if the real box has sloping walls as in the sketch, then the computed box volume may not ultimately agree with the actual box volume, in which case a modified length, LB, should be inserted as data so that the calculation operates with a realistic box volume. The reader will understand the importance of that statement more fully if the relevant acoustic theory is reexamined in Sect. 8.5.2 and Eqs. 8.5.2.1-5. That comment aside, the computed acoustic transmission loss is seen to peak at 475 Hz, which is sufficiently close to the target frequency of 500 Hz for maximum suppression.

The box B, a diffusing silencer, has 16 mm diameter pipes entering and leaving, each pipe being reentrant by 20 mm. The section length, LB, is 200 mm. The transmission loss is quite broad-banded, but the most important frequency of suppression is at 850 Hz, this lesson being the principal one learned from the work of Coates(8.3) with his SYSTEM 2 and discussed in Sect. 8.5.1. The first pass band is also of direct interest, and it is seen to exist at 1300 Hz.

If further design of this silencing system was left at this point, and the removal of the pass-band hole assigned to the absorption silencer segment, there would exist the strong possibility that this could turn out to be a rather noisy frequency. With the first two boxes, A and B, there is a good transmission loss for the silencer up to 1300 Hz, and the evidence from Sect. 8.5.1 is that the pass-band hole at that frequency would exist. Thus, a third small box of the side-resonant type, box C, is added and is targeted to have high attenuation at that specific frequency. It can be seen that a short box, just 50 mm long, with 16 holes of 2.6 mm diameter in the 16 mm diameter through-pipe with a 1.5 mm thick wall, has a transmission loss which peaks at 1275 Hz. The addition of this third box, C, making three silencing segments before the absorption silencer portion, would provide strong attenuation of the noise up to about 3500 Hz, with no visible pass-band holes in that transmission loss. The absorption silencer could then be designed to remove the remainder of the high-frequency content, either in the conventional manner or by using the annular flow system recommended in Sect. 8.5.3.2.

The reader should recall the discussion regarding the sound radiation possibilities from the outside surface of silencer and pipe walls and, in a real design situation, ensure that the wall is double-skinned or suitably damped acoustically along the entire length of the system.

8.6.3 Silencing the untuned exhaust system

The design of an untuned exhaust system for a two-stroke engine is discussed in Chapter 5, particularly in connection with the example of the chainsaw engine, where the small bulk required of the entire powerplant precludes the availability of space for adequate silencers, be they intake or exhaust. The problems in this regard are sketched in Fig. 8.25. There are other two-stroke engines with untuned or relatively untuned exhaust pipes, and these are to be found on agricultural and electricity generating equipment where powerplant space may not be at the same premium as it is on a chainsaw or a brushcutter. In this case the design approach is identical to that already described for the silencing section of the tuned pipe in the previous section. Another relatively straightforward example is the outboard motor, where the virtually unsilenced exhaust gas is directed underwater into the propeller wash which provides very effective exhaust silencing without any particular skill being employed by the designer. Whether this underwater racket is offensive to fish is not known.

The most difficult engine for which to design an adequate silencing characteristic is the chainsaw type of device. It is self-evident that, as noise is a function of the square of the dP, or pressure, fluctuation as seen in Eq. 8.1.2, then the larger the silencer volume into which the exhaust pulses are "dumped," the lesser fundamentally will be the dP value transmitted into the atmosphere. The designer of the chainsaw is continually looking for every available cubic centimeter of space to

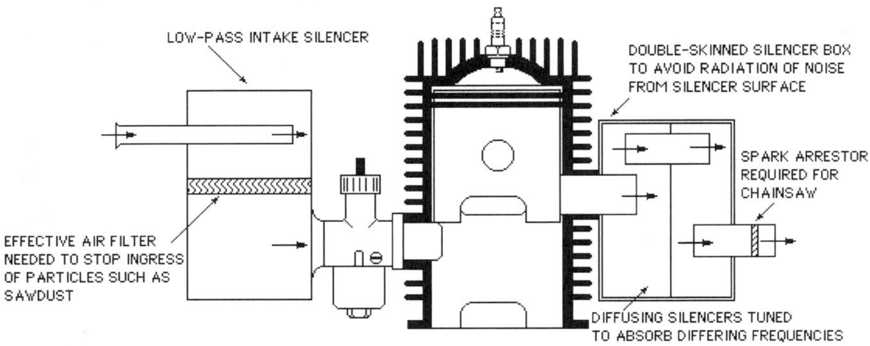

Fig. 8.25 Compact silencer designs for engines such as chainsaws where the bulk of the entire power unit must be minimized.

become part of the exhaust silencer. The space may be so minimal that the two-box design shown in Fig. 8.25 becomes an impossible luxury. The designer is left with the basic option of trimming out the first fundamental firing frequency using a diffuser type or side-resonant type of silencer as described in Sect. 8.5.

As a rule of thumb, the designer of a silencer for a chainsaw type of engine will know that there are going to be real silencing difficulties if there is not a total volume available for a silencer which is at least 15 times greater than the cylinder swept volume. If that is not the case, then the only design methodology left open is to choke the exhaust system by a restrictive silencer, thereby reducing the delivery ratio, and accept the consequential loss of power. Further complications in design terms arise from the legal necessity in many applications for the incorporation of a spark arrestor in the final tail pipe before entry to the atmosphere(8.22). By definition, a spark arrestor is a form of area restriction, as it has to inhibit, and more importantly extinguish, any small glowing particles of carbon leaving the silencer. This need is obvious for engines such as chainsaws and brushcutters, where the device is being used in an environment with a known fire hazard.

In any event, of equal importance is the necessity to damp the vibrations of the skin of the silencer, for this compact device is exposed to the normal "impact" of the unattenuated exhaust pulses. Consequently, double-skinning of the silencer surface is almost essential.

8.7 Concluding remarks on noise reduction

The principal concepts which the designer or developer has to retain at the forefront of the thought process when silencing the two-stroke engine are:

(a) Identify the principal source of noise from the engine be it mechanical, induction or exhaust noise and suppress that as the first priority. The other sources of noise could be completely eliminated and the total noise from the engine would be virtually unaffected if the single noisiest source remains unmuffled.

(b) Within any given noise source, the technique for suppressing it is to determine the frequency spectrum of that noise and to tackle, as a first priority, the frequency band with the highest noise content. Should that remain unmuffled, then the other frequency bands could be completely silenced and the total noise content would virtually remain at the original level.

(c) In the identification of noise sources, it is useful to remember that noise level, or loudness, is a function of the square of the distance from any given source(8.3) to the microphone recording that noise level. Thus placing the microphone close to the inlet tract end, or the exhaust pipe end, or a gearbox, in a manner which differentially distances the other sources allows for an initial and approximate identification of the relative contributions of the several sources of noise to the total noise output.

NOTATION for CHAPTER 8

NAME	SYMBOL	SI UNIT	OTHER
Reynolds Number	Re		
Density	D	kg/m^3	
Frequency of sound	FR	Hz	cpm
Length	L	m	mm
Mass flow rate	q	kg/s	
Pressure	P	Pa	bar
Pressure increment	dP	Pa	
Section area	S	m^2	mm^2
Sound Intensity	E	W/m^2	
Sound Intensity Level	EL	dB	
Sound Pressure Level	Lp	dB	
Speed of sound	A	m/s	
Temperature	T	K	ºC
Transmission Loss or Attenuation	Tr	dB	
Viscosity	μ	kg/ms	
Volume	V	m^3	cm^3
Wavelength of sound	Ω	m	

REFERENCES for CHAPTER 8

8.1 G.P. Blair, J.A. Spechko, "Sound Pressure Levels Generated by Internal Combustion Engine Exhaust Systems," SAE Automotive Congress, Detroit, January, 1972, SAE Paper No.720155.

8.2 G.P. Blair, S.W. Coates, "Noise Produced by Unsteady Exhaust Efflux from an Internal Combustion Engine," SAE Automotive Congress, Detroit, January, 1973, SAE Paper No.730160.

8.3 S.W. Coates, G.P. Blair, "Further Studies of Noise Characteristics of Internal Combustion Engines," SAE Farm, Construction and Industrial Machinery Meeting, Milwaukee, Wisconsin, September, 1974, SAE Paper No.740713.

8.4 D.D. Davis, G.M. Stokes, D. Moore, G.L. Stevens, "Theoretical and Experimental Investigation of Mufflers," NACA Report No.1192, 1954.

8.5 P.O.A.L. Davies, R.J. Alfredson, "The Radiation of Sound from an Engine Exhaust," *J.Sound Vib.*, Vol.13, p.389, 1970.

8.6 P.O.A.L. Davies, "The Design of Silencers for Internal Combustion Engines," *J.Sound Vib.,* Vol.1, p.185, 1964.

8.7 P.O.A.L. Davies, "The Design of Silencers for Internal Combustion Engine Exhaust Systems," Conference on Vibration and Noise in Motor Vehicles, *Proc.I.Mech.E.,* 1972.

8.8 G.H. Trengrouse, F.K. Bannister, "The Reduction of Noise Generated by Pressures Waves of Finite Amplitude–Attenuation due to Rows of Holes in Pipes," University of Birmingham, Research Report No. 40, 1964.

8.9 M. Fukuda, H. Izumi, "Kuudougata Shouonki no Tokusei ni Kansuru Kenkyuu," *Trans.Japan Soc.Mech.E,* Vol.34, No.263, p. 1294, 1968.

8.10 P.M. Nelson, "Some Aspects of Motorcycle Noise and Annoyance," SAE Paper No.850982.

8.11 R. Taylor, Noise, Penguin, London, 1975.

8.12 C.M. Harris, Handbook of Noise Control, McGraw-Hill, New York, 1957.

8.13 L.L. Beranek, Noise Reduction, McGraw-Hill, New York, 1960.

8.14 M.B. Johnston, "Exhaust Port Shapes for Sound and Power," SAE Farm, Construction and Industrial Machinery Meeting, Milwaukee, Wisconsin, September, 1973, SAE Paper No.730815.

8.15 K. Groth, N. Kania, "Modifications on the Intake Ports with the Aim to Reduce the Noise of a Two-Stroke Crankcase-Scavenged Engine," SAE International Off-Highway Meeting and Exposition, Milwaukee, Wisconsin, September, 1982, SAE Paper No.821069.

8.16 SAE Noise Standards:

SAE J47-Maximum Sound Level Potential for Motorcycles

SAE J331a-Sound Levels for Motorcycles

SAE J192-Exterior Sound Level Measurement for Snowmobiles

SAE J1174-Operator Ear Sound Level Measurement Procedure for Small Engine Powered Equipment

SAE J1046a-Exterior Sound Level Measurement Procedure for Small Engine Powered Equipment

SAE J1175-Bystander Sound Level Measurement Procedure for Small Engine Powered Equipment

SAE J1074-Engine Sound Level Measurement Procedure

SAE J1207-Measurement Procedure for Determination of Silencer Effectiveness in Reducing Engine Intake or Exhaust Sound

8.17 S.W. Coates, "The Prediction of Exhaust Noise Characteristics of Internal Combustion Engines," Doctoral Thesis, The Queen's University of Belfast, April, 1974.

8.18 E. Kato, R. Ishikawa, "Motorcycle Noise," Dritte Grazer Zweiradtagung, Technische Universitat, Graz, Austria, 13-14 April, 1989.

8.19 N.A. Hall, Thermodynamics of Fluid Flow, Longmans, London, 1957.

8.20 G.M. Jenkins, D.G. Watts, Spectral Analysis and its Applications, Holden-Day, San Francisco, 1968.

8.21 G.P. Blair, "Computer-Aided Design of Small Two-Stroke Engines for both Performance Characteristics and Noise Levels," I.Mech.E Automobile Division Conference on Small Internal Combustion Engines, Isle of Man, May 31-June 2, 1978, Paper No.C120/78.

8.22 SAE Standards, "Multiposition Small Engine Exhaust System Fire Ignition Suppression," SAE Standard J335b.

Postscript

The reader, having struggled through the book to this point, will be relieved to know that no further mathematical equations are to be thrust upon him or her. In this postscript it is intended to emphasize some of the salient points which this book has attempted to stress at various intervals within it, and finally to pass comment on the professional arena in which designers must operate and on the education to prepare them for it.

(a) The design process

The first point to be made is that a real design process for an internal combustion engine is one where the designer will specify every detail of the physical geometry of the engine, and when that engine is built, it will do three things: first, it will produce the target performance characteristics required of it; second, it will continue to produce those performance characteristics for the desired life of the powerplant without any mechanical failure; third, it will be capable of being produced and manufactured efficiently within the structure of the particular organization involved. This book, as the reader by now will understand, concentrates on that first criterion. For wider advice and guidance on the second and third of these approaches to design, the reader will have to peruse elsewhere.

Having said that, however, it is not possible to design for performance characteristics alone, nor for mechanical reliability alone, nor solely for manufacturing efficiency, as the resulting power unit may be inadequate on any of these grounds if attention has not been paid equally to all approaches to its design.

The technology of mechanical design and design for manufacture is as profound, in a mathematical and technological sense, as that for design for performance characteristics. How then is the reader to acquire these three technologies if they are inseparably linked for the successful prosecution of any design? The answer is straightforward: practice them in earnest, continually study them, design, followed by the almost inevitable failure first time around, practice again, study some more, design again, fail again, until eventually the design methodologies fit into place, success is achieved, and a new designer has arrived. Any notion that there is a short-cut to this procedure is a fallacy. Any notion that one can be a successful designer by being an expert in just one of those design technologies, or being even more of a specialized expert in part of just one of those technologies, is also a fallacy. The key words in the sentences above are "practice" and "study." For those who wish to be successful that means every waking minute, not just from "nine to five." If that prospect seems to be too draconian, then mentally start settling for second place or another career.

On the "study" front it is hoped that this book will provide insights into the technology of performance design which did not exist before in the mind of the reader. On the "practice" front it is hoped that the use of the computer programs associated with this book will assist the designer to rapidly examine many designs, and many more aspects of particular designs, before settling on the development of a particular physical geometry.

(b) Further study by the designer and engineer

For any engineer, but particularly the young engineer, association with other engineering designers is an extremely useful part of the learning curve. Apart from serious study of the technical papers which emanate from learned societies around the world, it is more than useful to go to Conferences where such engineers congregate, such as SAE at the Detroit and Milwaukee Conferences, or IMechE in the United Kingdom, and from discussions with those of a like mind acquire the distilled experience of others in minutes what would take years to accumulate by personal effort. Of course, one should not believe everything one hears!

(c) Some brief thoughts on engineering education

The advancement of technology is now proceeding at a pace which makes it difficult for any one engineer to be in command of all of its facets. However, it is vital to remember that all technological advancement is rooted in the sciences of mathematics, physics and chemistry. Therefore, the education of the engineer must leave the graduate with a fundamental understanding of these subjects and their engineering applications in thermodynamics, fluid and gas dynamics, mechanics, engineering design, manufacturing technology, and materials science. Without that fundamental understanding the young engineer will not be able to participate in the advancement of technology, much less comprehend it. In many countries, the author notes with chagrin that the engineering science fundamentals within undergraduate engineering education are being reduced to the bare minimum to permit that education to be expanded to include subject material relating to business, management, finance, law and politics. This philosophy of approach leads to graduates who may manage more efficiently the people who manufacture a product but who will have virtually no comprehension of its future development in a highly competitive world. For example, this book was not written after much study and practice of a generalist nature, but by the application of mathematics and engineering science. If this book is found to assist with the onward development of the two-stroke engine, in terms of study, practice and education, then it is contended that the philosophical criticism being made above deserves serious consideration.

(d) Concluding comments

The foreword to this book set out the aims and objectives for it. It is hoped that the readers, be they a designer in industry, an undergraduate or postgraduate engineering student, or any interested party in the research and development of the two-stroke engine, will feel that their understanding of this seemingly simple, but in reality very complex, power unit has been enhanced. If that is the end result then the author will feel that the toil inherent in the writing of this book has been a labor well spent. There is little doubt that the two-stroke engine is at a threshold of new development and is emerging from the mold of being thought of as merely the inexpensive manufacturing engine option. Serious consideration is being given to it as the future automobile powerplant, and this implies a more extensive application of funding for research and development than has been the case hitherto. Whatever else happens, this will have a knock-on effect on the knowledge base for its use in the more traditional applications for the engine type. For those of us who have been party to this research and development process, truly these have been exciting times in which to work.

COMPUTER PROGRAM APPENDIX

These programs are written in Microsoft QuickBASIC for the Apple Macintosh computer. In particular they have been developed on the Macintosh II, but they will run on any Macintosh machine. However, some of the earlier Macintosh machines have a smaller screen size than the Macintosh II and not all of the graphics for all of the programs will fit within the screen space of those computers. The QuickBASIC language produced by Microsoft Inc. is available for the Macintosh and the IBM personal computers, and the language is identical. However, some of the graphics statements are different for the IBM and Macintosh computers.

The primary intention is that the reader of this book will type in the programs from the listings below, and be immediately able to carry out the calculations which illustrate the book and use them for the design purpose. The programs will be available from SAE in disk form, which will bypass the entire typing process and ensure the accurate transmission of the programs into the computer used by the reader, whether it be an IBM type machine or a Macintosh II.

For those not familiar with the BASIC language, some useful advice is dispensed to get the reader started with the computational use of these programs. Should the reader be expert in such matters, the following simplistic paragraphs may be bypassed. The first program, Prog.1.1, is a good example of the type of prompts which the BASIC language uses to make the user respond to a request for data or direction for the program. The first prompt in Prog.1.1, upon getting the program running, is that the screen shows the following:

RUNNING THE PROGRAM(Y or N?)

The Y or the N requested is a "yes" or a "no" and the program needs, and will <u>only</u> recognize, a capital, i.e., uppercase Y or uppercase N, typed in from the keyboard and the "return" key depressed. (Clearly the user is going to type in a Y, as the program is started for that purpose in the first place!) The next prompt the screen will show is:

enter the BORE in mm?

The machine will do nothing until the reader types in the cylinder bore dimension in millimeter units, although, as computers are non-discriminatory, it will recognize any old number but it will not recognize a letter or a word for this type of prompt. The "return" key is depressed after the appropriate data number has been inserted. The same procedure is followed for the subsequent data values, whereupon the results appear automatically as shown in Fig. 1.11. Indeed, all of the previous discussion is shown in Fig. 1.11.

Finally, the program types upon the screen:

WANT A PRINT-OUT(Y OR N?)?

If the user wants the results printed, then a Y is typed; remember, not a y but a Y, and ensure that the printer is switched on and the paper inserted!

All of these commands and queries can be seen within the listing below and some simple study will reveal the ease with which programs are formulated.

EDITOR'S NOTE: If you intend to use the programs presented in this appendix, the following are the hardware and software requirements.

The printed appendix addresses the Apple platform and is written to support the Apple Imagewriter printer. If you are using an Apple Laserwriter, you will need to change some of the print statements to also support a Laserwriter. If you are using the Apple programs available from SAE, the print statements have been changed to also support a Laserwriter. These programs are identified by an "L" prefix.

Please note that the IBM programs available from SAE are compiled and independent of the Microsoft QuickBASIC interpreter. The Apple programs are compiled but require the use of the Microsoft QuickBASIC libraries.

If you are using the IBM programs, and are having difficulties printing the graphs with GRAPHICS.COM, it is recommended that you use GRAFPLUS screen dump utility software (manufactured by Jewel Technologies).

System Requirements:

1. Apple Platform
   Hardware: Apple Macintosh II, Macintosh IIX, or Macintosh IICX system
   - Extended memory – RAM above 1 MB
   - System Version 4.1 or later and Finder 5.5 or later
   - Hard disk drive is recommended
   - Access to the 68881 math coprocessor (compiler versions)

   Software: Microsoft QuickBASIC for Apple Macintosh systems
   - Some working knowledge of MS QuickBASIC
   - Binary version

2. IBM Platform
   Hardware: IBM PC or 100% compatible system
   - 640K RAM
   - MS or PC DOS 2.0 or later
   - Hard disk drive is recommended – at least 2 MB free storage
   - EGA or VGA color interface and display
   - IBM/Epson compatible printer

   Software: Microsoft QuickBASIC for IBM PCs or compatibles
   - Some working knowledge of MS QuickBASIC

   GRAPHICS.COM – External DOS command
   - Allows screen dump to support hard copy printing of graphic images

ProgList1.0 Program listings from Chapter 1

COMPUTER PROGRAM PROG.1.1
PROGRAM NAME "PISTON POSITION"
ProgList.1.1

```
REM Prog.1.1
REM ProgList.1.1
REM Program PISTON POSITION
REM***crank-angle piston movement calculation***
DIM ADEG(360),HTA(360),HBA(360),VB(360),VT(360)
100 INPUT"RUNNING THE PROGRAM(Y OR N?)";ANS$
IF  ANS$="N"  THEN  END
INPUT"enter BORE in mm",BO
INPUT"enter STROKE in mm",ST
INPUT"enter CON-ROD LENGTH  in mm",CRL
INPUT"enter  CRANK-ANGLE  after tdc at  start  of calculation",ANG1
INPUT"enter  CRANK-ANGLE  after tdc at end of calculation",ANG2
INPUT"enter  CRANK-ANGLE interval for each step of the calculation from start to finish",DANG
PI=3.141592654#
DTOR=PI/180
RTOD=1/DTOR
PA=PI*BO*BO/4
SVOL=PA*ST/1000
PRINT
PRINT  "Swept volume,CM3,=" ;USING"####.#";SVOL
CRNK=ST/2
RANG1=ANG1*DTOR
RANG2=ANG2*DTOR
N=0
PRINT "crank-angle after tdc      height from tdc        height from bdc"
FOR ANG=ANG1 TO ANG2 STEP DANG
N=N+1
ANGR=ANG*DTOR
CALL PISTPOS(ST,CRNK,CRL,ANGR,HT,HB)
HTA(N)=HT
HBA(N)=HB
ADEG(N)=ANG
VT(N)=PA*HT/1000
VB(N)=PA*HB/1000
PRINT USING"############.##";ANG ,HT,HB
NEXT  ANG
INPUT"WANT A PRINT-OUT(Y OR N?)";ANS$
IF ANS$="N" THEN GOTO 100
```

```
INPUT "WANT TO NAME THE DATA(Y OR N?)";ANS$
IF ANS$="Y" THEN INPUT HANDLE$
LPRINT HANDLE$
LPRINT" BORE STROKE CON-ROD SWEPT VOLUME PISTON
AREA
LPRINT" MM MM MM CM3 MM2
LPRINT USING"###########.#";BO,ST,CRL,SVOL,PA
LPRINT
LPRINT "CRANK-ANGLE HTDC HBDC VTDC VBDC"
LPRINT"DEG.ATDC MM MM CM3 CM3"
FOR N=1 TO N STEP 1
LPRINT USING"#######.#";ADEG(N),HTA(N),HBA(N),VT(N),VB(N)
NEXT N
GOTO 100
END
SUB PISTPOS(S,R,C,A,HT,HB)STATIC
HT=R+C-R*COS(A)-SQR(C*C-(R*SIN(A))^2)
HB=S-HT
END SUB
```

COMPUTER PROGRAM PROG.1.2

PROGRAM NAME "LOOP ENGINE DRAW"

ProgList.1.2

```
REM computer listing ProgList.1.2
REM program LOOP ENGINE DRAW
REM computer program PROG.1.2
REM computer program to give basic design parameters for a loop scavenged
two-stroke
REM click in any circle in the data box until all data inserted
REM after you have CLICKED in a box a third screen will come up and demand
a DATA value
REM do not RUN the program until all data values inserted
REM the data can be filed on hard disk by clicking the FILING box
REM if you wish to use FILED data type a F when prompted
REM if you wish to use NEW data type a N when prompted
REM this form of data and program operation in CLICK boxes uses the MAC
at high speed
REM this program will give a print-out to an Apple Imagewriter
REM the engine rotates for a complete engine cycle from tdc to tdc and
REM the print-out on the printer is exactly as seen on the screen
REM do not forget to CLICK on "tall adjusted" on the printer demand or circles
REM will appear as ellipses!
REM many program statements are common for PROG.1.2 and PROG.1.3 so
REM a shrewd typist can use the MAC very speedily to load both programs
```

```
WINDOW 1
REM  A TWO STROKE ENGINE
WINDOW 2,"CLICK BOX",(201,50)-(400,450),1
PRINT "SELECT BY BUTTON"
WINDOW 3,"CURRENT DATA",(10,50)-(200,450),1
WINDOW 4,"DATA INPUT",(400,50)-(600,450),1
WINDOW 5,"ACTION BOX",(10,430)-(600,480),1
WINDOW 6,"EXHAUST TYPE",(400,50)-(600,450),1
WINDOW 1
100 INPUT"NEW DATA OR FILE DATA (TYPE N OR F)?";D$
IF D$="N" THEN GOTO 110
IF D$="F" THEN GOSUB FILED
110 WINDOW 3
CLS
GOSUB CURRENTDATA
WINDOW 2
BUTTON 1,1,"BORE",(5,25)-(100,45),3
BUTTON 2,1,"STROKE",(5,45)-(100,65),3
BUTTON 3,1,"CON-ROD",(5,65)-(100,85),3
BUTTON 4,1,"EXHAUST TIMING",(5,85)-(200,105),3
BUTTON 5,1,"TRANSFER TIMING",(5,105)-(200,125),3
BUTTON 6,1,"INLET TIMING",(5,125)-(200,145),3
BUTTON 7,1,"COMPRESSION RATIO",(5,145)-(200,165),3
BUTTON 8,1,"SQUISH CLEARANCE",(5,165)-(200,185),3
BUTTON 9,1,"WRIST PIN TO CROWN LENGTH",(5,185)-(200,205),3
BUTTON 10,1,"WRIST PIN TO SKIRT LENGTH",(5,205)-(200,225),3
BUTTON 30,1,"RUN",(5,325)-(100,350),3
BUTTON 31,1,"QUIT",(5,350)-(100,375),3
ACTIVITY=DIALOG(0)
WHILE ACTIVITY<>1:ACTIVITY=DIALOG(0):WEND
BUTTONPUSHED=DIALOG(1)
IF BUTTONPUSHED=30 THEN GOTO 130
IF BUTTONPUSHED=31 THEN GOTO 556
ON BUTTONPUSHED GOSUB BORE,STROKE,CONROD,EXTIMING,
TRTIMING,INTIMING, COMPRAT, SQUISH,PINCROWN,PINSKIRT
GOTO 110
130 WINDOW 1
CLS
PI=3.1415927#
DTOR=PI/180
RTOD=180/PI
CRK=ST/2
DHEAD=.7*BO
SIGN=1
```

```
GOSUB CRANKFACT
PZTDC=CRK+CRL+GPC
PZBDC=PZTDC-ST
DEG=EO
GOSUB PIST
PZE=PZTDC-HTDC
YEX=YTDC+HTDC*SCAL
SV=PI*BO*BO*ST/4
TSV=PI*BO*BO*HTDC/4
CVOL=TSV/(TRAPCR-1)
SQVOL=PI*BO*BO*SQCL/4
DEG=TRO
GOSUB PIST
PZT=PZTDC-HTDC
YTR=YTDC+HTDC*SCAL
DEG=INO
GOSUB PIST
PZIO=PZTDC-GPC-GPS-HTDC
PZIFO=PZTDC-GPC-GPS
YIN=YITDC+HTDC*SCAL
GOSUB CRANKFACT
GOSUB CYLDRAW
DEG=0
150 GOSUB PIST
GOSUB CRANKDRAW
DEG=DEG+2
GOTO 150
557 GOSUB SUBPRINT
556 END
CYLDRAW:
R=RC1
X=XCL-XC1
Y=SQR(R*R-X*X)
AC3=ATN(Y/X)
YC3=YCL-Y
X=XCL-XL1
Y=SQR(R*R-X*X)
AL2=ATN(Y/X)
YL2=YCL-Y
TRDEEP=.5*(YBDC-YTR)
XTD1=XL1-TRDEEP
XTD2=XTD1-LINER
X=XCL-XTD1
Y=SQR(R*R-X*X)
```

```
AT1=ATN(Y/X)
YT1=YCL-Y
R=RC2
X=XCL-XTD2
Y=SQR(R*R-X*X)
AT2=ATN(Y/X)
YT2=YCL-Y
X=XL2-XCL
Y=SQR(R*R-X*X)
AL4=ATN(Y/X)
YL4=YCL-Y
CIRCLE(XCL,YCL),RC1,,PI-AC3,PI-AL2
CIRCLE(XCL,YCL),RC1,,PI-AT1,AC3
CIRCLE(XCL,YCL),RC2,,PI-AT2,AL4
YD=YTR+TRDEEP
CIRCLE(XL1,YD),TRDEEP,,PI/2,PI
CIRCLE(XL1,YD),(TRDEEP+LINER),,PI/2,PI
CIRCLE(XC1,YBDC+LINER),LINER,,PI/2,PI
LINE(XC1,YTOP)-(XC1,YC3)
LINE(XL1,YTOP1)-(XL1,YTR-LINER)
LINE(XTD2,YD)-(XTD2,YT2)
LINE(XTD1,YD)-(XTD1,YT1)
LINE(XL1,YBDC+LINER)-(XL1,YL2)
LINE(XC1,YTR)-(XL1,YTR)
LINE(XC2,YTOP)-(XC2,YC3)
LINE(XL2,YTOP1)-(XL2,YEX)
LINE(XL2,YBDC)-(XL2,YITDC)
LINE(XL2,YIN)-(XL2,YL4)
XPIP=XL2+25
LINE(XC2,YEX)-(XPIP,YEX)
LINE(XC2,YBDC)-(XPIP,YBDC)
LINE(XC2,YITDC)-(XPIP,YITDC)
LINE(XC2,YIN)-(XPIP,YIN)
CIRCLE(XPIP,(YEX+YBDC)/2),(YBDC-YEX)/2,,,,2
CIRCLE(XPIP,(YIN+YITDC)/2),(YIN-YITDC)/2,,,,2
LINE(XH1,YTOP)-(XH1,YRH)
LINE(XH2,YTOP)-(XH2,YRH)
LINE(XH1+RH,YHEAD)-(XH2-RH,YHEAD)
CIRCLE(XH1+RH,YRH),RH,,PI/2,PI
CIRCLE(XH2-RH,YRH),RH,,0,PI/2
LINE(XH1,YTOP)-(XH1,YRH)
LINE(XH2,YTOP)-(XH2,YRH)
LINE(XH1+RH,YHEAD)-(XH2-RH,YHEAD)
CIRCLE(XH1+RH,YRH),RH,,PI/2,PI
```

```
CIRCLE(XH2-RH,YRH),RH,,0,PI/2
YTOP2=YHEAD-LINER
LINE(XH3,YTOP1)-(XH3,YRH)
LINE(XH4,YTOP1)-(XH4,YRH)
LINE(XH1+RH,YTOP2)-(XH2-RH,YTOP2)
CIRCLE(XH1+RH,YRH),RH+LINER,,PI/2,PI
CIRCLE(XH2-RH,YRH),RH+LINER,,0,PI/2
LINE(XC1,YTOP)-(XH1,YTOP)
LINE(XC2,YTOP)-(XH2,YTOP)
LINE(XL1,YTOP1)-(XH3,YTOP1)
LINE(XL2,YTOP1)-(XH4,YTOP1)
RETURN
PIST:
DR=DEG*DTOR
HTDC=CRK+CRL-CRK*COS(DR)-SQR(CRL*CRL-(CRK*SIN(DR))^2)
HBDC=ST-HTDC
RETURN
CRANKFACT:
CRT=45
BEYE=10
SEYE=.75*BEYE
SCAL=CRT/CRK
SQX=SQCL*SCAL
BOX=BO*SCAL
RB=CRT/3
RBC=RB+BEYE
RF=CRT+RBC
RS=.75*RB
WR=2*RB
CR=CRL*SCAL
QQ=4
DQ=20
PLU=GPC*SCAL
PLD=GPS*SCAL
PL=PLU+PLD
XCL=360
YCL=350
YTDC=YCL-CR-CRT-PLU
YBDC=YTDC+2*CRT
YITDC=YTDC+PL
YTOP=YTDC-SQX
XT1=XCL-BOX/2
XT2=XCL+BOX/2
XC1=XT1-2
```

```
XC2=XT2+2
XR1=XC1+1
XR2=XC2-1
HRW=RB
RD=3
LINER=6
RC1=RF+6
RC2=RC1+LINER
XL1=XC1-LINER
XL2=XC2+LINER
YTOP1=YTOP-LINER
HHED=(CVOL-SQVOL)/(PI*DHEAD*DHEAD/4)
YHEAD=YTOP-HHED*SCAL
VALU=.35*BOX
XH1=XCL-VALU
XH2=XCL+VALU
XH3=XH1-LINER
XH4=XH2+LINER
RH=5
YRH=YHEAD+RH
RETURN
CRANKDRAW:
HTOP=HTDC*SCAL
HBOT=HBDC*SCAL
YPT=YTDC+HTOP
YPB=YPT+PL
YS=YPT+PLU
XB=XCL+CRT*SIN(DR)
YB=YCL-CRT*COS(DR)
CRZ=SQR(CR*CR-(XB-XCL)^2)
DRZ=ATN((XB-XCL)/CRZ)
XINC=HRW*COS(DRZ)
YINC=HRW*SIN(DRZ)
DYQ=DQ*COS(DR)
DXQ=DQ*SIN(DR)
XRD1=XCL-XINC
YRD1=YS+YINC
XRD4=XCL+XINC
YRD4=YS-YINC
XRD3=XB+XINC
YRD3=YB-YINC
XRD2=XB-XINC
YRD2=YB+YINC
YP=YCL-YPB
XT=XRD1-XCL
```

```
YT=YCL-YRD1
XTT=XRD2-XCL
YTT=YCL-YRD2
GOSUB INTERCEPT
X6=XCL+XM
Y6=YCL-YM
X5=XCL+XP
Y5=YPB
XT=XRD4-XCL
YT=YCL-YRD4
XTT=XRD3-XCL
YTT=YCL-YRD3
GOSUB INTERCEPT
X8=XCL+XM
Y8=YCL-YM
X7=XCL+XP
Y7=YPB
XQ1=DXQ+XCL
YQ1=YCL-DYQ
XQ2=-DXQ+XCL
YQ2=YCL+DYQ
XQ3=-DYQ+XCL
YQ3=YCL-DXQ
XQ4=DYQ+XCL
YQ4=YCL+DXQ
GOSUB CRANKERASE
LINE (XQ1,YQ1)-(XQ2,YQ2),33
LINE (XQ3,YQ3)-(XQ4,YQ4),33
LINE (X5,Y5)-(X6,Y6),33
LINE (X7,Y7)-(X8,Y8),33
CIRCLE(XCL,YCL),RF,33
CIRCLE(XCL,YCL),DQ,33
CIRCLE(XB,YB),RB,33
LINE(XT1,YPT)-(XT2,YPB),30,BF
LINE(XT1,YPT)-(XT2,YPB),,B
IF YPT<YTR THEN GOTO 922
LINE((XT1+XCL)/2,YTR)-((XCL+XT2)/2,YPT),,B
922 CIRCLE(XCL,YS),RS,33
LINE(XR1,YPT+4)-(XR2,YPT+4+RD),33,BF
LINE(XR1,YPT+11)-(XR2,YPT+11+RD),33,BF
EXQ1=XQ1
EYQ1=YQ1
EXQ2=XQ2
EYQ2=YQ2
```

```
EXQ3=XQ3
EYQ3=YQ3
EXQ4=XQ4
EYQ4=YQ4
EX5=X5
EY5=Y5
EX6=X6
EY6=Y6
EX7=X7
EY7=Y7
EX8=X8
EY8=Y8
EYB=YB
EXB=XB
EYPT=YPT
EYPB=YPB
EYS=YS
RETURN
CRANKPRINT:
LINE (XQ1,YQ1)-(XQ2,YQ2),33
LINE (XQ3,YQ3)-(XQ4,YQ4),33
LINE (X5,Y5)-(X6,Y6),33
LINE (X7,Y7)-(X8,Y8),33
CIRCLE(XCL,YCL),RF,33
CIRCLE(XCL,YCL),DQ,33
CIRCLE(XB,YB),RB,33
LINE(XT1,YPT)-(XT2,YPB),30,BF
LINE(XT1,YPT)-(XT2,YPB),,B
CIRCLE(XCL,YS),RS,33
LINE(XR1,YPT+4)-(XR2,YPT+4+RD),33,BF
LINE(XR1,YPT+11)-(XR2,YPT+11+RD),33,BF
RX=XCL-140
RY=YCL-2
GOSUB DARROW
RY=YTDC+2
GOSUB UARROW
LINE (RX,YCL-2)-(RX,YTDC+2)
LINE (RX-5,YTDC)-(XL1-5,YTDC)
LINE (RX-5,YCL)-(XCL-RC2-5,YCL)
RX=XCL-110
RY=YCL-2
GOSUB DARROW
RY=YTR+2
```

```
GOSUB UARROW
LINE (XCL-110,YCL-2)-(XCL-110,YTR+2)
LINE (RX-5,YTR)-(XTD2-5,YTR)
RX=XCL+120
RY=YCL-2
GOSUB DARROW
RY=YIN+2
GOSUB UARROW
LINE (XCL+120,YCL-2)-(XCL+120,YIN+2)
LINE(XPIP+5,YIN)-(XCL+125,YIN)
RX=XCL+180
RY=YCL-2
GOSUB DARROW
RY=YEX+2
GOSUB UARROW
LINE (XCL+180,YCL-2)-(XCL+180,YEX+2)
LINE (RX+5,YCL)-(XCL+RC2+5,YCL)
LINE(XPIP+5,YEX)-(XCL+185,YEX)
RX=XCL+150
RY=YCL-2
GOSUB DARROW
RY=YITDC+2
GOSUB UARROW
LINE (XCL+150,YCL-2)-(XCL+150,YITDC+2)
LINE(XPIP+5,YITDC)-(XCL+155,YITDC)
LOCATE 8,27:PRINT USING "###.#";PZTDC
LOCATE 16,30:PRINT USING "###.#";PZT
LOCATE 18,58:PRINT USING "###.#";PZIO
LOCATE 16,62:PRINT USING "###.#";PZIFO
LOCATE 14,66:PRINT USING "###.#";PZE
RETURN
INTERCEPT:
REM THIS GIVES THE X,Y VALUE OF A LINE SLOPE M THROUGH A
CIRCLE RADIUS R
REM LINE GOES THROUGH TARGET POINT XT,YT REFERRED TO
REM CIRCLE CENTRE AT 0,0
REM RR AND XM REFERS TO THE FLYWHEEL
IF DEG=0 THEN GOTO 799
IF DEG=180 THEN GOTO 799
IF DEG=360 THEN GOTO 799
IF DEG >180 THEN SIGN=-1
IF DEG >360 THEN GOTO 557
M=(YTT-YT)/(XTT-XT)
A=M*M+1
```

```
B=2*M*YT-2*M*M*XT
R=RF
C=M*M*XT*XT+YT*YT-2*M*XT*YT-R*R
XM=(-B-SIGN*SQR(B*B-4*A*C))/(2*A)
YM=M*(XM-XT)+YT
XP=(YP-YT)/M+XT
GOTO 800
799 YM=SQR(RF*RF-XT*XT)
XM=XT
XP=XT
800 RETURN
CRANKERASE:
LINE (EXQ1,EYQ1)-(EXQ2,EYQ2),30
LINE (EXQ3,EYQ3)-(EXQ4,EYQ4),30
LINE (EX5,EY5)-(EX6,EY6),30
LINE (EX7,EY7)-(EX8,EY8),30
CIRCLE(XCL,YCL),RF,30
CIRCLE(XCL,YCL),DQ,30
CIRCLE(EXB,EYB),RB,30
LINE(XT1,EYPT)-(XT2,EYPB),30,BF
CIRCLE(XCL,EYS),RS,30
LINE(XR1,EYPT+4)-(XR2,EYPT+4+RD),30,BF
LINE(XR1,EYPT+11)-(XR2,EYPT+11+RD),30,BF
RETURN
SUBPRINT:
CLS
LOCATE 1,1:PRINT"BORE,mm=";BO
LOCATE 2,1:PRINT"STROKE,mm=";ST
LOCATE 3,1:PRINT"CON-ROD,mm=";CRL
LOCATE 4,1:PRINT"EXHAUST OPENS,ºatdc=";EO
LOCATE 5,1:PRINT"TRANSFER OPENS,ºatdc=";TRO
LOCATE 6,1:PRINT"INLET OPENS,ºbtdc=";INO
LOCATE 7,1:PRINT "TRAP COMPRESSION RATIO=";TRAPCR
LOCATE 8,1:PRINT"SQUISH CLEARANCE,mm=";SQCL
LOCATE 9,1:PRINT"WRIST PIN TO CROWN,mm=";GPC
LOCATE 10,1:PRINT"WRIST PIN TO SKIRT,mm=";GPS
LOCATE 11,1:PRINT"SWEPT VOLUME,cm3="; USING "###.#";SV/1000
LOCATE 12,1:PRINT"TRAP SWEPT VOLUME,cm3="; USING
"###.#";TSV/1000
LOCATE 13,1:PRINT"CLEARANCE VOLUME,cm3="; USING
"###.#";CVOL/1000
GOSUB CYLDRAW
GOSUB CRANKPRINT
WINDOW 5
```

```
BUTTON 1,1,"RECALCULATE",(5,5)-(125,50),3
BUTTON 2,1,"PRINT",(125,5)-(200,50),3
BUTTON 3,1,"FILING DATA",(200,5)-(325,50),3
BUTTON 4,1,"FILE INPUT",(325,5)-(450,50),3
BUTTON 31,1,"QUIT",(450,5)-(550,50),3
ACTIVITY=DIALOG(0)
WHILE ACTIVITY<>1:ACTIVITY=DIALOG(0):WEND
BUTTONPUSHED=DIALOG(1)
IF BUTTONPUSHED=31 THEN GOTO 556
ON BUTTONPUSHED GOSUB RECALC,SUBLPRINT,FILING,FILED
RETURN
RECALC:
GOTO 110
RETURN
LARROW:
REM LEFTWARDS
LINE(RX,RY)-(RX+3,RY-3)
LINE(RX,RY)-(RX+3,RY+3)
RETURN
RARROW:
REM RIGHTWARDS
LINE(RX,RY)-(RX-3,RY-3)
LINE(RX,RY)-(RX-3,RY+3)
RETURN
UARROW:
REM UPWARDS
LINE(RX,RY)-(RX-3,RY+3)
LINE(RX,RY)-(RX+3,RY+3)
RETURN
DARROW:
REM DOWNWARDS
LINE(RX,RY)-(RX-3,RY-3)
LINE(RX,RY)-(RX+3,RY-3)
RETURN
SUBLPRINT:
INPUT"TYPE IN A TITLE FOR THE PRINT-OUT";TITLE$
CLS
OPEN"LPT1:PROMPT" FOR OUTPUT AS  #1
WINDOW OUTPUT#1
LOCATE 1,1:PRINT"BORE,mm=";BO
LOCATE 2,1:PRINT"STROKE,mm=";ST
LOCATE 3,1:PRINT"CON-ROD,mm=";CRL
LOCATE 4,1:PRINT"EXHAUST OPENS,ºatdc=";EO
LOCATE 5,1:PRINT"TRANSFER OPENS,ºatdc=";TRO
LOCATE 6,1:PRINT"INLET OPENS,ºbtdc=";INO
```

```
LOCATE 7,1:PRINT "TRAP COMPRESSION RATIO=";TRAPCR
LOCATE 8,1:PRINT"SQUISH CLEARANCE,mm=";SQCL
LOCATE 9,1:PRINT"WRIST PIN TO CROWN,mm=";GPC
LOCATE 10,1:PRINT"WRIST PIN TO SKIRT,mm=";GPS
LOCATE 11,1:PRINT"SWEPT VOLUME,cm3="; USING "###.#";SV/1000
LOCATE 12,1:PRINT"TRAP SWEPT VOLUME,cm3="; USING "###.#";TSV/1000
LOCATE 13,1:PRINT"CLEARANCE VOLUME,cm3="; USING "###.#";CVOL/1000
IF TITLE$="N" THEN GOTO 913
LOCATE 25,1:PRINT  TITLE$
913 GOSUB CYLDRAW
GOSUB CRANKPRINT
CLOSE#1
GOSUB SUBPRINT
RETURN
FILING:
INPUT "TYPE IN TITLE NEEDED FOR THE FILE WITH '.PDRAW' END CODE";P$
OPEN P$ FOR OUTPUT AS #1
WRITE#1,BO,ST,CRL,EO,TRO,INO,TRAPCR,SQCL,GPC,GPS
FINISH:
CLOSE#1
GOTO 130
RETURN
FILED:
INPUT "TYPE IN THE FILENAME WITH A '.PDRAW'";FIL$
OPEN"I",#1,FIL$
INPUT#1,BO,ST,CRL,EO,TRO,INO,TRAPCR,SQCL,GPC,GPS
CLOSE#1
RETURN
BORE:
WINDOW 4
CLS
INPUT"BORE";BO
RETURN
STROKE:
WINDOW 4
CLS
INPUT"STROKE";ST
RETURN
CONROD:
WINDOW 4
CLS
```

```
INPUT"CON-ROD";CRL
RETURN
EXTIMING:
WINDOW 4
CLS
PRINT"EXHAUST PORT"
INPUT"OPENS, DEG ATDC";EO
RETURN
INTIMING:
WINDOW 4
CLS
PRINT"INLET PORT"
INPUT"OPENS, DEG BTDC";INO
RETURN
SQUISH:
WINDOW 4
CLS
INPUT"SQUISH CLEARANCE";SQCL
RETURN
COMPRAT:
WINDOW 4
CLS
PRINT "TRAPPED"
INPUT"COMPRESSION RATIO";TRAPCR
RETURN
TRTIMING:
WINDOW 4
CLS
PRINT "TRANSFER PORT"
INPUT"OPENS, DEG ATDC";TRO
RETURN
PINCROWN:
WINDOW 4
CLS
PRINT"DISTANCE FROM GUDGEON PIN"
PRINT"TO PISTON CROWN"
INPUT"LENGTH,mm";GPC
RETURN
PINSKIRT:
WINDOW 4
CLS
PRINT"DISTANCE FROM GUDGEON PIN"
PRINT"TO PISTON SKIRT"
INPUT"LENGTH,mm";GPS
RETURN
```

```
CURRENTDATA:
PRINT "BORE=";BO
PRINT "STROKE=";ST
PRINT "CON-ROD=";CRL
PRINT "EXHAUST OPENS=";EO
PRINT "TRANSFER OPENS=";TRO
PRINT "INLET OPENS=";INO
PRINT "COMPRESSION RATIO=";TRAPCR
PRINT "SQUISH CLEARANCE=";SQCL
PRINT"WRIST PIN TO CROWN=";GPC
PRINT"WRIST PIN TO SKIRT=";GPS
RETURN
```

COMPUTER PROGRAM PROG.1.3

PROGRAM NAME "QUB CROSS ENGINE DRAW"

ProgList.1.3

```
REM ProgList.1.3
REM program QUB CROSS ENGINE DRAW
REM program for a QUB style cross scavenged two-stroke engine
REM click in any circle in the data box until all data inserted
REM after you have CLICKED in a box a third screen will come up and demand a DATA value
REM do not RUN the program until all data values inserted
REM the data can be filed on hard disk by clicking the FILING box
REM if you wish to use FILED data type a F when prompted
REM if you wish to use NEW data type a N when prompted
REM this form of data and program operation in CLICK boxes uses the MAC at high speed
REM the engine rotates for a complete engine cycle from tdc to tdc and then prints
REM the print-out on the printer is exactly as seen on the screen
REM do not forget to CLICK on "tall adjusted" on the printer demand or circles
REM will appear as ellipses!
REM many program statements are common for PROG.1.2 and PROG.1.3 so
REM a shrewd typist can use the MAC very speedily to load both programs
WINDOW 1
WINDOW 2,"CLICK BOX",(201,50)-(400,450),1
PRINT "SELECT BY BUTTON"
WINDOW 3,"CURRENT DATA",(10,50)-(200,450),1
WINDOW 4,"DATA INPUT",(400,50)-(600,450),1
WINDOW 5,"ACTION BOX",(10,430)-(600,480),1
WINDOW 6,"EXHAUST TYPE",(400,50)-(600,450),1
WINDOW 1
100 INPUT"NEW DATA OR FILE DATA (TYPE N OR F)?";D$
IF D$="N" THEN GOTO 110
```

```
IF D$="F" THEN GOSUB FILED
110 WINDOW 3
CLS
GOSUB CURRENTDATA
WINDOW 2
BUTTON 1,1,"BORE",(5,25)-(100,45),3
BUTTON 2,1,"STROKE",(5,45)-(100,65),3
BUTTON 3,1,"CON-ROD",(5,65)-(100,85),3
BUTTON 4,1,"EXHAUST TIMING",(5,85)-(200,105),3
BUTTON 5,1,"TRANSFER TIMING",(5,105)-(200,125),3
BUTTON 6,1,"INLET TIMING",(5,125)-(200,145),3
BUTTON 7,1,"COMPRESSION RATIO",(5,145)-(200,165),3
BUTTON 8,1,"SQUISH CLEARANCE",(5,165)-(200,185),3
BUTTON 9,1,"WRIST PIN TO CROWN",(5,185)-(200,205),3
BUTTON 10,1,"WRIST PIN TO SKIRT",(5,205)-(200,225),3
BUTTON 11,1,"DEFLECTOR HEIGHT",(5,225)-(200,245),3
BUTTON 30,1,"RUN",(5,325)-(100,350),3
BUTTON 31,1,"QUIT",(5,350)-(100,375),3
ACTIVITY=DIALOG(0)
WHILE ACTIVITY<>1:ACTIVITY=DIALOG(0):WEND
BUTTONPUSHED=DIALOG(1)
IF BUTTONPUSHED=30 THEN GOTO 130
IF BUTTONPUSHED=31 THEN GOTO 556
ON BUTTONPUSHED GOSUB BORE,STROKE,CONROD,EXTIMING,
TRTIMING,INTIMING,COMPRAT,SQUISH,PINCROWN,PINSKIRT,
DEFLCTR
GOTO 110
130 WINDOW 1
CLS
PI=3.1415927#
DTOR=PI/180
RTOD=180/PI
CRK=ST/2
DHEAD=.7*BO
SIGN=1
GOSUB CRANKFACT
PZTDC=CRK+CRL+GPC
PZTOP=PZTDC+DEFH
PZBDC=PZTDC-ST
DEG=EO
GOSUB PIST
PZE=PZTOP-HTDC
YEX=YTDC+HTDC*SCAL-DEFHS
SV=PI*BO*BO*ST/4
```

```
TSV=PI*BO*BO*HTDC/4
CVOL=TSV/(TRAPCR-1)
SQVOL=PI*BO*BO*SQCL/4
DEG=TRO
GOSUB PIST
PZT=PZTDC-HTDC
YTR=YTDC+HTDC*SCAL
DEG=INO
GOSUB PIST
PZIO=PZTDC-GPC-GPS-HTDC
PZIFO=PZTDC-GPC-GPS
YIN=YITDC+HTDC*SCAL
GOSUB CRANKFACT
GOSUB CYLDRAW
DEG=0
150 GOSUB PIST
GOSUB CRANKDRAW
DEG=DEG+2
GOTO 150
557 GOSUB SUBPRINT
556 END
CYLDRAW:
R=RC1
X=XCL-XC1
Y=SQR(R*R-X*X)
AC3=ATN(Y/X)
YC3=YCL-Y
X=XCL-XL1
Y=SQR(R*R-X*X)
AL2=ATN(Y/X)
YL2=YCL-Y
TRDEEP=.5*(YBDC-YTR)
XTD1=XL1-TRDEEP
XTD2=XTD1-LINER
X=XCL-XTD1
Y=SQR(R*R-X*X)
AT1=ATN(Y/X)
YT1=YCL-Y
R=RC2
X=XCL-XTD2
Y=SQR(R*R-X*X)
AT2=ATN(Y/X)
YT2=YCL-Y
```

```
X=XL2-XCL
Y=SQR(R*R-X*X)
AL4=ATN(Y/X)
YL4=YCL-Y
CIRCLE(XCL,YCL),RC1,,PI-AC3,PI-AL2
CIRCLE(XCL,YCL),RC1,,PI-AT1,AC3
CIRCLE(XCL,YCL),RC2,,PI-AT2,AL4
YD=YTR+TRDEEP
CIRCLE(XL1,YD),TRDEEP,,PI/2,PI
CIRCLE(XL1,YD),(TRDEEP+LINER),,PI/2,PI
CIRCLE(XC1,YBDC+LINER),LINER,,PI/2,PI
LINE(XC1,YTOP)-(XC1,YC3)
LINE(XL1,YTOP1)-(XL1,YTR-LINER)
LINE(XTD2,YD)-(XTD2,YT2)
LINE(XTD1,YD)-(XTD1,YT1)
LINE(XL1,YBDC+LINER)-(XL1,YL2)
LINE(XC1,YTR)-(XL1,YTR)
LINE(XC2,YTOP)-(XC2,YC3)
LINE(XL2,YTOP1)-(XL2,YEX)
LINE(XL2,YEXBDC)-(XL2,YITDC)
LINE(XL2,YIN)-(XL2,YL4)
XPIP=XL2+25
LINE(XC2,YEX)-(XPIP,YEX)
LINE(XC2,YEXBDC)-(XPIP,YEXBDC)
LINE(XC2,YITDC)-(XPIP,YITDC)
LINE(XC2,YIN)-(XPIP,YIN)
CIRCLE(XPIP,(YEX+YEXBDC)/2),(YEXBDC-YEX)/2,,,,2
CIRCLE(XPIP,(YIN+YITDC)/2),(YIN-YITDC)/2,,,,2
LINE(XC1,YTOP)-(XC2,YTOP)
LINE(XL1,YTOP1)-(XL2,YTOP1)
RETURN
PIST:
DR=DEG*DTOR
HTDC=CRK+CRL-CRK*COS(DR)-SQR(CRL*CRL-(CRK*SIN(DR))^2)
HBDC=ST-HTDC
RETURN
CRANKFACT:
CRT=45
BEYE=10
SEYE=.75*BEYE
SCAL=CRT/CRK
SQX=SQCL*SCAL
BOX=BO*SCAL
RB=CRT/3
```

```
RBC=RB+BEYE
RF=CRT+RBC
RS=.75*RB
WR=2*RB
CR=CRL*SCAL
DEFHS=DEFH*SCAL
RDEF=.5*DEFHS
QQ=4
DQ=20
PLU=GPC*SCAL
PLS=GPS*SCAL
PLD=PLU+DEFHS
PL=PLU+PLS
XCL=360
YCL=350
YTDC=YCL-CR-PLU-CRT
YBDC=YTDC+2*CRT
YEXBDC=YBDC-DEFHS
YITDC=YTDC+PL
YTOP=YTDC-SQX-DEFHS
XT1=XCL-BOX/2
XT2=XCL+BOX/2
XDT=XT1+.3*(XT2-XT1)
XC1=XT1-2
XC2=XT2+2
XR1=XC1+1
XR2=XC2-1
HRW=RB
RD=3
LINER=6
RC1=RF+6
RC2=RC1+LINER
XL1=XC1-LINER
XL2=XC2+LINER
YTOP1=YTOP-LINER
VALU=.35*BOX
RETURN
CRANKDRAW:
HTOP=HTDC*SCAL
HBOT=HBDC*SCAL
YPD=YTDC+HTOP-DEFHS
YPT=YTDC+HTOP
YDT=YPT-DEFHS
YPB=YPT+PL
```

```
YS=YPT+PLU
XB=XCL+CRT*SIN(DR)
YB=YCL-CRT*COS(DR)
CRZ=SQR(CR*CR-(XB-XCL)^2)
DRZ=ATN((XB-XCL)/CRZ)
XINC=HRW*COS(DRZ)
YINC=HRW*SIN(DRZ)
DYQ=DQ*COS(DR)
DXQ=DQ*SIN(DR)
XRD1=XCL-XINC
YRD1=YS+YINC
XRD4=XCL+XINC
YRD4=YS-YINC
XRD3=XB+XINC
YRD3=YB-YINC
XRD2=XB-XINC
YRD2=YB+YINC
YP=YCL-YPB
XT=XRD1-XCL
YT=YCL-YRD1
XTT=XRD2-XCL
YTT=YCL-YRD2
GOSUB INTERCEPT
X6=XCL+XM
Y6=YCL-YM
X5=XCL+XP
Y5=YPB
XT=XRD4-XCL
YT=YCL-YRD4
XTT=XRD3-XCL
YTT=YCL-YRD3
GOSUB INTERCEPT
X8=XCL+XM
Y8=YCL-YM
X7=XCL+XP
Y7=YPB
XQ1=DXQ+XCL
YQ1=YCL-DYQ
XQ2=-DXQ+XCL
YQ2=YCL+DYQ
XQ3=-DYQ+XCL
YQ3=YCL-DXQ
XQ4=DYQ+XCL
YQ4=YCL+DXQ
```

```
GOSUB CRANKERASE
LINE (XQ1,YQ1)-(XQ2,YQ2),33
LINE (XQ3,YQ3)-(XQ4,YQ4),33
LINE (X5,Y5)-(X6,Y6),33
LINE (X7,Y7)-(X8,Y8),33
CIRCLE(XCL,YCL),RF,33
CIRCLE(XCL,YCL),DQ,33
CIRCLE(XB,YB),RB,33
LINE(XT1,YPT+4)-(XT2,YPB),30,BF
LINE(XT1,YPT+4)-(XT2,YPB),,B
LINE(XT1,YPT+4)-(XT1,YPT),33
LINE(XT1,YPT)-(XDT-RDEF,YPT),33
LINE(XDT,YPT-RDEF)-(XDT,YDT),33
LINE(XDT,YDT)-(XT2,YDT),33
LINE(XT2,YDT)-(XT2,YPT+4),33
CIRCLE(XDT-RDEF,YPT-RDEF),RDEF,33,3*PI/2,2*PI
CIRCLE(XCL,YS),RS,33
LINE(XR1,YPT+4)-(XR2,YPT+4+RD),33,BF
LINE(XR1,YPT+11)-(XR2,YPT+11+RD),33,BF
EXQ1=XQ1
EYQ1=YQ1
EXQ2=XQ2
EYQ2=YQ2
EXQ3=XQ3
EYQ3=YQ3
EXQ4=XQ4
EYQ4=YQ4
EX5=X5
EY5=Y5
EX6=X6
EY6=Y6
EX7=X7
EY7=Y7
EX8=X8
EY8=Y8
EYB=YB
EXB=XB
EYPT=YPT
EYPB=YPB
EYDT=YDT
EYS=YS
RETURN
CRANKPRINT:
LINE (XQ1,YQ1)-(XQ2,YQ2),33
```

```
LINE (XQ3,YQ3)-(XQ4,YQ4),33
LINE (X5,Y5)-(X6,Y6),33
LINE (X7,Y7)-(X8,Y8),33
CIRCLE(XCL,YCL),RF,33
CIRCLE(XCL,YCL),DQ,33
CIRCLE(XB,YB),RB,33
LINE(XT1,YPT+4)-(XT2,YPB),30,BF
LINE(XT1,YPT+4)-(XT2,YPB),,B
LINE(XT1,YPT+4)-(XT1,YPT),33
LINE(XT1,YPT)-(XDT-RDEF,YPT),33
LINE(XDT,YPT-RDEF)-(XDT,YDT),33
LINE(XDT,YDT)-(XT2,YDT),33
LINE(XT2,YDT)-(XT2,YPT+4),33
CIRCLE(XDT-RDEF,YPT-RDEF),RDEF,33,3*PI/2,2*PI
CIRCLE(XCL,YS),RS,33
LINE(XR1,YPT+4)-(XR2,YPT+4+RD),33,BF
LINE(XR1,YPT+11)-(XR2,YPT+11+RD),33,BF
RX=XCL-140
RY=YCL-2
GOSUB DARROW
RY=YTDC+2
GOSUB UARROW
LINE (RX,YCL-2)-(RX,YTDC+2)
LINE (RX-5,YTDC)-(XL1-5,YTDC)
LINE (RX-5,YCL)-(XCL-RC2-5,YCL)
RX=XCL-110
RY=YCL-2
GOSUB DARROW
RY=YTR+2
GOSUB UARROW
LINE (XCL-110,YCL-2)-(XCL-110,YTR+2)
LINE (RX-5,YTR)-(XTD2-5,YTR)
RX=XCL+120
RY=YCL-2
GOSUB DARROW
RY=YIN+2
GOSUB UARROW
LINE (XCL+120,YCL-2)-(XCL+120,YIN+2)
LINE(XPIP+5,YIN)-(XCL+125,YIN)
RX=XCL+180
RY=YCL-2
GOSUB DARROW
RY=YEX+2
GOSUB UARROW
```

```
LINE (XCL+180,YCL-2)-(XCL+180,YEX+2)
LINE (RX+5,YCL)-(XCL+RC2+5,YCL)
LINE(XPIP+5,YEX)-(XCL+185,YEX)
RX=XCL+150
RY=YCL-2
GOSUB DARROW
RY=YITDC+2
GOSUB UARROW
LINE (XCL+150,YCL-2)-(XCL+150,YITDC+2)
LINE(XPIP+5,YITDC)-(XCL+155,YITDC)
LOCATE 18,27:PRINT USING "###.#";PZTDC
LOCATE 16,30:PRINT USING "###.#";PZT
LOCATE 18,58:PRINT USING "###.#";PZIO
LOCATE 16,62:PRINT USING "###.#";PZIFO
LOCATE 14,66:PRINT USING "###.#";PZE
RETURN
INTERCEPT:
REM THIS GIVES THE X,Y VALUE OF A LINE SLOPE M THROUGH A
CIRCLE RADIUS R
REM LINE GOES THROUGH TARGET POINT XT,YT REFERRED TO
REM CIRCLE CENTRE AT 0,0
REM RR AND XM REFERS TO THE FLYWHEEL
IF DEG=0 THEN GOTO 799
IF DEG=180 THEN GOTO 799
IF DEG=360 THEN GOTO 799
IF DEG >180 THEN SIGN=-1
IF DEG >360 THEN GOTO 557
M=(YTT-YT)/(XTT-XT)
A=M*M+1
B=2*M*YT-2*M*M*XT
R=RF
C=M*M*XT*XT+YT*YT-2*M*XT*YT-R*R
XM=(-B-SIGN*SQR(B*B-4*A*C))/(2*A)
YM=M*(XM-XT)+YT
XP=(YP-YT)/M+XT
GOTO 800
799 YM=SQR(RF*RF-XT*XT)
XM=XT
XP=XT
800 RETURN
CRANKERASE:
LINE (EXQ1,EYQ1)-(EXQ2,EYQ2),30
LINE (EXQ3,EYQ3)-(EXQ4,EYQ4),30
LINE (EX5,EY5)-(EX6,EY6),30
```

```
LINE (EX7,EY7)-(EX8,EY8),30
CIRCLE(XCL,YCL),RF,30
CIRCLE(XCL,YCL),DQ,30
CIRCLE(EXB,EYB),RB,30
LINE(XT1,EYPT+4)-(XT2,EYPB),30,BF
LINE(XT1,EYPT+4)-(XT1,EYPT),30
LINE(XT1,EYPT)-(XDT-RDEF,EYPT),30
LINE(XDT,EYPT-RDEF)-(XDT,EYDT),30
LINE(XDT,EYDT)-(XT2,EYDT),30
LINE(XT2,EYDT)-(XT2,EYPT+4),30
CIRCLE(XDT-RDEF,EYPT-RDEF),RDEF,30,3*PI/2,2*PI
CIRCLE(XCL,EYS),RS,30
LINE(XR1,EYPT+4)-(XR2,EYPT+4+RD),30,BF
LINE(XR1,EYPT+11)-(XR2,EYPT+11+RD),30,BF
RETURN
SUBPRINT:
CLS
LOCATE 1,1:PRINT"BORE,mm=";BO
LOCATE 2,1:PRINT"STROKE,mm=";ST
LOCATE 3,1:PRINT"CON-ROD,mm=";CRL
LOCATE 4,1:PRINT"EXHAUST OPENS,ºatdc=";EO
LOCATE 5,1:PRINT"TRANSFER OPENS,ºatdc=";TRO
LOCATE 6,1:PRINT"INLET OPENS,ºbtdc=";INO
LOCATE 7,1:PRINT "TRAP COMPRESSION RATIO=";TRAPCR
LOCATE 8,1:PRINT"SQUISH CLEARANCE,mm=";SQCL
LOCATE 9,1:PRINT"WRIST PIN-CROWN,mm=";GPC
LOCATE 10,1:PRINT"WRIST PIN-SKIRT,mm=";GPS
LOCATE 11,1:PRINT"DEFLECTOR HEIGHT,mm=";DEFH
LOCATE 12,1:PRINT"SWEPT VOLUME,cm3="; USING "###.#";SV/1000
LOCATE 13,1:PRINT"TRAP SWEPT VOLUME,cm3="; USING
"###.#";TSV/1000
LOCATE 14,1:PRINT"CLEARANCE VOLUME,cm3="; USING
"###.#";CVOL/1000
GOSUB CYLDRAW
GOSUB CRANKPRINT
666 WINDOW 5
BUTTON 1,1,"RECALCULATE",(5,5)-(125,50),3
BUTTON 2,1,"PRINT",(125,5)-(200,50),3
BUTTON 3,1,"FILING DATA",(200,5)-(325,50),3
BUTTON 4,1,"FILE INPUT",(325,5)-(450,50),3
BUTTON 31,1,"QUIT",(450,5)-(550,50),3
ACTIVITY=DIALOG(0)
WHILE ACTIVITY<>1:ACTIVITY=DIALOG(0):WEND
BUTTONPUSHED=DIALOG(1)
```

```
IF BUTTONPUSHED=31 THEN GOTO 556
ON BUTTONPUSHED GOSUB RECALC,SUBLPRINT,FILING,FILED
GOTO 666
RETURN
RECALC:
GOTO 110
RETURN
LARROW:
REM LEFTWARDS
LINE(RX,RY)-(RX+3,RY-3)
LINE(RX,RY)-(RX+3,RY+3)
RETURN
RARROW:
REM RIGHTWARDS
LINE(RX,RY)-(RX-3,RY-3)
LINE(RX,RY)-(RX-3,RY+3)
RETURN
UARROW:
REM UPWARDS
LINE(RX,RY)-(RX-3,RY+3)
LINE(RX,RY)-(RX+3,RY+3)
RETURN
DARROW:
REM DOWNWARDS
LINE(RX,RY)-(RX-3,RY-3)
LINE(RX,RY)-(RX+3,RY-3)
RETURN
SUBLPRINT:
INPUT"TYPE IN A TITLE FOR THE PRINT-OUT";TITLE$
CLS
OPEN"LPT1:PROMPT" FOR OUTPUT AS  #1
WINDOW OUTPUT#1
LOCATE 1,1:PRINT"BORE,mm=";BO
LOCATE 2,1:PRINT"STROKE,mm=";ST
LOCATE 3,1:PRINT"CON-ROD,mm=";CRL
LOCATE 4,1:PRINT"EXHAUST OPENS,ºatdc=";EO
LOCATE 5,1:PRINT"TRANSFER OPENS,ºatdc=";TRO
LOCATE 6,1:PRINT"INLET OPENS,ºbtdc=";INO
LOCATE 7,1:PRINT "TRAP COMPRESSION RATIO=";TRAPCR
LOCATE 8,1:PRINT"SQUISH CLEARANCE,mm=";SQCL
LOCATE 9,1:PRINT"WRIST PIN-CROWN,mm=";GPC
LOCATE 10,1:PRINT"WRIST PIN-SKIRT,mm=";GPS
LOCATE 11,1:PRINT"DEFLECTOR HEIGHT,mm=";DEFH
```

```
LOCATE 12,1:PRINT"SWEPT VOLUME,cm3="; USING "###.#";SV/1000
LOCATE 13,1:PRINT"TRAP SWEPT VOLUME,cm3="; USING "###.#";TSV/
1000
LOCATE 14,1:PRINT"CLEARANCE VOLUME,cm3="; USING
"###.#";CVOL/1000
IF TITLE$="N" THEN GOTO 913
LOCATE 25,1:PRINT TITLE$
913 GOSUB CYLDRAW
GOSUB CRANKPRINT
CLOSE#1
RETURN
FILING:
INPUT "TYPE IN TITLE NEEDED FOR THE FILE WITH '.XDRAW' END
CODE";P$
OPEN P$ FOR OUTPUT AS #1
WRITE#1,BO,ST,CRL,EO,TRO,INO,TRAPCR,SQCL,GPC,GPS,DEFH
FINISH:
CLOSE#1
GOTO 130
RETURN
FILED:
INPUT "TYPE IN THE FILENAME WITH A '.XDRAW' END CODE";FIL$
OPEN"I",#1,FIL$
INPUT#1,BO,ST,CRL,EO,TRO,INO,TRAPCR,SQCL,GPC,GPS,DEFH
CLOSE#1
RETURN
BORE:
WINDOW 4
CLS
INPUT"BORE";BO
RETURN
STROKE:
WINDOW 4
CLS
INPUT"STROKE";ST
RETURN
CONROD:
WINDOW 4
CLS
INPUT"CON-ROD";CRL
RETURN
EXTIMING:
WINDOW 4
CLS
```

```
PRINT"EXHAUST PORT"
INPUT"OPENS, DEG ATDC";EO
RETURN
INTIMING:
WINDOW 4
CLS
PRINT"INLET PORT"
INPUT"OPENS, DEG BTDC";INO
RETURN
SQUISH:
WINDOW 4
CLS
INPUT"SQUISH CLEARANCE";SQCL
RETURN
COMPRAT:
WINDOW 4
CLS
PRINT "TRAPPED"
INPUT"COMPRESSION RATIO";TRAPCR
RETURN
TRTIMING:
WINDOW 4
CLS
PRINT "TRANSFER PORT"
INPUT"OPENS, DEG ATDC";TRO
RETURN
PINCROWN:
WINDOW 4
CLS
PRINT"DISTANCE FROM GUDGEON PIN"
PRINT"TO PISTON CROWN"
INPUT"LENGTH,mm";GPC
RETURN
PINSKIRT:
WINDOW 4
CLS
PRINT"DISTANCE FROM GUDGEON PIN"
PRINT"TO PISTON SKIRT"
INPUT"LENGTH,mm";GPS
RETURN
DEFLCTR:
WINDOW 4
CLS
```

```
PRINT"DISTANCE FROM GUDGEON PIN"
PRINT"TO DEFLECTOR TOP"
INPUT"LENGTH,mm";DEFH
RETURN
CURRENTDATA:
PRINT "BORE=";BO
PRINT "STROKE=";ST
PRINT "CON-ROD=";CRL
PRINT "EXHAUST OPENS=";EO
PRINT "TRANSFER OPENS=";TRO
PRINT "INLET OPENS=";INO
PRINT "COMPRESSION RATIO=";TRAPCR
PRINT "SQUISH CLEARANCE=";SQCL
PRINT"WRIST PIN-CROWN=";GPC
PRINT"WRIST PIN-SKIRT=";GPS
PRINT "DEFLECTOR HEIGHT=";DEFH
RETURN
```

COMPUTER PROGRAM PROG.1.4
PROGRAM NAME "EXHAUST GAS ANALYSIS"
ProgList.1.4

```
REM Prog.1.4 EXHAUST GAS ANALYSIS
REM ProgList.1.4
REM This program is intended for use with an engine test analysis program
REM and so the program formulation is for attachment to some major data analysis
REM So this program is intended for information or for direct use
100 INPUT"RUNNING THE PROGRAM(Y OR N?)";ANS$
IF ANS$="N" THEN END
PRINT "In this calculation the molecular weight of the (dry) exhaust gas, MWEX,"
PRINT"is assumed to be 29 kg/kgmol."
MWEX=29
PRINT "It is assumed that the oxides of nitrogen emission is recorded as NO equivalent"
PRINT "The sensitivity factor K , see Fig.1.19, between calculated brake specific hydrocarbon emission values"
PRINT "which have been measured by either NDIR or FID systems is assumed to be unity."
PRINT "The trapping efficiency value is meaningful only where the combustion process"
PRINT "has consumed all oxygen present and the exhaust oxygen value"
PRINT "is from scavenge air only."
PRINT
```

```
INPUT"enter air to fuel ratio, AF";AF
INPUT"enter brake specific fuel consumption, BSFC, as kg/kWh";BSFC
INPUT"enter oxygen concentration in the exhaust gas as % vol, O2VOL";O2VOL
INPUT"enter carbon monoxide emission as % vol, COVOL";COVOL
INPUT"enter carbon dioxide emission as % vol, CO2VOL";CO2VOL
INPUT"enter oxides of nitrogen emission as ppm, NO equivalent";NOX
PRINT "the unburned hydrocarbon emission values will have been measured
as-"
PRINT "either HC ppm by a NDIR system as hexane equivalent, C6H14, "
PRINT "or as HC ppm by a FID system as methane equivalent, CH4"
INPUT"type in the name of the type of measurement system, either 'NDIR' or
'FID'";EM$
IF EM$="NDIR" THEN INPUT"enter the HC as ppm measured by NDIR
system";HCNDIR
IF EM$="FID" THEN INPUT"enter the HC as ppm measured by FID
system";HCFID
PRINT
MWHEX=86
MWMETHANE=16
MWO2=32
MWNO=30
MWCO=28
MWCO2=44
BSHCNDIR=(1+AF)*BSFC*HCNDIR*MWHEX/(MWEX*1000)
BSHCFID=(1+AF)*BSFC*HCFID*MWMETHANE/(MWEX*1000)
BSNOX=(1+AF)*BSFC*NOX*MWNO/(MWEX*1000)
BSCO=(1+AF)*BSFC*COVOL*MWCO/(MWEX/10)
BSCO2=(1+AF)*BSFC*CO2VOL*MWCO2/(MWEX/10)
BSO2=(1+AF)*BSFC*O2VOL*MWO2/(MWEX/10)
TRAPEFF=100-((1+AF)*O2VOL/(.21*AF))
BSAC=AF*BSFC
PRINT
PRINT "OUTPUT DATA"
PRINT "The brake specific air consumption, BSAC, in kg/kWh units,
is";USING"####.#";BSAC
PRINT "The brake specific emission values printed are in g/kWh units"
PRINT "The brake specific carbon monoxide value, BSCO,
is";USING"####.#";BSCO
PRINT "The brake specific nitrogen oxide value, BSNOX,
is";USING"####.#";BSNOX
PRINT "The brake specific carbon dioxide value, BSCO2,
is";USING"####.#";BSCO2
PRINT "The brake specific oxygen value, BSO2, is";USING"####.#";BSO2
IF EM$="NDIR" THEN PRINT "The brake specific hydrocarbon value by
NDIR system, BSHC, is";USING"####.#"; BSHCNDIR
```

```
IF EM$="FID" THEN PRINT "The brake specific hydrocarbon value by FID,
BSHC, is";USING"####.#";BSHCFID
PRINT "The Trapping Efficiency, TE , as %, is";USING"####.#";TRAPEFF
INPUT"WANT A PRINT-OUT(Y OR N?)";ANS$
IF ANS$="Y" THEN GOSUB SUBLINEPRINTER
IF ANS$="N" THEN GOTO 100
END
SUBLINEPRINTER:
INPUT "WANT TO NAME THE DATA(Y OR N?)";A$
IF A$="Y" THEN INPUT HANDLE$
LPRINT HANDLE$
LPRINT"INPUT DATA"
LPRINT"air to fuel ratio, AF";AF
LPRINT"brake specific fuel consumption, BSFC, as kg/kWh";BSFC
LPRINT"oxygen concentration in the exhaust gas as % vol, O2VOL";O2VOL
LPRINT"carbon monoxide emission as % vol, COVOL";COVOL
LPRINT"carbon dioxide emission as % vol, CO2VOL";CO2VOL
LPRINT"oxides of nitrogen emission as ppm, NOX";NOX
IF EM$="NDIR" THEN LPRINT"the HC as ppm measured by NDIR
system";HCNDIR
IF EM$="FID" THEN LPRINT"the HC as ppm measured by FID system";HCFID
LPRINT
LPRINT "OUTPUT DATA"
LPRINT "The brake specific air consumption, BSAC, in kg/kWh units,
is";USING"####.#";BSAC
LPRINT "The brake specific emission values printed are in g/kWh units"
LPRINT "The brake specific carbon monoxide value, BSCO,
is";USING"####.#";BSCO
LPRINT "The brake specific nitrogen oxide value, BSNOX,
is";USING"####.#";BSNOX
LPRINT "The brake specific carbon dioxide value, BSCO2,
is";USING"####.#";BSCO2
LPRINT "The brake specific oxygen value, BSO2, is";USING"####.#";BSO2
IF EM$="NDIR" THEN LPRINT "The brake specific hydrocarbon value by
NDIR system, BSHC, is";USING"####.#";BSHCNDIR
IF EM$="FID" THEN LPRINT "The brake specific hydrocarbon value by FID,
BSHC, is";USING"####.#";BSHCFID
LPRINT "The Trapping Efficiency, TE , as %, is";USING"####.#";TRAPEFF
LPRINT
LPRINT
GOTO 100
RETURN
```

ProgList2.0 Program listings from Chapter 2

COMPUTER DATA BANK 2.0
DATA FILE "UFLO"
This File is an array ABX(I,J) of 21 by 41 elements
This file contains the sonic and subsonic solution for inflow and outflow for unsteady flow where g=1.4.
The following banks of 41 numbers are the J array elements and each bank in turn is an I set, starting at 1 and going up to 21
The AK value at the start of each bank is for information only and is not part of the data set.
The ABX(I,J) is the numeric value of (A1+B1)/XC in Fig.2.7.
The ABX(I=1,J) value is BX(J) in the program and is the value of B1/XC shown in Fig.2.7.
The AK, or port to pipe area ratio, are the numbers in front of each bank below.

AK=0 and BX(1,1 to 41)
.6,.62,.64,.66,.68
.7,.72,.74,.76,.78
.8,.82,.84,.86,.88
.9,.92,.94,.96,.98
1,1.02,1.04,1.06,1.08
1.1,1.12,1.14,1.16,1.18
1.2,1.22,1.24,1.26,1.28
1.3,1.32,1.34,1.36,1.38
1.4

AK=.05
.68173,.69321,.70526,.7179,.73114
.74498,.75942,.77443,.79,.80608
.82264,.83963,.85702,.87475,.89279
.9111,.92961,.94807,.96635,.98425
1,1.01626,1.03497,1.05426,1.07387
1.09367,1.11355,1.13344,1.15332,1.17321
1.19309,1.21298,1.23286,1.25275,1.27263
1.29252,1.3124,1.33229,1.35217,1.37206
1.39194

AK=.1
.71418,.72447,.73519,.74636,.75803
.77021,.78292,.79617,.80997,.82431
.83919,.85457,.87043,.88675,.90349
.92061,.93799,.95516,.97189,.98777

1,1.01315,1.0304,1.04882,1.0679
1.08741,1.10716,1.12693,1.1467,1.16647
1.18624,1.20601,1.22578,1.24556,1.26533
1.2851,1.30487,1.32464,1.34441,1.36418
1.38395

AK=.15
.73547,.74528,.75542,.76595,.77688
.78825,.80008,.81239,.82521,.83854
.85238,.86673,.88159,.89692,.91272
.92896,.94542,.96143,.97671,.99062
1,1.01061,1.02628,1.04369,1.06213
1.08124,1.1008,1.12046,1.14012,1.15977
1.17943,1.19909,1.21875,1.2384,1.25806
1.27772,1.29737,1.31703,1.33669,1.35634
1.376

AK=.2
.75146,.76102,.77086,.78102,.79153
.80241,.8137,.82543,.83761,.85027
.86341,.87705,.89117,.90579,.92088
.93642,.95207,.96699,.98084,.99287
1,1.00855,1.02261,1.0389,1.05656
1.07517,1.09446,1.114,1.13355,1.15309
1.17263,1.19218,1.21172,1.23126,1.25081
1.27035,1.2899,1.30944,1.32898,1.34853
1.36807

AK=.25
.7643,.77372,.78338,.79332,.80355
.81412,.82505,.83638,.84811,.86029
.87293,.88603,.8996,.91366,.92819
.94317,.95805,.9719,.98434,.99458
1,1.0069,1.01937,1.03444,1.05122
1.06922,1.08813,1.10754,1.12697,1.1464
1.16583,1.18526,1.20469,1.22412,1.24355
1.26298,1.28241,1.30184,1.32127,1.3407
1.36013

AK=.3
.77504,.78438,.79394,.80372,.81377
.82412,.8348,.84583,.85724,.86906
.88131,.894,.90714,.92075,.93482
.94935,.96343,.97619,.98724,.99586
1,1.00557,1.01652,1.03032,1.04612

1.06341,1.08184,1.10105,1.12036,1.13968
1.159,1.17831,1.19763,1.21695,1.23626
1.25558,1.2749,1.29421,1.31353,1.33285
1.35216

AK=.35
78427,.79358,.80306,.81275,.82267
.83286,.84335,.85416,.86532,.87686
.88881,.90117,.91396,.92721,.9409
.955,.96824,.97988,.98959,.9968
1,1.00451,1.01403,1.02654,1.04127
1.05775,1.0756,1.0945,1.1137,1.13291
1.15211,1.17131,1.19051,1.20971,1.22892
1.24812,1.26732,1.28652,1.30572,1.32492
1.34413

AK=.4
.79238,.80167,.81111,.82073,.83056
.84063,.85097,.86162,.87258,.8839
.8956,.90769,.9202,.93314,.94652
.96016,.97248,.98298,.99145,.99747
1,1.00365,1.01187,1.02308,1.03668
1.05224,1.06942,1.08788,1.10697,1.12605
1.14514,1.16422,1.18331,1.2024,1.22148
1.24057,1.25965,1.27874,1.29782,1.31691
1.33599

AK=.45
.79959,.80888,.8183,.82787,.83764
.84763,.85786,.86836,.87917,.89031
.90181,.91368,.92595,.93863,.95174
.96484,.97616,.98553,.99288,.99796
1,1.00296,1.00998,1.01993,1.03235
1.0469,1.06329,1.08121,1.10012,1.11908
1.13805,1.15702,1.17599,1.19496,1.21392
1.23289,1.25186,1.27083,1.28979,1.30876
1.32773

AK=.50
.8061,.81539,.8248,.83435,.84407
.85399,.86413,.87453,.88522,.89621
.90753,.91922,.93128,.94375,.95662
.96899,.97926,.98756,.99397,.9983
1,1.0024,1.00835,1.01707,1.02827
1.04174,1.05724,1.0745,1.09312,1.11197

1.13082,1.14966,1.16851,1.18736,1.2062
1.22505,1.2439,1.26274,1.28159,1.30044
1.31929

AK=.55
.81201,.82132,.83073,.84027,.84996
.85982,.8699,.88021,.89079,.90166
.91285,.92437,.93626,.94853,.9612
.97258,.98178,.98914,.99478,.99855
1,1.00193,1.00693,1.01448,1.02445
1.03675,1.05126,1.06775,1.08594,1.10466
1.12338,1.14211,1.16083,1.17955,1.19827
1.217,1.23572,1.25444,1.27317,1.29189
1.31061

AK=.60
.81744,.82677,.83619,.84572,.85538
.86521,.87524,.88548,.89597,.90674
.9178,.92919,.94093,.95304,.96532
.97555,.98375,.99033,.99537,.99873
1,1.00155,1.00569,1.01215,1.02089
1.03195,1.04534,1.06094,1.07851,1.0971
1.11569,1.13429,1.15288,1.17148,1.19007
1.20867,1.22726,1.24586,1.26445,1.28305
1.30164

AK=.65
.82244,.83181,.84124,.85077,.86042
.87022,.8802,.89039,.90081,.91149
.92245,.93372,.94533,.95729,.96881
.97789,.98525,.99121,.9958,.99885
1,1.00123,1.00462,1.01003,1.01756
1.02734,1.0395,1.05406,1.07086,1.08922
1.10768,1.12614,1.1446,1.16306,1.18153
1.19999,1.21845,1.23691,1.25537,1.27383
1.29229

AK=.70
.82709,.83648,.84593,.85547,.86512
.8749,.88485,.89499,.90535,.91595
.92682,.93799,.94948,.96131,.97156
.97964,.98634,.99184,.9961,.99894
1,1.00096,1.00368,1.00813,1.01446
1.02291,1.03373,1.04709,1.06295,1.08092

```
1.09924,1.11756,1.13588,1.1542,1.17252
1.19084,1.20916,1.22748,1.2458,1.26412
1.28245

AK=.75
.83143,.84085,.85032,.85987,.86952
.87929,.88921,.89931,.90962,.92016
.93095,.94203,.95342,.96472,.97355
.98088,.9871,.99228,.99631,.999
1,1.00073,1.00286,1.00641,1.01157
1.01866,1.02803,1.03999,1.05472,1.07206
1.09023,1.1084,1.12657,1.14474,1.16291
1.18108,1.19925,1.21742,1.23559,1.25376
1.27193

AK=.80
.83549,.84495,.85444,.864,.87365
.88342,.89332,.90339,.91365,.92414
.93487,.94587,.95716,.96703,.97489
.98171,.98762,.99258,.99645,.99904
1,1.00054,1.00213,1.00485,1.0089
1.01459,1.02237,1.03272,1.04604,1.06243
1.08043,1.09844,1.11645,1.13446,1.15246
1.17047,1.18848,1.20649,1.22449,1.2425
1.26051

AK=.85
83931,.8488,.85832,.8679,.87756
.88732,.89721,.90725,.91748,.92792
.93859,.94952,.96009,.9684,.97571
.98224,.98795,.99277,.99655,.99906
1,1.00037,1.0015,1.00345,1.00642
1.0107,1.01676,1.02523,1.03676,1.0518
1.06949,1.08731,1.10513,1.12296,1.14079
1.15861,1.17644,1.19426,1.21208,1.22991
1.24774

AK=.90
.84291,.85244,.86199,.87159,.88125
.89101,.90089,.91092,.92112,.93151
.94213,.95281,.96141,.96912,.97618
.98254,.98815,.99289,.9966,.99908
1,1.00023,1.00094,1.00218,1.00411
1.00697,1.01118,1.0174,1.02656,1.03966
```

1.05667,1.07428,1.09189,1.10951,1.12712
1.14473,1.16234,1.17995,1.19756,1.21517
1.23278

AK=.95
.84632,.85589,.86547,.87508,.88476
.89452,.9044,.91441,.92458,.93494
.94519,.95384,.96192,.96945,.9764
.98269,.98825,.99294,.99664,.99908
1,1.00011,1.00044,1.00104,1.00198
1.00341,1.0056,1.00908,1.01485,1.02464
1.04013,1.05747,1.0748,1.09214,1.10947
1.12681,1.14414,1.16148,1.17882,1.19615
1.21349

AK=1
.84891,.85852,.86812,.87776,.88744
.89721,.90708,.91708,.92724,.93663
.94554,.95403,.96204,.96954,.97646
.98273,.98827,.99296,.99664,.99909
1,1.00002,1.00008,1.00012,1.00038
1.00067,1.00112,1.0019,1.0034,1.00704
1.01807,1.03504,1.05201,1.06898,1.08595
1.10291,1.11988,1.13685,1.15382,1.17078
1.18775

COMPUTER PROGRAM PROG.2.1
PROGRAM NAME "EXHAUST"
ProgList.2.1

```
REM this is Prog.2.1
REM this is ProgList.2.1
REM this program includes subroutines common with Prog.2.2 "INDUCTION"
REM these subroutines are listed at the end of Proglists for Chapter 2
DIM ABX(21,45),BX(45),AKA(45)
DIM AE(50),BE(50),FE(50),DFE(50),DF(50),F(50)
DIM PS(50),PR(50),PL(50),MN(50)
DIM XAM(50),YAM(50),A2(50),B2(50)
DIM XM(50),YM(50),MM(50)
DIM XCM(50),YCM(50),XDM(50),YDM(50)
DIM XG(600),YGPC(600),YGPP(600)
OPEN"I",#1,"UFLO"
FOR I=1 TO 21 STEP 1
FOR J=1 TO 41 STEP 1
INPUT#1,ABX(I,J)
```

```
NEXT J
NEXT I
FINISH:
CLOSE #1
FOR J=1 TO 41
BX(J)=ABX(1,J)
NEXT J
AKA(1)=0
FOR I=2 TO 21
AKA(I)=AKA(I-1)+.05
NEXT I
50 INPUT "RESPONSE FOR NEW DATA IS 'N' AND FILED DATA IS 'F'";H$
IF H$="F" THEN GOSUB FILED
IF H$="F" THEN GOTO 502
CLS
GOSUB DATASET
INPUT"NEW DATA FOR CYLINDER VOLUME(Y/N)?";A$
IF A$="Y" THEN INPUT"TYPE IN CYLINDER VOLUME,CM3";VCY
INPUT"NEW DATA FOR INITIAL CYLINDER PRESSURE(Y/N)?";A$
IF A$="Y" THEN INPUT"TYPE IN INITIAL CYLINDER PRESSURE,BAR";PCY
INPUT"NEW DATA FOR INITIAL CYLINDER TEMPERATURE(Y/N)?";A$
IF A$="Y" THEN INPUT"TYPE IN INITIAL CYLINDER TEMPERATURE,DEG C";TCY
INPUT"NEW DATA FOR CYLINDER/EXHAUST CYCLING RATE(Y/N)?";A$
IF A$="Y" THEN INPUT"TYPE IN CYLINDER/EXHAUST CYCLING RATE,CYCLES/MIN";RPM
INPUT"NEW DATA FOR EXHAUST PIPE DIAMETER(Y/N)?";A$
IF A$="Y" THEN INPUT"TYPE IN EXHAUST PIPE DIAMETER,MM";DEP
INPUT"NEW DATA FOR EXHAUST PORT MAXIMUM DIAMETER(Y/N)?";A$
IF A$="Y" THEN INPUT"TYPE IN EXHAUST PORT MAXIMUM DIAMETER,MM";DPP
INPUT"NEW DATA FOR EXHAUST PORT OPENING PERIOD(Y/N)?";A$
IF A$="Y" THEN INPUT"TYPE IN EXHAUST PORT OPENING PERIOD,DEG.CRANKSHAFT";DEO
INPUT"NEW DATA FOR EXHAUST PIPE LENGTH(Y/N)?";A$
IF A$="Y" THEN INPUT"TYPE IN EXHAUST PIPE LENGTH,MM";LEP
INPUT"NEW DATA FOR TRIP ANGLE(Y/N)?";A$
IF A$="Y" THEN INPUT"TYPE IN TRIP ANGLE";DEGTRIP
INPUT"NEW DATA FOR TRIP CYCLE(Y/N)?";A$
IF A$="Y" THEN INPUT"TYPE IN TRIP CYCLE";CYCTRIP
502 CLS
```

```
GOSUB DATASET
INPUT"TYPE 'RUN' FOR CALCULATION OR 'NEW' FOR A DATA
CHANGE";Z$
IF Z$="NEW" THEN GOTO 50
CLS
CYC=1
NN=0
PI=3.141593
GE=1.4
GK=287
CCP=1005
CCV=718
GR=GE*GK
GF1=(GE-1)/GE
TC1=TCY+273
PA=101325!
TA=293
PC1=PCY*100000!
TREF=TC1*(PA/PC1)^GF1
TEXC=TREF-273
DREF=PA/(GK*TREF)
SDE=SQR(GR*TREF)
DAIR=PA/(GE*TA)
SDA=SQR(GR*TA)
VC1=VCY/1000000!
MC1=PC1*VC1/(GK*TC1)
MCYL=MC1
PCYL=PC1
TCYL=TC1
DEG=0
CALL NINT(KE,LEP,TREF,TREF)
KE1=1
A=1000
AKM=(DPP/DEP)^2
DEPM=DEP/A
LEM=LEP/(A*(KE-1))
CALL SETS(AE(),BE(),KE)
FEP=PI*DEPM*DEPM/4
CALL AREA(DEPM,DEPM,KE1,KE,DFE(),FE())
GOSUB SCALE
GOSUB ASET
GOSUB BSET
GOSUB CSET
GOSUB DSET
GOSUB ESET
```

```
COUNT=0
PCOUNT=0
FOR KK=1 TO 20000
NN=NN+1
CALL STAB(AE(),BE(),KE,SDE,LEM)
DZ=.99*DZ
DEG=DEG+DC
IF DEG>360 THEN GOSUB START
IF CYC>2 THEN GOTO 500
IF CYC=CYCTRIP THEN PCOUNT=PCOUNT+1
IF CYC=CYCTRIP THEN GOSUB CLIPDRAW
508 CALL PIPE(KE1,KE,AE(),BE(),DFE(),FE())
REM for a throttled pipe exit the following two lines are replaced by a statement which
REM gives a new value of AKO at the pipe exit where AKO=[(pipe exit ø)/(pipe ø)]^2
REM you need a data input statement for 'pipe exit ø'
REM you need a program statement converting the 'pipe exit ø' to metre units
IF AE(KE)>=1 THEN AKO=.94
IF AE(KE)<1 THEN AKO=.6
REM this next statement calls the PORT subroutine and finds the reflected ß from the incident ∂
REM from the port-to-pipe area ratio, AKO, while the cylinder pressure is PA in this case of outlet
CALL PORT(BE(KE),AE(KE),PA,AKO)
IF DEG>DEO THEN GOTO 150
CALL EXPORT(DEG,DEO,AKD,AKM)
CALL PORT(AE(1),BE(1),PC1,AKD)
CALL FLOW(AE(),BE(),KE1,KE,FEP,PS(),PR(),PL(),MN())
CALL BOX(PC1,PC2,TC1,TC2,VC1,MC1,MC2,MEP,CEP,TEP)
MRAT=MC2/MCYL
GOSUB NSET
GOTO 250
150 AE(1)=BE(1)
AKD=0
CALL FLOW(AE(),BE(),KE1,KE,FEP,PS(),PR(),PL(),MN())
250 CALL SHOK(AE(),BE(),KE1,KE)
PCYR=PC1/PA
IF NN>1 THEN GOSUB AWHITE
IF NN>1 THEN GOSUB BWHITE
IF NN>1 THEN GOSUB CWHITE
IF NN>1 THEN GOSUB DWHITE
GOSUB CLOCK
```

```
GOSUB ABLACK
GOSUB BBLACK
GOSUB CBLACK
GOSUB DBLACK
GOSUB EBLACK
NEXT KK
500 INPUT"RESPONSE FOR RECALCULATION IS R OR FILING DATA
IS (F) OR QUITTING IS (Q)";D$
IF D$="R" THEN GOTO 50
IF D$="F" THEN GOSUB FILING
IF D$="Q" THEN GOTO 501
GOTO 500
501 END
DATASET:
LOCATE 1+ZS,1:PRINT "CURRENT DATA FOR EXHAUST
CALCULATION"
LOCATE 2+ZS,1:PRINT "CYLINDER VOLUME, cc=";VCY
LOCATE 2+ZS,31:PRINT "CYLINDER PRESSURE, bar=";PCY
LOCATE 3+ZS,1:PRINT "CYLINDER TEMPERATURE, degC=";TCY
LOCATE 3+ZS,31:PRINT "EXHAUST PULSE RATE, rpm=";RPM
LOCATE 4+ZS,1:PRINT "EXHAUST PIPE DIAMETER, mm=";DEP
LOCATE 4+ZS,31:PRINT "EXHAUST PORT DIAMETER, mm=";DPP
LOCATE 5+ZS,1:PRINT "EXHAUST OPENING PERIOD, deg=";DEO
LOCATE 5+ZS,31:PRINT "EXHAUST PIPE LENGTH, mm=";LEP
LOCATE 6+ZS,1:PRINT "INSPECTION ANGLE, deg=";DEGTRIP
LOCATE 6+ZS,31:PRINT "INSPECTION CYCLE=";CYCTRIP
RETURN
FILING:
INPUT "TYPE IN TITLE NEEDED FOR THE FILE WITH '.EXHAUST'
END CODE";P$
OPEN P$ FOR OUTPUT AS #1
WRITE#1,VCY,PCY,TCY,RPM,DEP,DPP,DEO,LEP,DEGTRIP,CYCTRIP
CLOSE#1
RETURN
FILED:
INPUT "TYPE IN THE FILENAME WITH A '.EXHAUST' END CODE";FIL$
OPEN"I",#1,FIL$
INPUT#1,VCY,PCY,TCY,RPM,DEP,DPP,DEO,LEP,DEGTRIP,CYCTRIP
CLOSE#1
RETURN
SCALE:
LOCATE 2,1:PRINT"1.5"
LOCATE 5,1:PRINT"1.0"
LOCATE 8,1:PRINT"0.5"
LOCATE 2,36:PRINT"1.5"
```

```
LOCATE 5,36:PRINT"1.0"
LOCATE 8,36:PRINT"0.5"
LOCATE 17,20:PRINT "CYLINDER PRESSURE AND PORT PRESSURE
LOCATE 21,50:PRINT"360 deg CRANKANGLE"
LOCATE 23,50:PRINT"EXHAUST TEMP, ºC=";USING"###.";TEXC
LOCATE 18,1:PRINT"1.5"
LOCATE 22,1:PRINT"1.0"
LOCATE 25,1:PRINT"0.5"
LOCATE 1,5:PRINT"TOTAL PRESSURE RATIO ALONG PIPE"
LOCATE 1,42:PRINT"OUTWARDS PULSE PRESSURE RATIO"
LOCATE 9,42:PRINT"INWARDS PULSE PRESSURE RATIO"
LOCATE 9,5:PRINT" GAS PARTICLE MACH NUMBER"
RETURN
```

COMPUTER PROGRAM PROG.2.2
PROGRAM NAME "INDUCTION"
ProgList.2.2

```
REM this is Prog.2.2
REM this is ProgList.2.2
REM this program includes subroutines common with Prog.2.2 "EXHAUST"
REM these subroutines are listed at the end of Proglists for Chapter 2
DIM ABX(21,45),BX(45),AKA(45)
DIM AE(50),BE(50),FE(50),DFE(50),DF(50),F(50)
DIM PS(50),PR(50),PL(50),MN(50)
DIM XAM(50),YAM(50),A2(50),B2(50)
DIM XM(50),YM(50),MM(50)
DIM XCM(50),YCM(50),XDM(50),YDM(50)
DIM XG(600),YGPC(600),YGPP(600)
OPEN"I",#1,"UFLO"
FOR I=1 TO 21 STEP 1
FOR J=1 TO 41 STEP 1
INPUT#1,ABX(I,J)
NEXT J
NEXT I
FINISH:
CLOSE #1
FOR J=1 TO 41
BX(J)=ABX(1,J)
NEXT J
AKA(1)=0
FOR I=2 TO 21
AKA(I)=AKA(I-1)+.05
NEXT I
```

```
50 INPUT "RESPONSE FOR NEW DATA IS 'N' AND FILED DATA IS 'F'";H$
IF H$="F" THEN GOSUB FILED
IF H$="F" THEN GOTO 502
CLS
GOSUB DATASET
INPUT"NEW DATA FOR CYLINDER VOLUME(Y/N)?";A$
IF A$="Y" THEN INPUT"TYPE IN CYLINDER VOLUME,CM3";VCY
INPUT"NEW DATA FOR INITIAL CYLINDER PRESSURE(Y/N)?";A$
IF A$="Y" THEN INPUT"TYPE IN INITIAL CYLINDER PRESSURE, BAR";PCY
INPUT"NEW DATA FOR INITIAL CYLINDER TEMPERATURE(Y/N)?";A$
IF A$="Y" THEN INPUT"TYPE IN INITIAL CYLINDER TEMPERATURE, DEG.C";TCY
INPUT"NEW DATA FOR PIPE REFERENCE AIR TEMPERATURE(Y/N)?";A$
IF A$="Y" THEN INPUT"TYPE IN PIPE REFERENCE AIR TEMPERATURE, DEG C";TPIPE
INPUT"NEW DATA FOR INTAKE CYCLING RATE(Y/N)?";A$
IF A$="Y" THEN INPUT"TYPE IN INTAKE CYCLING RATE, CYCLES/MIN";RPM
INPUT"NEW DATA FOR INTAKE PIPE DIAMETER(Y/N)?";A$
IF A$="Y" THEN INPUT"TYPE IN INTAKE PIPE DIAMETER,MM";DEP
INPUT"NEW DATA FOR INTAKE PORT MAXIMUM DIAMETER(Y/N)?";A$
IF A$="Y" THEN INPUT"TYPE IN INTAKE PORT MAXIMUM DIAMETER, MM";DPP
INPUT"NEW DATA FOR INTAKE PORT OPENING PERIOD(Y/N)?";A$
IF A$="Y" THEN INPUT"TYPE IN TOTAL INTAKE PORT OPENING PERIOD, DEG.CRANKSHAFT";DEO
INPUT"NEW DATA FOR INTAKE PIPE LENGTH(Y/N)?";A$
IF A$="Y" THEN INPUT"TYPE IN INTAKE PIPE LENGTH, MM";LEP
INPUT"NEW DATA FOR INTAKE PIPE END TYPE(Y/N)?";A$
IF A$="Y" THEN INPUT"TYPE IN NAME OF PIPE END AS EITHER 'BELLMOUTH' OR 'PLAIN'";PEND$
INPUT"NEW DATA FOR TRIP ANGLE(Y/N)?";A$
IF A$="Y" THEN INPUT"TYPE IN TRIP ANGLE";DEGTRIP
INPUT"NEW DATA FOR TRIP CYCLE(Y/N)?";A$
IF A$="Y" THEN INPUT"TYPE IN TRIP CYCLE";CYCTRIP
502 CLS
GOSUB DATASET
INPUT"TYPE 'RUN' FOR CALCULATION OR 'NEW' FOR A DATA CHANGE";Z$
IF Z$="NEW" THEN GOTO 50
CLS
```

```
CYC=1
NN=0
PI=3.141593
GE=1.4
GK=287
CCP=1005
CCV=718
GR=GE*GK
GF1=(GE-1)/GE
TC1=TCY+273
PA=101325!
TA=293
PC1=PCY*100000!
TREF=TPIPE+273
DREF=PA/(GK*TREF)
SDE=SQR(GR*TREF)
DAIR=PA/(GK*TA)
SDA=SQR(GR*TA)
VC1=VCY/1000000!
MC1=PC1*VC1/(GK*TC1)
MREF=PA*VC1/(GK*TA)
MCYL=MC1
PCYL=PC1
TCYL=TC1
DEG=0
CALL NINT(KE,LEP,TREF,TREF)
KE1=1
A=1000
AKM=(DPP/DEP)^2
DEPM=DEP/A
LEM=LEP/(A*(KE-1))
CALL SETS(AE(),BE(),KE)
FEP=PI*DEPM*DEPM/4
CALL AREA(DEPM,DEPM,KE1,KE,DFE(),FE())
GOSUB SCALE
GOSUB ASET
GOSUB BSET
GOSUB CSET
GOSUB DSET
GOSUB ESET
COUNT=0
PCOUNT=0
FOR KK=1 TO 20000
NN=NN+1
CALL STAB(AE(),BE(),KE,SDE,LEM)
```

```
DZ=.99*DZ
DEG=DEG+DC
IF DEG>360 THEN GOSUB START
IF CYC>2 THEN GOTO 500
IF CYC=CYCTRIP THEN PCOUNT=PCOUNT+1
IF CYC=CYCTRIP THEN GOSUB CLIPDRAW
508 CALL PIPE(KE1,KE,AE(),BE(),DFE(),FE())
REM for a throttled pipe exit the following four lines are replaced by a statement
which
REM gives a new value of AKO at the pipe exit where AKO=[(pipe exit ø)/(pipe
ø)]^2
REM you need a data input statement for 'pipe exit ø'
REM you need a program statement converting the 'pipe exit ø' to metre units
REM you need a statemet to CALL the PORT subroutine as in the EXHAUST
program
REM this next statement calls the PORT subroutine and finds the reflected ß
from the incident ∂
REM from the port-to-pipe area ratio, AKO, while the cylinder pressure is PA
in this case of outlet
IF AE(KE)>=1 THEN AKO=.94
IF AE(KE)<1 THEN AKO=.6
IF PEND$="PLAIN" THEN CALL PORT(BE(KE),AE(KE),PA,AKO)
IF PEND$="BELLMOUTH" THEN CALL
BELMOUTH(BE(KE),AE(KE),PA)
IF DEG>DEO THEN GOTO 150
CALL EXPORT(DEG,DEO,AKD,AKM)
CALL PORT(AE(1),BE(1),PC1,AKD)
CALL FLOW(AE(),BE(),KE1,KE,FEP,PS(),PR(),PL(),MN())
CALL BOX(PC1,PC2,TC1,TC2,VC1,MC1,MC2,MEP,CEP,TEP)
MRAT=MC2/MREF
GOSUB NSET
GOTO 250
150 AE(1)=BE(1)
AKD=0
CALL FLOW(AE(),BE(),KE1,KE,FEP,PS(),PR(),PL(),MN())
250 CALL SHOK(AE(),BE(),KE1,KE)
PCYR=PC1/PA
IF NN>1 THEN GOSUB AWHITE
IF NN>1 THEN GOSUB BWHITE
IF NN>1 THEN GOSUB CWHITE
IF NN>1 THEN GOSUB DWHITE
GOSUB CLOCK
GOSUB ABLACK
GOSUB BBLACK
GOSUB CBLACK
```

```
GOSUB DBLACK
GOSUB EBLACK
NEXT KK
500 INPUT"RESPONSE FOR RECALCULATION IS R OR FILING DATA
IS (F) OR QUITTING IS (Q)";D$
IF D$="R" THEN GOTO 50
IF D$="F" THEN GOSUB FILING
IF D$="Q" THEN GOTO 501
GOTO 500
501 END
DATASET:
LOCATE 1+ZS,1:PRINT "CURRENT DATA FOR INDUCTION
CALCULATION"
LOCATE 2+ZS,1:PRINT "CYLINDER VOLUME, cc=";VCY
LOCATE 2+ZS,31:PRINT "CYLINDER PRESSURE, bar=";PCY
LOCATE 3+ZS,1:PRINT "CYLINDER TEMPERATURE, degC=";TCY
LOCATE 3+ZS,31:PRINT "INTAKE PULSE RATE, rpm=";RPM
LOCATE 4+ZS,1:PRINT "INTAKE PIPE DIAMETER, mm=";DEP
LOCATE 4+ZS,31:PRINT "INTAKE PORT DIAMETER, mm=";DPP
LOCATE 5+ZS,1:PRINT "INTAKE OPENING PERIOD, deg=";DEO
LOCATE 5+ZS,31:PRINT "INTAKE PIPE LENGTH, mm=";LEP
LOCATE 6+ZS,1:PRINT "PIPE END IS ";PEND$
LOCATE 6+ZS,31:PRINT "PIPE AIR TEMPERATURE, degC=";TPIPE
LOCATE 7+ZS,1:PRINT "INSPECTION ANGLE, deg=";DEGTRIP
LOCATE 7+ZS,31:PRINT "INSPECTION CYCLE=";CYCTRIP
RETURN
FILING:
INPUT "TYPE IN TITLE NEEDED FOR THE FILE WITH '.INTAKE' END
CODE";P$
OPEN P$ FOR OUTPUT AS #1
WRITE#1,VCY,PCY,TCY,TPIPE,RPM,DEP,DPP,DEO,LEP,PEND$,DEGTRIP,CYCTRIP
CLOSE#1
RETURN
FILED:
INPUT "TYPE IN THE FILENAME WITH A '.INTAKE' END CODE";FIL$
OPEN"I",#1,FIL$
INPUT#1,VCY,PCY,TCY,TPIPE,RPM,DEP,DPP,DEO,LEP,PEND$,DEGTRIP,CYCTRIP
CLOSE#1
RETURN
SCALE:
LOCATE 2,1:PRINT"1.5"
LOCATE 5,1:PRINT"1.0"
LOCATE 8,1:PRINT"0.5"
LOCATE 2,36:PRINT"1.5"
LOCATE 5,36:PRINT"1.0"
```

```
LOCATE 8,36:PRINT"0.5"
LOCATE 17,20:PRINT "CYLINDER PRESSURE AND PORT PRESSURE
LOCATE 21,50:PRINT"360 deg CRANKANGLE"
LOCATE 23,50:PRINT"PIPE REF. TEMP. ºC=";TPIPE
LOCATE 18,1:PRINT"1.5"
LOCATE 22,1:PRINT"1.0"
LOCATE 25,1:PRINT"0.5"
LOCATE 1,5:PRINT"TOTAL PRESSURE RATIO ALONG PIPE"
LOCATE 1,42:PRINT"OUTWARDS PULSE PRESSURE RATIO"
LOCATE 9,42:PRINT"INWARDS PULSE PRESSURE RATIO"
LOCATE 9,5:PRINT" GAS PARTICLE MACH NUMBER"
RETURN
```

COMPUTER PROGRAM SUBROUTINES FOR Prog.2.1 and 2.2

These common subroutines are to be attached to Prog.2.1 and Prog.2.2

```
REM common subroutines for Prog.2.1 and 2.2
REM common subroutines for "EXHAUST and "INDUCTION"
ASET:
XA1=30
XA2=280
XMH=INT((XA2-XA1)/(KE-1))
XA2=XA1+(KE-1)*XMH
FOR J=1 TO KE
XAM(J)=XA1+XMH*(J-1)
XM(J)=XAM(J)
NEXT J
PK=100
YA1=80
YA2=5
YA3=130
Y15=YA1-.5*PK
Y5=YA1+.5*PK
RETURN
ABLACK:
Y15=YA1-.5*PK
Y5=YA1+.5*PK
LINE(XA1,YA2)-(XA1,YA3)
LINE(XA1,YA1)-(XA2,YA1)
LINE(XA1,Y15)-(XA1-4,Y15)
LINE(XA1,Y5)-(XA1-4,Y5)
FOR J=1 TO KE
YAM(J)=INT(YA1-(PS(J)-1)*PK)
```

```
NEXT J
FOR J=1 TO KE-1
LINE(XAM(J),YAM(J))-(XAM(J+1),YAM(J+1)),33
NEXT J
RETURN
AWHITE:
FOR J=1 TO KE-1
LINE(XAM(J),YAM(J))-(XAM(J+1),YAM(J+1)),30
NEXT J
RETURN
BSET:
XB1=XA1
XB2=XB1-28
YB4=YA3+5
YB3=YB4
YB1=YB4+100
YB2=YB1
XB5=XA2
YB7=YB1-10
YB6=YB7
YB5=YB1
RETURN
BBLACK:
LINE(XB1,YB1)-(XB2,YB2)
LINE(XB2,YB2)-(XB2,YB3)
LINE(XB2,YB3)-(XB1,YB4)
LINE(XB1,YB7)-(XB1,YB4)
LINE(XB1-5,YB7)-(XB1,YB7-4),33,BF
LINE(XB1,YB7)-(XB5,YB7)
LINE(XB1,YB1)-(XB5,YB1)
LINE(XB5,YB4)-(XB5,YB6-3)
LINE(XB5,YB6-50)-(XB5-4,YB6-50)
YB8=YB1-(1-AKD)*(YB1-YB7)
LINE(XB1-5,YB1)-(XB1,YB8+1),33,BF
FOR J=1 TO KE
YM(J)=YB7-ABS(MN(J))*100
MM(J)=MN(J)
GOSUB ARROB
NEXT J
YBM=YB4+(1-MRAT)*100
LINE(XB2,YB3)-(XB1,YBM),33,BF
RETURN
BWHITE:
FOR J=1 TO KE
GOSUB ARROW
```

```
NEXT J
LINE(XB2,YB2)-(XB1,YBM),30,BF
LINE(XB1-5,YB1)-(XB1,YB8+1),30,BF
RETURN
ARROB:
ZM=XMH/2
IF MN(J)<0 THEN GOTO 216
LINE(XM(J),YM(J))-(XM(J)+ZM,YM(J))
LINE(XM(J)+ZM,YM(J))-(XM(J)+ZM-2,YM(J)-2)
LINE(XM(J)+ZM,YM(J))-(XM(J)+ZM-2,YM(J)+2)
GOTO 217
216 LINE(XM(J),YM(J))-(XM(J)+ZM,YM(J))
LINE(XM(J),YM(J))-(XM(J)+2,YM(J)-2)
LINE(XM(J),YM(J))-(XM(J)+2,YM(J)+2)
217 RETURN
ARROW:
IF MM(J)<0 THEN GOTO 218
LINE(XM(J),YM(J))-(XM(J)+ZM,YM(J)),30
LINE(XM(J)+ZM,YM(J))-(XM(J)+ZM-2,YM(J)-2),30
LINE(XM(J)+ZM,YM(J))-(XM(J)+ZM-2,YM(J)+2),30
GOTO 219
218 LINE(XM(J),YM(J))-(XM(J)+ZM,YM(J)),30
LINE(XM(J),YM(J))-(XM(J)+2,YM(J)-2),30
LINE(XM(J),YM(J))-(XM(J)+2,YM(J)+2),30
219 RETURN
CLOCK:
LOCATE 26,2:PRINT "CYCLE NO.=";USING"#";CYC
LOCATE 26,12:PRINT "CRANKANGLE=";USING"###.#";DEG
LOCATE 26,29:PRINT"CYLINDER MASS RATIO=";USING"#.###";MRAT
LOCATE 26,53:PRINT"CYLINDER P, atm=";USING"#.##";PCYR
RETURN
CSET:
XC1=310
XC2=560
XMHC=INT((XC2-XC1)/(KE-1))
XC2=XC1+(KE-1)*XMHC
FOR J=1 TO KE
XCM(J)=XC1+XMHC*(J-1)
NEXT J
PK=100
YC1=80
YC2=5
YC3=130
YC15=YC1-.5*PK
YC5=YC1+.5*PK
```

```
RETURN
CBLACK:
FOR J=1 TO KE
YCM(J)=INT(YA1-(PR(J)-1)*PK)
NEXT J
LINE(XC1,YC2)-(XC1,YC3)
LINE(XC1,YC1)-(XC2,YC1)
LINE(XC1,YC15)-(XC1-4,YC15)
LINE(XC1,YC5)-(XC1-4,YC5)
FOR J=1 TO KE-1
LINE(XCM(J),YCM(J))-(XCM(J+1),YCM(J+1)),33
NEXT J
RETURN
CWHITE:
FOR J=1 TO KE-1
LINE(XCM(J),YCM(J))-(XCM(J+1),YCM(J+1)),30
NEXT J
RETURN
DSET:
XD1=XC1
XD2=XC2
XMHD=XMHC
XD2=XD1+(KE-1)*XMHD
FOR J=1 TO KE
XDM(J)=XCM(J)
NEXT J
PK=100
YD1=200
YD2=YC3+5
YD3=250
YD15=YD1-.5*PK
YD5=YD1+.5*PK
RETURN
DBLACK:
LOCATE 10,36:PRINT"1.5"
LOCATE 13,36:PRINT"1.0"
LOCATE 16,36:PRINT"0.5"
LOCATE 11,31:PRINT "0.5"
FOR J=1 TO KE
YDM(J)=INT(YD1-(PL(J)-1)*PK)
NEXT J
LINE(XD1,YD2)-(XD1,YD3)
LINE(XD1,YD1)-(XD2,YD1)
LINE(XD1,YD15)-(XD1-4,YD15)
LINE(XD1,YD5)-(XD1-4,YD5)
```

```
FOR J=1 TO KE-1
LINE(XDM(J),YDM(J))-(XDM(J+1),YDM(J+1)),33
NEXT J
RETURN
DWHITE:
FOR J=1 TO KE-1
LINE(XDM(J),YDM(J))-(XDM(J+1),YDM(J+1)),30
NEXT J
RETURN
ESET:
XE1=30
XE2=400
PK=100
YE1=340
YE2=265
YE3=400
Y15=YE1-.5*PK
Y5=YE1+.5*PK
XEO=XE1
YSO=YE1
YCO=INT(YE1-(PCY-1)*PK)
LINE(XE1,YE2)-(XE1,YE3)
LINE(XE1,YE1)-(XE2,YE1)
LINE(XE1,Y15)-(XE1-4,Y15)
LINE(XE1,Y5)-(XE1-4,Y5)
XE90=XE1+90
LINE(XE90,YE1-4)-(XE90,YE1+4)
XE180=XE1+180
LINE(XE180,YE1-4)-(XE180,YE1+4)
XE270=XE1+270
LINE(XE270,YE1-4)-(XE270,YE1+4)
XE360=XE1+360
LINE(XE360,YE1-4)-(XE360,YE1+4)
RETURN
EBLACK:
PK=100
YSN=INT(YE1-(PS(1)-1)*PK)
YCN=INT(YE1-(PCYR-1)*PK)
XEN=XE1+DEG
XG(PCOUNT)=XEN
YGPC(PCOUNT)=YCN
YGPP(PCOUNT)=YSN
IF COUNT=1 THEN GOTO 511
LINE(XEO,YSO)-(XEN,YSN)
IF PCYR>2.5 THEN GOTO 503
```

```
LINE(XEO,YCO)-(XEN,YCN)
GOTO 503
511 FOR I= 1 TO PCOUNT-1
LINE(XG(I),YGPC(I))-(XG(I+1),YGPC(I+1))
LINE(XG(I),YGPP(I))-(XG(I+1),YGPP(I+1))
NEXT I
503 YCO=YCN
YSO=YSN
XEO=XEN
RETURN
NSET:
TC1=TC2
PC1=PC2
MC1=MC2
RETURN
START:
PC1=PCYL
TC1=TCYL
MC1=MCYL
XEO=XE1
YCO=INT(YE1-(PCY-1)*PK)
DEG=DEG-360
CYC=CYC+1
RETURN
CLIPDRAW:
IF DEG<DEGTRIP THEN GOTO 509
COUNT=COUNT+1
IF COUNT>1 THEN GOTO 509
INPUT"WANT A PRINTER OUTPUT?, TYPE A 'Y' OR A 'N', ONLY!";PR$
IF PR$="N" THEN GOTO 509
OPEN"LPT1:PROMPT" FOR OUTPUT AS  #1
WINDOW OUTPUT#1
ZS=27
GOSUB DATASET
GOSUB SCALE
GOSUB ASET
GOSUB BSET
GOSUB CSET
GOSUB DSET
GOSUB ESET
GOSUB CLOCK
GOSUB ABLACK
GOSUB BBLACK
GOSUB CBLACK
GOSUB DBLACK
```

```
GOSUB EBLACK
CLOSE #1
COUNT=COUNT+1
ZS=0
509 RETURN
SUB EXPORT(DEG,DEO,AKD,AKM)STATIC
SLOP=2*AKM/DEO
HALF=.5*DEO
PD=DEG
IF DEG>HALF THEN PD=DEO-DEG
AKD=SLOP*PD
END SUB
SUB STAB(A(),B(),K,S,E)STATIC
SHARED DZ,DTM,DC,RPM
C=60*S/(RPM*E)
DZ=1
FOR J=1 TO K
Z=2/((A(J)+B(J))+ABS(5*(A(J)-B(J))))
IF Z<DZ THEN DZ=Z
NEXT J
DC=360*DZ/C
DTM=DC/(6*RPM)
END SUB
SUB SETS(A(),B(),K)STATIC
FOR J=1 TO K
A(J)=1
B(J)=1
NEXT J
END SUB
SUB AREA(DA,DB,KA,KB,DF(),F())STATIC
C=(DB-DA)/(KB-KA)
FOR J=KA TO KB
A=J-KA
F(J)=((DA+C*A)^2)*.7854
NEXT J
KC=KB-1
FOR J=KA TO KC
DF(J)=F(J+1)-F(J)
NEXT J
END SUB
SUB PIPE(KA,KB,A(),B(),DF(),F())STATIC
SHARED DZ,A2(),B2()
G1=3
G2=2
KC=KB-1
```

```
FOR J=KA TO KC
AA=DZ*A(J+1)*B(J)
BB=DZ*A(J)*B(J+1)
Y=(1+DZ*(G1*(A(J+1)-A(J))-G2*(B(J+1)-B(J))))
Z=(1+DZ*(G1*(B(J)-B(J+1))-G2*(A(J)-A(J+1))))
WA=(A(J+1)+G2*AA-G2*BB)/Y
RA=(A(J)-G1*BB+G1*AA)/Z
WB=(B(J+1)+G1*AA-G1*BB)/Y
RB=(B(J)+G2*AA-G2*BB)/Z
AA=WA-WB
BB=RA-RB
CC=WA+WB
DD=RA+RB
IF AA=0 THEN GOTO 22
QA=-DZ*CC*AA*DF(J)/(4*F(J))
GOTO 33
22 QA=0
33 IF BB=0 THEN GOTO 44
QB=-DZ*BB*DD*DF(J)/(4*F(J))
GOTO 55
44 QB=0
55 A2(J+1)=WA+QA
B2(J)=RB+QB
NEXT J
FOR J=KA TO KC
A(J+1)=A2(J+1)
B(J)=B2(J)
NEXT J
END SUB
SUB SHOK(A(),B(),KA,KB)STATIC
FOR J=KA TO KB
R1=5*ABS(A(J)-B(J))/(A(J)+B(J))
IF R1<1 THEN GOTO 56
R2=SQR((R1*R1+5)/(7*R1*R1-1))
R3=(A(J)+B(J))*((7*R1*R1-1)/6)^(1/7)
IF A(J)>B(J) THEN GOTO 26
A(J)=(5-R2)*R3/10
B(J)=(5+R2)*R3/10
GOTO 56
26 A(J)=(5+R2)*R3/10
B(J)=(5-R2)*R3/10
56 R1=0
NEXT J
END SUB
SUB PORT(A,B,P,AK)STATIC
```

```
SHARED ABX(),BX(),AKA(),PA
X1=(P/PA)^(1/7)
BX1=B/X1
IF AK>.95 THEN AK=.95
IF BX1<.6 THEN BX1=.6
I=INT(AK/.05)+1
J=INT((BX1-.6)/.02)+1
RA=(AK-AKA(I))/(AKA(I+1)-AKA(I))
AB1=RA*(ABX(I+1,J+1)-ABX(I,J+1))+ABX(I,J+1)
AB2=RA*(ABX(I+1,J)-ABX(I,J))+ABX(I,J)
RB=(BX1-BX(J))/(BX(J+1)-BX(J))
ABX1=RB*(AB1-AB2)+AB2
A=2*X1*ABX1-B
END SUB
SUB NINT(KI,PL,TE,TP)STATIC
A=SQR(TE/TP)
RI=PL*A/30
KI=INT(RI)
BI=RI-KI
IF BI>.5 THEN KI=KI+1
KI=KI+1
END SUB
SUB FLOW(A(),B(),KA,KB,F,PS(),PR(),PL(),MN())STATIC
SHARED TREF,DREF,DA,SDE,TEP,MEP,CEP,GR
FOR J=KA TO KB
AB=(A(J)+B(J))/2
C=2.5*SDE*(A(J)-B(J))
IF J=KA THEN CEP=C
T=TREF*AB*AB
IF J=KA THEN TEP=T
M=C*F*DREF*AB^5
IF J=KA THEN MEP=M
MN(J)=C/SQR(GR*T)
PS(J)=AB^7
PR(J)=(.5*(A(J)+1))^7
PL(J)=(.5*(B(J)+1))^7
NEXT J
END SUB
SUB BOX(PB1,PB2,TB1,TB2,VB,MB1,MB2,MDE,CE,TE)STATIC
SHARED CCP,CCV,DTM
N=0
PB2=PB1
DMI=MDI*DTM
DME=MDE*DTM
MB2=MB1+DMI-DME
```

```
EI=DMI*(CCP*TI+.5*CI*CI)
EE=DME*(CCP*TE+.5*CE*CE)
103 N=N+1
TB2=(EI-EE+CCV*MB1*TB1)/(CCV*MB2)
PB2=(MB2*TB2*PB1*VB)/(MB1*TB1*VB)
IF N<2 THEN GOTO 103
END SUB
SUB BELMOUTH(B,A,PA)STATIC
REM this is strictly speaking a SUB for Induction only but is better kept with the SUB set
REM you never know when you'd like to program it in!
M=.5*(A+1)
N=((4*M+1)+SQR(1+20*M-20*M*M))/6
B=2*N-1
END SUB
```

COMPUTER PROGRAM PROG.2.3
PROGRAM NAME "CYLINDER-PIPE FLOW"
ProgList.2.3

```
REM this is Prog.2.3
REM this is program "CYLINDER-PIPE FLOW"
REM this is ProgList.2.3
DIM ABX(21,45),BX(45),AKA(45)
OPEN"I",#1,"UFLO"
FOR I=1 TO 21 STEP 1
FOR J=1 TO 41 STEP 1
INPUT#1,ABX(I,J)
NEXT J
NEXT I
FINISH:
CLOSE #1
FOR J=1 TO 41
BX(J)=ABX(1,J)
NEXT J
AKA(1)=0
FOR I=2 TO 21
AKA(I)=AKA(I-1)+.05
NEXT I
100 CLS
GOSUB DATASET
INPUT"NEW DATA FOR INITIAL CYLINDER PRESSURE(Y/N)?";A$
IF A$="Y" THEN INPUT"TYPE IN INITIAL CYLINDER PRESSURE,BAR";PCY
```

```
PA=101325!
PC1=PCY*100000!
PCYR=PC1/PA
X1=(PCYR)^(1/7)
INPUT”NEW DATA FOR INCIDENT PRESSURE RATIO AT THE PORT(Y/N)?”;A$
IF A$=”Y” THEN INPUT”TYPE IN INCIDENT PRESSURE RATIO AT THE PORT,”;PI1
XI1=PI1^(1/7)
BE(1)=2*XI1-1
XVAL=BE(1)/X1
IF XVAL<=1 THEN GOTO 200
INPUT”NEW DATA FOR PIPE REFERENCE TEMPERATURE(Y/N)?”;A$
IF A$=”Y” THEN INPUT”TYPE IN PIPE REFERENCE TEMPERATURE,DEG C”;TREFC
200 INPUT”NEW DATA FOR INITIAL CYLINDER TEMPERATURE(Y/N)?”;A$
IF A$=”Y” THEN INPUT”TYPE IN INITIAL CYLINDER TEMPERATURE,DEG C”;TCY
INPUT”NEW DATA FOR EXHAUST PIPE DIAMETER(Y/N)?”;A$
IF A$=”Y” THEN INPUT”TYPE IN EXHAUST PIPE DIAMETER,MM”;DEP
INPUT”NEW DATA FOR EXHAUST PORT MAXIMUM DIAMETER(Y/N)?”;A$
IF A$=”Y” THEN INPUT”TYPE IN EXHAUST PORT MAXIMUM DIAMETER,MM”;DPP
GE=1.4
GK=287
CCP=1005
CCV=718
GR=GE*GK
GF1=(GE-1)/GE
GF2=2/GF1
GF3=2/(GE-1)
TC1=TCY+273
TA=293
IF XVAL>1 THEN GOTO 300
TREF=TC1*(PA/PC1)^GF1
TREFC=TREF-273
300 TREF=TREFC+273
DREF=PA/(GK*TREF)
SDE=SQR(GR*TREF)
DAIR=PA/(GE*TA)
SDA=SQR(GR*TA)
AKM=DPP/DEP
AKD=AKM*AKM
```

```
CALL PORT(AE(1),BE(1),PC1,AKD)
XR1=(AE(1)+1)/2
PR1=XR1^GF2
XS1=XI1+XR1-1
PRS=XS1^GF2
FPIP=3.14159*DEP*DEP/4000000!
CEX=GF3*SDE*(XR1-XI1)
DEX=DREF*XS1^GF3
MFLOW=DEX*FPIP*CEX
CLS
GOSUB DATASET
INPUT"WANT A PRINTER OUTPUT?, TYPE A 'Y' OR A 'N', ONLY!";PR$
IF PR$="Y" THEN GOSUB DATAPRINT
GOTO 100
END
DATAPRINT:
LPRINT
IF XVAL>1 THEN GOTO 700
LPRINT "OUTFLOW CALCULATION FOR AIR WITH PIPE REFERENCE PRESSURE AT 1 atm"
GOTO 800
700 LPRINT "INFLOW CALCULATION FOR AIR WITH PIPE REFERENCE PRESSURE AT 1 atm"
800 LPRINT "INCIDENT WAVE PRESSURE RATIO=";PI1
LPRINT "CYLINDER PRESSURE, atm=";USING"##.##";PCYR
LPRINT "CYLINDER TEMPERATURE, deg.C=";TCY
LPRINT"PIPE REF. TEMP, deg.C=";USING"###.";TREFC
LPRINT "PIPE DIAMETER, mm=";DEP
LPRINT "CYLINDER PORT DIAMETER, mm=";DPP
LPRINT "GAS VELOCITY IN PIPE, m/s=";USING"####.#";CEX
LPRINT "PULSE PRESSURE RATIO IN PIPE=";USING"#.###";PR1
LPRINT "SUPERPOSITION PRESSURE RATIO=";USING"#.###";PRS
LPRINT "PIPE MASS FLOW RATE, kg/s,=";USING"##.####";MFLOW
LPRINT "GAS DENSITY IN PIPE, kg/m3,=";USING"#.###";DEX
RETURN
DATASET:
IF XVAL>1 THEN GOTO 400
LOCATE 1+ZS,1:PRINT "CURRENT RESULT OF OUTFLOW CALCULATION"
GOTO 500
400 LOCATE 1+ZS,1:PRINT "CURRENT RESULT OF INFLOW CALCULATION"
500 LOCATE 2+ZS,1:PRINT "INCIDENT WAVE PRESSURE RATIO=";PI1
LOCATE 2+ZS,32:PRINT "CYLINDER PRESSURE, atm=";USING"##.##";PCYR
```

```
LOCATE 3+ZS,1:PRINT "CYLINDER TEMPERATURE, degC=";TCY
LOCATE 3+ZS,32:PRINT"PIPE REF. TEMP, ºC=";USING"###.";TREFC
LOCATE 4+ZS,1:PRINT "PIPE DIAMETER, mm=";DEP
LOCATE 4+ZS,32:PRINT "CYLINDER PORT DIAMETER, mm=";DPP
LOCATE 5+ZS,1:PRINT "GAS VELOCITY IN PIPE, m/
s=";USING"####.#";CEX
LOCATE 5+ZS,32:PRINT "PULSE PRESSURE RATIO IN
PIPE=";USING"#.###";PR1
LOCATE 6+ZS,1:PRINT " PIPE MASS FLOW RATE, kg/
s,=";USING"##.####";MFLOW
LOCATE 6+ZS,32:PRINT "GAS DENSITY IN PIPE, kg/
m3,=";USING"#.###";DEX
LOCATE 7+ZS,1:PRINT "SUPERPOSITION PRESSURE
RATIO=";USING"#.###";PRS
RETURN
SUB PORT(A,B,P,AK)STATIC
SHARED ABX(),BX(),AKA(),PA
X1=(P/PA)^(1/7)
BX1=B/X1
IF AK>.95 THEN AK=.95
IF BX1<.6 THEN BX1=.6
I=INT(AK/.05)+1
J=INT((BX1-.6)/.02)+1
RA=(AK-AKA(I))/(AKA(I+1)-AKA(I))
AB1=RA*(ABX(I+1,J+1)-ABX(I,J+1))+ABX(I,J+1)
AB2=RA*(ABX(I+1,J)-ABX(I,J))+ABX(I,J)
RB=(BX1-BX(J))/(BX(J+1)-BX(J))
ABX1=RB*(AB1-AB2)+AB2
A=2*X1*ABX1-B
END SUB
```

ProgList3.0 Program listings from Chapter 3

COMPUTER PROGRAM PROG.3.1
PROGRAM NAME "BENSON-BRANDHAM SCAVENGING MODEL"
ProgList.3.1

```
REM PROGRAM Prog.3.1
REM ProgList.3.1
REM PROGRAM NAME 'BENSON-BRANDHAM SCAVENGING MODEL'
PRINT"THIS IS A PROGRAM FOR A TWO STROKE ENGINE IN WHICH "
PRINT"IT IS POSSIBLE TO VARY THE SCAVENGE MODEL TO SIMULATE THE EFFECTS OF"
PRINT"PERFECT DISPLACEMENT, PERFECT MIXING AND SHORT-CIRCUITING OF THE SCAVENGING FLOW."
PRINT"THE MODEL IS BASED ON THAT OF BENSON AND BRANDHAM IN REFERENCE(1.4)."
PRINT "THIS IS A TWO-PART MODEL."
PRINT "THE PERIOD OF PERFECT DISPLACEMENT SCAVENGING"
PRINT "IS FOLLOWED BY A PERFECT MIXING PROCESS FOR THE REMAINDER OF SCAVENGING."
PRINT "SHORT-CIRCUITING AS A PROPORTION OF THE SCAVENGE AIR, SHOULD YOU INSERT A VALUE"
PRINT "OCCURS IN BOTH PARTS OF THE MODEL AT THE LEVEL INSERTED AS DATA."
110 INPUT "NEW DATA FOR THE LOWER LIMIT OF SCAVENGE RATIO? (Y/N)";DAT$
IF DAT$="Y" THEN INPUT"TYPE IN NEW VALUE OF LOWER LIMIT OF SCAVENGE RATIO";SR1
INPUT "NEW DATA FOR THE HIGHER LIMIT OF SCAVENGE RATIO? (Y/N)";DAT$
IF DAT$="Y" THEN INPUT"TYPE IN NEW VALUE OF HIGHER LIMIT OF SCAVENGE RATIO";SR2
INPUT "NEW DATA FOR THE SCAVENGE RATIO STEP IN THE CALCULATION? (Y/N)";DAT$
IF DAT$="Y" THEN INPUT"TYPE IN NEW VALUE OF THE SCAVENGE RATIO STEP IN THE CALCULATION";SRI
INPUT "NEW DATA FOR THE PERFECT DISPLACEMENT PERIOD? (Y/N)";DAT$
IF DAT$="Y" THEN INPUT"TYPE IN NEW VALUE OF THE PERFECT DISPLACEMENT PERIOD";PS
INPUT "NEW DATA FOR THE SHORT-CIRCUIT PROPORTION? (Y/N)";DAT$
IF DAT$="Y" THEN INPUT"TYPE IN NEW VALUE OF THE SHORT-CIRCUIT PROPORTION";SC
SR=SR1
```

```
NSRLIMIT=1+((SR2-SR1)/SRI)
PRINT
PRINT"PERFECT SCAVENGING PROPORTION=";USING"#.##";PS
PRINT"SHORT CIRCUITING PROPORTION=";USING"#.##";SC
PRINT" SCAVENGE RATIO  SCAVENGING EFFICIENCY  TRAPPING
EFFICIENCY"
SR=SR-SRI
FOR NSR=1 TO NSRLIMIT
SR=SR+SRI
GOSUB SCAV
PRINT USING"###########.###";SR,SEFF,TEFF
NEXT NSR
INPUT "WANT A PRINTOUT?Y/N)";PR$
IF PR$="Y" THEN GOSUB PRINTOUT
INPUT "CALCULATING AGAIN?Y/N)";PR$
IF PR$="Y" THEN GOTO 110
END
SCAV:
IF SR>(PS/(1-SC)) THEN SEFF=1-(1-PS)*EXP(PS-(1-SC)*SR)
IF SR<=(PS/(1-SC)) THEN SEFF=SR*(1-SC)
TEFF=SEFF/SR
RETURN
PRINTOUT:
SR=SR1-SRI
LPRINT
LPRINT"PERFECT SCAVENGING PROPORTION=";USING"#.##";PS
LPRINT"SHORT CIRCUITING PROPORTION=";USING"#.##";SC
LPRINT" SCAVENGE RATIO  SCAVENGING EFFICIENCY  TRAPPING
EFFICIENCY"
FOR NSR=1 TO NSRLIMIT
SR=SR+SRI
GOSUB SCAV
LPRINT USING"#############.###";SR,SEFF,TEFF
NEXT NSR
RETURN
```

COMPUTER PROGRAM PROG.3.2
PROGRAM NAME "BLAIR SCAVENGING MODEL"
ProgList.3.2

```
REM ProgList.3.2
REM Prog.3.2
REM Program Name "BLAIR SCAVENGING MODEL"
REM The PSV value printed is the volume proportion of the total cylinder
volume
```

```
REM which contains fresh charge only-the perfect scavenge volume of Fig.3.15.
REM The exhaust purity value printed out is the charge purity entering the exhaust port.
DIM SRA(20),SEA(20),TEA(20),FSEA(20),YYA(20),SSA(20),VDISP(20), PUREX(20)
750 INPUT"NEW VALUE OF SCAVENGE RATIO WHERE PERFECT SCAVENGING CEASES, 'SRM'? (Y/N)";Y$
IF Y$="Y" THEN INPUT "TYPE IN NEW VALUE OF 'SRM'";SRM
INPUT"NEW VALUE OF INITIAL SCAVENGE RATIO PERIOD OF PERFECT SCAVENGING, 'SRPD'? (Y/N)";Y$
IF Y$="Y" THEN INPUT "TYPE IN NEW VALUE OF 'SRPD'";SRPD
INPUT"NEW VALUE OF PERFECT SCAVENGING PROPORTION, 'YMAX'? (Y/N)";Y$
IF Y$="Y" THEN INPUT "TYPE IN NEW VALUE OF 'YMAX'";YMAX
INPUT"NEW VALUE OF SHORT CIRCUITING PROPORTION AT SR='SRPD', 'SCPD'? (Y/N)";Y$
IF Y$="Y" THEN INPUT "TYPE IN NEW VALUE OF 'SCPD'";SCPD
INPUT"NEW VALUE OF SHORT CIRCUITING PROPORTION AT SR=1, 'SCAT1'? (Y/N)";Y$
IF Y$="Y" THEN INPUT "TYPE IN NEW VALUE OF 'SCAT1'";SCAT1
V=1000
DVA=1
YY=1
DSR=DVA/V
VM=V
VD=0
VAD=0
VAM=0
SEM=0
SED=0
SR=0
YM=1
IF SRPD=0 THEN SRPD=.0001
PRINT "SCAVENGE RATIO WHERE PERFECT DISPLACEMENT FINISHES, 'SRM'=";SRM
PRINT "SCAVENGE RATIO PERIOD OF INITIAL PERFECT SCAVENGING, 'SRPD'=";SRPD
PRINT "MAXIMUM VALUE OF PERFECT DISPLACEMENT PROPORTION, 'YMAX'=";YMAX
PRINT "VALUE OF SHORT CIRCUITING PROPORTION AT SR='SRPD', 'SCPD'=";SCPD
PRINT "VALUE OF SHORT CIRCUITING PROPORTION AT SR=1, 'SCAT1'=";SCAT1
PRINT" SR SE TE LOG(1-SE) YMODEL SCMODEL PSV PURITY EXHST"
```

```
NN=0
IF SRM<=SRPD THEN SRM=SRPD+.0001
MM=YMAX/(SRPD-SRM)
MP=SC/SRPD
MS=(SCAT1-SCPD)/(1-SRPD)
JC=1
SS=SC
100 IF SR<=SRPD THEN YY=YMAX
IF SR>SRPD THEN YY=MM*(SR-SRM)
IF SR>=SRM THEN YY=0
IF SR<=SRPD THEN SS=SCPD
IF SR>SRPD THEN SS=SCPD+MS*(SR-SRPD)
IF SS<0 THEN SS=0
DSC=SS*DVA
DVR=DVA-DSC
DVD=YY*DVR
DVAD=DVD
DVAM=(1-YY)*DVR
VM=VM-DVD
VD=VD+DVD
VAD=VAD+DVAD
VAM=VAM+DVAM-SEM*DVR
SEM=VAM/VM
SEEX=(SEM*DVR+DSC)/DVA
SE=(VAM+VAD)/V
SR=SR+DSR
FSE=LOG(1-SE)
TE=SE/SR
VDPRT=VD/V
VMPRT=VM/V
NN=NN+1
JN=NN/100
IF JN<JC THEN GOTO 350
PRINT USING"####.#####";SR,SE,TE,FSE,YY,SS,VDPRT,SEEX
SRA(JC)=SR
SEA(JC)=SE
TEA(JC)=TE
FSEA(JC)=FSE
YYA(JC)=YY
VDISP(JC)=VDPRT
PUREX(JC)=SEEX
SSA(JC)=SS
JC=JC+1
IF NN=1400 THEN GOTO 200
350 GOTO 100
```

```
200 GOSUB LINEFIT
PRINT
PRINT"BEST STRAIGHT LINE FIT THROUGH LOG(1-SE),SR POINTS
BETWEEN SR=0.3 TO 1.4"
PRINT "Y INTERCEPT,CC      SLOPE,MM      AVERAGE ERROR,DEV"
PRINT USING"######.####";CC,MM,DEV
751 INPUT "WANT A PRINTOUT?(P), OR RE-CALCULATING AGAIN?
(R)";P$
IF P$="P" THEN GOSUB PRINTING
IF P$="P" THEN GOTO 751
IF P$="R" THEN GOTO 750
END
PRINTING:
INPUT"NEED A TITLE? (Y/N)";PT$
IF PT$="Y" THEN INPUT"TYPE IN TITLE";TIT$
LPRINT
LPRINT
LPRINT TIT$
LPRINT"BLAIR'S MODEL OF SCAVENGING WHERE A PROPORTION
OF THE INCOMING FRESH CHARGE"
LPRINT "IS ALWAYS TREATED AS PERFECTLY TRAPPED IN THE
CYLINDER AND THAT PROPORTION"
LPRINT" IS BASED ON A NEARLY LINEAR FUNCTION OF THE SCAV-
ENGE RATIO WHICH IS"
LPRINT" CONSTANT AT 'YMAX' FROM SR=0, TO SR='SRPD', THENCE
LINEAR TO ZERO AT 'SRM'."
LPRINT"THIS MODEL ALSO INCORPORATES A MODEL OF DIRECT
SHORT CIRCUITING"
LPRINT"TO THE EXHAUST PORT, AS AN INITIAL VALUE 'SCPD' UP
TO SR=SRPD"
LPRINT"AND DECAYS LINEARLY, OR INCREASES, TO 'SCAT1' AT
SR=1"
LPRINT "SCAVENGE RATIO WHERE PERFECT DISPLACEMENT
SCAVENGING CEASES, 'SRM'=";SRM
LPRINT "SCAVENGE RATIO PERIOD OF INITIAL PERFECT SCAVENG-
ING, 'SRPD'=";SRPD
LPRINT "MAXIMUM VALUE OF PERFECT DISPLACEMENT PROPOR-
TION INITIALLY, 'YMAX'=";YMAX
LPRINT "VALUE OF SHORT CIRCUITING AT SR='SRPD',
'SCPD'=";SCPD
LPRINT "VALUE OF SHORT CIRCUITING AT SR=1, 'SCAT1'=";SCAT1
LPRINT" SR   SE   TE LOG(1-SE) YMODEL SCMODEL  PSV
EXHAUST PURITY"
FOR J=1 TO JC-1
```

```
   LPRINT        USING"####.####";SRA(J),SEA(J),TEA(J),FSEA(J),
YYA(J),SSA(J),VDISP(J),PUREX(J)
   NEXT J
   LPRINT
   LPRINT"BEST STRAIGHT LINE FIT THROUGH LOG(1-SE),SR POINTS
BETWEEN SR=0.3 TO 1.4"
   LPRINT "Y INTERCEPT,CC   SLOPE,MM    AVERAGE ERROR,DEV"
   LPRINT USING"#######.####";CC,MM,DEV
   RETURN
   LINEFIT:
   REM STRAIGHT LINE FIT FOR A JN SERIES OF Y,X POINTS ON A LINE
Y=CC+MM*X
   REM IT IS POSSIBLE TO START AT JSTART
   REM NUMBER OF POINTS NPT IS JN-JSTART+1
   JSTART=3
   JN=JC-1
   NPT=JN-JSTART+1
   SUMY=0
   SUMX=0
   SUMXY=0
   SUMXX=0
   FOR J= JSTART TO JN
   SUMY=SUMY+FSEA(J)
   SUMX=SUMX+SRA(J)
   SUMXX=SUMXX+SRA(J)*SRA(J)
   SUMXY=SUMXY+FSEA(J)*SRA(J)
   NEXT J
   DEN=NPT*SUMXX-SUMX*SUMX
   CC=(SUMY*SUMXX-SUMX*SUMXY)/DEN
   MM=(NPT*SUMXY-SUMX*SUMY)/DEN
   DEV=0
   FOR J=JSTART TO JN
   DEV=DEV+(CC+MM*SRA(J)-FSEA(J))^2
   NEXT J
   DEV=SQR(DEV/(NPT-1))
   RETURN
```

COMPUTER PROGRAM PROG.3.3
PROGRAM NAME "QUB CROSS PORTS"
ProgList.3.3

```
REM Prog.3.3
REM Program Name "QUB CROSS PORTS"
REM ProgList.3.3
REM***PROGRAM QUB CROSS PORTS***
```

```
100 INPUT"NEW DATA OR FILE DATA(N OR F)?";FF$
IF FF$="F" THEN GOSUB FILED
IF FF$="F" THEN GOTO 150
GOSUB DATAIN
107 CLS
IF ANS$="P" THEN CLS
IF ANS$="P" THEN OPEN"LPT1:PROMPT" FOR OUTPUT AS #1
IF ANS$="P" THEN WINDOW OUTPUT#1
LOCATE 1,1:PRINT"CURRENT INPUT DATA FOR DEFLECTOR PISTON
CROWN AND PORTING"
LOCATE 2,1:PRINT"(B) BORE,mm=";BO
LOCATE 2,26:PRINT"(S) STROKE,mm=";ST
LOCATE 2,50:PRINT"(C) CON-ROD,mm=";CRL
LOCATE 3,1:PRINT"(T) TRANSFER OPENS,deg atdc=";TRO
LOCATE 3,35:PRINT"(SM) SIDE/MIDDLE PORT WIDTH RATIO =";SM
LOCATE 4,1:PRINT"(CR) PORT CORNER RADII,mm=";PCR
LOCATE 5,1:PRINT"(WD) DEFLECTOR WIDTH,mm=";WD
LOCATE 5,35:PRINT"(DR) DEFLECTOR RADIUS,mm=";LD
LOCATE 4,35:PRINT"(BC) PORT BAR CHORD WIDTH,mm=";BA
LOCATE 7,1:PRINT "OUTPUT DATA"
150 PI=3.1415927#
DTOR=PI/180
RTOD=1/DTOR
PA=PI*BO*BO/4
RB=BO/2
SV=PA*ST
SVCC=SV/1000
CRK=ST/2
DEG=TRO
GOSUB PIST
THR=WD/BO
TH=ATN(THR/SQR(1-THR*THR))
RZ=SQR(LD*LD-WD*WD/4)
PP=ATN(WD/(2*RZ))
PQ=PI/2-PP
PH=PQ-TH
PHI=PH*RTOD
EF=LD*SIN(TH+PH)
BE=LD-EF
OE=WD/(2*TAN(TH))
OB=OE-BE
XB=RB+OB
YB=BO-XB
Z1=.5*WD*(OE+EF)
```

```
AL=PI/2-TH-PH
Z2=LD*LD*AL
Z3=RB*RB*TH
NS=Z2+Z3-Z1
SQAREA=PI*RB*RB-NS
PISTAREA=PI*RB*RB
SA=SQAREA/PISTAREA
FS=PA-NS
BBR=BA/BO
BB=2*ATN(BBR/SQR(1-BBR*BBR))
MM=2*(TH-BB)/(2*SM+1)
SS=MM*SM
WS=BO*SIN(SS/2)
WM=BO*SIN(MM/2)
WP=2*WS+WM
WR=WP/BO
PAREA=WP*HBDC-3*(PCR*PCR)*(4-PI)
SCEFF=NS/PAREA
BDEG=BB*RTOD
SDEG=SS*RTOD
MDEG=MM*RTOD
TDEG=2*TH*RTOD
GOSUB CYLDRAW
LOCATE 8,1:PRINT"SQUISH AREA RATIO=";USING"#.###";SA
LOCATE 9,1:PRINT"PORT WIDTH/BORE RATIO=";USING"#.##";WR
LOCATE 10,1:PRINT"DEFLECTION RATIO=";USING"#.##";SCEFF
LOCATE 11,1:PRINT"TRANSFER HEIGHT & AREA=";USING
"#####.#";HBDC,PAREA
LOCATE 12,1:PRINT"SIDE, MIDDLE PORT WIDTHS,mm=";USING
"###.#";WS,WM
LOCATE 13,1:PRINT"SIDE PORT ANGLE,deg=";USING"##.#";SDEG
LOCATE 14,1:PRINT"MIDDLE PORT ANGLE,deg=";USING"##.#";MDEG
LOCATE 15,1:PRINT"DEFLECTOR ANGLE,deg=";USING"###.#";TDEG
LOCATE 16,1:PRINT"DEFLECTION RATIO,DY/BORE=";USING
"#.###";YB/BO
LOCATE 17,1:PRINT"DEFLECTOR POSITION,mm,DX & DY=";USING
"###.#";BO-YB,YB
LOCATE 18,1:PRINT"PORT-DEFLECTOR ANGLE,deg=";USING
"##.#";PHI
LOCATE 19,1:PRINT"SQUISH AREA,mm2,=";USING"####";SQAREA
LOCATE 20,1:PRINT"DEFLECTOR FLOW AREA,mm2,=";USING
"####";NS
IF ANS$="P" THEN LOCATE 23,1:PRINT TIT$
```

```
IF ANS$="P" THEN CLOSE#1
108 LOCATE 25,1:PRINT"TYPE CODE FOR NEW DATA AND A RECALCULATION"
LOCATE 26,1:INPUT"OR TYPE CODE FOR PRINTING(P), FILING(F) 0R QUITTING(Q) ?";ANS$
IF ANS$="B" THEN INPUT"TYPE IN NEW VALUE OF CYLINDER BORE,mm";BO
IF ANS$="S" THEN INPUT"TYPE IN NEW VALUE OF CYLINDER STROKE,mm";ST
IF ANS$="C" THEN INPUT"TYPE IN NEW VALUE OF CON-ROD CENTRES,mm";CRL
IF ANS$="T" THEN INPUT"TYPE IN NEW VALUE OF TRANSFER PORT TIMING,deg atdc";TRO
IF ANS$="SM" THEN INPUT"TYPE IN NEW VALUE OF SIDE/MIDDLE PORT WIDTH RATIO";SM
IF ANS$="CR" THEN INPUT"TYPE IN NEW VALUE OF PORT CORNER RADII,mm";PCR
IF ANS$="WD" THEN INPUT"TYPE IN NEW VALUE OF DEFLECTOR WIDTH,mm";WD
IF ANS$="DR" THEN INPUT"TYPE IN NEW VALUE OF DEFLECTOR PLAN RADIUS,mm";LD
IF ANS$="BC" THEN INPUT"TYPE IN NEW VALUE OF PORT BAR CHORDAL WIDTH,mm";BA
IF ANS$="P" THEN INPUT"TYPE IN A PRINT-OUT TITLE";TIT$
IF ANS$="P" THEN GOTO 107
IF ANS$="F" THEN GOSUB FILING
IF ANS$="Q" THEN GOTO 257
INPUT" FOR MORE NEW DATA TYPE 'OK'";OK$
IF OK$="OK" THEN GOTO 108
GOTO 107
257 END
REM PISTON POSITION SUBROUTINE
PIST:
DR=DEG*DTOR
HTDC=CRK+CRL-CRK*COS(DR)-SQR(CRL*CRL-(CRK*SIN(DR))^2)
HBDC=ST-HTDC
RETURN
DATAIN:
BO=70
ST=70
CRL=140
TRO=118
SM=1
PCR=3
```

```
WD=65
LD=120
BA=5
RETURN
FILING:
INPUT "TYPE IN TITLE NEEDED FOR THE FILE WITHA '.3PORT' END
CODE";P$
OPEN P$ FOR OUTPUT AS #1
WRITE#1,BO,ST,CRL,TRO,SM,PCR,WD,LD,BA
FINISH:
CLOSE#1
RETURN
FILED:
INPUT "TYPE IN THE FILENAME WITH A '.3PORT' END CODE";F$
OPEN"I",#1,F$
INPUT#1,BO,ST,CRL,TRO,SM,PCR,WD,LD,BA
CLOSE#1
RETURN
CYLDRAW:
CX=400
CY=235
CR=100
CIRCLE(CX,CY),CR
LTH=.12*CR
CR2=CR+LTH
CIRCLE(CX,CY),CR2
PP=2*PI
ZP=25
CR3=CR2+ZP
AST=PP-TH
CALL PORTEDGE(CX,CY,CR,CR3,AST)
AF=AST+SS
CALL PORTEDGE(CX,CY,CR,CR3,AF)
CIRCLE(CX,CY),CR3,,AST,AF
AF=TH
CALL PORTEDGE(CX,CY,CR,CR3,AF)
AST=TH-SS
CALL PORTEDGE(CX,CY,CR,CR3,AST)
CIRCLE(CX,CY),CR3,,AST,AF
AST=PP-MM/2
CALL PORTEDGE(CX,CY,CR,CR3,AST)
AF=MM/2
CALL PORTEDGE(CX,CY,CR,CR3,AF)
CIRCLE(CX,CY),CR3,,AST,AF
```

```
X5=CX-CR2-20
X8=CX+CR3+20
Y6=CY-CR2-20
Y7=CY+CR2+20
LINE(X5,CY)-(X8,CY)
LINE(CX,Y6)-(CX,Y7)
BG=YB*CR/RB
DR=LD*CR/RB
CC=CX+CR+DR-BG
AST=PI-AL
AF=PI+AL
CIRCLE(CC,CY),DR,,AST,AF
RETURN
SUB PORTEDGE(X,Y,R1,R2,PH)STATIC
C=COS(PH)
S=SIN(PH)
X1=X+R1*C
X2=X+R2*C
Y1=Y+R1*S
Y2=Y+R2*S
LINE(X1,Y1)-(X2,Y2)
END SUB
```

COMPUTER PROGRAM PROG.3.4
PROGRAM NAME "LOOP SCAVENGE PORT DESIGN"
ProgList.3.4

```
REM Prog.3.4
REM Program 'LOOP SCAVENGE PORT DESIGN'
REM ProgList.3.4
REM This program designs the scavenge port plan geometry for
REM loop scavenged two-stroke engines which have
REM between two and five transfer ports.
PI=3.1415927#
DTOR=PI/180
RTOD=180/PI
CLS
GOSUB TWOPORT
GOSUB THREEPORT
GOSUB FOURPORT
GOSUB FIVEPORT
LOCATE 15,1:PRINT"WHICH TYPE OF PORT DESIGN DO YOU WISH TO EMPLOY?"
LOCATE 17,1:PRINT"TYPE 'TWO' FOR THE TWO-PORT DESIGN, 'THREE' FOR THE THREE-PORT DESIGN, ETC,....."
```

```
INPUT"TYPE IN HERE YOUR CHOICE OF PORT TYPE";PRT$
IF PRT$="TWO" THEN GOSUB TWOP
IF PRT$="THREE" THEN GOSUB THREEP
IF PRT$="FOUR" THEN GOSUB FOURP
IF PRT$="FIVE" THEN GOSUB FIVEP
200 END
DRAWCYL:
REM CYLINDER BORE
YC=100
RC=50
CIRCLE(XC,YC),RC
LINE(XC-70,YC)-(XC+70,YC)
LINE(XC,YC-70)-(XC,YC+70)
RETURN
DRAWEXHPORT:
REM EXHAUST PORT
LINE (XC-40,YC+30)-(XC-65,YC+30)
LINE (XC-40,YC-30)-(XC-65,YC-30)
LINE(XC-65,YC+30)-(XC-65,YC-30)
RETURN
DRAWMAINPORT:
REM MAIN TRANSFER PORT
LINE (XC-20,YC+45)-(XC-30,YC+63)
LINE(XC-30,YC+63)-(XC+3,YC+80)
LINE(XC+3,YC+80)-(XC+23,YC+44)
LINE (XC-20,YC-45)-(XC-30,YC-63)
LINE(XC-30,YC-63)-(XC+3,YC-80)
LINE(XC+3,YC-80)-(XC+23,YC-44)
RETURN
DRAWSIDEPORT:
REM SIDE TRANSFER PORTS
LINE(XC+25,YC+43)-(XC+25,YC+60)
LINE (XC+25,YC+60)-(XC+40,YC+60)
LINE(XC+40,YC+60)-(XC+40,YC+30)
LINE(XC+25,YC-43)-(XC+25,YC-60)
LINE (XC+25,YC-60)-(XC+40,YC-60)
LINE(XC+40,YC-60)-(XC+40,YC-30)
RETURN
DRAWREARPORT:
REM REAR TRANSFER PORT
LINE(XC+46,YC+20)-(XC+65,YC+20)
LINE(XC+46,YC-20)-(XC+65,YC-20)
LINE(XC+65,YC+20)-(XC+65,YC-20)
RETURN
```

```
ONESLOPEDSIDEPORT:
REM SIDE TRANSFER PORTS
LINE(XC+25,YC+43)-(XC+20,YC+60)
LINE (XC+20,YC+60)-(XC+40,YC+60)
LINE(XC+40,YC+60)-(XC+40,YC+30)
LINE(XC+25,YC-43)-(XC+20,YC-60)
LINE (XC+20,YC-60)-(XC+40,YC-60)
LINE(XC+40,YC-60)-(XC+40,YC-30)
RETURN
RADIALREARPORT:
REM RADIAL TRANSFER PORTS
LINE(XC+25,YC+43)-(XC+35,YC+60)
LINE (XC+35,YC+60)-(XC+55,YC+40)
LINE(XC+55,YC+40)-(XC+40,YC+30)
LINE(XC+25,YC-43)-(XC+35,YC-60)
LINE (XC+35,YC-60)-(XC+55,YC-40)
LINE(XC+55,YC-40)-(XC+40,YC-30)
RETURN
TWOPORT:
XC=75
GOSUB DRAWCYL
GOSUB DRAWEXHPORT
GOSUB DRAWMAINPORT
LOCATE 13,1:PRINT"TWO TRANSFER PORTS"
RETURN
THREEPORT:
XC=225
GOSUB DRAWCYL
GOSUB DRAWEXHPORT
GOSUB DRAWMAINPORT
GOSUB DRAWREARPORT
LOCATE 13,19:PRINT"THREE TRANSFER PORTS"
RETURN
FOURPORT:
XC=375
GOSUB DRAWCYL
GOSUB DRAWEXHPORT
GOSUB DRAWMAINPORT
GOSUB DRAWSIDEPORT
LOCATE 13,39:PRINT"FOUR TRANSFER PORTS"
RETURN
FIVEPORT:
XC=525
GOSUB DRAWCYL
```

```
GOSUB DRAWEXHPORT
GOSUB DRAWMAINPORT
GOSUB DRAWREARPORT
GOSUB DRAWSIDEPORT
LOCATE 13,59:PRINT"FIVE TRANSFER PORTS"
RETURN
REM THIS IS THE LISTING FOR TWO SIDE TRANSFER PORTS
TWOP:
GOSUB COMMON1
INPUT"TYPE IN THE TITLE FOR THE ENGINE PORTING DESIGN";TI$
INPUT"NEW DATA (N) OR FILED DATA (F)";FZ$
IF FZ$="F" THEN GOSUB FILED
100 CLS
IF F$="P" THEN INPUT"NEEDING A CHANGED TITLE FOR THE EN-
GINE PORTING DESIGN";AN$
IF AN$="Y" THEN INPUT"TYPE IN THE NEW TITLE";TI$
IF F$="P" THEN OPEN"LPT1:PROMPT" FOR OUTPUT AS #1
IF F$="P" THEN WINDOW OUTPUT#1
UPMR=UPM*DTOR
REM TWO PORTS
SF=2*RR/BO
RRL=RR+CYT*SF
EW=EWS*SF
RRE=RR-5
DP1=55
DP2=35
GOSUB MAINPRT
GOSUB COMMON2
LOCATE 19,1:PRINT"OUTPUT DATA"
LOCATE 20,1:PRINT"MAIN PORT PLAN WIDTH=";USING"###.#";WID1S
TTEW=2*WID1S*COS(UPMR)
MWR=TTEW/BO
LOCATE 21,1:PRINT"TOTAL EFFECTIVE PORT WIDTH=";USING
"###.#";TTEW
LOCATE 22,1:PRINT"TOTAL EFFECTIVE WIDTH/BORE RATIO
=";USING"#.##";MWR
IF F$="P" THEN GOSUB PAGE2
60 LOCATE 24,1:PRINT"                                        "
IF F$="P" THEN CLOSE #1
65 LOCATE 24,1:PRINT"                                        "
LOCATE 24,1:PRINT"                                        "
LOCATE 25,1:PRINT"                                        "
LOCATE 24,1:PRINT"TYPE CODE FOR NEW DATA FOR 'BORE', 'LINER',
'EXHAUST', 'UP', 'AM1', 'AM2', 'MT1', OR 'MT2'"
```

```
LOCATE 25,1:INPUT"TYPE 'R' FOR RECALCULATING, 'P' FOR PRINT-
OUT, 'F' FOR FILING DATA, OR 'Q' FOR QUITTING";F$
GOSUB COMMON3
IF F$="R" THEN GOTO 100
IF F$="P" THEN GOTO 100
IF F$="F" THEN GOSUB FILING
IF F$="Q" THEN GOTO 200
GOTO 65
RETURN
REM
REM THREE PORTS
REM THIS STARTS THE LISTING FOR A THREE PORT ENGINE
THREEP:
GOSUB COMMON1
UPR=50
RWS=25
INPUT"TYPE IN THE TITLE FOR THE ENGINE PORTING DESIGN";TI$
INPUT"NEW DATA (N) OR FILED DATA (F)";FZ$
IF FZ$="F" THEN GOSUB FILED
300 CLS
IF F$="P" THEN INPUT"NEEDING A CHANGED TITLE FOR THE EN-
GINE PORTING DESIGN";AN$
IF AN$="Y" THEN INPUT"TYPE IN THE NEW TITLE";TI$
IF F$="P" THEN OPEN"LPT1:PROMPT" FOR OUTPUT AS #1
IF F$="P" THEN WINDOW OUTPUT#1
UPMR=UPM*DTOR
UPRR=UPR*DTOR
SF=2*RR/BO
RRL=RR+CYT*SF
EW=EWS*SF
RW=RWS*SF
RRE=RR-5
DP1=55
DP2=35
GOSUB MAINPRT
GOSUB COMMON2
LOCATE 12,1:PRINT "REAR PORT, UPRº AT";UPR
LOCATE 13,1:PRINT"REAR PORT PLAN WIDTH=";RWS
LOCATE 19,1:PRINT"OUTPUT DATA"
LOCATE 20,1:PRINT"MAIN PORT PLAN WIDTH=";USING"###.#";WID1S
TTEW=2*WID1S*COS(UPMR)+RWS*COS(UPRR)
MWR=TTEW/BO
LOCATE 21,1:PRINT"TOTAL EFFECTIVE PORT WIDTH=";USING
"###.#";TTEW
```

```
LOCATE 22,1:PRINT"TOTAL EFFECTIVE WIDTH/BORE RATIO
=";USING"#.##";MWR
IF F$="P" THEN GOSUB PAGE2
360 LOCATE 24,1:PRINT"                                                "
IF F$="P" THEN CLOSE #1
365 LOCATE 23,1:PRINT"                                                "
REM THREE PORTS
LOCATE 24,1:PRINT"                                                "
LOCATE 25,1:PRINT"                                                "
LOCATE 23,1:PRINT"TYPE CODE FOR NEW DATA FOR 'BORE', 'LINER',
'EXHAUST', 'UPM', 'AM1', 'AM2', 'MT1', OR 'MT2'"
LOCATE 24,1:PRINT"OR TYPE CODE FOR REAR PORT DATA, 'UPR'
FOR ANGLE, OR 'REAR' FOR THE EFFECTIVE WIDTH"
LOCATE 25,1:INPUT"TYPE 'R' FOR RECALCULATING, 'P' FOR PRINT-
OUT, 'F' FOR FILING DATA, OR 'Q' FOR QUITTING";F$
GOSUB COMMON3
IF F$="UPR" THEN INPUT"TYPE IN NEW VALUE FOR UPRº";UPR
IF F$="REAR" THEN INPUT"TYPE IN NEW VALUE FOR REAR PORT
PLAN WIDTH";RWS
IF F$="R" THEN GOTO 300
IF F$="P" THEN GOTO 300
IF F$="F" THEN GOSUB FILING
IF F$="Q" THEN GOTO 200
GOTO 365
RETURN
FOURP:
NPRT$="FOUR"
CLS
LOCATE 1,1:PRINT"THE WALLS OF THE SIDE TRANSFER PORTS ARE
EITHER-"
XC=75
GOSUB DRAWCYL
GOSUB DRAWEXHPORT
GOSUB DRAWMAINPORT
GOSUB DRAWSIDEPORT
LOCATE 14,4:PRINT"BOTH STRAIGHT"
XC=300
GOSUB DRAWCYL
GOSUB DRAWEXHPORT
GOSUB DRAWMAINPORT
GOSUB ONESLOPEDSIDEPORT
LOCATE 14,33:PRINT"ONE SLOPED"
XC=525
GOSUB DRAWCYL
```

```
GOSUB DRAWEXHPORT
GOSUB DRAWMAINPORT
GOSUB RADIALREARPORT
LOCATE 14,60:PRINT" RADIAL PORTS"
LOCATE 15,1:INPUT"TYPE IN THE NAME OF THE SIDE PORT CON-
FIGURATION YOU WANT";PRT$
IF PRT$="BOTH STRAIGHT" THEN GOSUB FOURPST
IF PRT$="ONE SLOPED" THEN GOSUB FOURP1S
IF PRT$="RADIAL" THEN GOSUB FOURPR
RETURN
REM
REM FOUR PORTS RADIAL WALLED SIDE TRANSFER PORTS
FOURPR:
GOSUB COMMON1
UPRS=20
MBRS=4
SBRS=30
INPUT"TYPE IN THE TITLE FOR THE ENGINE PORTING DESIGN";TI$
INPUT"NEW DATA (N) OR FILED DATA (F)";FZ$
IF FZ$="F" THEN GOSUB FILED
4400 CLS
IF F$="P" THEN INPUT"NEEDING A CHANGED TITLE FOR THE EN-
GINE PORTING DESIGN";AN$
IF AN$="Y" THEN INPUT"TYPE IN THE NEW TITLE";TI$
IF F$="P" THEN OPEN"LPT1:PROMPT" FOR OUTPUT AS #1
IF F$="P" THEN WINDOW OUTPUT#1
UPMR=UPM*DTOR
UPRSR=UPRS*DTOR
SF=2*RR/BO
RRL=RR+CYT*SF
EW=EWS*SF
SBR=SBRS*SF
RRE=RR-5
MBR=MBRS*SF
DP1=55
DP2=35
DPS=35
GOSUB MAINPRT
GOSUB RADIALSIDEPORTS
GOSUB COMMON2
LOCATE 12,1:PRINT "SIDE PORT, UPRSº AT";UPRS
LOCATE 13,1:PRINT"MAIN-SIDE BAR, MBAR=";MBRS
LOCATE 14,1:PRINT"SIDE-SIDE BAR, SBAR=";SBRS
LOCATE 18,1:PRINT"OUTPUT DATA"
```

```
LOCATE 19,1:PRINT"MAIN PORT PLAN WIDTH=";USING"###.#";WID1S
LOCATE 20,1:PRINT"SIDE PORT PLAN WIDTH=";USING"###.#";WIDSS
TTEW=2*WID1S*COS(UPMR)+2*WIDSS*COS(UPRSR)
MWR=TTEW/BO
LOCATE 21,1:PRINT"TOTAL EFFECTIVE PORT WIDTH=";USING
"###.#";TTEW
LOCATE 22,1:PRINT"TOTAL EFFECTIVE WIDTH/BORE RATIO
=";USING"#.##";MWR
IF F$="P" THEN GOSUB PAGE2
4460 LOCATE 24,1:PRINT"                                        "
IF F$="P" THEN CLOSE #1
4465 LOCATE 23,1:PRINT"                                        "
REM FOUR PORTS WITH RADIAL SIDE PORTS
LOCATE 23,1:PRINT"                                        "
LOCATE 24,1:PRINT"                                        "
LOCATE 25,1:PRINT"                                        "
LOCATE 26,1:PRINT"                                        "
LOCATE 23,1:PRINT"TYPE CODE FOR NEW DATA FOR 'BORE', 'LINER',
'EXHAUST', 'UPM', 'AM1', 'AM2', 'MT1', OR 'MT2'"
LOCATE 24,1:PRINT"OR TYPE CODE FOR SIDE PORT DATA, 'UPRS'
FOR ANGLE, OR 'MBAR' OR 'SBAR' FOR THE BAR WIDTHS"
LOCATE 25,1:INPUT"TYPE 'R' FOR RECALCULATING, 'P' FOR PRINT-
OUT, 'F' FOR FILING DATA, OR 'Q' FOR QUITTING";F$
GOSUB COMMON3
IF F$="UPRS" THEN INPUT"TYPE IN NEW VALUE FOR UPRSº";UPRS
IF F$="MBAR" THEN INPUT"TYPE IN NEW VALUE FOR NOMINAL
MAIN-SIDE PORT BAR";MBRS
IF F$="SBAR" THEN INPUT"TYPE IN NEW VALUE FOR SIDE-SIDE
PORT BAR";SBRS
IF F$="R" THEN GOTO 4400
IF F$="P" THEN GOTO 4400
IF F$="F" THEN GOSUB FILING
IF F$="Q" THEN GOTO 200
GOTO 4465
RETURN
REM
REM FOUR PORTS BOTH STRAIGHT SIDE-WALLED SIDE TRANSFER
PORTS
FOURPST:
GOSUB COMMON1
UPS=20
MBRS=4
SBRS=30
INPUT"TYPE IN THE TITLE FOR THE ENGINE PORTING DESIGN";TI$
```

```
INPUT"NEW DATA (N) OR FILED DATA (F)";FZ$
IF FZ$="F" THEN GOSUB FILED
400 CLS
IF F$="P" THEN INPUT"NEEDING A CHANGED TITLE FOR THE ENGINE PORTING DESIGN";AN$
IF AN$="Y" THEN INPUT"TYPE IN THE NEW TITLE";TI$
IF F$="P" THEN OPEN"LPT1:PROMPT" FOR OUTPUT AS #1
IF F$="P" THEN WINDOW OUTPUT#1
UPMR=UPM*DTOR
UPSR=UPS*DTOR
SF=2*RR/BO
RRL=RR+CYT*SF
EW=EWS*SF
SBR=SBRS*SF
RRE=RR-5
MBR=MBRS*SF
DP1=55
DP2=35
DPS=35
GOSUB MAINPRT
GOSUB STRAIGHTSIDEPORTS
GOSUB COMMON2
LOCATE 12,1:PRINT "SIDE PORT, UPSº AT";UPS
LOCATE 13,1:PRINT"MAIN-SIDE BAR, MBAR=";MBRS
LOCATE 14,1:PRINT"SIDE-SIDE BAR, SBAR=";SBRS
LOCATE 18,1:PRINT"OUTPUT DATA"
LOCATE 19,1:PRINT"MAIN PORT PLAN WIDTH=";USING"###.#";WID1S
LOCATE 20,1:PRINT"SIDE PORT PLAN WIDTH=";USING"###.#";WIDSS
TTEW=2*WID1S*COS(UPMR)+2*WIDSS*COS(UPSR)
MWR=TTEW/BO
LOCATE 21,1:PRINT"TOTAL EFFECTIVE PORT WIDTH=";USING "###.#";TTEW
LOCATE 22,1:PRINT"TOTAL EFFECTIVE WIDTH/BORE RATIO =";USING"#.##";MWR
IF F$="P" THEN GOSUB PAGE2
460 LOCATE 24,1:PRINT"                                        "
IF F$="P" THEN CLOSE #1
465 LOCATE 23,1:PRINT"                                        "
REM FOUR PORTS BOTH STRAIGHT
LOCATE 23,1:PRINT"                                        "
LOCATE 24,1:PRINT"                                        "
LOCATE 25,1:PRINT"                                        "
LOCATE 26,1:PRINT"                                        "
LOCATE 23,1:PRINT"TYPE CODE FOR NEW DATA FOR 'BORE', 'LINER', 'EXHAUST', 'UPM', 'AM1', 'AM2', 'MT1', OR 'MT2'"
```

```
LOCATE 24,1:PRINT"OR TYPE CODE FOR SIDE PORT DATA, 'UPS' FOR ANGLE, OR 'MBAR' OR 'SBAR' FOR THE BAR WIDTHS"
LOCATE 25,1:INPUT"TYPE 'R' FOR RECALCULATING, 'P' FOR PRINT-OUT, 'F' FOR FILING DATA, OR 'Q' FOR QUITTING";F$
GOSUB COMMON3
IF F$="UPS" THEN INPUT"TYPE IN NEW VALUE FOR UPSº";UPS
IF F$="MBAR" THEN INPUT"TYPE IN NEW VALUE FOR NOMINAL MAIN-SIDE PORT BAR";MBRS
IF F$="SBAR" THEN INPUT"TYPE IN NEW VALUE FOR SIDE-SIDE PORT BAR";SBRS
IF F$="R" THEN GOTO 400
IF F$="P" THEN GOTO 400
IF F$="F" THEN GOSUB FILING
IF F$="Q" THEN GOTO 200
GOTO 465
RETURN
REM
REM FOUR PORTS ONE STRAIGHT AND ONE SLOPED WALL SIDE TRANSFER PORTS
FOURP1S:
GOSUB COMMON1
UPS=20
MBRS=4
SBRS=30
INPUT"TYPE IN THE TITLE FOR THE ENGINE PORTING DESIGN";TI$
INPUT"NEW DATA (N) OR FILED DATA (F)";FZ$
IF FZ$="F" THEN GOSUB FILED
4100 CLS
IF F$="P" THEN INPUT"NEEDING A CHANGED TITLE FOR THE ENGINE PORTING DESIGN";AN$
IF AN$="Y" THEN INPUT"TYPE IN THE NEW TITLE";TI$
IF F$="P" THEN OPEN"LPT1:PROMPT" FOR OUTPUT AS #1
IF F$="P" THEN WINDOW OUTPUT#1
UPMR=UPM*DTOR
UPSR=UPS*DTOR
SF=2*RR/BO
RRL=RR+CYT*SF
EW=EWS*SF
SBR=SBRS*SF
RRE=RR-5
MBR=MBRS*SF
DP1=55
DP2=35
DPS=30
GOSUB MAINPRT
```

```
GOSUB SLOPESIDEPORTS
GOSUB COMMON2
LOCATE 12,1:PRINT "SIDE PORT, UPSº AT";UPS
LOCATE 13,1:PRINT"MAIN-SIDE BAR, MBAR=";MBRS
LOCATE 14,1:PRINT"SIDE-SIDE BAR, SBAR=";SBRS
LOCATE 15,1:PRINT"SIDEWALL ANGLE, SM1=";PS2
LOCATE 18,1:PRINT"OUTPUT DATA"
LOCATE 19,1:PRINT"MAIN PORT PLAN WIDTH=";USING"###.#";WID1S
LOCATE 20,1:PRINT"SIDE PORT PLAN WIDTH=";USING"###.#";WIDSS
TTEW=2*WID1S*COS(UPMR)+2*WIDSS*COS(UPSR)
MWR=TTEW/BO
LOCATE 21,1:PRINT"TOTAL EFFECTIVE PORT WIDTH=";USING
"###.#";TTEW
LOCATE 22,1:PRINT"TOTAL EFFECTIVE WIDTH/BORE RATIO
=";USING"#.##";MWR
IF F$="P" THEN GOSUB PAGE2
4160 LOCATE 24,1:PRINT"                                         "
IF F$="P" THEN CLOSE #1
4165 LOCATE 23,1:PRINT"                                         "
REM FOUR PORTS ONE SLOPED
LOCATE 23,1:PRINT"                                         "
LOCATE 24,1:PRINT"                                         "
LOCATE 25,1:PRINT"                                         "
LOCATE 26,1:PRINT"                                         "
LOCATE 23,1:PRINT"TYPE CODE FOR NEW DATA FOR 'BORE', 'LINER',
'EXHAUST', 'UPM', 'AM1', 'AM2', 'MT1', OR 'MT2'"
LOCATE 24,1:PRINT"OR FOR SIDE PORT DATA, 'UPS' FOR ANGLE,
'MBAR' OR 'SBAR' WIDTHS, OR 'SM1' FOR ANGLE"
LOCATE 25,1:INPUT"TYPE 'R' FOR RECALCULATING, 'P' FOR PRINT-
OUT, 'F' FOR FILING DATA, OR 'Q' FOR QUITTING";F$
GOSUB COMMON3
IF F$="UPS" THEN INPUT"TYPE IN NEW VALUE FOR UPSº";UPS
IF F$="MBAR" THEN INPUT"TYPE IN NEW VALUE FOR NOMINAL
MAIN-SIDE PORT BAR,MBAR";MBRS
IF F$="SBAR" THEN INPUT"TYPE IN NEW VALUE FOR SIDE-SIDE
PORT BAR, SBAR";SBRS
IF F$="SM1" THEN INPUT"TYPE IN NEW VALUE FOR SIDEWALL
ANGLE,SM1";PS2
IF F$="R" THEN GOTO 4100
IF F$="P" THEN GOTO 4100
IF F$="F" THEN GOSUB FILING
IF F$="Q" THEN GOTO 200
GOTO 4165
RETURN
```

```
FIVEP:
NPRT$="FIVE"
CLS
LOCATE 1,1:PRINT"THE WALLS OF THE SIDE TRANSFER PORTS ARE
EITHER-"
XC=75
GOSUB DRAWCYL
GOSUB DRAWEXHPORT
GOSUB DRAWMAINPORT
GOSUB DRAWSIDEPORT
GOSUB DRAWREARPORT
LOCATE 14,5:PRINT"BOTH STRAIGHT"
XC=300
GOSUB DRAWCYL
GOSUB DRAWEXHPORT
GOSUB DRAWMAINPORT
GOSUB ONESLOPEDSIDEPORT
GOSUB DRAWREARPORT
LOCATE 14,34:PRINT"ONE SLOPED"
XC=525
GOSUB DRAWCYL
GOSUB DRAWEXHPORT
GOSUB DRAWMAINPORT
GOSUB RADIALREARPORT
GOSUB DRAWREARPORT
LOCATE 14,65:PRINT"RADIAL"
LOCATE 15,1:INPUT"TYPE IN THE NAME OF THE SIDE PORT CON-
FIGURATION YOU WANT";PRT5$
IF PRT5$="BOTH STRAIGHT" THEN GOSUB FIVEPST
IF PRT5$="ONE SLOPED" THEN GOSUB FIVEP1S
IF PRT5$="RADIAL" THEN GOSUB FIVEPR
RETURN
REM
REM FIVE PORTS RADIAL WALLED SIDE TRANSFER PORTS
FIVEPR:
GOSUB COMMON1
UPRS=20
UPB=50
RWS=20
MBRS=4
SBRS=30
INPUT"TYPE IN THE TITLE FOR THE ENGINE PORTING DESIGN";TI$
INPUT"NEW DATA (N) OR FILED DATA (F)";FZ$
IF FZ$="F" THEN GOSUB FILED
```

```
5400 CLS
IF F$="P" THEN INPUT"NEEDING A CHANGED TITLE FOR THE ENGINE PORTING DESIGN";AN$
IF AN$="Y" THEN INPUT"TYPE IN THE NEW TITLE";TI$
IF F$="P" THEN OPEN"LPT1:PROMPT" FOR OUTPUT AS #1
IF F$="P" THEN WINDOW OUTPUT#1
UPMR=UPM*DTOR
UPRSR=UPRS*DTOR
UPBR=UPB*DTOR
SF=2*RR/BO
RW=RWS*SF
RRL=RR+CYT*SF
EW=EWS*SF
SBR=SBRS*SF
RRE=RR-5
MBR=MBRS*SF
DP1=55
DP2=35
DPS=35
GOSUB MAINPRT
GOSUB RADIALSIDEPORTS
GOSUB COMMON2
LOCATE 11,1:PRINT "SIDE PORT, UPRSº AT";UPRS
LOCATE 12,1:PRINT"MAIN-SIDE BAR, MBAR=";MBRS
LOCATE 13,1:PRINT"SIDE-SIDE BAR, SBAR=";SBRS
LOCATE 15,1:PRINT"REAR PORT, UPBº AT";UPB
LOCATE 16,1:PRINT"REAR PORT WIDTH, RWS=";RWS
LOCATE 18,1:PRINT"OUTPUT DATA"
LOCATE 19,1:PRINT"MAIN PORT PLAN WIDTH=";USING"###.#";WID1S
LOCATE 20,1:PRINT"SIDE PORT PLAN WIDTH=";USING"###.#";WIDSS
TTEW=2*WID1S*COS(UPMR)+2*WIDSS*COS(UPRSR)+RWS*COS(UPBR)
MWR=TTEW/BO
LOCATE 21,1:PRINT"TOTAL EFFECTIVE PORT WIDTH=";USING "###.#";TTEW
LOCATE 22,1:PRINT"TOTAL EFFECTIVE WIDTH/BORE RATIO =";USING"#.##";MWR
IF F$="P" THEN GOSUB PAGE2
5460 LOCATE 24,1:PRINT"                                                  "
IF F$="P" THEN CLOSE #1
5465 LOCATE 23,1:PRINT"                                                  "
REM FIVE PORTS WITH RADIAL SIDE PORTS
LOCATE 23,1:PRINT"                                                  "
LOCATE 24,1:PRINT"                                                  "
LOCATE 25,1:PRINT"                                                  "
```

```
LOCATE 26,1:PRINT”                                                              “
LOCATE 23,1:PRINT”TYPE CODE FOR NEW DATA FOR ‘BORE’, ‘LINER’, ‘EXHAUST’, ‘UPM’, ‘AM1’, ‘AM2’, ‘MT1’, OR ‘MT2’”
LOCATE 24,1:PRINT”OR SIDE PORT-’UPRS’, ‘MBAR’, OR ‘SBAR’, AND FOR REAR PORT-’UPB’, OR ‘RWS’”
LOCATE 25,1:INPUT”TYPE ‘R’ FOR RECALCULATING, ‘P’ FOR PRINT-OUT, ‘F’ FOR FILING DATA, OR ‘Q’ FOR QUITTING”;F$
GOSUB COMMON3
IF F$=”UPRS” THEN INPUT”TYPE IN NEW VALUE FOR UPRSº”;UPRS
IF F$=”MBAR” THEN INPUT”TYPE IN NEW VALUE FOR NOMINAL MAIN-SIDE PORT BAR”;MBRS
IF F$=”SBAR” THEN INPUT”TYPE IN NEW VALUE FOR SIDE-SIDE PORT BAR”;SBRS
IF F$=”UPB” THEN INPUT”TYPE IN NEW VALUE FOR UPBº”;UPB
IF F$=”RWS” THEN INPUT”TYPE IN NEW VALUE FOR REAR PORT WIDTH”;RWS
IF F$=”R” THEN GOTO 5400
IF F$=”P” THEN GOTO 5400
IF F$=”F” THEN GOSUB FILING
IF F$=”Q” THEN GOTO 200
GOTO 5465
RETURN
REM
REM FIVE PORTS BOTH STRAIGHT SIDE-WALLED SIDE TRANSFER PORTS
FIVEPST:
GOSUB COMMON1
UPS=20
UPB=50
RWS=20
MBRS=4
SBRS=30
INPUT”TYPE IN THE TITLE FOR THE ENGINE PORTING DESIGN”;TI$
INPUT”NEW DATA (N) OR FILED DATA (F)”;FZ$
IF FZ$=”F” THEN GOSUB FILED
500 CLS
IF F$=”P” THEN INPUT”NEEDING A CHANGED TITLE FOR THE ENGINE PORTING DESIGN”;AN$
IF AN$=”Y” THEN INPUT”TYPE IN THE NEW TITLE”;TI$
IF F$=”P” THEN OPEN”LPT1:PROMPT” FOR OUTPUT AS  #1
IF F$=”P” THEN WINDOW OUTPUT#1
UPMR=UPM*DTOR
UPSR=UPS*DTOR
UPBR=UPB*DTOR
```

```
SF=2*RR/BO
RRL=RR+CYT*SF
EW=EWS*SF
RW=RWS*SF
SBR=SBRS*SF
RRE=RR-5
MBR=MBRS*SF
DP1=55
DP2=35
DPS=35
GOSUB MAINPRT
GOSUB STRAIGHTSIDEPORTS
GOSUB COMMON2
LOCATE 11,1:PRINT "SIDE PORT, UPSº AT";UPS
LOCATE 12,1:PRINT"MAIN-SIDE BAR, MBAR=";MBRS
LOCATE 13,1:PRINT"SIDE-SIDE BAR, SBAR=";SBRS
LOCATE 15,1:PRINT"REAR PORT, UPBº AT";UPB
LOCATE 16,1:PRINT"REAR PORT WIDTH, RWS=";RWS
LOCATE 18,1:PRINT"OUTPUT DATA"
LOCATE 19,1:PRINT"MAIN PORT PLAN WIDTH=";USING"###.#";WID1S
LOCATE 20,1:PRINT"SIDE PORT PLAN WIDTH=";USING"###.#";WIDSS
TTEW=2*WID1S*COS(UPMR)+2*WIDSS*COS(UPSR)+RWS*COS(UPBR)
MWR=TTEW/BO
LOCATE 21,1:PRINT"TOTAL EFFECTIVE PORT WIDTH=";USING
"###.#";TTEW
LOCATE 22,1:PRINT"TOTAL EFFECTIVE WIDTH/BORE RATIO
=";USING"#.##";MWR
IF F$="P" THEN GOSUB PAGE2
560 LOCATE 24,1:PRINT"                                            "
IF F$="P" THEN CLOSE #1
565 LOCATE 23,1:PRINT"                                            "
REM FIVE PORTS BOTH STRAIGHT SIDEWALL SIDE PORTS
LOCATE 23,1:PRINT"                                            "
LOCATE 24,1:PRINT"                                            "
LOCATE 25,1:PRINT"                                            "
LOCATE 26,1:PRINT"                                            "
LOCATE 23,1:PRINT"TYPE CODE FOR NEW DATA FOR 'BORE', 'LINER',
'EXHAUST', 'UPM', 'AM1', 'AM2', 'MT1', OR 'MT2'"
LOCATE 24,1:PRINT "FOR SIDE PORT-'UPS', 'MBAR', OR 'SBAR'; FOR
REAR PORT-'UPB' OR 'RWS' "
LOCATE 25,1:INPUT"TYPE 'R' FOR RECALCULATING, 'P' FOR PRINT-
OUT, 'F' FOR FILING DATA, OR 'Q' FOR QUITTING";F$
GOSUB COMMON3
IF F$="UPS" THEN INPUT"TYPE IN NEW VALUE FOR UPSº";UPS
```

```
IF F$="MBAR" THEN INPUT"TYPE IN NEW VALUE FOR NOMINAL MAIN-SIDE PORT BAR";MBRS
IF F$="SBAR" THEN INPUT"TYPE IN NEW VALUE FOR SIDE-SIDE PORT BAR";SBRS
IF F$="UPB" THEN INPUT"TYPE IN NEW VALUE FOR UPBº";UPB
IF F$="RWS" THEN INPUT"TYPE IN NEW VALUE FOR REAR PORT WIDTH";RWS
IF F$="R" THEN GOTO 500
IF F$="P" THEN GOTO 500
IF F$="F" THEN GOSUB FILING
IF F$="Q" THEN GOTO 200
GOTO 565
RETURN
REM
REM FIVE PORTS ONE STRAIGHT AND ONE SLOPED WALL SIDE TRANSFER PORTS AND REAR TRANSFER
FIVEP1S:
GOSUB COMMON1
UPS=20
UPB=50
RWS=20
MBRS=4
SBRS=30
INPUT"TYPE IN THE TITLE FOR THE ENGINE PORTING DESIGN";TI$
INPUT"NEW DATA (N) OR FILED DATA (F)";FZ$
IF FZ$="F" THEN GOSUB FILED
5100 CLS
IF F$="P" THEN INPUT"NEEDING A CHANGED TITLE FOR THE ENGINE PORTING DESIGN(Y/N?)";AN$
IF AN$="Y" THEN INPUT"TYPE IN THE NEW TITLE";TI$
IF F$="P" THEN OPEN"LPT1:PROMPT" FOR OUTPUT AS #1
IF F$="P" THEN WINDOW OUTPUT#1
UPMR=UPM*DTOR
UPSR=UPS*DTOR
UPBR=UPB*DTOR
SF=2*RR/BO
RW=RWS*SF
RRL=RR+CYT*SF
EW=EWS*SF
SBR=SBRS*SF
RRE=RR-5
MBR=MBRS*SF
DP1=55
DP2=35
```

```
DPS=30
GOSUB MAINPRT
GOSUB SLOPESIDEPORTS
GOSUB COMMON2
LOCATE 11,1:PRINT "SIDE PORT, UPSº AT";UPS
LOCATE 12,1:PRINT"MAIN-SIDE BAR, MBAR=";MBRS
LOCATE 13,1:PRINT"SIDE-SIDE BAR, SBAR=";SBRS
LOCATE 14,1:PRINT"SIDEWALL ANGLE, SM1=";PS2
LOCATE 16,1:PRINT"REAR PORT, UPBº AT";UPB
LOCATE 17,1:PRINT"REAR PORT WIDTH, RWS=";RWS
LOCATE 18,1:PRINT"OUTPUT DATA"
LOCATE 19,1:PRINT"MAIN PORT PLAN WIDTH=";USING"###.#";WID1S
LOCATE 20,1:PRINT"SIDE PORT PLAN WIDTH=";USING"###.#";WIDSS
TTEW=2*WID1S*COS(UPMR)+2*WIDSS*COS(UPSR)+RWS*COS(UPBR)
MWR=TTEW/BO
LOCATE 21,1:PRINT"TOTAL EFFECTIVE PORT WIDTH=";USING
"###.#";TTEW
LOCATE 22,1:PRINT"TOTAL EFFECTIVE WIDTH/BORE RATIO
=";USING"#.##";MWR
IF F$="P" THEN GOSUB PAGE2
5160 LOCATE 24,1:PRINT"                                    "
IF F$="P" THEN CLOSE #1
5165 LOCATE 23,1:PRINT"                                    "
REM FIVE PORTS ONE SLOPED
LOCATE 23,1:PRINT"                                     "
LOCATE 24,1:PRINT"                                     "
LOCATE 25,1:PRINT"                                     "
LOCATE 26,1:PRINT"                                     "
LOCATE 23,1:PRINT"TYPE CODE FOR NEW DATA FOR 'BORE', 'LINER',
'EXHAUST', 'UPM', 'AM1', 'AM2', 'MT1', OR 'MT2'"
LOCATE 24,1:PRINT"FOR SIDE PORT-'UPS', 'MBAR', 'SBAR', OR 'SM1';
FOR REAR PORT-'UPB' OR 'RWS'"
LOCATE 25,1:INPUT"TYPE 'R' FOR RECALCULATING, 'P' FOR PRINT-
OUT, 'F' FOR FILING DATA, OR 'Q' FOR QUITTING";F$
IF F$="BORE" THEN INPUT"TYPE IN NEW VALUE FOR BORE,MM";BO
GOSUB COMMON3
IF F$="UPS" THEN INPUT"TYPE IN NEW VALUE FOR UPSº";UPS
IF F$="MBAR" THEN INPUT"TYPE IN NEW VALUE FOR NOMINAL
MAIN-SIDE PORT BAR,MBAR";MBRS
IF F$="SBAR" THEN INPUT"TYPE IN NEW VALUE FOR SIDE-SIDE
PORT BAR, SBAR";SBRS
IF F$="SM1" THEN INPUT"TYPE IN NEW VALUE FOR SIDEWALL
ANGLE,SM1";PS2
IF F$="UPB" THEN INPUT"TYPE IN NEW VALUE FOR UPBº";UPB
```

```
IF F$="RWS" THEN INPUT"TYPE IN NEW VALUE FOR REAR PORT
WIDTH";RWS
IF F$="R" THEN GOTO 5100
IF F$="P" THEN GOTO 5100
IF F$="F" THEN GOSUB FILING
IF F$="Q" THEN GOTO 200
GOTO 5165
RETURN
MAINPRT:
GOSUB BOREL
TG1=TG1S*SF
XT1=XX+TG1
YT1=YY
PH=PH1
XT=XT1-XX
YT=YT1-YY
RRP=RRL+DP1
GOSUB INTERCEPT
M1=M
PH1R=PHR
XM1=XM+XX
YM1D=YM+YY
YM1U=YY-YM
XE1=XE+XX
XM1S=ABS(XM/SF)
YE1D=YE+YY
YE1U=YY-YE
XL1=XL+XX
XL1S=XL/SF
YL1D=YL+YY
YL1U=YY-YL
TG2=TG2S*SF
XT2=XX+TG2
YT2=YY
PH=PH2
XT=XT2-XX
YT=YT2-YY
RRP=RRL+DP2
GOSUB INTERCEPT
M2=M
PH2R=PHR
XM2=XM+XX
YM2D=YM+YY
YM2U=YY-YM
```

```
XE2=XE+XX
XM2S=ABS(XM/SF)
YE2D=YE+YY
YE2U=YY-YE
XL2=XL+XX
XL2S=ABS(XL/SF)
YL2D=YL+YY
YL2U=YY-YL
XP2=XP+XX
YP2D=YP+YY
YP2U=YY-YP
REM
REM ***ALL Y VALUES SHOULD BE TAKEN FROM THE YTD VALUE
FOR ZERO REFERENCE***********
REM
REM CALCULATING INTERCEPT FROM LINE2 TO 90 DEG TO LINE1
FOR THE PORTBACK
M3=-1/M1
XP=XP2-XX
YP=YP2D-YY
XT=XT1-XX
YT=YT1-YY
YPI=(XP-XT+YT/M1-YP/M3)/(1/M1-1/M3)
XPI=(YPI-YT)/M1+XT
XP1=XPI+XX
YP1D=YPI+YY
YP1U=YY-YPI
REM TWO PORTS
REM CALCULATING INTERCEPT FROM LINE2 TO 90 DEG TO LINE1
FOR THE PORTWIDTH
XM=XM2-XX
YM=YM2D-YY
YPM=(XM-XT+YT/M1-YM/M3)/(1/M1-1/M3)
XPM=(YPM-YT)/M1+XT
WID1=SQR((XM-XPM)^2+(YM-YPM)^2)
WID1S=WID1/SF
RX=XX+3
RY=YT1-65
GOSUB LARROW
RX=XT1-3
RY=YT1-65
GOSUB RARROW
LINE (XT1,YT1-5)-(XT1,YT1-70)
LINE (XT1-3,YT1-65)-(XX+3,YT1-65)
```

```
LINE (XT2,YT2-5)-(XT2,YT2-40)
RX=XX+3
RY=YT2-35
GOSUB LARROW
RX=XT2-3
RY=YT2-35
GOSUB RARROW
LINE (XT2-3,YT1-35)-(XX+3,YT2-35)
CIRCLE (XT1,YT1),25,,PI-PH1R,PI
CIRCLE (XT2,YT2),25,,PI-PH2R,PI
LOCATE 11,47:PRINT USING"##";PH2
LOCATE 11,57:PRINT USING"##";PH1
LOCATE 12,26:PRINT EWS
LINE(XT1,YT1)-(XE1,YE1U)
LINE(XM1,YM1U)-(XP1,YP1U)
LINE(XM1,YM1D)-(XP1,YP1D)
LINE(XT2,YT2)-(XE2,YE2U)
LINE(XM2,YM2U)-(XP2,YP2U)
LINE(XM2,YM2D)-(XP2,YP2D)
LINE(XP1,YP1U)-(XP2,YP2U)
LINE(XP1,YP1D)-(XP2,YP2D)
RETURN
RADIALSIDEPORTS:
REM CALCULATING AND DRAWING RADIAL SIDE PORTS
XS=XM1-XX+MBR
YS=SQR(RR*RR-XS*XS)
YS2U=YY-YS
YS2D=YY+YS
XS2=XS+XX
YS=SBR/2
XS=SQR(RR*RR-YS*YS)
YS1U=YY-YS
YS1D=YY+YS
XS1=XX+XS
MBA=SQR((XS2-XM1)^2+(YS2U-YM1U)^2)
MBAS=MBA/SF
RBA=SQR((XRP-XS1)^2+(YRPU-YS1U)^2)
RBAS=RBA/SF
RS1=ATN((YY-YS1U)/(XS1-XX))
RS2=ATN((YY-YS2U)/(XS2-XX))
R3=RR+CYT+40
XS3=XX+R3*COS(RS2)
XS4=XX+R3*COS(RS1)
YS3=R3*SIN(RS2)
```

```
YS4=R3*SIN(RS1)
YS3U=YY-YS3
YS4U=YY-YS4
YS3D=YY+YS3
YS4D=YY+YS4
LINE (XS2,YS2U)-(XS3,YS3U)
LINE (XS1,YS1U)-(XS4,YS4U)
LINE (XS2,YS2D)-(XS3,YS3D)
LINE (XS1,YS1D)-(XS4,YS4D)
CIRCLE(XX,YY),R3,,RS1,RS2
CIRCLE(XX,YY),R3,,2*PI-RS2,2*PI-RS1
RE3=RR-5
XES3=XX+RE3*COS(RS2)
XES4=XX+RE3*COS(RS1)
YES3=RE3*SIN(RS2)
YES4=RE3*SIN(RS1)
YES3U=YY-YES3
YES4U=YY-YES4
YES3D=YY+YES3
YES4D=YY+YES4
LINE (XX,YY)-(XES3,YES3D)
LINE(XX,YY)-(XES4,YES4D)
CIRCLE(XX,YY),30,,2*PI-RS2,2*PI-RS1
RSD=(RS2-RS1)*RTOD
LOCATE 14,54:PRINT USING"##.#";RSD
WIDS=SQR((XS1-XS2)^2+(YS1U-YS2U)^2)
WIDSS=WIDS/SF
YYY=YS3U-14
LINE(XM1,YM1U-5)-(XM1,YYY-3)
LINE(XS2,YS2U-3)-(XS2,YYY-3)
LINE(XM1-3,YYY)-(XM1-20,YYY)
LINE(XS2+3,YYY)-(XS2+20,YYY)
RX=XM1-3
RY=YYY
GOSUB RARROW
RX=XS2+3
GOSUB LARROW
LOCATE 2,58:PRINT USING"#.#";MBRS
XRP=XS1+5
LINE(XRP,YS1U)-(XRP+60,YS1U)
LINE(XRP,YS1D)-(XRP+60,YS1D)
LINE(XRP+58,YS1D+3)-(XRP+58,YS1D+25)
LINE(XRP+58,YS1U-3)-(XRP+58,YS1U-25)
RX=XRP+58
```

```
RY=YS1D+3
GOSUB UARROW
RX=XRP+58
RY=YS1U-3
GOSUB DARROW
LOCATE 6,67:PRINT SBRS
LINE (XT2-3,YT1-35)-(XX+3,YT2-35)
CIRCLE (XT1,YT1),25,,PI-PH1R,PI
CIRCLE (XT2,YT2),25,,PI-PH2R,PI
LOCATE 11,47:PRINT USING"##";PH2
LOCATE 11,57:PRINT USING"##";PH1
LOCATE 12,26:PRINT EWS
RETURN
STRAIGHTSIDEPORTS:
REM CALCULATING AND DRAWING STRAIGHT SIDE PORTS
XS=XM1-XX+MBR
YS=SQR(RR*RR-XS*XS)
YS2U=YY-YS
YS2D=YY+YS
XS2=XS+XX
YS=SBR/2
XS=SQR(RR*RR-YS*YS)
YS1U=YY-YS
YS1D=YY+YS
XS1=XX+XS
MBA=SQR((XS2-XM1)^2+(YS2U-YM1U)^2)
MBAS=MBA/SF
RBA=SQR((XRP-XS1)^2+(YRPU-YS1U)^2)
RBAS=RBA/SF
WIDS=XS1-XS2
WIDSS=WIDS/SF
YS3U=YS2U-45
YS4U=YS3U
YS3D=YS2D+45
YS4D=YS3D
LINE (XS2,YS2U)-(XS2,YS3U)
LINE (XS1,YS1U)-(XS1,YS4U)
LINE(XS2,YS3U)-(XS1,YS4U)
LINE (XS2,YS2D)-(XS2,YS3D)
LINE (XS1,YS1D)-(XS1,YS4D)
LINE(XS2,YS3D)-(XS1,YS4D)
YYY=YS3U-14
LINE(XM1,YM1U-5)-(XM1,YYY-3)
LINE(XS2,YS3U-3)-(XS2,YYY-3)
```

```
LINE(XM1-3,YYY)-(XM1-20,YYY)
LINE(XS2+3,YYY)-(XS2+20,YYY)
RX=XM1-3
RY=YYY
GOSUB RARROW
RX=XS2+3
GOSUB LARROW
XRP=XS1+5
LINE(XRP,YS1U)-(XRP+60,YS1U)
LINE(XRP,YS1D)-(XRP+60,YS1D)
LINE(XRP+58,YS1D+3)-(XRP+58,YS1D+25)
LINE(XRP+58,YS1U-3)-(XRP+58,YS1U-25)
RX=XRP+58
RY=YS1D+3
GOSUB UARROW
RX=XRP+58
RY=YS1U-3
GOSUB DARROW
LOCATE 6,67:PRINT SBRS
LOCATE 2,58:PRINT USING"#.#";MBRS
LINE (XT2-3,YT1-35)-(XX+3,YT2-35)
CIRCLE (XT1,YT1),25,,PI-PH1R,PI
CIRCLE (XT2,YT2),25,,PI-PH2R,PI
LOCATE 11,47:PRINT USING"##";PH2
LOCATE 11,57:PRINT USING"##";PH1
LOCATE 12,26:PRINT EWS
RETURN
SLOPESIDEPORTS:
REM CALCULATING AND DRAWING SINGLE SLOPED SIDE PORTS
XS=XM1-XX+MBR
YS=SQR(RR*RR-XS*XS)
YS2U=YY-YS
YS2D=YY+YS
XS2=XS+XX
YS=SBR/2
XS=SQR(RR*RR-YS*YS)
YS1U=YY-YS
YS1D=YY+YS
XS1=XX+XS
MBA=SQR((XS2-XM1)^2+(YS2U-YM1U)^2)
MBAS=MBA/SF
WIDS=XS1-XS2
WIDSS=WIDS/SF
REM INSERTING THE CALCULATION FOR THE SLOPING WALL
```

```
XM=XS2-XX
YM=YS2D-YY
PH=PS2
YT=0
RRP=RRL+DPS
GOSUB INTERCEPTS
MS2=M
PS2R=PHR
YTS2=YT+YY
XTS2=XT+XX
XES2=XE+XX
XMS2S=ABS(XM/SF)
YES2D=YE+YY
YES2U=YY-YE
XLS2=XL+XX
XLS2S=XL/SF
YLS2D=YL+YY
YLS2U=YY-YL
XS3=XP+XX
YS3D=YP+YY
YS3U=YY-YP
YS4U=YS3U
YS4D=YS3D
LINE (XS2,YS2U)-(XS3,YS3U)
LINE (XS1,YS1U)-(XS1,YS4U)
LINE(XS3,YS3U)-(XS1,YS4U)
LINE (XS2,YS2D)-(XS3,YS3D)
LINE (XS1,YS1D)-(XS1,YS4D)
LINE(XS3,YS3D)-(XS1,YS4D)
YYY=YS3U-14
LINE(XM1,YM1U-5)-(XM1,YYY-3)
LINE(XS2,YS2U-5)-(XS2,YYY-3)
LINE(XM1-3,YYY)-(XM1-20,YYY)
LINE(XS2+3,YYY)-(XS2+20,YYY)
RX=XM1-3
RY=YYY
GOSUB RARROW
RX=XS2+3
GOSUB LARROW
LOCATE 2,58:PRINT USING"#.#";MBRS
XRP=XS1+5
LINE(XRP,YS1U)-(XRP+60,YS1U)
LINE(XRP,YS1D)-(XRP+60,YS1D)
RX=XRP+58
RY=YS1D+3
```

```
GOSUB UARROW
LINE(RX,RY)-(RX,RY+25)
RX=XRP+58
RY=YS1U-3
GOSUB DARROW
LINE(RX,RY)-(RX,RY-25)
LOCATE 7,66:PRINT SBRS
LINE(XTS2,YTS2)-(XES2,YES2D)
CIRCLE (XTS2,YTS2),25,,PI,PI+PS2R
LOCATE 13,55:PRINT USING"##";PS2
LINE (XT2-3,YT1-35)-(XX+3,YT2-35)
CIRCLE (XT1,YT1),25,,PI-PH1R,PI
CIRCLE (XT2,YT2),25,,PI-PH2R,PI
LOCATE 11,47:PRINT USING"##";PH2
LOCATE 11,57:PRINT USING"##";PH1
LOCATE 12,26:PRINT EWS
RETURN
COMMON1:
CLS
EW=120
GOSUB BOREL
ZZ=2
YDN=0
GOSUB UPSWEEP
LOCATE 12,28:PRINT"PORT"
LOCATE 13,28:PRINT"WIDTH"
LOCATE 20,1: PRINT"EFFECTIVE PORT WIDTH IS THE CHORD AT 90º TO THE GAS FLOW DIRECTION"
LOCATE 21,1:PRINT"TRUE EFFECTIVE WIDTH FOR FURTHER CALCULATIONS IS THE PLAN CHORDAL WIDTH*COSINE(UP)"
LOCATE 22,1:PRINT"ANY TRANSFER PORT WIDTH RATIOS CALCULATED ARE TOTAL EFFECTIVE WIDTH÷CYLINDER BORE"
LOCATE 25,1:PRINT"                                                  "
LOCATE 26,1:INPUT"TO PROCEED PRESS ANY KEY";V$
TG1S=30
PS2=70
PH1=60
TG2S=5
PH2=50
EWS=40
BO=70
CYT=5
UPM=10
RETURN
```

```
COMMON2:
LOCATE 1,1:PRINT"CURRENT INPUT DATA FOR ";TI$
LOCATE 2,1:PRINT"BORE DIAMETER=";BO
LOCATE 3,1:PRINT"LINER THICKNESS=";CYT
LOCATE 4,1:PRINT"MAXIMUM EXHAUST WIDTH=";EWS
LOCATE 7,51:PRINT USING"###.#";TG1S
LOCATE  9,50:PRINT USING"###.#";TG2S
LOCATE 6,1:PRINT "MAIN PORT, UPº AT";UPM
LOCATE 7,1:PRINT"TARGET MT1=";USING"###.#";TG1S
LOCATE 8,1:PRINT"TARGET MT2=";USING"###.#";TG2S
LOCATE 9,1:PRINT"ANGLE AM1=";PH1
LOCATE 10,1:PRINT"ANGLE AM2=";PH2
RETURN
COMMON3:
IF F$="BORE" THEN INPUT"TYPE IN NEW VALUE FOR BORE,MM";BO
IF F$="LINER" THEN INPUT"TYPE IN NEW VALUE FOR LINER THICKNESS,MM";CYT
IF F$="EXHAUST" THEN INPUT"TYPE IN NEW VALUE FOR EXHAUST PORT CHORDAL WIDTH,MM";EWS
IF F$="UP" THEN INPUT"TYPE IN NEW VALUE FOR UPº";UPM
IF F$="AM1" THEN INPUT"TYPE IN NEW VALUE FOR AM1º";PH1
IF F$="AM2" THEN INPUT"TYPE IN NEW VALUE FOR AM2º";PH2
IF F$="MT1" THEN INPUT"TYPE IN NEW VALUE FOR MT1";TG1S
IF F$="MT2" THEN INPUT"TYPE IN NEW VALUE FOR MT2";TG2S
RETURN
REM 2ND HALF OF OUTPUT PAGE
PAGE2:
YINC=350
ZZ=ZZ+22
LINE(XM1,YM1D-3+YINC)-(XM1,YM1D-80+YINC)
LINE(XM2,YM2D-3+YINC)-(XM2,YM2D-80+YINC)
ZD=215
RX=XM2+3
RY=YINC+ZD
GOSUB LARROW
LINE (XM2+3,ZD+YINC)-(XX-3,ZD+YINC)
RX=XX-3
RY=ZD+YINC
GOSUB RARROW
RX=XM1-3
RY=ZD+YINC
GOSUB RARROW
LINE (XM1-3,ZD+YINC)-(XX+3,ZD+YINC)
RX=XX+3
```

```
RY=ZD+YINC
GOSUB LARROW
LINE(XL1,YL1U-3+YINC)-(XL1,YL1U-80+YINC)
LINE(XL2,YL2U-3+YINC)-(XL2,YL2U-80+YINC)
RX=XL2+3
RY=20+YINC
GOSUB LARROW
LINE (XL2+3,20+YINC)-(XX-3,20+YINC)
RX=XX-3
RY=20+YINC
GOSUB RARROW
RX=XL1-3
RY=20+YINC
GOSUB RARROW
LINE (XL1-3,20+YINC)-(XX+3,20+YINC)
RX=XX+3
RY=20+YINC
GOSUB LARROW
LOCATE 23,45:PRINT USING"###.#";XL2S
LOCATE 23,50:PRINT USING"###.#";XL1S
LOCATE 35,45:PRINT USING"###.#";XM2S
LOCATE 35,50:PRINT USING"###.#";XM1S
GOSUB UPSWEEP
GOSUB BOREL2
LINE(XM1,YM1U+YINC)-(XP1,YP1U+YINC)
LINE(XM1,YM1D+YINC)-(XP1,YP1D+YINC)
LINE(XM2,YM2U+YINC)-(XP2,YP2U+YINC)
LINE(XM2,YM2D+YINC)-(XP2,YP2D+YINC)
LINE(XP1,YP1U+YINC)-(XP2,YP2U+YINC)
LINE(XP1,YP1D+YINC)-(XP2,YP2D+YINC)
REM GETTING THE PORT WIDTH DIMENSION LINE
LOCATE 30,51:PRINT USING"##.#";WID1S
M1=-M1
M3=-M3
M2=-M2
XK=XE2
YK=YE2U+YINC
YL=YT1+YINC
XL=(YL-YK)/M1+XK
YM=YE1U+YINC
XM=XE1
XJ=(YK-YM+M3*XM-M1*XK)/(M3-M1)
YJ=M3*(XJ-XM)+YM
XI=XJ+3
```

```
YI=M3*(XI-XM)+YM
XH=XM-3
YH=M3*(XH-XM)+YM
LINE(XJ,YJ)-(XK,YK)
LINE(XI,YI)-(XH,YH)
RX=XI
RY=YI
GOSUB LRARROW
RX=XH
RY=YH
GOSUB RRARROW
YR=YM-1
YS=YM+8
XR=(YR-YM)/M1+XM
XS=(YS-YM)/M1+XM
LINE(XR,YR)-(XS,YS)
LOCATE 39,1: PRINT"PORT WIDTH IS THE CHORD"
LOCATE 40,1: PRINT"AT 90º TO THE GAS FLOW DIRECTION."
LOCATE 42,1:PRINT"THE EFFECTIVE WIDTH IS THE"
LOCATE 43,1:PRINT"PLAN CHORDAL WIDTH*COSINE(UP)."
LOCATE 45,1:PRINT"TRANSFER PORT WIDTH RATIOS ARE "
LOCATE 46,1:PRINT"TOTAL EFFECTIVE WIDTH÷CYLINDER BORE."
IF PRT$="TWO" THEN GOTO 600
IF PRT$="THREE" THEN GOTO 600
LOCATE 36,1:PRINT"TRUE MAIN-SIDE CHORDAL BAR=";USING
"#.#";MBAS
IF NPRT$="FOUR" THEN GOTO 580
LOCATE 37,1:PRINT"TRUE REAR-SIDE CHORDAL BAR=";USING
"#.#";RBAS
GOTO 600
580 LOCATE 37,1:PRINT"TRUE REAR CHORDAL BAR=";USING
"#.#";RBAR
600 YQ1=YQ1-YINC
ZZ=2
YINC=0
RETURN
INTERCEPT:
REM THIS GIVES THE X,Y VALUE OF A LINE SLOPE M THROUGH A
CIRCLE RADIUS R
REM LINE GOES THROUGH TARGET POINT XT,YT REFERRED TO
REM CIRCLE CENTRE AT 0,0
REM RR AND XM REFERS TO THE BORE
REM RRL AND XL REFERS TO THE LINER OD
REM RRE AND XE REFERS TO THE INDICATOR POINT
```

```
REM RRP AND XP REFERS TO THE PORT BACKWALL
PHR=PH*DTOR
M=-TAN(PHR)
A=M*M+1
B=2*M*YT-2*M*M*XT
R=RR
C=M*M*XT*XT+YT*YT-2*M*XT*YT-R*R
XM=(-B-SQR(B*B-4*A*C))/(2*A)
YM=M*(XM-XT)+YT
R=RRE
C=M*M*XT*XT+YT*YT-2*M*XT*YT-R*R
XE=(-B-SQR(B*B-4*A*C))/(2*A)
YE=M*(XE-XT)+YT
R=RRL
C=M*M*XT*XT+YT*YT-2*M*XT*YT-R*R
XL=(-B-SQR(B*B-4*A*C))/(2*A)
YL=M*(XL-XT)+YT
R=RRP
C=M*M*XT*XT+YT*YT-2*M*XT*YT-R*R
XP=(-B-SQR(B*B-4*A*C))/(2*A)
YP=M*(XP-XT)+YT
RETURN
INTERCEPTS:
REM THIS GIVES THE X,Y VALUE OF A LINE SLOPE M THROUGH A
CIRCLE RADIUS R
REM LINE GOES THROUGH  TARGET POINT XT ON X-AXIS
REM CIRCLE CENTRE AT 0,0
REM RR AND XM REFERS TO THE BORE, XM, YM BEING KNOWN FOR
SIDE PORTS
REM RRL AND XL REFERS TO THE LINER OD
REM RRE AND XE REFERS TO THE INDICATOR POINT
REM RRP AND XP REFERS TO THE PORT BACKWALL
PHR=PH*DTOR
M=-TAN(PHR)
XT=XM+(YT-YM)/M
A=M*M+1
B=2*M*YT-2*M*M*XT
R=RRE
C=M*M*XT*XT+YT*YT-2*M*XT*YT-R*R
XE=(-B-SQR(B*B-4*A*C))/(2*A)
YE=M*(XE-XT)+YT
R=RRL
C=M*M*XT*XT+YT*YT-2*M*XT*YT-R*R
XL=(-B-SQR(B*B-4*A*C))/(2*A)
```

```
YL=M*(XL-XT)+YT
R=RRP
C=M*M*XT*XT+YT*YT-2*M*XT*YT-R*R
XP=(-B-SQR(B*B-4*A*C))/(2*A)
YP=M*(XP-XT)+YT
RETURN
BORE:
XX=400
YY=180
RR=100
LINE(XX-RR-30,YY)-(XX+RR+30,YY)
LINE(XX,YY-RR-30)-(XX,YY+RR+30)
CIRCLE (XX,YY),RR
RETURN
BOREL2:
REM THIS DRAWS THE CYLINDER, LINER, EXHAUST PORT, SIDE
AND REAR PORT ON PAGE2
XX=400
YY=180+YINC
RR=100
LINE(XX-RR-30,YY)-(XX+RR+30,YY)
LINE(XX,YY-RR-80)-(XX,YY+RR+80)
CIRCLE (XX,YY),RR
CIRCLE (XX,YY),RRL
XEX=XX-SQR(RR*RR-EW*EW/4)
YEXU=YY-EW/2
YEXD=YY+EW/2
LINE(XEX,YEXU)-(XEX-60,YEXU)
LINE(XEX,YEXD)-(XEX-60,YEXD)
LINE(XEX-60,YEXD)-(XEX-60,YEXU)
IF PRT$="THREE" THEN GOSUB P2REAR
IF NPRT$="FIVE" THEN GOSUB P2REAR
IF PRT$="BOTH STRAIGHT" THEN GOSUB P2ST
IF PRT$="ONE SLOPED" THEN GOSUB P1SL
IF PRT5$="BOTH STRAIGHT" THEN GOSUB P2ST
IF PRT5$="ONE SLOPED" THEN GOSUB P1SL
IF PRT$="RADIAL" THEN GOSUB P2RAD
IF PRT5$="RADIAL" THEN GOSUB P2RAD
IF F$="P" THEN GOTO 153
REM DRAWING DIMENSIONS ON EXHAUST PORT
LINE(XEX-63,YEXU)-(XEX-100,YEXU)
LINE(XEX-63,YEXD)-(XEX-100,YEXD)
LINE(XEX-90,YEXD-3)-(XEX-90,YEXU+3)
RX=XEX-90
```

```
RY=YEXD-3
GOSUB DARROW
RX=XEX-90
RY=YEXU+3
GOSUB UARROW
153 RETURN
P2RAD:
REM DRAWING RADIAL SIDE PORTS ON PAGE 2
YS2U2=YS2U+YINC
YS2D2=YS2D+YINC
YS1U2=YS1U+YINC
YS1D2=YS1D+YINC
YS3U2=YS3U+YINC
YS4U2=YS4U+YINC
YS3D2=YS3D+YINC
YS4D2=YS4D+YINC
LINE (XS2,YS2U2)-(XS3,YS3U2)
LINE (XS1,YS1U2)-(XS4,YS4U2)
LINE (XS2,YS2D2)-(XS3,YS3D2)
LINE (XS1,YS1D2)-(XS4,YS4D2)
R3=RR+CYT+40
CIRCLE(XX,YY),R3,,RS1,RS2
CIRCLE(XX,YY),R3,,2*PI-RS2,2*PI-RS1
YES4U2=YES4U+YINC
RE4=RRL
XL1=XX+RE4*COS(RS1)
XL2=XX+RE4*COS(RS2)
YL2=RE4*SIN(RS2)
YL1=RE4*SIN(RS1)
YL2D=YY+YL2
YL1D=YY+YL1
RS1D=RS1*RTOD
LINE(XX,YY)-(XES4,YES4U2)
CIRCLE(XX,YY),30,,0,RS1
LOCATE 33,55:PRINT USING"##.#";RS1D
YUU=YL2D+57
YVV=YUU+18
LINE(XL1,YL1D+5)-(XL1,YVV+5)
LINE(XL2,YL2D+5)-(XL2,YUU+5)
LINE(XX+3,YUU)-(XL2-3,YUU)
RX=XX+3
RY=YUU
GOSUB LARROW
RX=XL2-3
```

```
GOSUB RARROW
LINE(XX+3,YVV)-(XL1-3,YVV)
RX=XX+3
RY=YVV
GOSUB LARROW
RX=XL1-3
GOSUB RARROW
XUU=(XL2-XX)/SF
XVV=(XL1-XX)/SF
IF NPRT$="FOUR" THEN GOTO 950
IF NPRT$="FIVE" THEN GOTO 950
LOCATE 43,52:PRINT USING"##.#";XUU
LOCATE 44,52:PRINT USING"##.#";XVV
GOTO 960
950 LOCATE 41,52:PRINT USING"##.#";XUU
LOCATE 42,52:PRINT USING"##.#";XVV
960 RETURN
P1SL:
YS1U2=YS1U+YINC
YS1D2=YS1D+YINC
YS2U2=YS2U+YINC
YS2D2=YS2D+YINC
YS3U2=YS3U+YINC
YS3D2=YS3D+YINC
YS4U2=YS4U+YINC
YS4D2=YS4D+YINC
YLS2D2=YLS2D+YINC
LINE (XS2,YS2U2)-(XS3,YS3U2)
LINE (XS1,YS1U2)-(XS1,YS4U2)
LINE(XS3,YS3U2)-(XS1,YS4U2)
LINE (XS2,YS2D2)-(XS3,YS3D2)
LINE (XS1,YS1D2)-(XS1,YS4D2)
LINE(XS3,YS3D2)-(XS1,YS4D2)
LINE(XS2,YS2U2-5)-(XS2,YS3U2-30)
LINE(XS1,YS4U2-3)-(XS1,YS4U2-30)
RY=YS4U2-20
LINE(XS2+3,RY)-(XS1-3,RY)
RX=XS2+3
GOSUB LARROW
RX=XS1-3
GOSUB RARROW
LINE(XLS2,YLS2D2+3)-(XLS2,YLS2D2+60)
RY=YLS2D2+57
LINE(XX+3,RY)-(XLS2-3,RY)
```

```
RX=XX+3
GOSUB LARROW
RX=XLS2-3
GOSUB RARROW
LOCATE 45,51:PRINT USING"##.#";XLS2S
LOCATE 23,57:PRINT USING"##.#";WIDSS
RETURN
P2ST:
YS1U2=YS1U+YINC
YS1D2=YS1D+YINC
YS2U2=YS2U+YINC
YS2D2=YS2D+YINC
YS3U2=YS3U+YINC
YS3D2=YS3D+YINC
YS4U2=YS4U+YINC
YS4D2=YS4D+YINC
LINE (XS2,YS2U2)-(XS2,YS3U2)
LINE (XS1,YS1U2)-(XS1,YS4U2)
LINE(XS2,YS3U2)-(XS1,YS4U2)
LINE (XS2,YS2D2)-(XS2,YS3D2)
LINE (XS1,YS1D2)-(XS1,YS4D2)
LINE(XS2,YS3D2)-(XS1,YS4D2)
LINE(XS2,YS3U2-3)-(XS2,YS3U2-30)
LINE(XS1,YS4U2-3)-(XS1,YS4U2-30)
RY=YS4U2-27
LINE(XS2+3,RY)-(XS1-3,RY)
RX=XS2+3
GOSUB LARROW
RX=XS1-3
GOSUB RARROW
LOCATE 23,57:PRINT USING"##.#";WIDSS
RETURN
P2REAR:
XRP=XX+SQR(RR*RR-RW*RW/4)
YRPU=YY-RW/2
YRPD=YY+RW/2
LINE(XRP,YRPU)-(XRP+50,YRPU)
LINE(XRP,YRPD)-(XRP+50,YRPD)
LINE(XRP+50,YRPD)-(XRP+50,YRPU)
IF F$="P" THEN GOTO 152
LINE(XRP+53,YRPU)-(XRP+65,YRPU)
LINE(XRP+53,YRPD)-(XRP+65,YRPD)
LINE(XRP+63,YRPD-3)-(XRP+63,YRPU+3)
RX=XRP+63
```

```
RY=YRPD-3
GOSUB DARROW
RX=XRP+63
RY=YRPU+3
GOSUB UARROW
152 RETURN
BOREL:
REM THIS DRAWS THE CYLINDER, LINER AND EXHAUST PORT ON
PAGE 1
XX=400
YY=180
RR=100
LINE(XX-RR-30,YY)-(XX+RR+30,YY)
LINE(XX,YY-RR-80)-(XX,YY+RR+80)
CIRCLE (XX,YY),RR
CIRCLE (XX,YY),RRL
XEX=XX-SQR(RR*RR-EW*EW/4)
YEXU=YY-EW/2
YEXD=YY+EW/2
LINE(XEX,YEXU)-(XEX-60,YEXU)
LINE(XEX,YEXD)-(XEX-60,YEXD)
LINE(XEX-60,YEXD)-(XEX-60,YEXU)
LINE(XEX-63,YEXU)-(XEX-100,YEXU)
LINE(XEX-63,YEXD)-(XEX-100,YEXD)
LINE(XEX-90,YEXD-3)-(XEX-90,YEXU+3)
RX=XEX-90
RY=YEXD-3
GOSUB DARROW
RX=XEX-90
RY=YEXU+3
GOSUB UARROW
IF PRT$="THREE" THEN GOTO 155
IF NPRT$="FIVE" THEN GOTO 155
GOTO 150
REM DRAWING THE REAR PORT ON PAGE 1
155 XRP=XX+SQR(RR*RR-RW*RW/4)
YRPU=YY-RW/2
YRPD=YY+RW/2
LINE(XRP,YRPU)-(XRP+50,YRPU)
LINE(XRP,YRPD)-(XRP+50,YRPD)
LINE(XRP+53,YRPU)-(XRP+65,YRPU)
LINE(XRP+53,YRPD)-(XRP+65,YRPD)
LINE(XRP+63,YRPD-3)-(XRP+63,YRPU+3)
RX=XRP+63
```

```
RY=YRPD-3
GOSUB DARROW
RX=XRP+63
RY=YRPU+3
GOSUB UARROW
LOCATE 12,68:PRINT RWS
LINE(XRP+50,YRPD)-(XRP+50,YRPU)
150 RETURN
LARROW:
REM LEFTWARDS
LINE(RX,RY)-(RX+3,RY-3)
LINE(RX,RY)-(RX+3,RY+3)
RETURN
RARROW:
REM RIGHTWARDS
LINE(RX,RY)-(RX-3,RY-3)
LINE(RX,RY)-(RX-3,RY+3)
RETURN
UARROW:
REM UPWARDS
LINE(RX,RY)-(RX-3,RY+3)
LINE(RX,RY)-(RX+3,RY+3)
RETURN
DARROW:
REM DOWNWARDS
LINE(RX,RY)-(RX-3,RY-3)
LINE(RX,RY)-(RX+3,RY-3)
RETURN
LRARROW:
REM DOWNWARDS
LINE(RX,RY)-(RX,RY-5)
LINE(RX,RY)-(RX+5,RY)
RETURN
RRARROW:
REM DOWNWARDS
LINE(RX,RY)-(RX-5,RY)
LINE(RX,RY)-(RX,RY+5)
RETURN
UPSWEEP:
LOCATE ZZ,1:PRINT"PORT SECTION THROUGH"
LOCATE ZZ+1,1:PRINT"INCLINED TRANSFER PORT"
LOCATE ZZ+2,1:PRINT"'UP' AT UPº"
LOCATE ZZ+7,4:PRINT"UPº"
XQ1=100
```

```
YQ1=50+YINC
YQ2=150+YINC
YQ3=YQ1+30
YQ4=YQ3+20
YQ5=YQ4+2
YQ6=YQ5+20
YQ7=YQ3+25
XQ2=XQ1-30
XQ5=XQ2-3
XQ6=XQ5-40
CIRCLE (XQ6,YQ6),30,,0,PI/7
LINE (XQ1,YQ1)-(XQ1+10,YQ1)
LINE (XQ1+10,YQ1)-(XQ1+10,YQ2)
LINE (XQ2,YQ2)-(XQ1+10,YQ2)
LINE (XQ1,YQ2)-(XQ1,YQ1)
LINE (XQ1,YQ7)-(XQ1+10,YQ7)
LINE (XQ2,YQ2)-(XQ2,YQ4)
LINE (XQ2,YQ4)-(XQ1+10,YQ3)
LINE (XQ6,YQ6)-(XQ5,YQ6)
LINE (XQ6,YQ6)-(XQ5,YQ5)
RETURN
FILING:
INPUT "TYPE IN TITLE NEEDED FOR THE FILE WITH '.LOOP' END
CODE";FL$
OPEN FL$ FOR OUTPUT AS #1
WRITE#1,BO,CYT,EWS,UPM,PH1,PH2,TG1S,TG2S,UPR,RWS,UPS,
MBRS,SBRS,PS2,UPRS,UPB
FINISH:
CLOSE#1
RETURN
FILED:
INPUT "TYPE IN THE FILENAME";FL$
OPEN"I",#1,FL$
INPUT#1,BO,CYT,EWS,UPM,PH1,PH2,TG1S,TG2S,UPR,RWS,UPS,
MBRS,SBRS,PS2,UPRS,UPB
CLOSE#1
RETURN
```

ProgList4.0 Program listings from Chapter 4

COMPUTER PROGRAM PROG.4.1
PROGRAM NAME "SQUISH VELOCITY"
ProgList.4.1

```
REM PROGRAM Prog.4.1
REM ProgList.4.1
REM PROGRAM NAME 'SQUISH VELOCITY'
REM***PROGRAM FOR SQUISH VELOCITY***
DIM X(181),Y(181)
14 INPUT "NEW DATA OR FILED DATA(N OR F)";G$
IF G$="F" THEN GOSUB FILED
IF G$="F" THEN GOTO 19
15 INPUT"NEW BASIC ENGINE DATA(Y/N?)";ANS$
IF ANS$="N" THEN GOTO 16
 INPUT"NEW DATA FOR CYLINDER BORE(Y/N?)";ANS$
IF ANS$="Y" THEN INPUT"TYPE IN CYLINDER BORE(B)";BO
INPUT"NEW DATA FOR CYLINDER STROKE(Y/N?)";ANS$
IF ANS$="Y" THEN INPUT"TYPE IN CYLINDER STROKE(S)";ST
INPUT"NEW DATA FOR CONNECTING ROD LENGTH(Y/N?)";ANS$
IF ANS$="Y" THEN INPUT"TYPE IN CONNECTING ROD LENGTH(CRL)";CRL
INPUT"NEW DATA FOR ENGINE SPEED(Y/N?)";ANS$
IF ANS$="Y" THEN INPUT"TYPE IN ENGINE SPEED(RPM)";RPM
PRINT
16 INPUT"NEW DATA FOR EXHAUST PORT (Y/N?)";ANS$
IF ANS$="N" THEN GOTO 17
INPUT"NEW DATA FOR EXHAUST PORT OPENING TIMING(Y/N?)";ANS$
IF ANS$="Y" THEN INPUT"TYPE IN EXHAUST PORT OPENING,DEG.ATDC";EPO
PRINT
17 INPUT"NEW DATA FOR CYLINDER HEAD(Y/N?)";ANS$
IF ANS$="N" THEN GOTO 19
INPUT"NEW DATA FOR CYLINDER HEAD TYPE(Y/N?)";ANS$
IF ANS$="Y" THEN PRINT "NAMED TYPES, FROM BOOK FIG.4.10, ARE 'CENTRAL', 'OFFSET', 'TOTAL OFFSET', OR 'DEFLECTOR'"
IF ANS$="Y" THEN INPUT"TYPE IN NAMED TYPE PRECISELY AS ONE OF THE ABOVE NAMES";HEAD$
IF HEAD$="DEFLECTOR" THEN INPUT"TYPE IN DEFLECTOR WIDTH";DEFWD
INPUT"NEW DATA FOR TRAPPED COMPRESSION RATIO(Y/N?)";ANS$
IF ANS$="Y" THEN INPUT"TYPE IN TRAPPED COMPRESSION RATIO";TCR
```

```
INPUT"NEW DATA FOR SQUISH AREA RATIO(Y/N?)";ANS$
IF ANS$="Y" THEN INPUT"TYPE IN SQUISH AREA RATIO";SAR
INPUT"NEW DATA FOR SQUISH CLEARANCE(Y/N?)";ANS$
IF ANS$="Y" THEN INPUT"TYPE IN SQUISH CLEARANCE, MM";SQ
19 PI=3.141592654#
SUMENERGY=0
PMAX=1
SQVMAX=0
DTOR=PI/180
RTOD=1/DTOR
BOM=BO/1000
STM=ST/1000
CRLM=CRL/1000
DEFWDM=DEFWD/1000
SQM=SQ/1000
PA=PI*BOM*BOM/4
SV=PA*STM
CRNK=ST/2
CRNKM=STM/2
EPOR= (360-EPO)*DTOR
REM CALCULATING THE RELEVANT HEIGHT DATA FOR THE EX-
HAUST
DEG=EPOR
GOSUB PISTPOS
HEOT=HT*1000
HEOT1=HT
HH=HB*1000
HOB=HH
DEG=EPOR
ANGD=360-EPO
TSV=PA*HEOT1
CVOL=TSV/(TCR-1)
ASQ=PA*SAR
ABL=PA-ASQ
VSQBAND=SQM*ASQ
VBS=ABL*SQM
VBOWL=CVOL-VBS-VSQBAND
DBL=SQR(4*ABL/PI)
VS1=HEOT1*ASQ+VSQBAND
VB1=HEOT1*ABL+VBS+VBOWL
VCYL1=HEOT1*PA+CVOL
VTRAP=VCYL1
GASR=287
PTRAP=101325!
```

```
TTRAP=293
PCYL1=PTRAP
TCYL1=TTRAP
GAMMA=1.4
MTRAP=PTRAP*VTRAP/(GASR*TTRAP)
PS1=PCYL1
PB1=PCYL1
TS1=TCYL1
TB1=TCYL1
RHOS1=PS1/(GASR*TS1)
RHOB1=PB1/(GASR*TB1)
MS1=MTRAP*VS1/VCYL1
MB1=MTRAP-MS1
DC=1
DT=DC/(6*RPM)
GOSUB CRANKFACT
CLS
GOSUB CYLDRAW
50 DEG=DEG+DTOR*DC
NDEG=NDEG+DC
ANGD=ANGD+DC
IF ANGD>360 THEN GOTO 500
GOSUB PISTPOS
HEOT=HT*1000
HEOT2=HT
DH=HEOT1-HEOT2
HH=HB*1000
HOB=HH
VS2=HEOT2*ASQ+VSQBAND
VB2=HEOT2*ABL+VBS+VBOWL
VCYL2=HEOT2*PA+CVOL
PCYL2=PTRAP*(PTRAP/VCYL2)^GAMMA
TCYL2=PCYL2*VCYL2/(MTRAP*GASR)
RHOCYL2=PCYL2/(GASR*TCYL2)
AREF=SQR(GAMMA*GASR*TCYL2)
PS2=PS1*(VS1/VS2)^GAMMA
PB2=PB1*(VB1/VB2)^GAMMA
MS2=MTRAP*VS2/VCYL2
MB2=MTRAP-MS2
DELMS=MS1-MS2
HSQ=HEOT1+SQM-.5*DH
IF HEAD$="CENTRAL" THEN SQL=PI*DBL
IF HEAD$="OFFSET" THEN SQL=.75*PI*DBL
IF HEAD$="TOTAL OFFSET" THEN SQL=BOM
```

```
IF HEAD$="DEFLECTOR" THEN SQL=DEFWDM
ASV=HSQ*SQL
SQV=DELMS/(RHOCYL2*ASV*DT)
PRATIO=PS2/PB2
IF PRATIO>PMAX THEN PMAX=PRATIO
KENERGY=DELMS*SQV*SQV/2
SUMENERGY=SUMENERGY+KENERGY
GOSUB CRANKDRAW
HEOT1=HEOT2
VS1=VS2
VB1=VB2
VCYL1=VCYL2
PCYL1=PCYL2
TCYL1=TCYL2
MS1=MS2
MB1=MB2
PS1=PCYL2
PB1=PCYL2
IF SQV>SQVMAX THEN DEGMAX=360-ANGD
IF SQV>SQVMAX THEN SQVMAX=SQV
GOTO 50
500 COMPWORK=(PCYL2*VCYL2-PTRAP*VTRAP)/(1-GAMMA)
TURBRATIO=-100*SUMENERGY/COMPWORK
GOSUB SUBPRINT
105 LOCATE 25,1:INPUT"CALCULATING AGAIN(Y OR N), OR PRINTING(P), FILING THE DATA(F), OR QUITTING(Q)";ANS$
IF ANS$="Y" THEN GOTO 14
IF ANS$="N" THEN GOTO 105
IF ANS$="Q" THEN GOTO 166
IF ANS$="P" THEN GOSUB SUBLPRINT
IF ANS$="F" THEN GOSUB FILING
GOTO 105
166 END
FILING:
INPUT "TYPE IN TITLE NEEDED FOR THE FILE WITH '.SQV' END CODE";P$
OPEN P$ FOR OUTPUT AS #1
WRITE#1,BO,ST,CRL,RPM,EPO,TCR,SAR,SQ,DEFWD,HEAD$
FINISH:
CLOSE#1
RETURN
FILED:
INPUT "TYPE IN THE FILENAME WITH '.SQV' TAIL-CODE";FIL$
OPEN"I",#1,FIL$
```

```
INPUT#1,BO,ST,CRL,RPM,EPO,TCR,SAR,SQ,DEFWD,HEAD$
CLOSE#1
RETURN
PISTPOS:
HT=CRNKM+CRLM-CRNKM*COS(DEG)-SQR(CRLM*CRLM-
(CRNKM*SIN(DEG))^2)
HB=ST-HT
RETURN
CRANKFACT:
CRT=50
XCL=330
SCAL=CRT/CRNK
BOR=BO*SCAL/2
GPR=.22*BOR
STS=2*CRT
Y1=50
QQ=4
PL=STS+QQ
Y2=Y1+2*STS+QQ
Y3=Y1+STS
YET=Y1+HEOT*SCAL
YTT=Y1+MHOT*SCAL
LT=7
XL1=XCL+BOR
XL2=XCL-BOR
XL3=XL1+LT
XL4=XL2-LT
CLER=2
XP1=XL1-CLER
XP2=XL2+CLER
RLT=10*SCAL
XL5=XL4-RLT
XL6=XL3+RLT
YR=(YET+Y3)/2
XM1=(XCL+XL2)/2
XM2=XM1+MW*SCAL
CBAR=5*SCAL
XM3=XM2+CBAR
XM4=XM3+SW*SCAL
XM5=XL1-.25*RW*SCAL
MRTS=MRT*SCAL
MRBS=MRB*SCAL
SRTS=SRT*SCAL
SRBS=SRB*SCAL
```

```
RRTS=RRT*SCAL
RRBS=RRB*SCAL
XMT1=XM1+MRTS
XMT2=XM2-MRTS
XMB1=XM1+MRBS
XMB2=XM2-MRBS
XMT3=XM3+SRTS
XMT4=XM4-SRTS
XMB3=XM3+SRBS
XMB4=XM4-SRBS
XMT5=XM5+RRTS
XMB5=XM5+RRBS
XME=XL2+.3*(XM1-XL2)
RCE=6*SCAL
XMRE=XME-RCE
ANQ=(90-RTSLOP)*DTOR
YQ=YR+RLT/SIN(ANQ)
XQ2=XL3+RLT*COS(ANQ)
YQ2=YR+(XQ2-XL3)/TAN(ANQ)
REM VALUES FOR THE HEAD
DBLS=DBL*1000*SCAL
SQS=SQ*SCAL
XCH1=XCL-DBLS/2
XCH2=XCL+DBLS/2
YSQ=Y1-SQS
YCH=YSQ-LT
HEAD=(CVOL-PI*BOM*BOM*SQM/4)/(PI*DBL*DBL/4)
HEAD=HEAD*1000
HEADS=HEAD*SCAL
YCHTI=YSQ-HEADS
YCHTO=YCHTI-LT
XCH3=XCH1-LT
XCH4=XCH2+LT
REM VALUES FOR THE SQ VEL PLOT
XPLOT=XL3+5
XSMAX=XPLOT+150
XSQ1=XPLOT
X(1)=XPLOT
YSQ1=YET
Y(1)=YET
J=1
XV10=XPLOT+10*(XSMAX-XPLOT)/25
XV20=XPLOT+20*(XSMAX-XPLOT)/25
RETURN
```

```
PORTDRAWR:
LINE(XMT5,YR)-(XL3,YR)
LINE(XMB5,Y3)-(XL3,Y3)
LINE(XM5,YR+RRTS)-(XM5,Y3-RRBS)
CIRCLE(XMT5,YR+RRTS),RRTS,,PI/2,PI
CIRCLE(XMB5,Y3-RRBS),RRBS,,PI,3*PI/2
RETURN
HEADDRAW:
LINE(XL4,YSQ)-(XL2,Y1),33,BF
LINE(XL1,YSQ)-(XL3,Y1),33,BF
IF HEAD$="CENTRAL" THEN GOTO 302
IF HEAD$="OFFSET" THEN XSH=XL1-XCH2
IF HEAD$="TOTAL OFFSET" THEN GOSUB TOHEAD
IF HEAD$="DEFLECTOR" THEN GOSUB DEFHEAD
GOTO 301
302 LINE(XL4,YCH)-(XCH1,YSQ),33,BF
LINE(XL3,YCH)-(XCH2,YSQ),33,BF
LINE(XCH3,YCHTO)-(XCH4,YCHTI),33,BF
LINE(XCH3,YCHTI)-(XCH1,YCH),33,BF
LINE(XCH2,YCHTI)-(XCH4,YCH),33,BF
GOTO 305
301 LINE(XL4,YCH)-(XCH1+XSH,YSQ),33,BF
LINE(XL3,YCH)-(XCH2+XSH,YSQ),33,BF
LINE(XCH3+XSH,YCHTO)-(XCH4+XSH,YCHTI),33,BF
LINE(XCH3+XSH,YCHTI)-(XCH1+XSH,YCH),33,BF
LINE(XCH2+XSH,YCHTI)-(XCH4+XSH,YCH),33,BF
305 RETURN
TOHEAD:
XCH1=XCL
XCH2=XL1
XCH4=XL3
XCH3=XCL-LT
XSH=0
RETURN
DEFHEAD:
XCH1=XCL+LT
XCH2=XL1
XCH4=XL3
XCH3=XCL
XSH=0
RETURN
CYLDRAW:
LINE(XPLOT,YET)-(XPLOT,Y1)
LINE(XPLOT,Y1)-(XSMAX,Y1)
```

```
LINE(XV10,Y1)-(XV10,Y1-3)
LINE(XV20,Y1)-(XV20,Y1-3)
LOCATE 3,56:PRINT"10"
LOCATE 3,64:PRINT"20"
LOCATE 1,54:PRINT"SQUISH"
LOCATE 2,54:PRINT"VELOCITY, m/s"
GOSUB HEADDRAW
LINE(XL4,Y1)-(XL2,Y2),33,BF
LINE(XL1,Y1)-(XL3,Y2),33,BF
LINE(XL5,YET)-(XL2,Y3),30,BF
LINE(XL5,YET)-(XMRE,YET)
LINE(XL5,Y3)-(XMRE,Y3)
LINE (XL4,Y2+1)-(XL3,Y2+1)
LINE(XL4,Y2)-(XL3,Y2)
IF NTP=2 THEN GOTO 675
IF NTP=4 THEN GOTO 675
LINE(XL1,YR)-(XL3,Y3),30,BF
LINE(XL6,Y2)-(XL3,Y2)
IF RTSLOP=0 THEN GOTO 676
CIRCLE(XL3,YQ),RLT,,0,ANQ
LINE(XL3,YR)-(XQ2,YQ2)
LINE(XL6,YQ)-(XL6,Y2)
GOTO 675
676 CIRCLE(XL3,YR+RLT),RLT,,0,PI/2
LINE(XL6,YR+RLT)-(XL6,Y2)
675 RETURN
CRANKDRAW:
REM THIS DRAWS THE EXHAUST PORT OTHERWISE IT WOULD GET
RUBBED EACH TIME
LINE (XL5,YET)-(XL5,Y3)
LINE(XME,YET+RCE)-(XME,Y3-RCE)
CIRCLE(XMRE,YET+RCE),RCE,,0,PI/2
CIRCLE(XMRE,Y3-RCE),RCE,,3*PI/2,0
GOSUB PORTDRAWR
GOSUB PISTERASE
666 GOSUB PISTDRAW
REM DRAWING THE SQUISH VELOCITY PLOT
J=J+1
XSQ2=XPLOT+SQV*(XSMAX-XPLOT)/25
X(J)=XSQ2
YSQ2=YP
Y(J)=YSQ2
LINE(XSQ1,YSQ1)-(XSQ2,YSQ2)
XSQ1=XSQ2
```

```
YSQ1=YSQ2
RETURN
PISTDRAW:
HTS=1000*HT*SCAL
YP=Y1+HTS
YPB=YP+PL
YGP=(YP+YPB)/2
LD1=5
LD2=LD1
RTH=3
YR1=YP+LD1
YR2=YR1+RTH+LD2
LINE(XP2,YP)-(XP1,YPB),30,BF
LINE(XP2,YP)-(XP1,YPB),,B
LINE(XL2+1,YR1)-(XL1-1,YR1+RTH),33,BF
LINE(XL2+1,YR2)-(XL1-1,YR2+RTH),33,BF
CIRCLE(XCL,YGP),GPR,33
EYP=YP
EYPB=YPB
RETURN
PISTERASE:
CIRCLE(XCL,YGP),GPR,30
LINE(XP2,EYP)-(XP1,EYPB),30,BF
LINE(XL2+1,YR1)-(XL1-1,YR1+RTH),30,BF
LINE(XL2+1,YR2)-(XL1-1,YR2+RTH),30,BF
RETURN
SUBPRINT:
GOSUB PRINTCONTENT
RETURN
SUBLPRINT:
INPUT"TYPE IN A TITLE FOR THE PRINT-OUT";TITLE$
OPEN"LPT1:PROMPT" FOR OUTPUT AS  #1
WINDOW OUTPUT#1
GOSUB PRINTCONTENT
LOCATE 20,1:PRINT TITLE$
CLOSE#1
RETURN
PRINTCONTENT:
LOCATE 1,1:PRINT "CYLINDER HEAD TYPE IS ";HEAD$
LOCATE 2,1:PRINT"BORE,mm=";BO
LOCATE 3,1:PRINT"STROKE,mm=";ST
LOCATE 4,1:PRINT"CON-ROD,mm=";CRL
LOCATE 5,1:PRINT"ENGINE SPEED, rev/min=";RPM
LOCATE 6,1:PRINT"EXHAUST OPENS,ºatdc=";EPO
```

```
LOCATE 8,1:PRINT"CYLINDER HEAD GEOMETRY"
LOCATE 9,1:PRINT"TRAPPED COMPRESSION RATIO=";TCR
LOCATE 10,1:PRINT"SQUISH AREA RATIO=";SAR
LOCATE 11,1:PRINT "SQUISH CLEARANCE,mm=";SQ
IF HEAD$="DEFLECTOR" THEN LOCATE 12,1:PRINT "DEFLECTOR WIDTH,mm=";DEFWD
LOCATE 14,1:PRINT "OUTPUT DATA"
LOCATE 15,1:PRINT"MAXIMUM SQUISH VELOCITY, m/s=";USING "###.#";SQVMAX
LOCATE 16,1:PRINT"MAXIMUM SQUISH PRESSURE RATIO=";USING "#.###";PMAX
LOCATE 17,1:PRINT"BOTH OCCURRED AT, deg.btdc=";DEGMAX
LOCATE 18,1:PRINT"SQUISHED KINETIC ENERGY, mJ=";USING "##.##";SUMENERGY*1000
GOSUB CYLDRAW
GOSUB CRANKDRAW
FOR I=1 TO J-1
LINE(X(I),Y(I))-(X(I+1),Y(I+1))
NEXT I
RETURN
```

COMPUTER PROGRAM PROG.4.2

PROGRAM NAME "HEMI-SPHERE CHAMBER"

ProgList.4.2

```
REM PROGRAM Prog.4.2
REM ProgList.4.2
REM PROGRAM NAME 'HEMI-SPHERE CHAMBER'
REM PROGRAM FOR COMBUSTION CHAMBER CLEARANCE VOLUME
REM FOR HEMISPHERICAL SEGMENT COMBUSTION CHAMBER
REM WITH PISTON CROWN WITH SPHERICAL RADIUS
PI=3.141593
DTOR=PI/180
RTOD=180/PI
INPUT"NEW DATA OR FILED DATA? (TYPE 'N' OR 'F')";D$
IF D$="F" THEN GOSUB FILED
IF D$="F" THEN GOTO 500
N=0
IF N=0 THEN GOSUB DATAIN
500 CLS
IF F$="P" THEN CLS
IF F$="P" THEN OPEN"LPT1:PROMPT" FOR OUTPUT AS #1
IF F$="P" THEN WINDOW OUTPUT#1
LOCATE 1,1:PRINT"CURRENT INPUT DATA FOR 2-STROKE 'SPHERICAL' BOWL"
```

```
LOCATE 2,1:PRINT"(B) BORE,mm=";BO
LOCATE 3,1:PRINT"(S) STROKE,mm=";ST
LOCATE 4,1:PRINT"(C) CON-ROD,mm=";CRL
LOCATE 5,1:PRINT"(E) EXHAUST CLOSES,deg btdc=";EO
LOCATE 7,1:PRINT"BOWL DETAILS,mm"
LOCATE 8,1:PRINT"(SH) PISTON TO HEAD CLEARANCE=";SQH
LOCATE 9,1:PRINT"(RP) PISTON CROWN RADIUS=";RPC
LOCATE 10,1:PRINT"(RS) SQUISH BAND RADIUS=";RSB
LOCATE 11,1:PRINT"(RC) CHAMBER RADIUS=";RCH
LOCATE 12,1:PRINT"(RB) BLENDING RADIUS=";RSQ
LOCATE 13,1:PRINT"(SR) SQUISH AREA RATIO=";SAR
CRK=ST/2
DEG=EO
BOR=BO/2
SQD=BO*SQR(1-SAR)
FPIST=PI*BO*BO/4
SV=FPIST*ST
SVCC=SV/1000
GOSUB PIST
TSV=VTDC
TSVCC=TSV/1000
GOSUB BOWLFACTS
GOSUB CIRCLEDATA
GOSUB BOWL
LOCATE 15,1:PRINT"OUTPUT DATA"
LOCATE 16,1:PRINT"SWEPT VOLUME,cc=";USING"###.#";SVCC
LOCATE 17,1:PRINT"TRAPPED SWEPT VOLUME,cc=";USING
"###.#";TSVCC
LOCATE 18,1:PRINT"TRAPPED STROKE,mm=";USING"##.#";HTDC
LOCATE 19,1:PRINT"CLEARANCE VOLUME,cc=";USING"##.#";CVCC
LOCATE 20,1:PRINT"GEOMETRIC COMPRESSION RATIO=";USING
"##.#";GCR
LOCATE 21,1:PRINT"TRAPPED COMPRESSION RATIO=";USING
"##.#";TCR
LOCATE 22,1:PRINT"MIN & MAX SQUISH CLEARANCE,mm=";USING
"####.#";SCS,SCB
GOSUB PISTDRAW
IF F$="P" THEN LOCATE 25,35:PRINT T$
IF F$="P" THEN CLOSE #1
LOCATE 23,1:PRINT"                                        "
LOCATE 24,1:PRINT"                                        "
LOCATE 25,1:PRINT"                                        "
LOCATE 26,1:PRINT"                                        "
LOCATE 23,1:PRINT"TYPE CODE FOR NEW DATA AND A
RECALCULATION"
```

```
LOCATE 24,1:INPUT"OR TYPE 'P' FOR PRINT-OUT, 'F' FOR FILING DATA, OR 'Q' FOR QUITTING";F$
IF F$="B" THEN INPUT"TYPE IN NEW VALUE FOR BORE,mm";BO
IF F$="S" THEN INPUT"TYPE IN NEW VALUE FOR STROKE,mm";ST
IF F$="C" THEN INPUT"TYPE IN NEW VALUE FOR CON-ROD CENTRES,mm";CRL
IF F$="E" THEN INPUT"TYPE IN NEW VALUE FOR EXHAUST CLOSING,deg.btdc";EO
IF F$="SH" THEN INPUT"TYPE IN NEW VALUE FOR HEAD CLEARANCE,mm";SQH
IF F$="RP" THEN INPUT"TYPE IN NEW VALUE FOR PISTON CROWN RADIUS,mm";RPC
IF F$="RS" THEN INPUT"TYPE IN NEW VALUE FOR SQUISH BAND RADIUS,mm";RSB
IF F$="RB" THEN INPUT"TYPE IN NEW VALUE FOR SQUISH CORNER RADIUS,mm";RSQ
IF F$="RC" THEN INPUT"TYPE IN NEW VALUE FOR COMBUSTION BOWL RADIUS,mm";RCH
IF F$="SR" THEN INPUT"TYPE IN NEW VALUE FOR SQUISH AREA RATIO";SAR
IF F$="P" THEN INPUT"TYPE IN A TITLE FOR THE PRINT-OUT";T$
IF F$="P" THEN GOTO 500
IF F$="F" THEN GOSUB FILING
IF F$="Q" THEN GOTO 200
GOTO 500
200 END
REM PISTON POSITION SUBROUTINE
PIST:
DR=DEG*DTOR
HTDC=CRK+CRL-CRK*COS(DR)-SQR(CRL*CRL-(CRK*SIN(DR))^2)
VTDC=HTDC*FPIST
RETURN
BOWLFACTS:
SA2=BOR/RPC
AN2=ATN(SA2/SQR((1-SA2*SA2)))
HPT=RPC*COS(AN2)
HHT=HPT+SQH
SA3=BOR/RSB
AN3=ATN(SA3/SQR((1-SA3*SA3)))
HST=RSB*COS(AN3)
HST2=HHT-HST
SA4=SQD/(2*RSB)
AN4=ATN(SA4/SQR((1-SA4*SA4)))
XH=(RSB+RSQ)*SA4
YH=SQR((RCH+RSQ)*(RCH+RSQ)-XH*XH)
```

```
HCT2=(RSB+RSQ)*COS(AN4)-YH
HCT=HCT2+HST2
AN5=ATN(XH/YH)
XCE=RCH*SIN(AN5)
XSQ=SQD/2
XBE=BOR
RETURN
BOWLXY:
A=1
B=-2*Q
C=Q*Q+((XD-P)*(XD-P))-R*R
SQU=SQR(B*B-4*A*C)
YPLUS=(-B+SQU)/(2*A)
YMINUS=(-B-SQU)/(2*A)
RETURN
CIRCLEDATA:
P1=0
Q1=0
R1=RPC
P2=0
Q2=HST2
R2=RSB
P3=XH
Q3=HCT+YH
R3=RSQ
P4=0
Q4=HCT
R4=RCH
RETURN
BOWL:
AYBAR=0
AA=0
XD=0
DR=.5
P=P1
Q=Q1
R=R1
GOSUB BOWLXY
YB1=YPLUS
P=P4
Q=Q4
R=R4
GOSUB BOWLXY
YT1=YPLUS
```

```
958 XD=XD+DR
P=P1
Q=Q1
R=R1
GOSUB BOWLXY
YB2=YPLUS
IF XD>XCE THEN GOTO 955
P=P4
Q=Q4
R=R4
GOSUB BOWLXY
YT2=YPLUS
DA=(YT1+YT2-YB1-YB2)*DR/2
GOTO 957
955 IF XD>XSQ THEN GOTO 956
P=P3
Q=Q3
R=R3
GOSUB BOWLXY
YT2=YMINUS
DA=(YT1+YT2-YB1-YB2)*DR/2
SCB=(YT2-YB2)*COS(AN2)
GOTO 957
956 IF XD>XBE THEN GOTO 959
P=P2
Q=Q2
R=R2
GOSUB BOWLXY
YT2=YPLUS
DA=(YT1+YT2-YB1-YB2)*DR/2
SCS=(YT1-YB1)*COS(AN2)
957 AA=AA+DA
AYBAR=AYBAR+DA*(XD-DR/2)
YB1=YB2
YT1=YT2
GOTO 958
959 YBAR=AYBAR/AA
CVOL=2*PI*YBAR*AA
CVCC=CVOL/1000
GCR=(CVOL+SV)/CVOL
TCR=(CVOL+TSV)/CVOL
RETURN
DATAIN:
BO=72
```

```
ST=60
CRL=120
EO=103
SQH=2
RPC=150
RCH=27
RSQ=4
RSB=152
SAR=.45
RETURN
FILING:
INPUT "TYPE IN TITLE NEEDED FOR THE FILE WITH '.HEMISPH' END
CODE";FIL$
OPEN FIL$ FOR OUTPUT AS #1
WRITE#1,BO,ST,CRL,EO,SQH,SAR,RPC,RSQ,RSB,RCH
FINISH:
CLOSE#1
RETURN
FILED:
INPUT "TYPE IN THE FILENAME WITH '.HEMISPH' END CODE";FIL$
OPEN"I",#1,FIL$
INPUT#1,BO,ST,CRL,EO,SQH,SAR,RPC,RSQ,RSB,RCH
CLOSE#1
RETURN
PISTDRAW:
Y1=80
X1=315
X2=515
XCL=(X1+X2)/2
SCAL=(X2-X1)/BO
Y2=Y1+SQH*SCAL
Y3=Y2+ST*SCAL
Y5=Y3+8
Y4=Y2+HTDC*SCAL
C1=Y2+HPT*SCAL
C2=Y1+HST*SCAL
C3=C1-HCT*SCAL
C4=C3-YH*SCAL
XC4R=XCL+XH*SCAL
XC4L=XCL-XH*SCAL
MH=HCT+RCH-HHT
YB=Y1-MH*SCAL
BDS=SQD*SCAL/2
XB1=XCL-BDS
```

```
XB2=XCL+BDS
X3=X1-30
REM DRAWS CYLINDER
LINE(X1-5,Y1)-(X1,Y5),33,BF
LINE(X1,Y4)-(X1,Y1)
LINE(X2+5,Y1)-(X2,Y5),33,BF
LINE(X3,Y4)-(X1,Y3),30,BF
LINE(X3,Y4)-(X1,Y3),,B
REM DRAWS PISTON
LINE(X1,Y5-1)-(X1,Y2)
LINE(X2,Y2)-(X2,Y5-1)
LINE(X1,Y5-1)-(X2,Y5-1)
CIRCLE(XCL,C1),RPC*SCAL,,PI/2-AN2,PI/2+AN2
REM DRAWS HEAD
CIRCLE(XCL,C2),RSB*SCAL,,PI/2-AN3,PI/2-AN4
CIRCLE(XCL,C2),RSB*SCAL,,PI/2+AN4,PI/2+AN3
CIRCLE(XCL,C3),RCH*SCAL,,PI/2-AN5,PI/2+AN5
AN6=PI/2-AN5
AN1=AN5-AN4
CIRCLE(XC4R,C4),RSQ*SCAL,,PI+AN6,PI+AN6+AN1
CIRCLE(XC4L,C4),RSQ*SCAL,,2*PI-AN6-AN1,2*PI-AN6
REM DRAWS PISTON RINGS
YR=Y2+4
LINE(X1,YR)-(X1+6,YR+4),33,BF
YR=YR+9
LINE(X1,YR)-(X1+6,YR+4),33,BF
YR=Y2+4
LINE(X2,YR)-(X2-6,YR+4),33,BF
YR=YR+9
LINE(X2,YR)-(X2-6,YR+4),33,BF
REM DRAWS DIMENSION LINES
X4=X3-32
X5=X3-10
Y6=Y5+20
LINE(X4-3,Y1)-(X1-8,Y1)
LINE(X4-3,Y2)-(X1-8,Y2)
LINE(X4-3,Y3)-(X3-3,Y3)
LINE(X5-3,Y4)-(X5+3,Y4)
LINE(X1,Y5+3)-(X1,Y5+25)
LINE(X2,Y5+3)-(X2,Y5+25)
LINE(X2+8,Y1)-(X2+28,Y1)
LINE(XCL+20,YB)-(X2+28,YB)
Y7=Y2+23
LINE(XB1,Y2+3)-(XB1,Y7+2)
```

```
LINE(XB2,Y2+3)-(XB2,Y7+2)
RX=XB1+2
RY=Y7
GOSUB LARROW
RX=XB2-2
GOSUB RARROW
LINE(RX,RY)-(XB1+2,RY)
RX=X1+2
RY=Y6
GOSUB LARROW
RX=X2-2
GOSUB RARROW
LINE(RX,RY)-(X1+2,RY)
RX=X4
RY=Y2+2
GOSUB UARROW
RY=Y3-2
GOSUB DARROW
LINE(RX,RY)-(RX,Y2+2)
RX=X5
RY=Y2+2
GOSUB UARROW
RY=Y4-2
GOSUB DARROW
LINE(RX,RY)-(RX,Y2+2)
RX=X5
RY=Y1-2
GOSUB DARROW
LINE(RX,RY)-(RX,Y1-25)
RX=X5
RY=Y2+2
GOSUB UARROW
RY=Y4-2
GOSUB DARROW
LINE(RX,RY)-(RX,Y2+2)
X6=X2+15
RX=X6
RY=Y1-2
GOSUB DARROW
RY=YB+2
GOSUB UARROW
LINE(RX,RY)-(RX,Y1-2)
REM WRITES THE DIMENSIONS ON THE DRAWING
ZL=18+1.6*ST/BO
```

```
LOCATE ZL,51:PRINT "Ø";BO
LOCATE 11,28:PRINT ST
LOCATE 9,36:PRINT USING"##.#";HTDC
LOCATE 8,47:PRINT "SQUISH DIA.";USING"##.#";SQD
LOCATE 4,68:PRINT USING"##.#";MH
LOCATE 3,33:PRINT USING"##.#";SQH
RETURN
LARROW:
REM LEFTWARDS
LINE(RX,RY)-(RX+3,RY-3)
LINE(RX,RY)-(RX+3,RY+3)
RETURN
RARROW:
REM RIGHTWARDS
LINE(RX,RY)-(RX-3,RY-3)
LINE(RX,RY)-(RX-3,RY+3)
RETURN
UARROW:
REM UPWARDS
LINE(RX,RY)-(RX-3,RY+3)
LINE(RX,RY)-(RX+3,RY+3)
RETURN
DARROW:
REM DOWNWARDS
LINE(RX,RY)-(RX-3,RY-3)
LINE(RX,RY)-(RX+3,RY-3)
RETURN
```

COMPUTER PROGRAM PROG.4.3
PROGRAM NAME "HEMI-FLAT CHAMBER"
ProgList.4.3

```
REM PROGRAM Prog.4.3
REM ProgList.4.3
REM PROGRAM NAME 'HEMI-FLAT CHAMBER'
REM PROGRAM FOR COMBUSTION CHAMBER CLEARANCE VOLUME
REM FOR HEMISPHERICAL SEGMENT COMBUSTION CHAMBER
REM WITH FLAT TOPPED PISTON CROWN
PI=3.141593
DTOR=PI/180
RTOD=180/PI
INPUT"NEW DATA OR FILED DATA? (TYPE 'N' OR 'F')";D$
IF D$="F" THEN GOSUB FILED
IF D$="F" THEN GOTO 500
```

```
N=0
IF N=0 THEN GOSUB DATAIN
500 CLS
IF F$="P" THEN CLS
IF F$="P" THEN OPEN"LPT1:PROMPT" FOR OUTPUT AS #1
IF F$="P" THEN WINDOW OUTPUT#1
LOCATE 1,1:PRINT"CURRENT INPUT DATA FOR 2-STROKE 'SPHERI-
CAL' BOWL"
LOCATE 2,1:PRINT"(B) BORE,mm=";BO
LOCATE 3,1:PRINT"(S) STROKE,mm=";ST
LOCATE 4,1:PRINT"(C) CON-ROD,mm=";CRL
LOCATE 5,1:PRINT"(E) EXHAUST CLOSES,deg btdc=";EO
LOCATE 7,1:PRINT"BOWL DETAILS,mm"
LOCATE 8,1:PRINT"(SQ) SQUISH CLEARANCE=";SQ
LOCATE 9,1:PRINT"(BR) BOWL RADIUS=";BR
LOCATE 10,1:PRINT"(BH) BOWL DEPTH=";BH
LOCATE 11,1:PRINT"(SR) SQUISH CORNER RADIUS=";SR
CRK=ST/2
DEG=EO
FPIST=PI*BO*BO/4
SV=FPIST*ST
SVCC=SV/1000
GOSUB PIST
TSV=VTDC
TSVCC=TSV/1000
GOSUB BOWL
LOCATE 13,1:PRINT"OUTPUT DATA"
LOCATE 14,1:PRINT"SWEPT VOLUME,cc=";USING"###.#";SVCC
LOCATE 15,1:PRINT"TRAPPED SWEPT VOLUME,cc=";USING
"###.#";TSVCC
LOCATE 16,1:PRINT"TRAPPED STROKE,mm=";USING"##.#";HTDC
LOCATE 17,1:PRINT"CLEARANCE VOLUME,cc=";USING"##.#";CVCC
LOCATE 18,1:PRINT"COMPRESSION RATIO=";USING"##.#";GCR
LOCATE 19,1:PRINT"TRAPPED COMPRESSION RATIO=";USING
"##.#";TCR
LOCATE 20,1:PRINT"SQUISH AREA RATIO=0";USING".###";SAR
GOSUB PISTDRAW
IF F$="P" THEN LOCATE 23,20:PRINT T$
IF F$="P" THEN CLOSE #1
LOCATE 23,1:PRINT"                                        "
LOCATE 24,1:PRINT"                                        "
LOCATE 25,1:PRINT"                                        "
LOCATE 26,1:PRINT"                                        "
LOCATE 23,1:PRINT"TYPE CODE FOR NEW DATA AND A
RECALCULATION"
```

```
LOCATE 24,1:INPUT"OR TYPE 'P' FOR PRINT-OUT, 'F' FOR FILING
DATA, OR 'Q' FOR QUITTING";F$
IF F$="B" THEN INPUT"TYPE IN NEW VALUE FOR BORE,mm";BO
IF F$="S" THEN INPUT"TYPE IN NEW VALUE FOR STROKE,mm";ST
IF F$="C" THEN INPUT"TYPE IN NEW VALUE FOR CON-ROD
CENTRES,mm";CRL
IF F$="E" THEN INPUT"TYPE IN NEW VALUE FOR EXHAUST
CLOSING,deg.btdc";EO
IF F$="SQ" THEN INPUT"TYPE IN NEW VALUE FOR SQUISH
CLEARANCE,mm";SQ
IF F$="BH" THEN INPUT"TYPE IN NEW VALUE FOR BOWL
DEPTH,mm";BH
IF F$="BR" THEN INPUT"TYPE IN NEW VALUE FOR UPPER BOWL
RADIUS,mm";BR
IF F$="SR" THEN INPUT"TYPE IN NEW VALUE FOR SQISH CORNER
RADIUS,mm";SR
IF F$="P" THEN INPUT"TYPE IN A TITLE FOR THE PRINT-OUT";T$
IF F$="P" THEN GOTO 500
IF F$="F" THEN GOSUB FILING
IF F$="Q" THEN GOTO 200
GOTO 500
200 END
REM PISTON POSITION SUBROUTINE
PIST:
DR=DEG*DTOR
HTDC=CRK+CRL-CRK*COS(DR)-SQR(CRL*CRL-(CRK*SIN(DR))^2)
VTDC=HTDC*FPIST
RETURN
BOWL:
SQVOL=FPIST*SQ
CH=BR-BH
BD=SQR((BR+SR)^2-(SR+CH)^2)
PHI=ATN((SR+CH)/BD)
THET=PI/2-PHI
BAREA=PI*BD*BD
SAR=(FPIST-BAREA)/FPIST
ZZ=COS(THET)
BR2=BR*COS(PHI)
BRH=BR*SIN(PHI)-CH
AYBAR=0
AA=0
XD=0
DR=.5
R1=BH
958 XD=XD+DR
```

```
IF XD>=BR2 THEN GOTO 955
R2=SQR(BR*BR-XD*XD)-CH
GOTO 956
955 ZRC=BD-XD
R2=SR-SQR(SR*SR-ZRC*ZRC)
956 DA=(R2+R1)*DR/2
AA=AA+DA
AYBAR=AYBAR+DA*(XD-DR/2)
IF XD>=BD THEN GOTO 957
R1=R2
GOTO 958
957 YBAR=AYBAR/AA
BVOL1=2*PI*YBAR*AA
CVOL=SQVOL+BVOL1
CVCC=CVOL/1000
GCR=(CVOL+SV)/CVOL
TCR=(CVOL+TSV)/CVOL
RETURN
DATAIN:
BO=72
ST=60
CRL=120
EO=103
SQ=1.5
BH=19
BR=23
SR=4
RETURN
FILING:
INPUT "TYPE IN TITLE NEEDED FOR THE FILE WITH '.HEMIFLAT'
END CODE";FIL$
OPEN FIL$ FOR OUTPUT AS #1
WRITE#1,BO,ST,CRL,EO,SQ,BH,BR,SR
FINISH:
CLOSE#1
RETURN
FILED:
INPUT "TYPE IN THE FILENAME WITH '.HEMIFLAT' END CODE";FIL$
OPEN"I",#1,FIL$
INPUT#1,BO,ST,CRL,EO,SQ,BH,BR,SR
CLOSE#1
RETURN
PISTDRAW:
Y1=80
```

```
X1=315
X2=515
SCAL=(X2-X1)/BO
Y2=Y1+SQ*SCAL
Y3=Y2+ST*SCAL
Y5=Y3+8
Y4=Y2+HTDC*SCAL
YB=Y1-BH*SCAL
BDS=BD*SCAL
XCL=(X1+X2)/2
XB1=(X1+X2)/2-BDS
XB2=(X1+X2)/2+BDS
BRS=BR*SCAL
YCH=Y1+CH*SCAL
SRS=SR*SCAL
YRS=Y1-SRS
XB3=XB1+BRS
XB4=XB2-BRS
X3=X1-30
REM DRAWS CYLINDER
LINE(X1,Y5)-(X1,Y3)
LINE(X1,Y4)-(X1,Y1)
LINE(X1,Y1)-(X2,Y1)
LINE(X2,Y1)-(X2,Y5)
LINE(X3,Y4)-(X1+2,Y3),,B
REM DRAWS PISTON
LINE(X1+2,Y5-1)-(X1+2,Y2)
LINE(X1+2,Y2)-(X2-2,Y2)
LINE(X2-2,Y2)-(X2-2,Y5-1)
LINE(X1+2,Y5-1)-(X2-2,Y5-1)
CIRCLE(XCL,YCH),BRS,,PHI,PI-PHI
CIRCLE(XB1,YRS),SRS,,3*PI/2,2*PI-PHI
CIRCLE(XB2,YRS),SRS,,PI+PHI,3*PI/2
REM DRAWS PISTON RINGS
YR=Y2+4
LINE(X1,YR)-(X1+6,YR+4),33,BF
YR=YR+9
LINE(X1,YR)-(X1+6,YR+4),33,BF
YR=Y2+4
LINE(X2,YR)-(X2-6,YR+4),33,BF
YR=YR+9
LINE(X2,YR)-(X2-6,YR+4),33,BF
REM DRAWS DIMENSION LINES
X4=X3-32
```

```
X5=X3-10
Y6=Y5+20
LINE(X4-3,Y1)-(X1-3,Y1)
LINE(X4-3,Y2)-(X1-3,Y2)
LINE(X4-3,Y3)-(X3-3,Y3)
LINE(X5-3,Y4)-(X5+3,Y4)
LINE(X1,Y5+3)-(X1,Y5+25)
LINE(X2,Y5+3)-(X2,Y5+25)
LINE(X2+3,Y1)-(X2+28,Y1)
LINE(XCL+20,YB)-(X2+28,YB)
RX=XCL
RY=YB+3
GOSUB UARROW
LINE(RX,RY+25)-(RX,RY)
Y7=Y2+23
LINE(XB1,Y2+3)-(XB1,Y7+2)
LINE(XB2,Y2+3)-(XB2,Y7+2)
RX=XB1+2
RY=Y7
GOSUB LARROW
RX=XB2-2
GOSUB RARROW
LINE(RX,RY)-(XB1+2,RY)
RX=X1+2
RY=Y6
GOSUB LARROW
RX=X2-2
GOSUB RARROW
LINE(RX,RY)-(X1+2,RY)
RX=X4
RY=Y2+2
GOSUB UARROW
RY=Y3-2
GOSUB DARROW
LINE(RX,RY)-(RX,Y2+2)
RX=X5
RY=Y2+2
GOSUB UARROW
RY=Y4-2
GOSUB DARROW
LINE(RX,RY)-(RX,Y2+2)
RX=X5
RY=Y1-2
GOSUB DARROW
```

```
LINE(RX,RY)-(RX,Y1-25)
RX=X5
RY=Y2+2
GOSUB UARROW
RY=Y4-2
GOSUB DARROW
LINE(RX,RY)-(RX,Y2+2)
X6=X2+15
RX=X6
RY=Y1-2
GOSUB DARROW
RY=YB+2
GOSUB UARROW
LINE(RX,RY)-(RX,Y1-2)
REM WRITES THE DIMENSIONS ON THE DRAWING
ZL=18+1.6*ST/BO
LOCATE ZL,51:PRINT "Ø";BO
LOCATE 11,28:PRINT ST
LOCATE 9,36:PRINT USING"##.#";HTDC
LOCATE 8,51:PRINT "Ø";USING"##.#";2*BD
LOCATE 4,68:PRINT BH
LOCATE 3,33:PRINT SQ
LOCATE 5,50:PRINT "rad";BR
LOCATE 5,38:PRINT "rad";SR
RETURN
LARROW:
REM LEFTWARDS
LINE(RX,RY)-(RX+3,RY-3)
LINE(RX,RY)-(RX+3,RY+3)
RETURN
RARROW:
REM RIGHTWARDS
LINE(RX,RY)-(RX-3,RY-3)
LINE(RX,RY)-(RX-3,RY+3)
RETURN
UARROW:
REM UPWARDS
LINE(RX,RY)-(RX-3,RY+3)
LINE(RX,RY)-(RX+3,RY+3)
RETURN
DARROW:
REM DOWNWARDS
LINE(RX,RY)-(RX-3,RY-3)
LINE(RX,RY)-(RX+3,RY-3)
RETURN
```

COMPUTER PROGRAM PROG.4.4
PROGRAM NAME "BATHTUB CHAMBER"
ProgList.4.4

```
REM PROGRAM Prog.4.4
REM ProgList.4.4
REM PROGRAM NAME 'BATHTUB CHAMBER'
REM PROGRAM FOR COMBUSTION CHAMBER CLEARANCE VOLUME
REM FOR BATHTUB COMBUSTION CHAMBER
REM WITH FLAT TOPPED PISTON CROWN
PI=3.141593
DTOR=PI/180
RTOD=180/PI
INPUT"NEW DATA OR FILED DATA? (TYPE 'N' OR 'F')";D$
IF D$="F" THEN GOSUB FILED
IF D$="F" THEN GOTO 500
N=0
IF N=0 THEN GOSUB DATAIN
500 CLS
IF F$="P" THEN CLS
IF F$="P" THEN OPEN"LPT1:PROMPT" FOR OUTPUT AS #1
IF F$="P" THEN WINDOW OUTPUT#1
LOCATE 1,1:PRINT"CURRENT INPUT DATA FOR 2-STROKE 'BATH-
TUB' BOWL"
LOCATE 2,1:PRINT"(B) BORE,mm=";BO
LOCATE 3,1:PRINT"(S) STROKE,mm=";ST
LOCATE 4,1:PRINT"(C) CON-ROD,mm=";CRL
LOCATE 5,1:PRINT"(E) EXHAUST CLOSES,deg btdc=";EO
LOCATE 7,1:PRINT"BOWL DETAILS,mm"
LOCATE 8,1:PRINT"(SQ) SQUISH CLEARANCE=";SQ
LOCATE 9,1:PRINT"(BD) BOWL DIAMETER=";BD
LOCATE 10,1:PRINT"(BH) BOWL DEPTH=";BH
LOCATE 11,1:PRINT"(BR) BOWL CORNER RADIUS=";BR
LOCATE 12,1:PRINT"(SR) SQUISH CORNER RADIUS=";SR
CRK=ST/2
DEG=EO
FPIST=PI*BO*BO/4
SV=FPIST*ST
SVCC=SV/1000
GOSUB PIST
TSV=VTDC
TSVCC=TSV/1000
GOSUB BOWL
LOCATE 14,1:PRINT"OUTPUT DATA"
LOCATE 15,1:PRINT"SWEPT VOLUME,cc=";USING"###.#";SVCC
```

```
LOCATE 16,1:PRINT"TRAPPED SWEPT VOLUME,cc=";USING
"###.#";TSVCC
LOCATE 17,1:PRINT"TRAPPED STROKE,mm=";USING"##.#";HTDC
LOCATE 18,1:PRINT"CLEARANCE VOLUME,cc=";USING"##.#";CVCC
LOCATE 19,1:PRINT"COMPRESSION RATIO=";USING"##.#";GCR
LOCATE 20,1:PRINT"TRAPPED COMPRESSION RATIO=";USING
"##.#";TCR
LOCATE 21,1:PRINT"SQUISH AREA RATIO=0";USING".###";SAR
GOSUB PISTDRAW
IF F$="P" THEN LOCATE 23,1:PRINT T$
IF F$="P" THEN CLOSE #1
LOCATE 23,1:PRINT"                                        "
LOCATE 24,1:PRINT"                                        "
LOCATE 25,1:PRINT"                                        "
LOCATE 26,1:PRINT"                                        "
LOCATE 23,1:PRINT"TYPE CODE FOR NEW DATA AND A RECALCU-
LATION"
LOCATE 24,1:INPUT"OR TYPE 'P' FOR PRINT-OUT, 'F' FOR FILING
DATA, OR 'Q' FOR QUITTING";F$
IF F$="B" THEN INPUT"TYPE IN NEW VALUE FOR BORE,mm";BO
IF F$="S" THEN INPUT"TYPE IN NEW VALUE FOR STROKE,mm";ST
IF F$="C" THEN INPUT"TYPE IN NEW VALUE FOR CON-ROD
CENTRES,mm";CRL
IF F$="E" THEN INPUT"TYPE IN NEW VALUE FOR EXHAUST
CLOSING,deg.btdc";EO
IF F$="SQ" THEN INPUT"TYPE IN NEW VALUE FOR SQUISH
CLEARANCE,mm";SQ
IF F$="BD" THEN INPUT"TYPE IN NEW VALUE FOR BOWL
DIAMETER,mm";BD
IF F$="BH" THEN INPUT"TYPE IN NEW VALUE FOR BOWL
DEPTH,mm";BH
IF F$="BR" THEN INPUT"TYPE IN NEW VALUE FOR UPPER BOWL
RADIUS,mm";BR
IF F$="SR" THEN INPUT"TYPE IN NEW VALUE FOR SQISH CORNER
RADIUS,mm";SR
IF F$="P" THEN INPUT"TYPE IN A TITLE FOR THE PRINT-OUT";T$
IF F$="P" THEN GOTO 500
IF F$="F" THEN GOSUB FILING
IF F$="Q" THEN GOTO 200
GOTO 500
200 END
REM PISTON POSITION SUBROUTINE
PIST:
DR=DEG*DTOR
```

```
HTDC=CRK+CRL-CRK*COS(DR)-SQR(CRL*CRL-(CRK*SIN(DR))^2)
VTDC=HTDC*FPIST
RETURN
BOWL:
SQVOL=FPIST*SQ
BAREA=PI*BD*BD/4
SD=BD+2*SR
SRAREA=PI*(SD*SD-BD*BD)/4
SAR=(FPIST-BAREA-SRAREA)/FPIST
BVOL1=BAREA*(BH-BR-SR)
RM=(BD-2*BR)/2
FMID=PI*RM*RM
BVOL2=FMID*BR
RC=RM+4*BR/(3*PI)
BVOL3=PI*PI*RC*BR*BR/2
BVOL4=PI*SD*SD*SR/4-(PI*SR*SR/4)*PI*(SD-8*SR/(3*PI))
CVOL=SQVOL+BVOL1+BVOL2+BVOL3+BVOL4
CVCC=CVOL/1000
GCR=(CVOL+SV)/CVOL
TCR=(CVOL+TSV)/CVOL
RETURN
DATAIN:
BO=72
ST=60
CRL=120
EO=103
SQ=1.5
BD=45
BH=14
BR=8
SR=3
RETURN
FILING:
INPUT "TYPE IN TITLE NEEDED FOR THE FILE WITH '.BATHTUB' END CODE";FIL$
OPEN FIL$ FOR OUTPUT AS #1
WRITE#1,BO,ST,CRL,EO,SQ,BD,BH,BR,SR
FINISH:
CLOSE#1
RETURN
FILED:
INPUT "TYPE IN THE FILENAME WITH 'BATHTUB' END CODE";FIL$
OPEN"I",#1,FIL$
INPUT#1,BO,ST,CRL,EO,SQ,BD,BH,BR,SR
```

```
CLOSE#1
RETURN
PISTDRAW:
Y1=80
X1=315
X2=515
SCAL=(X2-X1)/BO
Y2=Y1+SQ*SCAL
Y3=Y2+ST*SCAL
Y5=Y3+8
Y4=Y2+HTDC*SCAL
YB=Y1-BH*SCAL
BDS=BD*SCAL/2
XCL=(X1+X2)/2
XB1=(X1+X2)/2-BDS
XB2=(X1+X2)/2+BDS
BRS=BR*SCAL
YR=YB+BRS
SRS=SR*SCAL
YRS=Y1-SRS
XB3=XB1+BRS
XB4=XB2-BRS
X3=X1-30
REM DRAWS CYLINDER
LINE(X1,Y5)-(X1,Y3)
LINE(X1,Y4)-(X1,Y1)
LINE(X1,Y1)-(X2,Y1)
LINE(X2,Y1)-(X2,Y5)
LINE(X3,Y4)-(X1+2,Y3),,B
REM DRAWS PISTON
LINE(X1+2,Y5-1)-(X1+2,Y2)
LINE(X1+2,Y2)-(X2-2,Y2)
LINE(X2-2,Y2)-(X2-2,Y5-1)
LINE(X1+2,Y5-1)-(X2-2,Y5-1)
CIRCLE(XB3,YR),BRS,,PI/2,PI
CIRCLE(XB4,YR),BRS,,0,PI/2
CIRCLE(XB1-SRS,YRS),SRS,,3*PI/2,2*PI
CIRCLE(XB2+SRS,YRS),SRS,,PI,3*PI/2
LINE(XB3,YB)-(XB4,YB)
LINE(XB1,YRS)-(XB1,YR)
LINE(XB2,YRS)-(XB2,YR)
REM DRAWS PISTON RINGS
YR=Y2+4
LINE(X1,YR)-(X1+6,YR+4),33,BF
```

```
YR=YR+9
LINE(X1,YR)-(X1+6,YR+4),33,BF
YR=Y2+4
LINE(X2,YR)-(X2-6,YR+4),33,BF
YR=YR+9
LINE(X2,YR)-(X2-6,YR+4),33,BF
REM DRAWS DIMENSION LINES
X4=X3-32
X5=X3-10
Y6=Y5+20
LINE(X4-3,Y1)-(X1-3,Y1)
LINE(X4-3,Y2)-(X1-3,Y2)
LINE(X4-3,Y3)-(X3-3,Y3)
LINE(X5-3,Y4)-(X5+3,Y4)
LINE(X1,Y5+3)-(X1,Y5+25)
LINE(X2,Y5+3)-(X2,Y5+25)
LINE(X2+3,Y1)-(X2+28,Y1)
LINE(XB2-3,YB)-(X2+28,YB)
Y7=YB-23
LINE(XB1,YB+3)-(XB1,Y7-2)
LINE(XB2,YB+3)-(XB2,Y7-2)
RX=XB1+2
RY=Y7
GOSUB LARROW
RX=XB2-2
GOSUB RARROW
LINE(RX,RY)-(XB1+2,RY)
RX=X1+2
RY=Y6
GOSUB LARROW
RX=X2-2
GOSUB RARROW
LINE(RX,RY)-(X1+2,RY)
RX=X4
RY=Y2+2
GOSUB UARROW
RY=Y3-2
GOSUB DARROW
LINE(RX,RY)-(RX,Y2+2)
RX=X5
RY=Y2+2
GOSUB UARROW
RY=Y4-2
GOSUB DARROW
```

```
LINE(RX,RY)-(RX,Y2+2)
RX=X5
RY=Y1-2
GOSUB DARROW
LINE(RX,RY)-(RX,Y1-25)
RX=X5
RY=Y2+2
GOSUB UARROW
RY=Y4-2
GOSUB DARROW
LINE(RX,RY)-(RX,Y2+2)
X6=X2+15
RX=X6
RY=Y1-2
GOSUB DARROW
RY=YB+2
GOSUB UARROW
LINE(RX,RY)-(RX,Y1-2)
REM WRITES THE DIMENSIONS ON THE DRAWING
LOCATE 19,51:PRINT "Ø";BO
LOCATE 11,28:PRINT ST
LOCATE 9,36:PRINT USING"##.#";HTDC
LOCATE 1,51:PRINT "Ø";BD
LOCATE 4,68:PRINT BH
LOCATE 3,33:PRINT SQ
LOCATE 4,53:PRINT "rad";BR
LOCATE 5,46:PRINT "rad";SR
RETURN
LARROW:
REM LEFTWARDS
LINE(RX,RY)-(RX+3,RY-3)
LINE(RX,RY)-(RX+3,RY+3)
RETURN
RARROW:
REM RIGHTWARDS
LINE(RX,RY)-(RX-3,RY-3)
LINE(RX,RY)-(RX-3,RY+3)
RETURN
UARROW:
REM UPWARDS
LINE(RX,RY)-(RX-3,RY+3)
LINE(RX,RY)-(RX+3,RY+3)
RETURN
```

```
DARROW:
REM DOWNWARDS
LINE(RX,RY)-(RX-3,RY-3)
LINE(RX,RY)-(RX+3,RY-3)
RETURN
```

COMPUTER PROGRAM PROG.4.5
PROGRAM NAME "TOTAL OFFSET CHAMBER"
ProgList.4.5

```
REM PROGRAM Prog.4.5
REM ProgList.4.5
REM PROGRAM FOR COMBUSTION CHAMBER CLEARANCE VOLUME
REM FOR TOTAL OFFSET COMBUSTION CHAMBER
REM WITH FLAT TOPPED PISTON CROWN
PI=3.141593
DTOR=PI/180
RTOD=180/PI
INPUT"NEW DATA OR FILED DATA? (TYPE 'N' OR 'F')";D$
IF D$="F" THEN GOSUB FILED
IF D$="F" THEN GOTO 500
N=0
IF N=0 THEN GOSUB DATAIN
500 CLS
IF F$="P" THEN CLS
IF F$="P" THEN OPEN"LPT1:PROMPT" FOR OUTPUT AS #1
IF F$="P" THEN WINDOW OUTPUT#1
LOCATE 1,1:PRINT"CURRENT INPUT DATA FOR 2-STROKE ' TOTAL
OFFSET' BOWL"
LOCATE 2,1:PRINT"(B) BORE,mm=";BO
LOCATE 3,1:PRINT"(S) STROKE,mm=";ST
LOCATE 4,1:PRINT"(C) CON-ROD,mm=";CRL
LOCATE 5,1:PRINT"(E) EXHAUST CLOSES,deg btdc=";EO
LOCATE 7,1:PRINT"BOWL DETAILS,mm"
LOCATE 8,1:PRINT"(SQ) SQUISH CLEARANCE=";SQ
LOCATE 9,1:PRINT"(BH) BOWL DEPTH=";BH
LOCATE 10,1:PRINT"(CR) CORNER RADIUS=";CR
LOCATE 11,1:PRINT"(SR) SQUISH CORNER RADIUS=";SR
LOCATE 12,1:PRINT"(TR) TOP CORNER RADIUS=";TR
CRK=ST/2
DEG=EO
FPIST=PI*BO*BO/4
SV=FPIST*ST
SVCC=SV/1000
GOSUB PIST
```

```
TSV=VTDC
TSVCC=TSV/1000
GOSUB BOWL
LOCATE 14,1:PRINT"OUTPUT DATA"
LOCATE 15,1:PRINT"SWEPT VOLUME,cc=";USING"###.#";SVCC
LOCATE 16,1:PRINT"TRAPPED SWEPT VOLUME,cc=";USING
"###.#";TSVCC
LOCATE 17,1:PRINT"TRAPPED STROKE,mm=";USING"##.#";HTDC
LOCATE 18,1:PRINT"CLEARANCE VOLUME,cc=";USING"##.#";CVCC
LOCATE 19,1:PRINT"COMPRESSION RATIO=";USING"##.#";GCR
LOCATE 20,1:PRINT"TRAPPED COMPRESSION RATIO=";USING
"##.#";TCR
LOCATE 21,1:PRINT"SQUISH AREA RATIO=0";USING".###";SAR
GOSUB PISTDRAW
GOSUB BOWLDRAW
IF F$="P" THEN LOCATE 23,1:PRINT T$
IF F$="P" THEN CLOSE #1
LOCATE 23,1:PRINT"                              "
LOCATE 24,1:PRINT"                              "
LOCATE 25,1:PRINT"                              "
LOCATE 26,1:PRINT"                              "
LOCATE 23,1:PRINT"TYPE CODE FOR NEW DATA "
LOCATE 24,1:PRINT"OR TYPE 'P' FOR PRINT-OUT"
LOCATE 25,1:PRINT"OR TYPE 'F' FOR FILING";F$
LOCATE 26,1:INPUT"OR TYPE 'Q' FOR QUITTING";F$
IF F$="B" THEN INPUT"TYPE IN NEW VALUE FOR BORE,mm";BO
IF F$="S" THEN INPUT"TYPE IN NEW VALUE FOR STROKE,mm";ST
IF F$="C" THEN INPUT"TYPE IN NEW VALUE FOR CON-ROD
CENTRES,mm";CRL
IF F$="E" THEN INPUT"TYPE IN NEW VALUE FOR EXHAUST
CLOSING,deg.btdc";EO
IF F$="SQ" THEN INPUT"TYPE IN NEW VALUE FOR SQUISH
CLEARANCE,mm";SQ
IF F$="CR" THEN INPUT"TYPE IN NEW VALUE FOR CORNER
RADIUS,mm";CR
IF F$="BH" THEN INPUT"TYPE IN NEW VALUE FOR BOWL
DEPTH,mm";BH
IF F$="TR" THEN INPUT"TYPE IN NEW VALUE FOR TOP CORNER
RADIUS,mm";TR
IF F$="SR" THEN INPUT"TYPE IN NEW VALUE FOR SQISH CORNER
RADIUS,mm";SR
IF F$="P" THEN INPUT"TYPE IN A TITLE FOR THE PRINT-OUT";T$
IF F$="P" THEN GOTO 500
IF F$="F" THEN GOSUB FILING
```

```
IF F$="Q" THEN GOTO 200
GOTO 500
200 END
REM PISTON POSITION SUBROUTINE
PIST:
DR=DEG*DTOR
HTDC=CRK+CRL-CRK*COS(DR)-SQR(CRL*CRL-(CRK*SIN(DR))^2)
VTDC=HTDC*FPIST
RETURN
BOWL:
SQVOL=FPIST*SQ
REM THE BOWL VOLUME IS CONSIDERED AS BH HIGH x BOWL AREA(BAREA) LESS +VSQ-VTR
PR1=BO/2-SR
PR2=CR
YC=SQR(PR1*PR1-2*PR1*PR2)
AN1=ATN(PR2/YC)
AN2=PI/2-AN1
AA=0
XD=0
DR=.5
R1=PR1
YC2=YC+PR2*COS(AN1)
YC3=YC+PR2
958 XD=XD+DR
IF XD>YC3 THEN GOTO 957
IF XD<=YC THEN GOTO 954
IF XD<=YC2 THEN GOTO 956
R2=2*SQR(PR2*PR2-XD2*XD2)
GOTO 955
956 XD2=XD-YC
R2=SQR(PR1*PR1-XD*XD)-(PR2-SQR(PR2*PR2-XD2*XD2))
GOTO 955
954 R2=SQR(PR1*PR1-XD*XD)
955 DA=(R1+R2)*DR/2
R1=R2
AA=AA+DA
GOTO 958
957 BAREA=2*AA
BVOL1=BH*BAREA
REM ATR IS THE AREA OF THE TOP RADIUS EXCESS FROM THE 'SQUARE'
REM YTR IS THE COG OF THAT AREA WRT THE VERTICAL WALL
REM LTR IS THE COG LENGTH SWEPT AROUND THE BOWL BY THAT AREA
```

```
REM VOLUME 'VTR' IS LTR x ATR
ATR=TR*TR-PI*TR*TR/4
YTR=(TR*TR*TR/2-PI*TR*TR*(TR-4*TR/(3*PI))/4)/ATR
Y=YTR
LTR=2*YC+(PI/2+AN1)*2*(PR2-Y)+(PI-2*AN1)*(PR1-Y)
VTR=LTR*ATR
REM ASQ, YSQ, LSQ AND VSQ ARE THE EQUIVALENT VALUES FOT
THE SQUISH RADIUS
ASQ=SR*SR-PI*SR*SR/4
YSQ=(SR*SR*SR/2-PI*SR*SR*(SR-4*SR/(3*PI))/4)/ASQ
Y=YSQ
LS=2*YC+(PI/2+AN1)*2*(PR2+Y)+(PI-2*AN1)*(PR1+Y)
LSQ=LS
VSQ=LSQ*ASQ
REM ASQ IS THE AREA OF THE SQUISH BAND FOR SQUISH AREA
RATIO PURPOSES
Y=SR/2
LS=2*YC+(PI/2+AN1)*2*(PR2+Y)+(PI-2*AN1)*(PR1+Y)
ASQ=LS*SR
SQAREA=BAREA+ASQ
SAR=(FPIST-SQAREA)/FPIST
REM VSQ IS EXTRA AND VTR IS EXCESS TO BE TAKEN OFF
CVOL=SQVOL+BVOL1+VSQ-VTR
CVCC=CVOL/1000
GCR=(CVOL+SV)/CVOL
TCR=(CVOL+TSV)/CVOL
RETURN
DATAIN:
BO=72
ST=60
CRL=120
EO=103
SQ=1.5
BH=14
CR=12
SR=4
TR=6
RETURN
FILING:
INPUT "TYPE IN TITLE NEEDED FOR THE FILE WITH '.TOTOFF' END
CODE";FIL$
OPEN FIL$ FOR OUTPUT AS #1
WRITE#1,BO,ST,CRL,EO,SQ,BH,BR,SR
FINISH:
CLOSE#1
```

```
RETURN
FILED:
INPUT "TYPE IN THE FILENAME WITH 'TOTOFF' END CODE";FIL$
OPEN"I",#1,FIL$
INPUT#1,BO,ST,CRL,EO,SQ,BH,BR,SR
CLOSE#1
RETURN
PISTDRAW:
Y1=80
X1=315
X2=515
SCAL=(X2-X1)/BO
Y2=Y1+SQ*SCAL
Y3=Y2+ST*SCAL
Y5=Y3+8
Y4=Y2+HTDC*SCAL
YB=Y1-BH*SCAL
BDS=BD*SCAL
BR=BO/2
BRS=BR*SCAL
PR1S=SCAL*PR1
PR2S=SCAL*PR2
SRS=SR*SCAL
TRS=TR*SCAL
YRS=Y1-SRS
XCL=(X1+X2)/2
XB1=XCL
XB2=X2-SRS
XB3=XB1+TRS
XB4=XB2-TRS
YB=Y1-BH*SCAL
YR=YB+TRS
SRS=SR*SCAL
YRS=Y1-SRS
X3=X1-30
REM DRAWS CYLINDER
LINE(X1,Y5)-(X1,Y3)
LINE(X1,Y4)-(X1,Y1)
LINE(X1,Y1)-(X2,Y1)
LINE(X2,Y1)-(X2,Y5)
LINE(X3,Y4)-(X1+2,Y3),,B
REM DRAWS PISTON
LINE(X1+2,Y5-1)-(X1+2,Y2)
LINE(X1+2,Y2)-(X2-2,Y2)
```

```
LINE(X2-2,Y2)-(X2-2,Y5-1)
LINE(X1+2,Y5-1)-(X2-2,Y5-1)
CIRCLE(XB3,YR),TRS,,PI/2,PI
CIRCLE(XB4,YR),TRS,,0,PI/2
CIRCLE(XB1-SRS,YRS),SRS,,3*PI/2,2*PI
CIRCLE(XB2+SRS,YRS),SRS,,PI,3*PI/2
LINE(XB3,YB)-(XB4,YB)
LINE(XB1,YRS)-(XB1,YR)
LINE(XB2,YRS)-(XB2,YR)
REM DRAWS PISTON RINGS
YR=Y2+4
LINE(X1,YR)-(X1+6,YR+4),33,BF
YR=YR+9
LINE(X1,YR)-(X1+6,YR+4),33,BF
YR=Y2+4
LINE(X2,YR)-(X2-6,YR+4),33,BF
YR=YR+9
LINE(X2,YR)-(X2-6,YR+4),33,BF
REM DRAWS DIMENSION LINES
X4=X3-32
X5=X3-10
Y6=Y5+20
LINE(X4-3,Y1)-(X1-3,Y1)
LINE(X4-3,Y2)-(X1-3,Y2)
LINE(X4-3,Y3)-(X3-3,Y3)
LINE(X5-3,Y4)-(X5+3,Y4)
LINE(X1,Y5+3)-(X1,Y5+25)
LINE(X2,Y5+3)-(X2,Y5+25)
LINE(X2+3,Y1)-(X2+28,Y1)
LINE(XB2-3,YB)-(X2+28,YB)
Y7=YB-23
LINE(XB1,YB+3)-(XB1,Y7-2)
LINE(XB2,YB+3)-(XB2,Y7-2)
RX=XB1+2
RY=Y7
GOSUB LARROW
RX=XB2-2
GOSUB RARROW
LINE(RX,RY)-(XB1+2,RY)
RX=X1+2
RY=Y6
GOSUB LARROW
RX=X2-2
GOSUB RARROW
```

```
LINE(RX,RY)-(X1+2,RY)
RX=X4
RY=Y2+2
GOSUB UARROW
RY=Y3-2
GOSUB DARROW
LINE(RX,RY)-(RX,Y2+2)
RX=X5
RY=Y2+2
GOSUB UARROW
RY=Y4-2
GOSUB DARROW
LINE(RX,RY)-(RX,Y2+2)
RX=X5
RY=Y1-2
GOSUB DARROW
LINE(RX,RY)-(RX,Y1-25)
RX=X5
RY=Y2+2
GOSUB UARROW
RY=Y4-2
GOSUB DARROW
LINE(RX,RY)-(RX,Y2+2)
X6=X2+15
RX=X6
RY=Y1-2
GOSUB DARROW
RY=YB+2
GOSUB UARROW
LINE(RX,RY)-(RX,Y1-2)
REM WRITES THE DIMENSIONS ON THE DRAWING
LOCATE 18,51:PRINT "Ø";BO
LOCATE 11,28:PRINT ST
LOCATE 9,36:PRINT USING"##.#";HTDC
LOCATE 1,56:PRINT USING"##.#";BR-SR
LOCATE 4,68:PRINT BH
LOCATE 3,33:PRINT SQ
LOCATE 4,55:PRINT "rad";TR
LOCATE 5,46:PRINT "rad";SR
LOCATE 9,45:PRINT"ELEVATION"
LOCATE 10,45:PRINT"ON CYLINDER AXIS"
IF F$="P" THEN LOCATE 36,40:PRINT"PLAN VIEW ON CYLINDER HEAD"
RETURN
```

```
BOWLDRAW:
CX2=415
CY2=420
CIRCLE(CX2,CY2),BRS
YCS=YC*SCAL
Y21=CY2-YCS
Y22=CY2+YCS
CX3=CX2+PR2S
CIRCLE(CX2,CY2),PR1S,,2*PI-AN2,AN2
CIRCLE(CX3,Y21),PR2S,,AN2,PI
CIRCLE(CX3,Y22),PR2S,,PI,2*PI-AN2
CIRCLE(CX3,Y21),PR2S+SRS,,AN2,PI
CIRCLE(CX3,Y22),PR2S+SRS,,PI,2*PI-AN2
LINE(CX2-SRS,Y21)-(CX2-SRS,Y22)
LINE(CX2,Y21)-(CX2,Y22)
LINE(X1-25,CY2)-(X2+25,CY2)
LINE(CX2,Y21-15)-(CX2,Y21-75)
LINE(CX2,Y22+15)-(CX2,Y22+75)
RY=Y21-PR2S+3
LINE(CX3,Y21)-(CX3,RY)
RX=CX3
GOSUB UARROW
LOCATE 25,54:PRINT CR
RETURN
LARROW:
REM LEFTWARDS
LINE(RX,RY)-(RX+3,RY-3)
LINE(RX,RY)-(RX+3,RY+3)
RETURN
RARROW:
REM RIGHTWARDS
LINE(RX,RY)-(RX-3,RY-3)
LINE(RX,RY)-(RX-3,RY+3)
RETURN
UARROW:
REM UPWARDS
LINE(RX,RY)-(RX-3,RY+3)
LINE(RX,RY)-(RX+3,RY+3)
RETURN
DARROW:
REM DOWNWARDS
LINE(RX,RY)-(RX-3,RY-3)
LINE(RX,RY)-(RX+3,RY-3)
RETURN
```

COMPUTER PROGRAM PROG.4.6
PROGRAM NAME "BOWL IN PISTON"
ProgList.4.6

```
REM PROGRAM Prog.4.6
REM ProgList.4.6
REM PROGRAM FOR COMBUSTION CHAMBER CLEARANCE VOLUME
REM FOR BOWL IN PISTON COMBUSTION CHAMBER
REM WITH FLAT TOPPED PISTON CROWN
PI=3.141593
DTOR=PI/180
RTOD=180/PI
INPUT"NEW DATA OR FILED DATA? (TYPE 'N' OR 'F')";D$
IF D$="F" THEN GOSUB FILED
IF D$="F" THEN GOTO 500
N=0
IF N=0 THEN GOSUB DATAIN
500 CLS
IF F$="P" THEN CLS
IF F$="P" THEN OPEN"LPT1:PROMPT" FOR OUTPUT AS #1
IF F$="P" THEN WINDOW OUTPUT#1
LOCATE 1,1:PRINT"CURRENT INPUT DATA FOR 2-STROKE BOWL"
LOCATE 2,1:PRINT"(B) BORE,mm=";BO
LOCATE 3,1:PRINT"(S) STROKE,mm=";ST
LOCATE 4,1:PRINT"(C) CON-ROD,mm=";CRL
LOCATE 5,1:PRINT"(E) EXHAUST CLOSES,deg btdc=";EO
LOCATE 7,1:PRINT"BOWL DETAILS,mm"
LOCATE 8,1:PRINT"(SQ) SQUISH CLEARANCE=";SQ
LOCATE 9,1:PRINT"(BD) BOWL DIAMETER=";BD
LOCATE 10,1:PRINT"(BH) BOWL DEPTH=";BH
LOCATE 11,1:PRINT"(BR) BOWL CORNER RADIUS=";BR
CRK=ST/2
DEG=EO
FPIST=PI*BO*BO/4
SV=FPIST*ST
SVCC=SV/1000
GOSUB PIST
TSV=VTDC
TSVCC=TSV/1000
GOSUB BOWL
LOCATE 13,1:PRINT"OUTPUT DATA"
LOCATE 14,1:PRINT"SWEPT VOLUME,cc=";USING"###.#";SVCC
LOCATE 15,1:PRINT"TRAPPED SWEPT VOLUME,cc=";USING
"###.#";TSVCC
LOCATE 16,1:PRINT"TRAPPED STROKE,mm=";USING"##.#";HTDC
```

```
LOCATE 17,1:PRINT"CLEARANCE VOLUME,cc=";USING"##.#";CVCC
LOCATE 18,1:PRINT"COMPRESSION RATIO=";USING"##.#";GCR
LOCATE 19,1:PRINT"TRAPPED COMPRESSION RATIO=";USING
"##.#";TCR
LOCATE 20,1:PRINT"SQUISH AREA RATIO=0";USING".###";SAR
GOSUB PISTDRAW
IF F$="P" THEN LOCATE 22,1:PRINT T$
IF F$="P" THEN CLOSE #1
LOCATE 23,1:PRINT"                                        "
LOCATE 24,1:PRINT"                                        "
LOCATE 25,1:PRINT"                                        "
LOCATE 26,1:PRINT"                                        "
LOCATE 23,1:PRINT"TYPE CODE FOR NEW DATA AND A
RECALCULATION"
LOCATE 24,1:INPUT"OR TYPE 'P' FOR PRINT-OUT, 'F' FOR FILING
DATA, OR 'Q' FOR QUITTING";F$
IF F$="B" THEN INPUT"TYPE IN NEW VALUE FOR BORE,mm";BO
IF F$="S" THEN INPUT"TYPE IN NEW VALUE FOR STROKE,mm";ST
IF F$="C" THEN INPUT"TYPE IN NEW VALUE FOR CON-ROD
CENTRES,mm";CRL
IF F$="E" THEN INPUT"TYPE IN NEW VALUE FOR EXHAUST
CLOSING,deg.btdc";EO
IF F$="SQ" THEN INPUT"TYPE IN NEW VALUE FOR SQUISH
CLEARANCE,mm";SQ
IF F$="BD" THEN INPUT"TYPE IN NEW VALUE FOR BOWL
DIAMETER,mm";BD
IF F$="BH" THEN INPUT"TYPE IN NEW VALUE FOR BOWL
DEPTH,mm";BH
IF F$="BR" THEN INPUT"TYPE IN NEW VALUE FOR INNER BOWL
RADIUS,mm";BR
IF F$="P" THEN INPUT"TYPE IN A TITLE FOR THE PRINT-OUT";T$
IF F$="P" THEN GOTO 500
IF F$="F" THEN GOSUB FILING
IF F$="Q" THEN GOTO 200
GOTO 500
200 END
REM PISTON POSITION SUBROUTINE
PIST:
DR=DEG*DTOR
HTDC=CRK+CRL-CRK*COS(DR)-SQR(CRL*CRL-(CRK*SIN(DR))^2)
VTDC=HTDC*FPIST
RETURN
BOWL:
SQVOL=FPIST*SQ
```

```
BAREA=PI*BD*BD/4
SAR=(FPIST-BAREA)/FPIST
BVOL1=BAREA*(BH-BR)
RM=(BD-2*BR)/2
FMID=PI*RM*RM
BVOL2=FMID*BR
RC=RM+4*BR/(3*PI)
BVOL3=PI*PI*RC*BR*BR/2
CVOL=SQVOL+BVOL1+BVOL2+BVOL3
CVCC=CVOL/1000
GCR=(CVOL+SV)/CVOL
TCR=(CVOL+TSV)/CVOL
RETURN
DATAIN:
BO=90
ST=90
CRL=180
EO=108
SQ=1.2
BD=40
BH=15
BR=6
RETURN
FILING:
INPUT "TYPE IN TITLE NEEDED FOR THE FILE WITH '.BOWLINP' END
CODE";FIL$
OPEN FIL$ FOR OUTPUT AS #1
WRITE#1,BO,ST,CRL,EO,SQ,BD,BH,BR
FINISH:
CLOSE#1
RETURN
FILED:
INPUT "TYPE IN THE FILENAME WITH '.BOWLINP' END CODE";FIL$
OPEN"I",#1,FIL$
INPUT#1,BO,ST,CRL,EO,SQ,BD,BH,BR
CLOSE#1
RETURN
PISTDRAW:
Y1=75
X1=315
X2=515
SCAL=(X2-X1)/BO
Y2=Y1+SQ*SCAL
Y3=Y2+ST*SCAL
```

```
Y5=Y3+11
Y4=Y2+HTDC*SCAL
YB=Y2+BH*SCAL
BDS=BD*SCAL/2
XCL=(X1+X2)/2
XB1=(X1+X2)/2-BDS
XB2=(X1+X2)/2+BDS
BRS=BR*SCAL
YR=YB-BRS
XB3=XB1+BRS
XB4=XB2-BRS
X3=X1-30
REM DRAWS CYLINDER
LINE(X1,Y5)-(X1,Y3)
LINE(X1,Y4)-(X1,Y1)
LINE(X1,Y1)-(X2,Y1)
LINE(X2,Y1)-(X2,Y5)
LINE(X3,Y4)-(X1+2,Y3),,B
REM DRAWS PISTON
LINE(X1+2,Y5-1)-(X1+2,Y2)
LINE(X1+2,Y2)-(X2-2,Y2)
LINE(X2-2,Y2)-(X2-2,Y5-1)
LINE(X1+2,Y5-1)-(X2-2,Y5-1)
CIRCLE(XB3,YR),BRS,,PI,3*PI/2
CIRCLE(XB4,YR),BRS,,3*PI/2,2*PI
LINE(XB3,YB)-(XB4,YB)
LINE(XB1,Y2)-(XB1,YR)
LINE(XB2,Y2)-(XB2,YR)
REM DRAWS PISTON RINGS
YR=Y2+4
LINE(X1,YR)-(X1+6,YR+4),33,BF
YR=YR+9
LINE(X1,YR)-(X1+6,YR+4),33,BF
YR=Y3+3
LINE(X1,YR)-(X1+6,YR+4),33,BF
YR=Y2+4
LINE(X2,YR)-(X2-6,YR+4),33,BF
YR=YR+9
LINE(X2,YR)-(X2-6,YR+4),33,BF
YR=Y3+3
LINE(X2,YR)-(X2-6,YR+4),33,BF
REM DRAWS DIMENSION LINES
X4=X3-32
X5=X3-10
```

```
Y6=Y1-20
LINE(X4-3,Y1)-(X1-3,Y1)
LINE(X4-3,Y2)-(X1-3,Y2)
LINE(X4-3,Y3)-(X3-3,Y3)
LINE(X5-3,Y4)-(X5+3,Y4)
LINE(X1,Y1-3)-(X1,Y1-25)
LINE(X2,Y1-3)-(X2,Y1-25)
LINE(X2+3,Y2)-(X2+28,Y2)
LINE(XB2+1,YB)-(X2+28,YB)
RX=X1+2
RY=Y6
GOSUB LARROW
RX=X2-2
GOSUB RARROW
LINE(RX,RY)-(X1+2,RY)
RX=X4
RY=Y2+2
GOSUB UARROW
RY=Y3-2
GOSUB DARROW
LINE(RX,RY)-(RX,Y2+2)
RX=X5
RY=Y2+2
GOSUB UARROW
RY=Y4-2
GOSUB DARROW
LINE(RX,RY)-(RX,Y2+2)
RX=X5
RY=Y1-2
GOSUB DARROW
LINE(RX,RY)-(RX,Y1-25)
RX=X5
RY=Y2+2
GOSUB UARROW
RY=Y4-2
GOSUB DARROW
LINE(RX,RY)-(RX,Y2+2)
X6=X2+15
RX=X6
RY=Y2+2
GOSUB UARROW
RY=YB-2
GOSUB DARROW
LINE(RX,RY)-(RX,Y2+2)
```

```
Y7=YB+35
LINE(XB1,YB+1)-(XB1,Y7+2)
LINE(XB2,YB+1)-(XB2,Y7+2)
RX=XB1+2
RY=Y7
GOSUB LARROW
RX=XB2-2
GOSUB RARROW
LINE(RX,RY)-(XB1+2,RY)
REM WRITES THE DIMENSIONS ON THE DRAWING
LOCATE 3,51:PRINT "Ø";BO
LOCATE 11,28:PRINT ST
LOCATE 9,36:PRINT USING"##.#";HTDC
LOCATE 10,51:PRINT "Ø";BD
LOCATE 6,68:PRINT BH
LOCATE 3,33:PRINT SQ
LOCATE 7,51:PRINT "rad";BR
RETURN
LARROW:
REM LEFTWARDS
LINE(RX,RY)-(RX+3,RY-3)
LINE(RX,RY)-(RX+3,RY+3)
RETURN
RARROW:
REM RIGHTWARDS
LINE(RX,RY)-(RX-3,RY-3)
LINE(RX,RY)-(RX-3,RY+3)
RETURN
UARROW:
REM UPWARDS
LINE(RX,RY)-(RX-3,RY+3)
LINE(RX,RY)-(RX+3,RY+3)
RETURN
DARROW:
REM DOWNWARDS
LINE(RX,RY)-(RX-3,RY-3)
LINE(RX,RY)-(RX+3,RY-3)
RETURN
```

COMPUTER PROGRAM PROG.4.7
PROGRAM NAME "QUB DEFLECTOR"
ProgList.4.7

```
REM PROGRAM Prog.4.7
REM ProgList.4.7
```

```
REM PROGRAM FOR COMBUSTION CHAMBER CLEARANCE VOLUME
REM FOR QUB TYPE DEFLECTOR PISTON COMBUSTION CHAMBER
REM QUB TYPE CROSS SCAVENGED ENGINE
100 INPUT"NEW DATA OR FILED DATA(N OR F?)";A$
IF A$="F" THEN GOSUB FILED
IF A$="F" THEN GOTO 99
INPUT"NEW DATA FOR CYLINDER BORE(Y/N?)";A$
IF A$="Y" THEN  INPUT"TYPE IN CYLINDER BORE,B, IN MM";BO
INPUT"NEW DATA FOR CYLINDER STROKE(Y/N?)";A$
IF  A$="Y" THEN INPUT"TYPE IN CYLINDER STROKE,S, IN MM";ST
INPUT"NEW DATA FOR TRAPPED STROKE ABOVE EXHAUST
CLOSURE(Y/N?)";A$
IF A$="Y" THEN INPUT"TYPE IN TRAPPED STROKE,T, IN MM";TS
INPUT"NEW DATA FOR DEFLECTOR HEIGHT(Y/N?)";A$
IF A$="Y" THEN INPUT"TYPE IN DEFLECTOR HEIGHT,H, IN MM";DH
INPUT"NEW DATA FOR SQUISH CLEARANCE(Y/N?)";A$
IF A$="Y" THEN INPUT"TYPE IN SQUISH CLEARANCE,Q, IN MM";SQ
INPUT"NEW DATA FOR DEFLECTOR RADIUS(Y/N?)";A$
IF A$="Y" THEN INPUT"TYPE IN DEFLECTOR RADIUS,R, IN MM";DR
INPUT"NEW DATA FOR DEFLECTION RADIUS(Y/N?)";A$
IF A$="Y" THEN INPUT"TYPE IN DEFLECTION RADIUS,D, IN MM";RD
INPUT"NEW DATA FOR DEFLECTOR WIDTH(Y/N?)";A$
IF A$="Y" THEN INPUT"TYPE IN DEFLECTOR WIDTH,W, IN MM";DW
99 PI=3.141592654#
SA=PI*BO*BO/4
SV=SA*TS
GSV=SA*ST
R2=BO/2
YY=SQR(DR*DR-DW*DW /4)
ZZ=SQR((BO*BO-DW*DW )/4)
A1=ATN(DW/(2*YY))
A2=ATN(DW/(2*ZZ))
F1=A1*DR*DR
F2=A2*BO*BO/4
FP= F1+F2-DW*(YY+ZZ)/2
RV =2*A1*DR*RD*RD*(1-PI/4)
CV=SA*SQ+FP*DH-RV
TCR=(SV+CV)/CV
GCR=(GSV+CV)/CV
Z3=(SA-FP)/SA
Z1=FP*(DH+SQ)/(SQ*(SA-FP))
RP=(3*CV/(4*PI))^(1/3)
Z2=(PI*BO*SQ+2*SA+2*DH*(A2*R2+A1*DR))*RP/(3*CV)
W1=YY+ZZ-DR+R2
```

```
W2=BO-W1
CVP=CV/1000
TSVP=SV/1000
GSVP=GSV/1000
CLS
GOSUB SUBPRINT
266 LOCATE 23,1:INPUT"RECALCULATING(R), FILING(F),
PRINTING(P), OR QUITTING(Q)";R$
IF R$="R" THEN GOTO 100
IF R$="P" THEN GOSUB SUBLPRINT
IF R$="F" THEN GOSUB FILING
IF R$="Q" THEN GOTO 267
GOTO 266
267 END
FILING:
INPUT "TYPE IN TITLE NEEDED FOR THE FILE WITH '.QUBDEF' END
CODE";P$
OPEN P$ FOR OUTPUT AS #1
WRITE#1,BO,ST,TS,DH,SQ,DR,RD,DW
FINISH:
CLOSE#1
RETURN
FILED:
INPUT "TYPE IN THE FILENAME WITH '.QUBDEF' END CODE";FIL$
OPEN"I",#1,FIL$
INPUT#1,BO,ST,TS,DH,SQ,DR,RD,DW
CLOSE#1
RETURN
SUBPRINT:
GOSUB DATABLOCK
GOSUB CHAMDRAW
RETURN
SUBLPRINT:
INPUT"REMEMBER TO PROMPT 'TALL ADJUSTED' IN THE DIALOG
BOX,OK!,(TYPE-OK)";OK$
INPUT"WANT A TITLE FOR THE OUTPUT(Y/N?)";A$
IF A$="Y" THEN INPUT"TYPE IN TITLE";TI$
OPEN"LPT1:PROMPT"FOR OUTPUT AS#1
WINDOW OUTPUT#1
IF A$="Y" THEN LOCATE 22,1:PRINT TI$
GOSUB DATABLOCK
GOSUB CHAMDRAW
CLOSE#1
777 RETURN
```

```
CHAMDRAW:
CY1=20
CX1=400
CX2=400
CY2=250
CR=65
PBO=2*CR
PP=PBO/BO
DHP=DH*PP
RDP=RD*PP
DRP=DR*PP
WYP=W2*PP
STP=ST*PP
SQP=SQ*PP
WDP=DW*PP
TSP=TS*PP
PC=2
LTH=6
RLP=5
RTP=3
DX1=CR+PC+LTH
X1=CX1-DX1
X2=CX1+DX1
PL=STP+5
LEX=20
Y2=CY1+LTH+SQP+PL
LINE(X1,Y2)-(X1,CY1)
LINE(X1,CY1)-(X2,CY1)
LINE(X2,CY1)-(X2,Y2)
LINE(X1,Y2)-(X1,Y2)
Y1=CY1+LTH
X3=X1+LTH
X4=X2-LTH
LINE(X1,Y2)-(X3,Y2)
LINE(X4,Y2)-(X2,Y2)
LINE(X3,Y2)-(X3,Y1)
LINE(X1,Y1)-(X2,Y1)
LINE(X4,Y1)-(X4,Y2)
Y3=Y1+SQP
X5=X3+PC
X6=X4-PC
Y4=Y3+PL
LINE(X5,Y4)-(X5,Y3)
X7=X6-WYP
```

```
Y5=Y3+DHP
X8=X6-WYP+RDP
Y6=Y5-RDP
LINE(X5,Y3)-(X7,Y3)
LINE(X7,Y3)-(X7,Y6)
LINE(X8,Y5)-(X6,Y5)
LINE(X6,Y5)-(X6,Y4)
LINE(X6,Y4)-(X5,Y4)
Y7=(Y4+Y5)/2
YB=Y3+STP
YEX=Y3+TSP
X0=X1-30
LINE(X0,YEX)-(X0,YB)
LINE(X0,YEX)-(X3,YEX)
LINE(X0,YB)-(X3,YB)
HTP=DHP/1.5
LINE(CX1,CY1-20)-(CX1,Y2)
CLP=.5*CR
GPR=.22*CR
LINE(CX1-CLP,Y7)-(CX1+CLP,Y7)
CIRCLE(CX1,Y7),GPR
AST=PI
AF=3*PI/2
CIRCLE(X8,Y6),RDP,,PI,AF
Y8=Y5+RLP
LINE(X3,Y8)-(X4,Y8+RTP),33,BF
Y8=Y8+RLP+RTP
LINE(X3,Y8)-(X4,Y8+RTP),33,BF
CIRCLE(CX2,CY2),CR
CXD=CX2+CR+DRP-WYP
WDZ=WDP/2
WW=SQR(DRP*DRP-WDZ*WDZ)
THP=ATN(WDZ/WW)
AST=PI-THP
AF=PI+THP
CIRCLE(CXD,CY2),DRP,,AST,AF
QQ=CR+20
LINE(CX2-QQ,CY2)-(CX2+QQ,CY2)
LINE(CX2,CY2-QQ)-(CX2, CY2+QQ)
RETURN
DATABLOCK:
LOCATE 1,1:PRINT "QUB DEFLECTOR PISTON DESIGN"
LOCATE 2,1:PRINT "INPUT DATA"
LOCATE 3,1:PRINT "BORE, mm=";USING"###.#";BO
```

```
LOCATE 4,1:PRINT "STROKE, mm=";USING"###.#";ST
LOCATE 5,1:PRINT "TRAPPED STROKE, mm=";USING"###.#";TS
LOCATE 6,1:PRINT "DEFLECTOR HEIGHT, mm=";USING"###.#";DH
LOCATE 7,1:PRINT "SQUISH CLEARANCE, mm=";USING"###.#";SQ
LOCATE 8,1:PRINT "DEFLECTOR RADIUS, mm=";USING"###.#";DR
LOCATE 9,1:PRINT "DEFLECTION RADIUS, mm=";USING"###.#";RD
LOCATE 10,1:PRINT "DEFLECTOR WIDTH, mm=";USING"###.#";DW
LOCATE 12,1:PRINT "OUTPUT DATA"
LOCATE 13,1:PRINT "TRAPPED COMPRESSION RATIO=";USING
"###.#";TCR
LOCATE 14,1:PRINT "GEOMETRIC COMPRESSION RATIO=";USING
"###.#";GCR
LOCATE 15,1:PRINT "CLEARANCE VOLUME, cc=";USING"###.#";CVP
LOCATE 16,1:PRINT "DEF. CENTRE TO EXH. BORE EDGE, mm=";USING
"###.#";W1
LOCATE 17,1:PRINT "DEF. CENTRE TO SCAV. BORE EDGE,
mm=";USING"###.#";W2
LOCATE 18,1:PRINT "SQUISH AREA RATIO=";USING"##.###";Z3
LOCATE 19,1:PRINT "CHAMBER TO SQUISH VOLUME RATIO=";USING
"##.##";Z1
LOCATE 20,1:PRINT "SURFACE TO VOLUME RATIO=";USING
"##.##";Z2
RETURN
```

ProgList5.0 Program listings from Chapter 5

COMPUTER PROGRAM PROG.5.1
PROGRAM NAME "ENGINE MODEL NO.1"
ProgList.5.1

```
REM PROGRAM Prog.5.1
REM ProgList.5.1
DIM ABX(21,45),BX(45),AKA(45),DF(20),F(20)
DIM AE(20),BE(20),FE(20),DFE(20)
DIM AI(20),BI(20),FI(20),DFI(20)
DIM AT(20),BT(20),FT(20),DFT(20)
DIM AET(20),BET(20),FET(20),DFET(20)
DIM A2(20),B2(20)
DIM XA(200),YC(200),YS(200),YE(200),YI(250)
DIM YF(13)
DIM YV(250),YP(250),XXA(250),YB(250)
PI=3.141593
RTOD=180/PI
DTOR=1/RTOD
TAF=13
KO=1
OPEN"I",#1,"UFLO"
FOR I=1 TO 21 STEP 1
FOR J=1 TO 41 STEP 1
INPUT#1,ABX(I,J)
NEXT J
NEXT I
FINISH:
CLOSE #1
FOR J=1 TO 41
BX(J)=ABX(1,J)
NEXT J
AKA(1)=0
FOR I=2 TO 21
AKA(I)=AKA(I-1)+.05
NEXT I
50 INPUT "RESPONSE FOR NEW DATA IS 'N' AND FILED DATA IS 'F'";H$
IF H$="F" THEN GOSUB FILED
IF H$="F" THEN GOSUB FILESET
IF H$="F" THEN GOTO 502
CLS
INPUT"NEW BASIC ENGINE DATA(Y/N?)";ANS$
IF ANS$="Y" THEN GOSUB ENGINEDAT
```

```
INPUT"NEW EXHAUST PORT DATA (Y/N?)";ANS$
IF ANS$="Y" THEN GOSUB EXPORTDAT
INPUT"NEW TRANSFER PORT DATA (Y/N?)";ANS$
IF ANS$="Y" THEN GOSUB TRPORTDAT
INPUT"NEW INLET PORT DATA (Y/N?)";ANS$
IF ANS$="Y" THEN GOSUB INPORTDAT
INPUT"NEW EXHAUST PIPE DATA (Y/N?)";ANS$
IF ANS$="Y" THEN GOSUB EXPIPDAT
INPUT"NEW TRANSFER PIPE DATA (Y/N?)";ANS$
IF ANS$="Y" THEN GOSUB TRPIPDAT
INPUT"NEW INLET PIPE DATA (Y/N?)";ANS$
IF ANS$="Y" THEN GOSUB INPIPDAT
INPUT"NEW DATA FOR CLOSED CYCLE(Y/N?)";ANS$
IF ANS$="Y" THEN GOSUB COMBDAT
502 GOSUB DATASET
GOSUB PORTSTART
INPUT"TYPE 'RUN' FOR CALCULATION OR 'NEW' FOR A DATA
CHANGE";Z$
IF Z$="NEW" THEN GOTO 50
CALL PAREA(NEP,WEPM,EXHT,RTEM,RBEM,FEPRT,DEPRT)
CALL PAREA(NTP,WTPM,TRHT,RTTM,RBTM,FTPRT,DTPRT)
CALL PAREA(NIP,WIPM,INHT,RTIM,RBIM,FIPRT,DIPRT)
FTDUCT=TRAR*FTPRT
DTDUCT=SQR(4*FTDUCT/PI)
TSV=FPISTON*HEOT
CYCV=TSV/(TCR-1)
VC1=TSV+CYCV
VCC1=CCCV+FPISTON*HEOB
CLS
CYC=1
NN=0
JJ=O
CCMEP=0
CYMEP=0
SEFF=0
CEFF=0
TEFF=0
MBACK=0
DRATIO=0
SCRATIO=0
EXRATIO=0
DRAT=1
SR=0
GE=1.4
```

```
GK=287
CCP=1005
CCV=718
GR=GE*GK
N27=1/3.5
N17=1/7
PA=101325!
TA=293
GOSUB CYLSTART
DA=PA/(GK*TA)
MREF=SV*DA
PC1=TCR*PA/2.5
TC1=1900
DCYL=PC1/(GK*TC1)
PCC1=.9*CCR*PA
DCC=PCC1/(GK*TTREF)
TEXC=TEREF-273
MC1=VC1*DCYL
MCC1=VCC1*DCC
TC1=PC1*VC1/(GK*MC1)
TCC1=PCC1*VCC1/(GK*MCC1)
PB1=1.1*PA
TB1=TEREF
MB1=PB1*VBOXM/(GK*TB1)
CVAL=4.3E+07
DEG=EPO
LMESH=30
CALL NINT(KE1,LEP1,TEREF,TEREF)
CALL NINT(KE2,LEP2,TEREF,TEREF)
CALL NINT(KT,LTP,TEREF,TTREF)
CALL NINT(KI,LIP,TEREF,TIREF)
LEM1=LE1M/(KE1-1)
LEM2=LE2M/(KE2-1)
LIMESH=LIM/(KI-1)
LTMESH=LTM/(KT-1)
CALL SETS(AE(),BE(),KE1)
CALL SETS(AET(),BET(),KE2)
CALL SETS(AT(),BT(),KT)
CALL SETS(AI(),BI(),KI)
CALL AREA(DE2M,DE2M,KO,KE2,DFET(),FET())
CALL AREA(DE1M,DE1M,KO,KE1,DFE(),FE())
CALL AREA(DTDUCT,DTDUCT,KO,KT,DFT(),FT())
CALL AREA(DIPRT,DIPM,KO,KI,DFI(),FI())
COUNT=0
```

```
PCOUNT=0
ANGD=EPO
CYCLE=1
CYCLESTOP=7
CYCLEP=CYCLESTOP+10
FOR KK=1 TO 20000
DZ=1
NN=NN+1
CALL STAB(AET(),BET(),KE2,SDE,LEM2)
CALL STAB(AT(),BT(),KT,SDT,LTMESH)
CALL STAB(AI(),BI(),KI,SDI,LIMESH)
CALL STAB(AE(),BE(),KE1,SDE,LEM1)
CALL STAB(AE(),BE(),KE1,SDE,LEM1)
DZ=.99*DZ
DEG=DEG+DC
ANGD=ANGD+DC
IF ANGD>360+EPO THEN CYCLE=CYCLE+1
IF CYCLE=CYCLEP THEN GOTO 430
IF CYCLE=CYCLEP+1 THEN GOTO 431
IF CYCLE=CYCLESTOP THEN GOTO 500
430 IF ANGD>360+EPO THEN ANGD=ANGD-360
IF DEG>540 THEN GOSUB RESETBDC
IF DEG>360+DC THEN GOTO 25
IF DEG>360 THEN GOSUB RESETTDC
25 CALL PISTPOS(STM,CRNK,CRLM,DEG,HTDC,HBDC)
VC2=CYCV+HTDC*FPISTON
VCC2=CCCV+HBDC*FPISTON
IF CYC=CYCTRIP THEN PCOUNT=PCOUNT+1
CALL PIPE(KO,KE1,AE(),BE(),DFE(),FE())
CALL PIPE(KO,KE2,AET(),BET(),DFET(),FET())
CALL PIPE(KO,KT,AT(),BT(),DFT(),FT())
CALL PIPE(KO,KI,AI(),BI(),DFI(),FI())
IF AE(KE1)>=BE(KE1) THEN AK=.9
IF AE(KE1)<BE(KE1) THEN AK=.75
IF SIL$="ABSORPTION" THEN AK=.6
CALL PORT(BE(KE1),AE(KE1),PB1,AK)
IF AET(1)>=BET(1) THEN AK=.75
IF AET(1)<BET(1) THEN AK=.9
CALL PORT(AET(1),BET(1),PB1,AK)
IF AET(KE2)>=BET(KE2) THEN AK=.9
IF AET(KE2)<BET(KE2) THEN AK=.75
CALL PORT(BET(KE2),AET(KE2),PA,AK)
AK=THAR
CALL PORT(BI(KI),AI(KI),PA,AK)
```

```
IF AT(KT)>=BT(KT) THEN AK=.9
IF AT(KT)<BT(KT) THEN AK=.75
CALL PORT(BT(KT),AT(KT),PCC1,AK)
IF HTDC<=HEOT THEN GOSUB EXPORTSHUT
IF HTDC>HEOT THEN GOSUB EXPORTOPEN
IF HTDC<=HTOT THEN GOSUB TRPORTSHUT
IF HTDC>HTOT THEN GOSUB TRPORTOPEN
IF HTDC<HIOT THEN GOSUB INPORTOPEN
IF HTDC>HIOT THEN GOSUB INPORTSHUT
CALL SHOK(AE(),BE(),KO,KE1)
CALL SHOK(AET(),BET(),KO,KE2)
CALL SHOK(AT(),BT(),KO,KT)
CALL SHOK(AI(),BI(),KO,KI)
IF AT(KT)>=BT(KT) THEN CD=.9
IF AT(KT)<BT(KT) THEN CD=.75
CALL OUTFLOW(BT(KT),AT(KT),FTDUCT,TCC1,HTKT,CTKT,
MTKT,D,CD)
IF AE(KE1)>=BE(KE1) THEN CD=1
IF AE(KE1)<BE(KE1) THEN CD=.75
CALL OUTFLOW(AE(KE1),BE(KE1),FEP1,TC1,HEK1,CEK1,MEK1,D,CD)
IF AET(1)>=BET(1) THEN CD=.75
IF AET(1)<BET(1) THEN CD=1
CALL OUTFLOW(AET(1),BET(1),FEP2,TB1,HET1,CET1,MET1,D,CD)
CALL BOX(PB1,PB2,TB1,TB2,VBOXM,VBOXM,MB1,MB2,MEK1,MET1,
CEK1,CET1,HEK1,HET1,BOXW)
CALL BOX(PCC1,PCC2,TCC1,TCC2,VCC1,VCC2,MCC1,MCC2,MI1,
MTKT,CI1,CTKT,HI1,HTKT,CCWORK)
IF HTDC<=HEOT THEN CALL CYLINDER(PC1,PC2,TC1,TC2,
VC1,VC2,MC2,CYWORK)
IF HTDC>HEOT THEN CALL BOX(PC1,PC2,TC1,TC2,VC1,VC2,MC1,
MC2,MT1,ME1,CT1,CE1,HT1,HE1,CYWORK)
IF CYCLE>1 THEN GOTO 426 ELSE GOTO 427
427 IF DEG<180 THEN GOTO 428
426 CCMEP=CCMEP+CCWORK/SV
CYMEP=CYMEP+CYWORK/SV
428 PCR=PC2/PA
PCCR=PCC2/PA
PBR=PB2/PA
PEX=((AE(1)+BE(1))/2)^7
PIR=((AI(1)+BI(1))/2)^7
PTR=((AT(1)+BT(1))/2)^7
PTRO=((AT(KT)+BT(KT))/2)^7
PEXP=((AE(KE1)+BE(KE1))/2)^7
PEXT=((AET(1)+BET(1))/2)^7
```

```
IF ANGD>EPC THEN GOTO 310
IF ANGD>EPO THEN GOTO 305 ELSE GOTO 307
305 IF ANGD<=EPO+DC THEN GOSUB SPLOT1
307 IF ANGD>EPO THEN GOSUB SPLOT2
310 IF ANGD>EPC THEN GOTO 405 ELSE GOTO 407
405 IF ANGD<=EPC+DC THEN GOSUB SPLOT3
407 IF ANGD>EPC THEN GOSUB SPLOT4
NEXT KK
500 INPUT”RESPONSE FOR RECALCULATION IS ‘R’, PRINTING IS ‘P’, FILING DATA IS ‘F’, OR QUITTING IS ‘Q’”;D$
IF D$=”R” THEN GOTO 50
IF D$=”F” THEN GOSUB FILING
IF D$=”Q” THEN GOTO 501
IF D$=”P” THEN GOSUB PRINTOUT
GOTO 500
501 END
DATOUT:
KON=100000&
LPRINT
LPRINT”OUTPUT DATA”
LPRINT”POWER,kW     BMEP,bar   BSFC,kg/kWh IMEP,bar   PMEP,bar FMEP,bar”
FOR JJJ=2 TO CYCLESTOP-2
LPRINT USING”######.###”;POWER(JJJ),BMEP(JJJ)/KON,BSFC(JJJ), CMEP(JJJ)/KON,-PMEP(JJJ)/KON,FMEP/KON
NEXT JJJ
LPRINT
LPRINT”DELIVERY RATIO SEFF    TEFF    CEFF    SR(mass) SR(vol) exFLOWratio”
FOR JJJ=2 TO CYCLESTOP-2
LPRINT USING”######.###”;DELRAT(JJJ),SEFFP(JJJ),TEFFP(JJJ), CEFFP(JJJ),SRMP(JJJ),SRVP(JJJ),EXRP(JJJ)
NEXT JJJ
LPRINT
LPRINT”PEAK PRESSURE, bar  and TEMPERATURE, K, from the combustion conditions”
FOR JJJ=2 TO CYCLESTOP-2
LPRINT USING”###########.#”;PEAKP(JJJ),PEAKT(JJJ)
NEXT JJJ
LPRINT
RETURN
PRINTOUT:
CYCLE=CYCLEP
INPUT”TYPE IN A TITLE FOR THE PRINTOUT”;TI$
```

```
LPRINT TI$;" on Prog.5.1, ENGINE WITH BOX SILENCER"
GOSUB DATASETL
GOSUB DATOUT
FOR NNN=1 TO 20
LPRINT
NEXT NNN
OPEN"LPT1:PROMPT"FOR OUTPUT AS#2
WINDOW OUTPUT#2
GOTO 430
431 CLOSE #2
RETURN
FILESET:
BOM=BO/1000
STM=ST/1000
CRLM=CRL/1000
RPS=RPM/60
FPISTON=PI*BOM*BOM/4
SV=FPISTON*STM
CCCV=SV/(CCR-1)
CRNK=STM/2
FMEP=100*STM*RPM
WEPM=WEP/1000
RTEM=RTE/1000
RBEM=RBE/1000
WTPM=WTP/1000
RTTM=RTT/1000
RBTM=RBT/1000
WIPM=WIP/1000
RTIM=RTI/1000
RBIM=RBI/1000
LE1M=LEP1/1000
LE2M=LEP2/1000
DE1M=DEP1/1000
DE2M=DEP2/1000
VBOXM=VBOX/1000000!
FEP1=PI*DE1M*DE1M/4
FEP2=PI*DE2M*DE2M/4
LTM=LTP/1000
LIM=LIP/1000
DIPM=DIP/1000
FIPM=PI*DIPM*DIPM/4
TEREF=TEXC+273
RETURN
RESETBDC:
DEG=DEG-360
```

```
JJ=JJ+1
DELRAT(JJ)=DRATIO
PEAKP(JJ)=PMAX
PEAKT(JJ)=TKMAX
PMEP(JJ)=CCMEP
CMEP(JJ)=CYMEP
BMEP(JJ)=CMEP(JJ)+PMEP(JJ)-FMEP
POWER(JJ)=(CMEP(JJ)+PMEP(JJ)-FMEP)*SV*RPS/1000
FUEL(JJ)=DRATIO*MREF/TAF
BSFC(JJ)=FUEL(JJ)*RPS*3600/POWER(JJ)
LOCATE 21,1:PRINT "ENGINE SPEED, rpm=";RPM
LOCATE 22,1:PRINT "POWER, kW=";USING"##.#";POWER(JJ)
LOCATE 22,20:PRINT "BSFC,kg/kWh=";USING"#.###";BSFC(JJ)
LOCATE 22,40:PRINT "BMEP, bar=";USING"##.##";BMEP(JJ)/(100000!)
LOCATE 23,1:PRINT "IMEP, bar=";USING"##.##";CMEP(JJ)/(100000!)
LOCATE 23,20:PRINT "PMEP, bar=";USING"##.##";-PMEP(JJ)/(100000!)
LOCATE 23,40:PRINT "FMEP, bar=";USING"##.##";FMEP/(100000!)
LOCATE 24,1:PRINT "DELIVERY RATIO=";USING"#.###";DRATIO
LOCATE 25,1:PRINT "PEAK CYLINDER PRESS., bar=";USING
"###.#";PMAX
LOCATE 25,28:PRINT "and TEMP., K=";USING"####.";TKMAX
LOCATE 25,44:PRINT "at deg.ATDC=";USING"##.#";MAXDEG
IF CYCLE=CYCLEP THEN GOTO 433
LOCATE 26,1:INPUT "PRESS 'RETURN' TO CONTINUE";A$
433 CCMEP=0
CYMEP=0
DRAT=DRATIO
DRATIO=0
RETURN
RESETTDC:
SEFFP(JJ)=SEFF
TEFFP(JJ)=TEFF
CEFFP(JJ)=CEFF
SRMP(JJ)=SCRATIO
SRVP(JJ)=SR
EXRP(JJ)=EXRATIO
LOCATE 22,1:PRINT "SCAVENGING EFF.=";USING"#.###";SEFF
LOCATE 22,23:PRINT "TRAPPING EFF.=";USING"#.###";TEFF
LOCATE 22,45:PRINT "CHARGING EFF.=";USING"#.###";CEFF
LOCATE 23,1:PRINT "SCAV. RATIO(mass)=";USING"#.###";SCRATIO
LOCATE 23,23:PRINT "SCAV. RATIO(vol)=";USING"#.###";SR
LOCATE 23,45:PRINT "EX. FLOW RATIO=";USING"#.###";EXRATIO
IF CYCLE=CYCLEP THEN GOTO 434
LOCATE 25,1:INPUT "PRESS 'RETURN' TO CONTINUE";A$
434 MBACK=0
```

```
SCRATIO=0
EXRATIO=0
SR=0
RETURN
EXPIPDAT:
19 INPUT"NEW DATA FOR INITIAL EXHAUST PIPE LENGTH(Y/N?)";ANS$
IF ANS$="Y" THEN INPUT"TYPE IN INITIAL EXHAUST PIPE LENGTH(LEP1)";LEP1
INPUT"NEW DATA FOR INITIAL EXHAUST PIPE DIAMETER(Y/N?)";ANS$
IF ANS$="Y" THEN INPUT"TYPE IN INITIAL EXHAUST PIPE DIAMETER(DEP1)";DEP1
INPUT"NEW DATA FOR OUTLET EXHAUST PIPE LENGTH(Y/N?)";ANS$
IF ANS$="Y" THEN INPUT"TYPE IN OUTLET EXHAUST PIPE LENGTH(LEP2)";LEP2
INPUT"NEW DATA FOR OUTLET EXHAUST PIPE DIAMETER(Y/N?)";ANS$
IF ANS$="Y" THEN INPUT"TYPE IN OUTLET EXHAUST PIPE DIAMETER(DEP2)";DEP2
INPUT"NEW DATA FOR EXHAUST BOX VOLUME(Y/N?)";ANS$
IF ANS$="Y" THEN INPUT"TYPE IN EXHAUST BOX VOLUME(VBOX AS cm3)";VBOX
INPUT"IS THE MUFFLER OF THE ABSORPTION TYPE(Y/N?)";ANS$
SIL$="PLAIN BOX"
IF ANS$="Y" THEN INPUT"TYPE IN THE WORD 'ABSORPTION'";SIL$
INPUT"MORE DATA,(Y/N?)";ANS$
IF ANS$="Y" THEN GOTO 19
LE1M=LEP1/1000
LE2M=LEP2/1000
DE1M=DEP1/1000
DE2M=DEP2/1000
VBOXM=VBOX/1000000!
FEP1=PI*DE1M*DE1M/4
FEP2=PI*DE2M*DE2M/4
RETURN
DATASET:
PRINT"PROGRAM FOR PISTON-PORTED ENGINE WITH EXHAUST BOX SILENCER"
PRINT"Prog.5.1, ENGINE WITH BOX SILENCER"
PRINT"ALL DATA IN mm AND deg UNITS"
PRINT"   BORE  STROKE  CON-ROD REV/MIN  TRAP-CR   CRANK-CASE-CR   TEXC"
PRINT USING"######.##";BO,ST,CRL,RPM,TCR,CCR,TEXC
PRINT
```

```
PRINT"EXHAUST PORT DATA"
PRINT"    OPENS  FULL-OPEN  NUMBER-OFF  WIDTH-EACH  TOP-
RAD  BOTTOM-RAD
PRINT USING"#########.#";EPO,EPFO,NEP,WEP,RTE,RBE
PRINT
PRINT"TRANSFER PORT DATA WITH SCAVENGING DESCRIBED AS
";SCAV$
PRINT"    OPENS  FULL-OPEN  NUMBER-OFF  WIDTH-EACH  TOP-
RAD  BOTTOM-RAD
PRINT USING"#########.#";TPO,TPFO,NTP,WTP,RTT,RBT
PRINT
PRINT"INLET PORT DATA"
PRINT"    OPENS  FULL-OPEN  NUMBER-OFF  WIDTH-EACH  TOP-
RAD  BOTTOM-RAD
PRINT USING"#########.#";IPO,IPFO,NIP,WIP,RTI,RBI
PRINT
PRINT "EXHAUST PIPE DATA WITH MUFFLER TYPE AS ";SIL$
PRINT"LENGTH.1  DIAMETER.1  LENGTH.2  DIAMETER.2  BOX
VOLUME, cm3
PRINT USING"######.##";LEP1,DEP1,LEP2,DEP2,VBOX
PRINT
PRINT "TRANSFER AND INLET PIPE DATA"
PRINT"LENGTH.T  TRANSFER.AR  LENGTH.I  DIAMETER.I
THROTTLE.AR
PRINT USING"######.##";LTP,TRAR,LIP,DIP,THAR
PRINT
PRINT"COMBUSTION MODEL PARAMETERS"
PRINT"   INDECES, COMP. AND EXP.  COMBUSTION.EFF  BURN
PERIOD AIR-TO-FUEL"
PRINT USING"########.##";NC,NE,BEFF,BDEG,TAF
PRINT
INPUT"OK?,(Y/N)";S$
RETURN
DATASETL:
LPRINT"   BORE   STROKE  CON-ROD  REV/MIN   TRAP-CR
CRANKCASE-CR TEXC"
LPRINT USING"######.##";BO,ST,CRL,RPM,TCR,CCR,TEXC
LPRINT
LPRINT"EXHAUST PORT DATA"
LPRINT"   OPENS  FULL-OPEN  NUMBER-OFF  WIDTH-EACH  TOP-
RAD  BOTTOM-RAD"
LPRINT USING"#########.#";EPO,EPFO,NEP,WEP,RTE,RBE
LPRINT
LPRINT"TRANSFER PORT DATA WITH SCAVENGING DESCRIBED AS
";SCAV$
```

```
LPRINT"   OPENS FULL-OPEN NUMBER-OFF WIDTH-EACH TOP-
RAD BOTTOM-RAD"
LPRINT USING"#########.#";TPO,TPFO,NTP,WTP,RTT,RBT
LPRINT
LPRINT"INLET PORT DATA"
LPRINT"   OPENS FULL-OPEN NUMBER-OFF WIDTH-EACH TOP-
RAD BOTTOM-RAD"
LPRINT USING"#########.#";IPO,IPFO,NIP,WIP,RTI,RBI
LPRINT
LPRINT "EXHAUST PIPE DATA WITH MUFFLER TYPE AS ";SIL$
LPRINT"   LENGTH.1 DIAMETER.1 LENGTH.2 DIAMETER.2 BOX
VOLUME, cm3
LPRINT USING"########.##";LEP1,DEP1,LEP2,DEP2,VBOX
LPRINT
LPRINT "TRANSFER AND INLET PIPE DATA"
LPRINT"   LENGTH.T TRANSFER.AR LENGTH.I DIAMETER.I
THROTTLE.AR
LPRINT USING"########.##";LTP,TRAR,LIP,DIP,THAR
LPRINT
LPRINT"COMBUSTION MODEL PARAMETERS"
LPRINT" COMBUSTION.EFF BURN PERIOD AIR-TO-FUEL COMP.
AND EXP. INDECES"
LPRINT USING"#########.##";BEFF,BDEG,TAF,NC,NE,
LPRINT
RETURN
SPLOT1:
NCOUNT=1
IF NCOUNT>=2 THEN GOTO 410
M=1
MG2=1
X0=50
Y0=250
YT=20
YZ=30
X2=460
X1=(X2+X0)/2
EP=2*(180-EPO)
DR=(X2-X0)/EP
BD=DR*(TPO-EPO)
X3=X0+BD
X4=X2-BD
GPS=1!
GSS=1.2
GTS=1500
```

```
GYS=(Y0-YZ)/GPS
YEF=(Y0-YZ)/GSS
YTT=(Y0-YZ)/GTS
IF D$="P" THEN GOTO 439
CLS
439 LOCATE 19,4
PRINT"CRANK-ANGLE from EXHAUST PORT OPENING to EXHAUST
PORT CLOSING"
LOCATE 17,6
PRINT "EPO"
LOCATE 17,14
PRINT"TPO"
LOCATE 17,31
PRINT"BDC"
LOCATE 17,48
PRINT"TPC"
LOCATE 17,57
PRINT"EPC"
LOCATE 4,11
PRINT "CYLINDER"
LOCATE 9,9
PRINT "EXHAUST"
LOCATE 9,33
PRINT"CRANKCASE"
LOCATE 4,2
PRINT"P."
LOCATE 5,2
PRINT"A"
LOCATE 6,2
PRINT"T"
LOCATE 7,2
PRINT"M"
LOCATE 2,2
PRINT"2.0"
LOCATE 9,2
PRINT"1.5"
LOCATE 16,2
PRINT"1.0"
YY1=Y0-.1*GYS
YY2=Y0-.2*GYS
YY3=Y0-.3*GYS
YY4=Y0-.4*GYS
YY5=Y0-.5*GYS
YY6=Y0-.6*GYS
```

```
YY7=Y0-.7*GYS
YY8=Y0-.8*GYS
YY9=Y0-.9*GYS
YYU=Y0-GYS
YT1=Y0-1500*YTT
YT2=Y0-1000*YTT
YT3=Y0-500*YTT
LINE(X0,YY1)-(X0+5,YY1)
LINE(X0,YY2)-(X0+5,YY2)
LINE(X0,YY3)-(X0+5,YY3)
LINE(X0,YY4)-(X0+5,YY4)
LINE(X0,YY5)-(X0+5,YY5)
LINE(X0,YY6)-(X0+5,YY6)
LINE(X0,YY7)-(X0+5,YY7)
LINE(X0,YY8)-(X0+5,YY8)
LINE(X0,YY9)-(X0+5,YY9)
LINE(X0,YYU)-(X0+5,YYU)
LINE(X0,Y0)-(X2,Y0)
LINE(X0,Y0)-(X0,YT)
LINE(X2,Y0)-(X2,YT)
LINE(X3,Y0)-(X3,YT)
LINE(X1,Y0)-(X1,YT)
LINE(X4,Y0)-(X4,YT)
FOR K=1 TO 13
YF(K)=Y0-K*YEF/10
NEXT K
410 RETURN
SPLOT2:
IF PCR>2! THEN PCR=2!
ZC=(PCR-1)*GYS
ZS=(PCCR-1)*GYS
ZE=(PEX-1)*GYS
XA(M)=INT(DR*(ANGD-EPO))+X0
YC(M)=Y0-ZC
YS(M)=Y0-ZS
YE(M)=Y0-ZE
IF M=1 THEN GOTO 633
LINE(XA(M),YS(M))-(XA(M-1),YS(M-1))
LINE(XA(M),YC(M))-(XA(M-1),YC(M-1))
LINE(XA(M),YE(M))-(XA(M-1),YE(M-1))
633 M=M+1
MG=MG+1
RETURN
SPLOT3:
XX0=50
```

```
YY0=180
YYB=260
YYT=5
YYZ=10
XX2=460
XX1=(XX2+XX0)/2
EP2=2*EPO
DR2=(XX2-XX0)/EP2
BD2=DR2*IPO
XX3=XX1+BD2
XX4=XX1-BD2
YPS=GYS/10
YVS=YEF/10
CLS
LOCATE 19,4
PRINT"CRANK-ANGLE from EXHAUST PORT CLOSING to EXHAUST
PORT OPENING"
LOCATE 17,7
PRINT "EPC"
LOCATE 17,16
PRINT"IPO"
LOCATE 17,33
PRINT"TDC"
LOCATE 17,46
PRINT"IPC"
LOCATE 17,56
PRINT"EPO"
LOCATE 6,36
PRINT "CYLINDER"
LOCATE 7,15
PRINT "DELIVERY RATIO"
LOCATE 14,9
PRINT "CRANKCASE"
LOCATE 13,51
PRINT"INLET"
LOCATE 13,59
PRINT"E"
LOCATE 14,59
PRINT"F"
LOCATE 15,59
PRINT"F."
LOCATE 7,2
PRINT"P."
LOCATE 8,2
PRINT"A"
```

```
LOCATE 9,2
PRINT"T"
LOCATE 10,2
PRINT"M"
LOCATE 12,59
PRINT"0.0"
LOCATE 11,59
PRINT"0.1"
LOCATE 10,59
PRINT"0.2"
LOCATE 8,59
PRINT"0.3"
LOCATE 7,59
PRINT"0.4"
LOCATE 6,59
PRINT"0.5"
LOCATE 5,59
PRINT"0.6"
LOCATE 4,59
PRINT"0.7"
LOCATE 3,59
PRINT"0.8"
LOCATE 2,59
PRINT"0.9"
LOCATE 5,2
PRINT"1.5"
LOCATE 12,2
PRINT"1.0"
LOCATE 16,2
PRINT"0.7"
LINE(XX0,YY0)-(XX2,YY0)
LINE(XX0,YYB)-(XX0,YYT)
LINE(XX2,YYB)-(XX2,YYT)
LINE(XX3,YYB)-(XX3,YYT)
LINE(XX1,YYB)-(XX1,YYT)
LINE(XX4,YYB)-(XX4,YYT)
Z=INT((YYB-YY0)/YPS)
Y=YY0+Z*YPS
YBL=Y
LINE(XX0,YBL)-(XX2,YBL)
631 LINE(XX0,Y)-(XX0+5,Y)
Y=Y-YPS
IF Y<YYZ THEN GOTO 632
GOTO 631
```

```
632 Y=YY0
635 LINE(XX2,Y)-(XX2-5,Y)
Y=Y-YVS
IF Y<YYZ THEN GOTO 634
GOTO 635
634 RETURN
SPLOT4:
M=MG2
ZB=(PCR-1)*GYS/40
ZI=(PIR-1)*GYS
ZP=(PCCR-1)*GYS
ZV=DRATIO*YEF
XXA(M)=INT(DR2*(ANGD-EPC))+XX0
YI(M)=YY0-ZI
YP(M)=YY0-ZP
YV(M)=YY0-ZV
IF YV(M)>YY0 THEN YV(M)=YY0
YB(M)=YBL-ZB
IF MG2=1 THEN GOTO 639
LINE(XXA(M),YP(M))-(XXA(M-1),YP(M-1))
LINE(XXA(M),YB(M))-(XXA(M-1),YB(M-1))
LINE(XXA(M),YI(M))-(XXA(M-1),YI(M-1))
IF XXA(M)>=XX3 THEN GOTO 639
LINE(XXA(M),YV(M))-(XXA(M-1),YV(M-1))
639 MG2=MG2+1
RETURN
FILING:
INPUT "TYPE IN TITLE NEEDED FOR THE FILE WITH '.BOX' END CODE";P$
OPEN P$ FOR OUTPUT AS #1
WRITE#1,BO,ST,CRL,RPM,TCR,CCR,TEXC
WRITE#1,EPO,EPFO,NEP,WEP,RTE,RBE
WRITE#1,TPO,TPFO,NTP,WTP,RTT,RBT,SCAV$
WRITE#1,IPO,IPFO,NIP,WIP,RTI,RBI
WRITE#1,LEP1,DEP1,LEP2,DEP2,VBOX,SIL$,LTP,TRAR,LIP,DIP,THAR
WRITE#1,NC,NE,BEFF,BDEG,TAF
CLOSE#1
RETURN
FILED:
INPUT "TYPE IN THE FILENAME WITH A '.BOX' END CODE";FIL$
OPEN"I",#1,FIL$
INPUT#1,BO,ST,CRL,RPM,TCR,CCR,TEXC
INPUT#1,EPO,EPFO,NEP,WEP,RTE,RBE
INPUT#1,TPO,TPFO,NTP,WTP,RTT,RBT,SCAV$
```

```
INPUT#1,IPO,IPFO,NIP,WIP,RTI,RBI
INPUT#1,LEP1,DEP1,LEP2,DEP2,VBOX,SIL$,LTP,TRAR,LIP,DIP,THAR
INPUT#1,NC,NE,BEFF,BDEG,TAF
CLOSE#1
RETURN
```

COMPUTER PROGRAM PROG.5.2
PROGRAM NAME "ENGINE MODEL NO.2"
ProgList.5.2

```
REM PROGRAM Prog.5.2
REM ProgList.5.2
DIM ABX(21,45),BX(45),AKA(45),DF(60),F(60)
DIM AE(60),BE(60),FE(60),DFE(60)
DIM AI(20),BI(20),FI(20),DFI(20)
DIM AT(20),BT(20),FT(20),DFT(20)
DIM A2(60),B2(60)
PI=3.141593
RTOD=180/PI
DTOR=1/RTOD
TAF=13
KO=1
OPEN"I",#1,"UFLO"
FOR I=1 TO 21 STEP 1
FOR J=1 TO 41 STEP 1
INPUT#1,ABX(I,J)
NEXT J
NEXT I
FINISH:
CLOSE #1
FOR J=1 TO 41
BX(J)=ABX(1,J)
NEXT J
AKA(1)=0
FOR I=2 TO 21
AKA(I)=AKA(I-1)+.05
NEXT I
50 INPUT "RESPONSE FOR NEW DATA IS 'N' AND FILED DATA IS 'F'";H$
IF H$="F" THEN GOSUB FILED
IF H$="F" THEN GOSUB FILESET
IF H$="F" THEN GOTO 502
CLS
INPUT"NEW BASIC ENGINE DATA(Y/N?)";ANS$
IF ANS$="Y" THEN GOSUB ENGINEDAT
```

```
INPUT"NEW EXHAUST PORT DATA (Y/N?)";ANS$
IF ANS$="Y" THEN GOSUB EXPORTDAT
INPUT"NEW TRANSFER PORT DATA (Y/N?)";ANS$
IF ANS$="Y" THEN GOSUB TRPORTDAT
INPUT"NEW INLET PORT DATA (Y/N?)";ANS$
IF ANS$="Y" THEN GOSUB INPORTDAT
INPUT"NEW EXHAUST PIPE LENGTH DATA (Y/N?)";ANS$
IF ANS$="Y" THEN GOSUB EXLENDAT
INPUT"NEW EXHAUST PIPE DIAMETER DATA (Y/N?)";ANS$
IF ANS$="Y" THEN GOSUB EXDIADAT
INPUT"NEW TRANSFER PIPE DATA (Y/N?)";ANS$
IF ANS$="Y" THEN GOSUB TRPIPDAT
INPUT"NEW INLET PIPE DATA (Y/N?)";ANS$
IF ANS$="Y" THEN GOSUB INPIPDAT
INPUT"NEW DATA FOR CLOSED CYCLE(Y/N?)";ANS$
IF ANS$="Y" THEN GOSUB COMBDAT
502 GOSUB DATASET
GOSUB PORTSTART
INPUT"TYPE 'RUN' FOR CALCULATION OR 'NEW' FOR A DATA CHANGE";Z$
IF Z$="NEW" THEN GOTO 50
CALL PAREA(NEP,WEPM,EXHT,RTEM,RBEM,FEPRT,DEPRT)
CALL PAREA(NTP,WTPM,TRHT,RTTM,RBTM,FTPRT,DTPRT)
CALL PAREA(NIP,WIPM,INHT,RTIM,RBIM,FIPRT,DIPRT)
FTDUCT=TRAR*FTPRT
DTDUCT=SQR(4*FTDUCT/PI)
TSV=FPISTON*HEOT
CYCV=TSV/(TCR-1)
VC1=TSV+CYCV
VCC1=CCCV+FPISTON*HEOB
CLS
CYC=1
NN=0
JJ=O
CCMEP=0
CYMEP=0
SEFF=0
CEFF=0
TEFF=0
MBACK=0
DRATIO=0
SCRATIO=0
EXRATIO=0
DRAT=1
```

```
SR=0
GE=1.4
GK=287
CCP=1005
CCV=718
GR=GE*GK
N27=1/3.5
N17=1/7
PA=101325!
TA=293
GOSUB CYLSTART
DA=PA/(GK*TA)
MREF=SV*DA
PC1=TCR*PA/2
TC1=1900
DCYL=PC1/(GK*TC1)
PCC1=.9*CCR*PA
DCC=PCC1/(GK*TTREF)
MC1=VC1*DCYL
MCC1=VCC1*DCC
TC1=PC1*VC1/(GK*MC1)
TCC1=PCC1*VCC1/(GK*MCC1)
CVAL=4.3E+07
DEG=EPO
LMESH=30
LEPM=LP01M+LP12M+LP23M+LP34M+LP45M+LP56M
CALL NINT(K1,LP01,TEREF,TEREF)
CALL NINT(KQ2,LP12,TEREF,TEREF)
K2=K1+KQ2-1
CALL NINT(KQ3,LP23,TEREF,TEREF)
K3=K2+KQ3-1
CALL NINT(KQ4,LP34,TEREF,TEREF)
K4=K3+KQ4-1
CALL NINT(KQ5,LP45,TEREF,TEREF)
K5=K4+KQ5-1
CALL NINT(KQ6,LP56,TEREF,TEREF)
K6=K5+KQ6-1
CALL NINT(KT,LTP,TEREF,TTREF)
CALL NINT(KI,LIP,TEREF,TIREF)
LEMESH=LEPM/(K6-1)
LIMESH=LIM/(KI-1)
LTMESH=LTM/(KT-1)
CALL SETS(AE(),BE(),K6)
CALL SETS(AT(),BT(),KT)
```

```
CALL SETS(AI(),BI(),KI)
CALL AREA(DE1M,DE1M,KO,K1,DFE(),FE())
CALL AREA(DE1M,DE2M,K1,K2,DFE(),FE())
CALL AREA(DE2M,DE3M,K2,K3,DFE(),FE())
CALL AREA(DE3M,DE4M,K3,K4,DFE(),FE())
CALL AREA(DE4M,DE5M,K4,K5,DFE(),FE())
CALL AREA(DE5M,DE6M,K5,K6,DFE(),FE())
CALL AREA(DTDUCT,DTDUCT,KO,KT,DFT(),FT())
CALL AREA(DIPRT,DIPM,KO,KI,DFI(),FI())
COUNT=0
PCOUNT=0
ANGD=EPO
CYCLE=1
CYCLESTOP=9
CYCLEP=CYCLESTOP+10
FOR KK=1 TO 20000
DZ=1
NN=NN+1
CALL STAB(AT(),BT(),KT,SDT,LTMESH)
CALL STAB(AI(),BI(),KI,SDI,LIMESH)
CALL STAB(AE(),BE(),K6,SDE,LEMESH)
DZ=.9*DZ
DEG=DEG+DC
ANGD=ANGD+DC
IF ANGD>360+EPO THEN CYCLE=CYCLE+1
IF CYCLE=CYCLEP THEN GOTO 430
IF CYCLE=CYCLEP+1 THEN GOTO 431
IF CYCLE=CYCLESTOP THEN GOTO 500
430 IF ANGD>360+EPO THEN ANGD=ANGD-360
IF DEG>540 THEN GOSUB RESETBDC
IF DEG>360+DC THEN GOTO 25
IF DEG>360 THEN GOSUB RESETTDC
25 CALL PISTPOS(STM,CRNK,CRLM,DEG,HTDC,HBDC)
VC2=CYCV+HTDC*FPISTON
VCC2=CCCV+HBDC*FPISTON
IF CYC=CYCTRIP THEN PCOUNT=PCOUNT+1
CALL PIPE(KO,K6,AE(),BE(),DFE(),FE())
CALL PIPE(KO,KT,AT(),BT(),DFT(),FT())
CALL PIPE(KO,KI,AI(),BI(),DFI(),FI())
IF AE(K6)>=BE(K6) THEN AK=.9
IF AE(K6)<BE(K6) THEN AK=.75
CALL PORT(BE(K6),AE(K6),PA,AK)
AK=THAR
CALL PORT(BI(KI),AI(KI),PA,AK)
```

```
IF AT(KT)>=BT(KT) THEN AK=.9
IF AT(KT)<BT(KT) THEN AK=.75
CALL PORT(BT(KT),AT(KT),PCC1,AK)
IF HTDC<=HEOT THEN GOSUB EXPORTSHUT
IF HTDC>HEOT THEN GOSUB EXPORTOPEN
IF HTDC<=HTOT THEN GOSUB TRPORTSHUT
IF HTDC>HTOT THEN GOSUB TRPORTOPEN
IF HTDC<HIOT THEN GOSUB INPORTOPEN
IF HTDC>HIOT THEN GOSUB INPORTSHUT
CALL SHOK(AE(),BE(),KO,K6)
CALL SHOK(AT(),BT(),KO,KT)
CALL SHOK(AI(),BI(),KO,KI)
IF AT(KT)>=BT(KT) THEN CD=1
IF AT(KT)<BT(KT) THEN CD=.75
CALL OUTFLOW(BT(KT),AT(KT),FTDUCT,TCC1,HTKT,CTKT,
MTKT,D,CD)
CALL BOX(PCC1,PCC2,TCC1,TCC2,VCC1,VCC2,MCC1,MCC2,MI1,
MTKT,CI1,CTKT,HI1,HTKT,CCWORK)
IF HTDC<=HEOT THEN CALL CYLINDER(PC1,PC2,TC1,TC2,
VC1,VC2,MC2,CYWORK)
IF HTDC>HEOT THEN CALL BOX(PC1,PC2,TC1,TC2,VC1,VC2,MC1,
MC2,MT1,ME1,CT1,CE1,HT1,HE1,CYWORK)
IF CYCLE>1 THEN GOTO 426 ELSE GOTO 427
427 IF DEG<180 THEN GOTO 428
426 CCMEP=CCMEP+CCWORK/SV
CYMEP=CYMEP+CYWORK/SV
428 PCR=PC2/PA
PCCR=PCC2/PA
PEX=((AE(1)+BE(1))/2)^7
PIR=((AI(1)+BI(1))/2)^7
PTR=((AT(1)+BT(1))/2)^7
PTRC=((AT(KT)+BT(KT))/2)^7
IF ANGD>EPO THEN GOTO 305 ELSE GOTO 307
305 IF ANGD<=EPO+DC THEN GOSUB SPLOT1
307 IF ANGD>EPO THEN GOSUB SPLOT2
NEXT KK
500 INPUT"RESPONSE FOR RECALCULATION IS 'R', PRINTING IS 'P',
FILING DATA IS 'F', OR QUITTING IS 'Q'";D$
IF D$="R" THEN GOTO 50
IF D$="F" THEN GOSUB FILING
IF D$="Q" THEN GOTO 501
IF D$="P" THEN GOSUB PRINTOUT
GOTO 500
501 END
```

```
DATOUT:
KON=100000&
LPRINT
LPRINT"OUTPUT DATA"
LPRINT"POWER,kW    BMEP,bar  BSFC,g/kWh  IMEP,bar  PMEP,bar
FMEP,bar"
FOR JJJ=2 TO CYCLESTOP-2
LPRINT  USING"######.###";POWER(JJJ),BMEP(JJJ)/KON,BSFC(JJJ),
CMEP(JJJ)/KON,-PMEP(JJJ)/KON,FMEP/KON
NEXT JJJ
LPRINT
LPRINT"DELIVERY RATIO SEFF   TEFF   CEFF   SR(mass)  SR(vol)
exFLOWratio"
FOR JJJ=2 TO CYCLESTOP-2
LPRINT  USING"######.###";DELRAT(JJJ),SEFFP(JJJ),TEFFP(JJJ),
CEFFP(JJJ),SRMP(JJJ),SRVP(JJJ),EXRP(JJJ)
NEXT JJJ
LPRINT
LPRINT"PEAK PRESSURE, bar, and TEMPERATURE, K, from the combus-
tion conditions"
FOR JJJ=2 TO CYCLESTOP-2
LPRINT USING"###########.#";PEAKP(JJJ),PEAKT(JJJ)
NEXT JJJ
LPRINT
RETURN
PRINTOUT:
CYCLE=CYCLEP
INPUT"TYPE IN A TITLE FOR THE PRINTOUT";TI$
LPRINT TI$;" on Prog.5.2, ENGINE WITH TUNED EXHAUST"
GOSUB DATASETL
GOSUB DATOUT
FOR NNN=1 TO 20
LPRINT
NEXT NNN
OPEN"LPT1:PROMPT"FOR OUTPUT AS#2
WINDOW OUTPUT#2
GOTO 430
431 CLOSE #2
RETURN
FILESET:
BOM=BO/1000
STM=ST/1000
CRLM=CRL/1000
RPS=RPM/60
```

```
FPISTON=PI*BOM*BOM/4
SV=FPISTON*STM
CCCV=SV/(CCR-1)
CRNK=STM/2
FMEP=100*STM*RPM
WEPM=WEP/1000
RTEM=RTE/1000
RBEM=RBE/1000
WTPM=WTP/1000
RTTM=RTT/1000
RBTM=RBT/1000
WIPM=WIP/1000
RTIM=RTI/1000
RBIM=RBI/1000
LP01M=LP01/1000
LP12M=LP12/1000
LP23M=LP23/1000
LP34M=LP34/1000
LP45M=LP45/1000
LP56M=LP56/1000
DE1M=DEP1/1000
DE2M=DEP2/1000
DE3M=DEP3/1000
DE4M=DEP4/1000
DE5M=DEP5/1000
DE6M=DEP6/1000
FEP1=PI*DE1M*DE1M/4
FEP2=PI*DE2M*DE2M/4
FEP3=PI*DE3M*DE3M/4
FEP4=PI*DE4M*DE4M/4
FEP5=PI*DE5M*DE5M/4
FEP6=PI*DE6M*DE6M/4
LTM=LTP/1000
LIM=LIP/1000
DIPM=DIP/1000
FIPM=PI*DIPM*DIPM/4
RETURN
RESETBDC:
DEG=DEG-360
JJ=JJ+1
PMEP(JJ)=CCMEP
CMEP(JJ)=CYMEP
BMEP(JJ)=CMEP(JJ)+PMEP(JJ)-FMEP
POWER(JJ)=(CMEP(JJ)+PMEP(JJ)-FMEP)*SV*RPS/1000
```

```
FUEL(JJ)=DRATIO*MREF/TAF
BSFC(JJ)=FUEL(JJ)*RPS*3600/POWER(JJ)
DELRAT(JJ)=DRATIO
PEAKP(JJ)=PMAX
PEAKT(JJ)=TKMAX
IF D$="P" THEN GOTO 433
LOCATE 21,1:PRINT "ENGINE SPEED, rpm=";RPM
LOCATE 21,27:PRINT "DELIVERY RATIO=";USING"#.###";DRATIO
LOCATE 22,1:PRINT "POWER, kW=";USING"##.#";POWER(JJ)
LOCATE 22,20:PRINT "BSFC,kg/kWh=";USING"#.###";BSFC(JJ)
LOCATE 22,40:PRINT "BMEP, bar=";USING"##.##";BMEP(JJ)/(100000!)
LOCATE 23,1:PRINT "IMEP, bar=";USING"##.##";CMEP(JJ)/(100000!)
LOCATE 23,20:PRINT "PMEP, bar=";USING"##.##";-PMEP(JJ)/(100000!)
LOCATE 23,40:PRINT "FMEP, bar=";USING"##.##";FMEP/(100000!)
LOCATE 24,1:PRINT "SEFF=";USING"#.###";SEFFP(JJ-1)
LOCATE 24,20:PRINT "TEFF=";USING"#.###";TEFFP(JJ-1)
LOCATE 24,40:PRINT "CEFF=";USING"#.###";CEFFP(JJ-1)
LOCATE 25,1:PRINT "PEAK CYLINDER PRESS., bar=";USING"###.#";PMAX
LOCATE 25,28:PRINT "and TEMP., K=";USING"####.";TKMAX
LOCATE 25,44:PRINT "at deg.ATDC=";USING"##.#";MAXDEG
433 CCMEP=0
CYMEP=0
DRAT=DRATIO
DRATIO=0
RETURN
RESETTDC:
SEFFP(JJ)=SEFF
TEFFP(JJ)=TEFF
CEFFP(JJ)=CEFF
SRMP(JJ)=SCRATIO
SRVP(JJ)=SR
EXRP(JJ)=EXRATIO
MBACK=0
SCRATIO=0
EXRATIO=0
SR=0
RETURN
EXLENDAT:
19 INPUT"NEW DATA FOR EXHAUST PIPE LENGTH, LP01(Y/N?)";ANS$
IF ANS$="Y" THEN INPUT"TYPE IN EXHAUST PIPE LENGTH, LP01";LP01
INPUT"NEW DATA FOR EXHAUST PIPE LENGTH, LP12(Y/N?)";ANS$
IF ANS$="Y" THEN INPUT"TYPE IN EXHAUST PIPE LENGTH, LP12";LP12
```

```
INPUT"NEW DATA FOR EXHAUST PIPE LENGTH, LP23(Y/N?)";ANS$
IF ANS$="Y" THEN INPUT"TYPE IN EXHAUST PIPE LENGTH,
LP23";LP23
INPUT"NEW DATA FOR EXHAUST PIPE LENGTH, LP34(Y/N?)";ANS$
IF ANS$="Y" THEN INPUT"TYPE IN EXHAUST PIPE LENGTH,
LP34";LP34
INPUT"NEW DATA FOR EXHAUST PIPE LENGTH, LP45(Y/N?)";ANS$
IF ANS$="Y" THEN INPUT"TYPE IN EXHAUST PIPE LENGTH,
LP45";LP45
INPUT"NEW DATA FOR EXHAUST PIPE LENGTH, LP56(Y/N?)";ANS$
IF ANS$="Y" THEN INPUT"TYPE IN EXHAUST PIPE LENGTH,
LP56";LP56
INPUT"MORE DATA,(Y/N?)";ANS$
IF ANS$="Y" THEN GOTO 19
LP01M=LP01/1000
LP12M=LP12/1000
LP23M=LP23/1000
LP34M=LP34/1000
LP45M=LP45/1000
LP56M=LP56/1000
RETURN
EXDIADAT:
119 INPUT"NEW DATA FOR EXHAUST PIPE DIAMETER, DEP1(Y/
N?)";ANS$
IF ANS$="Y" THEN INPUT"TYPE IN EXHAUST PIPE DIAMETER,
DEP1";DEP1
INPUT"NEW DATA FOR EXHAUST PIPE DIAMETER, DEP2(Y/N?)";ANS$
IF ANS$="Y" THEN INPUT"TYPE IN EXHAUST PIPE DIAMETER,
DEP2";DEP2
INPUT"NEW DATA FOR EXHAUST PIPE DIAMETER, DEP3(Y/N?)";ANS$
IF ANS$="Y" THEN INPUT"TYPE IN EXHAUST PIPE DIAMETER,
DEP3";DEP3
INPUT"NEW DATA FOR EXHAUST PIPE DIAMETER, DEP4(Y/N?)";ANS$
IF ANS$="Y" THEN INPUT"TYPE IN EXHAUST PIPE DIAMETER,
DEP4";DEP4
INPUT"NEW DATA FOR EXHAUST PIPE DIAMETER, DEP5(Y/N?)";ANS$
IF ANS$="Y" THEN INPUT"TYPE IN EXHAUST PIPE DIAMETER,
DEP5";DEP5
INPUT"NEW DATA FOR EXHAUST PIPE DIAMETER, DEP6(Y/N?)";ANS$
IF ANS$="Y" THEN INPUT"TYPE IN EXHAUST PIPE DIAMETER,
DEP6";DEP6
INPUT"MORE DATA,(Y/N?)";ANS$
IF ANS$="Y" THEN GOTO 119
DE1M=DEP1/1000
```

```
DE2M=DEP2/1000
DE3M=DEP3/1000
DE4M=DEP4/1000
DE5M=DEP5/1000
DE6M=DEP6/1000
FEP1=PI*DE1M*DE1M/4
FEP2=PI*DE2M*DE2M/4
FEP3=PI*DE3M*DE3M/4
FEP4=PI*DE4M*DE4M/4
FEP5=PI*DE5M*DE5M/4
FEP6=PI*DE6M*DE6M/4
RETURN
DATASET:
PRINT"PROGRAM FOR PISTON-PORTED ENGINE WITH EXHAUST
BOX SILENCER"
PRINT"Prog.5.2, ENGINE WITH TUNED EXPANSION CHAMBER PIPE"
PRINT"ALL DATA IN mm AND deg UNITS"
PRINT"     BORE   STROKE   CON-ROD  REV/MIN  EXHAUST,degC
TRAP-CR   CRANKCASE-CR "
PRINT USING"######.##";BO,ST,CRL,RPM,TEXC,TCR,CCR
PRINT
PRINT"EXHAUST PORT DATA"
PRINT"     OPENS  FULL-OPEN  NUMBER-OFF  WIDTH-EACH  TOP-
RAD  BOTTOM-RAD
PRINT USING"#########.#";EPO,EPFO,NEP,WEP,RTE,RBE
PRINT
PRINT"TRANSFER PORT DATA WITH SCAVENGING DESCRIBED AS
";SCAV$
PRINT"    OPENS   FULL-OPEN   NUMBER-OFF  WIDTH-EACH  TOP-
RAD  BOTTOM-RAD
PRINT USING"#########.#";TPO,TPFO,NTP,WTP,RTT,RBT
PRINT
PRINT"INLET PORT DATA"
PRINT"    OPENS   FULL-OPEN   NUMBER-OFF  WIDTH-EACH  TOP-
RAD  BOTTOM-RAD
PRINT USING"#########.#";IPO,IPFO,NIP,WIP,RTI,RBI
PRINT
PRINT "TUNED EXHAUST EXPANSION CHAMBER DATA, LENGTHS
AND DIAMETERS"
PRINT"LP01  LP12  LP23  LP34  LP45  LP56"
PRINT USING"######.";LP01,LP12,LP23,LP34,LP45,LP56
PRINT"DEP1  DEP2  DEP3  DEP4  DEP5  DEP6"
PRINT USING"######.";DEP1,DEP2,DEP3,DEP4,DEP5,DEP6
PRINT
```

PRINT “TRANSFER AND INLET PIPE DATA”
PRINT”LENGTH.T TRANSFER.AR LENGTH.I DIAMETER.I THROTTLE.AR
PRINT USING”######.##”;LTP,TRAR,LIP,DIP,THAR
PRINT
PRINT”COMBUSTION MODEL PARAMETERS”
PRINT” INDECES, COMP. AND EXP. COMBUSTION.EFF BURN PERIOD AIR-TO-FUEL”
PRINT USING”########.##”;NC,NE,BEFF,BDEG,TAF
PRINT
INPUT”OK?,(Y/N)”;S$
RETURN
DATASETL:
LPRINT” BORE STROKE CON-ROD REV/MIN EXHAUST,degC TRAP-CR CRANKCASE-CR “
LPRINT USING”######.##”;BO,ST,CRL,RPM,TEXC,TCR,CCR
LPRINT
LPRINT”EXHAUST PORT DATA”
LPRINT” OPENS FULL-OPEN NUMBER-OFF WIDTH-EACH TOP-RAD BOTTOM-RAD”
LPRINT USING”#########.#”;EPO,EPFO,NEP,WEP,RTE,RBE
LPRINT
LPRINT”TRANSFER PORT DATA WITH SCAVENGING DESCRIBED AS “;SCAV$
LPRINT” OPENS FULL-OPEN NUMBER-OFF WIDTH-EACH TOP-RAD BOTTOM-RAD”
LPRINT USING”#########.#”;TPO,TPFO,NTP,WTP,RTT,RBT
LPRINT
LPRINT”INLET PORT DATA”
LPRINT” OPENS FULL-OPEN NUMBER-OFF WIDTH-EACH TOP-RAD BOTTOM-RAD”
LPRINT USING”#########.#”;IPO,IPFO,NIP,WIP,RTI,RBI
LPRINT
LPRINT “TUNED EXHAUST EXPANSION CHAMBER DATA, LENGTHS AND DIAMETERS”
LPRINT” LP01 LP12 LP23 LP34 LP45 LP56"
LPRINT USING”######.”;LP01,LP12,LP23,LP34,LP45,LP56
LPRINT” DEP1 DEP2 DEP3 DEP4 DEP5 DEP6"
LPRINT USING”######.”;DEP1,DEP2,DEP3,DEP4,DEP5,DEP6
LPRINT
LPRINT “TRANSFER AND INLET PIPE DATA”
LPRINT” LENGTH.T TRANSFER.AR LENGTH.I DIAMETER.I THROTTLE.AR
LPRINT USING”########.##”;LTP,TRAR,LIP,DIP,THAR

```
LPRINT
LPRINT"COMBUSTION MODEL PARAMETERS"
LPRINT" COMBUSTION.EFF  BURN PERIOD  AIR-TO-FUEL  COMP.
AND EXP. INDECES"
LPRINT USING"#########.##";BEFF,BDEG,TAF,NC,NE,
LPRINT
RETURN
SPLOT1:
NCOUNT=1
IF NCOUNT>=2 THEN GOTO 410
M=1
MG2=1
X0=50
Y0=130
YY0=230
YYB=280
YT=20
YZ=30
X2=460
DR=(X2-X0)/360
EP=2*(180-EPO)
IP=2*IPO
TP=2*(180-TPO)
XEC=X0+EP*DR
XBDC=(X0+XEC)/2
XTDC=(X2+XEC)/2
XTO=XBDC-TP*DR/2
XTC=XBDC+TP*DR/2
XIO=XTDC-IPO*DR
IF IPO<=EPO THEN XIC=XTDC+IP*DR/2
IF IPO>EPO THEN XIC=X0+(IPO-EPO)*DR
GPS=1
GYS=(Y0-YZ)/GPS
IF D$="P" THEN GOTO 466
CLS
466 LOCATE 1,8:PRINT"CRANK-ANGLE from EXHAUST PORT OPEN-
ING for one cycle"
LOCATE 19,5
PRINT "EO"
LOCATE 19,11
PRINT"TO"
LOCATE 19,27
PRINT"TC"
LOCATE 19,33
```

```
PRINT"EC"
LOCATE 19,18
PRINT"BDC"
LOCATE 19,44
PRINT"TDC"
LOCATE 19,58
PRINT"EO"
IF IPO<=EPO THEN LOCATE 19,36
IF IPO>EPO THEN LOCATE 19,30
PRINT"IO"
IF IPO<=EPO THEN LOCATE 19,54
IF IPO>EPO THEN LOCATE 19,8
PRINT "IC"
LOCATE 2,3
PRINT"2.0"
LOCATE 5,3
PRINT"1.5"
LOCATE 9,3
PRINT"1.0"
LOCATE 12,3
PRINT"0.5"
LOCATE 12,60
PRINT"1.5"
LOCATE 15,60
PRINT"1.0"
LOCATE 18,60
PRINT"0.5"
FOR JN= 1 TO 10
PIP=0
IF JN=5 THEN PIP=4
IF JN=10 THEN PIP=4
LINE(X0,Y0-.1*JN*GYS)-(X0-PIP-3,Y0-.1*JN*GYS)
NEXT JN
FOR JN= 1 TO 5
PIP=0
IF JN=5 THEN PIP=4
LINE(X0,Y0+.1*JN*GYS)-(X0-PIP-3,Y0+.1*JN*GYS)
NEXT JN
FOR JN= 1 TO 10
PIP=0
IF JN=5 THEN PIP=4
IF JN=10 THEN PIP=4
LINE(X2,YY0-.1*JN*GYS)-(X2+PIP+3,YY0-.1*JN*GYS)
NEXT JN
```

```
FOR JN= 1 TO 5
PIP=0
IF JN=5 THEN PIP=4
LINE(X2,YY0+.1*JN*GYS)-(X2+PIP+3,YY0+.1*JN*GYS)
NEXT JN
LINE(X0,YY0)-(X2,YY0)
LINE(X0,Y0)-(X2,Y0)
LINE(X0,YYB)-(X0,YT)
LINE(X2,YYB)-(X2,YT)
LINE(XEC,YYB)-(XEC,YT)
LINE(XTO,YYB)-(XTO,YT)
LINE(XTC,YYB)-(XTC,YT)
LINE(XIO,YYB)-(XIO,YT)
LINE(XIC,YYB)-(XIC,YT)
LINE(XTDC,YYB)-(XTDC,YT)
LINE(XBDC,YYB)-(XBDC,YT)
410 RETURN
SPLOT2:
IF PCR>2 THEN PCR=2
ZC=(PCR-1)*GYS
ZS=(PCCR-1)*GYS
ZE=(PEX-1)*GYS
ZI=(PIR-1)*GYS
ZT=(PTR-1)*GYS
ZTC=(PTRC-1)*GYS
XAN=INT(DR*(ANGD-EPO))+X0
YCN=Y0-ZC
YSN=Y0-ZS
YEN=Y0-ZE
YIN=YY0-ZI
YCCN=YY0-ZS
IF M=1 THEN GOTO 633
LINE(XAO,YSO)-(XAN,YSN)
LINE(XAO,YCO)-(XAN,YCN)
LINE(XAO,YEO)-(XAN,YEN)
LINE(XAO,YIO)-(XAN,YIN)
LINE(XAO,YCCO)-(XAN,YCCN)
633 XAO=XAN
YCO=YCN
YSO=YSN
YEO=YEN
YIO=YIN
YCCO=YCCN
M=M+1
```

```
RETURN
FILING:
INPUT "TYPE IN TITLE NEEDED FOR THE FILE WITH '.BOX' END
CODE";P$
OPEN P$ FOR OUTPUT AS #1
WRITE#1,BO,ST,CRL,RPM,TEXC,TCR,CCR
WRITE#1,EPO,EPFO,NEP,WEP,RTE,RBE
WRITE#1,TPO,TPFO,NTP,WTP,RTT,RBT,SCAV$
WRITE#1,IPO,IPFO,NIP,WIP,RTI,RBI
WRITE#1,LP01,LP12,LP23,LP34,LP45,LP56
WRITE#1,DEP1,DEP2,DEP3,DEP4,DEP5,DEP6
WRITE#1,LTP,TRAR,LIP,DIP,THAR
WRITE#1,NC,NE,BEFF,BDEG,TAF
CLOSE#1
RETURN
FILED:
INPUT "TYPE IN THE FILENAME WITH A '.BOX' END CODE";FIL$
OPEN"I",#1,FIL$
INPUT#1,BO,ST,CRL,RPM,TEXC,TCR,CCR
INPUT#1,EPO,EPFO,NEP,WEP,RTE,RBE
INPUT#1,TPO,TPFO,NTP,WTP,RTT,RBT,SCAV$
INPUT#1,IPO,IPFO,NIP,WIP,RTI,RBI
INPUT#1,LP01,LP12,LP23,LP34,LP45,LP56
INPUT#1,DEP1,DEP2,DEP3,DEP4,DEP5,DEP6
INPUT#1,LTP,TRAR,LIP,DIP,THAR
INPUT#1,NC,NE,BEFF,BDEG,TAF
CLOSE#1
RETURN
```

COMPUTER PROGRAMS PROG.5.1 and 5.2

PROGRAM NAMEs "ENGINE MODEL NO.1 and ENGINE MODEL NO.2"

ProgLists.5.1 and 5.2

<u>COMMON SUBROUTINES FOR BOTH PROGRAMS</u>

```
TRPIPDAT:
20 INPUT"NEW DATA FOR TRANSFER PIPE LENGTH(Y/N?)";ANS$
IF ANS$="Y" THEN INPUT"TYPE IN TRANSFER PIPE
LENGTH(LTP)";LTP
INPUT"NEW DATA FOR TRANSFER PORT TO PIPE AREA RATIO(Y/
N?)";ANS$
IF ANS$="Y" THEN INPUT"TYPE IN TRANSFER PORT TO PIPE AREA
RATIO(TRAR)";TRAR
INPUT"MORE DATA,(Y/N?)";ANS$
IF ANS$="Y" THEN GOTO 20
```

```
LTM=LTP/1000
RETURN
INPIPDAT:
21 INPUT"NEW DATA FOR INLET PIPE LENGTH(Y/N?)";ANS$
IF ANS$="Y" THEN INPUT"TYPE IN INLET PIPE LENGTH(LIP)";LIP
INPUT"NEW DATA FOR INLET PIPE DIAMETER(Y/N?)";ANS$
IF ANS$="Y" THEN INPUT"TYPE IN INLET PIPE DIAMETER(DEP1)";DIP
INPUT"NEW DATA FOR THROTTLE AREA RATIO(Y/N?)";ANS$
IF ANS$="Y" THEN INPUT"TYPE IN THROTTLE AREA RATIO(THAR)";THAR
INPUT"MORE DATA,(Y/N?)";ANS$
IF ANS$="Y" THEN GOTO 21
LIM=LIP/1000
DIPM=DIP/1000
FIPM=PI*DIPM*DIPM/4
RETURN
COMBDAT:
22 INPUT"NEW DATA FOR POLYTROPIC COMPRESSION INDEX(Y/N?)";ANS$
IF ANS$="Y" THEN INPUT"TYPE IN POLYTROPIC COMPRESSION INDEX(NC)";NC
INPUT"NEW DATA FOR POLYTROPIC EXPANSION INDEX(Y/N?)";ANS$
IF ANS$="Y" THEN INPUT"TYPE IN POLYTROPIC EXPANSION INDEX(NE)";NE
INPUT"NEW DATA FOR COMBUSTION EFFICIENCY(Y/N?)";ANS$
IF ANS$="Y" THEN INPUT"TYPE IN COMBUSTION EFFICIENCY(BEFF)";BEFF
INPUT"NEW DATA FOR COMBUSTION PERIOD(Y/N?)";ANS$
IF ANS$="Y" THEN INPUT"TYPE IN COMBUSTION PERIOD(BDEG)";BDEG
INPUT"NEW DATA FOR TRAPPED AIR-FUEL RATIO(default value is 13)(Y/N?)";ANS$
IF ANS$="Y" THEN INPUT"TYPE IN TRAPPED AIR-FUEL RATIO(TAF)";TAF
INPUT"MORE DATA,(Y/N?)";ANS$
IF ANS$="Y" THEN GOTO 22
RETURN
PORTSTART:
EPC=360-EPO
IPC=360+IPO
IPOD=360-IPO
TPC=360-TPO
PFI=0
```

```
PFE=0
PFT=0
PFIM=0
PFEM=0
PFTM=0
PWE= WEPM-2*RTEM
PWT= WTPM-2*RTTM
PWI= WIPM-2*RTIM
PHE=0
PHT=0
PHI=0
CALL PISTPOS(STM,CRNK,CRLM,EPO,HEOT,HEOB)
CALL PISTPOS(STM,CRNK,CRLM,EPFO,HEFT,HEFB)
EXHT=HEFT-HEOT
HME=EXHT-RTEM-RBEM
CALL PISTPOS(STM,CRNK,CRLM,TPO,HTOT,HTOB)
CALL PISTPOS(STM,CRNK,CRLM,TPFO,HTFT,HTFB)
TRHT=HTFT-HTOT
HMT=TRHT-RTTM-RBTM
CALL PISTPOS(STM,CRNK,CRLM,IPO,HIOT,HIOB)
CALL PISTPOS(STM,CRNK,CRLM,IPFO,HIFT,HIFB)
INHT=HIOT-HIFT
HMI=INHT-RTIM-RBIM
IF SCAV$="UNIFLOW" THEN MSC=-1.7827
IF SCAV$="UNIFLOW" THEN CSC=.2094
IF SCAV$="SCRE" THEN MSC=-1.6709
IF SCAV$="SCRE" THEN CSC=.1899
IF SCAV$="YAM1" THEN MSC=-1.6993
IF SCAV$="YAM1" THEN CSC=.3053
IF SCAV$="YAM6" THEN MSC=-1.3516
IF SCAV$="YAM6" THEN CSC=.1435
IF SCAV$="CD" THEN MSC=-1.0104
IF SCAV$="CD" THEN CSC=-.117
IF SCAV$="QUBCR" THEN MSC=-1.6325
IF SCAV$="QUBCR" THEN CSC=.1397
RETURN
EXPORTSHUT:
AKE=0
ME1=0
AE(1)=BE(1)
RETURN
EXPORTOPEN:
PC1R=PC1/PA
IF PEX<=PC1R THEN CCE=.8
```

```
IF PEX>PC1R THEN CCE=.8
CALL FPORT(HEOT,HTDC,EXHT,FEPRT,FEP1,AKE,CCE)
CALL PORT(AE(1),BE(1),PC1,AKE)
IF PEX<=PC1R THEN CDE=.75
IF PEX>PC1R THEN CDE=.75
IF PEX<=PC1R THEN CALL OUTFLOW(AE(1),BE(1),FEP1,
TC1,HE1,CE1,ME1,D,CDE)
IF PEX>PC1R THEN CALL OUTFLOW(AE(1),BE(1),FEP1,
TC1,HE1,CE1,ME1,D,CDE)
EXRATIO=EXRATIO+ME1*DTM/MREF
RETURN
TRPORTSHUT:
AKT=0
MT1=0
AT(1)=BT(1)
RETURN
```

SCAVENGING MODEL NO.1 IN SUBROUTINE 'TRPORTOPEN'

```
REM Eqn.5.1.11 is being programmed here
TRPORTOPEN:
IF PTR<=PC1R THEN CCT=.8
IF PTR>PC1R THEN CCT=.8
CALL FPORT(HTOT,HTDC,TRHT,FTPRT,FTDUCT,AKT,CCT)
CALL PORT(AT(1),BT(1),PC1,AKT)
IF PTR<=PC1R THEN CDT=.75
IF PTR>PC1R THEN CDT=.75
IF PTR<=PC1R THEN CALL OUTFLOW(BT(1),AT(1),FTDUCT,TC1,
HT1,CT1,MT1,DSCAV,CDT)
IF PTR>PC1R THEN CALL OUTFLOW(BT(1),AT(1),FTDUCT,TCC1,
HT1,CT1,MT1,DSCAV,CDT)
EXPAND=PTR/PC1R
VOLINC=CT1*FTDUCT*DTM*EXPAND/VC1
SCRATIO=SCRATIO+MT1*DTM/MREF
IF SCRATIO<0 THEN GOTO 911
SR=SR+VOLINC
SEFF=1-EXP(MSC*SR+CSC)
911 RETURN
```

SCAVENGING MODEL NO.2 IN SUBROUTINE 'TRPORTOPEN'

```
REM Eqn.5.1.13 is being programmed here
TRPORTOPEN:
IF PTR<=PC1R THEN CCT=.8
IF PTR>PC1R THEN CCT=.8
CALL FPORT(HTOT,HTDC,TRHT,FTPRT,FTDUCT,AKT,CCT)
```

```
CALL PORT(AT(1),BT(1),PC1,AKT)
IF PTR<=PC1R THEN CDT=.75
IF PTR>PC1R THEN CDT=.75
IF PTR<=PC1R THEN CALL OUTFLOW(BT(1),AT(1),FTDUCT,TC1,
HT1,CT1,MT1,DSCAV,CDT)
IF PTR>PC1R THEN CALL OUTFLOW(BT(1),AT(1),FTDUCT,TCC1,
HT1,CT1,MT1,DSCAV,CDT)
DCY=PC1/(GK*TC1)
EXPAND=DSCAV/(.6*DSCAV+.4*DCY)
VOLINC=CT1*FTDUCT*DTM*EXPAND/VC1
SCRATIO=SCRATIO+MT1*DTM/MREF
IF SCRATIO<0 THEN GOTO 911
SR=SR+VOLINC
SEFF=1-EXP(MSC*SR+CSC)
911 RETURN
INPORTSHUT:
AKI=0
MI1=0
AI(1)=BI(1)
RETURN
INPORTOPEN:
PCC1R=PCC1/PA
IF PIR<=PCC1R THEN CCI=.8
IF PIR>PCC1R THEN CCI=.9
CALL FPORT(HTDC,HIOT,INHT,FIPRT,FIPRT,AKI,CCI)
CALL PORT(AI(1),BI(1),PCC1,AKI)
IF PIR<=PCC1R THEN CDI=.75
IF PIR>PCC1R THEN CDI=.9
IF PIR<=PCC1R THEN CALL OUTFLOW(BI(1),AI(1),FIPRT,TCC1,
HI1,CI1,MI1,D,CDI)
IF PIR>PCC1R THEN CALL INFLOW(BI(1),AI(1),FIPRT,TIREF,
HI1,CI1,MI1,D,CDI)
DRATIO=DRATIO+MI1*DTM/MREF
RETURN
ENGINEDAT:
15 INPUT"NEW DATA FOR CYLINDER BORE(Y/N?)";ANS$
IF ANS$="Y" THEN INPUT"TYPE IN CYLINDER BORE(B)";BO
INPUT"NEW DATA FOR CYLINDER STROKE(Y/N?)";ANS$
IF ANS$="Y" THEN INPUT"TYPE IN CYLINDER STROKE(S)";ST
INPUT"NEW DATA FOR CONNECTING ROD LENGTH(Y/N?)";ANS$
IF ANS$="Y" THEN INPUT"TYPE IN CONNECTING ROD
LENGTH(CRL)";CRL
INPUT"NEW DATA FOR ENGINE SPEED(Y/N?)";ANS$
IF ANS$="Y" THEN INPUT"TYPE IN ENGINE SPEED(RPM)";RPM
```

```
INPUT"NEW DATA FOR CYLINDER COMPRESSION RATIO(Y/N?)";ANS$
IF ANS$="Y" THEN INPUT"TYPE IN CYLINDER COMPRESSION RATIO(TCR)";TCR
INPUT"NEW DATA FOR CRANKCASE COMPRESSION RATIO(Y/N?)";ANS$
IF ANS$="Y" THEN INPUT"TYPE IN CRANKCASE COMPRESSION RATIO(CCR)";CCR
INPUT"NEW DATA FOR EXHAUST TEMPERATURE(Y/N?)";ANS$
IF ANS$="Y" THEN INPUT"TYPE IN EXHAUST PIPE REFERENCE TEMPERATURE, ºC";TEXC
INPUT"MORE DATA,(Y/N?)";ANS$
IF ANS$="Y" THEN GOTO 15
BOM=BO/1000
STM=ST/1000
CRLM=CRL/1000
RPS=RPM/60
FPISTON=PI*BOM*BOM/4
SV=FPISTON*STM
CCCV=SV/(CCR-1)
CRNK=STM/2
FMEP=100*STM*RPM
TEREF=TEXC+273
RETURN
EXPORTDAT:
16 INPUT"NEW DATA FOR EXHAUST PORT OPENING TIMING(Y/N?)";ANS$
IF ANS$="Y" THEN INPUT"TYPE IN EXHAUST PORT OPENING,DEG. ATDC";EPO
INPUT"NEW DATA FOR EXHAUST PORT FULLY OPEN TIMING(Y/N?)";ANS$
IF ANS$="Y" THEN INPUT"TYPE IN EXHAUST PORT FULLY OPEN TIMING,DEG. ATDC";EPFO
INPUT"NEW DATA FOR NUMBER OF EXHAUST PORTS(Y/N?)";ANS$
IF ANS$="Y" THEN INPUT"TYPE IN NUMBER OF EXHAUST PORTS";NEP
INPUT"NEW DATA FOR MAXIMUM WIDTH OF EACH EXHAUST PORT(Y/N?)";ANS$
IF ANS$="Y" THEN INPUT"TYPE IN MAXIMUM WIDTH OF EACH EXHAUST PORT(WE)";WEP
INPUT"NEW DATA FOR EXHAUST PORT TOP RADIUS(Y/N?)";ANS$
IF ANS$="Y" THEN INPUT"TYPE IN EXHAUST PORT TOP RADIUS(RTE)";RTE
INPUT"NEW DATA FOR EXHAUST BOTTOM RADIUS(Y/N?)";ANS$
```

```
IF ANS$="Y" THEN INPUT"TYPE IN EXHAUST PORT BOTTOM
RADIUS(RBE)";RBE
INPUT"MORE DATA,(Y/N?)";ANS$
IF ANS$="Y" THEN GOTO 16
WEPM=WEP/1000
RTEM=RTE/1000
RBEM=RBE/1000
RETURN
TRPORTDAT:
17 INPUT"NEW DATA FOR TRANSFER PORT OPENING TIMING(Y/
N?)";ANS$
IF ANS$="Y" THEN INPUT"TYPE IN TRANSFER PORT OPENING,DEG.
ATDC";TPO
INPUT"NEW DATA FOR TRANSFER PORT FULLY OPEN TIMING(Y/
N?)";ANS$
IF ANS$="Y" THEN INPUT"TYPE IN TRANSFER PORT FULLY OPEN
TIMING,DEG. ATDC";TPFO
INPUT"NEW DATA FOR NUMBER OF TRANSFER PORTS(Y/N?)";ANS$
IF ANS$="Y" THEN INPUT"TYPE IN NUMBER OF TRANSFER
PORTS";NTP
INPUT"NEW DATA FOR MAXIMUM WIDTH OF EACH TRANSFER
PORT(Y/N?)";ANS$
IF ANS$="Y" THEN INPUT"TYPE IN MAXIMUM WIDTH OF EACH
TRANSFER PORT(WE)";WTP
INPUT"NEW DATA FOR TRANSFER PORT TOP RADIUS(Y/N?)";ANS$
IF ANS$="Y" THEN INPUT"TYPE IN TRANSFER PORT TOP
RADIUS(RTE)";RTT
INPUT"NEW DATA FOR TRANSFER PORT BOTTOM RADIUS(Y/
N?)";ANS$
IF ANS$="Y" THEN INPUT"TYPE IN TRANSFER PORT BOTTOM
RADIUS(RBT)";RBT
INPUT"NEW DATA FOR SCAVENGING QUALITY(Y/N?)";ANS$
IF ANS$="Y" THEN PRINT"TYPE IN 'UNIFLOW', OR 'SCRE', OR 'YAM1',
OR 'YAM6', OR 'CD', OR 'QUBCR'"
IF ANS$="Y" THEN INPUT"AS DESCRIBED IN Fig.3.14 of Sect.3.3.1 of
Chapter 3";SCAV$
INPUT"MORE DATA,(Y/N?)";ANS$
IF ANS$="Y" THEN GOTO 17
WTPM=WTP/1000
RTTM=RTT/1000
RBTM=RBT/1000
RETURN
INPORTDAT:
18 INPUT"NEW DATA FOR INLET PORT OPENING TIMING(Y/N?)";ANS$
```

```
IF ANS$="Y" THEN INPUT"TYPE IN INLET PORT OPENING,DEG.
BTDC";IPO
INPUT"NEW DATA FOR INLET PORT FULLY OPEN TIMING(Y/
N?)";ANS$
IF ANS$="Y" THEN INPUT"TYPE IN INLET PORT FULLY OPEN
TIMING,DEG. ATDC";IPFO
INPUT"NEW DATA FOR NUMBER OF INLET PORTS(Y/N?)";ANS$
IF ANS$="Y" THEN INPUT"TYPE IN NUMBER OF INLET PORTS";NIP
INPUT"NEW DATA FOR MAXIMUM WIDTH OF EACH INLET PORT(Y/
N?)";ANS$
IF ANS$="Y" THEN INPUT"TYPE IN MAXIMUM WIDTH OF EACH
INLET PORT(WIP)";WIP
INPUT"NEW DATA FOR INLET PORT TOP RADIUS(Y/N?)";ANS$
IF ANS$="Y" THEN INPUT"TYPE IN INLET PORT TOP RADIUS(RTI)";RTI
INPUT"NEW DATA FOR INLET PORT BOTTOM RADIUS(Y/N?)";ANS$
IF ANS$="Y" THEN INPUT"TYPE IN INLET PORT BOTTOM
RADIUS(RBI)";RBI
INPUT"MORE DATA,(Y/N?)";ANS$
IF ANS$="Y" THEN GOTO 18
WIPM=WIP/1000
RTIM=RTI/1000
RBIM=RBI/1000
RETURN
CYLSTART:
TEREF=TEXC+273
TTREF=320
TIREF=300
SDE=SQR(GE*GK*TEREF)
SDT=SQR(GE*GK*TTREF)
SDI=SQR(GE*GK*TIREF)
DEREF=PA/(GK*TEREF)
DTREF=PA/(GK*TTREF)
DIREF=PA/(GK*TIREF)
RETURN
SUB FPORT(HOT,HTDC,PHT,FPRT,FPIP,AK,CC)STATIC
FINST=(HTDC-HOT)/PHT
IF FINST>1 THEN FINST=1
AK=CC*FINST*FPRT/FPIP
END SUB
SUB STAB(A(),B(),K,S,E)STATIC
SHARED DZ,DTM,DC,RPM
C=60*S/(RPM*E)
FOR J=1 TO K
Z=2/((A(J)+B(J))+ABS(5*(A(J)-B(J))))
```

```
IF Z<DZ THEN DZ=Z
NEXT J
DC=360*DZ/C
DTM=DC/(6*RPM)
END SUB
SUB SETS(A(),B(),K)STATIC
FOR J=1 TO K
A(J)=1
B(J)=1
NEXT J
END SUB
SUB AREA(DA,DB,KA,KB,DF(),F())STATIC
C=(DB-DA)/(KB-KA)
FOR J=KA TO KB
A=J-KA
F(J)=((DA+C*A)^2)*.7854
NEXT J
KC=KB-1
FOR J=KA TO KC
DF(J)=F(J+1)-F(J)
NEXT J
END SUB
SUB PIPE(KA,KB,A(),B(),DF(),F())STATIC
SHARED DZ,A2(),B2()
G1=3
G2=2
KC=KB-1
FOR J=KA TO KC
AA=DZ*A(J+1)*B(J)
BB=DZ*A(J)*B(J+1)
Y=(1+DZ*(G1*(A(J+1)-A(J))-G2*(B(J+1)-B(J))))
Z=(1+DZ*(G1*(B(J)-B(J+1))-G2*(A(J)-A(J+1))))
WA=(A(J+1)+G2*AA-G2*BB)/Y
RA=(A(J)-G1*BB+G1*AA)/Z
WB=(B(J+1)+G1*AA-G1*BB)/Y
RB=(B(J)+G2*AA-G2*BB)/Z
AA=WA-WB
BB=RA-RB
CC=WA+WB
DD=RA+RB
IF AA=0 THEN GOTO 23
QA=-DZ*CC*AA*DF(J)/(4*F(J))
GOTO 33
23 QA=0
```

```
33 IF BB=0 THEN GOTO 44
QB=-DZ*BB*DD*DF(J)/(4*F(J))
GOTO 55
44 QB=0
55 A2(J+1)=WA+QA
B2(J)=RB+QB
NEXT J
FOR J=KA TO KC
A(J+1)=A2(J+1)
B(J)=B2(J)
NEXT J
END SUB
SUB SHOK(A(),B(),KA,KB)STATIC
FOR J=KA TO KB
R1=5*ABS(A(J)-B(J))/(A(J)+B(J))
IF R1<1 THEN GOTO 56
R2=SQR((R1*R1+5)/(7*R1*R1-1))
R3=(A(J)+B(J))*((7*R1*R1-1)/6)^(1/7)
IF A(J)>B(J) THEN GOTO 26
A(J)=(5-R2)*R3/10
B(J)=(5+R2)*R3/10
GOTO 56
26 A(J)=(5+R2)*R3/10
B(J)=(5-R2)*R3/10
56 R1=0
NEXT J
END SUB
SUB PORT(A,B,P,AK)STATIC
SHARED ABX(),BX(),AKA(),PA
X1=(P/PA)^(1/7)
BX1=B/X1
IF AK>.95 THEN AK=.95
IF BX1<.6 THEN BX1=.6
I=INT(AK/.05)+1
J=INT((BX1-.6)/.02)+1
RA=(AK-AKA(I))/(AKA(I+1)-AKA(I))
AB1=RA*(ABX(I+1,J+1)-ABX(I,J+1))+ABX(I,J+1)
AB2=RA*(ABX(I+1,J)-ABX(I,J))+ABX(I,J)
RB=(BX1-BX(J))/(BX(J+1)-BX(J))
ABX1=RB*(AB1-AB2)+AB2
A=2*X1*ABX1-B
END SUB
SUB NINT(KI,PL,TE,TP)STATIC
SHARED LMESH
```

```
A=SQR(TE/TP)
RI=PL*A/LMESH
KI=INT(RI)
BI=RI-KI
IF BI>.5 THEN KI=KI+1
KI=KI+1
END SUB
SUB OUTFLOW(A,B,F,TR,H,C,M,D,CD)STATIC
SHARED GK,GE,PA
CP=481.17*TR^.125
GAM=1.7847/(TR^.0417)
AB=(A+B)/2
TREF=TR/AB^2
REM TR IS REGARDED AS THE SUPERPOSITION TEMPERATURE
SD=SQR(GAM*GK*TREF)
C=SD*(A-B)/(GAM-1)
H=CP*TR
D=AB^5*PA/(GK*TREF)
M=CD*C*F*D
MN=C/SQR(GAM*GK*TR)
P1=(.5*(A+1))^7
P2=(.5*(B+1))^7
END SUB
SUB INFLOW(A,B,F,TR,H,C,M,D,CD)STATIC
SHARED GK,GE,PA
CP=481.17*TR^.125
GAM=1.7847/(TR^.0417)
AB=(A+B)/2
SD=SQR(GAM*GK*TR)
C=2.5*SD*(A-B)
T=TR*AB*AB
H=CP*T
D=AB^5*PA/(GK*TR)
M=CD*C*F*D
MN=C/SQR(GAM*GK*T)
PR=(.5*(A+1))^7
PL=(.5*(B+1))^7
END SUB
SUB BOX(PB1,PB2,TB1,TB2,VB1,VB2,MB1,MB2,MIN,MO,CIN,CO,HIN,
HO,WORK)STATIC
SHARED CCV,DTM,GK,GE
CV1=269.49*TB1^0.1667
DMI=MIN*DTM
DME=MO*DTM
```

```
MB2=MB1+DMI-DME
EIN=DMI*(HIN+.5*CIN*CIN)
EO=DME*(HO+.5*CO*CO)
DW=PB1*(VB2-VB1)
TB2=(EIN-EO+CV1*MB1*TB1-DW)/(CV1*MB2)
CV2=269.49*TB2^0.1667
TB2=(EIN-EO+CV1*MB1*TB1-DW)/(CV2*MB2)
PB2=MB2*GK*TB2/VB2
WORK=(PB1+PB2)*(VB2-VB1)/2
PB1=PB2
TB1=TB2
MB1=MB2
VB1=VB2
END SUB
SUB BELMOUTH(B,A,PA)STATIC
M=.5*(A+1)
N=((4*M+1)+SQR(1+20*M-20*M*M))/6
B=2*N-1
END SUB
SUB PISTPOS(S,R,C,D,HT,HB)STATIC
SHARED DTOR
A=D*DTOR
HT=R+C-R*COS(A)-SQR(C*C-(R*SIN(A))^2)
HB=S-HT
END SUB
SUB PAREA(NP,WP,HP,RT,RB,FP,DP)STATIC
SHARED PI
FP=NP*(WP*HP-(2-PI/2)*(RT*RT+RB*RB))
DP=SQR(4*FP/PI)
END SUB
SUB CYLINDER(P1,P2,T1,T2,V1,V2,MC,WORK)STATIC
SHARED GK,GE,NE,NC,DEG,SEFF,DC,BEFF,BDEG,TAF,CVAL,EPC,PA
SHARED PMAX,TKMAX,MAXDEG,MREF,CEFF,TEFF,DRAT
DEGC=DEG
IF DEGC<=360 THEN POLYINDEX=NC
IF DEGC>360 THEN POLYINDEX=NE
IF DEGC>EPC+5 THEN GOTO 30
MAIR=SEFF*MC
CEFF=MAIR/MREF
TEFF=MAIR/(DRAT*MREF)
IGN=360-BDEG/6
BURNEND=IGN+BDEG
TAILDEG=BURNEND-BDEG/3
PEAK=IGN+BDEG/3
```

```
TAFB=13
IF TAF>TAFB THEN TAFB=TAF
HEAT=BEFF*CVAL*MAIR/TAFB
QMAX=36*HEAT/(14*BDEG)
S1=3*QMAX/BDEG
S2=5*QMAX/(2*BDEG)
S3=QMAX/(2*BDEG)
30 IF DEGC<=IGN THEN GOTO 90
IF DEGC>=BURNEND THEN GOTO 90
IF DEGC>IGN THEN GOTO 20
IF DEGC<=BURNEND THEN GOTO 20
90 P2=P1*(V1/V2)^POLYINDEX
T2=P2*V2/(MC*GK)
GOTO 50
20 IF DEGC>PEAK THEN GOTO 60
HT2=S1*(DEGC-IGN)
GOTO 80
60 IF DEGC<=TAILDEG THEN GOTO 70
HT2=(BURNEND-DEGC)*S3
GOTO 80
70 HT2=(TAILDEG-DEGC)*S2
80 QHT=DC*(HT1+HT2)/2
VR=V1/V2
VRN=(VR)^POLYINDEX
REM RASSWEILER AND WITHROW THEORY
TRM1=(GE+1)*V2/(2*(GE-1))-VR
TRM1=V2/(GE-1)
P2=(QHT/TRM1)+P1*VRN
T2=P2*V2/(MC*GK)
50 WORK=(P1+P2)*(V2-V1)/2
IF P2>P1 THEN DELTAP=(P2-P1)/(100000!*DC)
IF P2>P1 THEN PMAX=P2/100000!
IF P2>P1 THEN MAXDEG=DEGC-360
IF T2>T1 THEN TKMAX=T2
IF DELTAP>DPMAX THEN DPMAX=DELTAP
P1=P2
T1=T2
V1=V2
HT1=HT2
END SUB
```

ProgList.6.0 These program listings emanate from Chapter 6

COMPUTER PROGRAM PROG.6.1
PROGRAM NAME "TIMEAREA TARGETS"
ProgList.6.1

```
REM Prog.6.1
REM ProgList.6.1
REM Program TIMEAREA TARGETS
K1=600000!
100 PRINT"YOU WILL HAVE AVAILABLE SOME INFORMATION ON THE PROPOSED PERFORMANCE TARGETS"
PRINT
PRINT"FOR INSTANCE, YOU MUST KNOW ONE OF THREE SETS OF INFORMATION"
PRINT "IN OTHER WORDS TO BE ABLE TO PROCEED, YOU MUST KNOW,"
PRINT "EITHER-"
PRINT"A TARGET POWER LEVEL FROM"
PRINT "A GIVEN SWEPT VOLUME AT A DESIRED ENGINE SPEED-THIS IS GROUP A"
PRINT"OR-"
PRINT"A TARGET BMEP LEVEL ONLY-THIS IS GROUP B"
PRINT "OR-"
PRINT "A TARGET BMEP LEVEL FROM A GIVEN SWEPT VOLUME AT A DESIRED ENGINE SPEED-THIS IS GROUP C"
PRINT
INPUT"TYPE IN THE GROUP OF INFORMATION YOU KNOW, AS 'A' OR 'B' OR 'C'";GROUP$
IF GROUP$="A" THEN INPUT"TYPE IN THE TARGET POWER OUTPUT AT THE CRANKSHAFT IN kW UNITS";KW
IF GROUP$="A" THEN INPUT"TYPE IN THE DESIRED TARGET ENGINE SPEED IN rpm UNITS";RPM
IF GROUP$="A" THEN INPUT"TYPE IN THE SWEPT VOLUME OF THE PROPOSED ENGINE IN cm3 UNITS";SVCC
IF GROUP$="B" THEN INPUT"TYPE IN THE DESIRED BMEP IN bar UNITS";BMEP
IF GROUP$="C" THEN INPUT"TYPE IN THE DESIRED BMEP IN bar UNITS";BMEP
IF GROUP$="C" THEN INPUT"TYPE IN THE DESIRED TARGET ENGINE SPEED IN rpm UNITS";RPM
IF GROUP$="C" THEN INPUT"TYPE IN THE SWEPT VOLUME OF THE PROPOSED ENGINE IN cm3 UNITS";SVCC
IF GROUP$="A" THEN GOSUB PERFORM2
IF GROUP$="C" THEN GOSUB PERFORM1
```

```
GOSUB TIMEAREA
IF GROUP$="A" THEN GOSUB PRINT1
IF GROUP$="C" THEN GOSUB PRINT1
IF GROUP$="B" THEN GOSUB PRINT2
300 PRINT
PRINT
PRINT "YOU CAN PRINT THIS INFORMATION TO THE LINE PRINTER
BY TYPING A 'P'"
PRINT "OR YOU CAN RECALCULATE BY TYPING AN 'R'"
INPUT"OR YOU CAN QUIT BY TYPING A 'Q'";ACTION$
IF ACTION$="P" THEN GOSUB PRINTOUT
IF ACTION$="R" THEN GOTO 100
IF ACTION$="Q" THEN GOTO 200
GOTO 300
200 END
TIMEAREA:
EXTA=(BMEP+5.975)/1050
BLOWTA=(BMEP-1.75)/8187
TRTA1=(BMEP+9.66)/2400
TRTA2=(BMEP-.128)/587
INTA=(BMEP+1.528)/774
RETURN
PERFORM1:
KW=BMEP*SVCC*RPM/K1
RETURN
PERFORM2:
BMEP=K1*KW/(SVCC*RPM)
RETURN
PRINT1:
PRINT
PRINT "POWER, kW=";KW
PRINT "BMEP, bar=";BMEP
PRINT "SWEPT VOLUME, cc=";SVCC
PRINT "ENGINE SPEED, rpm=";RPM
GOSUB PRINTTA
PRINT
RETURN
PRINT2:
PRINT
PRINT "BMEP, bar=";BMEP
GOSUB PRINTTA
PRINT
RETURN
```

```
LPRINT1:
LPRINT "POWER, kW=";KW
LPRINT "BMEP, bar=";BMEP
LPRINT "SWEPT VOLUME, cc=";SVCC
LPRINT "ENGINE SPEED, rpm=";RPM
GOSUB LPRINTTA
PRINT
RETURN
LPRINT2:
LPRINT "BMEP, bar=";BMEP
GOSUB LPRINTTA
PRINT
RETURN
LPRINTTA:
LPRINT "TOTAL EXHAUST TIME AREA, s/m=";USING"##.####";EXTA
LPRINT "EXHAUST BLOWDOWN TIME AREA, s/m=";USING
"##.#####";BLOWTA
LPRINT "UPPER TARGET TOTAL TRANSFER TIME AREA, s/m=";USING
"##.####";TRTA1
LPRINT "LOWER TARGET TOTAL TRANSFER TIME AREA, s/m=";USING
"##.####";TRTA2
LPRINT "TOTAL INLET TIME AREA, s/m=";USING"##.####";INTA
RETURN
PRINTTA:
PRINT "TOTAL EXHAUST TIME AREA, s/m=";USING"##.####";EXTA
PRINT "EXHAUST BLOWDOWN TIME AREA, s/m=";USING
"##.#####";BLOWTA
PRINT "UPPER TARGET TOTAL TRANSFER TIME AREA, s/m=";USING
"##.####";TRTA1
PRINT "LOWER TARGET TOTAL TRANSFER TIME AREA, s/m=";USING
"##.####";TRTA2
PRINT "TOTAL INLET TIME AREA, s/m=";USING"##.####";INTA
RETURN
PRINTOUT:
LPRINT
INPUT"WANT A TITLE FOR THE PRINTOUT(Y/N?)";A$
IF A$="Y" THEN INPUT"TYPE IN YOUR TITLE";TIT$
LPRINT TIT$
IF GROUP$="A" THEN GOSUB LPRINT1
IF GROUP$="C" THEN GOSUB LPRINT1
IF GROUP$="B" THEN GOSUB LPRINT2
RETURN
```

COMPUTER PROGRAM PROG.6.2
PROGRAM NAME “EXPANSION CHAMBER”
ProgList.6.2

```
REM Prog.6.2
REM ProgList.6.2
REM Program EXPANSION CHAMBER
REM***PROGRAM TUNED EXPANSION CHAMBER FOR HIGH SPECIFIC OUTPUT 2-STROKE ENGINES***
WINDOW 1
REM A TUNED EXHAUST PIPE PROGRAM FOR A TWO STROKE ENGINE
WINDOW 2,"CLICK BOX",(201,50)-(400,450),1
PRINT "SELECT BY BUTTON"
WINDOW 3,"CURRENT DATA",(10,50)-(200,450),1
WINDOW 4,"DATA INPUT",(400,50)-(600,450),1
WINDOW 5,"ACTION BOX",(10,400)-(600,450),1
WINDOW 6,"EXHAUST TYPE",(400,50)-(600,450),1
WINDOW 1
100 INPUT"NEW DATA OR FILE DATA (TYPE N OR F)?";D$
IF D$="N" THEN GOTO 110
IF D$="F" THEN GOSUB FILED
110 WINDOW 3
CLS
GOSUB CURRENTDATA
WINDOW 2
BUTTON 1,1,"BORE",(5,25)-(100,45),3
BUTTON 2,1,"STROKE",(5,45)-(100,65),3
BUTTON 3,1,"CON-ROD",(5,65)-(100,85),3
BUTTON 4,1,"EXHAUST TIMING",(5,85)-(200,105),3
BUTTON 5,1,"NUMBER OF PORTS",(5,105)-(200,125),3
BUTTON 6,1,"WIDTH OF EACH PORT",(5,125)-(200,145),3
BUTTON 7,1,"TOP CORNER RADIUS",(5,145)-(200,165),3
BUTTON 8,1,"BOTTOM CORNER RADIUS",(5,165)-(200,185),3
BUTTON 9,1,"DOWN PIPE DIAMETER",(5,185)-(200,205),3
BUTTON 10,1,"MID-SECTION DIAMETER",(5,205)-(200,225),3
BUTTON 11,1,"RPM",(5,225)-(200,245),3
BUTTON 12,1,"EXHAUST TEMPERATURE",(5,245)-(200,265),3
BUTTON 13,1,"EXHAUST TYPE?",(5,265)-(200,285),3
BUTTON 30,1,"RUN",(5,325)-(100,350),3
BUTTON 31,1,"QUIT",(5,350)-(100,375),3
ACTIVITY=DIALOG(0)
WHILE ACTIVITY<>1:ACTIVITY=DIALOG(0):WEND
BUTTONPUSHED=DIALOG(1)
IF BUTTONPUSHED=30 THEN GOTO 130
```

```
IF BUTTONPUSHED=31 THEN GOTO 556
ON BUTTONPUSHED GOSUB BORE,STROKE,CONROD,EXTIMING,
PORTNO,PORTWID,TOPR,BOTTOMR,DOWNDIA,MIDDIA,SPEED,EXTEMP,
EXHTYPE
GOTO 110
130 WINDOW 1
CLS
PI=3.1415927#
DTOR=PI/180
RTOD=180/PI
CRK=ST/2
DEG=EO
GOSUB PIST
GASR=287
GAM=1.4
TEXK=TEXC+273
AO=SQR(GAM*GASR*TEXK)
REM THE 2 FACTOR IS FOR THE TAILPIPE ENTRY REFLECTION TO
RETURN AT EXH CLOSURE
REM THE TAILCONE SENDS + REFLECTIONS BACK EARLIER
THETA=2*(180-EO)
LT=1000*AO*THETA/(12*RPM)
X10=.1*LT
X12=.41*LT
X24=.14*LT
X20=X10+X12
X45=.11*LT
X60=LT
X56=.24*LT
X67=X56
X40=X20+X24
X50=X40+X45
X70=X60+X67
X14=X12+X24
REM DID IS THE 'IDEAL' EXHAUST PIPE DIAMETER
DID=1.125*EXD
D10=DI
D40=DM
D50=DM
REM THIS WAS THE ORIGINAL STATEMENT D60=DID*SQR(DID/D50)
D60=EXK*EXD
D70=D60
REM INTERIM STATEMENTS UNTIL MICROSOFT SUB CALL STATE-
MENTS SORTED OUT
```

```
LHORN=X14
LBIT=X12
DEND=D40
DSTART=D10
GOSUB BESSEL
D20=DBIT
REM END OF INTERIM STATEMENTS
GOSUB SUBPRINT
INPUT"WANT A PRINT-OUT OF THE PIPE CONE MANUFACTURING
DIMENSIONS?,(Y/N)";C$
IF C$="Y" THEN GOSUB CONEDRAW
GOTO 130
556 END
EXHTYPE:
WINDOW 6
LOCATE 1,1:PRINT "CLICK WHICH TYPE"
LOCATE 2,1:PRINT "OF EXHAUST APPLICATION"
BUTTON 1,1,"MOTOCROS",(5,85)-(200,105),3
BUTTON 2,1,"ENDURO",(5,105)-(200,125),3
BUTTON 3,1,"GP ROAD RACE",(5,125)-(200,145),3
ACTIVITY=DIALOG(0)
WHILE ACTIVITY<>1:ACTIVITY=DIALOG(0):WEND
BUTTONPUSHED=DIALOG(1)
ON BUTTONPUSHED GOSUB MOTOX,ENDUR,GPROAD
RETURN
MOTOX:
EXH$="MOTOCROS PIPE"
EXK=.65
RETURN
ENDUR:
EXH$="ENDURO PIPE"
EXK=.7
RETURN
GPROAD:
EXH$="GRAND PRIX PIPE"
EXK=.6
RETURN
SUBPRINT:
CLS
GOSUB CADRAW
LOCATE 18,1:PRINT"BORE=";BO
LOCATE 18,10:PRINT"STROKE=";ST
LOCATE 18,21:PRINT"CON-ROD=";CRL
LOCATE 18,34:PRINT"EXHAUST OPENS,ºatdc=";EO
```

```
LOCATE 18,57:PRINT"EXHAUST PORTS=";PNO
LOCATE 19,1:PRINT"WIDTH EACH PORT=";EWID
LOCATE 19,20:PRINT "EXHAUST PORT TOP RADIUS=";TR
LOCATE 19,46:PRINT"AND BOTTOM RADIUS=";BR
LOCATE 20,1:PRINT"ENGINE SPEED=";RPM
LOCATE 20,20: PRINT"EXHAUST TEMPERATURE,ºC=";TEXC
LOCATE 3,1: PRINT EXH$
LOCATE 21,1:PRINT"AREA RATIOS FOR PORT TO PIPE=";USING
"#.###";PTOPR
LOCATE 21,32:PRINT"AND PORT TO MID-SECTION=";USING
"##.##";PTOMR
WINDOW 5
BUTTON 1,1,"RECALCULATE",(5,5)-(125,50),3
BUTTON 2,1,"PRINT",(125,5)-(200,50),3
BUTTON 3,1,"FILING DATA",(200,5)-(325,50),3
BUTTON 4,1,"FILE INPUT",(325,5)-(450,50),3
BUTTON 31,1,"QUIT",(450,5)-(550,50),3
ACTIVITY=DIALOG(0)
WHILE ACTIVITY<>1:ACTIVITY=DIALOG(0):WEND
BUTTONPUSHED=DIALOG(1)
IF BUTTONPUSHED=31 THEN GOTO 556
ON BUTTONPUSHED GOSUB RECALC,SUBLPRINT,FILING,FILED
RETURN
RECALC:
GOTO 110
RETURN
CONEDRAW:
DX=D10
DY=D20
LC=X12
GOSUB CONE
RX1=RX
RY1=RY
TH1=THD
CHX1=CHX
CHY1=CHY
DX=D20
DY=D40
LC=X24
GOSUB CONE
RX3=RX
RY3=RY
TH3=THD
CHX3=CHX
```

```
CHY3=CHY
DX=D60
DY=D50
LC=X56
GOSUB CONE
RX4=RX
RY4=RY
TH4=THD
CHX4=CHX
CHY4=CHY
CLS
OPEN"LPT1:PROMPT" FOR OUTPUT AS #1
WINDOW OUTPUT#1
LOCATE 8,35:PRINT USING"####";D10
LOCATE 9,35:PRINT USING"####";D20
LOCATE 10,35:PRINT USING"####";X12
LOCATE 12,35:PRINT USING"####";RX1
LOCATE 13,35:PRINT USING"####";RY1
LOCATE 14,35:PRINT USING"####";CHX1
LOCATE 15,35:PRINT USING"####";CHY1
LOCATE 16,35:PRINT USING"##.#";TH1
LOCATE 8,44:PRINT USING"####";D20
LOCATE 9,44:PRINT USING"####";D40
LOCATE 10,44:PRINT USING"####";X24
LOCATE 12,44:PRINT USING"####";RX3
LOCATE 13,44:PRINT USING"####";RY3
LOCATE 14,44:PRINT USING"####";CHX3
LOCATE 15,44:PRINT USING"####";CHY3
LOCATE 16,44:PRINT USING"####";TH3
LOCATE 8,51:PRINT USING"####";D60
LOCATE 9,51:PRINT USING"####";D50
LOCATE 10,51:PRINT USING"####";X56
LOCATE 12,51:PRINT USING"####";RX4
LOCATE 13,51:PRINT USING"####";RY4
LOCATE 14,51:PRINT USING"####";CHX4
LOCATE 15,51:PRINT USING"####";CHY4
LOCATE 16,51:PRINT USING"####";TH4
LOCATE 8,30:PRINT "DX"
LOCATE 9,30:PRINT"DY"
LOCATE 10,30:PRINT"LC"
LOCATE 12,30:PRINT"RX"
LOCATE 13,30:PRINT"RY"
LOCATE 14,30:PRINT"CHX"
LOCATE 15,30:PRINT"CHY"
```

```
LOCATE 16,30:PRINT"Øº"
LOCATE 6,35:PRINT"CONE1"
LOCATE 6,44:PRINT"CONE2"
LOCATE 6,51:PRINT"CONE"
LOCATE 4,51:PRINT"TAIL"
LOCATE 5,51:PRINT"PIPE"
LOCATE 4,35:PRINT"DIFFUSER CONES"
LOCATE 9,58:PRINT"CONE"
LOCATE 10,58:PRINT"DIMENSIONS"
LOCATE 14,58:PRINT"FLAT PLATE"
LOCATE 15,58:PRINT"DIMENSIONS"
X5=100
Y5=20
Y6=100
X6=X5
R1=20
R2=50
CIRCLE (X5,Y5),R1,,,,.2
CIRCLE (X6,Y6),R2,,,,.2
X1=X5-R1
X2=X5+R1
X3=X6-R2
X4=X6+R2
LINE(X1,Y5)-(X3,Y6)
LINE(X2,Y5)-(X4,Y6)
LINE(X1-3,Y5)-(X1-35,Y5)
RX=X1-3
RY=Y5
GOSUB RARROW
LINE(X2+3,Y5)-(X2+R2-R1+25,Y5)
RX=X2+3
RY=Y5
GOSUB LARROW
LOCATE 1,6:PRINT"DX"
LINE(X3-3,Y6)-(X3-35,Y6)
RX=X3-3
RY=Y6
GOSUB RARROW
LINE(X4+3,Y6)-(X4+25,Y6)
RX=X4+3
RY=Y6
GOSUB LARROW
LOCATE 6,3:PRINT"DY"
LINE (X4+20,Y6-3)-(X4+20,Y5+3)
```

```
RX=X4+20
RY=Y6-3
GOSUB DARROW
RY=Y5+3
GOSUB UARROW
LOCATE 4,23:PRINT"LC"
YA=Y6+20
XA=X5
RCX=150
PD=20
S=(270-PD)*DTOR
E=(270+PD)*DTOR
CIRCLE (XA,YA),RCX,,S,E
CIRCLE (XA,YA),60,,S,E
LOCATE 11,13:PRINT"Øº"
LOCATE 15,13:PRINT"RX"
LOCATE 18,7:PRINT"FLAT SHEET TO"
LOCATE 19,7:PRINT"ROLL INTO CONE"
LOCATE 21,13:PRINT"RY"
RCY=250
S=(270-PD)*DTOR
E=(270+PD)*DTOR
CIRCLE (XA,YA),RCY,,S,E
PH=PD*DTOR
XX=SIN(PH)
YY=COS(PH)
XB=XA-RCX*XX
XC=XA+RCX*XX
YB=YA+RCX*YY
YC=YB
XD=XA-RCY*XX
XE=XA+RCY*XX
YD=YA+RCY*YY
YE=YD
LINE(XB,YB)-(XD,YD)
LINE(XC,YC)-(XE,YE)
LINE(XA,YA)-(XB+1,YB-3)
LINE(XA,YA)-(XC-1,YC-3)
XZ=XA
YZ=YA+RCY-3
LINE(XZ,YZ)-(XZ,YZ-25)
RX=XZ
RY=YZ
GOSUB DARROW
```

```
YZ=YA+RCX-3
LINE(XZ,YZ)-(XZ,YZ-25)
RX=XZ
RY=YZ
GOSUB DARROW
RX=XB-3
RY=YB
LINE (RX,RY)-(RX-20,RY)
GOSUB RARROW
RX=XD-3
RY=YD
LINE (RX,RY)-(RX-20,RY)
GOSUB RARROW
RX=XC+3
RY=YC
LINE (RX,RY)-(RX+20,RY)
GOSUB LARROW
RX=XE+3
RY=YE
LINE (RX,RY)-(RX+20,RY)
GOSUB LARROW
LOCATE 16,21:PRINT"CHX"
LOCATE 22,25:PRINT"CHY"
KI=16
KY=94
FOR J= 1 TO 10
LINE (220,KY+KI*J)-(455,KY+KI*J)
NEXT J
LINE (265,30)-(265,254)
LINE (325,70)-(325,254)
LINE (220,110)-(220,254)
LINE (455,30)-(455,254)
LINE (265,30)-(265,254)
LINE (265,30)-(455,30)
LINE (265,70)-(395,70)
LINE (220,110)-(220,254)
LINE (395,30)-(395,254)
CLOSE#1
RETURN
CONE:
CX=PI*DX
CY=PI*DY
RY=LC/(1-(CX/CY))
RX=RY-LC
```

```
TH=CY/RY
THD=TH*180/PI
SINE=SIN(TH/2)
CHX=2*RX*SINE
CHY=2*RY*SINE
RETURN
LARROW:
REM LEFTWARDS
LINE(RX,RY)-(RX+3,RY-3)
LINE(RX,RY)-(RX+3,RY+3)
RETURN
RARROW:
REM RIGHTWARDS
LINE(RX,RY)-(RX-3,RY-3)
LINE(RX,RY)-(RX-3,RY+3)
RETURN
UARROW:
REM UPWARDS
LINE(RX,RY)-(RX-3,RY+3)
LINE(RX,RY)-(RX+3,RY+3)
RETURN
DARROW:
REM DOWNWARDS
LINE(RX,RY)-(RX-3,RY-3)
LINE(RX,RY)-(RX+3,RY-3)
RETURN
SUBLPRINT:
PRINT "REMEMBER-DO NOT PROMPT 'TALL ADJUSTED' IN THE
DIALOG BOX"
INPUT"DOES THE OUTPUT DATA NEED A TITLE(Y/N?)";T$
IF T$="Y" THEN INPUT"TYPE IN TITLE";TI$
CLS
OPEN"LPT1:PROMPT" FOR OUTPUT AS #1
WINDOW OUTPUT#1
GOSUB CADRAW
LOCATE 18,1:PRINT"BORE=";BO
LOCATE 18,10:PRINT"STROKE=";ST
LOCATE 18,21:PRINT"CON-ROD=";CRL
LOCATE 18,34:PRINT"EXHAUST OPENS,ºatdc=";EO
LOCATE 18,57:PRINT"EXHAUST PORTS=";PNO
LOCATE 19,1:PRINT"WIDTH EACH PORT=";EWID
LOCATE 19,20:PRINT"EXHAUST PORT TOP RADIUS=";TR
LOCATE 19,46:PRINT"AND BOTTOM RADIUS=";BR
LOCATE 20,1:PRINT"ENGINE SPEED=";RPM
```

```
LOCATE 20,20: PRINT"EXHAUST TEMPERATURE,ºC=";TEXC
LOCATE 3,1: PRINT EXH$
LOCATE 21,1:PRINT"AREA RATIOS FOR PORT TO PIPE=";USING
"#.###";PTOPR
LOCATE 21,32:PRINT"AND PORT TO MID-SECTION=";USING
"##.##";PTOMR
IF T$="Y" THEN LOCATE 23,1:PRINT TI$
CLOSE#1
RETURN
CADRAW:
XA=90
XH=490
X2=XA
XE=XA
X3=XA
X1=X2-80
X4=X1
XZ=XA+10
XD=XZ
YB=100
YC=120
YA=YB
Y2=YA-30
Y3=YA+50
Y4=Y3
Y1=Y2
Y8=Y1+40
X8=X1+40
REM SETS UP PIPE DIMENSIONS
YY0=(YC+YB)/2
YSC=(YC-YB)/DI
XSC=(XH-XA)/LT
CX1=X10*XSC+XA
CX2=X20*XSC+XA
CX4=X40*XSC+XA
CX5=X50*XSC+XA
CX6=X60*XSC+XA
CX7=X70*XSC+XA
XB=CX1
XC=XB
SD2=D20*YSC/2
SD4=D40*YSC/2
SD5=D50*YSC/2
SD6=D60*YSC/2
```

```
SD7=D70*YSC/2
Y2U=YY0-SD2
Y2D=YY0+SD2
Y4U=YY0-SD4
Y4D=YY0+SD4
Y5U=YY0-SD5
Y5D=YY0+SD5
Y6U=YY0-SD6
Y6D=YY0+SD6
Y7U=YY0-SD7
Y7D=YY0+SD7
REM DRAWS TOP HALF OF PIPE
LINE (XB,YB)-(CX2,Y2U)
LINE (CX2,Y2U)-(CX4,Y4U)
LINE (CX4,Y4U)-(CX5,Y5U)
LINE (CX5,Y5U)-(CX6,Y6U)
LINE (CX6,Y6U)-(CX7,Y7U)
LINE (CX7,Y7U)-(CX7,Y7D)
REM DRAWS BOTTOM HALF OF PIPE
LINE (XC,YC)-(CX2,Y2D)
LINE (CX2,Y2D)-(CX4,Y4D)
LINE (CX4,Y4D)-(CX5,Y5D)
LINE (CX5,Y5D)-(CX6,Y6D)
LINE (CX6,Y6D)-(CX7,Y7D)
REM DRAWS CONE DIVIDERS
LINE (CX2,Y2U)-(CX2,Y2D)
LINE (CX4,Y4U)-(CX4,Y4D)
LINE (CX5,Y5U)-(CX5,Y5D)
LINE (CX6,Y6U)-(CX6,Y6D)
LINE (XB,YB)-(XC,YC)
REM DRAWS PISTON
REM
CIRCLE(X8,Y8),10
LINE(X8-20,Y8)-(X8+20,Y8)
LINE(X8,Y8+20)-(X8,Y8-20)
LINE(X1,Y1)-(X2,Y2)
LINE(X2,Y2)-(X3,Y3)
LINE(X3,Y3)-(X4,Y4)
LINE(X4,Y4)-(X1,Y1)
X6=X1
X7=X2
Y6=Y1+5
Y7=Y6
LINE(X6,Y6)-(X7,Y7)
```

```
LINE(X6,Y6+1)-(X7,Y7+1)
LINE(X6,Y6+4)-(X7,Y7+4)
LINE(X6,Y6+5)-(X7,Y7+5)
REM DRAWING FIRST DOWNPIPE
YZ=YB
YD=YC
YE=YD-5
LINE(XA,YA)-(XE,YE)
LINE(XA,YA)-(XB,YB)
LINE(XE,YE)-(XD,YD)
LINE(XD,YD)-(XZ,YZ)
LINE(XD,YD)-(XC,YC)
REM FINISHES DRAWING FIRST DOWNPIPE
REM DRAWS SCALE LINES
YN=200
YR=5
RX=X3
RY=Y3+3
LINE (RX,YN)-(RX,RY)
RX=XC
RY=YC+3
GOSUB UARROW
LINE (RX,YN)-(RX,RY)
RX=CX2
RY=Y2D+3
GOSUB UARROW
LINE (RX,YN)-(RX,RY)
RX=CX4
RY=Y4D+3
GOSUB UARROW
LINE (RX,YN)-(RX,RY)
RX=CX5
RY=Y5D+3
GOSUB UARROW
LINE (RX,YN)-(RX,RY)
RX=CX6
RY=Y6D+3
GOSUB UARROW
LINE (RX,YN)-(RX,RY)
RX=CX7
RY=Y7D+3
GOSUB UARROW
LINE (RX,YN)-(RX,RY)
REM DRAWS TOP SCALE LINES
```

```
RX=XB
RY=YB-3
GOSUB DARROW
LINE (RX,YR)-(RX,RY)
RX=CX2
RY=Y2U-3
GOSUB DARROW
LINE (RX,YR)-(RX,RY)
RX=CX4
RY=Y4U-3
GOSUB DARROW
LINE (RX,YR)-(RX,RY)
RX=CX5
RY=Y5U-3
GOSUB DARROW
LINE (RX,YR)-(RX,RY)
RX=CX6
RY=Y6U-3
GOSUB DARROW
LINE (RX,YR)-(RX,RY)
RX=CX7
RY=Y7U-3
GOSUB DARROW
LINE (RX,YR)-(RX,RY)
REM DRAWS HORIZONTAL SCALE LINES
YN=YN-5
RY=YN
RX=X3+3
GOSUB LARROW
RX=XC-3
GOSUB RARROW
LINE (X3+3,YN)-(XC-3,YN)
RX=CX1+3
GOSUB LARROW
RX=CX2-3
GOSUB RARROW
LINE (CX1+3,YN)-(CX2-3,YN)
RX=CX2+3
GOSUB LARROW
RX=CX4-3
GOSUB RARROW
LINE (CX2+3,YN)-(CX4-3,YN)
RX=CX4+3
GOSUB LARROW
```

```
RX=CX5-3
GOSUB RARROW
LINE (CX4+3,YN)-(CX5-3,YN)
RX=CX5+3
GOSUB LARROW
RX=CX6-3
GOSUB RARROW
LINE (CX5+3,YN)-(CX6-3,YN)
RX=CX6+3
GOSUB LARROW
RX=CX7-3
GOSUB RARROW
LINE (CX6+3,YN)-(CX7-3,YN)
YN=YN+50
RX=X3+3
RY=YN-5
GOSUB LARROW
RX=CX6-3
GOSUB RARROW
LINE (X3+3,RY)-(CX6-3,RY)
LINE (X3,YN)-(X3,YN-50)
LINE (CX6,YN)-(CX6,YN-50)
REM PRINTS LABELS ON DRAWING
LOCATE 1,13:PRINT"Ø"; USING"##";DI
LOCATE 1,33:PRINT"Ø"; USING"##";D20
LOCATE 1,45:PRINT"Ø"; USING"###";D40
LOCATE 1,58:PRINT"Ø"; USING"##";D60
LOCATE 1,70:PRINT"Ø"; USING"##";D70
LOCATE 12,13:PRINT USING"###";X10
LOCATE 12,24:PRINT USING"###";X12
LOCATE 12,39:PRINT USING"###";X24
LOCATE 12,45:PRINT USING"###";X45
LOCATE 12,54:PRINT USING"###";X56
LOCATE 12,66:PRINT USING"###";X67
LOCATE 15,15:PRINT"TUNED LENGTH FROM PISTON TO TAIL PIPE ENTRY IS ";USING"####";LT
LOCATE 17,1:PRINT"DESIGN LENGTHS AND DIAMETERS OF AN EXPANSION CHAMBER FOR A TWO-STROKE ENGINE"
RETURN
BESSEL:
DBIT=DSTART*(DEND/DSTART)*((LBIT/LHORN)^1.33)
RETURN
FILING:
INPUT "TYPE IN TITLE NEEDED FOR THE FILE WITH '.TUNEP' END CODE";P$
```

```
OPEN P$ FOR OUTPUT AS #1
WRITE#1,BO,ST,CRL,EO,PNO,EWID,TR,BR,DI,DM,RPM,TEXC,EXH$
FINISH:
CLOSE#1
GOTO 130
RETURN
FILED:
INPUT "TYPE IN THE FILENAME WITH A '.TUNEP'";FIL$
OPEN"I",#1,FIL$
INPUT#1,BO,ST,CRL,EO,PNO,EWID,TR,BR,DI,DM,RPM,TEXC,EXH$
CLOSE#1
RETURN
REM PISTON POSITION SUBROUTINE
PIST:
DR=DEG*DTOR
HTDC=CRK+CRL-CRK*COS(DR)-SQR(CRL*CRL-(CRK*SIN(DR))^2)
EXAREA=PNO*(ST-HTDC)*EWID-2*PNO*(TR*TR-PI*TR*TR/4)-
2*PNO*(BR*BR-PI*BR*BR/4)
EXD=SQR(4*EXAREA/PI)
PTOPR=(DI/EXD)^2
PTOMR=(DM/EXD)^2
RETURN
BORE:
WINDOW 4
CLS
INPUT"BORE";BO
RETURN
STROKE:
WINDOW 4
CLS
INPUT"STROKE";ST
RETURN
CONROD:
WINDOW 4
CLS
INPUT"CON-ROD";CRL
RETURN
EXTIMING:
WINDOW 4
CLS
PRINT"EXHAUST PORT"
INPUT"OPENS, DEG ATDC";EO
RETURN
PORTNO:
WINDOW 4
```

```
CLS
PRINT"NUMBER OF"
INPUT"EXHAUST PORTS";PNO
RETURN
PORTWID:
WINDOW 4
CLS
PRINT"WIDTH OF EACH"
INPUT"EXHAUST PORT";EWID
RETURN
SPEED:
WINDOW 4
CLS
INPUT"ENGINE SPEED, REV/MIN";RPM
RETURN
BOTTOMR:
WINDOW 4
CLS
PRINT"BOTTOM RADIUS OF EACH"
INPUT"EXHAUST PORT";BR
RETURN
TOPR:
WINDOW 4
CLS
PRINT"TOP RADIUS OF EACH"
INPUT"EXHAUST PORT";TR
RETURN
DOWNDIA:
WINDOW 4
CLS
PRINT"EXHAUST FIRST DOWN-"
INPUT"PIPE DIAMETER";DI
RETURN
MIDDIA:
WINDOW 4
CLS
PRINT"EXHAUST MID-SECTION"
INPUT"PIPE DIAMETER";DM
RETURN
EXTEMP:
WINDOW 4
CLS
PRINT"EXHAUST GAS TEMPERATURE"
INPUT"AT MID-SECTION, DEG.C";TEXC
RETURN
```

```
CURRENTDATA:
PRINT "BORE=";BO
PRINT "STROKE=";ST
PRINT "CON-ROD=";CRL
PRINT "EXHAUST OPENS=";EO
PRINT "NO. PORTS=";PNO
PRINT "WIDTH EACH PORT=";EWID
PRINT "TOP RADIUS=";TR
PRINT "BOTTOM RADIUS=";BR
PRINT "DOWN PIPE DIA=";DI
PRINT "MID SECTION DIA=";DM
PRINT "RPM=";RPM
PRINT "EXHAUST TEMP,ºC=";TEXC
PRINT EXH$
IF EXH$="ENDURO PIPE" THEN EXK=.675
IF EXH$="GRAND PRIX PIPE" THEN EXK=.6
IF EXH$="MOTOCROS PIPE" THEN EXK=.625
RETURN
```

COMPUTER PROGRAM PROG.6.3
PROGRAM NAME "TIME-AREAS"
ProgList.6.3

```
REM Prog.6.3
REM ProgList.6.3
REM Program TIME-AREAS
REM***TIME AREA PROGRAM FOR RECTANGULAR PORTS WITH
RADIUSED CORNERS***
14 INPUT "NEW DATA OR FILED DATA(N OR F)";G$
IF G$="F" THEN GOSUB FILED
IF G$="F" THEN GOTO 19
15 INPUT"NEW BASIC ENGINE DATA(Y/N?)";ANS$
IF ANS$="N" THEN GOTO 16
INPUT"NEW DATA FOR CYLINDER BORE(Y/N?)";ANS$
IF ANS$="Y" THEN INPUT"TYPE IN CYLINDER BORE(B)";BO
INPUT"NEW DATA FOR CYLINDER STROKE(Y/N?)";ANS$
IF ANS$="Y" THEN INPUT"TYPE IN CYLINDER STROKE(S)";ST
INPUT"NEW DATA FOR CONNECTING ROD LENGTH(Y/N?)";ANS$
IF ANS$="Y" THEN INPUT"TYPE IN CONNECTING ROD
LENGTH(CRL)";CRL
INPUT"NEW DATA FOR ENGINE SPEED(Y/N?)";ANS$
IF ANS$="Y" THEN INPUT"TYPE IN ENGINE SPEED(RPM)";RPM
PRINT
16 INPUT"NEW DATA FOR EXHAUST PORT (Y/N?)";ANS$
IF ANS$="N" THEN GOTO 17
```

```
INPUT"NEW DATA FOR EXHAUST PORT OPENING TIMING(Y/N?)";ANS$
IF ANS$="Y" THEN INPUT"TYPE IN EXHAUST PORT OPENING,DEG. ATDC";EPO
INPUT"NEW DATA FOR EXHAUST PORT FULLY OPEN TIMING(Y/N?)";ANS$
IF ANS$="Y" THEN INPUT"TYPE IN EXHAUST PORT FULLY OPEN TIMING,DEG. ATDC";EPFO
INPUT"NEW DATA FOR NUMBER OF EXHAUST PORTS(Y/N?)";ANS$
IF ANS$="Y" THEN INPUT"TYPE IN NUMBER OF EXHAUST PORTS";NEP
INPUT"NEW DATA FOR MAXIMUM WIDTH OF EACH EXHAUST PORT(Y/N?)";ANS$
IF ANS$="Y" THEN INPUT"TYPE IN MAXIMUM WIDTH OF EACH EXHAUST PORT(WE)";WEP
INPUT"NEW DATA FOR EXHAUST PORT TOP RADIUS(Y/N?)";ANS$
IF ANS$="Y" THEN INPUT"TYPE IN EXHAUST PORT TOP RADIUS(RTE)";RTE
INPUT"NEW DATA FOR EXHAUST BOTTOM RADIUS(Y/N?)";ANS$
IF ANS$="Y" THEN INPUT"TYPE IN EXHAUST PORT BOTTOM RADIUS(RBE)";RBE
PRINT
17 INPUT"NEW DATA FOR TRANSFER PORT(Y/N?)";ANS$
IF ANS$="N" THEN GOTO 18
INPUT"NEW DATA FOR TRANSFER PORT OPENING TIMING(Y/N?)";ANS$
IF ANS$="Y" THEN INPUT"TYPE IN TRANSFER PORT OPENING,DEG. ATDC";TPO
INPUT"NEW DATA FOR TRANSFER PORT FULLY OPEN TIMING(Y/N?)";ANS$
IF ANS$="Y" THEN INPUT"TYPE IN TRANSFER PORT FULLY OPEN TIMING,DEG. ATDC";TPFO
INPUT"NEW DATA FOR NUMBER OF TRANSFER PORTS(Y/N?)";ANS$
IF ANS$="Y" THEN INPUT"TYPE IN NUMBER OF TRANSFER PORTS";NTP
INPUT"NEW DATA FOR MAXIMUM WIDTH OF EACH TRANSFER PORT(Y/N?)";ANS$
IF ANS$="Y" THEN INPUT"TYPE IN MAXIMUM WIDTH OF EACH TRANSFER PORT(WE)";WTP
INPUT"NEW DATA FOR TRANSFER PORT TOP RADIUS(Y/N?)";ANS$
IF ANS$="Y" THEN INPUT"TYPE IN TRANSFER PORT TOP RADIUS(RTE)";RTT
INPUT"NEW DATA FOR TRANSFER PORT BOTTOM RADIUS(Y/N?)";ANS$
```

```
IF ANS$="Y" THEN INPUT"TYPE IN TRANSFER PORT BOTTOM RADIUS(RBT)";RBT
PRINT
18 INPUT"NEW DATA FOR INLET PORT(Y/N?)";ANS$
IF ANS$="N" THEN GOTO 19
INPUT"NEW DATA FOR INLET PORT OPENING TIMING(Y/N?)";ANS$
IF ANS$="Y" THEN INPUT"TYPE IN INLET PORT OPENING,DEG. BTDC";IPO
INPUT"NEW DATA FOR INLET PORT FULLY OPEN  TIMING(Y/N?)";ANS$
IF ANS$="Y" THEN INPUT"TYPE IN INLET PORT FULLY OPEN TIMING,DEG. BTDC";IPFO
INPUT"NEW DATA FOR NUMBER OF INLET PORTS(Y/N?)";ANS$
IF ANS$="Y" THEN INPUT"TYPE IN NUMBER OF INLET PORTS";NIP
INPUT"NEW DATA FOR MAXIMUM WIDTH OF EACH INLET PORT(Y/N?)";ANS$
IF ANS$="Y" THEN INPUT"TYPE IN MAXIMUM WIDTH OF EACH INLET PORT(WI)";WIP
INPUT"NEW DATA FOR INLET PORT TOP RADIUS(Y/N?)";ANS$
IF ANS$="Y" THEN INPUT"TYPE IN INLET PORT TOP RADIUS(RTI)";RTI
INPUT"NEW DATA FOR INLET PORT BOTTOM RADIUS(Y/N?)";ANS$
IF ANS$="Y" THEN INPUT"TYPE IN INLET PORT BOTTOM RADIUS(RBI)";RBI
19 PI=3.141592654#
DTOR=PI/180
RTOD=1/DTOR
PA=PI*BO*BO/4
SV=PA*ST
CRNK=ST/2
EPOR= EPO*DTOR
TPOR= TPO*DTOR
IPOR= 2*PI-IPO*DTOR
EPFR=EPFO*DTOR
TPFR= TPFO*DTOR
IPFR= 2*PI-IPFO*DTOR
PFE=0
PFT=0
PFI=0
PWE= WEP-2*RTE
PWT= WTP-2*RTT
PWI=WIP-2*RBI
PHI=0
PHE=0
PHT=0
```

```
EXTA=0
BLOW=0
SCAV=0
SUCK=0
CALL PISTPOS(ST,CRNK,CRL,EPOR,HEOT,HEOB)
CALL PISTPOS(ST,CRNK,CRL,EPFR,HEFT,HEFB)
EXHT=HEFT-HEOT
HME=EXHT-RTE-RBE
CALL PISTPOS(ST,CRNK,CRL,TPOR,HTOT,HTOB)
CALL PISTPOS(ST,CRNK,CRL,TPFR,HTFT,HTFB)
TRHT=HTFT-HTOT
HMT=TRHT-RTT-RBT
CALL PISTPOS(ST,CRNK,CRL,IPOR,HIOT,HIOB)
CALL PISTPOS(ST,CRNK,CRL,IPFR,HIFT,HIFB)
INHT= HIOT-HIFT
HMI=INHT-RTI-RBI
REM***IT IS ONLY 1E7 BECAUSE PRINT OF TIME AREA IS 'BY 10^4'***
KDT=1E+07/(6*RPM*SV)
GOSUB BAPRINT
PRINT
INPUT"DOING THE EXHAUST AND SCAVENGE PORT CALCULATION FIRST(Y/N)";ANS$
IF ANS$="N" THEN GOTO 80
ANGC=EPOR
50 ANGC=ANGC+DTOR
IF ANGC>PI THEN GOTO 500
ANGD=ANGC*RTOD
CALL PISTPOS(ST,CRNK,CRL,ANGC,HTP,HBP)
CALL TAREA(HEOT,EXHT,RTE,RBE,HME,WEP,PWE,PHE,PFE,EXTA,HTP)
IF ANGC<=TPOR THEN GOTO 60
CALL TAREA(HTOT,TRHT,RTT,RBT,HMT,WTP,PWT,PHT,PFT,SCAV,HTP)
60 IF ANGC<=TPOR THEN BLOW=EXTA*KDT*NEP
GOTO 50
500 SCAV=2*SCAV*KDT*NTP
EXTA=2*EXTA*KDT*NEP
PFE=PFE*NEP
PFT=PFT*NTP
GOSUB ETPRINT
80 INPUT"DOING THE INLET PORT CALCULATION NOW(Y/N?)";ANS$
IF ANS$="N" THEN GOTO 105
ANGC=IPOR
```

```
66 ANGC=ANGC+DTOR
ANGD=ANGC*RTOD
IF ANGC>2*PI THEN GOTO 70
CALL PISTPOS(ST,CRNK,CRL,ANGC,HTP,HBP)
CALL TAREA(HTP,INHT,RBI,RTI,HMI,WIP,PWI,PHI,PFI,SUCK,HIOT)
IF ANGC<2*PI THEN GOTO 66
70 PFI=PFI*NIP
SUCK=SUCK*KDT*NIP*2
GOSUB INPRINT
INPUT"WANT A PRINT-OUT(Y/N?)";ANS$
IF ANS$="N" THEN GOTO 105
106 GOSUB LBAPRINT
INPUT"IS THERE EXHAUST AND TRANSFER DATA FOR THE LINE-PRINTER(Y/N?)";ANS$
IF ANS$="Y" THEN GOSUB LETPRINT
INPUT"IS THERE INLET DATA FOR THE LINE-PRINTER(Y/N?)";ANS$
IF ANS$="Y" THEN GOSUB LINPRINT
105 INPUT"CALCULATING AGAIN(Y OR N), OR PRINTING(P), FILING THE DATA(F), OR QUITTING(Q)";ANS$
IF ANS$="Y" THEN GOTO 14
IF ANS$="N" THEN GOTO 105
IF ANS$="Q" THEN GOTO 166
IF ANS$="P" THEN GOTO 106
IF ANS$="F" THEN GOSUB FILING
GOTO 105
166 END
FILING:
INPUT "TYPE IN TITLE NEEDED FOR THE FILE WITH '.TIM' END CODE";P$
OPEN P$ FOR OUTPUT AS #1
WRITE#1,BO,ST,CRL,RPM
WRITE#1,EPO,EPFO,NEP,WEP,RTE,RBE
WRITE#1,TPO,TPFO,NTP,WTP,RTT,RBT
WRITE#1,IPO,IPFO,NIP,WIP,RTI,RBI
FINISH:
CLOSE#1
RETURN
FILED:
INPUT "TYPE IN THE FILENAME";FIL$
OPEN"I",#1,FIL$
INPUT#1,BO,ST,CRL,RPM
INPUT#1,EPO,EPFO,NEP,WEP,RTE,RBE
INPUT#1,TPO,TPFO,NTP,WTP,RTT,RBT
INPUT#1,IPO,IPFO,NIP,WIP,RTI,RBI
```

```
CLOSE#1
RETURN
BAPRINT:
PRINT"PROGRAM FOR TIME-AREAS FOR PISTON PORTED ENGINE
WITH RECTANGULAR PORTS AND CORNER RADII"
PRINT"PROGRAM'TIMAREA.1'"
PRINT"ALL DATA IN MM,DEG,AND TIME-AREAS IN S/M UNITS"
PRINT"      BORE   STROKE   CON-ROD  REV/MIN"
PRINT USING"#######.#";BO,ST,CRL,RPM
RETURN
ETPRINT:
PRINT
PRINT"EXHAUST PORT DATA"
PRINT"      OPENS  FULL-OPEN  NUMBER-OFF WIDTH-EACH  TOP-
RAD  BOTTOM-RAD
PRINT USING"#########.#";EPO,EPFO,NEP,WEP,RTE,RBE
PRINT
PRINT"  TOTAL-AREA  HTDC-OPEN      HEIGHT  TOTAL-TIME-
AREA &  BLOWDOWN*10^4"
PRINT USING"###########.#";PFE,HEOT,EXHT,EXTA,BLOW
PRINT
PRINT"TRANSFER PORT DATA"
PRINT"     OPENS   FULL-OPEN   NUMBER-OFF WIDTH-EACH  TOP-
RAD  BOTTOM-RAD
PRINT USING"#########.#";TPO,TPFO,NTP,WTP,RTT,RBT
PRINT
PRINT"TOTAL-AREA     HTDC-OPEN        HEIGHT    TOTAL-TIME-
AREA*10^4 "
PRINT USING"#########.#";PFT,HTOT,TRHT,SCAV
RETURN
INPRINT:
PRINT
PRINT"INLET PORT DATA"
PRINT"     OPENS   FULL-OPEN   NUMBER-OFF WIDTH-EACH  TOP-
RAD  BOTTOM-RAD
PRINT USING"########.#";IPO,IPFO,NIP,WIP,RTI,RBI
PRINT
PRINT"TOTAL-AREA       HTDC-OPEN       HEIGHT    TOTAL-TIME-
AREA*10^4 "
PRINT USING"#########.#";PFI,HIOT,INHT,SUCK
RETURN
LBAPRINT:
INPUT"NEED A TITLE FOR THE PRINT-OUT(Y/N?)";ANS$
IF ANS$="Y" THEN INPUT"TYPE IN TITLE FOR THE PRINT-OUT";TIT$
```

```
LPRINT
LPRINT
IF ANS$="Y" THEN LPRINT TIT$
LPRINT"PROGRAM FOR TIME-AREAS FOR PISTON PORTED ENGINE
WITH RECTANGULAR PORTS AND CORNER RADII"
LPRINT"ALL DATA IN MM,DEG,AND TIME-AREAS IN S/M UNITS"
LPRINT"   BORE   STROKE  CON-ROD  REV/MIN"
LPRINT USING"#######.#";BO,ST,CRL,RPM
RETURN
LETPRINT:
LPRINT
LPRINT"EXHAUST PORT DATA"
LPRINT"   OPENS  FULL-OPEN  NUMBER-OFF  WIDTH-EACH  TOP-
RAD  BOTTOM-RAD
LPRINT USING"#########.#";EPO,EPFO,NEP,WEP,RTE,RBE
LPRINT
LPRINT" TOTAL-AREA   HTDC-OPEN   HEIGHT TOTAL-TIME-AREA
& BLOWDOWN*10^4"
LPRINT USING"##########.#";PFE,HEOT,EXHT,EXTA,BLOW
LPRINT
LPRINT"TRANSFER PORT DATA"
LPRINT"   OPENS  FULL-OPEN  NUMBER-OFF  WIDTH-EACH  TOP-
RAD  BOTTOM-RAD
LPRINT USING"#########.#";TPO,TPFO,NTP,WTP,RTT,RBT
LPRINT
LPRINT"  TOTAL-AREA    HTDC-OPEN    HEIGHT   TOTAL-TIME-
AREA*10^4 "
LPRINT USING"###########.#";PFT,HTOT,TRHT,SCAV
RETURN
LINPRINT:
LPRINT
LPRINT"INLET PORT DATA"
LPRINT"   OPENS  FULL-OPEN  NUMBER-OFF  WIDTH-EACH  TOP-
RAD  BOTTOM-RAD
LPRINT USING"#########.#";IPO,IPFO,NIP,WIP,RTI,RBI
LPRINT
LPRINT"  TOTAL-AREA   HTDC-OPEN    HEIGHT   TOTAL-TIME-
AREA*10^4 "
LPRINT USING"###########.#";PFI,HIOT,INHT,SUCK
RETURN
SUB PISTPOS(S,R,C,A,HT,HB)STATIC
 HT=R+C-R*COS(A)-SQR(C*C-(R*SIN(A))^2)
 HB=S-HT
END SUB
```

```
SUB TAREA(HOT,HP,RT,RB,HM,WP,PW,PH,PF,PT,HTP)STATIC
PH2=HTP-HOT
DH=PH2-PH
IF PH2>HP  THEN FI=0
IF PH2>HP THEN GOTO 300
IF PH2>=RT THEN  GOTO 100
PW2=2*SQR(2*RT*PH2-PH2*PH2)+WP-2*RT
GOTO  400
100 IF PH2>(RT+HM) THEN GOTO 200
PW2=WP
GOTO 400
200 BH=HP-PH2
PW2=2*SQR(2*RB*BH-BH*BH)+WP-2*RB
GOTO 400
400 FI=(PW2+PW)*DH/2
300 PT=(PF+FI/2)+PT
PF=PF+FI
PW=PW2
PH=PH2
END SUB
```

COMPUTER PROGRAM PROG.6.4
PROGRAM NAME "REED VALVE DESIGN"
ProgList.6.4

```
REM Prog.6.4
REM ProgList.6.4
REM Program REED VALVE DESIGN
50 INPUT "RESPONSE FOR NEW DATA IS 'N' AND FILED DATA IS
'F'";H$
IF H$="F" THEN GOSUB FILED
IF H$="F" THEN GOSUB FILESET
IF H$="F" THEN GOTO 502
CLS
INPUT"NEW REED PETAL DIMENSIONS(Y/N?)";ANS$
IF ANS$="Y" THEN GOSUB PETALDAT
INPUT"NEW REED MATERIAL DATA (Y/N?)";ANS$
IF ANS$="Y" THEN GOSUB MATDAT
INPUT"NEW REED BLOCK DATA (Y/N?)";ANS$
IF ANS$="Y" THEN GOSUB BLOCKDAT
INPUT"NEW ENGINE DATA (Y/N?)";ANS$
IF ANS$="Y" THEN GOSUB ENGINEDAT
502 GOSUB DATINPUT
INPUT"DIFFERENT DATA NEEDED?(Y/N)";A$
IF A$="Y" THEN GOTO 50
```

```
   GOSUB FILESET
   INTA=(BMEP+1.528)/774
   THETA=200
   TPERDEG=1/(6*RPM)
   SVM=SVCC*.000001
   TADEG=INTA*SVM/TPERDEG
   PAREA=2000000!*TADEG/THETA
   RESTRICT=1.45
   PORTAREA=(RDPL*RDPW-RDPRAD*RDPRAD*(1-.25*PI))*RDPNO
*SIN(RDBAR)
   CARB=SQR(4*PORTAREA/(PI*RESTRICT))
   INERTIA=RDWDM*RDTHM^3/12
   MASS=RDWDM*RDTHM*RDRH
   TERM=SQR(YMOD*INERTIA/(MASS*RDLNM^4))
   FREQ1=1.758*TERM/PI
   FREQ2=11.015*TERM/PI
   LOADING=.18*(CCR-1)*101325!
   DEFLECTION=LOADING*RDWDM*RDLNM^4/(8*YMOD*INERTIA)
   TIPMOVE=DEFLECTION*1000!
   TIPLIFTRATIO=TIPMOVE/RDLN
   TIPQ=.3
   STOPRAD=(1-TIPQ*TIPQ)*RDLN/(2*TIPQ)
   CLS
   GOSUB CAD
   900 LOCATE 22,1:INPUT "TYPE A 'R' FOR RECALCULATION, 'P' FOR
A PRINTOUT, OR A 'F' FOR FILING, OR 'Q' FOR QUITTING" ;N$
   IF N$="Q" THEN GOTO 911
   IF N$="R" THEN GOTO 50
   IF N$="P" THEN GOSUB PRINTOUT
   IF N$="F" THEN GOSUB FILING
   GOTO 900
   911 END
   CAD:
   KP=4
   IF N$="P" THEN KP=3
   LR=KP*RDLN
   X1=250
   Y1=150
   SCAL=LR/RDLN
   RAD=STOPRAD*SCAL
   LP2=RDPS*SCAL
   RBK=RDBAR
   LP1=(RDPL+RDPS)*SCAL
   LC=LR/4
```

```
LE=LC
LN=.1*LR
LB1=LR+LN
LB=LB1+LC+LE
T2=1*SCAL
T1=3*SCAL
T3=2*SCAL
C=COS(RBK)
S=SIN(RBK)
T=TAN(RBK)
X2=X1+LB*C
Y2=Y1-LB*S
REM DRAWS THE BLOCK
LINE(X1,Y1)-(X2,Y2)
LINE(X1,Y1)-(X2,Y1)
LINE(X2,Y1)-(X2,Y2)
DX=T1/S
X11=X1+DX
DY=T1/C
Y21=Y2+DY
REM DRAWS THE REED PORT
LINE(X11,Y1)-(X2,Y21)
X12=X1+(LB1-LP1)*C
Y12=Y1-(LB1-LP1)*S
LINE(X12,Y12)-(X12+DX,Y12)
X13=X1+(LB1-LP2)*C
Y13=Y1-(LB1-LP2)*S
LINE(X13,Y13)-(X13+DX,Y13)
REM DRAWS THE REED
X14=X1+LN*C-T2*S
Y14=Y1-LN*S-T2*C
X15=X1+(LB1+LC)*C-T2*S
Y15=Y1-(LB1+LC)*S-T2*C
LINE(X14,Y14)-(X15,Y15)
X14P=X1+LN*C
Y14P=Y1-LN*S
LINE(X14,Y14)-(X14P,Y14P)
REM DRAWS THE STOP-PLATE
XC=X14+LR*C-RAD*S
YC=Y14-LR*S-RAD*C
CIRCLE (XC,YC),RAD,,3*PI/2,3*PI/2+RBK
CIRCLE (XC,YC),RAD-T3,,3*PI/2,3*PI/2+RBK
IF XC<X14 THEN LINE(XC,YC+RAD)-(XC,YC+RAD-T3)
IF XC>=X14 THEN LINE(XC,YC+RAD)-(X14,YC+RAD)
```

```
IF XC>=X14 THEN LINE(XC,YC+RAD-T3)-(X14,YC+RAD-T3)
IF XC>=X14 THEN LINE(X14,YC+RAD)-(X14,YC+RAD-T3)
REM DRAWS THE REST OF THE STOP PLATE
X5=X14+LR*C-T3*S
Y5=Y14-LR*S-T3*C
X6=X5+LC*C
Y6=Y5-LC*S
X7=X1+(LB1+LC)*C
Y7=Y1-(LB1+LC)*S
LINE(X7,Y7)-(X6,Y6)
LINE(X5,Y5)-(X6,Y6)
REM DRAWS THE SCREW
XCS=(X5+X6)/2
YCS=(Y5+Y6)/2
CIRCLE (XCS,YCS),LC/3,33,RBK,PI+RBK,1
REM STARTING TO DRAW THE PLAN VIEW
REM SETTING SCALE VALUES
WR=RDWD*SCAL
WP=RDPW*SCAL
CRP=RDPRAD*SCAL
CRR=.5*CRP
Y1L=180
X2L=X1+LB
WB=1.2*WR
Y2L=Y1L+WB
REM DRAWS BLOCK
LINE(X1,Y1L)-(X2L,Y2L),,B
YCL=(Y1L+Y2L)/2
X40=X1+LB1+LC
X41=X1+LB1
Y40=YCL-WR/2
Y41=YCL+WR/2
REM DRAWS CLAMP FOOTPRINT
LINE(X41,Y40)-(X40,Y41),33,BF
REM DRAWS THE REED
X42=X1+LN+CRR
Y42=Y40+CRR
Y43=Y41-CRR
LINE(X42,Y40)-(X41,Y40)
LINE(X42,Y41)-(X41,Y41)
LINE(X1+LN,Y42)-(X1+LN,Y43)
CIRCLE (X42,Y42),CRR,,PI/2,PI
CIRCLE (X42,Y43),CRR,,PI,3*PI/2
REM DRAWS THE REED PORT
```

```
X30=X1+LB1-LP1+CRP
X32=X1+LB1-LP2-CRP
Y30=YCL-WP/2+CRP
Y31=YCL+WP/2-CRP
YTE=Y30-CRP
YBE=Y31+CRP
XIE=X30-CRP
XOE=X32+CRP
LINE(X30,YTE)-(X32,YTE)
LINE(X30,YBE)-(X32,YBE)
LINE(XIE,Y30)-(XIE,Y31)
LINE(XOE,Y30)-(XOE,Y31)
CIRCLE (X30,Y30),CRP,,PI/2,PI
CIRCLE (X30,Y31),CRP,,PI,3*PI/2
CIRCLE (X32,Y30),CRP,,0,PI/2
CIRCLE (X32,Y31),CRP,,3*PI/2,0
REM DRAWS SCREW HOLE
XSC=(X40+X41)/2
CIRCLE (XSC,YCL),LC/5,30
REM DRAWS PORT INSIDE EDGE
X30=X30+T1/T
XIE=X30-CRP
CIRCLE (X30,Y30),CRP,,PI/2,PI
CIRCLE (X30,Y31),CRP,,PI,3*PI/2
LINE(XIE,Y30)-(XIE,Y31)
LOCATE 1,1:PRINT "CYLINDER SV, cc=";SVCC
LOCATE 1,21:PRINT "SPEED, rpm=";RPM
LOCATE 1,39:PRINT "BMEP, bar=";BMEP
LOCATE 1,52:PRINT "CCR=";CCR
LOCATE 2,1:PRINT "PETAL MATERIAL IS ";MAT$
LOCATE 3,1:PRINT "PETAL THICKNESS 'RDTH', mm=";RDTH
LOCATE 4,1:PRINT "REED BLOCK ANGLE 'RDBA', deg=";RDBA
LOCATE 5,1:PRINT "PETAL NUMBER 'RDNO'=";RDNO
LOCATE 6,1:PRINT "PORTS NUMBER 'RDPNO'=";RDPNO
LOCATE 7,1:PRINT "PETAL WIDTH 'RDWD', mm=";RDWD
LOCATE 8,1:PRINT "PORT WIDTH 'RDPW', mm=";RDPW
LOCATE 9,1:PRINT "CORNER RADIUS 'CRAD', mm=";RDPRAD
LOCATE 10,1:PRINT "PETAL LENGTH 'RDLN', mm=";RDLN
LOCATE 11,1:PRINT "PORT LENGTH 'RDPL', mm=";RDPL
LOCATE 12,1:PRINT "LENGTH FROM CLAMP 'RDPS', mm=";RDPS
LOCATE 13,1:PRINT "OUTPUT DATA"
LOCATE 14,1:PRINT "REQUIRED FLOW AREA, mm2=";USING
"####.";PAREA
LOCATE 15,1:PRINT "DESIGNED FLOW AREA, mm2=";USING
"####.";PORTAREA
```

```
LOCATE 16,1:PRINT "CARBURETTER DIA., mm=";USING"###.";CARB
LOCATE 17,1:PRINT "REED NATURAL FREQUENCY, Hz=";USING
"####.";FREQ1
LOCATE 18,1:PRINT "ENGINE NATURAL FREQUENCY, Hz=";USING
"####.";RPM/60
LOCATE 19,1:PRINT "TIP LIFT RATIO (0.15-0.25)=";USING "#.##";TI-
PLIFTRATIO
LOCATE 20,1:PRINT "STOP PLATE RADIUS 'SPRD', mm=";USING
"###.";STOPRAD
LOCATE 11,35:PRINT "SECTION ON HALF BLOCK"
LOCATE 20,35:PRINT "PLAN ON PORT AND PETAL OUTLINE"
LOCATE 21,35:PRINT "DARK AREA IS CLAMP FOOTPRINT"
RETURN
FILING:
INPUT "TYPE IN TITLE NEEDED FOR THE FILE WITH '.REED' END
CODE";P$
OPEN P$ FOR OUTPUT AS #1
WRITE#1,RDNO,RDTH,RDWD,RDLN,RDBA,RDPNO,RDPL,RDPS,
RDPW,RDPRAD,MAT$,SVCC,BMEP,CCR,RPM
CLOSE#1
RETURN
FILED:
INPUT "TYPE IN THE FILENAME WITH A '.REED' END CODE";FIL$
OPEN"I",#1,FIL$
INPUT#1,RDNO,RDTH,RDWD,RDLN,RDBA,RDPNO,RDPL,RDPS,RDPW,
RDPRAD,MAT$,SVCC,BMEP,CCR,RPM
CLOSE#1
RETURN
PETALDAT:
INPUT"NEW NUMBER OF REED PETALS, 'RDNO'(Y/N?)";ANS$
IF ANS$="Y" THEN INPUT"TYPE IN NUMBER OF REED PETALS";RDNO
INPUT"NEW REED PETAL WIDTH, 'RDWD'(Y/N?)";ANS$
IF ANS$="Y" THEN INPUT"TYPE IN REED WIDTH, mm";RDWD
INPUT"NEW REED PETAL FREE LENGTH, 'RDLN'(Y/N?)";ANS$
IF ANS$="Y" THEN INPUT"TYPE IN REED LENGTH, mm";RDLN
INPUT"NEW REED PETAL THICKNESS, 'RDTH'(Y/N?)";ANS$
IF ANS$="Y" THEN INPUT"TYPE IN REED PETAL THICKNESS,
mm";RDTH
RETURN
MATDAT:
INPUT"NEW TYPE OF REED PETAL MATERIAL(Y/N?)";ANS$
IF ANS$="Y" THEN PRINT "YOU CAN SELECT FROM 'STEEL' OR
'GLASS-FIBRE' OR 'CARBON-FIBRE'
IF ANS$="Y" THEN INPUT"TYPE IN THE WORD 'STEEL' OR 'GLASS-
FIBRE' OR 'CARBON-FIBRE'";MAT$
```

```
IF MAT$="STEEL" THEN YMOD=2.07E+11
IF MAT$="GLASS-FIBRE" THEN YMOD=2.15E+10
IF MAT$="CARBON-FIBRE" THEN YMOD=2.08E+10
IF MAT$="STEEL" THEN RDRH=7800
IF MAT$="GLASS-FIBRE" THEN RDRH=1850
IF MAT$="CARBON-FIBRE" THEN RDRH=1380
RETURN
BLOCKDAT:
PRINT "REMEMBER THAT THIS DATA IS FOR A REED BLOCK FOR
ONE CYLINDER"
PRINT
INPUT"NEW DATA FOR NUMBER OF REED BLOCK PORTS 'RDPNO'(Y/
N?)";ANS$
IF ANS$="Y" THEN INPUT"TYPE IN NUMBER OF REED BLOCK
PORTS";RDPNO
INPUT"NEW REED BLOCK PORT WIDTH, 'RDPW'(Y/N?)";ANS$
IF ANS$="Y" THEN INPUT"TYPE IN REED BLOCK PORT WIDTH,
mm";RDPW
INPUT"NEW REED BLOCK PORT LENGTH, 'RDPL'(Y/N?)";ANS$
IF ANS$="Y" THEN INPUT"TYPE IN REED BLOCK PORT LENGTH,
mm";RDPL
INPUT"NEW REED BLOCK PORT LENGTH POSITION, 'RDPS'(Y/
N?)";ANS$
IF ANS$="Y" THEN INPUT"TYPE IN REED BLOCK PORT LENGTH
POSITION, mm";RDPS
INPUT"NEW REED BLOCK PORT CORNER RADII, 'CRAD'(Y/N?)";ANS$
IF ANS$="Y" THEN INPUT"TYPE IN REED BLOCK PORT CORNER
RADII, mm";RDPRAD
INPUT"NEW REED BLOCK HALF ANGLE, 'RDBA'(Y/N?)";ANS$
IF ANS$="Y" THEN INPUT"TYPE IN REED BLOCK HALF ANGLE,
deg";RDBA
RETURN
ENGINEDAT:
PRINT "REMEMBER TO QUOTE DATA HERE FOR ONE CYLINDER
ONLY, EVEN IF IT IS A MULTI-CYLINDER UNIT"
PRINT "REMEMBER THAT IT IS ASSUMED THAT EACH CYLINDER
HAS A SEPARATE REED BLOCK AND PETALS"
PRINT
INPUT"NEW DATA FOR ENGINE CYLINDER SWEPT VOLUME(Y/
N?)";ANS$
IF ANS$="Y" THEN INPUT"TYPE IN SWEPT VOLUME OF EACH CYL-
INDER, cc";SVCC
INPUT"NEW DATA FOR ENGINE SPEED AT PEAK POWER OUTPUT(Y/
N?)";ANS$
```

```
IF ANS$="Y" THEN INPUT"TYPE IN ENGINE SPEED FOR PEAK POWER,
rpm";RPM
INPUT"NEW DATA FOR THE BMEP LEVEL EXPECTED, KNOWN, OR
ESTIMATED AT THAT ENGINE SPEED(Y/N?)";ANS$
IF ANS$="Y" THEN INPUT"TYPE IN THE BMEP LEVEL, bar";BMEP
INPUT"NEW DATA FOR CRANKCASE COMPRESSION RATIO, 'CCR'(Y/
N?)";ANS$
IF ANS$="Y" THEN INPUT"TYPE IN CRANKCASE COMPRESSION
RATIO";CCR
RETURN
FILESET:
PI=3.1415926#
DTOR=PI/180
RTOD=1/DTOR
IF MAT$="STEEL" THEN YMOD=2.07E+11
IF MAT$="GLASS-FIBRE" THEN YMOD=2.15E+10
IF MAT$="CARBON-FIBRE" THEN YMOD=2.08E+10
IF MAT$="STEEL" THEN RDRH=7800
IF MAT$="GLASS-FIBRE" THEN RDRH=1850
IF MAT$="CARBON-FIBRE" THEN RDRH=1380
RDBAR=RDBA*DTOR
A=1000
RDTHM=RDTH/A
RDLNM=RDLN/A
RDWDM=RDWD/A
RETURN
DATINPUT:
PRINT "DATA FOR REED PETAL, THE MATERIAL FOR WHICH IS
";MAT$
PRINT "   REEDS  THICKNESS  WIDTH  LENGTH"
PRINT USING"######.##";RDNO,RDTH,RDWD,RDLN
PRINT
PRINT "REED BLOCK AT HALF ANGLE, deg,=";RDBA
PRINT "   PORTS  WIDTH  LENGTH  CORNER RADII  LENGTH FROM
CLAMP POINT"
PRINT USING"######.##";RDPNO,RDPW,RDPL,RDPRAD,RDPS
PRINT
PRINT "ENGINE DATA"
PRINT "   CYLINDER,cc  SPEED, rpm    BMEP, bar   CRANKCASE
COMPRESSION RATIO"
PRINT USING"########.##";SVCC,RPM,BMEP,CCR
RETURN
PRINTOUT:
LOCATE 23,1:INPUT"NEED A PRINTOUT TITLE(Y/N)";T$
```

```
IF T$="Y" THEN LOCATE 24,1:INPUT"TYPE IN TITLE";TIT$
LOCATE 25,1:INPUT"*****INDEX 'TALL-ADJUSTED' FOR SKETCH
TO BE SCALED CORRECTLY***** PRESS RETURN";AAAAAA
CLS
OPEN"LPT1:PROMPT" FOR OUTPUT AS #1
WINDOW OUTPUT#1
GOSUB CAD
LOCATE 22,1:PRINT TIT$
CLOSE#1
RETURN
DATOUTPUT:
PRINT "REQUIRED FLOW AREA  DESIGNED FLOW AREA  RECOM-
MENDED CARBURETTER DIAMETER"
PRINT USING"############.#";PAREA,PORTAREA,CARB
PRINT
PRINT "NATURAL FREQUENCIES OF THE REED PETALS AND OF THE
ENGINE SUCTION, Hz"
PRINT "FIRST ORDER   SECOND ORDER  THE ENGINE"
PRINT USING"########.#";FREQ1,FREQ2,RPM/60
PRINT
PRINT "RECOMMENDED TIP LIFT RATIO IS BETWEEN 0.15 AND O.25"
PRINT "THE DESIGNED TIP LIFT RATIO FOR THE PETAL
IS=";USING"#.##";TIPLIFTRATIO
PRINT
PRINT "THE RECOMMENDED STOP PLATE RADIUS FOR THIS DE-
SIGN IS, mm,=";USING"###.";STOPRAD
PRINT
RETURN
```

COMPUTER PROGRAM PROG.6.5

PROGRAM NAME "DISC VALVE DESIGN"

ProgList.6.5

```
REM Prog.6.5
REM ProgList.6.5
REM Program DISC VALVE DESIGN
50 INPUT "RESPONSE FOR NEW DATA IS 'N' AND FILED DATA IS
'F'";H$
IF H$="F" THEN GOSUB FILED
IF H$="F" THEN GOTO 502
CLS
INPUT"NEW ENGINE DATA (Y/N?)";ANS$
IF ANS$="Y" THEN GOSUB ENGINEDAT
INPUT"NEW DISC TIMING DATA(Y/N?)";ANS$
IF ANS$="Y" THEN GOSUB TIMINGDAT
```

```
INPUT"NEW DISC DIMENSION DATA (Y/N?)";ANS$
IF ANS$="Y" THEN GOSUB DIMDAT
502 GOSUB DATINPUT
INPUT"DIFFERENT DATA NEEDED?(Y/N)";A$
IF A$="Y" THEN GOTO 50
KEXP=1.44
PI=3.1415926#
DTOR=PI/180
RTOD=1/DTOR
INTA=(BMEP+1.528)/774
PERIOD=DVO+DVC
ANGD=PERIOD-ANGH
TPERDEG=1/(6*RPM)
SVM=SVCC*.000001
TADEG=INTA*SVM/TPERDEG
DPA=1000000!*TADEG/(PERIOD-ANGH)
CARB=SQR(4*DPA*SQR(KEXP)/PI)
RMAX=SQR((360*(KEXP*DPA+(RCRN*RCRN*(1-PI/4)))/(PI*ANGH))
+RMIN*RMIN)
R=(RMAX+RMIN)/2
HEIGHT=RMAX-RMIN
WIDE=R*SIN(ANGH*DTOR)
ASPECT=WIDE/HEIGHT
CLS
GOSUB CAD
900 LOCATE 18,1:INPUT "TYPE A 'R' FOR RECALCULATION, 'P' FOR
A PRINTOUT, OR A 'F' FOR FILING, OR 'Q' FOR QUITTING" ;N$
IF N$="Q" THEN GOTO 911
IF N$="R" THEN GOTO 50
IF N$="P" THEN GOSUB PRINTOUT
IF N$="F" THEN GOSUB FILING
GOTO 900
911 END
PRINTOUT:
LOCATE 21,1:INPUT"NEED A PRINTOUT TITLE(Y/N)";T$
IF T$="Y" THEN LOCATE 22,1:INPUT"TYPE IN TITLE";TIT$
LOCATE 25,1:INPUT"*****INDEX 'TALL-ADJUSTED' FOR SKETCH
TO BE SCALED CORRECTLY***** PRESS RETURN";AAAAAA
CLS
OPEN"LPT1:PROMPT" FOR OUTPUT AS #1
WINDOW OUTPUT#1
GOSUB CAD
LOCATE 16,1:PRINT TIT$
CLOSE#1
```

```
RETURN
CAD:
XC=360
YC=140
RD=100
SCAL=RD/RMAX
RMINS=RMIN*SCAL
ANGHR=ANGH*DTOR
ANGDR=ANGD*DTOR
RHOL=.5*RMINS
RDISC=RD+5
RHOUS=RDISC+5
CIRCLE (XC,YC),RHOUS
CIRCLE (XC,YC),RHOL
STRT=PI/2+ANGDR/2
FIN=PI/2-ANGDR/2
CIRCLE (XC,YC),RDISC,,STRT,FIN
STRT=PI/2-ANGDR/2
FIN=PI/2+ANGDR/2
CIRCLE (XC,YC),RMINS,,STRT,FIN
RC=RCRN*SCAL
AC=RD-RC
RB=SQR(AC*AC-RC*RC)
AN2=ATN(RC/RB)
AN1=ANGHR/2
AN3=AN1-AN2
FIN=PI/2+AN3
STRT=PI/2-AN3
CIRCLE (XC,YC),RD,,STRT,FIN
XA=XC-AC*SIN(AN3)
XA1=XC+AC*SIN(AN3)
YA=YC-AC*COS(AN3)
YA1=YA
STRT=PI/2+AN3
FIN=PI+AN1
CIRCLE (XA,YA),RC,,STRT,FIN
STRT=2*PI-AN1
FIN=PI/2-AN3
CIRCLE (XA1,YA1),RC,,STRT,FIN
XR=XC-RB*SIN(AN1)
XR1=XC+RB*SIN(AN1)
YR=YC-RB*COS(AN1)
YR1=YR
REM LOWER CORNER RADII
```

```
PC=RMINS+RC
QC=SQR(PC*PC-RC*RC)
PN2=ATN(RC/QC)
XQ=XC-QC*SIN(AN1)
XQ1=XC+QC*SIN(AN1)
YQ=YC-QC*COS(AN1)
YQ1=YQ
PN3=AN1-PN2
XP=XC-PC*SIN(PN3)
XP1=XC+PC*SIN(PN3)
YP=YC-PC*COS(PN3)
YP1=YP
STRT=PI+AN1
FIN=3*PI/2+PN3
CIRCLE (XP,YP),RC,,STRT,FIN
STRT=3*PI/2-PN3
FIN=2*PI-AN1
CIRCLE (XP1,YP1),RC,,STRT,FIN
X1=XR
X3=XR1
X2=XQ
X4=XQ1
Y1=YR
Y3=YR
Y2=YQ
Y4=Y2
LINE(X1,Y1)-(X2,Y2)
LINE(X3,Y3)-(X4,Y4)
X5=XC-RDISC*SIN(ANGDR/2)
X8=XC+RDISC*SIN(ANGDR/2)
X6=XC-RMINS*SIN(ANGDR/2)
X7=XC+RMINS*SIN(ANGDR/2)
Y5=YC-RDISC*COS(ANGDR/2)
Y8=Y5
Y6=YC-RMINS*COS(ANGDR/2)
Y7=Y6
LINE(X5,Y5)-(X6,Y6)
LINE(X7,Y7)-(X8,Y8)
LOCATE 1,1:PRINT "SWEPT VOLUME, cc=";SVCC
LOCATE 1,23:PRINT "ENGINE BMEP, bar=";BMEP
LOCATE 1,42:PRINT "SPEED, rpm= ";RPM
LOCATE 2,1:PRINT "DISC VALVE OPENS, deg.btdc=";DVO
LOCATE 3,1:PRINT "DISC VALVE CLOSES, deg.atdc=";DVC
LOCATE 6,1:PRINT "PORT ANGLE, 'ANGH', deg.=";ANGH
```

```
LOCATE 4,1:PRINT "DISC MINIMUM RADIUS, 'RMIN', mm=";RMIN
LOCATE 5,1:PRINT "PORT CORNER RADII, 'RCRN', mm=";RCRN
LOCATE 8,1:PRINT "OUTPUT DATA"
LOCATE 9,1:PRINT "MAXIMUM RADIUS, 'RMAX', mm=";USING
"###.";RMAX
LOCATE 10,1:PRINT"DISC ANGLE, 'ANGD', deg=";USING"###.";ANGD
LOCATE 11,1:PRINT "CARBURETTER DIA., mm=";USING"###.";CARB
LOCATE 12,1:PRINT "PORT HEIGHT, mm=";USING"###.";HEIGHT
LOCATE 13,1:PRINT "PORT WIDTH at mid-height, mm=";USING
"###.";WIDE
LOCATE 14,1:PRINT "ASPECT RATIO, width/height=";USING
"#.##";ASPECT
LOCATE 12,37:PRINT "DISC VALVE AND PORT"
LOCATE 13,40:PRINT "AS DESIGNED"
RETURN
FILING:
INPUT "TYPE IN TITLE NEEDED FOR THE FILE WITH '.DISC' END
CODE";P$
OPEN P$ FOR OUTPUT AS #1
WRITE#1,DVO,DVC,ANGH,RMIN,RCRN,SVCC,BMEP,RPM
CLOSE#1
RETURN
FILED:
INPUT "TYPE IN THE FILENAME WITH A '.DISC' END CODE";FIL$
OPEN"I",#1,FIL$
INPUT#1,DVO,DVC,ANGH,RMIN,RCRN,SVCC,BMEP,RPM
CLOSE#1
RETURN
TIMINGDAT:
INPUT"NEW VALUE FOR DISC VALVE OPENING, 'DVO'(Y/N?)";ANS$
IF ANS$="Y" THEN INPUT"TYPE IN VALUE FOR DISC VALVE OPEN-
ING, deg.btdc";DVO
INPUT"NEW VALUE FOR DISC VALVE CLOSING, 'DVC'(Y/N?)";ANS$
IF ANS$="Y" THEN INPUT"TYPE IN VALUE FOR DISC VALVE CLOS-
ING, deg.atdc";DVC
RETURN
DIMDAT:
INPUT"NEW DATA FOR DISC PORT OPENING ANGLE, 'ANGH'(Y/
N?)";ANS$
PRINT
IF ANS$="Y" THEN PRINT"IF YOU ARE STUCK FOR AN INITIAL
GUESS, TRY 50 deg !"
IF ANS$="Y" THEN INPUT"TYPE IN VALUE FOR DISC PORT OPENING
ANGLE, deg";ANGH
```

```
INPUT"NEW DATA FOR MINIMUM DISC RADIUS, 'RMIN'(Y/N?)";ANS$
IF ANS$="Y" THEN INPUT"TYPE IN VALUE FOR MINIMUM DISC RADIUS, mm";RMIN
INPUT"NEW DATA FOR DISC PORT CORNER RADII, 'RCRN'(Y/N?)";ANS$
IF ANS$="Y" THEN INPUT"TYPE IN VALUE FOR DISC PORT CORNER RADII, mm";RCRN
RETURN
ENGINEDAT:
PRINT "REMEMBER TO QUOTE DATA HERE FOR ONE CYLINDER ONLY, EVEN IF IT IS A MULTI-CYLINDER UNIT"
PRINT "REMEMBER THAT IT IS ASSUMED THAT EACH CYLINDER HAS A SEPARATE DISC VALVE AND PORT"
PRINT
INPUT"NEW DATA FOR ENGINE CYLINDER SWEPT VOLUME(Y/N?)";ANS$
IF ANS$="Y" THEN INPUT"TYPE IN SWEPT VOLUME OF EACH CYLINDER, cc";SVCC
INPUT"NEW DATA FOR ENGINE SPEED AT PEAK POWER OUTPUT(Y/N?)";ANS$
IF ANS$="Y" THEN INPUT"TYPE IN ENGINE SPEED FOR PEAK POWER, rpm";RPM
INPUT"NEW DATA FOR THE BMEP LEVEL EXPECTED, KNOWN, OR ESTIMATED AT THAT ENGINE SPEED(Y/N?)";ANS$
IF ANS$="Y" THEN INPUT"TYPE IN THE BMEP LEVEL, bar";BMEP
RETURN
DATINPUT:
PRINT "DATA FOR DISC VALVE"
PRINT " DVO    DVC    ANGH    RMIN    RCRN"
PRINT USING"####.##";DVO,DVC,ANGH,RMIN,RCRN
PRINT
PRINT "ENGINE DATA"
PRINT "  CYLINDER,cc  SPEED, rpm    BMEP, bar"
PRINT USING"########.##";SVCC,RPM,BMEP
RETURN
```

ProgList.8.0 These program listings emanate from Chapter 8

COMPUTER PROGRAM PROG.8.1
PROGRAM NAME "DIFFUSING SILENCER"
ProgList.8.1

```
REM Prog.8.1
REM ProgList.8.1
REM Program DIFFUSING SILENCER
REM DIFFUSING SILENCER DESIGN PROGRAM
DIM XD(300),YD(300)
PI=3.1415926#
100 INPUT"DESIGNING A DIFFUSING SILENCER? (Y/N)";Q$
IF Q$="Y" THEN GOSUB DIFFUSING
CLS
GOSUB DATASET
GOSUB GRAPHAXIS
FR=0
J=1
YD(J)=YB
XD(J)=50*XSCAL+XL
200 FR=25+FR
J=J+1
GOSUB FUKUDA
IF FR>XHZ THEN GOTO 300
GOTO 200
300 LOCATE 26,1:INPUT"IF YOU WANT A PRINT ON THE PRINTER
THEN TYPE A 'P'";A$
IF A$="P" THEN GOSUB SUBLPRINT
GOTO 100
END
SUBLPRINT:
INPUT"TYPE IN A TITLE FOR THE PRINT-OUT";TITLE$
CLS
OPEN"LPT1:PROMPT" FOR OUTPUT AS #1
WINDOW OUTPUT#1
GOSUB GRAPHAXIS
FR=0
J=1
YD(J)=YB
XD(J)=50*XSCAL+XL
500 FR=25+FR
J=J+1
GOSUB FUKUDA
```

```
IF FR>XHZ THEN GOTO 600
GOTO 500
600 LOCATE 25,1:PRINT  TITLE$
CLOSE#1
RETURN
GRAPHAXIS:
YT=45
YB=345
XR=475
XL=75
XHZ=4000
YDB=50
XSCAL=(XR-XL)/XHZ
YSCAL=(YB-YT)/YDB
YSTEP=10*YSCAL
XSTEP1=100*XSCAL
XSTEP2=1000*XSCAL
LINE(XL,YB)-(XL,YT)
LINE(XL,YB)-(XR,YB)
FOR I=1 TO 5
LINE(XL,YB-YSTEP*I)-(XL-5,YB-YSTEP*I)
NEXT I
FOR I=1 TO 40
LINE(XL+XSTEP1*I,YB)-(XL+XSTEP1*I,YB+3)
NEXT I
FOR I=1 TO 4
LINE(XL+XSTEP2*I,YB)-(XL+XSTEP2*I,YB+8)
NEXT I
LOCATE 2,10:PRINT "DESIGN FOR A DIFFUSING SILENCER"
LOCATE 5,62:PRINT "INPUT DATA"
LOCATE 6,62:PRINT "LB=";LB
LOCATE 7,62:PRINT "DSB=";DSB
LOCATE 8,62:PRINT "DS1=";DS1
LOCATE 9,62:PRINT "DS2=";DS2
LOCATE 10,62:PRINT "L1=";L1
LOCATE 11,62:PRINT "L2=";L2
LOCATE 12,62:PRINT "LT=";LT
LOCATE 13,62:PRINT "TºC=";TEXC
LOCATE 5,5:PRINT "A"
LOCATE 6,5:PRINT "T"
LOCATE 7,5:PRINT "T"
LOCATE 8,5:PRINT "E"
LOCATE 9,5:PRINT "N"
LOCATE 10,5:PRINT "U"
```

```
LOCATE 11,5:PRINT "A"
LOCATE 12,5:PRINT "T"
LOCATE 13,5:PRINT "I"
LOCATE 14,5:PRINT "O"
LOCATE 15,5:PRINT "N"
LOCATE 17,5:PRINT "dB"
LOCATE 3,7:PRINT "50"
LOCATE 7,7:PRINT "40"
LOCATE 11,7:PRINT "30"
LOCATE 15,7:PRINT "20"
LOCATE 18,7:PRINT "10"
LOCATE 22,7:PRINT " 0"
LOCATE 23,10:PRINT "0"
LOCATE 23,20:PRINT "1000"
LOCATE 23,32:PRINT "2000"
LOCATE 23,45:PRINT "3000"
LOCATE 23,57:PRINT "4000"
LOCATE 24,35:PRINT "FREQUENCY, Hz"
RETURN
FUKUDA:
K=2*PI*FR/AO
TERM=MFUKUDA*((SIN(K*LBM)*SIN(K*LTM))/(COS(K*L1M)
*COS(K*L2M)))
ATTFUKUDA=10*(LOG(TERM*TERM))/LOG(10)
IF ATTFUKUDA<0 THEN ATTFUKUDA=0
IF ATTFUKUDA>50 THEN ATTFUKUDA=50
YD(J)=YB-ATTFUKUDA*YSCAL
XD(J)=FR*XSCAL+XL
LINE(XD(J-1),YD(J-1))-(XD(J),YD(J))
RETURN
DIFFUSING:
INPUT"NEW DATA FOR DIFFUSING SILENCER BOX LENGTH, LB (Y/
N?)";A$
IF A$="Y" THEN INPUT"TYPE IN SILENCER BOX LENGTH, mm";LB
INPUT"NEW DATA FOR DIFFUSING SILENCER BOX DIAMETER, DSB
(Y/N?)";A$
IF A$="Y" THEN INPUT"TYPE IN SILENCER BOX DIAMETER, mm";DSB
INPUT"NEW DATA FOR INLET PIPE DIAMETER, DS1 (Y/N?)";A$
IF A$="Y" THEN INPUT"TYPE IN INLET PIPE DIAMETER, mm";DS1
INPUT"NEW DATA FOR OUTLET PIPE DIAMETER, DS2 (Y/N?)";A$
IF A$="Y" THEN INPUT"TYPE IN OUTLET PIPE DIAMETER, mm";DS2
INPUT"NEW DATA FOR INLET PIPE INSERTION LENGTH, L1 (Y/
N?)";A$
IF A$="Y" THEN INPUT"TYPE IN INLET PIPE INSERTION LENGTH,
mm";L1
```

```
INPUT"NEW DATA FOR OUTLET PIPE INSERTION LENGTH, L2 (Y/
N?)";A$
IF A$="Y" THEN INPUT"TYPE IN OUTLET PIPE INSERTION LENGTH,
mm";L2
INPUT"NEW DATA FOR OUTLET PIPE LENGTH, LT (Y/N?)";A$
IF A$="Y" THEN INPUT"TYPE IN OUTLET PIPE LENGTH, mm";LT
INPUT"NEW DATA FOR EXHAUST GAS TEMPERATURE, TEX (Y/
N?)";A$
IF A$="Y" THEN INPUT"TYPE IN EXHAUST GAS TEMPERATURE,
ºC";TEXC
RETURN
DATASET:
LBM=LB/1000
L1M=L1/1000
L2M=L2/1000
LTM=LT/1000
DSBM=DSB/1000
DS1M=DS1/1000
DS2M=DS2/1000
TEXK=TEXC+273
AO=SQR(1.4*287*TEXK)
S1=PI*DS1M*DS1M/4
S2=PI*DS2M*DS2M/4
SB=PI*DSBM*DSBM/4
MROE=SB/S1
MFUKUDA=SB/S2
RETURN
```

COMPUTER PROGRAM PROG.8.2
PROGRAM NAME "SIDE-RESONANT SILENCER"
ProgList.8.2

```
REM Prog.8.2
REM ProgList.8.2
REM Program SIDE-RESONANT SILENCER
REM SIDE-RESONANT SILENCER DESIGN PROGRAM
DIM XD(300),YD(300)
PI=3.1415926#
100 INPUT"DESIGNING A SIDE-RESONANT SILENCER? (Y/N)";Q$
IF Q$="Y" THEN GOSUB RESONANT
CLS
GOSUB DATASET
GOSUB GRAPHAXIS
FR=0
J=1
YD(J)=YB
```

```
XD(J)=50*XSCAL+XL
200 FR=25+FR
J=J+1
GOSUB DAVIS
IF FR>XHZ THEN GOTO 300
GOTO 200
300 LOCATE 26,1:INPUT"IF YOU WANT A PRINT ON THE PRINTER
THEN TYPE A 'P'";A$
IF A$="P" THEN GOSUB SUBLPRINT
GOTO 100
END
SUBLPRINT:
INPUT"TYPE IN A TITLE FOR THE PRINT-OUT";TITLE$
CLS
OPEN"LPT1:PROMPT" FOR OUTPUT AS #1
WINDOW OUTPUT#1
GOSUB GRAPHAXIS
FR=0
J=1
YD(J)=YB
XD(J)=50*XSCAL+XL
500 FR=25+FR
J=J+1
GOSUB DAVIS
IF FR>XHZ THEN GOTO 600
GOTO 500
600 LOCATE 25,1:PRINT TITLE$
CLOSE#1
RETURN
GRAPHAXIS:
YT=45
YB=345
XR=475
XL=75
XHZ=4000
YDB=50
XSCAL=(XR-XL)/XHZ
YSCAL=(YB-YT)/YDB
YSTEP=10*YSCAL
XSTEP1=100*XSCAL
XSTEP2=1000*XSCAL
LINE(XL,YB)-(XL,YT)
LINE(XL,YB)-(XR,YB)
FOR I=1 TO 5
```

```
LINE(XL,YB-YSTEP*I)-(XL-5,YB-YSTEP*I)
NEXT I
FOR I=1 TO 40
LINE(XL+XSTEP1*I,YB)-(XL+XSTEP1*I,YB+3)
NEXT I
FOR I=1 TO 4
LINE(XL+XSTEP2*I,YB)-(XL+XSTEP2*I,YB+8)
NEXT I
LOCATE 2,10:PRINT "DESIGN FOR A SIDE-RESONANT SILENCER"
LOCATE 5,62:PRINT "INPUT DATA"
LOCATE 6,62:PRINT "LB=";LB
LOCATE 7,62:PRINT "DSB=";DSB
LOCATE 8,62:PRINT "DS3=";DS3
LOCATE 9,62:PRINT "TH=";TH
LOCATE 10,62:PRINT "NH=";NH
LOCATE 11,62:PRINT "DSH=";DSH
LOCATE 12,62:PRINT "TºC=";TEXC
LOCATE 5,5:PRINT "A"
LOCATE 6,5:PRINT "T"
LOCATE 7,5:PRINT "T"
LOCATE 8,5:PRINT "E"
LOCATE 9,5:PRINT "N"
LOCATE 10,5:PRINT "U"
LOCATE 11,5:PRINT "A"
LOCATE 12,5:PRINT "T"
LOCATE 13,5:PRINT "I"
LOCATE 14,5:PRINT "O"
LOCATE 15,5:PRINT "N"
LOCATE 17,5:PRINT "dB"
LOCATE 3,7:PRINT "50"
LOCATE 7,7:PRINT "40"
LOCATE 11,7:PRINT "30"
LOCATE 15,7:PRINT "20"
LOCATE 18,7:PRINT "10"
LOCATE 22,7:PRINT " 0"
LOCATE 23,10:PRINT "0"
LOCATE 23,20:PRINT "1000"
LOCATE 23,32:PRINT "2000"
LOCATE 23,45:PRINT "3000"
LOCATE 23,57:PRINT "4000"
LOCATE 24,35:PRINT "FREQUENCY, Hz"
RETURN
DAVIS:
TERM=(2*S3*(FR/NATFR-NATFR/FR))^2
```

```
ATTDAVIS=10*(LOG(1+CO*VBM/TERM))/LOG(10)
IF ATTDAVIS<0 THEN ATTDAVIS=0
IF ATTDAVIS>50 THEN ATTDAVIS=50
YD(J)=YB-ATTDAVIS*YSCAL
XD(J)=FR*XSCAL+XL
LINE(XD(J-1),YD(J-1))-(XD(J),YD(J))
RETURN
RESONANT:
INPUT"NEW DATA FOR RESONANT SILENCER BOX LENGTH, LB (Y/N?)";A$
IF A$="Y" THEN INPUT"TYPE IN SILENCER BOX LENGTH, mm";LB
INPUT"NEW DATA FOR RESONANT SILENCER BOX DIAMETER, DSB (Y/N?)";A$
IF A$="Y" THEN INPUT"TYPE IN SILENCER BOX DIAMETER, mm";DSB
INPUT"NEW DATA FOR INLET PIPE DIAMETER, DS3 (Y/N?)";A$
IF A$="Y" THEN INPUT"TYPE IN INLET PIPE DIAMETER, mm";DS3
INPUT"NEW DATA FOR THICKNESS OF INNER PIPE WALL, TH (Y/N?)";A$
IF A$="Y" THEN INPUT"TYPE IN THICKNESS OF INNER PIPE WALL, mm";TH
INPUT"NEW DATA FOR NUMBER OF HOLES, NH (Y/N?)";A$
IF A$="Y" THEN INPUT"TYPE IN NUMBER OF HOLES";NH
INPUT"NEW DATA FOR DIAMETER OF HOLES, DSH (Y/N?)";A$
IF A$="Y" THEN INPUT"TYPE IN DIAMETER OF HOLES";DSH
INPUT"NEW DATA FOR EXHAUST GAS TEMPERATURE, TEX (Y/N?)";A$
IF A$="Y" THEN INPUT"TYPE IN EXHAUST GAS TEMPERATURE, ºC";TEXC
RETURN
DATASET:
LBM=LB/1000
DSBM=DSB/1000
DS3M=DS3/1000
DSBM=DSB/1000
DSHM=DSH/1000
THM=TH/1000
TEXK=TEXC+273
AO=SQR(1.4*287*TEXK)
S3=PI*DS3M*DS3M/4
SB=PI*DSBM*DSBM/4
SH=PI*DSHM*DSHM/4
VBM=PI*LBM*(DSBM*DSBM-(DS3M+THM)*(DS3M+THM))/4
CO=NH*SH/(THM+.8*SH)
NATFR=(AO/(2*PI))*SQR(CO/VBM)
RETURN
```

INDEX